HANDBOOK OF
RELIABILITY ENGINEERING
AND MANAGEMENT

Other McGraw-Hill Quality Books of Interest

Carrubba • PRODUCT ASSURANCE PRINCIPLES: INTEGRATING DESIGN AND QUALITY ASSURANCE

Crosby • LET'S TALK QUALITY

Crosby • QUALITY IS FREE

Crosby • QUALITY WITHOUT TEARS

Feigenbaum • TOTAL QUALITY CONTROL, 3D ED., REVISED (FORTIETH ANNIVERSARY EDITION)

Grant, Leavenworth • STATISTICAL QUALITY CONTROL

Hradesky • TOTAL QUALITY MANAGEMENT HANDBOOK

Johnson • ISO 9000: THE NEW INTERNATIONAL STANDARD

Juran, Gryna • JURAN'S QUALITY CONTROL HANDBOOK, 4TH ED.

Juran, Gryna • QUALITY PLANNING AND ANALYSIS

Kazmierski • STATISTICAL PROBLEM SOLVING IN QUALITY ENGINEERING

Menon • TQM IN NEW PRODUCT MANUFACTURING

Mills • THE QUALITY AUDIT

Ott, Schilling • PROCESS QUALITY CONTROL

Ross • TAGUCHI TECHNIQUES FOR QUALITY ENGINEERING

Saylor • TQM SIMPLIFIED

Slater • INTEGRATED PROCESS MANAGEMENT: A QUALITY MODEL

Taylor • OPTIMIZATION AND VARIATION REDUCTION IN QUALITY

Vani • THE MCGRAW-HILL CERTIFIED QUALITY ENGINEER EXAMINATION GUIDE

To order or receive additional information on these or any other McGraw-Hill titles, please call 1-800-822-8158 in the United States. In other countries, contact your local McGraw-Hill representative.

HANDBOOK OF RELIABILITY ENGINEERING AND MANAGEMENT

W. Grant Ireson

Clyde F. Coombs, Jr.

Richard Y. Moss

Second Edition

McGraw-Hill

New York San Francisco Washington, D.C. Auckland Bogotá
Caracas Lisbon London Madrid Mexico City Milan
Montreal New Delhi San Juan Singapore
Sydney Tokyo Toronto

Library of Congress Cataloging-in-Publication Data

Handbook of reliability engineering and management / [edited by] W.
    Grant Ireson, Clyde F. Coombs, Jr., Richard Y. Moss—2nd ed.
            p.   cm.
      Includes index.
      ISBN 0-07-012750-6
      1. Reliability (Engineering)   I. Ireson, William Grant, date.
    II. Coombs, Clyde F.   III. Moss, Richard Y.
    TA169.H36   1995
    620'.00452—dc20   95-43962
                                                       CIP

McGraw-Hill

A Division of The McGraw·Hill Companies

1 2 3 4 5 6 7 8 9 0   QBM/QBM   9 0 0 9 8 7 6 5

ISBN 0-07-012750-6

*The sponsoring editor for this book was Harold Crawford, the editing supervi-
sor was Fred Bernardi, and the production supervisor was Pamela A. Pelton.
It was set in Times Roman by Priscilla Beer of McGraw-Hill's Professional
Book Group composition unit.*

Printed and bound by Quebecor/Martinsburg.

McGraw-Hill books are available at special quantity discounts to use as premi-
ums and sales promotions, or for use in corporate training programs. For more
information, please write to the Director of Special Sales, McGraw-Hill, 11
West 19th Street, New York, NY 10011; or contact your local bookstore.

CONTENTS

Chapter 10. Reliability Information Collection and Analysis **10.1**

Chapter 11. Designing Experiments to Measure and Improve Reliability **11.1**

Chapter 12. Accelerated Testing **12.1**

Chapter 13. Failure Analysis System—Root Cause and Corrective Action **13.1**

Chapter 14. Physics of Failure **14.1**

Chapter 15. Maintainability and Reliability   15.1

Chapter 16. Component Reliability   16.1

Chapter 17. Thermal Management and Reliability of Electronics   17.1

Chapter 18. Mechanical Stress and Analysis   18.1

CONTRIBUTORS

John R. Adams (CHAPTER 13). Mr. Adams has over 30 years experience in the Reliability field, during which he was a charter member and Chapter Chairman of the Denver IEEE Reliability Society, elected to National ADCOM for IEEE Reliability Society from 1992 to 1995, and published papers on software and hardware reliability.

Necip Doganaksoy, Ph.D. (CHAPTER 27). Dr. Doganaksoy is a statistician at GE Corporate R&D Center and an adjunct professor at Union College, Schenectady, where he received his Ph.D. in Administrative and Engineering Systems. He also serves as Associate Editor for *Technometrics*. His research interests are in reliability, life testing, and quality control.

Richard L. Doyle (CHAPTERS 18, 19, 20). Mr. Doyle is a Registered Electrical and Civil Engineer, with over 25 years experience in the Reliability field. His recent emphasis has been on Ocean Mining, Ship Building, and Nuclear Power. He teaches a three-day seminar on Mechanical Stress Analysis, and has served as Chairman of the IEEE Reliability Society.

Thomas L. Fagan (CHAPTER 2). Mr. Fagan is a Fellow of the IEEE and a past President of the IEEE Reliability Society. He was General Chairman of the 1983 "Reliability and Maintainability Symposium." He has also served in the United States Senate as an IEEE Congressional Science and Engineering Fellow.

Spencer Graves, Ph.D. (CHAPTER 11). Dr. Graves is a Professional Engineer with a Ph.D. in Statistics. He has more than 13 years experience teaching and consulting in quality and productivity improvement, emphasizing engineering trouble shooting and experimentation.

Dennis R. Hoffman (CHAPTER 8). Mr. Hoffman is a Senior Member of the Technical Staff and a Strategy Director within Texas Instrument Corporation Defense Systems and Electronic Group. He is a Registered Professional Engineer in Electrical Engineering, and a past Chairman of the Dallas Chapter of the IEEE Reliability Society.

Kailash C. Kapur, Ph.D. (CHAPTERS 24, 25). Dr. Kapur is Chairman of the Industrial Engineering Department at the University of Washington, Seattle. He has consulted, taught, and written extensively in Reliability, including papers and books. In 1987 he shared the Allan Chop Technical Advancement Award from the Reliability Division of ASQC.

Samuel J. Keene, Ph.D. (CHAPTER 22). Dr. Keene is a Senior Member of the IEEE, and a past President of the IEEE Reliability Society. He has planned hardware and software reliability, as well as software development. He has published over 50 papers in the reliability field, and taught tutorials on software reliability.

Henry J. Kohoutek (CHAPTERS 4, 7, 9). Mr. Kohoutek is a researcher, consultant, author, and lecturer in engineering management with many years experience in high-tech industries. His academic background is in engineering and humanities from John Hus University and the Bohemian Technological Institute, both in Prague.

Richard A. Kowalski, Ph.D. (CHAPTER 15). Dr. Kowalski is the Director, Product Assurance for ARINC's Research Division, where he is responsible for hardware and software quality program policy, planning, and execution. He is a member of the IEEE Reliability Society Administrative Committee and a past editor of the IEEE *Transactions* on Reliability.

J. David Lazor (CHAPTER 6). Mr. Lazor has over 25 years experience in Managing the Quality and Reliability sciences. His main product focus has been automotive electrical and electronic component design and maintenance.

Henry A. Malec (CHAPTER 21). Mr. Malec is the author of over fifty technical papers and a Chief Editor for *Quality and Reliability International*. He is an executive board member of the IEEE Reliability Society, a member of the Malcolm Baldrige National Quality Award Board of Examiners, and has co-edited a NASA reliability text.

Vivek Mansingh, Ph.D. (CHAPTER 17). Dr. Mansingh is a Senior Thermal Scientist and Marketing

Manager at Fujitsu Computer Products Technologies. He has been involved with R&D work in the area of Electronic Cooling for more than 12 years, and authored more than 25 technical papers, and taught related courses in Europe, the United States, and India.

Arthur A. McGill (Appendix C). Mr. McGill is a Staff Engineer at Lockheed Martin in Sunnyvale, California. He is a registered Professional Engineer (Quality) in California, and a member of the IEEE. In 1985 he received the Navy "Naval Material Command's Reliability, Maintainability, and Quality Assurance Award."

Thomas Menten, Ph.D. (Chapter 11). Dr. Menten has served as Statistician and Reliability Engineer for Hewlett-Packard Co. since 1981. In this role he has consulted in, and used, reliability analysis, DOE, and other probability and statistical methods, most recently for the Laser Printer Division.

Richard Y. Moss (Chapter 5). Mr. Moss is Reliability Engineering Manager in the Corporate Quality Department of Hewlett-Packard Co., and has more than 35 years experience in electronic product design and reliability. He holds degrees in electrical engineering from Princeton and Stanford universities.

David Nelson, Ph.D. (Chapter 23). Dr. Nelson is a manager for the Advanced Quality System in Procurement Quality Assurance, for the Boeing Company. Since joining Boeing in 1969 as a Statistician he has provided technical contributions in commercial airplane, defense, computing, and federal aeronautics and energy programs.

Wayne Nelson, Ph.D. (Chapters 26, 27). Dr. Nelson consults and gives training courses in Reliability Data Analysis and Accelerated Testing throughout industry. For outstanding contributions in these areas, he was elected a Fellow of the Institute of Electrical and Electronics Engineers, American Society for Quality in Control, and American Statistical Association.

Keith V. Rohrbach, Ph.D. (Chapter 3). Dr. Rohrbach has extensive experience in the development and manufacturing of state-of-the-art devices. He has served as Chairperson of the Biomedical Division, and has presented numerous papers and graduate courses in Reliability and related subjects.

Charles Schinner (Chapter 12). Mr. Schinner has over 15 years experience in hardware reliability at Hewlett-Packard Co., and is currently the Hardware Reliability Manager for the San Diego Division. He has authored several papers on "Accelerated Reliability Techniques" published by the Institute of Environmental Sciences and IEEE Groups.

Jack S. Smith (Chapter 14). Mr. Smith holds an MS in Physics from Syracuse University. He has spent more than 30 years in the study of failure mechanisms in semiconductor devices, and has authored over 30 articles, as well as serving on numerous Symposium Committees, and chaired the EOS/ESD Symposium.

Fred Watt (Chapter 16). Mr. Watt is President of Sav-Soft Products, a consulting firm in Milpitas, California, and has trained over 1500 component and reliability engineers in his Component Technology and Reliability Seminars. He has over 28 years experience in product development and reliability.

PREFACE TO THE SECOND EDITION

The importance of reliability as an element of customer satisfaction and product differentiation in all industries cannot be overstated. It is no longer an issue that is connected only to large systems purchased at premium prices. Reliability is now fundamental to the success of all consumer and industrial companies as well as military equipment suppliers. The customer now expects a product that functions to specification indefinitely, at no extra cost. The company that develops a reputation for poor reliability pays a heavy penalty in the marketplace. In addition, the cost of repair or replacement for failed products after delivery can make the difference between profit and loss. At the same time, the issue of legal liability for the effects of "unreliable" products can be catastrophic, and a detailed and documented process for ensuring and demonstrating that all measures to assure reliable operation have been defined and followed may be the difference between winning and losing in court. Therefore, it is critical that the tools of reliability engineering be available to, and used by, all members of the organization who have a role to play in any part of the product life cycle.

To improve an organization's ability to understand and implement the best tools and industry practices available for achieving the highest appropriate level of reliability, we have made major changes to the thrust of this second edition. Specifically, we have added material that concentrates attention on the areas of design for reliability, reliability as a process, and related issues concerning electronic equipment, mechanical equipment, systems, and software. We have also retained and updated the material on testing, prediction, and the mathematics of reliability. The result is a book that provides the tools, and the understanding, to make reliability a competitive advantage in the modern industrial environment.

It is axiomatic that a product cannot be manufactured to be more reliable than it is defined in its design. Therefore, the ability to design for reliability is a critical part of the overall life cycle of a new design. As a result, reliability is not solely the domain of the reliability engineers, but is a fundamental responsibility of all members of the new product development team. This requires that tools be available, and understood by all members of the team, for predicting, and thereby ensuring, the achievement of established reliability goals at each step, rather than testing the final design to discover the failure modes and rates, and then trying to fix them. As a result, a main new thrust of this book is the description of design for reliability—what it means, what its elements are, and how to achieve it. Specifically we have included chapters on "Design for Reliability," "Specification and Goal Setting," "Concurrent Engineering," "FMEA and FTA," "Thermal Management and Reliability," and "Design for Mechanical Reliability."

This material is directed at all those who are responsible for product development, regardless of their organizational function. At the same time, we have provided updates of the traditional tools used by the reliability professional for the detailed technical analysis of product performance.

It is also clear that reliability is a process that can be characterized, controlled, and improved as part of the overall life cycle of a product. This concept of "process" means that the tools, methodologies, and approaches developed for total quality control can generally be applied to the resolution of reliability issues as well. New chapters are included on "Managing Reliability as a Process," "Design of Experiments," and "Human-Centered Design." In addition, the revised chapters have included this concept as well, and, therefore we have described this area in some detail. Those already familiar with TQC will see the parallels between the two disciplines and the application of the general approach. For those who are not familiar with TQC, the material will stand alone as a description of tools for managing the reliability process.

We are confident that the result is a reference book that can significantly help an organization turn the reliability of its products into a competitive advantage, both externally with its customers and internally in dealing with the total cost of doing business.

We are indebted to the IEEE Reliability Society, New York, for the early and consistent support shown on this project. The content and structure of the text benefited from suggestions made by the Society leadership during its development, and many of the authors of this book are Society members.

We also thank the American Society for Quality Control, Milwaukee, Wisconsin, for permission to reproduce Appendixes B.3, B.4, B.5, B.6, and B.7, Charts of Confidence Limits on Reliability, from Lloyd and Liporo's *Reliability: Management, Methods, and Mathematics*, copyright 1984.

We are grateful to the literary executor of the late Sir Ronald A. Fisher, F.R.S., to Dr. Frank Yates, F.R.S., and to the Longman Group Ltd., London, for permission to reprint data from Table 3 from their book, *Statistical Table for Biological, Agricultural, and Medical Research* (6th edition, 1974).

W. Grant Ireson
Clyde F. Coombs, Jr.
Richard Y. Moss

HANDBOOK OF
RELIABILITY ENGINEERING
AND MANAGEMENT

CHAPTER 1
INTRODUCTION, DEFINITIONS, AND RELATIONSHIPS

W. Grant Ireson
PROFESSOR EMERITUS, STANFORD UNIVERSITY

1.1 PURPOSE OF SECTION

This section of the handbook describes in broad terms the entire process by which the quality characteristic known as *reliability* is developed in a product or service. Its purpose is to assist in the development of an effective reliability program plan by

1. Showing the total spectrum of activities needed
2. Defining the relationships of reliability engineers, quality assurance engineers, production engineers, purchasing, marketing, and management
3. Demonstrating the need for top management understanding and support of a well-planned and well-executed reliability program

1.2 CUSTOMER SATISFACTION

Quality and reliability are not free, but poor quality and reliability usually cost much more than good quality and reliability. Warranties, liabilities, recalls, and repairs cost millions of dollars each year because quality and reliability were not given enough emphasis during the design, manufacture, and use stages of product development to attain customer satisfaction. Just as in medicine, the cost of preventing poor quality and reliability is usually much less than the resulting costs of inferior quality and reliability.

Every producer of goods or services knows that the success of a business depends upon the customers' satisfaction with the product or the service, relative to the price charged. This commonly means that the quality of the product or service meets the customers' expectations at a price or cost considered reasonable. The differences in the expectations of different customers account for the fact that products that perform the same general functions are available in widely differing "qualities" and at widely differing prices. These differences are the result of different specifications and the conformance of the product to those specifications.

There are two general concepts of quality as regards specifications:

Quality of design
Quality of conformance

Automobiles offer excellent examples of differences in quality of design. The Rolls Royce, Mercedes, BMW, and Cadillac represent high quality of design when compared with less expensive cars such as the lower-priced Chevrolets, Fords, and Plymouths, and that superiority is represented in the prices charged. All the cars perform the same basic function: provide transportation for the customer. Specification of better materials, tighter tolerances, improved control systems, and many other factors make the higher-priced cars "better" cars.

Quality of conformance deals with how well the actual product conforms to the specifications. Therefore, most quality assurance activities deal with the measurement of the quality characteristics specified and procedures and programs to assure that each part that goes into the product complies with all the individual specifications. A product can be produced with total compliance with the specifications but still be inferior if the specifications are poorly prepared or inferior.

In most cases, the quality of conformance can be determined very quickly by inspection of the parts. Conformance to the specifications may or may not determine how long the product will function properly. The quality of design usually determines how well and how long the product will perform its designated function in an acceptable manner.

1.2.1 Reliability Is Just One Quality Characteristic

Reliability is defined in many different ways, but the most widely accepted definition states that it is the ability or capability of the product to perform the specified function in the designated environment for a minimum length of time or minimum number of cycles or events.

The "life" of an individual product cannot be determined except by running or operating it for the desired time (or number of cycles) or until it fails. Obviously, you cannot wear out all the products to prove that they meet the specifications, and so, just as in quality assurance, you must rely on data gained by testing samples of the product. This in turn means that the statements regarding reliability must be in terms of probability of surviving the specified life with satisfactory performance throughout. Thus, reliability is normally stated as one or more of the following:

MTTF	Mean time to failure.
MTBF	Mean time between failures.
MTBMA	Mean time between (or before) maintenance actions.
MTBR	Mean time between (or before) repairs.
Θ_0	Mean life in some units such as hours or cycles. Read "theta subzero."
λ	Failure rate in some specified time period. Read "lambda."

Contrary to ordinary quality assurance, reliability assurance requires testing over time, and frequently for very long times, in order to prove beyond a reasonable doubt that the product as designed and *built* will have an acceptable life. Since this will be a probability statement, the confidence level of the results will vary with the amount of testing. To attain a high confidence level of high reliability usually requires that the tests on a number of products be run over several months. The higher the confidence level and the reliability desired, the more testing that will be necessary.

A formal reliability program plan is as necessary as a quality assurance plan because, contrary to the quality assurance situation, it may be months before errors in design, manufacturing, and assembly that adversely affect reliability can be identified. This puts a great premium on DOING IT RIGHT THE FIRST TIME!

1.2.2 Reliability Engineering

Reliability engineering has become a recognized profession, and the American Society for Quality Control conducts examinations by which individuals can become Certified Reliability Engineers. The certification examinations usually cover about six subject areas, such as

Definitions

Analytical methods

Various distributions and their uses

Hazard functions

Step stress, life, and demonstration testing

Product life-cycle characteristics

Information regarding the examinations and other requirements, such as experience, for certification can be obtained from the Certification Department, American Society for Quality Control, 310 West Wisconsin Street, Milwaukee, WI 53203. Many of the local sections give short preparatory training programs before the examinations. The exams are given in a large number of locations in the United States as well as in some other countries.

The functions performed by the reliability engineer are probably best described by MIL-STD-785, *Reliability Program for Systems and Equipment Development and Production,* as follows:

> Tasks shall focus on the prevention, detection, and correction of reliability design deficiencies, weak parts, workmanship defects. Reliability engineering shall be an integral part of the item design process, including design changes. The means by which reliability engineering contributes to the design, and the level of authority and constraints on the engineering discipline, shall be identified in the reliability program plan. An efficient reliability program shall stress early investment in reliability engineering tasks to avoid subsequent costs and schedule delays.

The same military standard calls for concurrent reliability accounting:

> Tasks shall focus on the provision of information essential to acquisition, operation, and support management, including properly defined inputs for estimates of operational effectiveness and ownership costs...ensuring...efforts to obtain management data [that] is clearly visible and carefully controlled.

A review of the life cycle of a product will assist in explaining how and where reliability engineering enters into the total process of designing, developing, manufacturing, and marketing a product.

1.3 LIFE CYCLE OF A PRODUCT

Table 1.1 gives a detailed series of steps in the process of designing, manufacturing, and marketing a new product. These steps and the inputs of quality assurance, reliability engineering, and production engineering are described in detail in the following sections.

1.3.1 Design Phase

System Definition (Goals).   Reliability engineering and quality assurance usually will have extensive records of past experiences in producing a similar product. Those records will inform the design engineers of the kinds of problems encountered previously and the solutions, if any. This assists the design engineers in setting realistic objectives for the new product.

Concept.   The design engineers will base the fundamental design concept on their prior experiences, results of research and development activities, and marketing research regarding the need

TABLE 1.1   Life Cycle of a System

Reliability and quality assurance activities at each stage	
Life-cycle stages	Activities
Design phase	
System definition (goals)	R&D data inputs
Concepts for design	
Preliminary design	R&Q consultation, data
Design disclosure (drawings and specs)	
Preliminary design review	Carried out by designers, reliability engineers, quality engineers, production engineers, customers or marketing department
Redesign: new design disclosure	
Design review repeated (may be repeated several times until design approved)	Same as above
Build prototype	
Prototype testing	Planned and executed by designers, reliability engineers, quality engineers
Design review	Same as above
Redesign	Developed concurrently are quality inspection plans, reliability test program plan, providing of facilities
Production phase	
Release for production (complete drawings and specs)	
Production begins	R&Q perform production tests and verifications; quality sets up change and configuration control system
Continue production program	R&Q continue tests and inspections
Support phase	
Deliver product to customer	R&Q set up field failure report system
Customer starts using product and may find malfunctions and failures	
Field failure, malfunction, and service reports received	Reliability analyzes reports and suggests design changes and corrective actions
Corrective action decided	May require full design review and customer's approval
Corrective action implemented	R&Q follow-up to assure effectiveness
Retirement of product or system	
End-of-life cycle	

for the proposed new product. The concepts themselves are founded on the research results and engineering science that support the design engineer's belief that a new product can be designed that will meet the market needs at a price that will provide an attractive profit to the manufacturer.

Preliminary Design.   A design disclosure package of drawings, specifications, etc., that can be evaluated by all concerned parties is prepared. The reliability and quality assurance people will usually be called upon by the design engineers for data inputs and consulting services. The reliability engineering and quality assurance inputs help the design engineers avoid mistakes in the selection of parts, components, manufacturing processes, etc., by pointing out causes of prior difficulties on similar products.

Preliminary Design Review.   A design review committee is usually appointed for any major project. It consists of representatives of the customers (marketing), design engineering, reliability engineering, quality assurance, production engineering, and project management. The preliminary design review (PDR) is carried out after the design disclosure but before any prototypes are built. The purpose is to bring together the knowledge of a number of specialists to analyze the proposed design and to identify any potential problems which may be prevented by design changes.

Remember: The design engineer(s) have *final* authority for the product design. Quality assurance, reliability engineering, production engineering, and marketing are advisers. In some companies, reliability engineering may have final approval authority.

Redesign.   The design engineers will use the results of the PDR to make changes in the design disclosure package and will present it for a second design review. This step may be repeated two or three times until the design is approved for a prototype.

Prototype.   One or more prototypes of the design are built by hand, using highly skilled technicians, machinists, etc., to make the prototypes conform to all the specifications.

Prototype Testing.   It is common for reliability engineers to have responsibility for performing all functional tests, but quality assurance will be responsible for ascertaining that the components, etc., conform to specifications. The test results are then presented for a preproduction design review. The same design review committee conducts that review.

Redesign.   Prototype testing usually reveals weaknesses, errors, or other desirable changes. Existing prototypes may be modified in order to test the "fix," and another design review may be desirable. Concurrent with the redesign, the reliability engineers will be developing the reliability test program and selecting and procuring necessary test or measuring equipment. Quality assurance will be developing the quality assurance plan, designing the inspection program and forms, and procuring any necessary equipment. Production engineering will be setting up production facilities, obtaining tools, and developing the production plan. All of this takes place in preparation for the next phase.

1.3.2 Production Phase

Release for Production.   Complete production drawings, specifications, parts lists, etc., are turned over to production.

Production Begins.   The first few production units are submitted to reliability engineering for "production verification tests" and to quality assurance for complete and full inspection against the specifications. Whereas the prototypes were made by highly skilled technicians under very strict control, the production units are the result of production tooling, ordinary workers, and unsorted components and parts. It is to be expected that the quality of conformance and the reliability may have been degraded by the production process unless special provisions have been made in anticipation of this problem. Even when the incoming acceptance procedures and the production personnel have been trained and warned against passing substandard parts, materials, and workmanship, there is still the danger of degradation of reliability.

The inspection and test results are submitted to the design review committee for a complete postproduction design review. As a result of the design review, changes in design, specifications, and production processes may be proposed.

Design Changes.   After production has begun, it is necessary for quality assurance to establish a change and configuration control system. This system documents any changes that are made, the serial numbers of the unit on which the change first appeared, and the drawing numbers and/or specifications involved. This is a necessary procedure throughout the life of the product so that spare parts, maintenance procedures, etc., can always be provided for each specific version of the product.

Continue Production.   Reliability engineering and quality assurance will continue the test and inspection programs as planned earlier. Results will be recorded, and any discrepancies will be brought to the attention of the project manager for corrective action.

1.3.3 Support Phase

Delivered Product.   When the customers start using the product, other unanticipated problems may arise. All malfunctions and failures in the customers' hands should be recorded by the service organization and the information fed back to the project manager. Quality assurance and reliability engineering will review all reports and analyze the data to identify causes and propose corrective actions.

Corrective Actions.   Decision on corrective action may require a full design review and the customer's approval. Reliability engineering and quality assurance will be responsible for follow-up to see that corrective action is taken and that it is effective.

1.3.4 Retirement Phase

Retirement or Replacement.   This is the end of the life cycle, but all the knowledge gained during the design, development, production, and marketing of the product should be recorded so that it can be used later to improve the process for repeat production of the same product or for new products.

1.4 RELIABILITY PROGRAM PLAN

Each company will (or should) develop its own reliability program plan, tailored to the needs of the particular products or services which it produces. Companies with several divisions manufacturing products of significantly different complexity should have different plans for the different lines of products. It is advisable, however, to have only one plan in any given plant so that there will be no confusion regarding which plan applies to which lines of products.

This handbook has been designed to provide how-to information about all the aspects of designing and operating a reliability program, and in most of the chapters, the subject is addressed from the viewpoint of the production of a complex product, failure of which could cause catastrophic results, such as death or severe injury. For simpler products which do not present serious safety or health hazards, the reliability program can be simplified to provide the necessary protection without wasting money unnecessarily on elaborate programs.

The most common description of a reliability program is that contained in the previously quoted MIL-STD-785. The U.S. Department of Defense has spent many millions of dollars over three decades developing and revising guidelines, standards, and specifications to be used in procurement of military and weapon systems; however, these basic principles are applicable to any production facility. Short descriptions of a large number of these standards and specifications are given in Appendix C. Each company must tailor the implementation of the generalized program plan to its own specific needs.

MIL-STD-785 lists 17 different tasks that should be considered in the development of a reliability program (see Table 1.2). The first task, 101, says, "The purpose of this task is to develop a reliability program plan which identifies, and ties together, all program management tasks required to accomplish program requirements."

TABLE 1.2   Reliability Program Tasks

Task Section 100: Program Surveillance and Control
Task:

101   Reliability program plan
102   Monitor/control of subcontractors and suppliers
103   Program reviews
104   Failure reporting, analysis, and corrective action system (FRACAS)
105   Failure Review Board (FRB)

Task Section 200: Design and Evaluation

201   Reliability modeling
202   Reliability allocations
203   Reliability predictions
204   Failure modes, effects, and criticality analysis (FMECA)
205   Sneak circuit analysis (SCA)
206   Electronic parts/circuits tolerance analysis
207   Parts program
208   Reliability of critical items
209   Effects of functional testing, storage, handling, packaging, transportation, and maintenance

Task Section 300: Developing and Production Testing

301   Environmental stress screening (ESS)
302   Reliability development/growth test (RDGT) program
303   Reliability qualification test (RQT) program
304   Production reliability acceptance test (PRAT) program

Source: MIL-STD-785B, *Reliability Program for Systems and Equipment Development and Production.*

The following is a list of the functions that are normally assigned to the reliability engineers.

1. Reliability estimation, prediction, and growth plan
2. Participate in all design reviews
3. Reliability apportionment
4. Plan and conduct reliability tests
5. Perform statistical analysis of test data
6. Maintain reliability data system
7. Provide assistance to:
   a. Production
   b. Quality assurance
   c. Purchasing
8. Write reliability specifications for purchased items
9. Identify causes of reliability degradation

Each of these functions is briefly explained in the following paragraphs:

1. *Reliability estimation, prediction, and growth plan:* At each stage of the life cycle, the reliability engineer is expected to use the information gained from tests plus recorded information about the reliability of components, parts, and materials specified in the design to estimate the reliability that can be attained. Tests (as described in Table 1.1) usually show that the estimated reliability has not been attained and that it is necessary to predict what will happen if certain changes are made. In addition, there is a "learning curve" regarding reliability, especially when the design is pushing the state of the art. Therefore, a reliability growth plan is often required for a complex system. The growth plan sets reliability goals in the future, in terms of either time or the number of systems produced. Then the actual test results are compared with the goals in order to make decisions regarding programmatic actions. The mathematical and statistical techniques used to analyze test results and predict future reliability accomplishments are described in Chaps. 24, 25, 26, and 27.

2. *Participate in all design reviews:* The importance of design reviews cannot be overemphasized. The design review presents the opportunity for all the different viewpoints to be presented and evaluated. If design engineers attempt to obtain the input ideas regarding a design by contacting reliability engineers, quality assurance engineers, production engineers, customers or marketing departments, and management one at a time, there is no chance that the differences can be resolved satisfactorily. Each representative has certain knowledge about reliability consequences that must be weighed against any opposing knowledge from others. The opportunity to present all the reliability considerations at the design stage can save great amounts of money and time (see Chap. 4).

3. *Reliability apportionment:* Since the reliability of the overall system or product is basically the product of the reliabilities of the individual parts, and since it is easier to attain very high reliability on some kinds of parts than on others, the specification of reliability requirements for the different parts should consider the ease of accomplishment. Lower requirements should be set for difficult parts or components. This allocation of the total reliability among different parts and components is known as apportionment, and is usually the task of reliability engineering.

4. *Plan and conduct reliability tests:* Most frequently, the reliability test equipment serves for research testing as well as for reliability tests, saving duplication of investment. The reliability engineers are specialists in designing test procedures and setting criteria for equipment.

5. *Perform statistical analysis of test data:* The reliability engineer is better equipped to interpret the results of tests and to present the findings to the design engineer and project manager. This also works right into the matter of estimating and predicting reliability of a design.

6. *Maintain reliability data system:* Throughout the product life cycle, reliability information is being collected and must be maintained in a form that can be used in the management of the program. Since the reliability engineer must evaluate and analyze the information in order to make suggestions for changes in design, procedures, processes, etc., the information usually is routed directly to the reliability engineering department, where it can be computerized for rapid analysis and for permanent storage into the database (see Chap 10).

7. *Provide assistance to:* production, quality assurance, and purchasing.

8. *Write reliability specifications for purchased items:* The reliability of all parts, materials, components, subassemblies, etc., that go into the final product influence the overall reliability. Unreliable supplies can ruin an otherwise good product, and the reliability engineer is the one who is most likely to understand what should be included in the specifications (see Chap. 23).

9. *Identify causes of reliability degradation:* Degradation can occur because of poor parts, improper process control, inadequate tooling, poor workmanship, and many other causes. The reliability engineer must make judgments based on quality assurance information and reliability tests and recommend corrective action.

1.5   SOME SPECIAL RELIABILITY CONSIDERATIONS

While reliability normally is associated with the hardware and its ability to perform the prescribed function, there are some other factors that must be considered as part of the overall reliability program in order to attain the desired customer satisfaction. They may directly affect the measured performance and/or the probability of functioning for the specified time or cycles. None of these come under the direct control of the reliability engineer, but they should be evaluated by reliability engineering during the design reviews.

1.5.1   Maintainability and Availability

Practically all electromechanical products require some maintenance to keep them operating in a satisfactory manner. Preventive maintenance is employed to try to prevent unscheduled downtime and interruptions in use; however, it is practically impossible to prevent all interruptions in service. The ease with which the system or product can be restored to operating condition and the time required are primary considerations in the design stage. The information that has been collected by reliability engineers and quality assurance personnel permits the prediction of frequency of malfunction or failure, and, in turn, the need for planned maintenance and repair actions.

Quantitative measures of maintainability are

MTTR	Mean time to repair
MTTRS	Mean time to restore system
MTTRF	Mean time to restore function
DLH/MA	Direct labor-hours per maintenance action
TPCR	Total parts cost per removal
PFD	Probability of fault detection

Maintainability demonstration is required on complex systems by most customers. This is usually carried out by industrial engineers using regular repair and maintenance personnel with specialized tools, test equipment, etc., designed for use with the system. Time studies of several trials of each maintenance action are averaged to obtain the expected mean time (see Chap. 15).

Availability is the total time minus the time required for maintenance and repair actions. Obviously, for systems that need to be available full time, there is considerable pressure to reduce the maintenance time and cost of downtime. Typical examples of high-cost systems that justify special attention to maintainability, availability, and cost of downtime are airplanes, large computers, communication systems, and electric power systems. An unscheduled break in service due to some unexpected part failure can cost thousands of dollars per hour in lost revenues.

Part of the maintainability program requires that a maintenance plan be developed. As shown in Table 1.3, the maintenance plan assures that the provisions for routine and emergency maintenance are considered and provided as part of the overall reliability program and that maintenance data will be analyzed and used in setting up the maintenance plan.

1.5.2   Human Factors

Human beings are involved in practically every stage of the life cycle of a product, and over time, humans can and frequently do make mistakes. These mistakes can render an otherwise good design unacceptable. History has recorded a great many costly events that could have been avoided if more consideration had been given to the human factors during the design stage. Also, instructions regarding the performance of duties, or of operation, should be so carefully written

TABLE 1.3 Maintainability Program

Development of a maintainability program involves:

1. Identify phases of program: conceptual, validation, full engineering design, production
2. Tailoring of plan to needs
3. Qualitative requirements
4. Set tasks and classifications

Program surveillance and control tasks:

101   Maintainability program plan
102   Monitor/control of subcontractors and vendors
103   Program reviews:
          Preliminary design reviews (PDR)
          Critical design reviews (CDR)
          Test readiness and others
104   Data collection, analysis, and corrective action system

Design and analysis tasks:

201   Maintainability modeling
202   Maintainability allocation
203   Maintainability prediction
204   Failure modes, effects, and criticality analysis; maintainability information
205   Maintainability analysis
206   Maintainability design criteria
207   Preparation of inputs to the detailed maintainability plan and logistic support analysis (LSA)

Evaluation and test tasks:

301   Maintainability demonstration

Source: MIL-STD-470A, *Maintainability Program for Systems and Equipment.*

that they cannot be misunderstood. Specifically, design engineers and reliability engineers should consider how human factors affect and in turn are affected by

1. Equipment design characteristics
2. Operational and maintenance procedures
3. Work environment
4. Technical data needed for operation and maintenance
5. Communications
6. Logistics
7. System organization

A massive amount of research has been performed to develop excellent design guidelines for all aspects of human performance. Two of these have been developed by the Department of Defense. They are:

MIL-HDBK-759, *Human Factors Engineering Design for Army Materiel*

MIL-STD-1472, *Human Engineering Design Criteria for Military Systems, Equipment, and Facilities*

Consideration of human factors can improve the maintainability of a system, reduce the possibility of detrimental operating procedure, and reduce the chance of damage to the system from improper use. Customers' operators frequently are the cause of malfunctions as a result of inadequate instructions regarding maintenance and operation. Connectors that can be joined in two or more ways (involving polarity, for example) can cause catastrophic damage. Complex corrective actions when something goes wrong may put the operator under such stress that the correct procedure is forgotten or critical steps are omitted. All such possibilities should be recognized and eliminated or reduced by improvements in the design (Chap. 9).

1.5.3   Cost-Effectiveness

Cost-effectiveness is not easily defined, but in simple terms it means that the cost of making some improvement in the design, production, or maintenance of a system is exceeded by the prevention of costs resulting from short lives, high maintenance costs, human injuries or deaths, replacement costs, etc. In other words, cost-effectiveness means spending some money to prevent the expenditure of a larger amount of money, with due consideration to the time value of money.

Many organizations, including the Department of Defense, look at the "life-cycle costs" of a system. In life-cycle cost analysis, all the expected costs over the entire life of the system are reduced by use of discounted cash flow methods (net present value at some specified discount rate) to obtain the present value of all future costs of the system at a given level and amount of service. Obviously, reliability is probably the most important factor in the development of life-cycle costs.

Reliability can be engineered into the products at the design stage if management is willing to provide the resources as described in this handbook. That does not mean that the product must be "gold-plated," but that the reliability program is tailored to the needs of the product line with due consideration of the consequences of having an inferior product. A breakdown of the costs of quality and reliability can be made as follows:

Controllable costs: The costs of those activities that are planned and included in the quality and reliability budget. These costs include all the activities to assure quality and reliability and all the activities to inspect and test to find out what quality and reliability level has been accomplished.

Resultant costs: These are unplanned costs that result from *not* attaining the required levels of quality and reliability. They include *internal failures,* such as scrap, rework, repairs, etc., within the plant, and *external failures,* which are all the costs incurred as a result of failures and malfunctions after the product is delivered to the customer: field services, replacement, warranties, reduced billings, repairs, liabilities, and loss of reputation.

These costs are frequently categorized under three titles: prevention, appraisal, and failure costs. *Prevention costs* are roughly the same costs that I have described as those budgeted and incurred to assure acceptable quality and reliability. *Appraisal cost* is a term applied to the costs budgeted and incurred to inspect and test the products to ascertain the level of quality and reliability attained. *Failure costs* refer to all of those costs incurred as a result of failures that occur either before or after shipment (internal and external resultant costs).

Reliability engineers should be aware of the effects of their activities on the costs of a system and make a special effort to optimize the overall reliability costs. The design engineers normally are not very concerned about cost matters and strive for perfection regardless of the costs. Reliability, quality assurance, and production engineers can have a great influence on these costs through their participation in the design reviews as well as by providing information to the design engineers (see Chap. 8).

CHAPTER 2
THE ROLE OF MANAGEMENT IN RELIABILITY

Thomas L. Fagan
ITT DEFENSE & ELECTRONICS
McLEAN, VIRGINIA

2.1   CHANGES IN RELIABILITY FROM THE 1960S TO THE 1990S

There has been a vast change in the perception of reliability engineering by virtually every government agency, contractor, and consumer since the early 1990s. Whereas in the 1960s and 1970s reliability was a discrete unit of a matrix organization, today reliability is an integral part of an engineering design team. In the past, reliability, although not an afterthought, was an extra bell and whistle, a number-crunching "after-the-fact" analysis tool which at best gave relative trade-offs among alternative design approaches. When budgets got tight and programs got cut, reliability engineering was always one of the first to be affected, a nicety.

However, as industry relearned quality and reliability from the Japanese, who had been taught them by Dr. Deming in the early 1960s, it became clear to everyone that reliability and quality actually equated to customer/user satisfaction. Once that message hit home, it wasn't long before the *total quality management* (TQM) concept swept the United States, with all of its associated accouterments such as the Baldrige Award,* continuous process improvement, quality functional deployment (house of quality), quality circles, ring of quality, etc.

Today we have reached a point where reliability is truly considered a major performance parameter and the design engineers who are part of an overall design team have all of the tools required, especially CAD/CAM tools, to allow reliability to be optimally considered in a design.

The management of this reliability discipline has also gone from a specialty to an integral part of the engineering function. "Reliability" is a parameter in transition. Many believe that "reliability" will disappear altogether and will no longer appear on an organization chart. As you will see in later parts of this chapter, reliability management as practiced by successful organizations starts with a commitment from top management to consider the reliability parameter as a mutually important performance parameter, a parameter which actually brings a design team together and provides a focus for ultimate customer/user satisfaction.

---

*The Malcolm F. Baldrige Award is given annually by the U.S. Department of Commerce to both large and small business, manufacturing, and service companies for outstanding achievement in reliability and quality.

2.2   THE TEAM APPROACH

For a new product to be successful, the reliability practitioner must be a fully integrated member of a larger design team. In fact, this team, in reality, is a cross-functional team which incorporates design, test, and production and then follows the classic principles of concurrent engineering.

2.2.1   Concurrent Engineering and Reliability

The reliability manager must ensure that the members of the cross-functional team consider reliability in all aspects of the design, including adequate redundancy, parts and materials selection, thermal considerations, etc. The whole essence, then, of concurrent engineering is designing a system or product which is producible. It must also be testable and, as the system or product becomes more complex, have a satisfactory amount of automatic fault isolation included.

2.2.2   Top Management's Role in Reliability

Top management's role is, in reality, simply fostering an environment wherein every single member of the team consistently keeps reliability and quality in the forefront of consciousness and applies all of the tools and principles of reliability at his or her command in all design and/or programmatic considerations. Most top-level managers understand that reliability is perceived through the eyes of a user. A company that provides high product reliability coupled with dedicated life-cycle support ensures that that company will be known as a world-class supplier, furnishing products which are built to last.

Whether a company embraces the nomenclature of total quality management (TQM) or simply utilizes those tools and techniques which make up TQM, that firm's image will become synonymous with high reliability and long life. By fostering the use of these tools and processes, top management sets that company apart and lowers both the cost and the risk of ownership for a customer or consumer. By creating this environment, top management saves time and money and ultimately winds up with a long-lasting image as a high-reliability, world-class supplier.

2.3   RELIABILITY GOALS

When an organization knows where it wants to go, or at least what its ultimate endpoint needs to be, the task at hand is one of setting goals which provide the direction and also measurable intermediate milestones along the schedule path. Reliability policies, on the other hand, provide general guidance and remove uncertainty. Section 2.3.1 will discuss the establishment of reliability policies, while Sec. 2.3.2 discusses the reliability organization and Sec. 2.3.3 reliability goals.

Goals are used to focus organizational attention on desired results and to serve as standards against which achievements can be measured. They are established in various ways and at various levels to facilitate achievement of firm requirements. Where requirements are not yet firm, goals serve to specify desired results.

Depending upon the type of business, product reliability requirements or goals may be contractually specified by customers or established in-house by those who decide what level of reliability the product must have. In the latter case, the reliability level is based on customer and market considerations, balancing the cost of reliability and the risk of unreliability. In either case, the level is specified as a measurable quantity which, for complex products, must be divided over the major subsystems or units to provide meaningful goals for designers. These goals may be further subdivided down to the lowest level of design responsibility.

All goals have certain characteristics in common. The following criteria can be used to assist in developing and arranging goals:

1. *Relevance.* All goals relate to some higher purpose and should be well suited to the particular needs and circumstances. A good match with needs and circumstances leads to greater success by directing focus to the right things.

2. *Attainability.* Goals should be set at levels that are reasonably attainable within the available time span. Oftentimes, however, a desired level is first specified and then a target date for completion is selected. Whichever is the case, quality, quantity, and time are the interacting variables. For any particular type of goal (quality), goal level (quantity) and completion date (time) are tradable. Other things being equal, the higher the goal, the longer the time required to achieve it. However, to maintain interest and commitment, stringent goals over long periods of time should be avoided. Dividing goals into subgoals over shorter time spans makes them more attainable and individually more satisfying. In all cases, goals should present sufficient challenge to be worthy of pursuit.

3. *Supportability.* Goal achievement requires wherewithal. The necessary instruments and resources must be available at the time they are needed. It should be determined beforehand what the instruments and resources will be and the extent to which they can or will be provided. Stay with what can realistically be supported.

2.3.1 Reliability Policies

Organizations periodically face situations that require direction from management to ensure that decisions made and actions taken consistently support the overall mission and goals. Policies provide this direction by defining specific areas of concern and then indicating a desired outcome or specifying the direction that decisions and actions should take. In this way, policies increase decisiveness by removing uncertainty and thereby reduce the potential for inefficiency, counterproductivity, inappropriate risk taking, or conflict.

While policies are, in a certain sense, related to procedures and rules, they are neither of these. Procedures are precise, step-by-step instructions for required actions in a definite sequence. Rules are requirements to take or not to take specific actions. Both are characterized by a high degree of specialization and rigidity. Neither allows any discretion or deviation. Policies, on the other hand, admit flexibility of judgment within a framework of guidance and direction. The best policies admit the widest possible range of expression and scope of action.

The need for policy is indicated when an organization is faced with decision options that may be precedent-setting. Situations calling for policy include those where

- Opinions may differ over the best course of action in situations that affect the achievement of the overall mission and goals.

- Decisions may have significant consequences beyond the local level at which they are made.

- The choice of action may lead to unnecessary risk, counterproductivity, inefficiency, or conflict.

- Cooperation and reciprocal actions on the part of one organizational element are needed to enable another element to function effectively.

Policies may be established to clarify, define, guide, regulate, direct, establish, integrate, authorize, enable, empower, commit, support, provide, admit, ensure, inhibit, prohibit, restrict, or disallow.

Reliability policies are needed at a number of levels in the organization to ensure reliability creation, reliability maintenance, and reliability improvement. They are especially needed to smooth the way for key or critical matters. Firms establishing a new reliability function have a compelling need for them; firms engaged in the manufacture and sale of products or systems that can present a significant public safety hazard or that can affect national security or national prestige have a critical need for them.

To be effective, reliability policy must issue from top-level management. Management's attitude toward reliability, as expressed through policy, is the most important single ingredient in the success of reliability engineering and reliability assurance in any organization. If the deeds of management in fact support written policies, policies gain credibility and legitimacy and will be respected.

Reliability policy and, ultimately, the final responsibility for the reliability of products and services rests with top management, albeit through successive management levels. The top assurance executive is responsible for pursuing reliability in the manner and to the extent prescribed by general management policy and establishing his or her own policies in support of this function.

Attributes of Policies.   Policies, in general, should be

1. Action-oriented (as contrasted with mere statements of belief)
2. Supportive of organizational goals
3. Consistent with other policies
4. Authoritative, credible, and acceptable at the level of implementation
5. Inclusive to the extent of embracing all aspects of the intended application
6. Specific to the extent of providing unambiguous direction and focus
7. Admissive to the extent of allowing maximum flexibility of choice within the prescribed framework of guidance or direction
8. Concise and readily understandable
9. Relevant to the times and circumstances
10. Stable over relatively long periods of time

Types and Levels of Policies.   Policies are established at all levels of the organization, beginning at the corporate level and proceeding through divisional levels and successively finer levels of organizational structure down to departments and functional units. High-level policies such as those at the corporate level are broad and general so as to deal effectively with the broad concerns of top-level management. Corporate reliability policy provides for the establishment and promotion of reliability activity and achievement to fulfill obligations to customers and to society at large. It deals with internal matters pertaining to overall performance and with external matters pertaining to relationships with customers and the community, and to involvement with regulatory and other such organizations. It sponsors the reliability function by declaring its intentions to the organization at large, and it provides for review and evaluation of the overall reliability system.

Divisional policies respond to corporate policies and relate to the more specific issues encountered by departments and functional units. Typically they deal with administration, organizational interrelationships, operating methods, and the maintenance or improvement of organizational performance. Departmental policies deal with situations and conditions that are more apt to arise on a day-to-day basis. These situations typically include matters relating to suppliers, subcontractors, product design activities, parts and materials, manufacturing, testing, auditing, and reviewing.

Developing and Establishing Policies.   The formation of effective and lasting policies requires a comprehensive view of the issues and a full appreciation of the circumstances leading to the need for policy. If the policymaker is not in full possession of the facts and nuances, it is advisable that he or she enlist the views of others who may be deeply involved in the situation and have the breadth of view, knowledge, judgment, and experience to make constructive contributions. It is desirable that those managers and supervisors who would be affected by the policy or who would be expected to carry it out should be considered as potential contributors. Not only can their views be constructive, but their involvement in developing the policy will increase their acceptance and support when the policy goes into effect. However, while the participative approach is useful, the responsible manager must nonetheless use the power of the office and provide the benefit of experience to create appropriate policy.

It is especially important to the formation of policies for a new reliability function to win the acceptance and cooperation of long-established groups, particularly those that play prominent roles in the organization. Bringing key individuals from those groups into the definition and development phases of policy formation can help ensure a well-conceived policy in which key issues are addressed.

While the development of many policies is straightforward, some policies are more involved and may require advance planning and study. To assist in such cases, the following outline is presented as a guide.

General Guide for Policy Development

1. State the need. Describe the situation that created the need. Identify who or what is involved, how they are involved, and to what extent.
2. Identify and review any existing policies that relate to the situation.
3. Survey managers and supervisors who will be affected by the new policy. Obtain pros and cons.
4. Determine if a new policy is actually needed or if existing policies should be revised to accommodate the situation.
5. Draft a preliminary policy statement for review and comment by the departments affected. Include purpose and scope.
6. Integrate appropriate suggestions and prepare a revised statement for additional review. Add sections on responsibilities and actions, if appropriate.
7. With executive approval and signoff, release the document for distribution.

Policies are communicated by memoranda, letters, instructions, or directives and are included in program plans and various manuals. They are given visibility by way of meetings, workshops, lectures, and training sessions. New policies should be routed to all departments and units affected and acknowledged by signature and date.

Policy statements range in size from simple statements of a paragraph or two in length to comprehensive documents that may include some or all of the following topics:

Background

Purpose

Scope

Definitions

Policy

Responsibility

Actions

Although policies are essential to effective and efficient operations, they should be held to the minimum allowed by the size of the organization, complexity of operations, criticality of processes, management style, and self-responsibility, awareness, and professional level of employees. As guiding forces, they should admit the widest possible latitude for action any given situation allows.

Policies should be reviewed for possible revision or cancellation on a scheduled basis and whenever major changes are made in organization, management, practice, or overall organizational strategy.

Example of Policy. Formal policies and their applications vary considerably from industry to industry. Some firms maintain elaborate systems of policies, plans, goals, and methods; others have relatively few. The extremes may be characterized on the one hand by the small, entrepreneurial organization functioning in a highly competitive, volatile marketplace, and on the other hand by the very large organization functioning in a highly structured environment such as exists in defense contracting. The small specialty house, functioning "organically," has less need for formal policy because its people have frequent exposure to high-level managers and know their goals, philosophies, ground rules, and views. The employees are well informed about the company's business, are highly self-managed, and see their jobs in the light of the operation as a whole.

The large defense contractor is, by comparison, mechanistic, compartmentalized, and rigorously structured, and functions through an elaborate system of controls. The more critical the end use of the product and the more costly and complex the system produced, the greater the need for order, system, rigor of method, and formalization of policies. Because of these widespread differences, there is no model policy that can satisfy the needs of every organization. The example given in this section, therefore, is presented more for structure and style than for specific content.

This example of corporate policy is for a hypothetical company engaged in the design, development, and production of systems and equipment for critical applications where failures can result in serious safety hazards to the public or extraordinary economic loss to the user.

Policy for Upholding Organizationwide Reliability

    I. *Background.* Products of the kind produced by the corporation may, through loss of function or degradation in performance, present potentially serious hazards to public safety or incur extraordinary economic losses to users. The risk of these hazards is diminished by high product reliability, a result attributable not only to the efforts of product designers and other technical specialists who endow the product with an inherent level of reliability, but also to the many others throughout the organization who safeguard product reliability by maintaining strict vigilance over operations and processes in offices as well as on the production floor.

    II. *Purpose.* The purpose of this policy is to establish a forward-looking attitude whereby individuals throughout the organization more readily anticipate the potential effects of their actions on the reliability of products as ultimately experienced by customers. The objective is to avert potential reliability problems in their formative stages and thereby statistically reduce the incidence of latent defects in the production process. Success in this area will reduce the burden on the production screening process and materially reduce the chances of undetected problems passing on into fielded operational systems, where the cost of failure can be exorbitant.

   III. *Scope.* This policy applies to all direct and supporting organizational functions for all aspects of endeavor from product design through procurement of parts and materials, manufacturing, assembly, shipment, and installation.

   IV. *Policy.* Emphasis is to be placed on forward-looking, anticipative, and preventive activities to reduce the potential for latent reliability problems further downstream in the product realization cycle. In particular, attention is to be directed to the following situations:

      A. Where economies at local levels are realizable, but the impact on the future reliability of the product is uncertain

      B. Where departures or deviations of any kind are perceived as expedient for schedule maintenance or recovery

      C. Where there appears to be a demand for decisions under uncertainty

      It is firm policy that safety-related reliability is never tradable downward from the established product baselines. Success in achieving the objectives of this policy document requires that all organizational elements be at all times alert to situations that may lead to compromise of product reliability.

    V. *Responsibilities.*

      A. The director of program management is responsible for ensuring that this policy is applied by all program managers in all programs under their administration.

B. The director of engineering is responsible for ensuring that this policy is carried out by all elements of system engineering, design engineering, project engineering, and test engineering throughout all program phases.

C. The director of operations is responsible for ensuring that this policy is carried out by the materials function, industrial engineering, manufacturing engineering, manufacturing operations, and all other functions within the department having the potential for affecting product reliability.

D. The director of product assurance is responsible for the dissemination of this policy, for its annual review, and for making recommendations to the chief operating officer concerning its continued relevance, application, and effectiveness.

E. The director of employee relations is responsible for ensuring that all new employees hired for entry into any of the departments identified in parts A through D in the foregoing are made aware of the intent of this policy as a part of their familiarization with the mission, goals, operations, and priorities of the company.

The above document should be signed and dated by the chief operating officer of the corporation.

2.3.2 Reliability Organization

While there is some concern that reliability engineering as a separate function will disappear from the corporate organization chart, as discussed previously, reliability has become an intrinsic, ingrained, organic parameter for all members of a design or product team. With the continued expansion of computer-aided design (CAD) and computer-aided manufacturing (CAM), software algorithms now exist for real-time tradeoffs for the most complex systems. Today's reliability analysis software programs are becoming expert systems which are filled with interactive decision rules. The bottom line is that each member of an integrated design team or an integrated process team now has the facility to address all aspects of reliability.

2.3.3 Reliability Goal Setting

Working from preestablished requirements, reliability goal setting consists of reducing the overall requirements to a series of subgoals, related in such a way as to reinforce the net outcome. Reliability apportionment is a common, long-standing technique. The goal-setting process is more involved when customer needs and competitive factors in an open-market situation are being addressed or when goals for organizational performance improvement are being developed. The following general procedure is useful in establishing reliability goals:

1. Clarify what is needed or desired.
2. Review the organizational purpose and mission of the organizational unit establishing the goal and the purposes and missions of the units that will be pursuing the goal.
3. Identify desired key result areas.
4. Determine where the highest payoffs are likely.
5. Select the result areas that are most desirable to pursue.
6. Select goal candidates.
7. Consider the resources required to pursue each goal to a successful conclusion.
8. Identify any impediments or risks to goal achievement and determine how they can be overcome or offset.
9. Rank candidate goals according to ease of achievement in combination with degree of payoff.

10. Consider goal interdependencies and adjust goal candidates for maximum coordination and mutual reinforcement.

11. Examine goals for relevance, attainability, supportability, compatibility, acceptability, and measurability.

12. Make final selection of goals and establish deadlines for achievement.

13. Establish measures of success.

14. Develop action plans for key goals. Include provisions for any motivational initiatives needed for obtaining commitment and any provisions for supervisory and management support.

15. Communicate goals in writing.

16. Periodically review progress toward goal achievement and make any adjustments needed to ensure success.

Figure 2.1 depicts a sample goal network and the supporting activities.

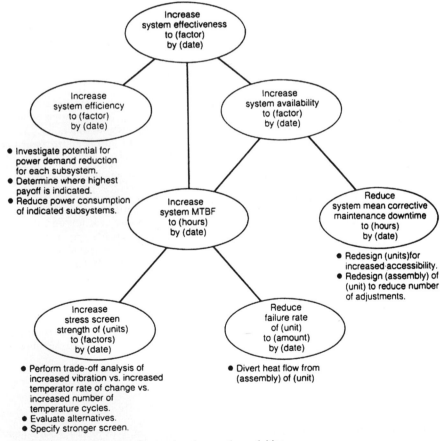

FIGURE 2.1   Example of goal network and supporting activities.

While goal setting is a straightforward, rational process which leads to statements with numerical values and associated due dates, it must be kept in mind that the successful pursuit of goals requires the leadership and active support of management. Once goals are established, managers must see that these goals are communicated, are clearly understood, and are accepted through individual commitment. They must ensure that a path toward these goals is indicated and that the resources and instruments needed to achieve them are made available. Indeed, because the goal process is fundamental to the planning and control functions of management, the effectiveness of managers rests to no small extent upon the degree to which they are sensitive to the needs of those pursuing goals and the degree to which they provide the leadership and the material and psychological support needed to achieve the goals.

CHAPTER 3
MANAGING RELIABILITY AS A PROCESS

Keith V. Rohrbach
MEDTRONIC CARDIORHYTHM
SAN JOSE, CALIF.

3.1 WHAT IS A PROCESS?

A simple definition of *process* is "a collection of successive steps that achieves a specific end." The steps in a process act on some specific inputs and result in specific outputs. A process is methodical, with purposeful steps in a particular sequence that eventually conclude at some specific endpoint. A process is a mechanism used to promote consistent change, whether in nature, business, education, or society at large.

In considering reliability as a process, the following definition will be applied:

> *Process:* "A collection of successive steps that takes inputs, adds value in customer satisfaction, and achieves a specific output."

A change occurs during a process. This change may add or remove value between the beginning inputs and the resulting outputs. Value is added if both of these conditions are met:

1. An action results in a change to something that the customer wants.
2. The desired action is done correctly the first time.

"Customer" can mean the person who is ultimately purchasing, paying for, and/or benefiting from the product *or* the next person in the process chain. If neither condition is met, then value will not be added and may actually be reduced. For example, the "homogenization process" applied to whole milk emulsifies the butterfat so that it will not easily separate into a cream layer. This process adds value to the milk, since today's consumers prefer to buy milk in which the butterfat remains evenly dispersed. Similarly, the "voter registration process" adds value to the "election process" and in turn the "democratic process" by including more eligible voters so that the population's preference is more fairly represented. However, if either the homogenization or the voter registration process is conducted poorly or without effective control, the contrary effect—lowered value of output—would result.

It is important to distinguish between natural and people-made processes. Natural processes, such as those associated with change over time, often result in lowered value (as perceived by people) between process inputs and outputs unless they are counteracted or controlled by people-made processes. Examples include the erosion of soil and beaches by wind and surf, the failure of paints and other protective coatings, the changing position of river beds, and the structural fracture associated with repeated freezing and thawing. The "aging process" exacts a high price in terms of costs of maintaining buildings, highways, utility systems, and even people themselves. The processes that people create to counteract these natural processes are all intended to preserve or sometimes increase the value that would be lost to natural processes.

In contrast to processes, *systems* are used to help control a process to ensure that it is operating effectively and efficiently. The homogenization process described above may include the machine control system that monitors and adjusts flow, pressure, and temperature and the quality control sampling system that periodically verifies the homogenizer's effectiveness.

Processes are used to conduct all of the fundamental aspects of business and achieve consistent outputs. Every business has a group of essential processes. Furthermore, each process must add value or the business is likely to fail. These essential processes are called business processes and are defined by H. James Harrington as "a group of logically related tasks that use the resources of the organization to provide defined results in support of the organization's objectives."[1]

For example, a successful used car dealer certainly would need business processes for customer needs analysis, supplier sourcing, inventory and materials management, billing and collections, and after-sale service. Depending on its size, a business may also need processes for hiring and training personnel. Similar processes would be required for such diverse businesses as an ice cream shop, an automobile factory, and a university.

There is one additional aspect that is critical to the success of modern businesses and business processes: the customer. Customer satisfaction is the ultimate determinant of whether a business process adds value and whether the organization's objectives are met. If the process adds value to its output in terms of customer satisfaction, then customer satisfaction should improve. Without processes that result in added customer satisfaction, the result for the business will be the same as that often associated with the natural aging process: deterioration.

3.2 HOW IS RELIABILITY A PROCESS?

The ways in which reliability is achieved meet the definition of process: "A collection of successive steps that takes inputs, adds value in customer satisfaction, and achieves a specific output." Creating a reliable product requires focused efforts to define, design, assess, and maintain reliability throughout the phases of the product's life cycle.

Figure 3.1 represents a conceptual view of the six phases of the product life cycle. This figure represents the product development and improvement process and encompasses the entire life cycle of a product from initial idea or concept to the conclusion of its commercial life. It shows that development does not occur only once in a product's life but should be part of a cyclical process of continuous improvement consistent with the demands of the marketplace and the objectives of the business.

The process of reliability is an integral part of these six phases of the product life cycle and of the overall product development and improvement process. The reliability process may achieve most of its essential outputs during the product design and development phase, but it derives essential feedback and controls from other phases of the life cycle.

Figures 3.2 through 3.7 depict the same six phases of the product life cycle as a process flow that fully encompasses product development and improvement. The reliability process, including the essential elements of a reliability program, is integrated into these flowcharts. Each life-cycle phase is described in terms of the essential process steps, decision points, documented outputs, and interaction with other phases. Consider this view of the reliability process: a collection of successive steps that takes inputs, adds value in customer satisfaction, and achieves a specific output.

Table 3.1 also lists the six phases of the product life cycle and the corresponding stages. For each stage, reliability process outputs or deliverables are also listed. This table summarizes the reliability deliverables that are shown in the process flowcharts (Figs. 3.2 through 3.7).

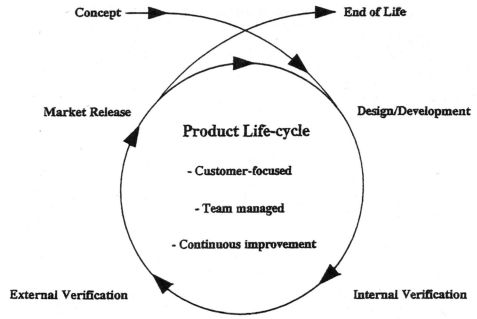

FIGURE 3.1   Six phases of the product life cycle.

Consider these phases and stages through which each product must pass to reach market and eventually conclude its commercial life. Consider how these phases relate to the creation, achievement, and maintenance of reliability. Notice that the outputs of each phase should represent value added to a product as defined by customer-specified, or customer-driven, reliability requirements.

Concept Phase

Summary.   Ideas, based on customer needs, are identified and defined into proven concepts worth development investment.

Idea: The "seed" of a product.

Idea definition: Determining the ideas and priorities that the organization wants to pursue.

Concept development: Basic development (sometimes "tinkering") to see if the idea is feasible.

Proof of concept: Evidence that specific product attributes (performance, reliability, maintainability, etc.) can be achieved at estimated cost levels for a generally defined market size and customer base.

Figure 3.2 is a flowchart of the concept phase. The inputs to the concept phase originate from the potential customers and markets, business trends, and the performance of current and competitive products. The minimum reliability required in a product is what the customer expects tempered by his or her willingness to pay and what the market and technology can provide. If the competition provides less than the customer expects, then better reliability is the obvious advantage in a new product concept, provided it is technologically feasible. If the competition is satisfying reliability expectations, then other product characteristics may offer a distinct competitive advantage. Whether or not reliability is among the distinguishing features of a product compared with its competition, it has to be assessed in this phase and given preliminary but realistic quantification.

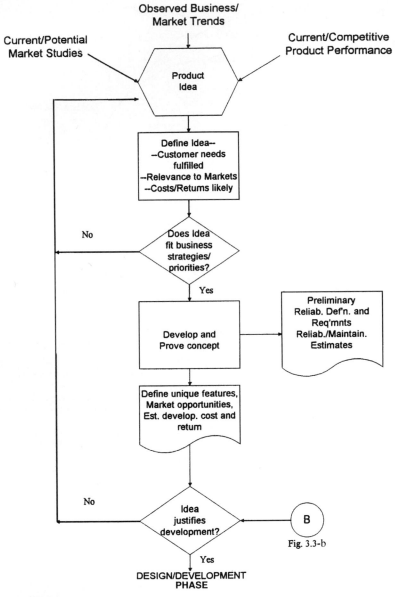

FIGURE 3.2   Concept phase.

The concept phase and its relationship to reliability rests on the business strategy and policies of the organization. This is apparent in the two key decision points shown in Fig. 3.2: "Does the idea fit business strategies and priorities?" and "Does the idea justify development?" Management policy, the organization's stated mission, and even its culture define the legal, financial, and marketing framework within which product development occurs. If reliability is not included in that framework as a core competency or even a concept, then the initial "activation energy" of the reliability process is not present. A framework or organizational infrastructure that

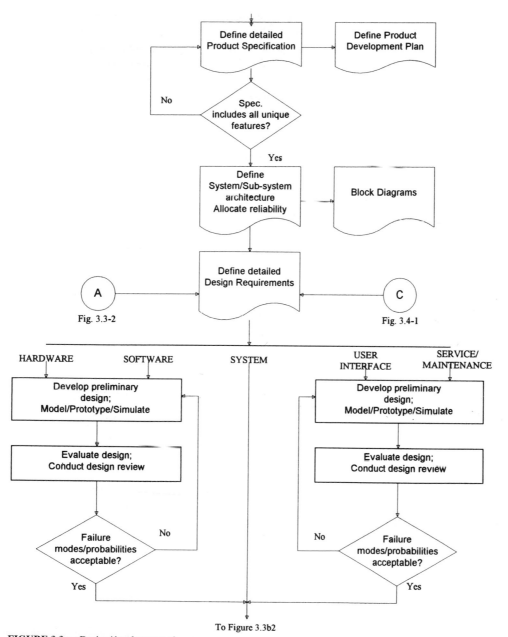

FIGURE 3.3a Design/development phase.

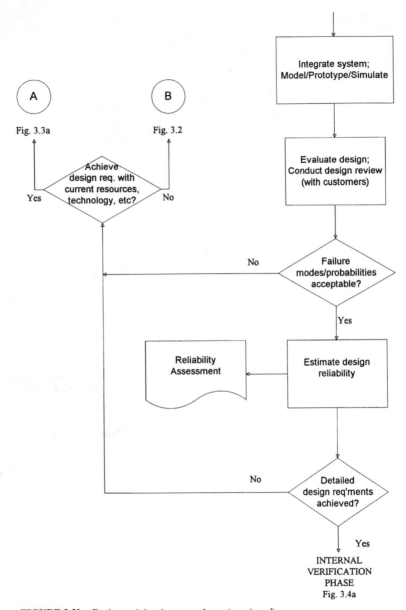

FIGURE 3.3*b* Design and development phase. (*continued*)

defines, promotes, and requires *defined* product reliability is a prerequisite if an organization is to have an effective reliability process.

The most substantial outputs of the reliability process in the concept phase are definition and initial estimation of reliability for the product concept under consideration. Although this definition will be preliminary, it should represent a quantification of system reliability for each of the key product performance requirements or unique features. If maintainability applies to the product, a design goal should be established along with an estimate of how the design concept will

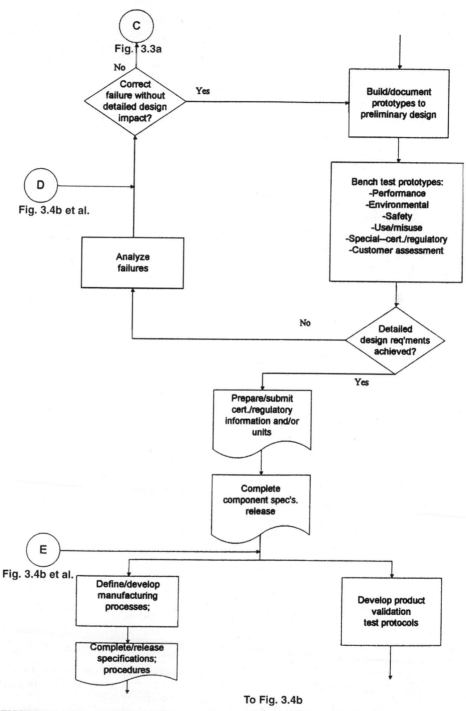

FIGURE 3.4a   Internal verification phase.

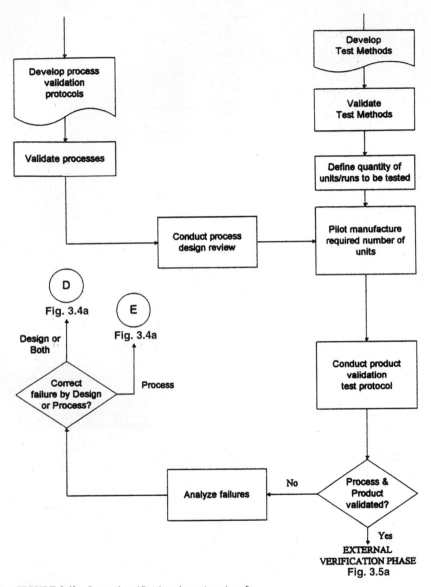

FIGURE 3.4*b*   Internal verification phase. (*continued*)

achieve it. These reliability targets and initial assessments are among the outputs that management considers in deciding whether the product concept justifies a substantial investment for further development.

Design/Development Phase

Summary

- Accepted concepts for new products and major changes to existing products are converted into designs/models that meet detailed requirements.

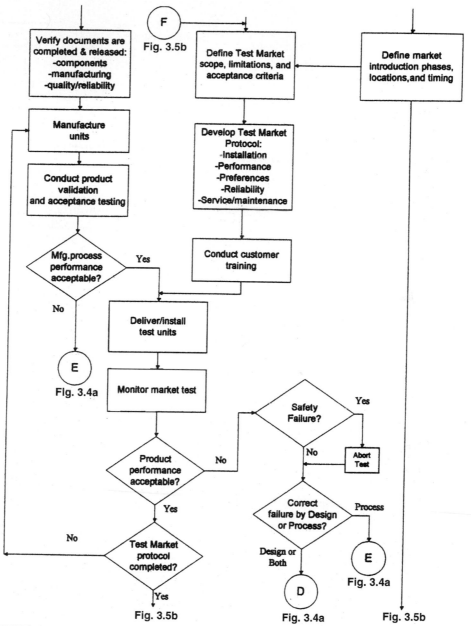

FIGURE 3.5a External verification phase.

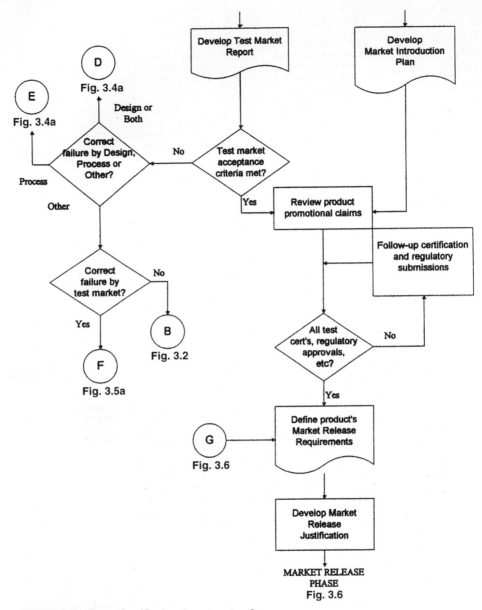

FIGURE 3.5b External verification phase. (*continued*)

- The phase concludes with qualification of the design and "design freeze."

  Product design: Product development project plan; serious specification definition and design work to achieve it.

  Prototype: A working model that, based on preliminary testing, meets essential design requirements, including hardware and software as applicable.

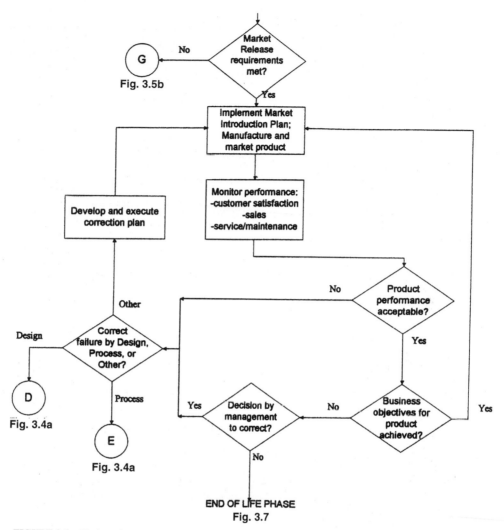

FIGURE 3.6   Market release phase.

Figure 3.3*a* and 3.3*b* shows the flowchart of the design/development phase. This phase of the process begins with the development of a detailed product specification and a product development plan. The product specification must be sufficiently detailed to clearly communicate to the designer the essential expectations of the customer, including reliability. These are design targets, product needs, features, or characteristics that have been generated in the concept phase. The product specification should go well beyond but not omit unique features. For example, a unique feature for a new refrigerator might be "lightweight and portable." The product specification should define the acceptable ranges for weight and dimensions. In complex products, the specification should reflect the apportionment of overall product reliability to each of its subsystems.

The product development plan is a schedule of milestones and responsibilities for the specific project. It should reflect the timing of subsequent steps of the product life cycle up to the market release phase. This is the framework in which the life-cycle phases are managed from design/development through external verification.

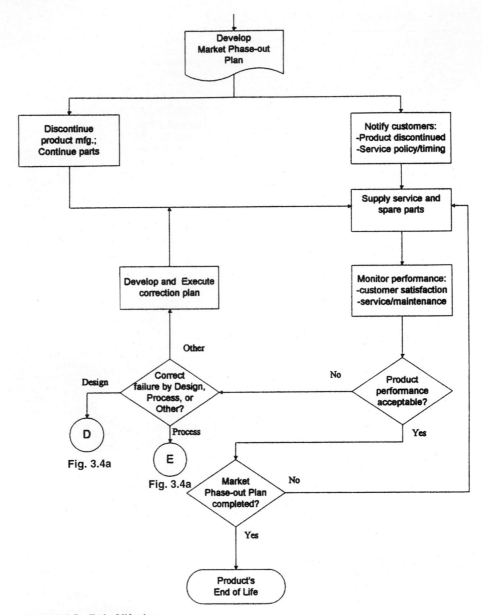

FIGURE 3.7 End-of-life phase.

The next major step in this phase is defining the subsystems of the product and their relationship and interactions to achieve the product specification. This is especially important for complex products where multiple designers are working simultaneously on various subsystems. Block diagrams can be used to define and communicate this architecture among the various people involved. Characteristics such as power source, extent and level of automation of controls, product accessories, and product inputs and outputs as a function of the user are examples of important elements of a product's architecture. For example, it is important to define whether an automobile is front-

TABLE 3.1 Elements of Reliability in the Product Life Cycle

Life-cycle phase	Development stage	Reliability process deliverables
Concept	Idea	
	Idea definition	
	Concept development	Definition of reliability levels
		Preliminary reliability requirements and specifications
		Preliminary reliability estimate
		Preliminary maintainability estimate
	Proof of concept	Reliability and cost trade studies
		Preliminary reliability assessment
		Product definition/unique features
Design/Development	Product design	Component/supplier selection
		Reliability prediction
		Reliability allocation/apportionment
		Failure mode studies (FMECA, FTA)
		Design reliability assessment
	Prototype/model/simulation	Preliminary reliability testing
		Failure analyses
		Initial reliability growth testing
Internal verification	Bench tests	Accelerated testing
		Life testing
	Process design	Stress screening test definition
		Process design review
	Pilot manufacturing	Failure analyses
		Process design review
		Reliability qualification/life testing
External verification	Field tests	Failure analyses
		Design review
		Test market report
	Market introduction planning	Promotional claims review
	Full-scale manufacturing	Reliability acceptance tests
		Failure analyses
Market release	Market release	Failure analysis reports
		Warranty cost analyses
	Service/improvement	Maintenance cost analyses
		Product/process changes or retrofits
End of life	Market discontinuation	Market phase-out plan

wheel, rear-wheel, or all-wheel drive before detailed design and development proceeds. Consider the difference in design implications if the choice is to be made by the car rather than the driver!

With the product specification and, as applicable, the definition of product architecture, detailed design requirements can be developed. As shown in Fig. 3.3a and 3.3b, this can be an iterative process. The detailed design requirements must be defined, and then a complex loop of preliminary design development, design evaluation, and resolution of failures to achieve the design requirements proceeds. Preliminary design development can be enhanced by the creation of engineering prototypes, models, or simulations and may proceed independently for the various subsystems involving hardware, software, user interface, and service/maintenance. Once the subsystem designs have achieved the design requirements, the system integration step (the product) is completed and evaluated. It is important that customers be included here because this is the first point in the process at which an actual design result is available for customer assessment.

Should the design fail to meet design requirements, a key decision must be made: Can the deficiency be corrected with current resources, technology, and priorities? If yes, the iterative process begins again with a return to design development. If no, a decision needs to be made whether to continue development of the product by redefining the product specification, i.e., what the product needs to be.

Notice throughout the design/development phase that following each design step, evaluations must be conducted and decisions made about whether the design meets the design requirements and whether its failure modes and probabilities for failure are acceptable. This design–evaluate–decide cycle is a fundamental concept in the reliability process and one that helps achieve the reliability (as well as other) objectives.

During the design/development phase, component selection and, in some instances, supplier selection are critical. Here the designer/reliability engineer consolidates experience, field failure analysis data from similar designs, component manufacturers' data, reliability estimation techniques, and derating practices to achieve the requirements.

Also, the use of failure mode studies such as FMECA (failure mode effects and criticality analysis) and FTA (fault-tree analysis) can be helpful to confirm that specific components are correctly specified and, as needed, derated in their applications. CASE (computer-aided software engineering) tools can also be used for comparable evaluations in software designs.

Once a prototype or model exists, tests should be performed to confirm at least some of the design requirements. Although tests are often limited by the number of units available, they should include at least functional performance, environmental and/or other stresses, and unexpected uses (misuse). Typically prototypes experience high failure rates and sometimes unexpected failure modes. Thorough failure analyses and root-cause corrections are essential. Again, where applicable and feasible, maintainability should also be evaluated when completing repairs.

The design/development phase is concluded when the design is shown to meet the detailed preliminary design requirements. This is achieved when the various subsystem designs have been completed and the system integrated to yield acceptable design reviews, acceptable failure modes and probabilities, acceptable reliability assessments, and the conclusion that all detailed design requirements have been achieved. Although this phase is based on the preliminary design, it can produce the final design if it is effectively done and is not affected by experiences in later phases. If it is done correctly the first time, it can maximize the potential for added value.

Internal Verification Phase

Summary

- In-house testing is done to verify the design against the specified requirements and to confirm/obtain third-party test certifications needed for commercial distribution.
- Process development and validation is done.
- Pilot manufacturing for product qualification and validation is done.

In-house testing: Qualification of the prototype and any subsequent iterations.

Process design: Design and specification development for the manufacturing and test processes.

Pilot manufacturing: Qualification of the product *and* processes.

Figure 3.4*a* and 3.4*b* contains a flowchart of the internal verification phase. This phase focuses on assuring that the design as it is manufactured achieves all applicable requirements, including reliability. This phase begins with the construction of prototypes according to the preliminary design. At least two series of prototypes should be constructed. The first should be produced as soon as possible after completion of the preliminary design. These would probably be the first ones to be constructed completely from the actual preliminary design and could provide early and important performance data to confirm the complete integrated design and satisfy certification and/or regulatory requirements. These first prototypes should be produced as a joint effort of the design and manufacturing personnel even though the manufacturing processes have not yet been fully defined and developed. Such involvement helps to reduce the length of time required to learn the new design and allows consideration of ways in which manufacturing can be done most efficiently.

The second series of prototypes should be built with developed and validated manufacturing processes. These prototypes will represent what the customer will see in the next phase, and probably in the marketplace. The most extensive and rigorous evaluations will be performed on these units to justify the qualification of *both* the manufacturing processes and the product resulting from them.

In order to obtain evidence that all design requirements have been achieved, as much testing as possible should be done. Tests should be conducted on units that are produced by standard manufacturing processes and that represent as closely as possible the design that will be supplied to customers. Bench testing of the first prototypes should be done to specific requirements contained in the product specifications and detailed design requirements. Testing should be performed to written protocols and include at least essential performance, environmental, safety, and use/misuse tests. Where product certification, regulatory requirements, or other standards apply, test results and/or units should be supplied. Other special tests such as accelerated life testing may also be performed. Customer participation and feedback is desirable, but it should *not* yet include release of products to customers for actual use.

Bench testing of the second series of prototypes should be conducted to a written product qualification/validation protocol containing the complete product requirements and rigorous tests based on validated methods. The development of this protocol and the definition and validation of test methods can occur in parallel with the development and validation of manufacturing processes. Test method validation establishes that the results of a test, obtained under defined and specified conditions, are reproducible and will achieve measured and acceptable limits of accuracy and precision. Similarly, process validation establishes that the resulting products will consistently meet their requirements under defined process conditions. Where possible, the quantity of units manufactured should provide a basis for a valid statistical conclusion for the qualification test results.

A key objective in developing and validating manufacturing processes is to demonstrate that they do not degrade design reliability. Subprocesses such as component supplies, environmental controls (ESD protection, peak temperature/heat history, foreign materials), materials handling, automation management, and personnel practices must be defined and controlled in this phase to assure that their outputs enhance reliability maintenance.

The internal verification phase is completed when the product and its processes are qualified and validated. This means that test protocols have been completed, and the results demonstrate that the product's requirements have been achieved, including those in the product specification and the detailed design requirements from the design/development phase.

If these requirements are not met, a key decision must be made. If correction can be achieved through manufacturing process design, then the next step is to return to the parallel path of process development and test protocol development. If correction can be achieved only through product design or through both product and process design, then the next step is to return to the beginning of the internal verification phase. Note, however, that it is possible for the next step to be to return to the concept phase if the failure to meet requirements has an impact on detailed design and the correction cannot be achieved with current resources, technology, and/or priorities.

External Verification Phase

Summary

- Place internally verified product in limited customer use: field tests, test markets, clinical studies, or customer preference tests.
- Collect and analyze external-use data for design verification and market introduction planning.
- Complete all test and standard certifications, regulatory clearances, or other approvals necessary for commercial distribution.
- Verify satisfactory achievement of all design objectives.
- Begin manufacturing scale-up.

- Develop market introduction plan.

    Field tests: Qualification of the product in the market environment.

    Market introduction planning: Identifying locations, methods, materials, and timing for product introduction.

    Full-scale manufacturing: Achieving ramp-up to throughput and yield requirements.

The external verification phase (Fig. 3.5*a* and 3.5*b*) is a continuation of the internal verification phase, except that much more should be known about the design's performance and reliability. Also, testing is conducted by actual product users with actual manufactured products. Ample evidence should already be available to show that the design meets the reliability specifications under the conditions of the qualification and validation methods. Any weaknesses in the product design or manufacturing methods should have been addressed already to avoid their potential contributions to unreliability.

All manufacturing in this phase should be conducted using fully defined, documented, and validated methods. Configuration control should be fully implemented. Evaluation of the products produced is both by the qualification and validation protocol (from the internal verification phase) and by standard acceptance testing. While this may seem like excessive testing, it serves to verify that the routine product acceptance tests are of sufficient scope and sensitivity to detect all relevant defects. The combination of these tests is used to assess whether the manufacturing processes are sufficiently developed and robust to achieve the unit volume needed for market release of the product.

Parallel to the initial manufacturing in this phase is the development of the test market protocol and the market introduction plan. Both are essential to confirm the product's market readiness and prepare for market release. Test market product evaluations must include actual customers, be representative of the actual market demographics, be objective and thorough, and, where possible, provide statistically significant results. A thorough study design in this step is necessary to assure valid results. Consideration should be given to the selection of the customer users, how their evaluation is to be conducted, the specific information sought, and the methods for collecting and analyzing the results.

The test market is complete only at the conclusion of the test market protocol requirements. Product performance in the test market must be continually monitored. Although it would be unusual, given the rigor of the process thus far, a safety failure of the product would clearly cause a decision to abort the test. Given this or any other type of failure, a decision would have to be made regarding how the failure could be corrected in the same way it would have been done in the internal verification phase.

All reliability-related aspects of the product should be evaluated in the test market. These include written product instructions and literature, performance, safety, ergonomics, maintainability, and compatibility with different users, peripheral equipment, accessories, and environments. A test market report should be complete and detailed, identifying all results of this phase's testing. The report's results should be included in a product design review, and actions should be taken on any deficiencies cited.

The conclusion of the external verification phase is determined by whether all test market acceptance criteria have been met and all test certifications and regulatory requirements have been satisfied. If so, the market release requirements and the compilation of evidence to justify product release can be completed and this phase concluded. If not, decisions and/or follow-up are necessary. Note again, as in the internal verification phase, that it is possible to return to the concept phase if the failure to meet test market requirements involves failure of the detailed design; inability to correct the design with current resources, technology, and/or priorities, or some other aspect outside the test market design.

Market Release Phase

Summary

- Management approves market release.

- Product and customers are serviced and appropriate improvements implemented throughout the product's commercial life.

    Market release: Completion of release reviews and approvals; implementation of market introduction plan.

    Service/improvement: Maintaining product support and implementing customer-driven improvement throughout the market phase.

The market release phase is depicted in Fig. 3.6. Hopefully this phase is the longest in the life cycle of the product! It starts with a management verification that all release requirements, including reliability, have been met. It continues with product launch via the market introduction plan and the subsequent loop of manufacturing, marketing, and performance monitoring of the product.

Performance monitoring involves all aspects of the product and its acceptance by customers. This means price, delivery, installation, quality, reliability, serviceability, and availability—all those aspects of product and service that influence the customers' perceptions of value and satisfaction.

If performance is not acceptable, the product development and improvement reverts to the internal verification phase for manufacturing process correction, or even to the design/development phase for product design correction. If this return to earlier phases is necessary, all of the process steps should be followed in their original order, but they may be abbreviated or otherwise modified to fit the extent of the change needed for correction. It is essential, however, that all changes be thoroughly developed, reviewed, and evaluated to minimize the opportunity for unintended iterations and prevent possible catastrophic results.

If performance is acceptable *and* the business objectives of the product are being met, then the loop through manufacturing, marketing, and performance monitoring continues. This is the desired "equilibrium" state sought for long periods of time.

Continuous improvement of a product is directly associated with the decision point where management determines, on an ongoing basis, whether a product in the market release phase is meeting its business objectives. A product may be showing excellent performance against its design requirements but not be meeting its sales or cost targets. Such a situation needs to be assessed by management. Under some circumstances, a correction to the marketing plan or a decision to pursue a process improvement or a cost reduction could be an effective approach. In other circumstances, a change in the product's design, e.g., addition of a feature, simplification, or multiple line extensions, could substantially improve the market acceptance of the product and extend its life. These are examples of a continuous improvement loop for marketed products wherein the process reverts to earlier phases of the product life cycle to gain improvements.

At this same decision point, it is also possible that management could conclude that further investment is not justified and that a product's commercial life should be ended. A product that faces competitive pressures from newer, more state-of-the-art designs or from less expensive versions produced with newer manufacturing technology would be a likely candidate for concluding commercial life.

The key reliability process outputs of this phase are measures of product performance: failure analyses, warranty replacement and service costs, product availability (uptime), and maintainability performance. The collection and effective use of these field data should provide early warning in the event of unexpected deficiencies in design or manufacturing-contributed reliability deficiencies. Regular reviews of field data should be conducted with product teams to determine if and when any actions should be taken to adjust designs to address deficiencies.

End-of-Life Phase

Summary

- Management decides to discontinue marketing of product.
- Manufacturing/sales/spare parts/service phase-out plan developed.

    Market discontinuation: Management decision to discontinue product.

The end-of-life phase (Fig. 3.7) begins with the development of a market phase-out plan. This plan needs to define when manufacturing of the product will cease, when customers will be notified, and what steps will be taken for products remaining in use between the end of manufacturing and the time when all stock is through the distribution channels and/or when spare parts and service are no longer provided. A typical plan, depending on the type of product, should include such elements as the final manufacturing date, estimates of spare parts usage (based on field data and installed base), spare parts manufacturing and distribution plan, and customer notification with a statement of service policy and timing. Since some market phase-out plans can span years, it is essential that the same product performance measurement system used in the market release phase remain in effect until service phase-out occurs.

The end-of-life phase is most significant in the product development and improvement process for repairable or serviceable products. As long as the product performance is acceptable and the market phase-out plan is not completed, service and spare parts continue to be supplied. If product performance is not acceptable, then correction is achieved via the same loops presented in earlier process phases.

3.3   WHY MANAGE RELIABILITY AS A PROCESS?

The view of reliability as a process that was described above is not a traditional view. The traditional view has been that the design engineer, with or without the contributions of a reliability engineering specialist, basically assured that a product satisfied market expectations. What this traditional view omitted was the contributions required throughout an organization to help define the reliability expectations of customers and to deliver these quickly and efficiently. A "process for reliability" as described is the only means likely to achieve success, especially with results that customers want and that are achieved correctly the first time!

Consider these additional reasons for managing reliability as a process:

Rapidly Changing Worldwide Markets.   Recent years have seen dramatic changes in the markets for virtually all types of products. Quality and reliability have been the unsurpassed focus of companies throughout the developed world. The consolidation of markets in Europe and North America to reduce trade barriers and the political reforms in Europe and Asia represent an expansion potential for virtually all industries. The growing environmental concerns, the aging population, changes in health-care delivery, the communications revolution, and other such forces all combine to produce rapid and substantial changes that pose business opportunities as well as challenges.

Only those businesses with internal processes that are as dynamic and flexible as the outside world will gain from the opportunities provided by a world of rapid change. To meet the expectations of these customers, with varied cultures, resources, education levels, and perceptions, businesses must constantly look at the world outside the business organization as well as inside.

Reliability as a process is not an exception. A reliability process must provide perpetual measurement of customer expectations and preferences, competitive developments, and existing product performance and then respond to those results with continuous improvement. This would be a reliability process meeting the opportunities of the world.

Customer-Focused and Minimum Time-to-Market Product Development and Evolution.   The successful businesses of the last decade have shifted product development from a technology-driven and series approach to a more integrated customer-driven and process approach. This was prompted by the need to develop high-quality products faster, more efficiently, and more effectively than the competition.

Common technological capabilities in the world had become so widespread that innovation and fierce competition were extensive in many industries. Technology alone could no longer drive product development and assure customer demand in the marketplace. Also, the series approach to product development, wherein each function makes its contribution to product development and delivery and then passes the product to the next function, was sluggish and expensive, especially in the high-technology industries. An alternative process approach was adopted to overlap and blend the transitions among groups. To minimize time and cost to market, the highest-risk technical barriers for a new product had to be investigated and resolved first. This approach provides for early development of contingency plans or, at the very least, recognition that an inevitable development or market failure is not worth the investment.

Customer feedback was recognized as the most effective way for a business to understand the market. This feedback was used to facilitate rapid adjustments or modifications in the product offering, leading to improved sales and sometimes entirely new products. It was also seen as a competitive edge. The more a company knew about its customers that the competition did not know, the better its opportunity to gain market share.

Cross-functional teams evolved as organizations within the organization that were given full authority to define, develop, evaluate, and launch new products. Various aspects of reliability, indeed the various process outputs described above, are the product of many members of the typical business organization—engineering, finance, reliability, quality, manufacturing, customer service, marketing, etc.—working together in a process.

Product Liability Incentives.   Product liability has gained considerable attention. Many businesses have experienced litigation involving claims that their products caused injuries or other user impairments. Examples of such claims abound, including the health effects of asbestos, the potential for crash-caused fires in certain vehicles, and failures or side effects of health care products.

The basic defense against such litigation has been for businesses to eliminate the potential causes. Efforts have been made to improve product safety and assure clear and consistent product claims and instructions. The product development process, including the reliability process as described above, is the means within an organization to define, design, verify, and document product safety. As a cross-functional process, the essential customer input, market experiences, and design/performance analyses can be viewed from multiple perspectives and yield a product with proven features and validated literature.

Regulatory Pressures.   Developed nations of the world have volumes of laws and regulations affecting the conduct of business. In the United States, at least fifteen federal commissions and agencies administer regulations. Similar organizations perform duplicate or additional regulatory functions at state and local levels. The regulations have continued to grow, along with the organizations that administer them.

Voluntary standards have also been developed to help define and control certain industries and their products and practices. The most significant recent example is the ISO 9000 series of quality standards developed and adopted by many of the nations of the world. These standards define quality management system practices for suppliers of goods and services. While these standards are voluntary, their widespread adoption among nations and industries is having the same effect as if they were regulations. Companies may be compelled to comply simply from competitive pressures.

These regulations and standards have significant impact on businesses. Because of the complexity of their requirements, specialists are often needed to maintain, interpret, and administer them within the organization. For business processes such as reliability, the regulations and standards for product performance, safety, and market registration, which often differ by nation, require contributions by these specialists. Successful recognition and achievement of regulatory requirements can be achieved only through planned and active participation in the process.

3.4 MANAGING THE RELIABILITY PROCESS

Defining the Process.   The prerequisite to achieving successful high-reliability products today is establishing a reliability process that fits the particular organization, its products, and its market environment. The reliability process is best implemented within the framework of the broader product development and improvement process as described in Figs. 3.1 through 3.7. Product reliability, quality, performance, and safety are the ultimate results of this process. All members of the organization, especially management and the participants in the process, must agree to a suitable definition and commit to its successful implementation and maintenance.

When developing this process, these objectives should be kept in mind: The product development and improvement process seeks to provide products to the market in the least time with the least overall cost that fully meet (and/or sometimes set) customer expectations. The overall process must be customer-focused and team-managed, and must seek continuous improvement of not only the products but the process itself.

A definition of the product development and improvement process should start with the identification of the overall process phases. Concept, design/development, internal verification, external verification, market release, and end of life were suggested above. However, depending on the business's organization and its products, other combinations with more or fewer phases may be suitable.

Recognize that each phase should apply to both new designs and changes to existing designs. The extent to which each phase applies to new or existing products, e.g., time required, resources needed, extent of documentation, etc., may vary depending on the nature of the products and the extent of experience with the design.

Next, define how each phase should be broken into stages, each with its own responsible person and defined outputs or deliverables. Consider these stages the essential steps in the process. The outputs of each stage should logically be the inputs to the next stage. Responsibility for each phase can be assigned to a team leader. A product team, with responsibility for design/development through market release, should execute the process for a particular new product or product family. The team leader should rotate within the team based on the phase of the process at the time. For example, in an early concept phase, the team leader would most likely be a design engineer, whereas the external verification phase might be the responsibility of a manufacturing engineer or marketing product manager. The team leader manages the stage, including all planning, coordination, and controlling, to meet the established schedule and outcome objectives in the project plan, including reliability.

Deliverables should also be defined consistent with the needs of the organization, type of product, and any associated regulatory demands. A typical deliverable is a written plan, specifications, or other record and report needed as evidence of achieving completion of that process stage. Usually, items such as project plans, customer preference test reports, design specifications, failure modes analyses, and design qualification bench test reports are listed as outputs at the appropriate stage(s).

Formal design reviews and release approvals should be considered at certain stages to help communicate project status and serve as "gates" so that management can assess the results of investment and need for resource reallocation. Design reviews should occur frequently during the process and should have varied participation outside the team, depending on the process stage.

At two particularly critical points there should be both design review and management approval: conclusion of the concept phase and the external verification phase. These points are crucial because each represents a substantial investment commitment. The concept phase ends with the decision to enter the full design and development phase, and the external verification phase ends with the decision to enter full market launch of the product. Management, along with the team and others involved in the process, should reach a formal consensus that financial investment in these subsequent phases is justified and that the organization as well as the market is ready.

Create a process flowchart that defines and sequences the phases and stages chosen. The chart should depict the process as "a collection of successive steps that takes inputs, adds value in customer satisfaction, and achieves a specific output." Here is a series of steps that can be followed to define a process flowchart:

1. Determine the scope of the process to be defined:

   - Specific purposes of the process and measures of their achievement (how does it contribute to customer satisfaction?)

- Beginning input(s)
- Ending output(s)

2. List the sequential steps needed from the beginning input to the ending output.

3. For each step listed, decide whether one or more of the following are needed to achieve the specific purposes of the process:

- Deliverable output from the step
- Person(s) responsible
- Decision before the next step (question to be answered) and alternative choices

4. Where applicable, add specific deliverable outputs, insert decision points reflecting alternatives, and identify the persons responsible.

As an example, one of the purposes of the product development and improvement process defined in Figs. 3.2 through 3.7 should be to "supply management information and require timely decisions." This purpose has value to both internal and external customers, and its achievement can be measured, e.g., time in concept phase decision queue. A process that achieves this purpose should clearly designate points where management decisions are required and identify what information the process should provide to enable timely decisions.

Implementing the Process.   Once the process is defined, use it to help train and develop the organization to effectively implement it. One way to adopt the process in the organization is to start implementation with selected pilot teams and projects. Identify a person, either within or outside the organization, to facilitate the learning and incorporation of the process by the teams. Where a process approach is new to the organization's culture, it is especially important that management be a participant in the implementation and demonstrate commitment to its success.

Several overall objectives need to transcend the design and development process implementation effort:

1. *Quantify external customer needs and expectations.* This is the foremost objective of the entire process and is, to a great extent, the source of eventual success in the marketplace.

2. *Understand the value-addition chain within the product development and improvement process and other interfacing business processes.* Recognize that each subsequent phase of the process should represent added value to the design and to the organization's potential in the marketplace. Decisions to allocate resources, add or delete product features, and delay or accelerate project schedules should be made in terms of value added in customer satisfaction. Again, "customer" can mean the person ultimately choosing the product, paying for it, and/or benefiting from it (the external customer) *or* the person next in the successive steps of the process chain (the internal customer). It is possible that a well-defined and effectively implemented design and development process can be hindered by other processes in the organization that are not effective. Here are some examples of other business processes that can affect product development and improvement:

- Production planning
- Materials management
- Hiring
- Customer needs analysis
- After-sales service (and its subprocesses)

    Repair and return

    Training of field repair personnel

    Field problem resolution

    Technical assistance

    Product servicing

    Warranty claims

3. *Translate customer needs/expectations into measurable characteristics that are evaluated* before *delivery to the customer.* This is typically done for external customers through detailed standards, specifications, and test methodologies. However, it may be overlooked for the internal customers in the organization—the team members, management, and the many others outside the design and development process circle that have impact and add value for one another as well as the outside customer.

Here are some steps to consider in meeting this particular objective:

- Define the internal customer(s) or clients.
- Determine their satisfaction expectations.
- Get agreement by both the supplier and the customer.
- Select measurements of customer satisfaction with highest improvement value.
- Measure objectively and reproducibly; avoid bias.
- Use what you measure or don't bother measuring it.

As an example, consider the manufacturing engineers who participate in the product development and improvement process. As internal customers of this process, what might be their most important expectations? It is likely that they would cite at least some of the following:

- Designers use as many materials as possible that have already been used in currently manufactured designs. In other words, avoid adding new materials unless they offer some unique advantages.
- The organization's design principles be consistently followed, e.g., avoiding use of ozone-reactive substances, incorporating only mechanical interlocking of parts rather than adhesives, developing completed part drawings only on one specific CAD (computer-aided design) system.
- Manufacturing engineer team member participates in the second and subsequent generation prototype construction.
- Material sources be selected and approved before freezing materials.

It is important that the suppliers for these examples verify that these do represent expectations of the manufacturing engineers and that more significant ones were not omitted. If, for example, the first expectation regarding the use of current materials was especially important to the organization, a goal could be established to achieve designs that included, say, 80 percent current materials. Each design then should be evaluated by the designer (supplier) against this goal before the design is submitted for prototyping. Focus on this important principle when developing a scheme for identifying customer expectations, selecting measurements, and evaluating achievement: If the measured process outputs relate to customer satisfaction and the measures are used to improve the process, then customer satisfaction should improve.

Maintaining the Process. Continuous improvement is necessary to prevent any process from becoming cumbersome and ineffective, even one that is well established and consistently achieving desired results. Persons participating in the process change, their interests and motivations change, and the organization can experience shifts in resources and priorities. Although change is inevitable, management and the process participants need to commit to the continuous cycle of definition, goal setting, measurement, and improvement action. Incorporating the following principles will help assure effective process maintenance:

1. Relate *every* measure, directly or indirectly, to the ultimate goal of customer satisfaction.

2. Define each measure operationally.

- Make certain that its units of measure are defined and that it is understood by all concerned.

3. Integrate all key measures in company business plans.

- Be sure that all measures are considered important.
- All parties should agree that the measures need to be watched closely and acted upon if performance is less than desirable.

- Measures monitored should be continuously improved.
- The benefit of the measurement should exceed the cost of taking it.

REFERENCE

1. H. James Harrington, *Business Process Improvement,* McGraw-Hill, New York, 1991, p. 9.

CHAPTER 4
ECONOMICS OF RELIABILITY

Henry J. Kohoutek
MANAGEMENT AND ENGINEERING STRATEGIES
SSA INTERNATIONAL
LOVELAND, COLORADO

4.1 INTRODUCTION

The task of economics is to devise optimal ways of allocating scarce resources among competing proposals for their use. One scheme that has been studied many times in detail, and that has historically proven to be effective in free markets, is allocation based on the price of a given product or service. The concept of price can represent buyers' and sellers' convictions, willingness to pay, personal preferences, efforts to obtain favor, etc., but, fundamentally, it refers to the worth of the item concerned as related to its market situation (described in terms of supply and demand) and manufacturing cost.

4.2 RELIABILITY AND VALUE

4.2.1 Preliminaries

Experience shows that customers select a particular product from the set of competing products not because it has the lowest price, but because it has the highest ratio of customer-perceived quality to price. This ratio represents a narrow empirical definition of product value:

$$\text{Value} = \frac{\text{customer-perceived quality}}{\text{price}}$$

Value includes all nonprice attributes the customer believes the product possesses. Value is, therefore, a measure of the product's relative desirability, and its ability to create customer satisfaction. In many situations, customers look beyond just the product they buy, and include in value every aspect of their interface with the supplier.

Given that value is the primary differentiator, it is necessary to understand its position in the business strategic situation the company faces. The strategic triangle in Fig. 4.1 illustrates that

- The company competes for customers in a given market by offering them products and services of superior customer-perceived value.

- The company competes directly with its competitors by better cost performance and higher levels of internal competencies.

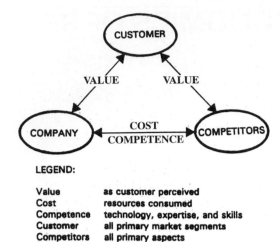

LEGEND:

Value	as customer perceived
Cost	resources consumed
Competence	technology, expertise, and skills
Customer	all primary market segments
Competitors	all primary aspects
Company	all primary functions

FIGURE 4.1   The strategic triangle.

Value must be measured to gauge the alignment between customers' expectations and the company's actual performance. Methods for quantification of the monetary aspects of value are found in economic analysis. To assess the nonmonetary value of a product, we must use cognitive socio-psychological methods, such as *sample surveys,* which are now widely accepted as a fundamental means of providing insight into customer valuation processes.[1]

4.2.2 Value of Reliability

Reliability contributes to the product value and must be included in value assessment and analysis. It should be considered as mission success probability in the traditional way, as well as in terms of its empirical attributes, such as failure rate, environmental ruggedness, cost of repair, level of fault tolerance, etc.

The most direct way to express and measure the value of reliability is by statements of how much reliability customers can obtain per some unit price. For example,

$$\text{Value of reliability} = \frac{\text{mean time to failure}}{\text{price}} \qquad \text{(years per thousand dollars)}$$

or

$$\text{Value of reliability} = \frac{\text{averaged failure rate in the first year of use}}{\text{price}} \qquad \begin{array}{l}\text{(percent per year per}\\\text{thousand dollars)}\end{array}$$

The advantages of these definitions are their simplicity, their ease of measurement and monitoring, and their frequently documented long-term validity for a given family of products. Knowledge of long-term trends in both reliability growth and price erosion is important, because it allows the point of diminishing return on investment in reliability improvement programs to be identified.

Using customer survey data about a product's acceptance and the relative importance of the primary attributes that drive the buying decision, it is possible to establish a *product value index.* This approach allows us to study changes in value due to reliability improvement, as shown by a simplified example in Fig. 4.2. To compensate for competitors' actions during the reliability

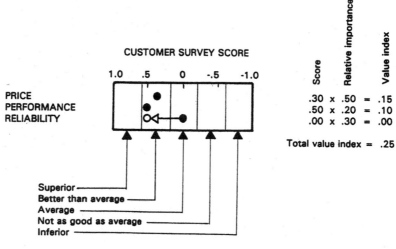

FIGURE 4.2   Product value change as a function of reliability improvement.

improvement project period, it is prudent to lower the expected new survey results by a fraction, reflecting industry-observed reliability growth rates.

It is important to recognize that superior product reliability also indicates the company's unique technical competence in reliability engineering and management.

4.2.3   Relative Importance of Reliability, Price, and Performance

The wealth of data about customer behavior, values, and attitudes often confirms that reliability is the most important product quality attribute, with the highest impact on the value in exchange, expressed by price, and value in use, expressed, for example, in terms of users' return on investment. Table 4.1 illustrates two extreme cases. Decisions in the space programs are dominated by

TABLE 4.1   Examples of Relative Importance of Price, Performance, and Reliability

	Relative importance of reliability for	
	Space program	Consumer electronics
Price	0.05	0.75
Performance	0.25	0.15
Reliability	0.70	0.10

reliability considerations (a weight of 0.7 on a scale of 0 to 1.0) because of the severity of the adverse consequences of a failure. In most purchase decisions in the area of consumer electronics, on the other hand, price is considered much more important than reliability, given the usually noncritical impact of failures, available protection by a warranty, seller's good will, etc. Knowledge of the relative position of reliability versus other product characteristics, expressed in terms of quantitative relative importance, is crucial for both formal and practical design considerations, and can significantly improve the rationality of many decision processes.

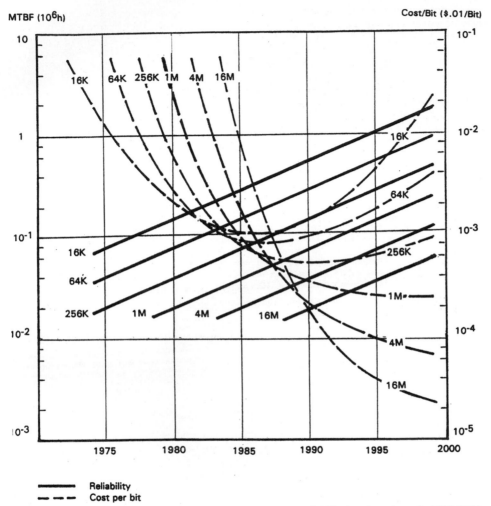

MTBF (10^6h)   Cost/Bit ($.01/Bit)

FIGURE 4.3   Example of long-term trends in cost, complexity, and reliability interdependence for MOS RAMs.

Balancing reliability, price, and performance can be constrained by many mutual interdependencies, often dominated by common technology or design principles. Understanding these interdependencies provides the insight needed for a realistic valuation of reliability in electronic components. Figure 4.3 suggests an example format for presenting the time dynamics of MOS RAMs' cost, complexity, and reliability interdependencies. In other cases, similar interdependencies may be expressed in very simple experience-based heuristics, such as, "At maturity, the infancy failure rate of any NMOS component will stabilize at around 0.05 percent per 1000 hours, independent of its complexity."

4.2.4   Reliability as a Capital Investment

Considering the relationship of reliability to value and its impact on price, product reliability can be analyzed in terms of its attractiveness as a capital investment. A primary measure of this attractiveness is *return on investment* (*ROI*), from which other measures can be easily derived.

ROI and payback period are only an approximation, because they do not consider the time value of money and they assume the project has an indefinite life. They can be used for "quick and dirty" analysis of proposed investments in reliability programs aimed at preventing or reducing the number of costly product field failures. It is highly recommended that more accurate methods, involving the use of present worth or equivalent uniform annual payments and returns, be used where the required rate of return is high or compensation for future uncertainties is necessary. These correct methods, as illustrated in the example below, can make a great difference in the estimates of ROI and the payback period.

- *Payback period* is simply the reciprocal of ROI.
- *Benefit/cost ratio* is ROI multiplied by the expected years of useful life.
- The *net return* is calculated by multiplying the benefit/cost ratio by the investment cost, in our case by the cost of a reliability improvement program, and subtracting from it again the investment cost.

The benefit/cost ratio and the net return can be discounted to reflect the time value of money by multiplying the ROI by the present worth factor, which is found in standard interest tables.

Example. A reliability improvement program under consideration has an expected cost C_R of $50,000 and will prevent $N = 250$ failures annually with an average repair cost C_{rep} of $850 each over an expected useful life of $L = 8$ years. Let us use different measures to assess the effectiveness of this investment assuming general and administrative overhead OH = 30 percent, profit coefficient $P = 10$ percent, and a 10 percent discounting schedule, so that the present worth factor for 8 years of service $F_8 = 5.335$.
Return on investment:

$$\text{ROI} = \frac{NC_{rep}}{C_R(1 + \text{OH})(1 + P)} = \frac{250 \times 850}{50,000(1 + 0.30)(1 + 0.10)} = 2.97$$

Payback period:

$$\text{PP} = \frac{1}{\text{ROI}} = \frac{1}{2.97} = 0.336 \text{ year}$$

Benefit/cost ratio:

$$\text{B/C} = \text{ROI} \times L = 2.97 \times 8 = 23.76$$

Discounted benefit/cost ratio:

$$\text{DB/C} = \text{ROI} \times F_8 = 2.97 \times 5.335 = 15.84$$

Net return:

$$\text{NR} = (\text{B/C} \times C_R) - C_R = (23.76 \times \$50,000) - \$50,000 = \$1,138,000$$

Discounted net return:

$$\text{DNR} = (\text{DB/C} \times C_R) - C_R = (15.84 \times \$50,000) - \$50,000 = \$742,000$$

All these terms can be expanded to take into account the particulars of a given reliability program, such as the impact of inflation or of changes in technology on cost of repair. With the aid of a personal computer, additional investigation is possible to assess the results' sensitivities to input variations or the impact of uncertainties, or to compare alternative programs. Similar studies can be made to assess the economic impact of unreliability on product value in terms of performance and availability degradation or increased cost of ownership.

This conventional approach compares only cost of investment against expected cost savings. It excludes any revenue enhancement or protection from the analysis, and it makes no allowance for lost opportunities. It unrealistically assumes that sales will remain constant with or without the contemplated investment. Fortunately, using the concept of value, it is possible to account for the impact of the investment on market share and for the risk of inaction by calculating the expected change in the product value index (see Fig. 4.2) and correlating it with the industry-specific information available in the *Profit Impact on Market Share* (PIMS) database. This data bank, formed initially by GE in 1972 and now managed by the Strategic Planning Institute in Cambridge, Massachusetts, contains the strategy experience of over 2000 business units for 20 years and covers close to 200 separate characteristics of each business experience. It has statistically proven a relationship between value improvement, which PIMS calls customer-perceived quality improvement, and market share increase for different types of business.[2] For example, in the capital goods business, a 15 percent improvement in product value, relative to the established competitive values, provides a potential for a 100 percent increase (i.e., doubling) in market share. Probable revenue and profit increases established in this way can be incorporated into the reliability improvement program financial justification.

4.3   RELIABILITY AND COST

4.3.1   Introduction

The concept of *product life cycle,* despite being a subject of continuing debate and even some controversy, forms the basis of all current views on product costing, and is applicable to reliability cost analysis. It translates directly into the concept of *life-cycle cost* (LCC), which combines cost and reliability information with the goal of approximating and evaluating the economic profiles of design alternatives. Because two major parties are involved in a product life cycle, LCC must accommodate two different perspectives:

- The vendor's perspective through *cost of reliability* (COR)
- The user's perspective by *cost of ownership* (COO)

Figure 4.4 indicates which primary cost elements are involved in each perspective, and provides an example of the actual accrued relative LCC cost structure for a typical consumer electronics product.

4.3.2   Manufacturer's Viewpoint: Cost of Reliability

The concept of cost of reliability (COR), the total cost the manufacturer incurs during the design, manufacture, and warranty period of a product of a given reliability, can be developed around the generally accepted notion of *cost of quality* (COQ). The principles of COQ, established in the 1950s, have been verified and validated in all segments of the manufacturing industry. COQ is applied to measure the economic state of quality, to identify opportunities for quality improvement, and to verify and document the effectiveness and impact of quality improvement programs. The accounting for COQ makes visible all the cost items associated with defects in products and processes and then sets them in contrast with the cost of doing and staying in business. The classic categorization of COQ applied to COR expresses as unique the costs associated with the following elements:

External failure	Cost of unreliability during the warranty period, cost of spare parts inventory, cost of failure analysis, etc.
Internal failure	Yield losses caused by reliability screens and tests; cost of failure-caused manufacturing equipment downtime; cost of redesign, rework,

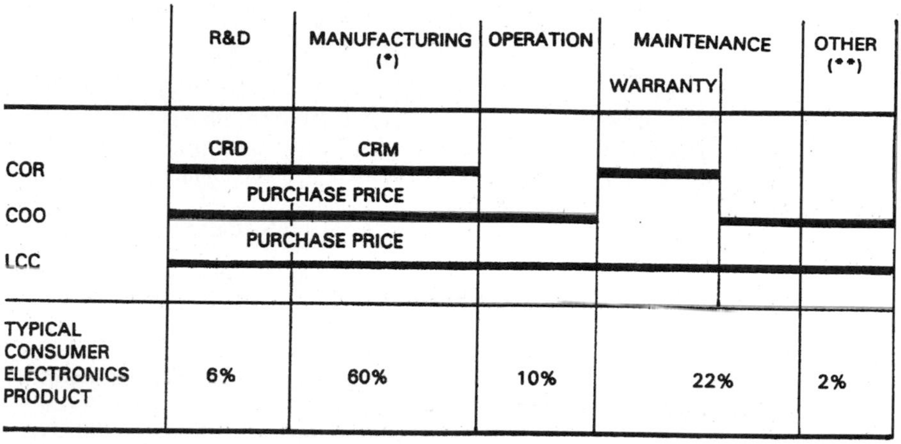

	R&D	MANUFACTURING (*)	OPERATION	MAINTENANCE WARRANTY	OTHER (**)
COR	CRD	CRM			
COO	PURCHASE PRICE				
LCC	PURCHASE PRICE				
TYPICAL CONSUMER ELECTRONICS PRODUCT	6%	60%	10%	22%	2%

(*)   includes installation
(**)   e.g., refurbishing, upgrades, obsolescence

FIGURE 4.4   Comparison of different views of the product total life-cycle cost.

	and retest with related extra material and work-in-process inventory, etc.
Reliability appraisal	Life testing, environmental ruggedness evaluation, abuse testing, failure data reporting and analysis, reliability modeling, etc.
Prevention	Design for reliability, reliability standards and guidelines development, customer requirements research, product qualification, reliability screens, design reviews, reliability training, fault-tree analysis, failure modes, effects, and criticality analysis, etc.

Understanding of these cost categories is a must for rational planning of reliability assurance resources, environmental and life-testing facilities, training programs, warranty policies, and other services needed for successful reliability programs.

For COR management, the cost information must be further restructured by product, process segment, or department to identify major contributors and therefore opportunities for improvement. The identified cost levels are usually compared with some measure of revenue or value added, to form ratios and management indexes. From the viewpoint of cost management, it is also important to recognize that prevention and appraisal costs are controllable with planning and budgetary mechanisms. Both types of failure cost are results of the inability to assure defect-free design, manufacturing, and distribution processes, and to control product users' application conditions and environment.

There are also some dangers associated with the use of both COQ and COR. Managers often forget that COQ and COR are dependent variables, reflecting successes and failures of the quality and reliability programs. By doing so, they run the risk of degrading reliability levels just to achieve short-term cost savings. Other possible dangers involve difficulties in the identification of defect causes, conflicts coming from cost charges transfer rules, preoccupation with reporting systems, tendencies toward perfectionism, and so on. To prevent these and other pitfalls, managerial prudence and knowledge are required in a search for facts, understanding, realistic objectives, priorities with rational execution plans, and progress monitors.

The model of total COR can be similarly developed as a tool for managing the costs and resources associated with the design for reliability, reliability manufacture, and warranty cost reflecting residual unreliability. Because of the different slopes in individual cost curves as functions of increased reliability, the applicability of the concept of cost optimum is self-evident.

In some industries, such as the semiconductor industry, enough experience-based information has been accumulated to rationally model the dependency of the final product reliability on the number of redesign cycles, the levels of screening and testing, or, in general, the level of reliability assurance. This information confirms the intuitive expectation that an increase in planned and controlled reliability assurance costs and activities significantly reduces unplanned failure cost. These models, combined with physical failure rate models and warranty cost calculations, allow optimization of the COR and, through that, rational planning of the project, manufacturing, and support resources very early in the design phase.[3,4]

Development of an optimal cost versus reliability strategy starts with a simple formula for total cost of reliability:

$$\text{COR}_{\text{total}} = \text{CRD} + \text{CRM} + \text{WC}$$

CRD is the *cost of reliability design,* modeled by an empirical relationship f_1 among the cost and number of design for reliability cycles, application environment stresses, complexity, and expected general level of quality, expressed, for example, through the quality coefficient π_Q. This coefficient, regularly used in MIL-HDBK-217, *Reliability Prediction of Electronic Equipment,* which contains reliability models, reflects the relative impact of reliability program actions on the base reliability defined by the physical nature of the components used.

$$\text{CRD} = f_1(\pi_Q)$$

The *cost of reliability manufacturing* (CRM) is a function f_2 of the effectiveness of manufacturing screens, expressed again by the values of the coefficient π_Q and associated fixed and variable costs:

$$\text{CRM} = f_2(\pi_Q)$$

The *warranty cost* (WC) is a function of the initial failure rate, but also takes into account the effect of the bathtub curve, the learning curve, and, of course, repair cost.

These individual expressions allow the study of the total cost of reliability

$$\text{COR}_{\text{total}} = f_1(\pi_Q) + f_2(\pi_Q) + \text{WC} = F(\pi_Q)$$

as a function F of the π_Q, and allow a search for an optimum (see Fig. 4.5), taking into consideration different warranty periods, learning factors, inflation rates, shapes of the bathtub curve, etc. The optimum conditions found must then be translated into resource levels and explained in technical terms of reliability engineering. This strategy for reliability, based on minimum total cost and its optimum apportioning among design, manufacturing, and warranty, is generally acceptable to all managers involved.

A word of caution is again in order. This classic approach to reliability optimization may not be valid when dealing with long-life system products, or when technology-based high MTTF exceeds the end product's useful life.

4.3.3 Combined Viewpoint: Life-Cycle Costing

According to reference 5, life-cycle cost is defined as follows:

> The Life Cycle Cost of a system is the total cost to the government of acquisition and ownership of that equipment over entire life.

One of the primary benefits of past applications of LCC methodology is the considerable evidence showing that by the time a product is ready to be released to manufacturing, a substantial portion of its eventual life-cycle cost is already locked in,[6] even though it has not been accrued yet:

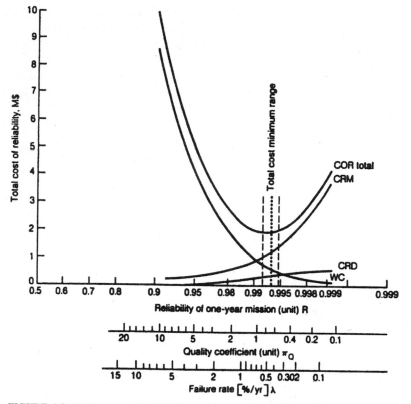

FIGURE 4.5 Optimum set of reliability cost curves. (Adapted from Henry J. Kohoutek, "Development of a Reliability Strategy for New IC Component Family and Process, *Microelectronics and Reliability,* vol. 23, no. 2, 1983, p. 388. Reproduced with permission.)

70 percent by concept review, i.e., by the end of the concept definition phase

85 percent by design review, i.e., by the end of the system definition phase

95 percent by production release, i.e., by the end of full-scale development

An important fundamental principle of LCC is the understanding that the lowest manufacturing cost does not necessarily yield the lowest life-cycle cost to either the manufacturer or the customer.

The customer has to find a balance between the contemplated mission requirements and the budget resources, taking into account the internal constraints given by

- Internal acquisition logistics—for example, status of incoming inspection and testing facilities, installation opportunity, asset management system and procedures.
- The support system, which will affect the actual application environment stresses, maintenance strategy, etc.

The vendor, who wants to satisfy the proposed mission requirements, must perform many analyses to gain insight into available implementation alternatives and find the one that guarantees the achievement of these requirements at minimum cost. Major constraints for this activity depend on available

- Cost-estimating skills and completeness of the historical cost database
- Tools, skills, information, and resources for accurate analysis
- Creative design and implementation alternatives

This process of finding the optimum tradeoff for both customer and vendor must be repeated at different levels of detail throughout the whole acquisition phase. If the vendor is an equipment manufacturer or system integrator, the LCC process is also repeated and refined for each new phase of the product's life.

- During the conceptual design phase, usually the cost of only one of the proposed alternatives is developed using parametric cost-estimating relationships. The cost of other alternatives is calculated from estimated cost differences. When design specifications and implementation strategies are defined, the project enters the validation phase. Here more detailed cost estimates are needed to justify design-to-cost objectives and to establish demonstrable LCC characteristics.
- In the phase of full-scale development, operating and support cost estimates become more accurate because reliability, maintainability, serviceability, and supportability characteristics have been demonstrated. The issues of warranty conditions and pricing are resolved also, and form a base for a new total LCC reassessment.
- In the phases of full-scale manufacturing, delivery, installation, application, and use, the actual cost is measured, and cost information is stored in the database to allow evaluation of the accuracy of previous estimates, adequacy of applied decision models, and their future improvement.

Effective local applications of the LCC methodology usually start by acceptance and experimentation with existing systems, many of which were described in past *Proceedings of Annual Reliability and Maintainability Symposium,* e.g., 1993 (pp. 305–310) or 1994 (pp. 166–171).

Continuously increasing requirements for LCC minimization motivate program managers to use computerized methods to achieve this goal by varying individual cost elements, considering more design alternatives, testing new concepts, evaluating implementation strategies, and accounting for uncertainties.[7] In profit-oriented business environments, the minimization of some cost is not necessarily the best solution available, and so new LCC models must be developed to assure strategies for net profit maximization. Profit is a random variable which depends on the amount of uptime logged during the period of product use. Key to the calculation of the total life-cycle profit is the profitability characteristic, which relates profit to the operating and support cost. It is a function of equipment age, and often exhibits three stages (early, useful, and wearout periods) similar to the stages of the bathtub curve.

In the majority of practical cases, the situations in which the user is facing decisions are much less complex than, for example, a major military procurement contract. Because the development and manufacturing costs have already been incurred, and are reflected in the equipment purchase price, the user's degrees of freedom in the search for minimum LCC are limited to evaluation of different support and maintenance strategies. The equipment's intrinsic reliability, which is also given, affects only the cost of maintenance M_C, which can be estimated assuming (each per some time period, usually a month or year)

- The cost of spare parts inventory C_{SPI}, reflecting the original manufacturing cost of spare parts C_M and the inventory cost rate I_{CR} (as a percentage), including depreciation, interest, handling cost, etc.
- Preventive maintenance cost C_{PM}
- Corrective maintenance cost C_{CM}

So,

$$M_C = C_{\text{SPI}} + C_{\text{PM}} + C_{\text{CM}}$$

$$= C_M I_{\text{CR}} + WH \, \frac{T_R^P + T_T^P}{T_I^P} + WH \, \frac{\text{MTTR} + T_T^C}{\text{MTBF}}$$

where W = hourly rate of service engineer, including prorated parts cost

$\quad\quad H$ = equipment usage in hours per time period considered (in-use time)

$\quad\quad T_R^P$ = scheduled time for preventive maintenance

$\quad\quad T_T^P$ = expected travel time for preventive maintenance

$\quad\quad T_I^P$ = scheduled preventive maintenance interval

$\quad$ MTTR = mean time to repair

$\quad\quad T_T^C$ = expected travel time for corrective maintenance

$\quad$ MTBF = mean time between failures, expressed in terms of in-use time, not calendar time

For example, typical values of these factors for personal computer system maintenance in the early 1990s are as follows:

C_M = $2000 at user's site	I_{CR} = 10% per year
W = $350 per hour	H = 4000 hours per year
T_R^P = 0.25 hour	T_T^P = 0.5 hour
T_I^P = 6 months	T_T^C = 0.5 hour
MTTR = 0.5 hour	MTBF = 6000 hours

For competitive analysis or evaluation and comparison of alternative strategies, it is customary to express yearly maintenance cost as a percent of the equipment's original purchase price or to standardize it, for example, per $1000 of price.

4.3.4 User's Viewpoint: Cost of Ownership

LCC analysis methodology is rarely used in the commercial world, with the automotive industry being an exception. When it is applied, it usually focuses on the manufacturer's portion of the life-cycle cost. Because the manufacturer's perspective may not lead to an optimum from the customers' viewpoint, a new approach is needed. Cost of ownership (COO) provides a viable alternative to LCC. The COO approach may be useful to manufacturers also by helping them to leverage sales and services through additional competitive information. A reasonable goal may be to try to close the gap between the two, LCC- and COO-based optima.

The primary cost elements of the COO model are

- Initial investment (bundled product price, cost of installation, cost of preuse training)
- Depreciation (initial investment, depreciation factor)
- Cost of operation (space, personnel, power, supplies, communication carrier charges)
- Cost of maintenance (downtime, repair cost, failure rates, cost of spares)
- Enhancement costs (hardware/software upgrades, training)

- Disposal cost at product's end of life
- Imposed costs (tax rate, inflation rate)

Considering the general character of these cost elements, it is clear that product reliability is a significant cost driver. COO analysis therefore provides a means of balancing and integrating reliability and maintenance strategies and cost, and allows additional insights that can be used in the buying decision. Because of the relative simplicity of the model, it can be executed in spreadsheet form on a personal computer. This provides the opportunity for various investigatory runs and sensitivity analyses.

An example of COO commercial application may be found in the semiconductor manufacturing equipment industry. The Sematech COO model[8] clearly indicates that in many cases, throughput, yield, reliability, training, and other factors contribute more to the total lifetime cost than the actual price paid for the equipment. For the particular needs of this yield-sensitive industry, COO is considered as the time-period cost associated with the equipment, divided by the number of product units manufactured by that equipment in that time period:

$$\text{COO} = \frac{\text{fixed costs} + \text{variable costs} + \text{yield costs}}{\text{throughput} \times \text{throughput yield} \times \text{equipment utilization}}$$

The three dominant cost drivers identified by the model applications are scrap, consumables, and equipment cost, and COO is expressed in terms of cost per good wafer out (an example of this calculation may be found in Ref. 8).

Equipment suppliers see benefits from this model in many areas: system design, communications with customers, competitive analysis, pricing, sensitivity analysis, and intelligent selling. Some problems may arise from inaccurate data, inflated equipment performance specifications, and lack of model validation as a result of limited customer feedback.

4.3.5 Design to Cost

The challenge of balancing reliability, maintainability, supportability, and availability specifications during the design planning phase is complicated by the ever-present cost and profit requirements. Design to cost is one possible approach to meet this challenge, useful especially when the design cycle time is under time-to-market constraints. To increase the desired success of the design team involved, careful cost planning is needed (Fig. 4.6).

First, the target cost must be determined by a rational link to the company business plan, namely its profit planning part. This is usually accomplished through the profitability index based on past experience with similar products. Then the target cost can be established from the product price by considering the planned cost reduction activities.[9] These, in turn, may reflect industry or technology trends or internal quality and productivity improvement initiatives. Since reliability, availability, maintainability, and supportability are LCC drivers, their analyses provide substantive inputs into the optimization of the product design specifications to be implemented by the design team.[10]

Even though target costing and design to cost are often practiced by Japanese manufacturers, they are encountering implementation difficulties in the United States:

- Radical changes in the overall design process, such as introduction of prototyping, concurrent engineering, or specialty engineering disciplines, are often required for their successful implementation.
- Optimization and simultaneous tradeoffs are hindered by lack of knowledge-based cost of design for reliability rules and validated heuristics.
- Easy-to-use computerization and an adequate database to assure automated repeatability of the planning process are seldom available.

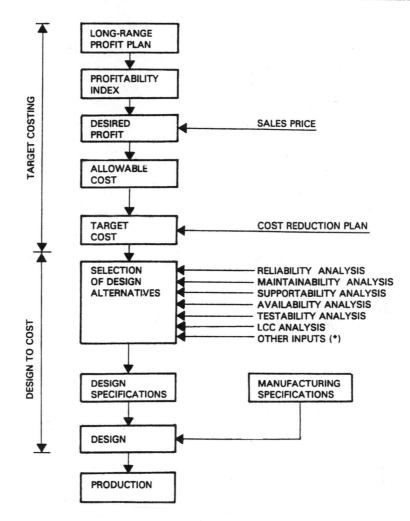

(*)   **Issues of product functionality, performance, etc.,
        are subject to concurrent planning process**

FIGURE 4.6   Design to cost planning process.

Increased competitive pressures and availability of decision support tools can make design to cost a viable and attractive alternative in the near future, because of its suitability to almost any product and application environment.

4.4   PARTICULAR ISSUES

The concepts of total COR, LCC, and COO discussed above provide an overall framework for cost optimization and for addressing and studying the economics of individual reliability-related issues, their impact on total cost and revenue, and selection of local implementation alternatives.

4.4.1 Economics of Warranty

In the narrow sense, defined by the Uniform Commercial Code, there are four basic types of warranties:

1. *Warranty of title and against infringement,* representing that the title is transferred and that the transfer is good, unencumbered, and free from patent, trademark, and similar claims
2. *Express warranty,* which is usually created to assure the buyer that the delivered products will conform with samples, models, and descriptions used by the seller during the bargaining process
3. *Implied warranty of fitness for use,* reflecting the assumption that the seller knows the intended end use of the goods by the buyer, who is relying on the seller's skills and judgment
4. *Implied warranty of merchantability,* assuring that the goods sold meet the minimum standards of merchantability as stated in the Uniform Commercial Code, paragraph 2-314 (1979)

These warranties are representations of an inducement to purchase for which the law will allow a remedy if it is proven not true. The key issues invoked by this concept involve disclaimers, coverage, limitation in time period and conditions of use, consistency, and the required disclosures. In the more general framework of the free market, however, warranties are viewed as an element of competitive strategy, and the economic issues associated with pricing, cost accounting, and determination of warranty reserves will dominate management's interest.

From the cost accounting point of view, which greatly influences company management thinking, warranties usually fall into three categories:

1. *Reimbursement warranty,* the simplest form, in which the manufacturer reimburses the dealership for any charges incurred to repair or replace defective merchandise
2. *Sales warranty,* in which charges are treated as an ordinary sale, except that the revenue account is reduced rather than the accounts receivable being increased
3. *Expense warranty,* the most popular, which involves a merger of warranty expenses with other operating costs or cost of goods sold

The lack of adequate costing figures as a satisfactory base for estimation, the complexity of the warranty reporting system, tax regulations, etc., are often stated as reasons for ignoring warranty costs despite the evidence that they should be controlled.

To assess the probable warranty cost, these cost elements and inputs must be taken into account:

1. Cost elements
   - Repair cost, including labor, material, travel, and overhead
   - Item transportation cost between the customer and repair center locations
   - Field service administration cost
   - Cost of spare parts inventory
   - Costs of warranty administration, including data tracking, monitoring, reporting, and analysis
   - Cost of settling warranty issues
2. Inputs
   - MTBF or failure rate
   - Number of installed product units under warranty
   - Warranty period (sometimes regulated by federal or local legislation)
   - Inflation rate

When warranty action is repair only, warranty cost calculation is simple, even if probability distributions of some of the inputs and cost elements are considered. These calculations may become difficult when

- The warranty period is restarted with each repair.
- A fleet warranty (a guaranteed mean life of a product population sold to a single customer) has been granted.

The warranty, as an expressly stated obligation and responsibility of the seller, can be perceived as an integral part of the value package, intended to promote and encourage sales by reducing potential risks for the buyer. This type of warranty increases the obvious expenses by the cost of promotion, which may be significant. Market analysis is necessary to evaluate the estimated ability of the warranty to increase sales and profits, especially in industries in which good warranty experience and repeat sales are not strongly correlated, or in which the rate of return on goodwill assets is limited by insufficient demand. If the market and competitive situations allow reasonable freedom in product pricing and in the length of the warranty period, an optimization model can be developed. In most market situations we see some evidence of tradeoffs between higher price and longer warranty versus lower price and shorter warranty, indicating a strong dependence of the cost optimum on the distribution of failures early in the life of the product.

4.4.2 Economics of Software Testing

In the current market situation, the software pricing strategy reflects, in most cases, the demand-dominated perceived product value. This unique but temporary pricing freedom focuses software reliability economics on developers' internal issues: dependencies among code size, complexity, productivity, schedule time, defect densities, and costs. The most frequently asked question about these relations is when to stop testing and debugging, which are considered the most critical elements of software reliability assurance and represent the final review of specifications, design, and coding. The search for a balance between cost of in-house testing and costs associated with field failures is the framework for both of the following methods.

Failure-Rate-Based Decision.   The objective of this approach is to minimize the total life-cycle cost associated with testing and fixing defects. Because the cost elements and their values are different before and after release, the time of the release t_{REL} is the critical decision parameter.

$$C_{BR} = \int_{t=0}^{t=t_{REL}} [c_{BF}\lambda(t) + c_{TEST}] \, dt$$

$$C_{AR} = \int_{t=t_{REL}}^{t=\infty} [c_{OF}\lambda(t)] \, dt$$

$$C_{TOTAL} = C_{BR} + C_{AR}$$

where C_{BR} = cost before release
   c_{BF} = average cost of defect fix during in-house testing
   c_{TEST} = testing cost per unit time
   C_{AR} = cost after release
   c_{OF} = average cost of defect fix in the field
   $\lambda(t)$ = instantaneous failure rate at time t
   C_{TOTAL} = total life-cycle cost associated with testing and fixing defects

To achieve our objective, C_{TOTAL} must be minimized, so we look for

$$\frac{d}{dt_{REL}} C_{TOTAL} = \frac{d}{dt_{REL}} \left\{ \int_{t=0}^{t=t_{REL}} [c_{BF}\lambda(t) + c_{TEST}] \, dt + \int_{t=t_{REL}}^{t=\infty} [c_{OF}\lambda(t)] \, dt \right\}$$

$$= c_{BF}\lambda(t = t_{REL}) + c_{TEST} - c_{OF}\lambda(t = t_{REL}) = 0$$

There we considered that the instantaneous cost at both $t = 0$ (testing start) and $t = \infty$ (approximation of obsolescence time) is zero.

The resulting optimum release time t_{REL} will be associated with the demonstrated failure rate:

$$\lambda(t = t_{REL}) = \frac{c_{TEST}}{c_{OF} - c_{BF}}$$

Because $\lambda(t)$ is assumed to be a continuous and monotonically decreasing function, $\lambda(t = t_{REL})$ is subject to the following constraint:

$$\lambda(t = 0) \geq \lambda\ (t = t_{REL}) \geq \lambda(t = \infty)$$

The primary benefits of this approach are

- It allows application of many software reliability models describing the failure rate $\lambda(t)$ analytically.
- It may be extended to situations where the failure rate is not continuous or monotonically decreasing.
- It may be extended by including additional cost elements.
- It may be extended to include present value correction of the cost elements involved.

Decision Based on Return on R&D Investment.   The overall objective of this approach is to protect ROI on the R&D cost invested in testing and fixing the software under development. Inadequate ROI means that the R&D test team will generate more benefits to the company by developing a completely new product than by continuing to test and debug the existing one. The following example illustrates the methodology.

- Number of expected customers for the product under consideration: 20,000
- Assumption: In the case of a serious defect

    25 percent of customers (5000) will demand an immediate update

    50 percent will wait for the next scheduled release

    25 percent will not ask for an update at all

- Estimated cost of an immediate update:

1 floppy disk	$ 1.00
Duplication	$ 5.00
Handling/postage	$ 5.00
Labor overhead 500 percent	$25.00
Total unit cost per update	$36.00
Total cost of update:	$180,000

- Considering 10:1 ROI, the required R&D cost savings is $18,000.
- At $100 per software R&D hour, it is necessary to save 180 R&D hours.
- Considering 33 percent effectiveness of testing and debugging, it is necessary to save quality assurance (QA) logged time of 60 hours.
- To ensure that there is no impact on the product's final reliability, it is prudent to discontinue testing only after MTBF>60 hours has been demonstrated.

This conclusion may be tested by heuristics based on past observation, such as:

- Past data indicate that a concentrated debugging effort by quality engineers places 50 to 100 times higher stress on the product than a typical customer application environment. This means that internally observed MTBF = 60 hours is an equivalent to 3000 to 6000 hours MTBF in the field, a performance comparable to system hardware reliability and therefore acceptable.
- MTBF = 60 hours must be compatible with the failure rate model describing the actually observed reliability growth.

4.4.3 Economic Aspects of Product Safety

The close relationship of reliability to issues of product safety is evident from the simple definition of a risk:

$$\text{Risk} = \text{probability of a failure} \times \text{exposure} \times \text{consequences}$$

Results obtained by fault-tree analysis techniques can be interpreted in terms of negative utility, event probabilities, and severity (criticality), which are standard subjects of cost-benefit studies. The cost factor, which is strongly dependent on the criticality level, must take into account costs of design and manufacturing for safety and losses caused by complaints, claims, suits, legal costs, unfavorable publicity, government intervention, etc. The best investment usually results from the lowest cost-benefit ratio, but implementation alternatives must be selected more carefully because cost may significantly increase at both probability extremes:

1. A low probability requirement could result in costly overdesign.
2. Accepting a high probability of occurrence can be perceived as negligence.

A cost-benefit analysis can be supplemented by a comparison of the cost of different alternatives for achieving a defined acceptable level of safety. An analytical approach to decision making about safety issues usually follows this simple sequence of steps:

- Identify potential hazards and resulting probable accidents associated with the current product design.
- Obtain credible data on accident rates for the product over time, e.g., from the National Emergency Injury Surveillance system or insurance companies.
- Propose a set of alternatives for providing additional increments of safety.
- Obtain cost data or cost assessments for these alternatives.
- Analyze the effectiveness and cost of the alternatives.

During the analysis, it is necessary to follow some practical principles, such as

- Analysis should be a support for judgment, not a substitute for it.
- Analysis must be open and explicit to be useful as a framework for constructive criticism and as a stimulant for improvement suggestions.
- There is seldom only one single best solution.
- Conclusions from the analysis should be simple.
- It is necessary to be realistic about improvement implementation prospects.

Consideration of economic aspects of product safety must include the imposed cost associated with government and industry regulations and considerations related to human reliability.[11]

4.4.4 Economics of Incoming Inspection

Testing objectives may differ from manufacturer to manufacturer (e.g., improved process control, minimized number of field failures, etc.), but they are always combined with the basic tendency to minimize production cost. Four basic incoming inspection (I.I.) strategies are available to implement those objectives:

1. *100 percent test-stress-retest.* This comprehensive I.I. technique is the most effective strategy, forcing infant failures to appear and failed parts to be removed before they are used in the assembly process. The correlation of test results prior to and after burn-in stress often provides important information for the vendor's process control and quality improvement. Implementation of this strategy requires investment in both testing and burn-in equipment, indicating that it is most suitable for automation at high product volumes.

2. *100 percent test only.* This approach assures the removal of defective parts delivered in particular batches, and therefore lower assembly rework cost, improved test effectiveness, and lower total cost of quality. Functional test is only partially effective as a reliability screen.

3. *Buying preburned and tested parts.* This variant is a very effective strategy that may bring benefits of economies of scale because of the expected difference in parts volume between the vendor's and the user's facilities. Sample testing for vendor's performance audit purposes and risk control will increase the total cost without any additional impact in improved quality or reliability.

4. *No I.I.* This is the best strategy assuming good vendor's process control and mutual trust in vendor-user relationships. In less favorable conditions, the cost of no I.I. will be recognized in process disruptions on subassembly or system levels and, potentially, also in the field.

To determine the most suitable alternative, numerous details of all strategies need to be examined and a composite picture formed. I.I. is often difficult to justify on purely economic terms, especially for low volumes. For high volumes, the cost-benefit ratio is favorable in most industrial situations. Also, let us not underestimate the importance of I.I.-generated information feedback to vendors.

A quick estimate of the breakeven component volume can be obtained by this simple reasoning: The breakeven point is defined by equality of the cost of not testing and the cost of testing per given period of time, usually one year. Cost of not testing is given by

$$C_{\text{NT}} = NR(P_I C_R^I + P_W C_R^W)$$

and the cost of testing is given by

$$C_{\text{T}} = P + \frac{NL}{n}$$

where N = number of component units under consideration
R = total fraction defective previously observed or estimated
P_I = fraction defective which fails in-plant
C_R^I = average cost of in-plant repair
P_W = fraction defective which fails during the warranty period
C_R^W = average cost of warranty repair
P = cost of the test equipment
n = number of units tested per hour
L = labor and overhead rate per test hour

So, the breakeven volume can be estimated from

$$C_{NT} = C_T$$

$$NR(P_I C_R^I + P_W C_R^W) = P + \frac{N}{n} L$$

$$N = \frac{P}{R(P_I C_R^I + P_W C_R^W) - L/n}$$

In all I.I. strategies the cost of discovery and removal of defectives is absorbed in the cost of acceptable units. Because the actual fraction defective is unknown before testing, many creative sampling and cost minimization schemes, with illustrative examples, have been proposed and described in studies available in the public domain, such as Ref. 12.

4.4.5 Economics of the Reliability Engineering Function

Funding the reliability engineering function presents a special challenge because it is defined as a service organization. Because there is currently no coherent theory of management that encompasses service organizations, all approaches to funding are pragmatic and empirical, and based on the assumption that all services within an enterprise are being paid for out of the surplus generated by the business's fundamental economic activities. The reliability engineering function, being a service organization, does not produce economic results directly or perform economically by itself. The central economic and managerial challenge is to assure the contribution of reliability engineering to the value delivered to customers and to the effectiveness of company internal competencies. This leads to the assumption that accumulating reliability engineering expertise, employing scientists, purchasing state-of-the-art test and analytical equipment, or producing literature cannot be considered as economic contributions.

Current economic justifications for funding of the reliability function fall into two categories:

1. *Deserved funding,* which is an allocation of funds based on providing normative services that the department's customers are required to subscribe to and consume.

2. *Earned funding,* where the reliability engineering function is being paid for its results and performance in an environment allowing dissatisfied or disinterested customers not to renew existing contracts.

Even though earned funding is generally preferable, the traditional allocation of funds is not necessarily bad or undesirable, because it reflects company management's recognition of long-term strategic needs related to staying in business. In both funding approaches, the key managerial issues are identical:

- How are contribution, performance, results, effectiveness, productivity, etc., to be measured?
- What makes the reliability engineering department capable of outstanding performance?
- How are the built-in psychological and organizational barriers to change in a knowledge-intensive professional service institution to be offset?

Progress toward resolution of these issues is a strong function of managerial ability to establish clear, specific, and highly focused objectives and targets, against which the actual performance can be measured.

4.4.6 Other Topics

Detailed optimization of the economic aspects of reliability over the whole product life cycle, and across the whole business organization as well, requires reliability management to engage many additional topics demanding serious economic consideration, such as

- Selection of effective reliability assurance technologies
- Evaluation of manufacturing reliability screening and assurance alternatives
- Identification of optimum maintenance strategies
- Calculation and management of the ROI on specialty instrumentation

All these topics may be addressed by applying tools and methods of economic evaluation.

4.5 REVIEW OF ECONOMIC TOOLS

The tools of economic analysis and related optimization methods are so extensive that they are the subject of many textbooks, monographs, and handbooks. For the purposes of this chapter, a simple overview is sufficient.

4.5.1 Methods of Economic Evaluation

In a majority of manufacturing industries, money will not be allocated to or spent on a project unless a good argument is presented to the management that this particular expenditure will assist the company in reaching its financial goals. To apply this criterion rationally, technical and economic evaluations of proposed projects are performed. Sometimes these evaluations are intuitive, based on experience; in other situations, formal detailed analyses are required.

The general concepts forming the framework of sound decisions are quite simple. In the case of economic decisions, they evolve around profit, growth rate, return on investment, cash flow, cost-benefit ratio, time value of money, etc.

The fundamental step in any economic analysis is the selection of decision criteria or figures of merit; these may differ significantly between the private sector (motivated by profit, increasing market share, rate of revenue growth, etc.) and the public sector (driven by perceived social benefits or desire for power). Situations driven strictly by economic benefits are much easier subjects for formal analyses, and in such cases the desired figure of merit is usually defined in monetary terms. But even in obvious situations of highly favorable cost-benefit ratios, it is necessary to assess possible changes in the assumed conditions to assure the stability of the expected benefits. It is necessary to understand the risks associated with inflation, changes in interest rates, business cycles, errors in cost estimates, effects of obsolescence, depreciation, taxation, social costs, and other uncertainties.

The basic methods of economic analysis most frequently used are:

- *Uniform cash flow method.* This requires conversion of all cash flows into a time-adjusted, equivalent, uniform annual amount.
- *Breakeven analysis.* This method is suitable for problem solving with incomplete data sets. Sensitivity analysis is required in order to minimize the consequences of errors in estimates.
- *Present worth method.* This is based on conversion of all cash flows to an equivalent amount discounted to time zero by application of appropriate interest rate formulas.
- *Rate of return method.* This method relates operating income to operating assets.
- *Benefit-cost method.* In this method, which is frequently used by federal and state government agencies to estimate the economic attractiveness of an investment, the ratio of probable annual benefits to the equivalent uniform cash flow is computed.

- *Cost-effectiveness analysis.* This allows comparison of alternatives on other than solely monetary measures, but very careful definition of effectiveness, preferably via quantitative terms such as performance, reliability, and relation to company objectives, is required. This presents many difficulties.

- *Replacement studies.* These studies are concerned with identification of the most economical time for renewal of existing assets; they utilize the concept of already incurred (sunk) cost.

In actual implementation of economic analyses and evaluations, frequent difficulties are encountered in creating consensus on figures of merit, decision criteria and their importance, cost estimation credibility, and the power of a wide variety of optimization methods. The major shortcoming of all methods of engineering and managerial economics is their complete disregard of the non-cash flow elements involved. This weakness must be compensated for by managerial experience and judgment.

4.5.2 Cost Estimation Methods

All cost estimating is done by means of analogies, and always reflects the future cost of a new system by relating it to some known past experience. Historical cost data incorporate experience with setbacks, design requirements changes, and other circumstances that are difficult to identify and control, in contrast to industrial engineering methods, which tend to be optimistic and do not allow for unforeseen problems. The role of analogy, and the methods of reasoning behind it, is crucial. The art of cost estimation is based on a seven-step process, described in detail, for example, by Barry W. Boehm.[13]

1. Establish objectives for the cost estimating activity, and reevaluate and modify them as the process progresses.
2. Assure adequate resources for this miniproject and prepare a simple plan.
3. Spell out reliability requirements and document all assumptions.
4. Explore as much detail as feasible, to assure a good understanding of the technical aspects of all parts of the product under consideration.
5. For actual estimation, use several independent methods (e.g., algorithmic models, expert judgment, Delphi technique, analogy, top-down and bottom-up approaches) and data sources.
6. Compare and iterate estimates to eliminate biases.
7. Follow up obtained estimates with regular comparison with actual cost data collected during the project implementation.

The most common form of estimating algorithms are:

- *Analytical models.*

$$\text{Cost} = f(X_1,\ldots,X_n)$$

where f is some mathematical function relating cost to cost variables $X_1,\ldots,X_n$ correlated with some physical or performance characteristics of the system.

For example, cost of spare parts procurement support is

$$C_{ps} = 0.037P_{rc}$$

where P_{rc} is parts repair cost.

- *Tabular (matrix form) models.* These provide relationships that are easy to understand, implement, and modify and that are difficult to express by explicit analytical formulas—for example, the relative cost of software qualification testing C as a function of required reliability.

C	Required reliability
$0.55C_n$	Very low
$0.75C_n$	Low
$1.00C_n$	Nominal
$1.25C_n$	High
$1.75C_n$	Very high

where C_n is qualification cost on a nominal reliability software product.

Note: C_n itself is usually a function of product size and complexity, design team skills and experience, programming language, etc.

The strength of algorithmic models is in their objectivity, repeatability, and computational efficiency in support of families of estimates or sensitivity analysis. They do not handle exceptional conditions, and they do not compensate for erroneous values for cost variables and model coefficients. Actual experience with cost estimation leads to a conclusion that no single method is substantially superior in all aspects, and that the strengths and weaknesses of many methods are complementary.

4.5.3 Calculation of Present Value

This method compares the present value of the cash flows expected from an investment to the initial cash outflow associated directly with that investment. In other words, the present value of X dollars spent at time t is the amount X_0 dollars spent at the start of the savings period.

$$\text{NPV} = \text{PV} - \text{CI}$$

$$\text{PV} = \frac{R_1}{(1 + i)} + \frac{R_2}{(1 + i)^2} + \cdots + \frac{R_n}{(1 + i)^n}$$

where NPV = net present value
   PV = present value
   CI = cash outflow resulting from the cost of the initial investment
   R_n = expected cash flow for the nth period
   i = required rate of return

If NPV is positive, the investment may be accepted.
   Advantages of this methodology are

- It reflects the time value of money.
- It allows calculations both when cash flows are an annuity and when they change from period to period.

The disadvantage of this method is that it requires a long-term forecast of incremental cash flows.

4.5.4   Cost Accounting

To assure a reasonable rate of diffusion of methods of economic analysis to support rational decision making about reliability, a feedback system of actual cost data is necessary for evaluation of the accuracy of prediction and cost estimates and the effectiveness of the analytical methods used. The first time an attempt is made to identify and measure the cost of reliability, it is highly probable that the cost data requirements will not match the established cost accounting system. In this situation, the decision to start with a study of a single project usually prevents deadlock and allows both the development of information for the reliability improvement program and identification of cost categories of key importance. Data collection will be manual, the necessary forms will be designed separately for this particular study, and data compression and interpretation will be done, most probably, on the reliability engineers' personal computers.

Only after a demonstrated success of reliability cost management in a single study or project environment can steps be taken to establish a reliability cost accounting system, with the objective of providing a continuing scoreboard. This continuing scoreboard should be based on a formal reliability cost accounting and reporting process which may parallel the general financial accounting system. This systematic approach requires at least

- A list of projects, products, and programs of interest
- A list of the functions or departments involved
- A list of accounts in which relevant charges are accumulated
- Cost categories (see Table 4.2 for an example suitable for an LCC model)
- Definitions of data entry requirements and formats
- Definitions of data process flow with control points
- Formats and frequency of reports and summaries
- Rules of data and results interpretation
- An established base for results comparison and evaluation
- A methodology for creation and improvement of cost standards

Collected data must be sorted and compressed to accommodate evaluation from many different viewpoints, such as

- Product, subassembly, part, component, material, etc.
- Organizational responsibility
- Place and time of occurrence or reporting
- Project or program association

In formatting the data and results, the general preference for graphical representations in the form of tables, Pareto-type charts, histograms, pie charts, trend lines, scattergrams, control charts, etc., is well established. Narratives by specialists or representatives from responsible teams can help in interpretation and assessment of the seriousness of data, especially when reports result in transfer of charges between departments and accounts.

4.5.5   Activity-Based Cost Management

Activity-based costing (ABC) and cost management reveals the links between company activities and consumption of company resources. Its emphasis on the distinction between value-adding and non-value-adding activities allows easier identification of cost reduction opportunities. The fundamental conceptual contribution of ABC is its recognition that cost elements are not intrinsically fixed or variable.

TABLE 4.2 Examples of Reliability Cost Categories

Prevention costs
   Hourly cost and overhead rates for design engineers, reliability engineers, materials engineers, technicians, test and evaluation personnel
   Hourly cost and overhead rates for reliability screens
   Cost of preventive maintenance program
   Cost of annual reliability training per capita

Appraisal costs
   Hourly cost and overhead rates for reliability evaluation, reliability qualification, reliability demonstration, environmental testing, life testing
   Average cost per part of assembly testing, screening, inspection, auditing, calibration
   Vendor assurance cost for new component qualification, new vendor qualification, vendor audit
   Cost of test results reports

Internal failure costs
   Hourly cost and overhead rates for troubleshooting and repair, retesting, failure analysis
   Replacement parts costs
   Spare parts inventory costs
   Cost of production changes administration
   Cost of reduced productivity

External failure costs
   Cost of repair and failure
   Service engineering hourly rate and overhead
   Replacement parts costs
   Cost of spare parts inventory
   Cost of failure analysis
   Warranty administration and reporting costs
   Incremental costs of sales
   Incremental installation costs
   Cost of liability insurance
   Incremental legal costs

Reliability-related activities may also be grouped into value-adding (e.g., design for reliability, reliability improvement programs, reliability-related certifications required and paid for by customers) and non-value-adding (e.g., warranty administration, design validation testing, corrective action system administration). The ABC approach, therefore, may supplement or replace the traditional cost of reliability system, provide additional insights and information that are valuable for reliability planning, and enable a search for lower-cost strategic alternatives in reliability management.

Examples of new insights provided by ABC are

- When data indicate that 1M DRAMs have higher failure rates than 256K DRAMs, there is no reason to discriminate against a vendor who supplies both and thus displays a higher overall failure rate than a 256K-only supplier.

- When some repair centers report higher no-defect-found returns on field-replaceable units, it is prudent to initiate analysis and corrective action.

- When some customers display higher warranty failure rates than others, this information should be considered during contract pricing and should stimulate investigation.

- Unless specific corrective actions are taking place, repetition of observed problems should be assumed and related failure costs included in forecasts and plans.

ABC-based cost management is forcing the responsible managers and teams to identify the areas causing the most significant problems, to explain which product or process characteristics drive the associated losses, and to suggest improvement actions. In general, ABC

- Supports the achievement of cost and reliability leadership.
- Contributes to providing superior value to customers.
- Compensates for the weaknesses of existing traditional cost accounting systems.

The greatest challenges in designing an ABC system for reliability cost management are

- Identification of reliability cost drivers and their impact on company behavior
- Separation of costs into product-driven and customer-driven categories
- Identification of individual activities and their correlation with resource consumption
- Understanding the customer-perceived value of reliability

4.5.6 Summary

Reliability engineering and management can greatly benefit from prudent application of tools from engineering economics and operations research, especially when facing decisions about costly investments and high-risk or complex situations. The power of those tools, with their rigorous methods and computational accuracy, must be understood in the context of their dependency on the quality and relevance of the underlying assumptions, historical data, cost estimates, and known unequal treatment of results from physical and human sciences. The growing computerization of the described methods, and the availability of computers for daily work, should allow management to concentrate more on the human and strategic aspects of decision making, while being assured of the rigor and accuracy of their dealing with the technical aspects.

4.6 REFERENCES AND SUGGESTED READING

4.6.1 References

1. Henry J. Kohoutek, "Methods for Valuing Quality," *Proceedings of the 30th Annual EOQC Conference,* 1986, pp. 125–136.
2. Robert Buzzell and Bradley Gale, *The PIMS Principles,* Freepress, New York, 1987.
3. Henry J. Kohoutek, "Establishing Reliability Goals for New Technology Products," *Proceedings, Reliability and Maintainability Symposium,* IEEE, 1982, pp. 460–465.
4. Henry J. Kohoutek, "Development of a Reliability Strategy for New IC Component Family and Process," *Microelectronics and Reliability,* vol. 23, no. 2, 1983, pp. 383–389.
5. M. Earls, *Factors, Formulas, and Structures for Life Cycle Costing* (from author's published notes), 89 Lee Drive, Concord, Mass. 01742.
6. D. N. Isaacson, *Life Cycle Cost Analysis,* SAE-RMS Guidebook (available from author), Warrendale, Pa., 1990.
7. E. Carrubba and M. Johnson, "Use of Life Cycle Cost in the Commercial Arena," *Proceedings, IASTED Symposium on Advances in Reliability and Quality Control,* 1988, pp. 72–76.
8. Richard L. LaFrance and Stephen B. Westrate, "Cost of Ownership: The Suppliers View," *Solid State Technology,* July 1993, pp. 33–37.
9. Michiharu Sakurai, "Target Costing and How To Use It," *Journal of Cost Management,* Summer 1989, pp. 39–50.
10. James R. Brennan and Jerrell T. Stracener, "Designing to Cost-Effectiveness: Enhancing Quality, " *Proceedings, Annual Reliability and Quality Symposium,* IEEE, 1992, pp. 44–52.
11. E. M. Dougherty, Jr. and J. R. Fragola, *Human Reliability Analysis,* John Wiley & Sons, New York, 1988.
12. Richard Y. Moss, "Cutting Component Inspection Cost," *Quality,* October 1983, pp. 35–36.
13. Barry W. Boehm, *Software Engineering Economics,* Prentice-Hall, Englewood Cliffs, N.J., 1981.

4.6.2 Suggestions for Further Reading

1. Many papers dealing with the subject of economics of reliability may be found in the proceedings of the annual reliability and maintainability symposia sponsored by IEEE, ASQC, and other professional organizations.

2. Anderson, Henry R., et al. *Management Accounting.* Houghton Mifflin Co., Boston, 1989.

3. Lee, John Y. *Management Accounting Changes for the 1990s.* McKay Business Systems, Artesia, Calif., 1987.

4. O'Guin, Michael C. *The Complete Guide to Activity-Based Costing.* Prentice-Hall, Englewood Cliffs, N.J., 1991.

5. Londeix, Bernard. *Cost Estimation for Software Development.* Addison-Wesley, Reading, Mass., 1987.

6. Shim, Jae K., et al. *The Vest-Pocket MBA.* Prentice-Hall, Englewood Cliffs, N.J., 1986.

CHAPTER 5
DESIGN FOR RELIABILITY

Richard Y. Moss II
HEWLETT-PACKARD COMPANY
PALO ALTO, CALIFORNIA

5.1 INTRODUCTION

In today's highly competitive markets, *successful* new products must satisfy three requirements:

- *Better*—they must have higher performance and reliability.
- *Faster*—they must get to market more quickly.
- *Cheaper*—they must have lower factory cost and prices.

The problem is that these three requirements seem to be mutually exclusive, particularly "better" and "faster"—it is very difficult to achieve high reliability at low cost with shortened design and life cycles that leave no time for testing and refinements. In the past, a three- to five-year design cycle was common, with a successful product expected to enjoy a market life of five more years. The reliability could be improved and the cost reduced gradually throughout the product's life, as manufacturing processes went through a learning curve and design improvements were made periodically.

Those days are gone forever! Design cycles now range from a few months to perhaps $1\frac{1}{2}$ years, with one year typical of all but the most complex systems. The market life of a new product is also only a year or two, and so production volume must ramp up steeply in the first few months, with no learning curve and few production changes. Preintroduction testing is cut short or eliminated entirely, and new products are rushed to market with defects.

The true costs of a product with defects are seldom reckoned. Despite the television advertisements for automobile maintenance which end with the phrase "Pay me now, or pay me later," schedule still gets put ahead of quality, and new products get introduced ready or not.

The argument justifying this is that the market window for a new product and technology has a fixed end date, so that a delay at introduction represents sales that are lost forever. What isn't mentioned is that disappointed customers are also lost forever, and studies have shown that the cost per order dollar to develop new customers is more than twice that of repeat orders by satisfied customers. Add the direct costs of poor quality—scrap and rework, troubleshooting and retest, warranty claims and customer support—and you have the reason why many new products are never profitable and their producers don't survive to try, try again.

What to do, then? High reliability and low cost traditionally have been achieved by evolutionary changes, not revolutionary ones, but now technology and customer expectations are changing too fast for slow-but-sure methods to be successful. The answer lies in process improvement: Adopt a design process that maximizes the use of the most effective methods and minimizes repetitious and ineffective methods. One such process has been developed; it's called *design for reliability,* or DFR for short.

5.2   THE DFR PROCESS

The development of this process was the result of market pressures and a challenge. In the early 1980s, the management of a large U.S. electronics firm became worried by a slow but steady increase in warranty costs and the erosion of its market share by competitors. The CEO issued the following statement:

> With above-average quality standards already well established…it would be difficult to ask for better results. Yet it was apparent that a major improvement was needed for us to retain a leadership position in the long run. Clearly, a bold approach was needed to convince people that a problem existed and to fully engage the entire organization in solving it. The proper place to start, we concluded, was with a startling goal—one that would get attention. ***The goal we chose was a tenfold reduction in the failure rates of our products in the 1980's.***[1]

About halfway through the decade of the 1980s, it was noted that the actual rate of improvement varied widely from one product line to another, and that the overall rate was not sufficient to achieve the goal. A task force studied the situation, and made several key recommendations:

- Management must have better visibility of progress, and conduct regular reviews.
- Management must develop specific action plans, assign ownership, and set or reaffirm high-priority improvement goals.
- A companywide survey of reliability improvement methods must be made, and a program developed to spread the most effective methods to all divisions.

There were other recommendations as well, but it was this last one that led directly to the discovery of the key design for reliability activities.

The survey methodology was fairly simple. A list of 37 activities that could be used to improve hardware reliability was made (Fig. 5.1). Then the engineering and quality staff at 12 divisions were asked to rate their usage of each activity, using a numerical scale (0 to 100 percent). Data were obtained which showed each division's reliability improvement in the previous three years, based primarily on warranty failure data. Finally, the usage and improvement data for each of the 37 improvement activities were compared and correlated.

Eight of the 37 activities were found to correlate well with reliability improvement; the organizations which used them extensively had consistently better results than those which did not. These eight activities are shown in Table 5.1, arranged in descending order of impact on improvement.

Impact is defined as the slope of the reliability improvement versus usage regression, and should be interpreted here as only a relative measure. The highest-impact activity, thermal design, has approximately four times the slope of the lowest-impact activity, FMEA.

The coefficient of determination (R^2) shows how well the data fit a simple straight-line model. If all the data points fell exactly on a straight line, R^2 would be 1.0; if the data did not correlate at all, R^2 would be 0. An R^2 of 0.5 or more is interesting, and more than 0.9 is practically a "sure thing."

From Table 5.1, it is apparent that there are really seven key DFR disciplines, the eighth being training in the other seven. Design for reliability is not part of the typical engineering curriculum, and so it is not surprising that the amount of DFR training correlated very well (0.95) with improvement.

The survey was then carried a step further. A second questionnaire was prepared, probing usage of the eight key activities in more depth, and every division which designed and manufac-

```
ENTITY: _____   PERSON: _____   DATE: _____

Scoring:   4 = 100%, top priority, always done
           3 = >75%, use normally expected
           2 = 25% - 75%, variable use
           1 = <25%, only occasional use
           0 = not done or discontinued

Management:
____ Goal setting for division
____ Priority of quality & reliability improvement
____ Management attention & follow up (ownership of goals)

Engineering:
____ Documented hardware design cycle
____ Reliability goal setting by product or module
____ Priority of reliability improvement goals
____ Ownership of reliability goals
____ Design For Reliability (DFR) training
____ Preferred technology selection (standardization)
____ Component qualification testing
____ OEM selection & qualification testing
____ Physical failure analysis of testing failures
____ Failure & root cause analysis
____ Statistically-designed engineering experiments
____ Design & stress derating rules
____ Design reviews & design rule checking
____ Failure rate estimation (prediction)
____ Thermal design & measurements
____ Worst case analysis
____ Failure Modes & Effects Analysis (FMEA)
____ Environmental (margin) testing
____ STRIFE (cyclical, multi-stress) testing
____ Design Defect Tracking (DDT)
____ Lessons-learned database

Manufacturing:
____ Design for manufacturability (DFM)
____ Priority of quality & reliability goals
____ Ownership of quality & reliability goals
____ Quality training programs
____ Statistical process control (SPC/SQC)
____ Internal process audits
____ Supplier process audits
____ Incoming inspection (100% or sampling)
____ Component-level burn-in
____ Assembly-level burn-in
____ Product-level burn-in
____ Manufacturing defect tracking
____ Corrective action reports
```

FIGURE 5.1   DFR survey checklist.

TABLE 5.1   The Eight Key DFR Activities

Activity	Impact	R^2
Thermal design and measurement	2.5	0.65
Worst-case analysis	1.9	0.76
Supplier process audits	1.4	0.48
Goals high priority	1.3	0.59
Supplier qualification testing	1.3	0.68
DFR training for designers	0.9	0.95
Component stress derating	0.9	0.98
Failure modes and effects analysis	0.6	0.61

tured hardware was asked to answer it. The DFR usage data from this survey were correlated with three success factors: warranty failure rate, rate of improvement toward the tenfold reduction goal, and operating profit. In each of these graphs, the data were grouped; that is, the warranty failure rates experienced by divisions reporting 20 percent to 40 percent usage were averaged and represented by a single data point at 30 percent, 40 percent to 60 percent usage at 50 percent, and so forth. The warranty rate axis has been normalized by a factor relating to the complexity of the products, both to allow combining of warranty data from vastly dissimilar products and to protect proprietary data.

Figure 5.2 tells a not unexpected story: products from divisions which used the eight key DFR methods 30 percent of the time had about three times higher failure rates than those that used them 90 percent of the time. Figure 5.3 shows that the 30 percent users were also about three times further behind the 10 times in a decade improvement rate goal. So, not only did the nonusers have a much higher hardware warranty failure rate, they were also improving more slowly!

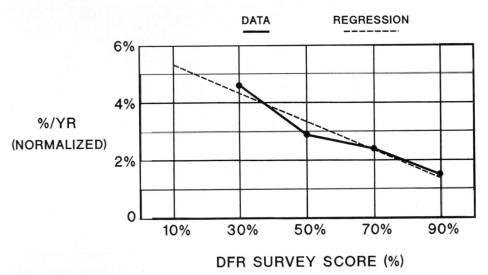

FIGURE 5.2   DFR use vs. warranty failure rate.

Figure 5.4 is the clincher, showing the correlation of pretax profit margin with DFR survey scores. While high DFR usage does not guarantee high profits, it is interesting that divisions with high DFR survey scores are enjoying much higher profits than those with low scores. High profit is the result of doing many things right, and DFR is certainly one of them.

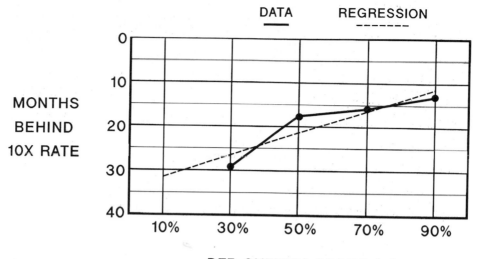

FIGURE 5.3   DFR use vs. rate of improvement.

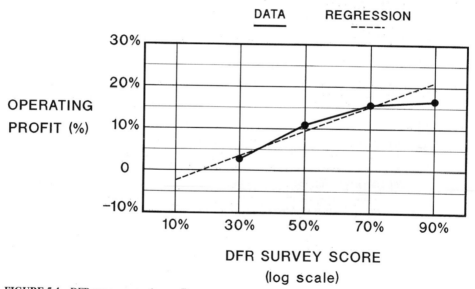

FIGURE 5.4   DFR use vs. operating profit.

At this point, it is appropriate to describe each of the eight key activities in more detail. This will be done by introducing a simple process model, and then describing each of the activities at the appropriate point in the process. New product development proceeds in phases, which have various names depending on the particular organization and technology. For discussion purposes, we will use a very simple structure, and assign all product development activities to just three phases, given in Fig. 5.5.

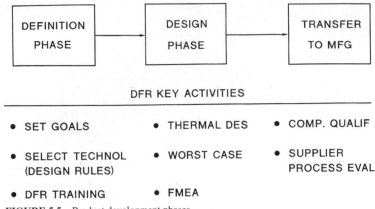

FIGURE 5.5   Product development phases.

5.3   DEFINITION PHASE

In the definition phase, a fairly complete description of the proposed product, both written and pictorial, is created. The market for the product is investigated, the investment necessary to bring it to market is estimated, and schedules and budgets are proposed. Three key DFR activities occur in this phase:

- Setting reliability improvement goals
- Technology selection, setting design and derating rules
- Teaching (and persuading) designers to use DFR

5.3.1   Goal Setting

This first phase is the time to set aggressive reliability improvement goals for the proposed product. To achieve a tenfold improvement in a decade requires a 21 percent improvement every year, or about double every three years. This means that the failure rate goal for a new product that will take $1\frac{1}{2}$ years to develop must be about 30 percent lower than the current product's rate, right at introduction. Coupled with more performance per dollar, that's a challenging task.

Committing to goals is just as important as setting them. It requires real courage and conviction to refuse to allow a new product to be sent to manufacturing when it's behind schedule and badly needed, but doesn't quite meet the reliability goals. This situation usually arises because too little attention is paid to reliability before the design is complete and scheduled to be turned over to manufacturing. If reliability improvement and assessment have been given a high priority and actively pursued from the start, this situation is much less probable.

When setting goals for reliability improvement, management's role should be to set general rates of improvement, like "a factor of two" or "tenfold in a decade." Engineering must break this down into a reliability budget, much as any other resource is allocated. The first step is to analyze a similar, current product: What are the contributions of the different parts of the product, and what fraction of the failures (and failure costs) can be allocated to each module or part? The next step is to determine the root causes of those failures. Then, some informed guesses can be made as to where the biggest opportunities for improvement lie. Some parts of the product can be improved a lot, and others very little. One important technique to use is "K.I.S.S."—"Keep It Simple, Stupid!" *The single most effective way to reduce failure rate is to reduce parts (and interconnection) count.* That's the biggest reliability benefit the electronics industry has enjoyed as discrete components were replaced by integrated circuits and parts counts were dramatically reduced.

Reliability goals can be expressed in several ways: failure rate, mean time between failures (the reciprocal of failure rate), or reliability (probability of operating for a given time in given conditions). All of these involve statistical inference, predicting the behavior of the whole population from data about a sample. Valid reliability specifications should therefore include information about the operating environment and confidence bounds.

One problem with setting reliability goals is that failure rate is not constant. It changes with the age of the product, as well as with its environment, producing the familiar "bathtub" curve in Fig. 5.6.

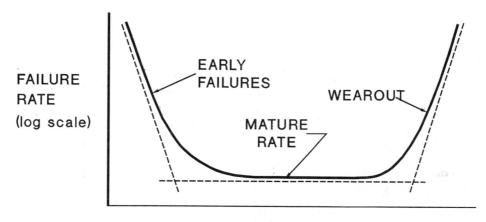

FIGURE 5.6   The reliability bathtub.

The initial period of decreasing failure rate is called early failures; these failures are caused by process damage and hidden defects in the products. If the starting failure rate is more than twice the mature rate, production screening or "burn-in" can help temporarily. (It's too expensive to continue forever!) The long-term cure for a high early failure rate is to improve process control, both within your own company and by suppliers.

The period of approximately constant failure rate is characterized by failures at random times. Despite this temporal randomness, there is no such thing as a random failure—they all have causes, which are a function of the design and the manufacturing process. The specific causes are combinations of application and environmental stress that infrequently occur, the cumulative effect of designed-in overstress, etc. Designing in plenty of margin (derating, worst-casing) and input/output protection circuits will help reduce failures caused by abuse and accidental overstress, and the conscientious use of component stress derating and worst-case analysis will prevent them.

The period of increasing failure rate is wearout, caused by failure mechanisms which create progressive, irreversible change (damage), such as corrosion, mechanical wear, and electromigration of metals. These failure mechanisms limit the service life of a product, the length of time it can economically be kept running. Many of the causes of wearout can be designed out by changing operating stress or material type, or the impact of wearout can be lessened by preventive maintenance. Wearout mechanisms *must* be discovered and counteracted, because otherwise they are time bombs. *Every* product (100 percent) containing a wearout mechanism will fail, whereas only a small fraction will experience an early or mature life failure.

An obvious implication of the bathtub concept is that a reliability specification cannot be one number. It depends on the age of the product and the environmental and application stresses present, and it must have a confidence bound around it to describe a band of uncertainty which results from sampling.

5.3.2 Technology Selection and Design Rules

The second definition phase of DFR activity is selecting technology and setting design rules for it. Every technology has its strengths and weaknesses; for example, aluminum electrolytic capacitors can tolerate high ripple current but have poor reliability at high altitude, at high temperature, or when there is no DC bias voltage applied. So, they are great for low-frequency (50- to 60-Hz) power supplies, but not so great for applications with long periods of nonoperating storage at high altitudes and high ambient temperatures. Selecting the best technology for each part of your product must be based on extensive knowledge of the physics of failure of each alternative, not just their price and performance.

Stress accelerates failures. That's why prototypes of new products are subjected to high stress testing—to increase the failure rate so that hidden failure mechanisms can be found and eliminated quickly. In the design itself, you want to do just the opposite. Instead of increasing it, you should *minimize* the stress on each component in an organized way, so that the failure rate is reduced. If you can't reduce the stress, use a component with a higher rating, so that a lower percentage of the component's strength is demanded.

The most destructive stress is heat, because it accelerates nearly all the failure mechanisms found in electronic components (more about that in the design phase discussion). Other stresses that must be controlled are electrical (voltage and current), chemical (particularly water, which dissolves and ionizes most chemicals, making them charged, more mobile, and more corrosive), and mechanical (shock, vibration, and thermal expansion).

There is another class of design rules that must be created and followed. These rules are dictated by the manufacturing processes, and include layout rules for printed wiring boards, temperature and chemical resistance that the components must have to survive the mounting and soldering process, and so forth. Most defects found in factory test are the result of damage done to components by the manufacturing processes, especially components which were never tested for ability to withstand the process stresses.

5.3.3 DFR Training

The third DFR activity which must occur at or before the start of the definition phase is DFR training. In the ideal case, the design team would attend a regular series of classes or seminars on reliability topics, starting with the eight key topics and then continuing with any others that are deemed important to your particular technologies and business goals. For example, the measurement of reliability involves the use of statistics—not all branches of that science, but certain parts, such as descriptive statistics and statistical inference. If these are taught with practical examples, such as how to plan and analyze an accelerated stress qualification test, they will be very helpful in achieving reliability improvement. The investment of one or two hours every two weeks in training will pay huge dividends in six months or so, in terms of a more knowledgeable and capable design team who can create new products that are far more reliable, in far less time.

There are two possible approaches to DFR training:

1. Use an in-house expert for both training and long-term technical support.
2. Hire an outside consultant to do the training, initially at least, and "grow" an in-house expert for long-term support.

As you can see, you need an in-house person in either case, to keep the flame burning and to answer users' questions after the consultant has left. If you do not invest in this level of support, the technique will not become widely adopted.

There are several advantages to the second approach. The first is the ability to get started more quickly—your future or "designated" expert can be learning at the same time as the users, and the technique can be deployed sooner than if the resident expert must become expert enough to teach the subject before its use can spread. Also, the future resident expert has someone to ask when he or she is unsure. The only disadvantage is the dollar cost of the consultant, but that should be more than offset by the savings from earlier reliability improvement.

Topics which *require real technical expertise,* and therefore are most likely to benefit from the use of the second approach, are

- Thermal design and measurement
- Failure and root cause analysis
- Statistical data analysis

The resident expert will require substantial education and training in both the theory and practice of the techniques involved. Complex software tools are likely to be involved, along with specialized and expensive hardware.

Topics that could benefit from some outside expertise at first, particularly to teach the use of tools and techniques, are

- Worst-case analysis
- FMEA and FTA
- Component qualification
- Design qualification testing

For these topics, one or more people could be sent to a seminar, with some consulting afterwards to gain the advantage of experience. The remaining topics,

- Goal setting
- Technology and design rule setting
- Design reviews
- Retrospective product reviews

are processes that are likely to be unique to each organization's culture and issues, and their theory can be learned by reading. The resident expert is more of a facilitator and instigator than a scientist or technician.

5.4   DESIGN PHASE

During the design phase, the new product is created. The process typically involves creating a circuit or logic diagram on a computer workstation, then using one or more software simulators to compute its expected performance. The next step is layout—of custom or application-specific integrated circuits (ASICs) as well as printed wiring boards—and then more simulation, often three-dimensional, of air flow and temperature rise. There are huge advantages to this software approach, mainly in reducing the time to make a change and evaluate it. Also, design software can include design rule checkers, which will help detect design errors and "enforce" the design rules necessary to manufacturability and reliability.

Three key DFR activities must happen in the design phase:

- Thermal design and measurement
- Worst-case analysis
- Failure modes and effects analysis (FMEA)

5.4.1   Thermal Design and Measurement

As was previously mentioned, high temperature is usually the greatest enemy of reliability. Increasing temperature almost always increases failure rate, and the increase is drastic for some components. Batteries and electrolytic capacitors contain liquids, whose vapor pressure increases with increasing temperature, causing them to leak or explode. Metals such as silver, copper, tin, and aluminum corrode, migrate, or grow dendrites (whiskers) faster when hot, causing open and short circuits. Semiconductor components with even a tiny amount of surface contamination experience increased leakage currents and inversions. Most failure rates increase exponentially with temperature (in °Kelvin), just as chemical reaction rates do. This behavior is called the Arrhenius relationship, in honor of the Swedish physicist and chemist Svante August Arrhenius (1859–1927), who publicized the exponential temperature behavior of chemical reactions.

Thermal design is simply the practice of designing a product so that every component experiences the smallest possible temperature rise and never exceeds a reasonable percentage of its maximum temperature rating. Not only does this design approach increase the reliability of the product, but it also can improve performance. Many components, particularly semiconductor devices, show a deterioration in performance (gain, bandwidth, noise, leakage, etc.) when hot, and work best near room temperature. Because the heat generated by one part can affect a neighboring one, thermal design is not an easy task. Also, heat is transferred by three mechanisms: conduction, convection, and radiation. They are not linear, and they are affected by ambient temperature, altitude, relative humidity, and air velocity, to name only a few factors. Fortunately, quite sophisticated thermal analysis and simulation software programs which can perform three-dimensional modeling rapidly and accurately are available. Their existence permits the achievement of high power densities while preserving reliability.

Thermal measurement can be done in several ways:

- Direct contact, by thermistors, thermocouples, or chemicals which change appearance with temperature
- Indirect means, by measuring the infrared emission

Direct contact measurements have the advantage of allowing the product to be oriented and operated exactly as it would be in use, while the infrared method usually requires that the product be uncovered, with the components to be measured visible, which may affect their temperature. The contact method may also affect the temperature of a very small device because the transducer acts as a heat sink and lowers the component temperature. The indirect method has the additional advantage that a scan or picture of a complete assembly can be made easily, while the contact method requires moving the transducers from component to component, or else skipping some that "couldn't possibly be hot." Whatever method is used, measurements are important to calibrate the simulations, to be sure the computed temperatures are accurate.

5.4.2   Worst-Case Analysis

Worst-case analysis obviously interacts with thermal design, since maximum input or output will generally result in maximum temperature rise. Other elements of worst-case analysis involve maximum variations in power supply voltage, timing or signal frequency, and component parameter tolerances. Trying to do this manually is time-consuming and tedious. Circuit simulators can do these computations quickly and thoroughly, and should be used regularly throughout the design phase. (Don't forget to repeat this if any design changes are made later, in production.) A design which is robust, which can tolerate a wide range of variations and still operate correctly, is a prerequisite for high reliability.

5.4.3   FMEA and FTA

FMEA (failure modes and effects analysis) is a systematic process to anticipate and prioritize failure modes and causes associated with the design and manufacturing processes of a product. The procedure is simple in concept: A list of possible failure modes of each component or subassembly is made, and then each is given a numeric rating for expected frequency of occurrence, criticality, and probability of causing a problem detectable by the user. Finally, these three numbers are multiplied together to obtain a risk priority number (RPN), which is used to steer design improvement effort to the most critical problems first.

Two aspects of FMEA are particularly important: a team approach and timeliness. A team approach helps because the broader the expertise that is brought to bear on the analysis, the more effective FMEA will be. Timeliness is important because FMEA is primarily a preventive tool that can help steer design decisions between alternatives before serious failure modes are designed in, avoiding delay and redesign after problems arise. FMEA is equally applicable to hardware and software, components and systems.

Another similar process is fault-tree analysis (FTA). While FMEA is a "bottom-up" approach, FTA is "top down." FTA starts at the highest (system) level with the assumption of a failure mode (such as "no output"), and then works down through the functional block diagram looking for possible causes of that kind of failure. Thus, FTA is most effective after the system is pretty well defined. FTA and FMEA are complementary; whenever possible, both should be used. For practical reasons, FTA should be limited to the really serious system-level failure modes, especially those involving safety or system damage. FTA can be used at the component, subassembly, and module levels, and is equally applicable to hardware and software.

5.5   RELEASE TO MANUFACTURING PHASE

As a new product design nears completion, the need for the final two key DFR activities accelerates:

- Component qualification
- Supplier process evaluation and monitoring

5.5.1   Component Qualification

Today's shorter design cycles and emphasis on high return on assets have resulted in more OEM (original equipment manufacturer) products and subassemblies being used, as well as complex and customized components. Their appeal is that the supplier has already made the initial investment in technology and manufacturing equipment for the device or component you need, which will save you time and money. The problem is that 70 percent of all hardware warranty claims result in components being replaced, and so the supplier's reliability becomes your reliability.

The first step in qualifying a component generally involves testing samples in your application or a test fixture. Once the performance has been measured and the leading candidate(s) identified, it is time to evaluate reliability. Obtain life and stress test data from each supplier, and calculate the impact on your design. Purchase an appropriate sample quantity, preferably from the normal distribution chain rather than carefully selected (and pretested) ones directly from the supplier, and subject them to your own accelerated stress tests. Where appropriate (ICs, for example), perform DPA (destructive physical analysis) on a few, looking for evidence of good (or bad) design practices and process control. Demand analysis by the supplier of any failures which occur during your testing, and consider doing your own parallel analysis of any repeat failures.

5.5.2   Supplier Process Evaluation and Monitoring

All of the above are arm's-length evaluation steps and, while important, are not as powerful as "arm-in-arm." The supplier's product is a result of its processes, from design to shipment. Determining which supplier has the best processes, or at least the willingness to continually improve its processes, is much better than exhaustively inspecting and testing the product, hoping to detect defects or process changes before they are installed in your product. Once you have selected your best supplier candidate, the surest way to achieve high reliability is to work *with* that supplier, starting with the design process, and then continuing through manufacturing, test, and shipping. You may discover that a supplier refuses to allow you access to its "secrets," either by overtly refusing or by hiding key parts. In such cases, examine what you can, looking for clues which suggest poor or excellent processes. Usually a supplier will answer general questions about its process, so long as it is not forced to reveal proprietary or trade secret information. Look for evidence of DFR and good manufacturing process control. For example, it is fairly easy to see at a glance whether ESD (electrostatic discharge) damage control is being practiced in a manufacturing area; lack of it implies a lack of the discipline necessary for good process quality control in general.

Once a supplier has been selected, it is a good idea to set up some sort of continuing monitor of its processes. Depending on the situation, you may want to request notification of any process changes and data on AOQL (average outgoing quality level) and ongoing life tests. You may need to repeat the most sensitive qualification tests periodically. Most important is to actually visit the supplier's facility regularly, to see what changes you can spot, and to renew your relationship face to face. Finally, you should track the factory and field failure rates that you experience with the supplier's product, and feed this information back to the supplier regularly. *Build relationships, but trust data.*

5.6   OTHER DFR ACTIVITIES

Several other reliability improvement activities are worth discussion, even though they weren't validated by the survey. Most of these didn't correlate because everyone was doing them 90 percent of the time, and so there were no 30 percent, 50 percent, or 70 percent users to survey; one was not on the list of 37 improvement activities. They are

- Design qualification testing
- Failure and root cause analysis
- Statistical data analysis
- Design reviews
- Retrospective product reviews

5.6.1   Design Qualification Testing

Just as a supplier's product should be tested, so should yours. The first kind of test is a margin test, where the product is subjected to maximum and minimum levels of input, output, load, and environmental conditions such as temperature, humidity, vibration, and shock. The test conditions are purposely 5 to 10 percent beyond the product's ratings, to ensure that there is some margin in the design. While these tests are traditionally conducted one variable at a time, it makes sense to vary two or more stresses at once, such as temperature and voltage, to increase the likelihood of finding the "corner case" where the product doesn't perform properly. Figures 5.7 and 5.8 show a simple model for which varying two stresses at once doubles the coverage of the test, compared with varying one stress at a time.

The object of a margin test is to determine if a product will function at 5 or 10 percent beyond its ratings. These tests don't last very long (minutes or hours), and generally involve only a small sample size. It is not unreasonable for such a test to be passed with no failures. Accelerated life testing, on the other hand, is *expected* to produce failures, for two purposes:

- To discover failure mechanisms that can be eliminated or minimized
- To allow statistical analysis and reliability prediction

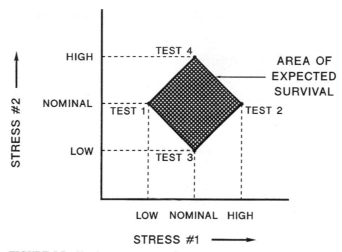

FIGURE 5.7   Varying one stress at a time.

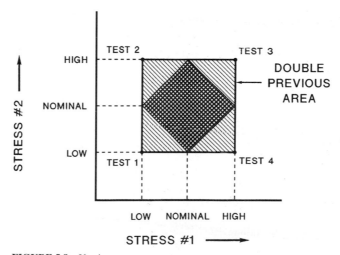

FIGURE 5.8   Varying two stresses at once.

A life test isn't successful until several failures have occurred. To make sure this happens, stress is increased in steps at predetermined time intervals until failures occur. These tests, called STRIFE (stress life) or HALT (highly accelerated life tests), have become so stressful that the test equipment is at risk for failure, not just the product. Designers usually object to these techniques once their product has begun failing, complaining that they are unrealistic and unreasonable. No real user, they reason, would subject the product to such stress extremes.

Those arguments miss the point. The object of these tests is to find the failure mechanisms inherent in the design so that they can be minimized or eliminated, not to prove that the product has adequate margin. To find a failure mechanism, you must first propagate failures. That's the real value of these tests. Generally, the first time this technique is used, even on existing products, many failures will occur at relatively low stress. After several generations of products have been subjected to this, it will become increasingly difficult to generate failures without extreme stress-es. Multiple, cyclical stresses are more effective than single, constant ones, particularly temperature cycles where the air temperature surrounding the product is changing at more than 10°C per minute.

5.6.2   Failure and Root Cause Analysis

The key to quality and reliability improvement is simple: Find and eliminate (or minimize) the root causes of failures. This is a two-step process. The first step is to force or obtain failures, then perform failure analysis on them to determine the cause. This process is a kind of "autopsy," in which a "dead" product is analyzed to find the failed part(s), and then those parts are carefully dissected to learn what caused them to fail. This requires the services of a well-equipped and expertly staffed failure laboratory, with chemical and mechanical sectioning capabilities, optical and electron microscopes, and other specialized test equipment. As a practical matter, not every failed part can be analyzed, nor can a proven cause be wrested from every analysis. To be really certain of cause, you should be able to prove the analysis by causing similar failures in new parts, particularly where electrical overstress is suspected.

Physical failure analysis is particularly important when dealing with suppliers. It is an article of faith in the world of the semiconductor manufacturer that most failures are caused by overstress applied after the part was shipped, while the electronic equipment manufacturer is equally certain that most component failures involve parts that were weak or defective when received from the supplier. This finger-pointing can usually be resolved by skilled failure analysis, and the root cause positively identified. More importantly, a reasonable plan to eliminate that cause in the future can be formulated, which will benefit all parties concerned.

The second part of the process is to be sure you really have the *root* cause. The way to do this is to ask "Why?" five times, or however many times it takes to get to the root of the situation. Your product failed. Why? A part failed. Why? The part suffered an internal open circuit. Why? A solder joint failed. Why? The solder fatigued. Why? The joint experienced repetitive shear strain each time the part operated. Why? There is no clamp or strain relief on the wire. Why? And so on, until no further information can be obtained and an effective solution can be proposed and tested. You can see that physical failure analysis is involved all the way to the determination of fatigue failure; after that, it requires design participation to answer the whys and propose a solution. More important still, the lesson learned needs to be added to the organization's memory, perhaps in the form of a new item on a design review checklist.

5.6.3   Statistical Data Analysis

Analyzing the failure data is just as important as analyzing the failures themselves. The most obvious reason is to prioritize effort, to identify the major or frequent failures whose elimination will have the greatest effect on reliability. One of the most powerful data analysis techniques is graphical, plotting the failures versus time on specialized graph paper, on which the result will be a straight line if the data fit the mathematical model on which the graph paper is based. A particularly simple and versatile model is the Weibull distribution, which can model data from all three portions of the reliability "bathtub" with equal ease. As well as having a simple mathematical equation, the Weibull distribution data plot as a straight line on ordinary log-log graph paper, which can be created by most graphics software. Specialized software is available which performs most of the data preparation, including estimating the coefficients of the equation and the confidence bounds.

A good rule of thumb is that five data points are needed to have reasonable assurance that the fitted line is valid; that means there must be at least five failures of the same type. Again, physical failure analysis is needed, to determine which failures came from which mechanism. If failures caused by different mechanisms are plotted together, the data are unlikely to fit any sort of simple model, and you will have a very hard time making sense of them.

Once you have a good plot, it has all sorts of uses. The quotation "A picture is worth a thousand words" is no exaggeration, especially when you are trying to persuade others to accept a particular analysis or plan of action. Plotting data on two different products or suppliers on the same graph can demonstrate a difference much more vividly than any words. One danger is the very human tendency to extend the line beyond the last data point. This is useful, so long as the obvious danger is not overlooked. Again, remember the bathtub. If you are in the early failure region, you really can't predict the flat portion; if you are in the flat region, you can't predict wearout.

5.6.4 Design Reviews

Since there are two general kinds of design rules, those necessary to assure manufacturability of the design and those aimed at high reliability, there are also two aspects to a review. The manufacturing design review can be partially accomplished by using computer-aided design (CAD) tools with built-in design rule checking, which assure that layouts have the correct minimum spacing, etc. Manufacturing process engineers can perform the rest, using their own experience and checklists to look for potential assembly difficulties, and verifying that all components chosen have been evaluated for their ability to withstand normal manufacturing process stresses. The other half of the design review should be an engineering peer review, with other designers invited to examine and comment on the design. They should also use checklists evolved from experience—the organizational memory—and ask why many times. With new components or circuits, the designer should be prepared with data to show how he or she has tested for failure mechanisms as well as performance deficiencies. Reliability engineers are part of this process, just as they have been part of the design team from the start. They are consultants to the designers, helping to anticipate problems, design tests, and analyze data. However, the reliability of the new product is design's responsibility, not reliability engineering's.

5.6.5 Retrospective Product Reviews

A common failing is for engineering to "throw the design over the wall" to manufacturing, never to see or hear about it again. Often design and manufacturing are physically miles apart, reinforcing the isolation. This lack of communication or feedback leads to several problems:

1. Designers never learn from their mistakes. Every new design incorporates new ideas and improvements, but engineering does not get feedback on how they worked out in practice.
2. Elitist attitudes develop, with each group thinking the other is inferior or doesn't understand the problem. Managers become defensive, and finger pointing is the order of the day.

The solution is to schedule regular reviews of products after they are released to manufacturing. The primary presenters are the people who have continuing responsibility for the product: the manufacturing engineer, the product marketing person and support persons, and the design engineer assigned continuing authority over any changes. As many of the design team as are still available should also participate, including the reliability engineer(s). These meetings should be held after three and six months of production, and every six months thereafter. Data should be presented on number and reasons for production changes, manufacturing yields and test times compared to goals, and standard costs. Product marketing and support should present data on customer reaction to the product, support costs, and any significant problems. Warranty data should be presented, although there will probably be little data for the first (three-month) meeting.

There are two purposes to the meeting:

- To determine if the product is succeeding in meeting its goals, and to look for further improvements that are needed and practical.
- To discover lessons learned which can be used to improve future products and avoid making the same mistakes again. This knowledge, often expensively gained, must be documented and communicated widely, so that it becomes part of the organizational memory.

5.7 MEASURING RESULTS

Unfortunately, there isn't any "magic wand" to wave over a new product to exorcise defects and make it reliable. The best plan is to be organized and thorough, to develop and document a process which will eliminate most mistakes, and then follow that process faithfully. Most reliability improvement activities are based on learning the causes of failure and designing them out. The most costly way to do this is to wait until near the end of the design cycle and then test-fix-test until the product is "good enough." In such a paradigm, you will typically run out of time before you have discovered all the defects, and end up fixing the rest in production.

A smarter, less costly approach is to start reliability improvement at the very beginning of product development, and pursue it relentlessly every step of the way. Use what has been learned in past projects to avoid repeating old mistakes, and communicate this lore by means of check-lists, design reviews, and seminars for design teams.

How do you know when it's good enough? This may be unanswerable, so instead, ask, "How do you know it's getting better?" Gather data from every possible source, and constantly compare them. Keep pushing on the second derivative, and if the trend is in the right direction, improvement will surely follow.

5.7.1 DFR Checklist

Here are some questions to help you decide if you are doing everything possible to improve your products. Have you:

- Set improvement goals for this product and made them top priority?
- Selected new technologies and a scale of integration that can improve product reliability?
- Developed or chosen manufacturing processes compatible with improved reliability?
- Developed design and derating rules for the chosen technologies and manufacturing processes?
- Provided DFR training for the design team?
- Provided thermal modeling and measurement tools and insisted on their use?
- Provided worst-case analysis tools and insisted on their use, including after manufacturing changes?
- Established an FMEA team for this project?
- Qualified every new part or OEM subassembly by stress testing?
- Evaluated suppliers' processes and set improvements in motion?
- Performed margin and accelerated stress life testing on each generation of prototypes of the new design?
- Analyzed every failure to determine the root cause, and taken appropriate steps to eliminate or minimize it?
- Held design and manufacturing reviews of the new design at every checkpoint (transition from one phase to the next)?

REFERENCE

1. *Wall Street Journal,* "Manager's Journal," July 25, 1983.

CHAPTER 6
FAILURE MODE AND EFFECTS ANALYSIS (FMEA) AND FAULT TREE ANALYSIS (FTA) (SUCCESS TREE ANALYSIS—STA)

J. D. Lazor
FORD MOTOR COMPANY

6.1 INTRODUCTION

6.1.1 Introduction to FMEA and FTA (STA)

The FMEA and FTA processes are problem prevention methodologies that can easily interface with many engineering and reliability methods. They have become two of the most all-encompassing problem prevention tools in a product development process (PDP). They can benefit a program from concept selection to the start of production. By their broad influence in representing potential product problems and planned responses to these problems, they establish an effective risk management environment.

As with many methods and procedures, FMEA and FTA offer great potential benefits to any product program when used wisely. This requires knowledge of both the rationale for their use and their detailed steps. This chapter will focus on FMEA and FTA and their application in reliability engineering.

FTA and FMEA are means of identifying or investigating potential failure modes and related causes. They can be

- Applied in the early concept selection or design phase and then progressively refined and updated as the design evolves
- Helpful in identification of all possible causes, including root causes in some cases, and also helpful in establishing the relationships between causes
- Used as a tool to aid in the improvement of the design of any given product or process

6.1.2 A Continuous Improvement Model

A continuous improvement model, as shown in Fig. 6.1, illustrates the flow of information through the PDP. Specific data from field, development testing, and/or service sources are input to the FMEA and FTA process. This process is described as a closed-loop model.

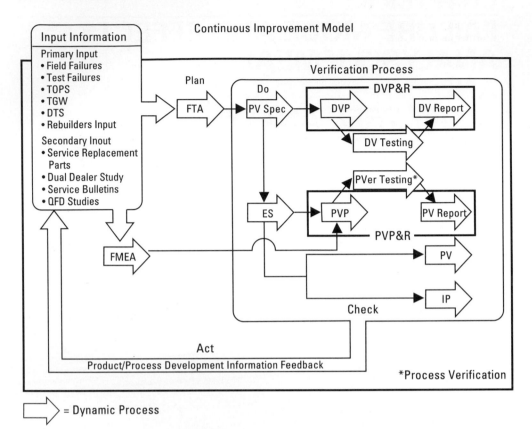

FIGURE 6.1 Continuous improvement model, where TOPS = Team-Oriented Problem-Solving Data Base; TGW = Things Gone Wrong (for the customer); DTS = Durability/High Lifetime Data; DVP & R = Design Verification Plan and Report; PV = Process Verification; PVP & R = Process Verification Plan and Report; DV = Design Verification.

There are two levels of input data in this model: primary and secondary input. Typical examples of primary input include

- Field and test failures
- Problem-solving records and case studies
- Customer surveys
- Rebuilder/remanufacturer input

Typical examples of secondary input include

- Service replacement parts
- Competitive product data
- Service information
- Quality function deployment (QFD) studies

Field data from these various sources are incorporated into the database and establish historical system performance levels, including failure information on various components. This historical information is one source of the information used as baseline data for the FMEA.

As shown in the continuous improvement model flowchart in Fig. 6.1, FMEA and FTA "feed" the product verification specification (PV spec). As the process illustrates, the PV spec feeds into the design verification plan and report (DVP&R) and the engineering specification (ES). This information provides

- A primary data source that will be processed through the FMEA database
- Continuous update and product development information for the analytical PDP

In this continuous closed-loop process, all "lessons learned" are incorporated into a historical database which returns as a primary input; this helps to create a dynamic process capable of assisting all engineers in their future design and reliability engineering activities. Therefore, part of the basic reliability documentation and analysis efforts, from concept through development, are FMEA and FTA.

These must be ongoing, evolving processes in order to be effective. This corresponds to the PDP itself. The approach taken with FMEA and FTA, and the level of sophistication used, will be dependent upon the nature and overall requirements of the individual program.

For this reason, it is necessary to tailor the requirements to each individual program. However, FMEA and FTA must contribute meaningfully to program decisions regardless of the degree of sophistication and tailoring performed. Properly performed, they are invaluable tools to those individuals responsible for program decisions concerning the feasibility and adequacy of a design approach. In short, FMEA and FTA document technical program and performance issues linked to functional intent so that the risk of nonperformance can be better managed.

6.2   DEFINITIONS AND PURPOSES

6.2.1   FMEA

An FMEA can be described as a systemized group of activities intended to

1. Recognize and evaluate the potential failure of a product/process and its effects.
2. Identify actions which could eliminate or reduce the chance of potential failure occurring.
3. Document the process.

It is complementary to the design process of defining positively what a design must do to satisfy a customer.

In the FMEA approach, the identification of potential failure modes of a system, product, or manufacturing/assembly operation is related directly to functions or attributes in either the design or the process.

FMEA is a key tool used to prevent problems from occurring and is a vital part of the up-front engineering disciplines. It also aids in identifying key design or process characteristics that require special controls (those above normal process controls) for manufacturing, as well as highlighting areas for improvement in characteristic control or performance.

6.2.2   FTA (STA)

Fault-tree analysis (FTA) can be described as a logical, graphical diagram that is used to determine all system, subsystem, assembly, module, and part/component faults and combinations of faults that can result in specific system symptoms or faults. This particular type of logic diagram tool has been proven to be extremely useful in reliability design and engineering activities.

The following definition will help introduce the subject of FTA:

Fault Tree Analysis is a graphical representation related to a particular product anomaly; a top-down type of analysis in which each of the events that contributes to a particular anomaly can be evaluated in both quantitative and qualitative terms.

Stated another way, the fault tree shows the cause-and-effect relationship between the top-level undesired event and the various contributing events. This is done by providing a logical statement of the cumulative effect of causes within a product or process.

Because this analysis method begins with a top-level undesired event and then proceeds through all known or possible causes that could lead to this top-level undesired event, the analysis process takes on a branching structure that is similar to that of a tree. For this reason, the analysis technique is termed a fault-tree analysis.

6.3 FMEA TYPES AND OBJECTIVES AND BENEFITS

6.3.1 FMEA Types

System-level FMEA   The system-level FMEA is the highest-level FMEA that can be performed. It is used to identify and prevent failures that are related to systems or subsystems in the early design concept stages. The system-level FMEA is performed to validate that the system design specifications minimize the risk of functional failure during operation.

Design-level FMEA   The design-level FMEA is used as a tool to help identify and prevent product failures that are related to the product design. This FMEA can be performed upon a system-level, subsystem-level, or component-level design proposal and is intended to validate the design parameters selected for a given functional performance requirement.

Process-level FMEA   The process-level FMEA is used to identify and prevent failures that are related to the manufacturing or assembly process for a specific component/assembly or for a family of components/assemblies.

6.3.2 System-Level FMEA

Benefits and objectives of the system-level FMEA:

- It identifies potential systemic failure modes caused by system interaction with other systems and/or by subsystem interactions, including those that may adversely affect safety or compliance with government regulations.
- It identifies potential system design parameters that may include deficiencies before hardware and/or software is released to production.
- It helps in selecting the optimum system design alternative.
- It enables actions to ensure that customer wants/expectations are satisfied to be initiated as early as possible in the development cycle and quality planning phases of the system design.
- It acts as the basis for developing system-level diagnostic and system fault management techniques.
- It provides an organized, systematic approach to identifying all potential effects of subsystem, assembly, and part failure modes for inclusion in design-level FMEAs.
- It serves as a historical record of the thought processes and the action taken in product development efforts.
- It helps engineers focus on eliminating product concerns and minimizing the probability of poorly performing products reaching the customer.
- It helps in determining, evaluating, and improving the system design verification (SV) test programs.
- It helps in generating failure mode occurrence ratings that can estimate a particular system design alternative's reliability target.

- It helps in determining if hardware redundancy is required in order to meet the reliability requirements.

6.3.3 Design-Level FMEA

Benefits and objectives of the design-level FMEA:

- It identifies potential design-related failure modes at a system, subsystem, or component level that may adversely effect safety or compliance with government regulations in the early stages (prior to hardware release) so that design actions to eliminate or mitigate the concerns can be identified.
- It increases the probability that potential failure modes and their effects on vehicle/system performance have been considered in the design/development process.
- It identifies key critical and significant characteristics of a design.
- It enables actions to ensure that customer wants/expectations are satisfied to be initiated as early as possible in the product development cycle and quality planning phases of the product design.
- It aids in the objective evaluation of design requirements and design alternatives and provides a reference to aid in analyzing field concerns to develop advanced designs in the future.
- It provides an organized, systematic approach to criticality reduction and risk reduction and establishes a priority for design improvement actions.
- It serves as a historical record of the thought processes and the action taken in product development efforts.
- It documents the rationale behind product design changes to guide the development of future product designs.
- It helps engineers focus on eliminating product concerns and minimizing the probability of poorly performing products reaching the customer.
- It increases the probability that potential failure modes and their effects on system/vehicle performance have been considered in the design/development process.
- It helps in determining, evaluating, and improving design verification (DV) test programs by providing information to help plan a thorough product design verification test program.
- It assists in the evaluation of product design requirements and alternatives.
- It enhances organizational learning by serving as a depository for valuable "lessons learned" to help organizations avoid making the same error repeatedly.

6.3.4 Process-Level FMEA

Benefits and objectives of the process-level FMEA:

- It identifies potential process-level failure modes at a system, subsystem, or operation level that may adversely affect safety or compliance with government regulations so that actions can be taken to eliminate the concern or mitigate its effects.
- It identifies key process critical and significant characteristics and aids in the development of thorough control plans.
- It identifies potential process deficiencies early in the process planning cycle, enabling engineers to focus on controls that will reduce the incidence of unacceptable products and the use of unacceptable methods and increase detection capability well before production begins.
- It enables actions to ensure that customer wants/expectations are satisfied to be initiated as early as possible in the process development cycle and quality planning phases of the process design.

- It eliminates or reduces product criticality through manufacturing and/or assembly process design improvements.
- It provides an organized, systematic approach to process change and process update prioritization.
- It establishes priorities for process improvement actions.
- It serves as a historical record of the thought processes and the action taken in process planning and development efforts.
- It helps engineers focus on eliminating product concerns caused by the manufacturing or assembly process, thus minimizing the probability of poorly performing products reaching the customer.
- It helps in determining, evaluating, and improving the production validation (PV) test programs.
- It documents the rationale behind process changes to guide the development of future manufacturing/assembly processes.

6.4 PREPARING TO DEVELOP THE FMEA

This section is intended to provide an introduction to the FMEA process. By focusing on the design-level FMEA, this section demonstrates and relates the methodologies and strategies applied to develop information for the FMEA. The same basic process can be applied in the development of either the system-level FMEA or the process-level FMEA by repeating the steps used in developing a design-level FMEA and adjusting the scope and nature of the analysis.

The overall purpose of the FMEA is to prevent a new design or process system (within a set scope or boundary conditions) from failing to meet any of its requirements. The logical approach and inherent thoroughness of the FMEA help the PD team to ensure that a new product or process has been thoroughly analyzed, validated, and tested prior to release. It is very important to begin the analysis as early in the development process as possible in order to effect changes at optimum points in the PD process.

In 1994, Ford, General Motors, and Chrysler, under the auspices of the USCAR (United States Council for Automotive Research), created and adopted Society of Automotive Engineers standard SAE J1739 for both their industry and their supply base. The ranking tables, forms, and definitions (and their methodologies/strategies) contained in this document are representative of that effort.

6.4.1 Introduction and General Objectives and Benefits

The proper development of an FMEA has several objectives and affords global benefits:

- The FMEA is an objective and cost-efficient tool used for the collection and documentation of failure modes, causes, and related failure effects for product/process functions. Applying FMEA methods improves the quality, reliability, and safety of products.
- The FMEA provides a logical method of identifying analysis and test requirements for the intended design.
- The FMEA is a method of establishing a risk assessment associated with product characteristics. It enhances prevention efforts and can reduce warranty costs associated with product performance.
- The FMEA acts as the basis for product design evaluation and for the evaluation of changes to the product design, the process design, or the materials.
- The FMEA serves as a depository of lessons learned from a historical perspective, documenting and passing along organizational learning from earlier-generation products and/or processes.

- The FMEA increases customer satisfaction toward best-in-class (BIC) levels.
- The FMEA reduces product development time and cost.

6.4.2   Using the Team Approach

Although an individual may be able to perform many of the steps in the development of an FMEA, it is more helpful to utilize the product development (PD) team concept. Using a team approach to the FMEA allows the members to share important process and product knowledge as well as ideas. Teams that are driven to analyze product performance are often the best and most complete source of information about product failure modes and their effects. Using the collective knowledge of the team helps to assure the development of both a reliable product and a reliable process. The FMEA should be a catalyst to stimulate the interchange of ideas between the functions affected. Thus, the team approach is essential.

The team should include any and all individual stakeholders needed to properly analyze a product or process design. The team should consist of those individuals who are aware of how a design change will affect the performance of the product or process and can effect changes in the product or process and deploy recommended actions. The core FMEA team should be cross-functional in nature and include

- Product engineers (component and system)
- Manufacturing or assembly process engineers
- Reliability engineers
- Quality engineers

Depending upon the type and level of analysis that is required, the FMEA core team may need the support of additional personnel. Typical examples include

- Materials engineer
- Purchasing
- Manufacturing supplier representatives
- Legal and safety staff
- Service engineers
- Test engineers
- Other subject matter experts as appropriate

Other team considerations are as follows:

- The system, product, or manufacturing engineer that owns the design or process usually leads the FMEA team.
- Team members will vary as the system, product, and process designs mature.
- For proprietary designs, suppliers are responsible. The responsible design engineer is required to evaluate the accuracy and thoroughness of the supplier's FMEAs, with assistance from the supplier's quality interface activity engineer and reliability engineer.

6.4.3   Timeliness and Timing

Timeliness is one of the most important factors required for the success of a project. As with other program development steps, the FMEA is meant to be developed before the fact, not performed after the fact as a corrective action.

To achieve the greatest value, the FMEA must be done before a design or process failure mode has been unknowingly designed into the product and released. The up-front time spent performing a comprehensive and thorough FMEA when most product/process changes can easily be incorporated will alleviate late change crises and reduce implementation costs.

Properly applied, the FMEA is an interactive, iterative process that is never-ending. The FMEA is a living document that should be updated continually as changes occur throughout the PDP. An FMEA is directly correlated with a given product/process design, so that as the product/process design changes, the FMEA must be updated to reflect those changes.

The FMEA should be started

- When new systems, products, and processes are being designed or new technologies launched
- When existing designs or processes are being changed
- When carryover designs or processes will be used in new applications or new environments
- For a system-level FMEA, after system functions (system design specifications or parameters) are defined, but usually before specific hardware is selected
- For a design-level FMEA, after product functions are defined, but before the design is approved and released to manufacturing
- For a process-level FMEA, when preliminary drawings of the product are available and the corresponding process flow is developed
- For new product programs, no later than the timing given in Table 6.1 (this table should be used as a general guideline only)

TABLE 6.1   FMEA Timing in the Product Development Cycle

Product development timing	System FMEA	Design FMEA	Process FMEA
Design concept	Initiate		
Design simulation	Complete*	Initiate	Initiate
Detailed design		Update	
Prototype test		Complete*	Update
Product launch			Complete*
Field usage	Update	Update	Update

*Since the FMEA is a living document, the term *complete* indicates the publication of the first iteration of the document and its release as input to the next level of FMEA to be performed or to be updated.

6.4.4 Preanalysis

Resource Requirements.   Because the FMEA procedure analyzes product designs, it may be very helpful for the PD team (or the FMEA team, if it is a subset) to assemble various resources and data before beginning the actual FMEA development. There are a great many items that are useful tools for the engineer in the development of the FMEA. These tools are examined prior to developing the FMEA and can serve as a basis for the FMEA. The resources that should be assembled include

- Historical documents describing similar products, materials, or processes that have been used; system drawings; and previous FMEAs and test reports
- Customer usage history, complaints, and field-diagnosed problem reports from similar designs
- Design specifications or parameters, checklists, guidelines, or rules

- Review of appropriate technological advance(s), in terms of previous or similar product development, relating to current design requirements and products
- Analysis of testing performed, results obtained, and standards to be met

Overall Requirements.   A comprehensive list of the overall requirements that describe the intended functionality of the design must be provided for the specific product and should be included in the preanalysis stage of the FMEA.

- It is important that the design be linked to the customer's requirements. The information developed to understand the customer's requirements (requirements analysis) forms part of the basis for the FMEA statement of function and the failure-mode-related effects that have an impact upon customer requirements.
- A comprehensive list of the basic functional specifications that must be met should be included.
- Usage environments and the conditions, extremes, and interactions that can be expected during product performance should also be provided as part of the preanalysis stage. Examples include temperature, humidity, atmosphere, vibration, shock, and EMI/RFI.
- In conjunction with the usage environment, requirements for shipping, handling, and storage for the particular design under analysis should be examined. This includes
  - Temperatures involved
  - Exposure to foreign materials, such as silicone, during shipment
  - Transportation loads
  - Shock to the product, such as being dropped onto a concrete floor

In addition to the above-listed basic design information to be controlled, various regulations, standards, and safety guidelines are part of the preanalysis phase of FMEA development. Examples include any or all of the following:

- Government and safety regulations (federal, state, and local) including
  - Federal/Canadian Motor Vehicle Safety Standards (F/CMVSS)
  - Clean Air Act [federal and California Air Resources Board (CARB)]
  - Occupational Safety and Health Act
  - Federal Communications Commission requirements
- Standards that include any internal organizational policies and/or objectives
- Other standards or requirements that may be needed for reference, such as those of the Society of Automotive Engineers (SAE) or ASTM
- Recyclability requirements or ultimate product service/disposal requirements which govern service-related standards for the design's repair or part replacement functions

Information Sources.   Other possible information sources include

- Engineering drawings/diagrams
- Design requirements/specifications
  System design specifications
  Engineering specifications
  Material specifications
  DV testing and highly accelerated life testing
  Manufacturing/process specifications
- Process flow diagrams
- Control plans

- Manufacturing feasibility studies
- Previous or similar data
  Warranty data
  Reliability data
  Field service data/customer returns
- Other studies
  Quality function deployment (QFD)
  Competitive quality studies
  Reliability analysis
- Reports
  Problem reports
  Service reports
  Laboratory test reports
- Other FMEAs
  Previous/similar designs
  Previous/similar processes

Block Diagrams.   Block diagrams are very helpful in the development of the FMEA because they can be used to

- Establish and present a logical order for the analysis process (hardware breakdown structure)
- Visually present the overall primary relationships between various parts in a system, subsystem, assembly, module, or part/component
- Encourage a thorough analysis of the parts within the system, subsystem, assembly, module, or part/component

Figure 6.2 shows typical examples of three types of block diagram.

- *Hierarchy block diagram.*   This breaks the product into natural and logical elements, becoming more detailed at each level down. (For example, if the top level of a hierarchy is the system, the next level down might be subsystems, the next level assemblies, and so on.)
- *Functional block diagram.*   This identifies the interrelationships between the elements and the "flow" from one element to another that is required in order to achieve successful operation.
- *Reliability block diagram.*   This is used to illustrate the dependence and/or independence of the elements in contributing to the reliability requirements. This type of block diagram yields the effects of the failures on the overall system performance. It provides a set of rules that defines the satisfactory or unsatisfactory operation of the system.

Block diagrams should be constructed as simply but as completely as possible. Comprehensive block diagrams should show the primary and secondary interrelationships between the blocks across system boundaries. Each type of block diagram described above will be referred to as required when performing the preanalysis phase of FMEA development.

System analysis may uncover design elements that include "gray box" or "black box" assemblies or components purchased from an outside source. These purchased assemblies or components may not require detailed analysis by the PD team if they fall outside of the FMEA project scope or are covered in detailed component FMEAs. However, the same level of preanalysis effort must be available and provided by the manufacturer (or supply source) to ensure that all parts of

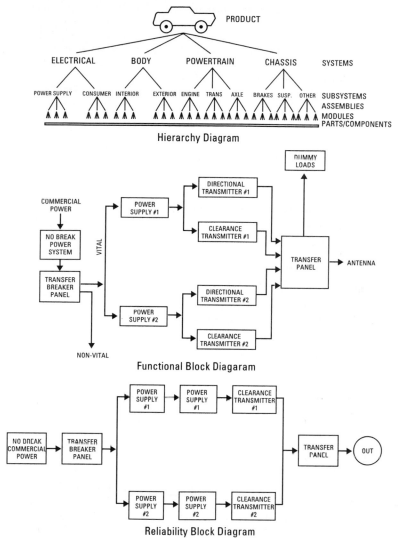

FIGURE 6.2   Three types of block diagram.

the system can be adequately analyzed during the FMEA. The PD team should always request these analyses from the manufacturer or source.

6.5   DEVELOPING THE FMEA

6.5.1   Introduction to Development

As discussed earlier, the FMEA can and should be developed continually throughout the PDP. Also, as knowledge of the system expands throughout development, refinements (usually in the form of updates) should be made to the FMEA. The FMEA becomes a living document that is continually refined and improved as the design is refined and improved.

Before the FMEA process begins, an important step for analysis success is to determine the reference point position, or scope of the project, that will be used in performing the FMEA. The reference point for the FMEA is usually the customer's standpoint, especially described in terms of product safety and/or performance.

Once the preanalysis for the design under study is complete, the appropriate information can be transferred to the FMEA worksheets. The FMEA provides the basic documentation of the reliability analysis efforts. It identifies

- The functions, failure modes, and causes of the failure modes at the various levels for the design, including part levels
- For each failure mode, the effects of the failure and its relative criticality to the overall system performance level and to the user

The credibility of the initial reliability prediction is easily determined by the level of documentation provided in the FMEA analysis and its completeness. As additional information from testing or field sources becomes available, the information on the FMEA documents must be continually updated to reflect the results. This, in turn, can be used to update the reliability prediction.

The FMEA analysis follows the following steps (see Fig. 6.3):

Start at the lowest level that is feasible for the analysis, including starting at part level.

Determine the functional specification for the part, including measurable parameters that describe the functionality.

Determine the failure modes for each function.

Determine the causes for each failure mode, including root causes for key design failure modes that relate to critical part function.

Determine the effects of each failure mode (consequences) on the next higher level up to the overall system and to the end user (the person who buys the product).

6.5.2 Documentation Requirements

A preferred FMEA form for the design-level FMEA is shown in Fig. 6.4. This same form can be used for system-level and process-level FMEAs with modifications to the header/title section of the form.

The first section of the FMEA form, the header/title information, involves information particular to the FMEA analysis scope, such as name, dates, and other aspects of identification. This section is also called the title block and runs across the top of the form. Items 1 through 8 are covered in this section. This section is primarily used for identification and traceability.

The various areas in Fig. 6.4 have been numbered, and the descriptions below are keyed to those numbers to aid in understanding where information should be placed and to give a typical example of the type of information needed.

1. FMEA Number.   Enter the FMEA document number, which may be used for tracking.

2. Type of FMEA (Component Name and Number).   Indicate the appropriate level of analysis and enter the name and other system classification number for the component, if relevant.

3. Design Responsibility.   Enter the department or group that has design responsibility. Also include other external design sources (suppliers) if known.

4. Prepared By.   Enter the name, telephone number, and company of the engineer responsible for preparing the FMEA.

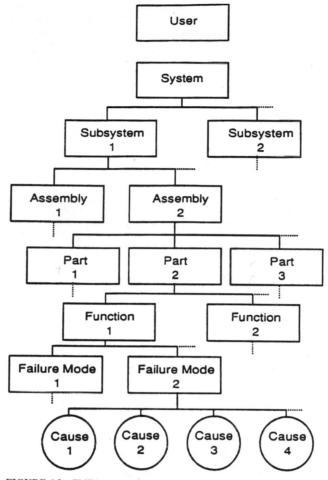

FIGURE 6.3 FMEA steps.

5. *Model Years/Vehicles.*   Enter the intended model year(s) or vehicle line(s) or design date for a component that will utilize and/or be affected by the design being analyzed (if known) on this FMEA.

6. *Key Date.*   Enter the initial FMEA due date, which should not be later than the scheduled production design release date.

7. *FMEA Date.*   Enter the original date FMEA being compiled (first interaction), followed by the latest revision date.

8. *Core Team.*   List the names of the responsible individuals and departments which have the authority to identify and/or perform tasks. It is recommended that all team members' names, departments, telephone numbers, addresses, etc., be included on a distribution list attached to the FMEA.

   To complete the form, all of the part-level information generated in the overall requirements section of preanalysis must be reviewed and consolidated for inclusion in the FMEA form.

POTENTIAL
FAILURE MODE AND EFFECTS ANALYSIS
(DESIGN FMEA)

Subsystem/Name _____ (1)

Design Responsibility _____ (2)

Other Areas Involved _____ (3)

Suppliers and Plants Affected _____ (1)

Model Year/Vehicle(s) _____ (2)

Engineering Release Date _____ (3)

Page __1__ Rows __1__ through __2__

Prepared By _____ (7)

FMEA Date (Orig.) _____ (8) (Rev.) _____ (8A)

Part Name & Number / Part Function	Potential Failure Mode	Potential Effects(s) of Failure	Sev	▷	Potential Cause(s) of Failure	Occur	Design Evaluation Technique	Detec	R.P.N.	Recommended Action(s)	Area/Individual Responsible & Completion Data	Action Results				
												Actions Taken	Sev	Occ	Det	R.P.N.
(9A) (9B)	(10)	(11)			(14)		(16)			(19)	(20)	(21)				

Part Function

Enter the Part Name and number.

Also, enter the part function(s) being analyzed. If the part has several functions, list each function separately. Use the verb-noun format to describe a Part function.

Potential Failure Mode (10)

Enter the potential failure mode(s).

Describe the failure mode as "loss of function" or in physical or engineering terms such as:
-fractured
-corroded
-short circuit
-loose

Potential Effects(s) of Failure (11)

For each Failure mode, list its consequences on:
-Part function/performance
-Next higher Assembly
-System
-Vehicle
-Customer
-Government Regulations

Severity (12)

For each failure mode, rate the most serious effect.

Enter rating in column 12.

Use Severity Rating Table for Design FMEA

If safety or compliance with government regulations are affected, enter a rating of 9 or 10

(13) Potential Critical Characteristic

Potential Cause(s) of Failure (14)

Enter the 1st level cause(s) for each failure mode. List separately. For high Severity, list root causes (describe as a part characteristic).

Assume:
1) Part is made correctly
2) Mfg/Assembly misbuild due to design deficiency (do not enter misbuilds due to process deficiency)

Occurence (15)

Estimate the number of cumulative failures that could occur for a given cause over the design life of the part.

Use Occurence Rating Table for Design FMEA.

Design Evaluation Technique (16)

Enter the methods, tests, or techniques to be used to detect the cause and/or failure mode before Engineering Release.

Some Techniques may affect the occurence of the cause. These are to be considered when estimating the Occurence rating.

Detection (17)

For each Technique, estimate the likelihood the cause of the failure mode, or the failure mode, will be detected. If several methods are listed, enter the lowest (best) rating.

Us the Detection Rating Table for Design FMEA.

(18) RPN

Risk Priority Number

Recommended Action(s) (19)

List design actions that can reduce the Severity, Occurence, then the Detection ratings.

If no action, enter None at this time

Area/Individual Responsible & Completion Data (20)

Enter:
-Design Department
-Design Engineer
-Target completion data

Actions Taken (21)

Enter a brief description of actions taken and their completion data.

Revised (22) RPN

After actions have been taken, re-estimate ratings for Severity, Occurence, and Detection, Enter revised ratings in columns to the right.

FIGURE 6.4 Design FMEA form.

Information sources researched in preparing for the overall FMEA project often include references or data on part functions and related failure modes. Enter the specific and descriptive part-level information by completing the following fields:

9(a) and 9(b).   Item Name/Function.   Enter the name and the number of the part or assembly that is being analyzed. Use the nomenclature and show the design level as indicated on the engineering drawing. Prior to initial release for a design, an experimental number should be used.

For the function, enter, as concisely as possible, the design intent of the function of the item being analyzed. Include information regarding the environment for the system and the design parameters that govern and describe that function in engineering terms. List all of the functions separately if doing more than one part number or component within the FMEA analysis.

This is the time to make use of the verb-noun functions developed in the function diagram.

6.5.3 Failure Modes

10.   Potential Failure Mode(s).   The purpose of this section is to identify any and all of the potential failures that might occur with a given part or assembly. A failure mode can be defined as the manner in which a component, subsystem, or system could potentially fail to meet the design intent, performance requirements, or even customer expectations. A potential failure mode may also be the cause of a potential failure mode in a higher-level subsystem or system, or the effect of one in a lower-level component.

List each potential failure mode for the particular part and function. (Refer to the information/resources acquired earlier, including test reports, previous FMEAs, and reports concerning quality, durability, warranty, and reliability.)

- Consider all the potential failures that might occur under certain operating conditions, such as hot, cold, wet, dry, or dusty.
- Consider the potential failures that might occur under certain component usage conditions, such as

  Above-average/below-average life cycle

  Harsh environment, such as vibration or usage pattern

- Consider the potential failure when the part is assembled or serviced incorrectly, such as installed upside down or backward:

  Can an incorrect part be substituted?

  Can the part be assembled or serviced incorrectly (contributing to functional failure)?

  Can the part be omitted?

FTA (fault-tree analysis, which will be covered later in this chapter) can be of assistance in determining the failure modes. As shown in Fig. 6.5, if the top-level event of a fault tree is listed as "failure to provide function," the first level of the fault tree (inputs to the topmost gate) identifies the failure modes. The next level of the fault tree identifies the failure causes. Subsequent levels determine intermediate causes and root causes.

6.5.4 Failure Effects

11.   Potential Effect(s) of Failure.   The potential effect(s) of failure are defined as the effects of the failure mode on the function, as perceived by the customer. Failure effects are described in terms of what the customer might experience or notice, remembering that the customer may be an internal customer (next assembly, component, or system) or an external customer (ultimate user). Any impact on safety or noncompliance with regulatory standards should be clearly stated.

The following acronyms can be used to identify the level of failure effect:

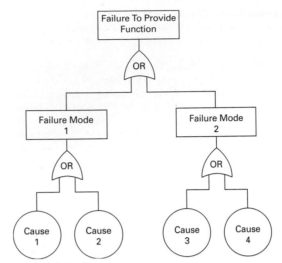

FIGURE 6.5   Fault-tree analysis structure.

LOF (*loss of function*): Loss of function implies that the part/component, assembly, or subsystem has lost the ability to provide basic functions as a result of a crippling failure mode.

PLOF (*possible loss of function*): Possible loss of function implies that the part/component, assembly, or subsystem has suffered a failure mode which may or may not prevent the item from performing basic and/or critical functions.

DF (*degraded function*): Degraded function implies that the part/component, assembly, or subsystem has suffered a failure mode which degrades intended performance levels to something less than completely operational.

MDF (*minor degraded function*): Minor degraded function implies that the part/component, assembly, or subsystem has suffered a failure mode which partially disables or minimally degrades the basic or critical functions.

NIE (*no immediate effect*): No immediate effect implies that the part/component, assembly, or subsystem failure has no immediate performance impact.

For each failure mode, the consequences for the part, next higher assembly, system, vehicle (or ultimate system design level), customer, and government should be identified.

6.5.5   Severity (Criticality)

12.   Severity   Severity is an assessment of the seriousness of the effect (listed in the previous column) for the potential failure mode under examination. Severity applies to the effect only. A possible severity ranking index scheme is shown in Table 6.2. A reduction in severity can be effected only through a design change that eliminates the failure modes and prevents the effect of the failure from occurring within the design. The highest index for severity applied to any effect in column 11 is carried into column 12. Each failure mode has only one overall severity ranking.

6.5.6   Process Steps

13.   Classification.   This column may be used to classify any special product characteristics (e.g., key, critical, or significant) for components, subsystems, or systems that may require addi-

TABLE 6.2 Suggested Severity Evaluation Criteria

Effect	Criteria: severity of effect	Ranking
Hazardous—without warning	Very high severity ranking when a potential failure mode affects safe vehicle operation and/or involves noncompliance with government regulation without warning.	10
Hazardous—with warning	Very high severity ranking when a potential failure mode affects safe vehicle operation and/or involves noncompliance with government regulation with warning.	9
Very high	Vehicle/item inoperable, with loss of primary function.	8
High	Vehicle/item operable, but at reduced level of performance. Customer dissatisfied.	7
Moderate	Vehicle/item operable, but comfort/convenience item(s) inoperable. Customer experiences discomfort.	6
Low	Vehicle/item operable, but comfort/convenience item(s) operable at reduced level of performance. Customer experiences some dissatisfaction.	5
Very low	Fit and finish/squeak and rattle item does not conform. Defect noticed by most customers.	4
Minor	Fit and finish/squeak and rattle item does not conform. Defect noticed by average customer.	3
Very minor	Fit and finish/squeak and rattle item does not conform. Defect noticed by discriminating customer.	2
None	No effect.	1

tional process controls to control manufacturing variability. It is used to flag those product characteristics that affect safe vehicle/product function or compliance with government regulations, or that are customer satisfaction–related.

Column 13 is normally addressed after Columns 9, 10, 11, 12, 14, and 15 are completed.

If the severity index table is applied (Table 6.2), then characteristics rated severity >8 (potential safety or governmental noncompliance effects) would be flagged in column 13. Mid-level severity-related items (those with a severity level between 4 and 8) might also be flagged as second-priority items on which to take recommended action. The use of this column is usually developed by the FMEA process leadership to reflect company standards or procedures.

14.   Potential Cause(s) and Mechanism(s) of Failure.   A potential cause of failure is defined as an indication of a design weakness, the consequence of which is the failure mode. Listed in this column, to the extent possible, is every conceivable failure cause and/or failure mechanism related to the failure mode shown in column 10.

Identify and list all possible causes of failure for each failure mode so that any remedial action can be directed to the pertinent cause. Look for the "root cause" for those characteristics that are key, critical, or significant so as to aim the remedial actions at the area where change will be most effective. Ask "what" and "why" questions. Try to be as concise, but as complete, as possible.

A design-level FMEA does not rely on process controls to overcome potential weaknesses in the design, but it does take the technical/physical limits of a manufacturing/assembly process into consideration in the way design parameters are specified and tolerances. The object is a robust design that can tolerate expected process variation.

Techniques such as cause-and-effect diagrams or fault-tree analysis can be used to help determine root causes.

15.   Occurrence.   Occurrence is the likelihood that a specific cause/mechanism listed in the previous column will occur. The likelihood of occurrence ranking number has a meaning rather than a value. Removing or controlling one or more of the causes/mechanisms of the failure mode through a design change is the only way a reduction in the occurrence ranking can be effected. This ranking is entered in column 15 on the form.

The likelihood of occurrence of the causes is usually ranked on a scale of 1 to 10. Table 6.3, or a similar table, can be used to determine the occurrence rating.

TABLE 6.3   Suggested Occurrence Evaluation Criteria

Probability of failure	Possible failure rates	Ranking
Very high: Failure is almost inevitable	≥1 in 2	10
	1 in 3	9
High: Repeated failures	1 in 8	8
	1 in 20	7
Moderate: Occasional failures	1 in 80	6
	1 in 400	5
	1 in 2000	4
Low: Relatively few failures	1 in 15,000	3
	1 in 150,000	2
Remote: Failure is unlikely	≤1 in 1,500,000	1

16.   Current Design Controls.   List current design controls or evaluation techniques aimed at preventing the design causes of the potential failure mode. Current design controls are prevention, design validation/verification (DV), or other activities which will assure the design adequacy for the failure mode and/or cause/mechanism under consideration.

Three types of controls are:

- Prevention types aimed at reducing or preventing the occurrence of the cause/mechanism or failure mode
- Detection types aimed at detecting the cause/mechanism and leading to corrective actions
- Those that detect the failure mode itself

Typical examples of current design controls include

Load testing

Prototype tests

Design reviews

Customer tests

Mathematical studies

Environmental tests

Feasibility studies

Lab tests

17.   Detection.   Detection is the assessment of the ability of the proposed current design control to detect the potential cause/mechanism or failure mode before the component, subsystem, or system is released for production. In order for a lower control ranking to be achieved, generally the planned design control method must be improved. The design control ranking relates to the control that is specifically applied to detect that cause/mechanism or failure, not to the overall effectiveness of the method.

TABLE 6.4   Suggested Detection Evaluation Criteria

Detection	Criteria: likelihood of detection by design control	Ranking
Absolute uncertainty	The design control will not and/or cannot detect a potential cause/mechanism and subsequent failure mode; or there is no design control.	10
Very remote	Very remote chance that the design control will detect a potential cause/mechanism and subsequent failure mode.	9
Remote	Remote chance that the design control will detect a potential cause/mechanism and subsequent failure mode.	8
Very low	Very low chance that the design control will detect a potential cause/mechanism and subsequent failure mode.	7
Low	Low chance that the design control will detect a potential cause/mechanism and subsequent failure mode.	6
Moderate	Moderate chance that the design control will detect a potential cause/mechanism and subsequent failure mode.	5
Moderately high	Moderately high chance that the design control will detect a potential cause/mechanism and subsequent failure mode.	4
High	High chance that the design control will detect a potential cause/mechanism and subsequent failure mode.	3
Very high	Very high chance that the design control will detect a potential cause/mechanism and subsequent failure mode.	2
Almost certain	The design control will almost certainly detect a potential cause/mechanism and subsequent failure mode.	1

The detection ranking system shown in Table 6.4 can be be used when developing the FMEA. As with the other ranking systems presented, the numbers represent suggested ranges only.

18.   Risk Priority Number (RPN).   The risk priority number (RPN) is the product of the severity (S), occurrence (O), and detection (D) ratings. The RPN is a measure of the design risk and should be used to rank or prioritize the concerns in the design (e.g., Pareto fashion). One technique found useful by some teams is to select a threshold RPN number, which, if exceeded, requires action. This RPN is dependent upon team consensus.

$$RPN = severity \times occurrence \times detection$$

The RPN will be between 1 and 1000 if 10-point scales are used for severity, occurrence, and detection. Where there are high calculated RPNs, the team must undertake efforts to reduce the risk through corrective actions. Special attention should always be focused on high-severity failure modes regardless of the RPN value.

19.   Recommended Action(s).   Note: It is extremely important to complete this section and assure meaningful follow-up. An FMEA without information in the right-hand section is an incomplete FMEA.

The intent of any recommended actions is to reduce the occurrence, severity, and/or detection rankings by improving the design. An improvement in design validation/verification will result in a reduction in the detection ranking only. A reduction in occurrence can be effected only by removing or controlling one or more causes/mechanisms of a failure mode through a design change. Severity is reduced only by elimination of the failure mode or a mitigation of the effects associated with the failure mode though design changes.

Corrective actions should be directed to the highest-ranked problem identified. Recommended actions should be considered in the following order:

- A failure mode has an effect with a severity rating of 9 or 10.
- A failure mode/cause combination has a high severity and occurrence ranking (based on team consensus).
- A failure mode/cause/design control combination has a high RPN ranking (based on team consensus).

The intent of design actions is to reduce the severity, occurrence, and detection ratings, in that order.

Note: One hundred percent test or inspection during manufacture cannot be used as a recommended action in a design-level or system-level FMEA. Recommended actions that involve manufacturing are not appropriate for an FMEA at these levels. Recommended actions listed must be able to be executed during the design phase, not after release of the design to production.

Quite often, specific engineering testing is required in order to determine the design concerns. Actions to be considered include

- Design of experiments (DOE)
- Revised test plan
- Revised design and/or material specification

When preparing recommended actions, include all available supporting details and facts in order to provide a traceable plan of action.

If there are no actions recommended for any specific failure cause, this should be noted in the section by entering "No action at this time."

20. Responsibility and Target Completion Date.   Enter the organization area and the individual who are responsible for any recommended action listed in the previous column. Items to include in this section might be the responsible individual's phone number, fax number, and department. This is solely dependent upon local requirements. Be sure to include the target completion date for the recommended action.

21. Actions Taken.   The individual responsible for the recommended action enters a brief description of the actual action taken and the date completed. This should include any change to recommended actions made and the results of the action.

Note: Failure to follow up on a corrective/recommended action puts the individual responsible at great risk in terms of liability.

22. Resulting RPN.   After corrective action has been implemented, re-estimate the severity rating, the occurrence rating, and the detection rating. Calculate a new value for the RPN and enter the number in this section of the form. If there were no recommended actions, leave the resulting RPN section and each of the rating columns blank.

Review all of the resulting RPNs and verify that each has been reduced to an acceptable level, as determined by the PD team or local requirements. If further action is necessary, repeat columns 19 through 22.

Follow-up.   The PD team is responsible for ensuring that all recommended actions have been implemented. Several typical examples of checks for design verification are:

Update of engineering drawings and specifications, showing design changes, key/critical characteristics, supplier/manufacturing test requirements, or other engineering actions related to the drawing specified on the FMEA

Sign-off of manufacturing installation drawings to verify that assembly concerns identified by the FMEA have been addressed by the installation drawings

Review of the process-level FMEA and resulting manufacturing control plans and strategies

Note: The FMEA is a living document and should always reflect the latest design level and relevant actions, including those occurring after production start-up or product launch.

6.5.7 FMEA Checklists

Proposed checklists for each type of FMEA are shown in Tables 6.5, 6.6, and 6.7.

TABLE 6.5   Concept (System) FMEA Checklist

Use the checklist below to help assure that the concept (system) FMEA is complete. All answers to the following questions should be yes.

Preliminaries	Are the system functions or wants listed?
	Is a system block diagram evident?
	Are cross-functional FMEA team members listed?
	Is there evidence that background information has been checked?
Header information	Are all the applicable entries in the header completed?
Purpose/function	Is the system function or purpose clear?
	Are functions described in verb-noun format?
Failure modes	Are failure modes identified as loss of function?
Failure effects	Are effects on safe vehicle operation and government regulations considered?
	Are effects on other systems, the vehicle, and the customer considered?
Failure causes/mechanisms	Are all causes for each failure mode identified?
	Are causes in terms of system element failure modes?
Design controls	Can methods listed detect the first-level causes (element failure modes) of the system failure modes?
Severity rating	Are ratings based upon the most serious effect of the system failure mode?
Occurrence rating	Are ratings based on the cumulative number of failures that could occur for each cause over the design life of the system?
Detection rating	Are ratings based on the likelihood of detecting the first-level causes (element failure modes) of the system failure mode?
Risk priority number (RPN)	Are the risk priority numbers ranked from high to low?
Recommended actions	Are remedial actions considered for the highest-ranked failure modes?
	Are responsibility and timing for the recommended actions listed?
	Are actions directed at eliminating or reducing the occurrence of the causes of the system failure modes?
Follow-up	Was the FMEA updated after recommended actions were implemented?

6.6 FTA

As discussed before, *fault-tree analysis* (FTA) is a term that has been applied to a logical, graphical diagram that is used to develop all system, subsystem, assembly, module, and part/component faults and combinations of faults that can result in specific system symptoms or higher-level

TABLE 6.6 Design FMEA Checklist

Use the checklist below to help assure that the design-level FMEA is complete. All answers to the following questions should be yes.

Preliminaries	Are the part functions listed?
	Has a cross-functional FMEA team been informed?
	Is there evidence that background information has been checked?
Header information	Are all the applicable entries in the header completed?
Purpose/function	Is the design intent or purpose clear?
	Are functions described clearly (verb-noun format)?
Failure modes	Do the failure modes relate to the part functions?
Failure effects	Are effects on safe vehicle operation and government regulations considered?
	Are effects on part, next higher assembly, system, vehicle, customer considered?
Failure causes/mechanisms	Are root causes identified?
	Are design deficiencies that may induce manufacturing/assembly variation considered?
	Are manufacturing/assembly causes excluded (but addressed in process FMEA)?
	Are causes described in terms of a part characteristic, where appropriate?
Design controls	Can design controls listed detect the cause(s) of failure modes before engineering release, or detect the failure mode(s) before engineering release?
	Are manufacturing/assembly detection methods excluded?
Severity rating	Are ratings based upon the most serious consequence of the failure mode?
Occurrence rating	Are ratings based on the occurrence of the first-level cause?
Detection rating	Are ratings based on the likelihood of detecting the failure mode prior to engineering release?
Classification	Do potential critical characteristics correspond to a part characteristic?
	Were potential critical characteristics communicated to the process FMEA team?
RPN	Are the risk priority numbers (RPN) ranked from high to low?
Recommended actions	Do actions address potential critical characteristics?
	Are remedial actions listed that reduce the RPN for the highest-ranked failure modes?
	Are responsibility and timing for recommended actions listed?
	Are preventive, instead of detection, actions listed where appropriate?
	Are actions aimed at making the design more robust?
	Are the actions listed design actions, not manufacturing/assembly controls?
Follow-up	Was the FMEA updated after recommended actions were implemented?
	Did the process FMEA team determine whether normal and customary or special controls were required for the identified potential critical characteristics?

TABLE 6.7 Process FMEA Checklist

Use the checklist below to help assure that the process FMEA is complete. All answers to the following questions should be yes.

Preliminaries	Was an FMEA team organized? Was background information checked?
Process flowchart	Was a process flowchart prepared? Are the part characteristics produced, or evaluated, at each operation listed? Are the process characteristics at each operation listed? Are the sources of incoming variation identified, where applicable?
Header information	Have all the applicable entries in the header been completed?
Description/purpose	Is the purpose or function of each operation listed?
Failure modes	Are failure modes listed in terms of "What would the part be rejected for?" Do failure modes list parts characteristics produced at the operation that would be rejected if the part characteristic were outside specification limits? Do failure modes include inspection/testing operations; i.e., accepting bad part or rejecting good part? If a failure mode adversely affects an operation downstream: • Is the failure mode listed as a cause for the downstream operation? • Is the downstream failure mode identified?
Failure effects	Are potentially hazardous effects to machinery, plant personnel, and the final customer considered? Are the effects described in terms of the impact of the failure mode on • The location operation, next operation, subsequent operation? • The next user (manufacturing or assembly facility)? The ultimate customer? • The vehicle? • Government regulations?
Failure causes/mechanisms	Are things that may go wrong in the process identified? Are causes described in terms of a characteristic that can be fixed or controlled? Are process characteristics considered? Are design weaknesses considered? Are material or parts incoming to each operation considered? Are operator actions considered?
Current process controls	Are the controls to be implemented to detect bad parts listed? Can the controls detect bad parts before they leave the manufacturing/assembly facility? Are controls identified according to whether they are used for detection or for prevention?
Severity rating	Are ratings based upon the most serious effect of the failure mode? Do the ratings for customer effects agree with the ratings shown in the design FMEA?
Occurrence rating	Do ratings consider the ability of prevention controls to reduce the occurrence of a failure mode?
Detection rating	Are ratings based on the ability of the current controls to detect a failure mode before the part leaves the manufacturing or assembly location?
Classification	Are critical characteristics and their special controls identified? Are critical characteristics identified with an inverted delta symbol? Are critical characteristics identified as a process (or part) characteristic? Were critical characteristics and their special controls communicated to the responsible design engineer?
Risk priority number	Are the risk priority numbers (RPNs) ranked from high to low?
Recommended actions	Are process actions considered to reduce criticality for critical characteristics? Are special manufacturing/assembly controls identified for critical characteristics? Are all critical characteristics addressed? Are remedial actions considered to reduce the RPN for the highest-ranked failure modes? Are responsibility and timing for the recommended actions listed? Are preventive, instead of detection, actions listed where appropriate? Are actions considered to eliminate/reduce occurrence of potentially hazardous failure modes, where applicable?
Follow-up	Are the risk priority numbers (RPNs) ranked from high to low?

faults. This particular type of logic diagram tool has been proven to be extremely useful in a variety of reliability design and engineering activities.

FTA is a means of identifying or investigating potential failure modes and related causes. FTA can be

- Applied in the early design phase and then progressively refined and updated as the design evolves
- Helpful in identification of all possible causes (at all possible levels) and relationships between causes
- Used as a tool to help improve the design of any given product or process

Construction and analysis of a fault tree is best done as a PD team activity. Even though FTA may be attempted by an individual, the trees that are developed by a team are generally more fully defined and complete. The reason for this is that a much broader sphere of information is presented and considered by a team. The collective knowledge of a team is more beneficial than that of an individual. This is analogous to the discussion in Sec. 6.4.2 on using the team approach for the FMEA.

6.6.1 Benefits of Fault-Tree Analysis

There are important benefits to be gained from learning precisely what can go wrong and how this will affect the system. The PD team gains new insight and sees new possibilities. The PD team can see what new data are needed for prevention purposes and can come up with better answers because they will be based on examination of the whole system rather than a single component.

FTA is of major value for its ability to

- Identify failures deductively.
- Point out the important aspects of a system with respect to failure.
- Provide a graphic aid for system management.
- Provide options for reliability analysis.
- Allow concentration on one particular system fault at a time.
- Provide genuine insight into system behavior.
- Better account for human error.

6.6.2 Reduced Fault Tree

If the same input is present at more than one place in the fault tree, it is possible to develop an equivalent "reduced" fault tree from the minimal cut sets. This reduced fault tree will not contain the duplicated inputs and can be used as the model for quantitative evaluation.

6.6.3 Success-Tree Analysis

A success tree is the complement or "dual" of the fault tree and focuses on what must happen for the top-level event to be a success.

The same methodology and logic symbols that are used for the fault tree are used to arrive at a success tree.

A fault tree can be converted to a success tree by

- Changing each OR gate to an AND gate
- Changing each AND gate to an OR gate

- Appropriately rewording statements within the blocks to their respective complements (success statements instead of fault statements)

The following example will illustrate the concept of changing a fault tree to a success tree. A section of a ceramic catalytic converter fault tree that asks the question, "What could cause catalyst deactivation?" is shown in Fig. 6.6. The causes for catalyst deactivation could be

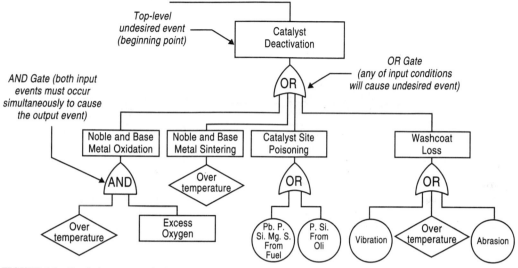

FIGURE 6.6   Catalytic converter fault tree.

- Noble and base metal oxidation

  OR

- Noble and base metal sintering

  OR

- Catalyst site poisoning

  OR

- Washcoat loss

In the illustration shown in Fig. 6.7, the fault tree has been converted to a success tree. Instead of the top-level event being a fault, it is shown as a success: "Catalyst remains active." Note that the AND gates and the OR gates have been switched, and the lower-level events have been reworded in such a way as to make this top-level event a success. The catalyst remains active if it

- Resists oxidation

  AND

- Resists sintering

Fault Tree Inverted To
A Success Tree

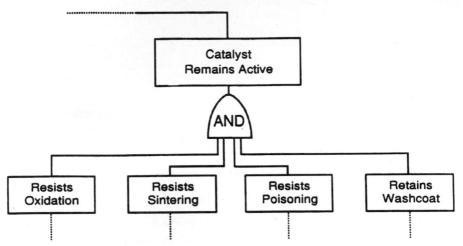

FIGURE 6.7   Catalytic converter fault tree inverted to a success tree.

AND

- Resists poisoning

AND

- Retains washcoat

In some situations, the PD team may find it helpful to develop a fault tree and a success tree in parallel. The change in perspective from one approach to the other will sometimes give the team new insight that might have been overlooked otherwise.

6.7   FAULT-TREE SYMBOLS

A fault tree may be viewed as a system of logic symbols, or gates, as they are called, that allow or inhibit the passage of a fault up the tree.

- A gate shows the relationships of logic events needed for the occurrence of a higher event.
- The higher events are the outputs of the gate.
- The lower events are the inputs to the gate.
- The gate symbol denotes the type of relationship between the input events and the output events.

Various symbols are used to aid interpretation of the fault tree and provide uniform representation. The symbols commonly used in fault-tree diagraming are shown in Figs. 6.8 through 6.10.

Two Common Logic Symbols

AND Gate

OR Gate

FIGURE 6.8 Two common logic symbols.

These symbols are used to show basic functional relationships in logic form. The same symbols are used in constructing success tree diagrams depicting successful operation.

6.7.1 Fault-Tree Logic Symbols

The gates are used to interconnect the lower-level events that contribute to the top-level event. The two logic gates used *most frequently* to develop the fault tree are shown in Fig. 6.8. They are

- The AND gate, which provides an output event only if all the input events occur
- The OR gate, which provides an output event if one or more of the input events are present

There are other logic symbols available, as illustrated in Fig. 6.9, but they are used less frequently:

INHIBIT gate: This is used with a conditional event. Input produces output directly only when the conditional input is satisfied.

PRIORITY (ORDERED) AND gate: This requires that the input events follow a specific "order of occurrence" in order for the output event to occur.

EXCLUSIVE OR gate: In order for the output event to occur, only one of the input events would have to occur. The output event will not occur if more than one input event occurs.

SAMPLING gate: This requires that at least m of the n possible input events occur (where $m \leq n - 1$) for the output event to occur.

6.7.2 Fault-Tree Event Symbols

Other symbols, called *event symbols*, are needed for the events themselves, to indicate whether they are normal, independent, or insignificant. These symbols are shown in Fig. 6.10 and listed below.

Rectangle: An event or a fault that results from the combination of more basic faults and that can be developed further.

OTHER LOGIC SYMBOLS

INHIBIT Gate

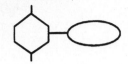

PRIORITY (ORDERED) AND Gate
(Alternate Symbols)

  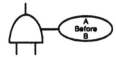

EXCLUSIVE OR Gate
(Alternate Symbols)

SAMPLING Gate
(Alternate Symbols)

 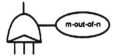

FIGURE 6.9 Other logic symbols.

Circle: A basic event or a fault that does not need to be developed any further. This type of event is independent of other events and indicates termination at that point. This event can be assigned a probability of occurrence.

Diamond: An undeveloped event or fault; an event that is not developed further either because further development is of insufficient consequence or because the necessary information is unavailable.

Triangle: Used as a transfer symbol to move or connect information from one part or page of the fault tree to another when constructing a lengthy or complex fault tree.

Oval: A conditional event. This usually functions in combination with a logic gate, generally an INHIBIT gate.

House: Input events that are not themselves faults that are expected to cause the output event to occur.

EVENT SYMBOLS

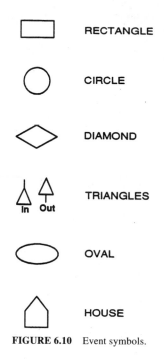

FIGURE 6.10   Event symbols.

The strength of the fault-tree symbolism lies in the fact that the symbols can readily be translated into algebraic terms, using boolean algebra, so that the fault tree can be simplified. It can be mathematically reduced so that all duplications are eliminated and the most important independent events are identified.

If the frequency of occurrence of the independent events is known or can be approximated, then the relative importance of the various independent events in producing the undesired event can be calculated.

6.8 *CONSTRUCTING A FAULT TREE*

6.8.1 General Rules of Construction

When constructing a fault tree, the scope of the analysis may need to be reduced in order to make it more manageable. This can be accomplished by using a block diagram such as the reliability block diagram for the system and equipment. A separate fault tree can then be constructed for each block. Conversely, a success tree could also be constructed for each block in order to identify what must occur for the block to be a success.

6.8.2 Basic Approach to FTA

Once the top-level fault event has been defined, there is a series of steps that the PD team should follow in order to properly analyze and construct the fault tree.

Note: In the following basic approach, the development, analysis, and eventual recommendation of corrective actions are presented as separate steps. In actual practice, there is a great deal of interaction between the steps listed. As a result, the fault tree that evolves includes additions and/or changes reflecting an improved understanding of the various faults.

1. Define the system and any assumptions to be used in the analysis. Also, define what constitutes a failure (i.e., limit, parametric shift, functionality, vibration, etc.).

2. If it is necessary to simplify the scope of the analysis, develop a simple block diagram of the system, showing input, output, and interfaces. *Note:* Early in the design phase, a design functional block diagram can be the basis for an FTA.

3. Identify and list the top-level fault events to be analyzed. If necessary, develop a separate fault tree for each top-level event. (This will depend upon how the top-level event is defined and the specificity of the event or scope of study.)

4. Using the fault-tree symbols presented earlier and a logic tree format, identify all of the contributing causes for the top-level event. In other words, using deductive reasoning, identify what events could cause the top-level event to occur.

5. Thinking of the causes of Step 4 as intermediate effects, continue the logic tree by identifying causes for these intermediate effects.

6. Develop the fault tree to the lowest level of detail needed for the analysis, typically with basic or undeveloped events.

7. Once the fault tree is completed, analyze it in order to understand the logic and the interrelations of the various fault paths, and to gain insight into the unique modes of product faults. Additionally, this analysis process should focus on those faults that, potentially, appear most likely to occur.

8. Determine where corrective action is dictated or a design change is required to eliminate fault paths wherever possible, or identify any control(s) that could possibly keep the fault from occurring.

9. Document the analysis process and then follow up to ensure that all appropriate action(s) have been taken.

6.8.3 Sample Fault Trees

The following examples are provided to help in understanding the method of an FTA. The examples illustrate the basic format of the fault tree. For each of the examples presented, the analysis method begins with the top-level event and then proceeds downward to the various input events or causes.

The diagrams also depict the various logic gates to independent or undeveloped events. Additionally, each example has been expanded to illustrate and explain the use of various symbols and logic for the analysis.

FTA Example 1: Potential Engine Damage.   Figure 6.11 shows an undesirable top-level event, potential engine damage relating to insufficient oil pressure. As illustrated in the figure, damage can result from a combination of three basic events. These are

1. Low oil pressure
2. Low oil level
3. Defective oil pressure indicator gage

If there is either low oil pressure or a low oil level and the operator fails to detect these faults because there is a defective oil pressure gage, then potentially the engine could be damaged.

POTENTIAL ENGINE DAMAGE RELATING TO INSUFFICIENT OIL PRESSURE

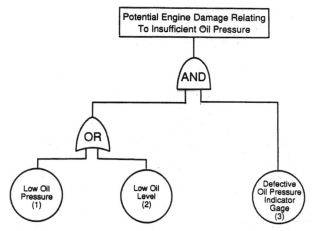

FIGURE 6.11   Potential engine damage resulting from insufficient oil pressure fault tree.

Note that the appropriate symbols are used in Fig. 6.11 to represent logical AND and OR gates. Once this step has been completed, a mathematical model can be developed describing the probability of occurrence for the top-level event. This subject will be covered in Sec. 6.9.2.

FTA Example 2: Coolant Supply to Heat Exchanger.   Figure 6.12 is a drawing of a coolant supply system. As depicted in the drawing, cooling water is pumped through a flow control valve. The flow control valve is linked to a controller on the heat exchanger. This device is used to divide the flow between the bypass line and the heat exchanger. The demand for cooling water by the heat exchanger will actually determine the direction of relative volume.

Figure 6.13 depicts the fault tree developed in response to the schematic shown in Fig. 6.12. It has as the top-level event being analyzed a loss of minimum flow to heat exchanger.

Figure 6.14 shows an additional portion of the FTA for Example 2. Note that this portion of the fault tree is linked to the complete fault tree (shown in Fig. 6.13) through transfer symbol 3. This figure depicts the use of an AND gate with a basic event, shown by the circle symbol, and an undeveloped event, which is depicted by the diamond symbol.

In this example, the fault event shown in the rectangle is described as "control valve closed beyond minimum flow position." This fault event might occur because failure of the valve stop occurs simultaneously with the valve going to the full closed position.

For this analysis, the cause depicted by the circle and identified as "valve stop fails when control valve moved to minimum position" is considered a primary event and requires no further development. This event is at a part level and is considered an independent event terminating the FTA process. However, if the FTA were actually focused on the failure of the valve, then this event would be expanded to a greater level of detail.

The other cause, shown in the figure as a diamond, is described as "valve to full closed position when valve stop fails." This event is referred to as an *undeveloped event* and falls beyond the scope of this analysis. The reason this event is not expanded further might be a lack of either available information or a need or desire to develop the detail of this event any further. Additionally, in terms of control functions, this event is related to the valve and interfacing control system, which falls beyond the analysis requirements of this fault tree.

COOLANT SUPPLY SYSTEM

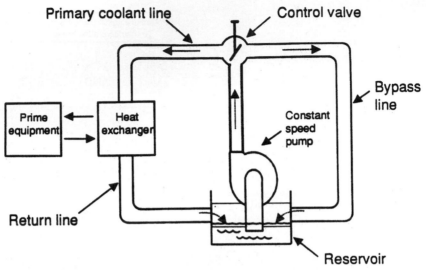

FIGURE 6.12   Coolant supply system diagram.

FTA Example 3: Motor Does Not Operate.   The top-level event describing the completed fault tree shown in Fig. 6.15 is listed as "motor does not operate." This fault-tree example illustrates the use of an INHIBIT gate. Note that the condition shown with the INHIBIT gate is that of a fuse that opens because of a current overload in the circuit, but is "inhibited" from opening until the current through the fuse exceeds the fuse rating.

Figure 6.16 shows the top portion of the FTA and illustrates the use of an OR gate with various input events. The purpose of the OR gate is to depict graphically the event(s) that need to occur, either individually or in combination, for the top-level fault event to occur.

This example illustrates the use of the "house" symbol. The event "switch opened" is not actually a fault event. This is an event that is normally expected to occur. The reason this event has been included in this example is to account for all of the events that would normally lead to the top-level event.

The event listed as "motor failure" and depicted by the circle in the figure is considered a basic or primary event for this analysis and does not require further development.

The event listed as "no current to motor" and depicted by the rectangle is a fault event that is the result of a cause or a combination of causes, as shown in Fig. 6.15.

6.9   ANALYSIS TECHNIQUES

6.9.1   Cut Sets and Path Sets

Cut sets are commonly used in FTA. A cut set is defined as any basic event or combination of basic events whose occurrence will cause the top-level event to occur. On the other hand, a path set is an event or combination of events whose nonoccurrence ensures the nonoccurrence of the top-level event.

Loss Of Minimum Flow
To Heat Exchanger

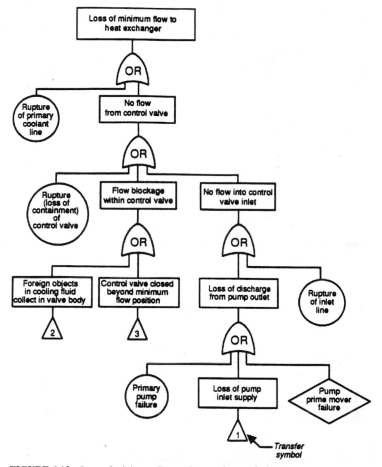

FIGURE 6.13   Loss of minimum flow to heat exchanger fault tree.

Determining the cut sets for a fault tree is a relatively simple task, but it can be a repetitious one. There are actually two simple rules that need to be followed:

1. An OR gate always increases the number of cut sets. (There will be a separate cut set for each OR gate input.)
2. An AND gate always increases the size of a cut set. (There will be one cut set for an AND gate, and each input will increase the size of the cut set.)

The cut sets that have been determined may not actually represent the simplest configuration. When this occurs, the next step is to eliminate all the repetitious items and reduce these cut sets to the *minimal cut set.*

A minimal cut set can be defined as the smallest combination of events that, if they all occur, will cause the top-level event to occur. This combination is the smallest combination of failures

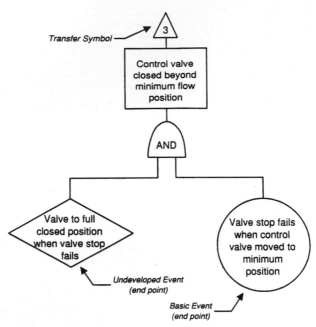

**Control Valve Closed Beyond
Minimum Flow Position**

FIGURE 6.14 Control valve closed beyond minimum position fault tree.

needed for the top-level event to occur. If any of the failures identified in the minimal cut set does not occur, then the top-level event will not occur (that is, the top-level event will not occur using this minimal cut set combination).

A complete set of minimal cut sets consists of all the failure causes for the given system fault. Minimal cut sets are important because they

- Point out the weakest links in the system.
- Depict which failures must be repaired in order for the top-level failure to be removed from the failed state.

Any fault tree will consist of a finite number of minimal cut sets that are unique for that top-level event.

After the minimal cut sets are determined, some idea of failure importance can be obtained by ordering the minimal cut sets according to their size. Single-component minimal cut sets are listed first, then double-component cut sets, then triple-component cut sets, and so on. After the minimal cut sets have been defined, qualitative and quantitative probability evaluations can be performed.

Figure 6.17 shows an example of the minimal cut sets on a reliability block diagram of a seven-component system. Each component is represented by a block. The cut sets are depicted by the vertical lines through the blocks indicating the failures that could occur. The path sets are depicted by the horizontal lines indicating the blocks that are needed for the event to be a success.

Cut sets and path sets can be determined using either a fault tree or a reliability block diagram. Only the FTA method will be examined here.

Motor Does Not Operate

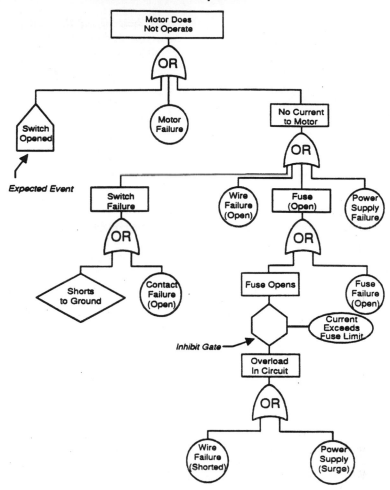

FIGURE 6.15   Motor does not operate fault tree.

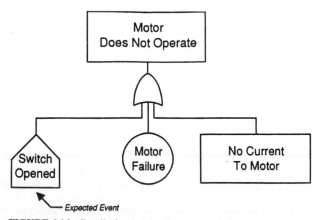

FIGURE 6.16   Detail of motor does not operate fault tree.

CUT SETS AND TIE SETS

CUT SETS

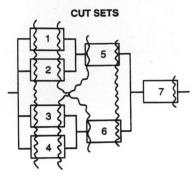

PATH SETS (TIE SETS)

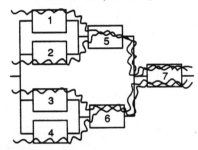

FIGURE 6.17   Cut sets and tie sets.

Cut Sets and Path Sets from a Fault Tree.   Figure 6.18 shows an example of a fault tree and its corresponding reliability block diagram for the top-level event described as "motor overheats."

Note from the block diagram in the figure that the OR gate represents events that are in series in the reliability block diagram and the AND gate represents events that are in parallel.

The minimal cut sets for this relatively simple fault tree can be seen to be the following events:

- (1) Primary motor failure (overheated)

    OR

- (2) Primary fuse failure (fuse fails to open) AND
  (3) Primary wiring failure (shorted)

    OR

- (2) Primary fuse failure (fuse fails to open) AND
  (4) Primary power supply failure (surge)

For more complicated fault trees, it will not be possible to determine the cut sets by observation. A more structured method will be necessary.

One method or algorithm for determining the cut sets directly from a fault tree involves numbering each gate and event in the fault tree. Remember that every input to an OR gate identifies a separate cut set and that every AND gate identifies just one cut set. Each input to an AND gate simply increases the size of the cut set.

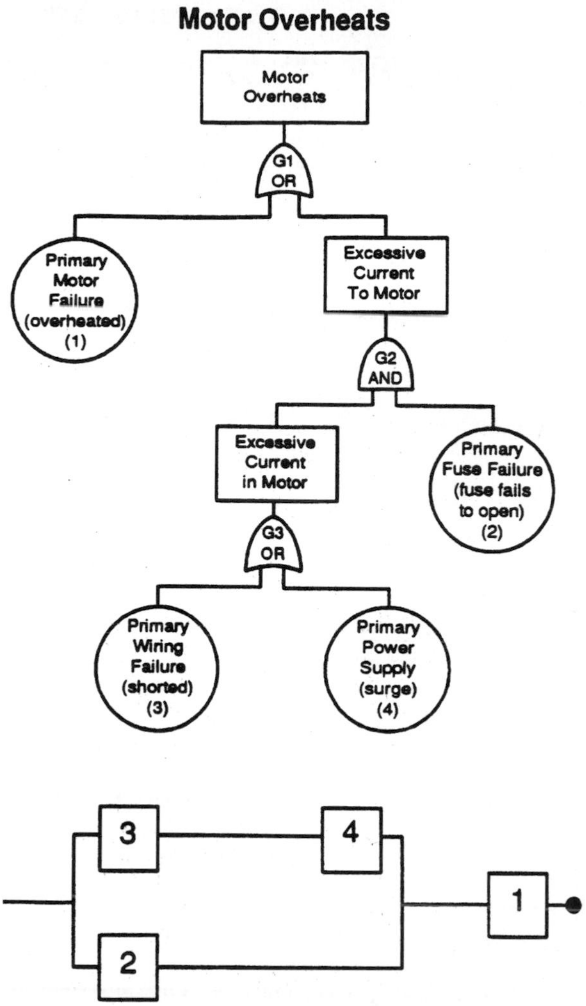

Motor Overheats

FIGURE 6.18   Motor overheats fault tree.

After numbering each gate and event, start with the topmost gate on the fault tree and replace it with its input events. In the example fault tree shown in Fig. 6.18, the topmost gate, labeled G1 for Gate 1, is an OR gate with two inputs, which identify two cut sets:

1

G2

Each cut set is written on a separate line. Each line that contains a gate must be expanded by replacing the gate with its respective inputs.

Gate 2 (G2) has two inputs. Since the gate is an AND gate, both of the inputs will be accounted for on the same line.

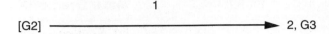

Using the same approach, replace Gate 3 (G3) with its inputs. Gate 3 is an OR gate with two inputs. This means that Gate 3 must be replaced with two separate lines—one line for each input.

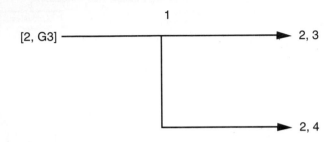

Now that all of the gates have been replaced by their inputs, a complete list of the cut sets for the example fault tree in Fig. 6.18 has been developed. The cut sets are

1

2, 3

2, 4

Note that this agrees with the results that we obtained by observation. This procedure can be used to obtain the cut sets for any size fault tree. For a large fault tree, you can save time by using a software program that uses the boolean algebra equations for the fault tree to determine the cut sets.

As mentioned earlier, a path set is an event or combination of events whose nonoccurrence ensures the nonoccurrence of the top-level event. A path set is minimal if it cannot be further reduced and still remain a path set.

Because of the dual, or complement, relationship between the fault tree and the success tree, the path set for the fault tree is the cut set for the "dual" success tree. Therefore, in order to find the path set in a fault tree, draw the corresponding success tree and find its cut sets using the previously described algorithm. The success-tree cut set will be a fault-tree path set.

6.9.2 Quantitative Evaluation

Once the appropriate logic gates, symbols, and event descriptions have been developed, the next level of complexity in FTA involves the calculation of the probability of occurrence of the top-level events. For this, the probability of occurrence values for the lowest-level events are required.

When the probabilities of occurrence of the lowest-level events have been determined, the probability of occurrence of the top-level events can be calculated, using boolean algebra probability techniques. (Note that fault-tree quantitative evaluation should be performed only on a reduced fault tree.) Probability equations for the AND gate and the OR gate are shown in Figs. 6.19 and 6.20.

In most cases, it will be more convenient to use computer software to perform these calculations. One such computer program is the Integrated Reliability And Risk Analysis System

FAULT TREE

PROBABILITY EQUATION

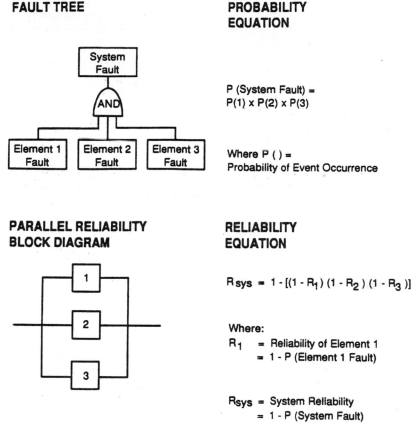

P (System Fault) =
P(1) x P(2) x P(3)

Where P ( ) =
Probability of Event Occurrence

PARALLEL RELIABILITY BLOCK DIAGRAM

RELIABILITY EQUATION

$R_{sys} = 1 - [(1 - R_1)(1 - R_2)(1 - R_3)]$

Where:
R_1 = Reliability of Element 1
= 1 - P (Element 1 Fault)

R_{sys} = System Reliability
= 1 - P (System Fault)

FIGURE 6.19   Probability and reliability equations for AND gate.

(IRRAS-PC) program, which was developed at the Idaho National Engineering Laboratory for the U.S. Nuclear Regulatory Commission.

6.9.3   Conversions to Reliability Block Diagrams

Fault-Tree Conversion to a Reliability Block Diagram.   An OR gate in a fault tree corresponds to a series reliability block diagram, as illustrated in Fig. 6.20. An AND gate in a fault tree corresponds to a parallel reliability block diagram, as illustrated in Fig. 6.19. To illustrate the concept just discussed, take another look at FTA Example 1, which is shown in Fig. 6.21. The probabilities of the lowest-level events are given in Table 6.8.

From the reliability block diagram, the system reliability is

$$R_{sys} = 0.99997$$

The probability of potential engine damage due to insufficient oil pressure, $P(D)$, can be found by subtracting R_{sys} from 1:

$$P(D) = 1 - R_{sys} = 1 - 0.99997 = 0.00003$$

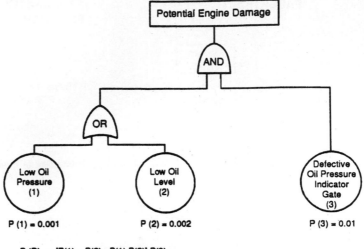

$$P(D) = [P(1) + P(2) - P(1) P(2)] P(3)$$
$$= [0.001 + 0.002 - (0.001) (0.002)] \, 0.01$$
$$= (0.002998) (0.01) = 0.00002998$$

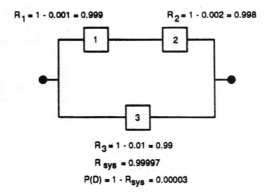

$$R_3 = 1 - 0.01 = 0.99$$
$$R_{sys} = 0.99997$$
$$P(D) = 1 - R_{sys} = 0.00003$$

FIGURE 6.20   Probability and reliability equations for OR gate.

As shown in Fig. 6.21, this agrees with the values of $P(D)$ obtained from the fault-tree probability equations.

Quantitative evaluation of fault trees enables the PD team to determine the

- Overall likelihood of the undesired top-level event
- Combination of input events most likely to lead to the undesired top-level event
- Single event contributing the most to this combination
- Most likely path(s) leading through the fault tree to the top-level event

Also, if the system under investigation is modified in any way, the fault tree can be changed easily and quickly to reflect this modification, and new calculations can be developed to determine the effect of the change. As a matter of fact, many modifications can be made to the system, and the effects of all of these can be evaluated at the same time.

FAULT TREE

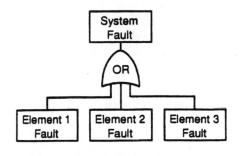

**PROBABILITY
EQUATION**

P (System Fault) =
1 - [1 - P(1)] [1 - P(2)] [1 - P(3)]

Where P ( ) =
Probability of Event Occurrence

**SERIES RELIABILITY
BLOCK DIAGRAM**

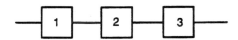

**RELIABILITY
EQUATION**

$R_{sys} = R_1 \times R_2 \times R_3$

Where:

R_1 = Reliability of Element 1

   = 1 - P (Element 1 Fault)

R_{sys} = System Reliability

   = 1 - P (System Fault)

FIGURE 6.21   Probability equations for engine damage.

TABLE 6.8   Probabilities for FTA Example 1

Event	Fault	Probability of fault occurrence, $P(F)$	Probability of success $R = 1 - P(F)$
1	Low oil pressure	0.001	0.999
2	Low oil level	0.002	0.998
3	Defective oil pressure indicator gage	0.01	0.99

Success-Tree Conversion to a Reliability Block Diagram. In addition to converting a fault tree to its corresponding reliability block diagram (as previously discussed), it is also possible to convert a success tree to a reliability block diagram. This conversion is shown in Fig. 6.22.

Success Tree Conversion To Reliability Block Diagram

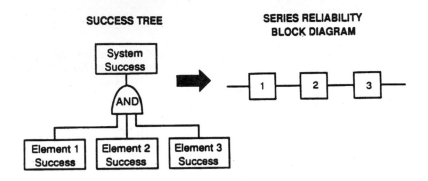

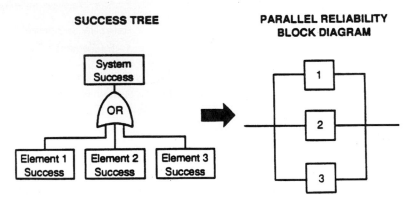

FIGURE 6.22 Success-tree conversion to reliability block diagram.

This success tree/reliability block diagram relationship follows from the fact that for a system success, the series reliability block diagram requires that Block 1 AND Block 2 AND Block 3 all be successes. Thus, the success-tree AND gate corresponds to a series reliability block diagram. The relationship between the success tree OR gate and a parallel reliability block diagram can also be seen in Fig. 6.22.

Quantitative Evaluation of a Success-Tree Reliability Prediction. Success trees can be quantitatively evaluated using the same procedure just discussed for fault trees.

The top-level probability of occurrence of a success tree is *probability of success,* which by definition is *reliability R.* Because of this fact, a success tree can be used as a reliability prediction model. If a success tree has been developed for a product and the probability of occurrence has been determined for each cause, the reliability of the product can be determined by calculating the top-level probability of success.

6.10 OTHER ASPECTS OF FTA

6.10.1 System Safety Analysis Feature

An additional feature of FTA is that the logical approach associated with it is very useful for system safety analysis to assess safety considerations or potential hazards. The following are a few of the reasons why fault trees are used in system safety analysis:

- Logical fault paths can be derived from FTA.
- The PD team is able to consider the effects of multiple causes or combinations of causes.
- The PD team is able to include external influences in the analysis, such as
  - Human error
  - Misuse or abuse
  - Judgment or decision error

By examining the hazard potential, the PD team may develop a better understanding of the safety issues associated with various fault paths. Using this approach may help in prioritizing areas for corrective action, if practical, or in assuring proper control methods, where safety or potential injury is a concern.

6.10.2 Troubleshooting Facilitation

The FTA approach of starting with a top-level fault and deductively determining the root causes makes the fault tree an excellent reference for developing service and maintenance manuals. The fault tree can provide a logical approach for service diagnostics. The analysis process is identical except that, instead of being performed prior to the event, service diagnostics are performed after the fault has occurred.

Because the fault trees are developed early in the design phase, they will be available to the individuals responsible for developing service diagnostics. This should result in substantial cost savings.

6.10.3 Use as a Reference Guide

The FTA can also be used throughout the PDP as a reference document for a number of other reliability activities, such as

- Failure analysis
- Design-level FMEA
- Process-level FMEA
- Design verification plan and report (DVP&R)
- Process verification plan and report (PVP&R)

6.11 COMPARISON TECHNIQUES

6.11.1 FTA and Cause-and-Effect Diagrams

A cause-and-effect diagram is a deductive analytical technique that uses a graphical "fishbone" to show the cause-and-effect relationships between an undesired event (failure mode) and the various contributing causes.

Usually, a cause-and-effect diagram is drawn up with the same orientation as a fault tree. That is, both techniques start with the end effect (or failure mode). From this, the next step is to develop all known or possible causes that lead to its occurrence.

The cause-and-effect diagram shown in Fig. 6.23 is developed by thinking of all the potential causes and then grouping these causes into five major categories: people, materials, machinery, environment, and methods. However, little or no attention is paid to any interactions or dependencies between the various categories.

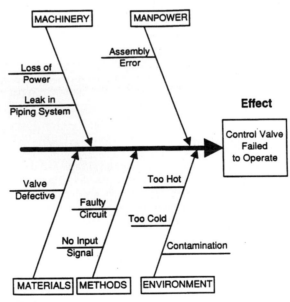

FIGURE 6.23 Cause-and-effect diagram.

In comparison to the cause-and-effect diagram, the FTA looks at the various causes that might lead to any given effect in a highly structured manner.

In contrast to the cause-and-effect diagram, the FTA focuses on any dependency or interaction that may exist between the cause and the effect. The FTA does not group causes into the five categories (except to look for any common cause that may apply to the various branches of the fault tree).

6.11.2 Failure Mode and Effects Analysis

FMEA and FTA are complementary analysis forms in many ways.
The classic FMEA, in its most common form and approach,

- Is a "bottom-up" type of analysis. This means that the analysis considers failure modes at the lowest level and works its way up to determine the effects at the highest level.
- Involves an inductive-analysis approach (as in Fig. 6.24) to consider how a low-level, single-point failure can lead to one or many effects at the top level.

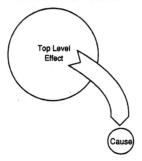

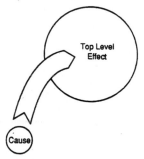

FIGURE 6.24 Deductive and inductive approaches.

FTA differs significantly from FMEA in approach and construction. The classic FTA

- Uses a "top-down" (rather than a bottom-up) analysis to determine the causes of the top-level effect.
- Utilizes deductive analysis (shown in Fig. 6.24) instead of the inductive analysis of the FMEA. The deductive analysis derives all of the possible causes, or combinations of causes, that may lead to the top-level event.

FTA involves an analysis of a given single top-level event. This analysis is used to identify all of the possible causes for the event. The probability that the top-level event will occur can be predicted by using the probability of the causes and relationships that lead to the top-level event.

In order to perform either FMEA or FTA, a PD team must possess considerable knowledge of the system. However, FTA does have an advantage in that

- The fault tree can be used as a design development tool. This tool does not require that all of the design details be complete before the analysis process can actually begin. These design details refer to such items as drawings, specifications, methods, and so forth.

- The analysis effort may begin earlier in the design development process; however, an understanding of how the system is expected or intended to operate must exist prior to beginning the FTA.

Table 6.9 depicts the various advantages of both FMEA and FTA, as well as the expected performance of the tool and the situations in which it is the best approach.

TABLE 6.9 FMEA—FTA Comparison Table

Characteristics	FTA	FMEA
Best approach for:		
Multiple failure description	X	
Analyzing single-point failures		X
Avoiding analysis of noncritical failures	X	
Identifying higher-level events liable to failure as a result of lower-level events	X	
Having a broader sphere in observation of failure mode phenomena		X
Having fewer restrictions and being easier to follow	X	
Identifying external influences	X	
Identifying critical characteristics		X
Providing a format for validation plans		X
Should be used when:		
A quantitative analysis is needed	X	
There is a need to ensure that each component/characteristic is examined for failure modes and effects		X
Information is limited to system schematics or basic functions	X	
Design information is available in detail drawings and specifications		X
Evaluating alternative designs/approaches	X	
Evaluating redundancy	X	
Evaluating design integrity including failure-safe, failure detection		X
The analysis objective is deductive (top-down) to derive failure causes	X	
The analysis objective is inductive (bottom-up) to derive failure effects		X

In reality, though, these techniques should be viewed as complementing each other. Both techniques focus on generating and taking action on the same information. That is, they both focus on failure modes, causes, and effects, but each tool approaches these from a different perspective.

These techniques are best used to support each other rather than being viewed as in competition or in conflict.

CHAPTER 7
RELIABILITY SPECIFICATIONS AND GOAL SETTING

Henry J. Kohoutek
MANAGEMENT AND ENGINEERING STRATEGIES
SSA INTERNATIONAL
LOVELAND, COLORADO

7.1 INTRODUCTION

Recognition of the importance of business objectives, and the need for their systematic treatment, goes back to Alfred P. Sloan, Jr. of General Motors (1875–1966) and Pierre S. du Pont (1870–1954). The fundamental driving force behind this recognition is the criticality of shared understanding by managers of what a company's business is and should be. Objectives represent a translation of the company's main purpose, mission, and vision of the future into a framework for defining, directing, and timing actions. Therefore, they should enable the company to

- Communicate principal business aspects in a small number of statements.
- Evaluate the validity of managerial decisions from the viewpoint of their match to business purpose.
- Predict future behavior and test its effectiveness.
- Accumulate experience, and use it systematically for performance improvement.

7.1.1 Forms of Objectives

Objectives are neither a description of fate nor commands that determine the future. They are directions and commitments that mobilize resources, energies, and efforts to shape the future. Because they must be deployed at all levels, objectives may take many different forms despite their common sources: market, customers, and business needs. For the purpose of this chapter, we will consider *objective* to be a generic concept, from which the following terms have been derived:

- A *policy* is actually a concentrated genuine high-level decision, characterized by many risks and therefore is frequently tested against market and technology dynamics.
- A *requirement* is a condition or capability which must be met or possessed by a product or system if it is to achieve an objective.
- *Product requirements* are requirements that are oriented toward solving customer problems, and toward support of company objectives.

- *Manufacturing requirements* address the objectives of manufacturability, product and process quality, cost, etc.
- A *specification* is usually a document that prescribes, in a complete, precise, and verifiable manner, the requirements, constraints, expected behavior, or other characteristics of a product or system.
- A *goal,* sometimes called a *target,* states a quantified performance level related to a specific measurable parameter, and is used to specify the desired operational achievement of an organization.

7.1.2 Desired Attributes

As a group, objectives must convey the set of priorities accepted by management and be balanced against profitability goals, current and distant demands and, quite importantly, each other. Successful objectives usually display a set of attributes, which may be arranged into two categories:

1. *Contextual,* such as the following:
   - Must be derived from the business purpose
   - Must be operational, i.e., capable of being converted into specific goals and actions
   - Must allow concentration of resources and efforts
   - Must be specified for all areas on which business survival and success depends

2. *Formal,* reflecting their role as a means of communication. These must be:
   - Relevant to the role of the organization charged with their implementation
   - Compatible with the existing reward and recognition system
   - Acceptable to those who will be actively involved in their pursuit
   - Attainable and supported by available resources
   - Clear, simple, measurable, and timed

This multiplicity of desired attributes indicates immediately the potential and actual conflicts that may be encountered during the process of developing and formulating objectives, and also leads to the pragmatic recognition that no design can fully satisfy all of them simultaneously.

7.1.3 Scope of the Need for Reliability Objectives

One characteristic illustrating the importance of reliability is its relevance to and pervasiveness throughout all the aspects of a business:

- From a potential customer's first impression to the level of satisfaction with long-term after-sale support of the product in the customer's application
- From the conceptual design of a new product to the legally required spare parts inventory to be maintained after its obsolescence
- From the overall business strategy of delivering to customers products of a superior value to the tactics of in-process handling of minute components

The scope of the reliability influence requires reliability managers to be familiar with the overall big picture, as well as with the individual goal-setting methods and their fit for particular situations. Table 7.1 attempts to outline this scope by pointing to the three types of reliability goals and their primary driving forces, and by listing their appropriate forms and the issues that will most probably be encountered during their development and deployment.

The width of the field calls for teamwork, especially between the top management and implementers, and its depth calls for familiarity with both economics of reliability and reliability assurance technologies. The time delay between the setting of goals and the availability of a

TABLE 7.1 Scope of the Reliability Goal-Setting Field

Reliability goal type	Primary topics to be considered	For details, refer to Sec.
Company level	Driving forces:	
	Market demands	
	Competitive pressure	
	Business economics	
	Forms:	
	Reliability policy statement	7.2.1
	Goals for companywide improvement program	7.2.2 and 7.2.3
	Goals for the reliability engineering function	7.2.4
	Issues:	
	Compatibility with business objectives and strategy	
	Value of reliability	4.2
	Credibility of management's commitment	
	Implementation control	
Product family	Driving forces:	
	Product price erosion	
	Technology progress	
	Regulations and standards	
	Forms:	
	Design requirements	
	Component derating specifications	7.3.1
	Design margins specifications	7.3.2
	Environmental ruggedness specifications	7.3.3
	Reliability growth maps	7.3.4
	Issues:	
	Customer feedback	
	Accuracy of reliability data	
	Experimental validation of reliability models	
Individual products	Driving forces:	
	Customer requirements	
	Competitive products' reliability	
	Cost of reliability	4.3 and 4.4
	Reliability assurance technology	
	Forms:	
	Direct:	7.4
	MTTF, MTBF requirements	
	Failure rate or warranty cost limits	
	Indirect:	7.5
	Testability requirements	
	Required production reliability screens	
	Environmental ruggedness specifications	7.3.3
	Issues:	
	Goal apportionment and deployment	7.4.6
	Implementation strategies	
	Product special considerations	7.6

high-confidence assessment of the actual field reliability performance puts an additional demand on reliability management's credibility and skills in justifying expenditures that frequently have the character of a long-term investment.

7.1.4 Goal-Setting Process

Because of its close link to the overall new product development or company-level quality improvement goals, the process of setting reliability goals is usually part of product or business planning. The technological dependence of design for reliability, and its assurance, on selected product design and manufacturing methods makes this link very strong, and demands concurrence and mutual support in planning and implementation.

The actual goal-setting process may have many variants, but in general, it goes through a few clearly identifiable stages:

1. Selection and outline of the domain to be responsible for the goal's successful accomplishment (company, function, project team, etc.)
2. Identification of the driving forces and reasoning behind the need to set this particular goal (customer demand, regulatory requirements, change in technology, etc.)
3. Identification of the most appropriate form of the goal statement (policy, design requirement, product specification, etc.)
4. Employment of past performance data, reliability models, economic evaluations, etc., to propose a range of goal values (from minimum acceptable level to preferable, or world-class performance)
5. Establishment and verification of the goal's linkage to higher-level objectives, defining their priority, and testing for desirable attributes
6. Decomposition of the overall goal into subgoals, and classification of those according to their importance (musts and wants)
7. Identification of major implementation issues, constraints, barriers to success, and proposed progress reporting and evaluation system
8. Assuring that the goal is well understood and accepted by the responsible implementation team

This list outlines the sequence of events only approximately because in actual industrial situations, restarts, repetition of steps, and participation of representatives of the implementation team may be required throughout the process.

Determining reliability specifications and goals is, in essence, a highly structured task requiring both engineering and economic reasoning. Major deviations from the dominance of the technical and methodological aspects of the process, although justifiable in some extraordinary situations, are usually harbingers of future difficulties or indicators of some weakness in management.

7.1.5 Primary Issues

Despite the fact that the overall approach to reliability objectives development, optimization, allocation and deployment, implementation, monitoring, etc., is based on established business and product planning practices, there are a few issues that deserve special attention.

Flexibility. Reliability objectives must be translated into a set of actions carrying some real level of uncertainty and therefore must not be too restrictive. They must be reasonably flexible because the expectations on which they are based may change with time, as a result of factors outside of company control. From the formal viewpoint, this flexibility can be expressed by an interval in which the performance level is expected to fall, or by a single specification limit: "Performance better than…," accompanied by a statement of the desired minimum confidence level.

Goal Apportionment and Deployment.   This is one of the most difficult steps in the goal-setting process. To assure success and acceptance here requires a clear and agreed-upon goal allocation strategy, factual knowledge to support it, involvement of participants, and flexibility to deal with eventual difficulties (see Sec. 7.4.6).

Performance Validation.   Validation is a time- and resources-consuming activity, calling for new innovative, indirect, or continuous assessment methods to improve its effectiveness and to minimize its unfavorable impact on project schedule and cost.

Importance of Innovation.   Innovation, the primary strategy for business success, plays a similar role in the field of reliability. The known managerial approaches to and engineering methods for continuous improvement, as represented in contemporary literature, handbooks, consultants' preaching, and even the experience of successful practitioners, must be supplemented by an unrelenting systematic search for opportunities for breakthrough and substantive innovation.

7.1.6   Reliability Management by Objectives

The fact that every employee contributes something different and the need for all to contribute for the success of the whole organization are behind Peter F. Drucker's principle of "management by objectives" (MBO), which emphasizes self-control and participation.

Implementation of MBO in reliability management must take into account the many forces acting against achieving common objectives in any organization or institution, such as

- Specialization and competition among individual managers and their organizations
- Isolation tendencies inherent in hierarchical organizational structures
- Differences in interpretation of the company vision and in personal values
- Tendency toward professional perfection as an important but unstated personal objective
- Lack of effective tools for performance measurement and control

Since the early 1980s, MBO, as a results-oriented approach to management, has often been discussed in contrast to process-oriented management, that is stressing management's role in stimulating and supporting employees' efforts to improve their work processes. But even the process-orientation enthusiasts admit the need for undivided attention to results.

7.2   COMPANY-LEVEL GOALS

In some business situations, the reliability of the delivered product is essential for customer satisfaction and business success, or to minimize the damaging adverse consequences of a failure. Power plants, airlines, and the chemical, medical, and communication industries are highly visible examples. In other situations, affordable high reliability may provide a significant competitive advantage that can be translated into improved profitability or market share, or inadequate reliability may result in a loss. Here, the automotive and consumer electronics industries are well-known examples. In all these situations it is necessary, or at least prudent, to deal with reliability management at the business strategic level, and to develop companywide reliability objectives and goals.

7.2.1 Reliability Policy

One possible way to approach company-level reliability goals is to provide a strategic decision framework in the form of a company reliability policy. The main purpose of such policy statements is to give company customers, suppliers, strategic partners, and employees a clear indication of top management's position and intentions toward reliability of company products, and of the importance of reliability for successful and satisfying business relations.

To assure credibility, and avoid potential uncertainties about its significance, this policy must be signed by the top management, and be compatible with other company policies. It is also beneficial to publish supplements to this policy, such as

- Explanatory statements relating the reliability policy to business objectives and strategy, and to the company's own long-term vision
- Reliability management fundamentals, including important definitions, assigned responsibilities, fundamental performance metrics, principles guiding the development of goals for individual products, resources justification guidelines, preferred implementation strategies, performance monitoring and appraisal system, etc.
- Outline of the reliability system for policy implementation, and implementation support

All this information may be issued as a company internal standard or in the form of a reliability management manual. Because of the strategic importance of this information, the document containing it must be a controlled document, with full distribution and version controls institutionalized. A set of good practices dealing with company policies has been established by the quality management community in relation to the ISO 9000 registration requirements.

Even though the development of policies is a reasonably straightforward process, advance planning is required to assure:

- A comprehensive view of all important issues involved
- Availability of all decisive facts and knowledge
- Participation and contribution by all involved and affected, as well as reliability experts
- A clear, acceptable, and effective strategy for policy implementation and enforcement

The actual policy may be a simple and short statement of intent, such as, "Every new product will have reliability at least 10 percent higher than the product being replaced by it." Some companies may prefer a comprehensive document containing, besides the policy statement itself, background information, purpose, scope, and explicitly stated responsibilities. Keeping policy documents to the minimum is preferable, because they should be auditable, as should the system dedicated to their implementation.

7.2.2 Companywide Improvement Programs

Companywide reliability improvement programs and initiatives, despite sounding like a great idea and being fashionable, are high-risk endeavors. Their problems originate in the simple fact that real improvement usually takes a long time, stretching the patience of both managers and participants. Successful programs are usually down-to-earth initiatives, well justified and explained, and focused on a few key performance measures. They will be geared to deliver both short-term and long-term results and frequently monitored, and the required efforts will be aligned with the existing reward and recognition system.

A five-year program, with both internally and externally focused goals, is used as an example to illustrate the goal setting and justification logic (for numerical values, see Table 7.2).

TABLE 7.2   Simplified Justification of a Companywide 5-Year Reliability Improvement Program

Parameter	n^a	$n+1$	$n+2$	$n+3$	$n+4$	$n+5$
Goal 1: Eliminate the need for additional staffing of the service engineering function.						
A: Current trend						
Number of installed products (thousands)^b	100	120	144	173	207	248
Average product failure rate, percent per year^c	1.00	1.00	1.00	1.00	1.00	1.00
Total number of service calls needed per year	1000	1200	1440	1730	2070	2480
Calls per service person per year^d	100	105	110	116	122	128
Number of service persons needed	10	11	13	15	17	19
B: Desired trend						
Desired number of service persons^e	10	10	10	10	10	10
Calls per service person per year^d	100	105	110	116	122	128
Total number of service calls possible per year	1000	1050	1100	1160	1220	1280
Required average product failure rate, percent per year	1.00	0.875	0.764	0.671	0.590	0.516
Improvement rate, percent per year	—	12.5	12.7	12.2	12.1	12.5
Goal 2: Improve value of product reliability delivered by 5 percent per year.						
C: Expected trend						
Average product failure rate expected, percent per year	1.00	0.875	0.764	0.671	0.590	0.516
Trend in product average price, thousands of dollars^f	10.00	8.00	6.40	5.12	4.10	3.28
Reliability value expected, percent per year per thousand dollars^g	0.100	0.109	0.119	0.131	0.144	0.157
D: Trend needed						
Needed reliability value, percent per year per thousand dollars^h	0.100	0.095	0.090	0.086	0.081	0.077
Trend in product average price, thousands of dollars^f	10.00	8.00	6.40	5.12	4.10	3.28
Needed average product failure rate, percent per year	1.00	0.760	0.576	0.440	0.332	0.253
Needed improvement rate, percent per year	—	24.0	24.2	23.6	24.5	23.8

^{a}Current year situation
^{b}20 percent per year growth expected
^{c}No improvement assumed
^{d}5 percent per year productivity improvement assumed
^{e}Unchanged staffing (Goal 1)
^{f}20 percent per year price reduction expected
^{g}Unfavorable trend in value
^{h}5 percent per year improvement (Goal 2)

Program Overall Goals

1. Internal goal: Eliminate the need for additional staffing of the service engineering function.
2. External goal: Provide company customers with product reliability value that is increasing continuously at a rate of 5 percent annually, or better.

Strategy.   Continuous product reliability improvement through better design and manufacture.

Constraints

1. No fundamental change in the product line characteristics and mix.
2. No fundamental change in service strategy.
3. Program should be expandable beyond the five-year period, if needed.

Analysis.   From the consideration of current trends and stated program goals, derive the required reliability improvement rate goal.

Resulting Implementation Goal.   The average improvement in the annualized product failure rate must be 24 percent per year, at least.

Comment.   Monotonic rate change trend has been assumed for simplification of the analysis example.

7.2.3   Requirements for Reliability Assurance Programs

Well-established market requirements, powerful customers, industry standards, etc., may demand contractual visibility of the company's internal framework for assuring the expected or required reliability. This visibility is usually achieved through a reliability manual describing company reliability programs. Such a manual, or a similar document, is a management communication tool and does not provide any technical implementation detail or reveal any proprietary information. Current reliability management practice calls for descriptions of:

• Program elements, such as reliability organization, responsibilities, and management controls
• Reliability planning and reporting
• Testing, especially development validation and qualification testing
• Parts and production reliability assurance
• Reliability engineering tools and methods in use
• Reliability engineering and assurance training

MIL-STD-785B, *Reliability Program for Systems and Equipment, Development and Production,* may serve as an example of general requirements and expected specific tasks for reliability programs, and as a guide for program planning. It also points to related documents that address specific issues, such as procurement and environmental testing.

7.2.4   Goals for the Reliability Engineering Function

The service and staff character of the reliability engineering and assurance function justifies a closer look at its cost-effectiveness. Its contribution to cost savings and value creation is indirect, and therefore management metrics and goals are also indirect, and often stated in relation to the performance of the line functions supported. The following list represents an example of a reasonable set of variables allowing performance assessment.

Economic Indicators:

• Cost of reliability engineering as a percentage of R&D cost, totals yearly and for each individual project
• Cost of reliability engineering as a percentage of cost of quality
• Cost per reliability engineering service hour, actual and compared to charges by external service companies and laboratories
• Capital intensity in thousands of dollars invested per capita, actual and compared to R&D

Service quality indicators:

• Satisfaction ranking by department customers
• Percentage of designers trained in design for reliability
• Percentage of service requests fulfilled on time
• Number and percentage of service requests not accepted

Reliability value indicators:

- MTBF per thousand dollars of product price
- Product warranty failure rate per thousand dollars of product price
- Reliability value growth rate

Other indicators:

- Cost of reliability engineering as a percentage of warranty cost
- Product warranty failure rate
- Customer-perceived company product reliability rank relative to key competitive products

The value indicators represent the reliability a customer is purchasing per unit price, e.g., per thousand dollars of the product price. They are useful for comparisons across different products, products from different vendors, or products of different vintages. Because reliability is a function of product complexity, the price is used here as an approximate measure of this complexity (see Sec. 4.2 for a more detailed discussion of the value of reliability).

The process of setting numerical goals for these indicators is usually an overlapping mixture of trend analysis, a negotiated balance with the budget plans of the supported departments, and internal performance improvement initiatives. Simple optimization methodology and prudent application of past actual performance data may provide additional credibility to the justification of budgeting proposals, and smooth out negotiations (see Sec. 4.4.5 for discussion of the economics of the reliability engineering function).

7.3   GOALS FOR FAMILIES AND GENERATIONS OF PRODUCTS

One of the important activities of reliability management and engineering is to provide business strategic planners and new product developers with information about trends in customer needs and company technology capabilities related to reliability performance and its improvement. This information, which describes expected changes and trends in reliability profiles of company product families across future generations, displays two sets of principal characteristics:

1. One set contains requirements that will remain unchanged over the planning period of interest. These are usually based on imposed regulations or reflect physical characteristics of the application environment which do not change with time.

2. The second group reflects changes in the complexity of the problems customers plan to solve, leading to demands for higher reliability, and progress in underlying technologies that will allow the delivery of products that meet these more stringent requirements. It also includes the expected price erosion of the products.

The first group can be addressed directly by permanent design for reliability policies dealing with technical issues of derating, design margins, etc. The time-dependent requirements must be approached through focused reliability growth programs, derived from longer-term trend projections and documented, for example, in product family reliability maps.

7.3.1   Derating Policies

The idea of operating components below their basic rating is fundamental to all fields of engineering design. Derating is effective because the reliability of the majority of components increases as the applied stress levels are decreased. The opposite is also true. Derating, as a design for reliability strategy addressing application of existing components, relates to all four basic families of stresses: mechanical, thermal, chemical, and electrical stresses. Studies of derating effectiveness indicate a correlation with improved reliability at the correlation coefficient level $r > 0.95$, because of the reduced energy available for activation of degradation mechanisms and processes.

Actual derating procedures vary with different types of parts and their application, but the underlying mathematical models have a common generic form relating the derating level to the part's base failure rate. In design for reliability guidelines, they are presented in one of these graphical formats:

- Base failure rate as a function of level of a particular stress
- Maximum stress ratio allowable as a function of level of a particular stress
- Base failure rate multiplier as a function of level of a particular stress
- Operating life as a function of application to nominal stress ratio

These policies are published internally as

- Derating tables (examples may be found in Ref. 1) specifying the component type, the parameter of interest, the nominal derating factor (i.e., recommended practice), and the worst-case derating factor
- Graphs indicating recommended and forbidden application regions (see example in Fig. 7.1)

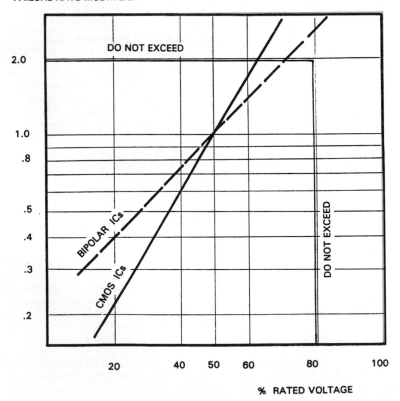

FIGURE 7.1   Example of a graphical representation of derating policy for a family of products.

- Database for reliability analysis supporting computer programs, such as RADC's ORACLE

Because it takes only one overstressed component to cause repetitive design-related product failures with potentially significant adverse business consequences, systematic enforcement of the derating policy is of great importance. Computer-aided reliability analysis tools, available to all designers for design evaluation, seem to be the best approach.

7.3.2 Design Margins as Reliability Specifications

Component and parts family designers face a similar problem, but their implementation strategy follows the logic of derating in the opposite direction. The reliability of the device under consideration must be assured by overdesigning to reliability specifications that significantly exceed the expected nominal stress level. The concepts underlying this strategy are stress–strength and cumulative damage models.

Stress-Tolerance Funnel.   During the design specification and goal-setting process for a family of general-purpose components, the actual range of application stresses and environments is seldom known. However, a reasonable set of design specifications can be derived from forecasted customer requirements by defining a variety of allowances for potential stresses in different phases of a given component's life. Because these allowances are considered additive, they can be presented in the form of a stress-tolerance funnel (see Fig. 7.2 for an example).

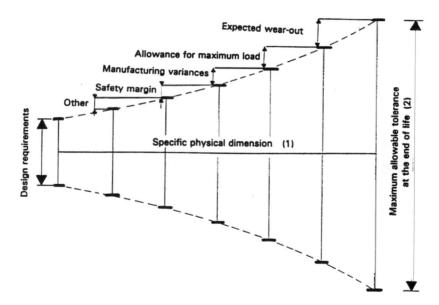

Notes: (1)   Reliability critical dimension, such as disk memory spindle diameter.
        (2)   Tolerance still assuring nominal performance.

FIGURE 7.2   Example of a physical dimension-tolerance funnel.

Since many components are designed as a family, based on well-defined technology and for some specified application domain, it is prudent to structure the existing experience with application stresses and their expected future changes into a validated stress-tolerance funnel, and use it as a guideline for setting design for reliability specifications.

Composite Failure-Line Graph (Schmoo Plot).   Knowing the variation range of failure-critical parameters of a device and their interdependencies forms a basis for development of reliability qualification requirements. This knowledge is often expressed as a composite failure-line graph, where each line represents device failure conditions in relation to a combination of given drift parameters and any other specified parameter. From these plots, tables of upper and lower limits for each parameter are derived and compared, and the most restrictive limits determined. These may serve as test conditions that must be met if the device is to meet the reliability qualification.

Figure 7.3 provides an example of a very simple Schmoo plot of the acceptable performance region of a consumer electronic product as a function of power line voltage and ambient temperature.

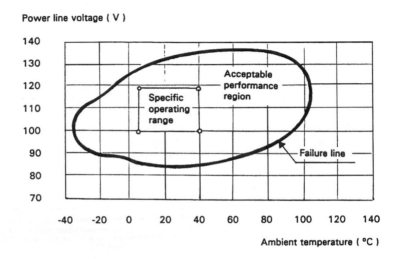

Note: Acceptable region boundary = Failure line

FIGURE 7.3   Example of a Schmoo plot for a consumer electronic product.

Safety Factors.   In the design of mechanical components for application in, for example, the housing, automotive, or civil engineering industry, legislated design codes often directly prescribe reliability requirements in the form of predefined multipliers (safety coefficients) of the nominal stress level, in order to assure the expected functionality even under overload conditions.

The National Electric Safety Code, which may serve as an example from the electric power industry, simultaneously addresses issues of system safety and reliability by specifying, for example,

- Line insulation level requirements (providing both safety and protection against corrosion)
- Vertical clearance of wires, conductors, cables, and live parts of equipment above ground, rails, and water
- Minimum clearance from live parts for low, medium, and high voltages

Here the application environment requirements have already been translated into design specifications for multiple families of products.

7.3.3   Specifications for Environmental Ruggedness

Reliability-oriented design considers the need for product environmental resistance early in the design planning cycle. Identification and detailed description of the environments in which the product is required to operate is the first step, which must be followed by identification of degradation processes affected by those environments. Fortunately, for the majority of products, the physical characteristics of the application environment are known and stable over long periods of time. This makes it possible to develop environmental specification standards that are applicable to many designs with no, or only minimal, changes. These standards usually are in tabular form, and describe, besides the environmental parameter of interest, the criteria level and associated qualification test. An example of a partial environmental specification is given in Table 7.3.

TABLE 7.3   Example of Environmental Specifications

Environment—mode	Criteria	Test conditions
Temperature		
Temperature—Operating	To −10°C	Stabilize at −10°C, 30 minutes performance test
Temperature—Survival	To −40°C	Stabilize at −40°C for 30 minutes, return to −10°C, 30 minutes performance test
High temperature—Operating	To +60°C	Stabilize at +60°C, 30 minutes performance test
High temperature—Survival	To +70°C	Stabilize at +70°C for 15 minutes, return to +60°C, 30 minutes performance test
Humidity		
Noncondensing—Operating	+40°C, 95% RH	Stabilize at +40°C, 95% RH, 30 minutes performance test
Condensation—Survival	+25°C, 50% RH	Stabilize at +25°C, 50% RH, transfer to +40°C, 95% RH, apply power, wait for 15 minutes, 30 minutes performance test
Altitude		
15,000 ft—Operating	+47°C, 8.29 psi	Stabilize, 60 minutes performance test

Notes: For additional environments, e.g., shock, vibration, transportation simulation.
       Number of units to be tested is 5 units minimum.

Special care is required when the equipment is expected to operate in unique environments characterized, for example, by high radiation intensity, electromagnetic fields, chemical corrosiveness, high dust and sand concentrations, etc. Accurate characterization and effective testing methodologies are needed for reliability assurance. The same reasoning applies to equipment that is expected to be misused.

7.3.4 Reliability Trend Maps

In industries where long-term trends, usually technology development–based, have been observed and documented, reliability requirements for future product families may be derived by projection of those trends. Trend maps of critical parameters may represent other assumptions also. In the case of product reliability, assumptions about product complexity and price erosion are the most important.

Example 1.   Long-term general reliability requirements for electronic products.

Critical parameters:

• Product annualized failure rate, percent per year.
• Product reliability value, percent per year per thousand dollars of price.

Constraints:

• 1990 situation: reliability value = 5 percent per year per thousand dollars of price.
• Maximum failure rate tolerable by customers: in 1990, 100 percent per year (based on survey data); in 1995, 50 percent per year (expected).

- Minimum failure rate economically feasible is 0.5 percent per year.
- Product price range: $100 through $100,000.

Assumptions:

- Progress in IC technology, levels of integration, and manufacturing automation indicates 35 percent per year.
- Reduction of component failure rate.
- Product list price can be used as an approximation of product complexity.
- Hardware only is taken into consideration.

For the resulting map, see Fig. 7.4.

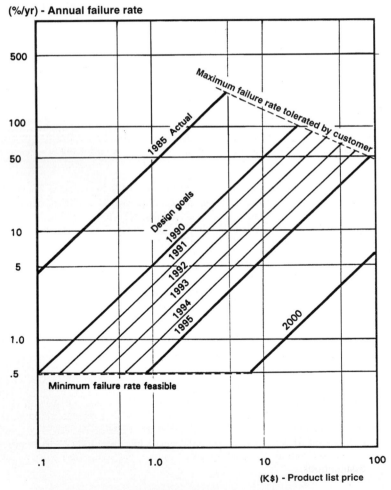

Note: Assumed linear improvement in failur rate: 35% per year

FIGURE 7.4   Reliability map for electronic products.

Example 2.   Long-term general reliability requirements for application software products.

Critical parameters:

- Total number of defects found during the first year after release
- Defect density, measured by the number of defects per 1000 lines of uncommented source code

Constraints:

- 1990 situation: Defect density observed is 0.05 defect per thousand lines.
- Maximum number of defects tolerable by customers: in 1990, 10 total per year (observed); in 1995, 6 total per year (expected). (*Note:* After 1995, the acceptable worst case is expected to be 0.02 defect per thousand lines *or* 2 defects in first three months.)
- Code size range: 5000 lines through 500,000 lines.
- Zero-defect products are considered to be outliers and are not included in calculation of averages or other statistics.
- Only new code is being considered.

Assumption:

- Improvement in defect density in released products is expected to be 20 percent per year as a result of new tools and better process controls.

For the resulting map, see Fig. 7.5.

Practical use of these maps starts with plotting the actual performance of past and existing product families, and calculating a few basic descriptive statistics. For the future requirements, four trends must be considered:

1. Reliability growth of existing products over their expected life, as a result of learning
2. Overall reliability growth over multiple generations of a given product family
3. Introductory performance and subsequent growth of new product families introduced
4. Price erosion of a typical product

Technical issues of assessing price reduction trends, estimating new products code size, and predicting potential discontinuities in product mix or changes in user requirements complicate the application of these maps as goal-setting tools for individual products. Experience is also needed to credibly translate past field performance data into future-oriented requirements for new products. In those situations, a particular product-oriented analysis is necessary.

7.4   RELIABILITY GOALS FOR INDIVIDUAL PRODUCTS

Developing reliability policies, planning improvement programs, and proposing long-term goals for product families taxes reliability management's skills in strategic thinking, insight into market and technology trends, and providing reliability leadership. Establishing reliability goals for individual products demands skills of a more analytical character and, often and preferably, domain-specific knowledge of the products, processes, and technologies involved.

For the actual goal-setting process, it is also necessary to understand the many unstated assumptions.

General:

- Preferences are transitory and mutually interrelated.

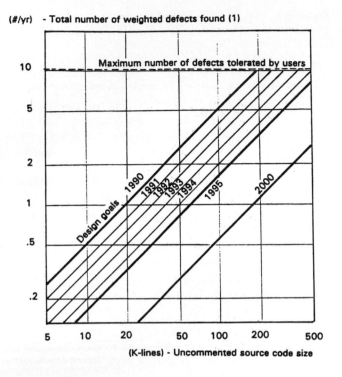

(#/yr) - Total number of weighted defects found (1)

Notes: (1) ... in the first 12 months after release
Assumed linear improvement in defect density: 20% per year

FIGURE 7.5 Reliability map for software products.

- Goals and results providing maximum possible satisfaction are preferable.
- Preference is given to results that are easier to achieve.
- There are many unimportant preference alternatives.

Reliability-specific:

- Higher levels of reliability are desirable.
- A lower-cost approach to achieving a given reliability level is preferable.
- No single goal, or design specification, can cover all the product reliability requirements.
- Stating goals in terms of a range of acceptable results stimulates self-control.

The wide variety of industrial situations has given rise to a variety of goal-setting approaches. Selection of the most suitable method from the manufacturer's perspective is governed by the dominant reliability driving force and by the availability of past performance data. Table 7.4 presents an overview of the options discussed later in this chapter, and may serve as an introductory selection guide as well.

Reliability requirements, goals, and overall objectives may also be established directly by the customer, and imposed on the vendor by contractual agreement. Despite the different motivation behind the customers' goal setting (application environment characteristics, task duration and

TABLE 7.4   Goal-Setting Approach Selection Guide

Customer's application	Driving forces		Company economy	Past performance data		Suitable goal-setting approach	Refer to example in Sec.
	Competitive pressure	Technology		Available	Not available		
	X				X	Arbitrary	7.4.1
	X			X		Economic analysis	7.4.2
			X	X		Economic analysis	7.4.2
		X		X		Comparison	7.4.3
		X			X	Reliability modeling	7.4.4
		X	X	X		Optimization	7.4.5
X				X		Economic analysis	7.4.2
X					X	Reliability modeling	7.4.4

complexity, cost of ownership, requirement of "no maintenance" strategy, etc.), the most frequently used methodologies are essentially the same.

In all goal-setting situations, it is necessary to keep in mind that what is a goal for management becomes directed and controlled activity for the teams and individuals involved and charged with successful implementation.

7.4.1 Arbitrary Goals

Arbitrary goals (which are not preferred but are still useful in some situations) can be both qualitative and quantitative:[2]

- Double the reliability of this particular product in the next two years.
- No product with a failure rate above 50 percent per year will be released to customer shipment.

They often take the form of indirectly stated goals, such as

- Let's establish a failure analysis laboratory.
- Let's eliminate burn-in.
- We need to include components reliability in supplier evaluation.

These self-explanatory goals are appropriate for initial planning, and usually reflect recognition of changes in the business environment, such as rising customer expectations, competitive pressures, reaction to highly publicized quality campaigns, or a breakthrough in managerial attitude.

The following set of statements, which represents the experience of the electronic industry in the early 1990s, may serve as an example of a reasonable set of quantitative arbitrary goals:

- Product warranty failure rate must be less than 5.0 percent per year per thousand dollars of product price.
- Warranty cost average should be less than 0.5 percent of the selling price for simple products and less than 2.0 percent of the selling price for complex products.
- Average active repair time must be less than 4 h.
- Average turn-around time must be less than 7 days.

Even though these goals are arbitrary, they reflect, to a strong degree, knowledge of the conditions under which customers' dissatisfaction becomes a clearly visible major business problem.

7.4.2   Goals Based on End Product Warranty Requirements

A company warranty policy and warranty cost control plans form a reasonable basis for new product reliability goals, as the following example illustrates.

Objective.   For the family of random-access memory (RAM) parts, propose a generic reliability goal that is consistent with the company goal of a warranty cost of 1.5 percent of the value-added shipped (VAS) at the time of the product maturity.

Model Development.   Assuming the independence of failures, the failure rate λ (percent per year) of the group of identical components in question is a simple function of the number of units n in the group:

$$\lambda = n\lambda_{comp}$$

where λ_{comp} is the failure rate of an individual RAM package. The relation K of the end product failure rate λ_P (percent per year) to the RAM component group failure rate can, for the purpose of a simple goal allocation, be assumed to be the same as the relation of the total end product's production cost to the cost contribution of the RAM component group.

$$K = \frac{\text{component group associated production cost (\$)}}{\text{end product total production cost (\$)}}$$

Therefore,

$$\lambda = K\lambda_P$$

The number of failures to be expected during the warranty period T (years) and the cost of warranty repair C_R (\$) then define the warranty cost related to hardware failure C_{HDWR} (\$):

$$C_{HDWR} = \lambda_P T C_R \frac{1}{100} = \lambda_{comp} \frac{n}{K} TC_R \frac{1}{100}$$

If

$$N = \frac{\text{hardware failure warranty cost}}{\text{total warranty cost}} = \frac{C_{HDWR}}{C_W}$$

then it is possible to express the total warranty cost C_W in relation to the component failure rate λ_{comp} as follows:

$$C_W = \frac{1}{N}\lambda_{comp}\frac{n}{K}TC_R\frac{1}{100}$$

In a ratio management environment, warranty cost C_W is expressed as a percentage of VAS, which for an individual product is given by its price P. Assuming that at the time of the new product's maturity, the numerical mix of new and old products under warranty will be close to the ratio M of shipment values

$$M = \frac{\text{new products VAS (\$)}}{\text{total VAS (\$)}}$$

we can predict the new level of the company warranty cost index as

$$W_{com} = MW_{new} + (1-M)W_{old}$$

where W_{old} = company current warranty cost index based on old products' warranty performance, %
W_{new} = warranty cost index for the new products, %, given by

$$W_{new} = \frac{C_W}{P} \times 100$$

In most situations, W_{com} is determined for the long term using other means. If values of n, K, T, C_R, N, P, M, and W_{old} are given by the current data and design considerations, then, combining the equations for C_W, W_{com}, and W_{new}, we have

$$W_{com} = \lambda_{comp} \frac{nTC_R M}{PNK} + (1-M)W_{old}$$

where λ_{comp} is the only unknown. So the desired goal for the component failure rate can be derived by a simple rearrangement of the previous formula:

$$\lambda_{comp} = \frac{PNK}{nTC_R M} [W_{com} - (1-M)W_{old}]$$

As a numerical example, let us assume these values to be given:

Company warranty cost as a percentage of VAS	W_{old} = 2.0 percent W_{com} = 1.5 percent
Warranty period	T_1 = 3 months for 75 percent of products shipped
	T_2 = 12 months for the remaining 25 percent of products
	Average warranty period T = 0.4375 year
Product mix	50 percent "old" products, 50 percent "new" products (M = 0.50)
Repair cost	C_R = \$500 per repair
Hardware-related warranty cost	N = 0.30 of total warranty cost
New product price	P = \$3000
RAM's cost contribution	K = 0.30 of total production cost
Total number of RAM packages	n = 12 in group

Then the annualized failure rate goal required to support the company warranty cost improvement goal can be calculated from

$$\lambda_{comp} = \frac{PNK}{nTC_R M} [W_{com} - (1-M)W_{old}]$$

$$= \frac{3000 \times 0.30 \times 0.30}{12 \times 0.4375 \times 500 \times 0.50} \; [1.5 - (1 - 0.50)2.0]$$

$$= 0.10 \text{ percent per year}$$

Result.   Assuming a duty cycle of 2000 hours of usage per year, the generic failure rate for an individual RAM during the warranty period must not exceed 0.05 percent per 1000 hours.

7.4.3 Reliability Goals Based on Past Performance

In the contemporary design environment, the task of replacing an existing hardware subassembly with a new one with a higher component integration level, made available as a result of technology advance, is quite routine. This situation provides an opportunity for increased reliability through a local performance breakthrough. A review of past warranty data, test results, and expected trends in technology provides the basic guidance for reliability goals proposals. A simple example of the reasoning behind this approach is presented in Table 7.5.

TABLE 7.5   Past Performance–Based Reliability Goal

Current version performance (based on warranty data)

Subassembly	Observed failure rate, %/yr[a]		
	Minimum	Average	Maximum
1	1.51	2.47	3.47
2	2.01	2.82	3.43
3	1.51	1.94	2.65
4	2.38	3.06	3.74
Total	7.41	10.29	13.29

Goals for new integrated version

Parts	Release to production	Early production	Mature production
New technology part[b]	5.00	4.00	2.50
Commercial parts[c]	2.00	1.90	1.75
Total	7.00[d]	5.90	4.25
Compared to current version[c]	10.29	9.78	9.29
Percent improvement[e]	32	40	54[f]

[a]Past 24 months.
[b]Technology-based estimate.
[c]Assumed 5 percent per year improvement.
[d]Slightly better than observed past minimum total.
[e]Against improving current version.
[f]Expected in two years after release (realistic, less than 25 percent per year improvement).

7.4.4 Technology Assessment–Based Goal

In many situations, the design of the new component or part is based on a quantitative extension of relatively well understood processes and technologies. If industry or company data and parameters of empirical reliability models are known, or can be estimated with reasonable accuracy and credibility, setting goals using a technology assessment methodology is a viable alternative. A wealth of related industry data and empirical reliability models can be found, for example, in military standards, handbooks, and special reports. These sources were used in the following example of goal-setting procedure for a thermal print head, using thin-film technology for the manufacture of 560 power resistors on a glazed ceramic substrate. Driving electronics resided in seven hybrid packages of NMOS drivers. A relatively new and unproven interconnect scheme was also proposed. This design strategy benefited from empirical reliability models published in MIL-HDBK-217, as presented in a condensed form in Figs. 7.6 and 7.7.[2]

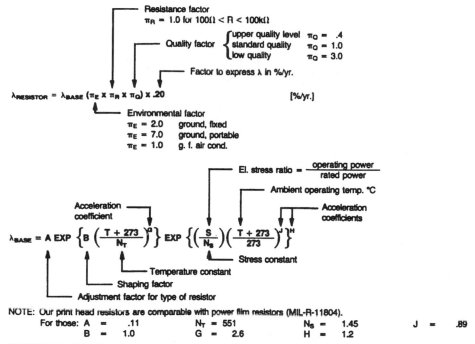

FIGURE 7.6   Thin-film power resistor failure rate model as per MIL-HDBK-217.

Actual calculations using internal design parameters and applicable empirical constants reflecting the state of the art documented in MIL-HDBK-217B are shown in Tables 7.6, 7.7, and 7.8. (The more current version, MIL-HDBK-217F, published in December 1991, documents the latest technological progress and gives different values for the empirical constants of interest.) These calculations follow the models mentioned above. The actual field reliability performance of this particular print head, as reported through the warranty system, indicated failure rates between 17.9 percent per year (worst case in early production) and 5.4 percent per year (best case), confirming the validity of the models used.

7.4.5 Optimal Total Cost of Reliability–Based Goal

The experience of the electronics industry, which correlates positively the amount of screening and testing with the reliability of semiconductor devices in actual field use, can be formalized to find the optimal balance between resources spent for reliability improvement and expected failure rate under product warranty.

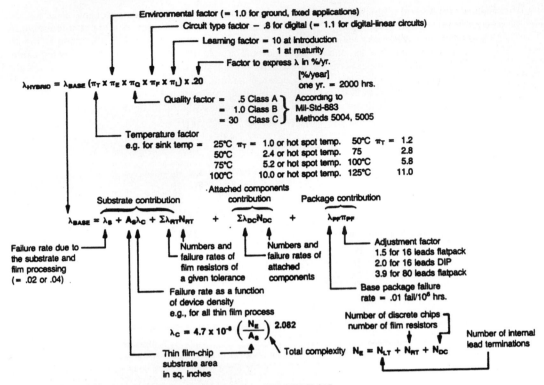

FIGURE 7.7   Hybrid microcircuit failure rate model as per MIL-HDBK-217.

The total cost of reliability is defined by three fundamental cost components:

1. Cost of design for reliability
2. Cost of reliability production
3. Warranty cost

All of these cost components could also include the cost of defect prevention, as it relates to reliability assessment and reporting. Relationships among the failure rate and application conditions, screening rate and level, etc., can be represented by empirical models based on, for example, military handbooks available in the public domain.

An outline of an application example may be found in Sec. 4.3.2 of this handbook, and a full description of the procedure is given in Ref. 3.

A more complicated situation arises when a design-to-cost strategy has been adopted for the supported project (see Sec. 4.3.5 of this handbook), so that verification of all proposed reliability goals for compliance is demanded. This is done by considering the impact of the cost of reliability manufacture on the product's total cost.

7.4.6   System Reliability Apportionment

The obvious step after development of product reliability specifications and goals is to translate them into goals for subsystems and lower-level assemblies. The key problem is to establish a rea-

TABLE 7.6   Thermal Print Head Hybrid Drivers Reliability Goal

Model parameters	Introduction	Early	Mature	Comment
N_{DC} Number of discrete chips	4	4	4	No attached comp.
N_{RT} Number of film resistors	—	—	—	—
N_{LT} Number of int. lead connects	128	128	128	4×32
N_E Complexity	132	132	132	
A_S Chip substrate area (in²)	0.0551	0.0551	0.0551	$(2840\mu \times 3180\mu)$ 9 mm²$\times 4$
N_E/A_S	2396	2396	2396	36.12 mm²
λ_C	0.5106	0.5106	0.5106	
λ_{PF} Base package failure rate	0.01	0.01	0.01	Modular package
π_{PF} Adjustment factor	4.0	4.0	4.0	Complex interconn.
π_T Temperature factor	3.3	3.3	3.3	Heat sink temp. 60°C
π_E Environmental factor	1.0	1.0	1.0	Ground, fixed, air cond.
π_F Circuit type factor	1.1	1.1	1.1	Lin. dig. comb.
π_Q Quality factor (Class C)	15	15	15	No screens
π_L Learning coefficient	2.0	1.5	1.0	Minimum learning
λ_S Substrate failure rate	0.0200	0.0200	0.0200	Thin film only
$A_S\lambda_C$	0.0281	0.0281	0.0281	
$\Sigma\lambda_{RT}N_{RT}$	—	—	—	Print res. separate
$\Sigma\lambda_{DC}N_{DC}$	—	—	—	No attached comp.
$\lambda_{PF}\pi_{PF}$	0.0400	0.0400	0.0400	
λ_{BASE}	0.0881	0.0881	0.0881	
λ_{HYBRID} (%/yr.)	1.7	1.4	0.9	
Hot spot temperature	N.A.	N.A.	N.A.	Not expected
Number of pins	N.A.	N.A.	N.A.	No pins
Failure rate increase for conn.	2.0	1.7	1.1	20 percent for spec. conn.
7 Hybrids/head (Class C)	14.0	11.9	7.7	Percent per year

TABLE 7.7   Thermal Print Head Resistors Reliability Goal

Model parameters	Introduction	Early	Mature	Comment
N_S Stress constant	1.45	1.45	1.45	
S Stress ratio	0.10	0.10	0.10	Based on duty cycle
T Ambient operating temperature	50	50	50	[°C]
J Acceleration coefficient	0.89	0.89	0.89	Power resistor
H Acceleration coefficient	1.30	1.30	1.30	Power resistor
A Adjustment factor	0.11	0.11	0.11	Power resistor
B Shape factor	1.0	1.0	1.0	
N_T Temperature constant	551	551	551	Power resistor
G Acceleration coefficient	2.6	2.6	2.6	Power resistor
λ_{BASE}	0.15	0.15	0.15	Failure 10^6 hrs.
π_E Environmental factor	1.0	1.0	1.0	Ground, fixed, air cond.
π_R Resistance factor	1.0	1.0	1.0	$R = 135\ \Omega$
π_Q Quality factor (Class B)	1.0	0.7	0.4	High quality a must
$\lambda_{RESISTOR}$ (percent per year)	0.030	0.021	0.012	Percent per year
$\lambda_{PRINT\ HEAD}$ (percent per year)	16.8	11.8	6.7	560 resistors

TABLE 7.8 Thermal Print Head Complete Assembly Reliability Goal and Performance

Part	At introduction	After 9 months	Mature	Comment
Substrate with resistors	16.8	11.8	6.7	See assessment
Hybrid drivers	14.0	11.9	7.7	See assessment
Interconnect system	3.0	1.0	0	10% impact
Total	33.8	24.7	14.4	Percent per year

Actual field performance				
Total head failure rate, %/yr.		Early	Mature	Comment
Worst case		17.9	—	5 months average
Best case			5.4	5 months average

sonable (from the viewpoint of design feasibility) and noncontroversial (i.e., consistent with design constraints) procedure for this goal allocation task. Six basic approaches to reliability apportioning have been established, all of them relying on some scheme of weighting the influence of the subsystems involved.[1]

1. *Simple Apportionment:*

$$0.9\lambda_{\text{SYSTEM}} = \sum_i \lambda_i$$

where λ = failure rate
i = individual subsystems, $i = 1,2,\ldots,n$

Failure rate estimates for the n subsystems are based on past data and experience. It has become a common practice to allocate only 90 percent of the system failure rate goal, in order to provide a design margin and to allow for later tradeoffs.

2. *Equal Apportionment:*

$$\lambda_{ia} = \lambda_{SR}\left(\frac{1}{N}\right)$$

where λ = failure rate
ia = allocated to individual item
SR = system requirement
N = number of subsystems

3. *Feasibility-of-objectives technique:* This scheme is a variant of equal apportionment in which the failure rates of individual subsystems λ_i are weighted by subsystem allocation factors a_i that are functions f_i of numerical ratings of system complexity, criticality, and reliability; technology state of the art; mission duration; and application environment conditions. The set of functions f_i is a mathematical symbol for a heuristic estimation process relying significantly on expert opinion and experience.

$$\lambda_{\text{SYSTEM}} = \sum_i a_i \lambda_i$$

where a_i = allocation factors, $a_i \equiv f_i$(complexity, criticality,...)

$$\lambda_i = \frac{\lambda_{\text{SYSTEM}}}{N}$$

i = individual subsystem, $i \leq 1 \leq N$.

4. *AGREE technique* (Advisory Group on Reliability of Electronic Equipment, Office of the Assistant Secretary of Defense): This technique takes into consideration both the complexity and the importance of each subsystem.

$$\Theta_i = \frac{N w_i t_i}{n_i[-\ln R(t)]}$$

$$R_i(t_i) = \exp\left\{\frac{-t_i}{\Theta_i}\right\}$$

$$N = \sum_{i=1}^{k} n_i$$

where Θ_i = allocated MTBF for subsystem i
$R(t)$ = system reliability goal
n_i = number of modules of the ith subsystem
w_i = importance factor for the ith subsystem
t_i = mission duration for the ith subsystem
N = total number of modules
$R_i(t_i)$ = reliability apportioned to the ith subsystem
t = required system mission time
k = number of subsystems

5. *ARINC technique* (Aeronautical Research Inc.): This technique makes the following assumptions:

- The system is made up of subsystems, with constant failure rates, which are in series.
- Any subsystem failure causes system failure.
- Subsystem mission time is equal to the system mission time.
- Predicted failure rates are based on historical data.

$$\lambda_{ia} = \lambda_{SR}\left(\frac{\lambda_{ip}}{\lambda_{Sp}}\right)$$

where λ = failure rate
p = predicted
S = System
R = Required
a = Allocated
i = the individual subsystem, $i = 1,2,...,n$

6. *Apportionment by dynamic programming:* This is a computerized optimization technique suitable for situations when subsystems are not equally difficult to develop.

Additional methodologies exist at individual companies where the allocation issue is of high importance. An approach which combines the EQUAL and ARINC methodologies has been proposed by James A. Boyd, Jr.[4] This combination allows the use of a single variable to weight the two methods according to user needs, and can be implemented in a simple spreadsheet format. It requires knowledge of the system block diagram and the predicted failure rates for both individual subassemblies and the total system. Its simple format allows sensitivity analysis and inclusion of a safety margin.

Accumulated experience with these approaches indicates a few limitations that should be considered when selecting a methodology for a particular project:

- Both AGREE and ARINC address product complexity, but provide no incentive for its reduction.
- Most of the approaches assume that subsystems are in series, that they operate for the same length of time as the total system, and that the failure rates are constant over time.

Allocation of system reliability among subsystems also neglects the impact of interfaces among the system components. To improve allocation results, the allocation must be agreed upon by both the system developers and the subsystem suppliers, and it must draw heavily on the experience of the reliability engineering teams of all parties involved. If some allocated goals appear too difficult to achieve, the underlying analysis may serve as a basis for the necessary tradeoffs.

7.5   INDIRECT GOALS FOR RELIABILITY

In contemporary design practice, there are many ways to achieve the desired product reliability. The designers may chose overdesign, simplicity, standardization, contribution of human engineering, environmental ruggedness, redundancy, design for testability, design for manufacturability, etc. Even though these design strategies do not address the reliability goal challenge directly, they are supportive of design for reliability and may allow more challenging reliability goals to be proposed. It is therefore desirable to have the reliability management and engineering function participate in design strategy selection and goal setting in reliability-related areas.

7.5.1   Goals for Reliability Production Assurance

The benefits of applied design for reliability engineering expertise can be realized only in actual production. No matter how theoretically sound the design is, the potential for degradation of the inherent reliability is always present because of

- Variability of manufacturing processes, materials, and available human skills
- Inadequate process quality control and reliability assurance
- Lack of understanding of the manufacturing technologies employed
- Physical and chemical changes taking place in parts and materials

This phenomenon is readily observable in burn-in and warranty failure data. The desired profile of a product's early warranty performance may well serve as a production reliability goal (see examples in Fig. 7.8). The maximum allowable degradation and the slope of the required learning curve are functions of the planned warranty economics and shipment rate.

7.5.2   Testability Requirements

Testability is a design attribute that allows the operational status of a component, a subassembly, or the whole product to be determined. It also aids in fault isolation and failure analysis. Adequacy of testability can be expressed pragmatically by

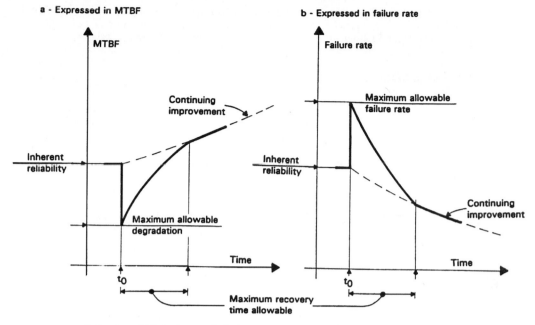

FIGURE 7.8   Example of goals for reliability production assurance.

- The level and structure of the product functional partitioning
- The number and density of test points
- The extent of the automatic test coverage

The main reliability concern is the proper definition of field replaceable units (FRUs), because they influence the maintenance strategy and therefore the cost of unreliability. There is no simple methodology to establish the optimal size of the FRU, which in real life may vary from a single component to the complete product, if a throw-away strategy has been chosen for replacement.

Reliability requirements for testability usually reflect the company's local experience, and are improved over time in an evolutionary fashion. MIL-STD-2165, *Testability Program for Electronic Systems and Equipment,* is a frequently consulted reference.

7.5.3   Reliability-Related Standards

Another indirect way to influence reliability specifications and goals is through stating requirements in terms of international, industrial, and other public domain standards. Reliability management practice cannot neglect consideration of key U.S. military standards, namely

- MIL-STD-721, *Definition of Terms for Reliability and Maintainability*
- MIL-STD-756, *Reliability Modeling and Predictions*
- MIL-STD-781, *Reliability Design Qualification and Production Acceptance Tests*
- MIL-STD-785, *Reliability Program for Systems and Equipment*
- MIL-STD-790, *Reliability Assurance Program for Electronic Parts Specification*

- MIL-STD-810, *Environmental Test Methods and Engineering Guidelines*

Note: Before applying individual standards, a check for the latest version is necessary.

7.6 SPECIAL CONSIDERATIONS

Defining reliability specifications and setting reliability goals is one of the more difficult tasks reliability management periodically faces. In some situations, generic approaches, based on reliability assessment and prediction methods, are quite sufficient and well established. In many other situations, the unique particularity of the market, product, or underlying technologies demands special considerations and requires innovative approaches outside the realm of established practices. The following paragraphs illustrate some of these unique situations.

7.6.1 Reliability Goals for Consumer Products

Setting reliability goals for consumer products and monitoring the actual field reliability performance are complicated by the discontinuity of information collection and flow that results from distribution channels. The accuracy, timeliness, and technical content of the available information are severely limited. Therefore, past performance and trend analyses are improper tools for goal setting for these products. Because the reliability of consumer products is reflected in customers' opinion more often than in quantitative failure data, marketing research methods are more suitable for this task.

Assessing market sensitivity to reliability, or lack of it, does not always require complicated statistically designed customer surveys.[2] In most situations there is enough knowledge available immediately in-house. It requires the skill of reliability engineering to translate this available, but frequently vague knowledge into the language of measurable reliability parameters, such as failure rate. As an example of a simple approach to assessing market sensitivity, let us consider a consumer electronic product that appeals to university students, and that has these basic product characteristics:

- Product price: $750
- Expected gross profit: 20 percent
- Warranty period: 1 year
- Repair cost: $150

From discussions with marketing and sales, a picture of probable customer discomfort emerged and was translated into a representative failure rate:

Disaster: Every second unit shipped fails in the first three months of life	200 percent per year
Severe difficulty: No profit—20 percent profit equals cost of one repair per warranty	100 percent per year
Difficulty: 5 percent profit only	25 percent per year
Acceptable situation: A university campus with 50 units sold where 5 customers will experience product failure in the same year	10 percent per year
Preferable situation: Warranty cost only 1 percent of the price (i.e., $7.50 per unit sold), which means that only 1 out of 20 units may fail	5 percent per year
Ideal solution: Less than one unit out of 50 units shipped will cause difficulties	2 percent per year

This reasoning may be represented graphically by defining a discomfort level scale (see Fig. 7.9).

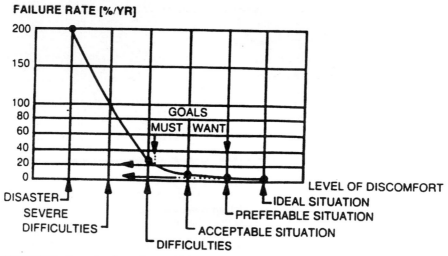

FIGURE 7.9   Example of a market sensitivity analysis.

This allows the full range of sensitivity to be represented by a continuous plot, and permits the selection of final reliability goals, e.g., 20 percent per year as a "must" goal, and 5 percent per year or better as a "want" goal.

7.6.2   Impact of Soft Failures

The explosion in the application of digital electronics accompanied by software has given rise to failure events that result in recoverable, temporary system malfunctions. These "soft" failures are transient, in contrast to "hard" failures, which are usually catastrophic. The current state of practice indicates that many mechanisms causing individual soft failures, such as function time delay, ground bounce, supply noise, input skew, output oscillations, software defects, etc., have been investigated, but no widely accepted general methodology to address this problem has yet emerged.

Because current experience indicates that up to 40 percent of observable and reported failure events are attributable to soft failures, these failures must be considered during reliability goal setting and specification development, as well as while planning the cost of reliability. The lack of structured experience limits the possible approaches to those employing subjective estimates of the overall reliability degradation due to soft failures. Work on more systematic and generalized approaches to soft failure events analysis and reliability modeling is in progress. An example of a proposal based on conditional probabilities, complemented by an extensive set of references, may be found in Ref. 5.

7.6.3   Impact of Redundancy

Redundancy, in a general sense, means creation of new parallel paths in the system structure to improve its reliability. The various implementation strategies, such as component, system, network, compromise, standby, etc., redundancy, are well understood and described by mathematical models, often based on the Markov process.

Even though redundancy is a strategy for design for reliability, i.e., implementation, it significantly influences the way reliability goals are stated. Examples of goal statements reflecting the practice of the aerospace industry are[6]

- Inherently critical components must have a reliability goal of less than one failure in the life of the total systems population.
- No single failure shall cause catastrophic loss.
- The system must stay operative after any two simultaneous subsystem failures.
- $R > 0.9999997$ for 1-h mission.

It is necessary to keep in mind that even highly redundant systems are susceptible to "common cause" failures, i.e., causes that affect all redundant paths simultaneously, such as power failure. This fact is most important for development of reliability specifications and design strategies for high-reliability safety protective systems. Their principal function is to interrupt the "domino effect" of a potentially dangerous sequence of events triggered by a failure or fault condition. This points to the need for independent safety and reliability assessments involving validation of the adequacy of the protective system's reliability under abnormal conditions. Here the issues of human reliability enter in. The frequent requirement of reliability $R > (1 - 10^{-6})$ is achieved through multiple separate and diverse automated protective systems. The importance of the underlying reliability policy, which must include assurance of continuous reliability growth, cannot be overestimated.[7]

7.6.4 Reliability Growth Models

Reliability growth is a concept that is helpful in planning reliability testing, in evaluation of reliability improvement proposals, and in reliability predictions. Initially introduced in the mid-1960s, it is still the subject of many discussions and controversies, especially concerning its proper application. Practically speaking, reliability growth means a systematic and permanent removal of failure mechanisms, regardless of their source. Its mathematical models are based on the concept of the learning curve (e.g., the Duane growth model) or on Kalman filtering (e.g., the dynamic linear model) employing a bayesian recursion process. New studies point to computerized models based on stochastic Petri nets.

Current experience admits general acceptance of growth models, reflecting their simplicity and conceptual intuitiveness, but it also recognizes potential controversy stemming from the inaccuracy of the data used to establish the values of growth curve parameters. The most successful applications reported involve

- Managing the reliability of long-life products that are subject to design changes and improvements
- Software development and testing

Despite the high number of published studies (e.g., close to 100 papers in the *Proceedings of the Annual Reliability and Maintainability Symposia* between 1980 and 1993) and many elegant mathematical models, the question of an adequate description of the reliability growth process is still open, and the overall approach has still only a heuristic character.

7.6.5 Importance of Intangibles

So far, all the discussions in this chapter have been oriented toward specifications and goals expressed in technical and economic terms and parameters, allowing reasonably straightforward measurement. But life, even in the highly rational field of reliability engineering and management, is more colorful and provides opportunities for contribution and satisfaction far beyond the appeal of numbers. So, while analyzing trends, assessing the power of new technologies, balancing costs and benefits, and searching for possibilities for innovation, let us not forget the personal and social needs of reliability program participants; let us also set goals for

- Adequate and fair recognition of contributors
- A social environment supportive of satisfaction from a job well done

- Creating attentive audiences for success stories
- Mutual work enrichment by sharing one another's expertise through multidisciplinary teams
- Discretionary time and resources for professional growth

Let us make visible our understanding that all management tasks are, in essence, tasks of dealing with people, the most valuable company resource.

7.6.6   Role of Ethics and Psychology

In a generic sense, goal setting is a part of the early phase of any problem-solving or decision-making process. In most industrial situations, this process is expected to be rational, with the concept of rationality formulated around physical and economic metaphors. Max Weber's concept of "substantive rationality," which includes ethical, political, altruistic, and hedonistic values, is gaining in general acceptance over the narrow economic rationality.

Failure of a product, whether it be something of trivial value or a major structure such as a bridge or an airplane, can potentially be traced to lack of reliability design or assurance, both of which are affected by the initial project goals. So, it may be prudent and ethical to accept a variation of the decision maker's practice of examining adverse consequences, and include a test of reliability goals against residual risks, those that are still present even when reliability goals are met, in a goal-setting process's internal loop.

Ethical aspects may be introduced into reliability management and goal setting both negatively, in conjunction with product liability concerns, and positively, through accepted codes of ethics adopted by various professional societies, such as IEEE, ASME, and ASQC. But differences of opinion still remain significant, with the largest disparity found between groups that believe that engineers have a public responsibility and those that hold that the engineer's primary responsibility is to the employer. Therefore, issues of ethics in reliability engineering and management do not have clear and universally accepted solutions and must be resolved on the personal level.

Psychological considerations may also enter the field through the problem-solving framework. For example,

- Reliability management can be interested in strengthening good design for reliability practices through modification of the existing reward and recognition system. Psychological aspects (habit forming, learning and unlearning, etc.) will have to be addressed.
- Selection of reliability goals may be influenced by the manager's personal preferences, and so the psychology of making choices, especially under risk or pressure, may be an issue.
- When reliability goals must be set without support from past performance data, the quality of human inference and social judgment, both with many psychological connotations, becomes the most important methodological aspect.

There may also be psychological problems associated with goal negotiation, apportionment, communication, and acceptance.

In a healthy industrial environment, the involvement of ethics or psychology is almost imperceptible and far in the background. When these issues start to dominate reliability management and goal setting, a paradigm shift, away from industrial rationality toward the metaphor of an Oriental bazaar, has already occurred, and major corrective action is due.

7.7   CONCLUSIONS

It is possible to conclude that the art and practice of reliability specifications development and goal setting are methodologically well supported and, in many cases, solidly based on a significant amount of credible data and experience. As in other areas of industrial engineering management, the significant challenges are at the frontiers of product and technology development. Uncertainties about valid approaches to goal setting are especially visible in

- Consumer products where there is a tendency toward a "sell and forget" strategy
- Products and components with extremely low failure rates
- Products with a high level of functional or structural complexity
- Products with significant human factor influence
- Autonomous systems
- Assessment of the impact of new computer-aided design tools on product reliability
- Management of reliability-related processes

The process of establishing reliability specifications and goals will be meaningless if it is not followed by design validation and final product qualification against those goals. Changes in the business environment that have elevated time to market to one of the most important strategic objectives for new product development exercise a strong influence on reliability planning and validation also. This creates a new demand to test the realism of proposed reliability goals against schedule constraints and against locally available validation methodologies.

New challenges are also created by the broadening of the concept of reliability, which now often includes consideration of operator errors and task definition incompleteness as new sources of potential system failures. This points to the increasing importance of the systems approach to reliability management, which bears in mind all three fundamental dimensions of the task: assuring desired product reliability, controlling the impact of design and manufacturing processes, and providing effective tools and support by reliability specialists. These trends indicate that not just reliability expertise but its combination with the application domain knowledge is necessary for world-class reliability.

7.8 REFERENCES AND SUGGESTED READING

7.8.1 References

1. N. B. Fuqua, *Reliability Engineering for Electronic Design,* Marcel Dekker, Inc., New York, 1987.
2. H. J. Kohoutek, "Establishing Reliability Goals for New Technology Products," *Proceedings of the Annual Reliability and Maintainability Symposium,* IEEE, 1982.
3. H. J. Kohoutek, "Development of a Reliability Strategy for New IC Component Family and Process," *Microelectronics and Reliability,* vol. 23, no. 2, 1983, pp. 383–389.
4. J. A. Boyd, Jr., "Allocation of Reliability Requirements: A New Approach," *Proceedings of the Annual Reliability and Maintainability Symposium,* IEEE, 1992.
5. H. B. Chenoweth, "Soft Failures and Reliability," *Proceedings of the Annual Reliability and Maintainability Symposium,* IEEE, 1990.
6. K. H. Eagle and A. S. Agarwala, "Redundancy: Design Philosophy for Catastrophic Loss Protection," *Proceedings of the Annual Reliability and Maintainability Symposium,* IEEE, 1992.
7. A. E. Green, *Safety Systems Reliability,* John Wiley & Sons Ltd., Chichester, 1983.

7.8.2 Suggestions for Further Reading

1. *Proceedings of the Annual Reliability and Maintainability Symposia,* IEEE.
2. P. F. Drucker, *Management: Tasks, Responsibilities, Practices,* Harper & Row, Publishers, New York, 1974 (or later edition, if available).

CHAPTER 8
CONCURRENT ENGINEERING

Dennis R. Hoffman
TEXAS INSTRUMENTS, INC.
DALLAS, TEXAS

8.1 BASIC APPLICABILITY AND PURPOSE

Concurrent engineering is a fundamentally different way of conceiving, specifying, and developing products, their enabling technologies, and their manufacturing and support processes. Conventional product development practices, characterized as "review-based" and "sequential," become progressively less efficient as product complexity and market demands increase.

Concurrent engineering (CE), or integrated product development (IPD), started as a Department of Defense (DoD) initiative for the defense industry, but it has also been used very successfully by commercial industries. Concurrent engineering is applicable to any industry that develops products and wants to remain competitive in today's marketplace.

Concurrent engineering involves a product development infrastructure that fosters a unified, collaborative approach that integrates inputs from business, engineering, and management specialists across the traditionally segregated phases of product development, starting with requirements definition and encompassing the design of manufacturing and support processes within product development. With this unified infrastructure, companies are able to achieve more competitive products and eliminate predictable cost, schedule, quality, and product functionality shortfalls that are inherent in sequential, review-based development practices.

Concurrent engineering is purported to yield a large competitive advantage to adopters of the methodology. Yet there is no widely accepted definition of the term or its concepts. However, the definitions used in this chapter are generally accepted in industry, and provide a working set of these terms and concepts.

Concurrent engineering supports many business objectives, including staying competitive and profitable, making quality product-oriented decisions that best meet the customer's total requirements, and complying with those requirements. This chapter is intended to clarify the intent of concurrent engineering as envisioned, provide insight into its objectives, and review the benefits derived from implementing it. It is also intended to provide an introduction to concurrent engineering principles, set the stage for broader concurrent engineering implementation, and accelerate the adoption of concurrent engineering in the reader's new product development process.

8.1.1 Reliability and Concurrent Engineering

Reliability is a design attribute just like any other performance attribute. Therefore, it is extremely important that the reliability perspective be represented during product development. Reliability engineering brings that perspective to the product development design team. That is why it is essential for the reliability engineering professional to understand concurrent engineering concepts and principles, what it takes to make a design team successful, and how reliability fits into and contributes to that design team.

8.1.2 Need for Concurrent Engineering

The Departments of Commerce and Defense have conducted numerous studies on the state of industry within the United States. What these studies found was often not encouraging. In a changing world marketplace in which sophisticated consumers are demanding more individualized products with steadily increasing quality, many U.S. companies are too rigid to adequately respond to the market and unable to match competitive quality levels at reasonable cost. The increased availability of cheap transportation and lowering of trade barriers have created a truly international marketplace containing not only more markets to compete in, but also more competitors.

The majority of approaches to developing a more competitive new product development process have involved optimizing specific areas of product design or fabrication. These point optimizations did result in incremental improvements in the process or product, but they segregated sections of product development and caused a loss of the view of the product as a whole.

The U.S. defense industry has been plagued by other manifestations of similar problems. For example, it is perceived that weapon systems cost too much in time and dollars (cost of acquisition and cost of ownership) for fielded capabilities. Defense contractors must deal with the need to constantly increase the capability and complexity of their product lines by incorporating new technology, and do so under unrealistic development time and budget constraints. The result has been a steady increase in time from program initiation to operational maturity, accompanied by a cost growth that has outpaced the relative growth in capability. The net result of delays in fielding new systems has been a shrinking of the effective technological advantage of developed weapons, in spite of vigorous technology development programs.

8.1.3 Benefits of Concurrent Engineering Defined

The Undersecretary of Defense for Acquisition directed the Institute for Defense Analysis (IDA) to evaluate the applicability of concurrent engineering concepts to the defense industry. In the resulting survey of U.S. companies that had applied CE concepts, concurrent engineering was found to have tangible benefits which were impressive and broadly based. For example:

1. Development and production lead times:

   Product development time reduced up to 60 percent

   Production time reduced 10 percent

   Total process time reduced up to 46 percent

2. Measurable quality improvements:

   Manufacturing defects reduced up to 87 percent

   Yield improvements of up to 4 times

   Field failure rates reduced up to 83 percent

3. Engineering process improvements:

   Engineering changes/drawing reduced up to 15 times

   Early production engineering changes reduced 50 percent

   Inventory items stocked reduced up to 60 percent

   Engineering prototype builds reduced up to 3 times

   Scrap and rework reduced up to 87 percent

The survey findings stated that where properly applied, concurrent engineering resulted in higher utility to the user, increased quality, and lower product cost with less development time. These benefits are possible within any resourceful company.

8.2 FUNDAMENTAL CONCEPTS OF CONCURRENT ENGINEERING

8.2.1 Definition of Concurrent Engineering

Concurrent engineering, as the term was originally coined, focuses on three major concepts:

1. The integrated product development process needs to be well enough understood and modeled to be repeatable, ensuring systematic success.
2. All relevant perspectives, from customer requirements through internal constraints, must be considered in the definition and design of the product.
3. All perspectives need to be integrated to yield a global optimum, i.e., a cost-effective, robust design that is tolerant of manufacturing and use variations.

Institute for Defense Analysis Definition.   The IDA report, mentioned earlier, provided the following definition of concurrent engineering:

> Concurrent Engineering is a systematic approach to the integrated concurrent design of products and their related processes, including manufacture and support. This approach is intended to cause the developers, from the offset, to consider all elements of the product life cycle from conception through disposal, including quality, cost, schedule, and user requirements.

Department of Defense Definition.   The Department of Defense issued a definition of concurrent engineering within MIL-HDBK-59 which states:

> Concurrent Engineering is a systematic approach to creating a product design that considers all elements of the product life cycle from conception through disposal. In doing so, Concurrent Engineering simultaneously defines the product, its manufacturing processes, and all other required life cycle processes, such as logistics support. Concurrent Engineering is not the arbitrary elimination of a phase of the existing, sequential, feed-forward engineering process, but rather the co-design of all downstream processes toward a more all encompassing, cost effective optimum. Nor does Concurrent Engineering entail simultaneous design of the product and execution of the production process, an approach which is demonstrably unsound. Concurrent Engineering is an integrated design approach that takes into account all desired downstream characteristics during upstream phases to produce a more robust design that is tolerant of manufacturing and use variation, at less cost than sequential design. It affects all system procurement activities from Milestone 0 (concept definition and exploration) to the start of Milestone III (the end of full scale development).

Another Definition.   The unofficial definition provided below was abstracted from an unpublished paper written within the DoD and captures the essence of concurrent engineering fairly concisely:

> Concurrent Engineering refers to the simultaneous, parallel design process that encompasses all aspects of a product's development in a top-down fashion by a multi-disciplinary team.
> The goal of Concurrent Engineering is to achieve mutual optimization of critical characteristics of the product and its related processes. This approach is intended to cause the product developers, from the outset, to consider all elements of the product life cycle from conception through disposal.

Concurrent Engineering is based on the realization that: "specialty" attributes of a product are more effective when designed into the product rather than installed after-the-fact; early design decisions must consider "specialty" attributes along with other performance characteristics; and quality products need a continuously evolving quality process.

8.2.2 Concurrent Engineering Concepts

DoD's concurrent engineering initiative has sparked quite a lot of interest. Concurrent engineering appeals to the DoD because it represents a way to get higher-quality products for less money. Concurrent engineering appeals to corporations because it can increase quality and broaden product acceptance in the marketplace. Concurrent engineering appeals to defense contractors, and industry in general, because it will increase competitive posture and internal profitability. Yet with all this appeal, few people have had or have taken the opportunity to understand the true components of concurrent engineering or identify what is required for its effective implementation.

Concurrent engineering concepts, as shown in Fig. 8.1, must encompass a balanced mixture of elements which include business requirements, human variables, and technical variables.

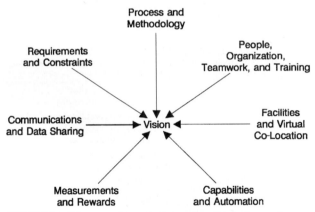

Effective Concurrent Engineering Implementation Must Assess and Potentially Affect These Key Elements:

FIGURE 8.1   Elements of a concurrent engineering environment.

Vision.   At the center of these elements is the long-term "vision," business goals that must be well defined and disseminated to all levels of management. The business goals must be distinct enough to allow development activities to leverage solutions in a cost-effective manner. The concept of change management is perhaps the most difficult to institute because of the diverse implementation roles of all personnel, from individual contributors through management. The combined efforts of all roles are necessary in order to plan, implement, and monitor the required changes in the process. When evaluating the implementation of concurrent engineering, the dominating common factor is the ability to manage change effectively.

The effective implementation of concurrent engineering requires commitment to changes which are business-driven and affect the elements of the work environment through a controlled change process. Commitment to the concurrent engineering process at all levels is a must. All levels of management must "buy in" through their words, actions, and expectations. Management must institute appropriate metrics and team recognition/reward systems.

Management commitment (active involvement) will be imperative to properly fund the up-

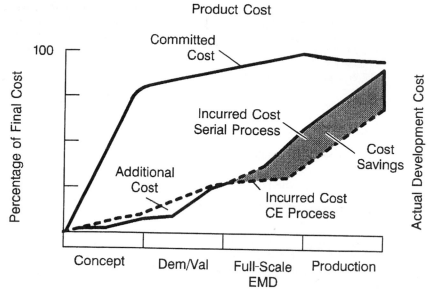

FIGURE 8.2   Early CE involvement yields potential savings.

front multidisciplinary tasks and analyses needed to reap the long-term payback and benefits. Although successful implementation requires additional and earlier involvement, the corresponding change in the funding profile is more than offset by the reduced cost of downstream (especially production-phase) design changes, as depicted in Fig. 8.2.

Use of Process.   Concurrent engineering is a seemingly simple idea that is based on a fundamentally different way of looking at how products are conceived, engineered, manufactured, and supported. The idea is that people can do a better job if they cooperate to achieve a common goal. To implement this concept successfully, members of management, engineering, test, manufacturing, and product support must develop a profoundly different insight into the nature of industrial activity, namely the "process" insight. The process insight is the realization that all the activities which transform a collection of inputs into a product satisfying a need are a single process. This process can be defined, measured, managed, and continually improved. Improvements include both the breakthroughs associated with new inventions and the small improvements that result from everyday suggestions.

Methodology.   One of the primary elements of a CE environment, as shown in Fig. 8.1, is the effective implementation of a CE process and methodology. This is a fundamental refocusing of the way products and their manufacturing and support processes are developed. It strives for globally efficient solutions to the satisfaction of customer requirements rather than tradeoffs between localized optima. It achieves design convergence through feedforward of relevant design knowledge. Design knowledge is most effectively utilized early in the design process. Concurrent engineering stands in contrast to traditional sequential and isolated development methods.

The fallacy underlying sequential engineering is its implied hypothesis that product characteristics are independent. Taken to the extreme, this postulate assumes that certain product parameters can be set at a given stage without consideration of the impact that other stages or other parameters might have on them. Engineers frequently make arbitrary or "rule of thumb" decisions when those decisions do not constrain their immediate work. However, such an arbitrary decision may constrain subsequent (downstream) engineering activities for years to come. These constraints force engineers downstream to make suboptimal engineering decisions and parameter assignments. Since downstream engineers have little visibility into upstream parameter assignments, it is difficult for them to challenge decisions made upstream. In addition, the interrelationships among these parameters are complex and nonlinear. Thus, the implicit constraints placed by one discipline or team member on another are difficult to assess by simple examination or analysis.

As stated, sequential engineering assumes that a problem can be fixed by adjusting one or two parameters while the others are held constant. This approach has worked in the relatively low-technology past. However, high technology demands a far greater consistency of materials, much higher complexity of interacting components, and tighter tolerances and margins. Nonlinear, interactive, and second-order effects play a much greater role in high-technology products. Sequential methods just break down.

On the other hand, concurrent engineering consists of simultaneous engineering (design) of the product, manufacturing processes, and support processes, as depicted in Fig. 8.3. This does not mean that the full design is accomplished near instantly. It does mean that

1. All aspects of the total engineering effort mature at the same relative pace as the product design evolves.

2. The risks of high-potential interdisciplinary dependency, or uncertainty, are resolved through experimentation or simulation.

3. The end-to-end development process is participatory, with all engineering disciplines contributing to one another's effort as well as their own areas.

4. The goal of all participants in the total engineering process is an engineering solution that is optimum with respect to all requirements [functional, schedule, life-cycle cost (LCC), etc.].

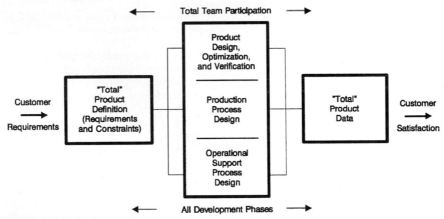

FIGURE 8.3   Concurrent engineering concept.

Concurrent engineering requires an emphasis on ongoing process development in parallel with the development of the product. In particular, development of the design process and interactive sharing of data, requirements, and constraints between processes (more of the key CE elements) form the basis for design optimization and build the knowledge base for continuous product and process improvement. Specifically, requirements and motivating criteria for decisions are seldom captured for use in later phases in the development cycle. Treating every engineering function as a process that can be formally described is the first technical step in concurrent engineering. This process-oriented communication and integration aspect of CE has led to its being called integrated product development (IPD) in some circles.

Multidiscipline Teams.   Interdisciplinary process improvement and multidisciplinary product development demands a new management approach—a different culture and viewpoint. What is required has come to be referred to as multidiscipline teams, another key CE element depicted in Fig. 8.1. Empowered teams (see Fig. 8.4) have been shown to be dramatically successful in shortening

A Concurrent Engineering team is
analogous to a football team:

- The team functions as a single unit.

- The team has a common goal.

- Each member has a specific assignment that must be completed

- Each member's contribution is crucial to success.

- The combined efforts of the team result in better solutions

- The team has a leader/facilitator

- The management organization is on the sidelines, supplying guidance and resources.

FIGURE 8.4   A concurrent engineering team is analogous to a football team.

development phases with fewer people. Characteristics of an effective team include involvement of the members who are essential for the completion of all aspects of product design and process development; a sense of responsibility among all team members for total success criteria of the product (product versus organizational focus); team authority and responsibility; a reward structure that emphasizes team success; and a team longevity that corresponds to the project's duration.

Process Development.   Concurrent engineering requires a long-term commitment to process development and optimization. This, in turn, requires a long-term commitment to the accumulation and application of knowledge. Lessons learned from testing and field experience must be evaluated to fix design problems and correct the process(es) in order to prevent future recurrence. Competitive benchmarking of the company's and competitors' products and processes is also a valuable source of design knowledge. This knowledge repository (information and tools) must be maintained and leveraged to ensure effective application to future generations of products. Communication and sharing of this information is another key CE element identified in Fig. 8.1.

Computer-Aided Engineering Application.   The application of computer-aided engineering (CAE) tools and techniques provides a vast capacity for knowledge generation and integration. It also can provide the informational and analytical capability to support complex design decisions that tend to optimize the total development effort (product and processes) when the decisions are timely. In addition, CAE can serve as a utility to structure the design process by capturing design decisions with (importantly) design intent, making explicit the engineering rationale behind them. Automation via computer-aided-engineering [including the related infrastructure of product data management (PDM), libraries, networks, etc.] is best viewed as a team enabler by linking dispersed team members and providing a vehicle for rapid communication and analysis of design concepts and ideas.

The role of CAE as a tool for multidiscipline parameter optimization and related design decision support is a novel one. Traditionally, CAE tools are applied sequentially in an analysis and verification role. Databases are structured to support individual tools. Progress is being made in converting CAE to play a more proactive role in concurrent engineering through open systems frameworks and design decision support tools.

8.3   THE FIRST PRINCIPLES OF CONCURRENT ENGINEERING

A DoD/industry team, sponsored by the NSIA/CALS Industry Steering Group, studied and defined what the concept of "concurrent engineering" means in practice. The report generated by this team defines the first principles of concurrent engineering, identifies influencing dimensions that affect the concurrent engineering environment, presents a self-assessment matrix, describes obstacles from the DoD and industry perspectives, and presents the benefits derived from concurrent engineering implementation. Highlights of that study report are captured in this section to provide an industry team's perspective and to present the team's findings, as the report does an excellent job of capturing the intent of CE as envisioned.*

8.3.1   A Common Concurrent Engineering Understanding

Successful implementation of concurrent engineering requires a commitment to the need for change on the part of the management and technical community. Once the commitment to concurrent engineering is made, organizations need an action plan to know what to implement. Without specific information on the immediate targets of change, time, energy, and resources will not be committed. Management must establish an atmosphere that is conducive to the formation and implementation of concurrent engineering within their corporations.

The difficulty has been in generating that clear set of targets. A broader "body of knowledge" or "common understanding" of how concurrent engineering is applied to the individual facets of a project has been missing. Without this common understanding, change remains risky and benefits cannot be systematically assessed. With a shared body of knowledge, concurrent engineering proponents become members of a larger community, with common ground rules and vocabulary that allow sharing of ideas and concepts. The level of knowledge or understanding can then rapidly increase as contributions of members are accepted into the body of common knowledge. The creation and dissemination of this common understanding is the goal of this chapter.

8.3.2   The "First Principles of Concurrent Engineering"

The first principles of concurrent engineering as outlined in the referenced report are as follows:

The craftsman model: Concurrent engineering's ideal model is either a craftsman or a collaborative multidiscipline product development team with the unity of purpose, integrated perception, and unified reasoning ability of a craftsman—depending upon product and market constraints.

---

*This author was fortunate enough to be involved in the subject study and made significant contributions to the content and the subject matter structure. The chair of this study group therefore granted this author permission to utilize any material from the study that may be useful in helping the reader to gain a greater appreciation and understanding of concurrent engineering. This chapter makes use of only selected portions of the study material, and it is not a direct reprint; there are modifications, deletions, and reformatting of the material to make it more compatible with the overall chapter content. Please refer to "First Principles of Concurrent Engineering: A Competitive Strategy for Product Development" for the complete report. This author wishes to acknowledge the contributions made to this study by all the team members who are identified in the subject report.

Discipline specialist organizations: Segregated discipline specialist organizations must be reserved as the incubators of new and evolving technologies, the reservoirs of specialist knowledge, and the advocates of unique discipline perspectives.

Optimum cross-discipline tradeoffs: Successful products depend upon focusing the diverse knowledge and unique perspectives of more specialized disciplines upon a broader scope of product development decisions.

Team infrastructure: A collaborative, team-centered infrastructure within and among all activities of complex product development is superior to any review-based sequential or nonteam approach.

Integrated development enterprises: All activities within a product development process must be considered one integrated enterprise, not independent activities that are the exclusive domain of particular disciplines.

The level playing field: Authority must always be equally shared among the members of a collaborative, multidiscipline product development team. Goals set by management determine a team's focus.

CE enabling tools and technology: The efficiency of a multidiscipline team can be substantially improved by a specific class of electronic tools and technologies that provide all members of the team complete visibility of evolving designs, support to identify and resolve conflicts, and the ability of each to influence decisions.

Total enterprise visibility: The effectiveness of product development teams is limited by the extent to which each team and team member are provided complete, continuous, and unambiguous visibility of the development enterprise as it evolves.

Management success standards: To implement concurrent engineering, the executive standards by which management success is measured, and accordingly career success is achieved, must reflect concurrent engineering principles.

Collaborative management: Discipline management must collaborate to define balanced, market-responsive product development team goals, empower team members to make tradeoffs, share responsibility in measuring team, not individual, performance toward these goals, and foster a unity of purpose within each team and among teams.

Continuous improvement: Continuous improvement of product development practices is essential to compete successfully in a global market.

8.3.3 Concurrent Engineering Assessment Technique—Implementation Road Map

Concurrent Engineering Influencing Dimensions Critical Self-examination—The "Should-Be" Environment.   The necessary and sufficient concurrent engineering capability is tightly associated with the nature of a particular program or project; therefore, influencing dimensions need to be considered in order to gauge the appropriate concurrent engineering capability. By assessing the program requirements as high or low on the influencing dimensions, an organization can assess the concurrent engineering capability that is appropriate for a program.

To illustrate this concept of influencing dimensions, consider a very complex high-technology program that involves many people spread across organizational and geographic boundaries. This type of program would naturally require more comprehensive concurrent engineering capabilities, whereas a smaller, less complex project could be accomplished with a simpler concurrent engineering approach. An attempt to move the required concurrent engineering environment beyond the capabilities that are necessary and sufficient to satisfy the needs of a program and project will not necessarily add value to that program or project. Of course, once an organization has achieved a concurrent engineering capability, it would not be prudent to purposely degrade its concurrent engineering environment. Such consideration within the self-assessment process is critical to the correct interpretation of the second matrix.

Since the concurrent engineering capability is so tightly coupled with the nature of a program, a set of influencing program and product dimensions are provided to aid in gauging the appropriate concurrent engineering approach. Each dimension deals with a specific aspect of program complexity. The specific dimensions itemized have an influence on the recommended approach to concurrent engineering. The aggregate of all influences determines the approach most appropriate for a specific program. The influencing dimensions are described here and listed in a comprehensive form in Table 8.1.

Product complexity

Product technology

Program structure

Program "futures"

Competition

Business relationships

Team scope

Resource tightness

Schedule tightness

TABLE 8.1   Concurrent Engineering Environment Influencing Dimensions

Influencing dimension	CE environment			
	A	B	C	D
Product complexity	"Catalog" items	Mostly common parts; limited number of state-of-the-art parts	State-of-the-art items; sensitive interfaces	Pushing state-of-the-art
Product technology	Available technology	New applications of existing technology and custom-built items	New capabilities from core technologies	New core technology
Program structure	Small staff and informal communications	Moderate size staff, layered structure	Multiple locations and formal communications	Large staff and deep reporting structure
Program "futures"	No follow-on planned	Investments made to minimize costs	Investments span contractual boundaries	Significant future opportunities
Competition	Minimal	Significant barriers to market entry exist	Competitive analysis and market expansion	Active competition and pressure to anticipate and react
Business relationships	Arm's length	Contractual	Teaming	Enterprisewide common goals
Team scope	Dominant perspective	Competing dominant perspectives	Competing perspectives, interrelated optimization	Aggressive optimization to meet requirements
Resource tightness	Not severely constrained	Limited in-process resolution	No in-process correction	Tightly constrained
Schedule tightness	Significant schedule slack time	Adequate for first-pass success	Aggressive; requires first-pass success	Severely constrained

These dimensions, when coupled together, define a program's attributes, its degree of execution difficulties, its limitations, and its expectations. All of these affect the concurrent engineering environment that is needed by the subject program. Each of these dimensions is discussed in the following sections to provide a better understanding of the viewpoint and the influence of the dimension on concurrent engineering.

Product Complexity.   Product complexity is inversely proportional to the number of people who fully understand how the product works. Complex products, for example, include those with electronic, software, mechanical, and optical functionality where few engineers truly understand the full spectrum of this functionality. Thus, concurrent engineering is essential. Complex products typically have many interrelated factors which make product design difficult. The identified levels of complexity are

A. Designs that are assembled using readily available "catalog" parts whose interfaces are standardized and robust.

B. Designs that are assembled using mostly common parts with a limited number of items representing state-of-the-art parts.

C. Designs that contain key elements which are state-of-the-art or have large numbers of state-of-the-art parts with many sensitive interfaces.

D. Designs that push the state-of-the-art envelope. Managing interdependencies is critical to product performance.

Product Technology.   Product technology refers to the availability of a capability or technology base which can be utilized in product design. The identified levels for technology are

A. Product and process designs utilize readily available technology.

B. Product and process designs require a new application of an existing technology, e.g., gears custom-built for the product or vapor phase reflow soldering for selected through-hole components.

C. Product and process designs require new capabilities from one or more core technologies, e.g., higher-speed integrated circuits (ICs) or surface mount technology.

D. Product and process designs require new core technology, e.g., gallium arsenide (GaAs) or multichip modules using TAB bonding.

Program Structure.   Program structure encompasses the number of people, layers of reporting hierarchy, role of formal and informal communication channels, and physical distribution of program staff. *Note:* The structure represents what is *needed* to execute the program, not necessarily the way the business is structured today (which tends to always look like category D in a large company). This relates to how you *want* to structure a program staff. The identified levels for program structure are

A. Program staff size is small, with informal reporting hierarchies and communication channels. Program objectives are broadly understood by all team members.

B. Program staff size is moderate, requiring layered reporting structures and more formal communications. Subgroups have specialized assignments. Informal communication channels are available.

C. Program staff size is moderately large and physically distributed across multiple locations within a building or across buildings or sites. Communication channels are typically more formal, with few informal means of communication.

D. Program staff size is large, with deep reporting hierarchies and structured communication channels, and physically distributed across multiple companies, often across numerous organizations. Typically, individual assignments are narrow in scope and highly focused.

Program "Futures."   Program "futures" refer to the follow-on opportunities for the program in the minds of all team members. "Futures" deal with how much incentive there is to invest in the current phase in order to optimize product success in later phases or the success of future products (in other words, requirements for long-range business decisions and investments). The levels for "futures" are as follows:

A. The program is stand-alone with no follow-on planned. No long-term investments are required.

B. Investments are made to minimize recurring (e.g., labor-saving devices/automation) and nonrecurring costs (e.g., hard tooling) and are aimed at reuse. Business base available to pay back investments.

C. Investments span contractual and business-base boundaries. Investment risks are shared across the enterprise. Program end-use criticality and life-cycle product cost call for investment in reuse and future improvements.

D. The program is strategically aligned with the enterprise, encouraging significant reuse in future generation programs and enabling significant future opportunities. Opportunities for midstream (or in-use) corrections are severely limited. Product has stringent end-use requirements, e.g., nuclear power plants, artificial organs.

Competition.   Competition refers to the level of activity in the relevant industry and the criticality of anticipating and reacting to competitors' moves. This dimension emphasizes the need for program flexibility and the ability to react quickly to competitive pressures. The levels of competition are as follows:

A. Competitive pressure is minimal because there are few competitors, there are close business relationships between established businesses and their customer base, or businesses have protected (product or strategy) niche market positions.

B. A significant portion of the available market is controlled by a few key competitors. Often significant barriers to entry exist.

C. Competing enterprises with significant resources channeled to competitive analysis and market expansion exist. Competitive benchmarking is extensive.

D. There is active competition with few barriers to entry, and so there are significant pressures to anticipate and react to competitors' actions. Product introduction schedules and costs are critical, as are features and other differentiators.

Business Relationships.   This refers to the formality of the relationships with customers, vendors, suppliers, teaming partners, and prime developers. The levels of relationships are as follows:

A. Arm's-length relationship. Commercial transactions (buying and selling of preexisting goods) are the primary form of interaction.

B. The relationships between business entities are formal and typically contractual. Directives governing business interaction are primarily unidirectional (one party has the leadership role and typically dictates requirements to others).

C. There is a teaming relationship to achieve joint or complementary goals. Companies selectively engage in bidirectional business relationships (e.g., consortia, strategic suppliers, joint development).

D. Customer, key suppliers, etc., are all working together as equals within the enterprise to satisfy relevant aspects of the program goals (establishing requirements and implementation approach).

Team Scope.   Team scope refers to the diversity of critical perspectives required for program execution. This relates to the dominant product requirements. The levels of team scope are as follows:

A. Small number of dominant perspectives (e.g., performance), with advice coming from numerous perspectives (e.g., test, packaging).

B. Small number of competing dominant perspectives which must be balanced to meet product requirements.

C. Large number of competing discipline perspectives involved in interrelated optimizations.

D. Aggressive optimization required to meet the total product requirement for the total life cycle.

Resource Tightness.   Resource tightness refers to limitations on the staffing or funding available to the program. In some areas, resources can be used to counteract deficiencies in the existing concurrent engineering methodology/environment. The levels of resource tightness are as follows:

A. Resources are not severely constrained and are available to be applied to correct a program weakness.

B. Resources are available for limited resolution of in-process problems.

C. Resources are not available for in-process correction.

D. Resources are tightly constrained. Inadequate resources to execute the program leads to creative changes in the development process.

Schedule Tightness.   Schedule tightness refers to the limited "schedule slack times" to counteract deficiencies in the existing concurrent engineering methodology/environment. The levels of schedule tightness are as follows:

A. Schedules include significant slack time on noncritical paths. Schedule is adequate for limited risk implementation. Trial/beta-test product introduction time is available.

B. Schedule is adequate for first-pass success. Schedule includes slack time on some noncritical paths.

C. Schedule is aggressive and requires first-pass success.

D. Schedules are severely constrained. Mistakes cause significant schedule slippage, cost overruns, and other negative business impacts. Product introduction is in a "catch-up" position.

Concurrent Engineering Influencing Dimensions Considerations.   From reviewing the material associated with the influencing dimensions matrix, it should be apparent that the appropriate concurrent engineering approach is dependent on program or product complexity. That is, there is no single implementation plan or step-by-step recipe for concurrent engineering. Concurrent engineering can be thought of as the mechanism for dealing with program complexity. It is a way to achieve "master craftsman" quality while meeting or exceeding business and management goals. As an illustration of these concepts, consider the following examples:

1. All of the elements of concurrent engineering are put into place, with the exception of data sharing. The program is large, so that it requires a single spacious facility occupied by the entire program staff. Teams are formed, but the only mechanism established to keep all team members current on activities and decisions is periodic meetings. Key inputs from members are lost or delayed, since information was not available to team members at the appropriate time. The program and product suffer.

2. Again, all elements of concurrent engineering are established except for team empowerment. Managers are included in each team as the mechanism for keeping everyone informed of the latest information. Decision making is slowed because each team has to wait until one or more managers review each recommendation and determine the appropriate action for every decision on the program. Development cycle time and labor costs are increased because of the idle time incurred while waiting for management decisions.

These are simplistic examples, but they should indicate the impact of not implementing concurrent engineering thoroughly and appropriately. The reader will gain more insight as the material of this chapter is comprehended.

Purpose of Differing Approaches.   The purpose of the differing approaches to concurrent engineering just described is to match the appropriate approach to concurrent engineering implementation for a particular program to the need of that program. The matrices can be used for several purposes: (1) determining the specific components of a concurrent engineering approach, (2) generating an implementation road map to enhance concurrent engineering capability, and (3) checking the consistency of the concurrent engineering approach currently in place. All of these activities involve critical self-examination, as described next.

Concurrent Engineering Self-Assessment—The "As-Is" Environment.   Concurrent engineering is a broad topic with numerous attributes. In order to examine the relevant aspects in detail, it is necessary to decompose the total concurrent engineering methodology into its integral components: organizational issues, requirements, communication issues, and product development methodology.

Organizational Issues.   Organizational issues refer to aspects of team dynamics, strategic business issues, and management and corporate culture that affect product development. The organization and its culture must support a concurrent engineering methodology for such a methodology to succeed. Existing cultural and organizational policies often counteract the intentions of concurrent engineering. The matrix focuses attention on several specific categories of corporate culture and management policies that are crucial to successful adoption of concurrent engineering methods.

Requirements.   Concurrent engineering has broadened the interpretation of requirements to include all product attributes that affect customer satisfaction. Adequately capturing and expressing the total set of these requirements is crucial to concurrent engineering. In addition, the matrix includes the need for planning, scheduling, and documentation for the product development team, along with validation of the total set of requirements. These topics must be worked in concert to ensure successful concurrent engineering.

Communication.   Communication is the next major category of critical concurrent engineering capabilities. Communication is the lifeblood of an enterprise. Strategies and common goals must flow out to every individual to mold the team into an efficient and productive unit. Feedback from knowledgeable individuals is essential in order to optimize design decisions and to improve the development, manufacturing, and support processes. The communications capabilities are broad organizational needs for data management and data sharing within and between departments (for example, logistics, manufacturing, and design) and between suppliers and customers. Next is "lessons learned," which come from various organizations but must be interpreted and analyzed by an individual engineer in order to influence a particular program. Next is decision traceability, which refers to the capture of an "audit trail" of decisions and tradeoffs that were considered during the development process, specifically the rationale for a decision, the other alternatives considered, and the rationale for their rejection. Finally, interpersonal communication is considered to be the single most important element of successful concurrent engineering today. Individuals in an enterprise must care deeply about the success of the team and be openly receptive to ideas for improvement and proactive in the dissemination of timely constructive assistance. Product development participants need to communicate several categories of information, such as working product data, lessons learned, decision rationale, and decision sequences. All are needed to track and optimize the process of product development. Interpersonal communication and interworkstation communication are crucial and are related to how data are acquired and shared within the project, program, and enterprise.

Product Development Methodology.   The process of concurrently enhancing the product and assessing its status in a concurrent engineering environment is quite novel. In particular, optimization, verification, and development processes are redefined for concurrent engineering. This affects the role of data libraries, reviews, and product architectures.

The Self-Assessment Matrix.   The self-assessment matrix was developed to provide industry with guidance on determining the level of the concurrent engineering environment presently implemented for comparison with the indicated level defined by the program's influencing dimensions. A concurrent engineering environment, as noted above, can be generalized into four main categories: organization, requirements, communications, and development methodology. These main topics are subdivided in the matrix to aid the evaluator in assessing the current environment. A summarized self-assessment matrix, Fig. 8.5, is provided to give the reader an overview insight

Attributes of CE		Theme
Organization		
Team membership	———	Team integration
Team leadership	———	Effectiveness
Team member contribution	———	Synergism
Business relationships	———	Participation
Training and education	———	Awareness
Responsibility and authority	———	Empowerment
Management decisions	———	Perspective
Requirements		
Definition	———	Completeness
Schedule types	———	Parallel
Planning style	———	Adaptability
Validation of specifications to requirements	———	Accuracy
Communication		
Management of working data	———	Control
Data acquisition and sharing	———	Accessibility
Lessons-learned feedback	———	Experience
Decision traceability	———	Legacy
Interpersonal	———	Equality
Development Methodology		
Optimization	———	Customer satisfaction
Data libraries	———	Consistency
Development process	———	Controllability
Reviews	———	Non-interruptive
Process measurements	———	Information content
Analysis architecture	———	Hierarchical
Verification	———	Compliance

FIGURE 8.5   CE self-assessment criteria summary.

into these key considerations. The complete self-assessment matrix, Table 8.2A through D, is presented to provide further detail and definition. The following paragraphs define the concurrent engineering categories and their subcategories.

Organizational (See Table 8.2A).   An enterprise's organizational maturity is defined by the structure and dynamics of its teams, its business relationships, and its decision-making apparatus. As product complexity increases, the enterprise must seek a tightly knit structure that includes team members across the enterprise, including both internal and external resources.

Critical team membership (*team integration*). In product development, a collection of individual experts must combine their efforts as a team. For increasing levels of product and process complexity, the critical dimension is a tighter integration of the team and not just its co-location (real or virtual).

A. Members have task perspective
B. Members have multidiscipline perspective
C. Members have product perspective
D. Members have strategic perspective

Team leadership (*effectiveness*). The effectiveness of a development team depends on its leadership structure, from one imposed by management to one selected by the team members themselves.

TABLE 8.2A Concurrent Engineering Self-Assessment Criteria: Organizational

	CE Environment				
	A	**B**	**C**	**D**	**N**
Team membership (critical members)	*Team Integration* →				D y n a m i c
	Members have task perspective	Members have multi-discipline perspective	Members have product perspective	Members have strategic perspective	
Team leadership	*Effectiveness* →				
	Management-appointed team leader	Management-selected team facilitator	Team-selected facilitator	Natural emergence of temporary, most knowledgeable leader	c o n t i n u o u s
Team member contributions	*Synergism* →				
	Segmented	Leveraged	Cooperative	Collaborative	
Business (key) relationships	*Participation* →				
	Transaction-based	Contractual	Joint venture	"Partnership"	i m p r o v e m e n t
Training and education	*Awareness* →				
	Team concepts	Multidiscipline understanding	Cooperative decision process	Synergistic knowledge discovery	
Responsibility and authority	*Empowerment* →				
	Member responsibilty and rewards	Multi-disciplinary group responsibility and rewards	Team decisions and responsibility	Team autonomy and rewards	
Management decisions	*Perspective* →				
	Profit-based decisions and planning	Single-phase planning and investments	Multi-phase planning and investments	Life-cycle-based decisions	

A. Management-appointed team leader
B. Management-selected team facilitator
C. Team-selected facilitator
D. Natural emergence of temporary, most knowledgeable leader

Team member contributions (synergism). Concurrent engineering environments are characterized by synergy in the interaction among the individual members of the team. Without synergy, interaction tends to occur between relatively isolated domains of expertise; with synergy, the isolation is eliminated.

A. Segmented (discipline-oriented contribution)
B. Leveraged (expert consultants provide advice)
C. Cooperative (team member equality)
D. Collaboration (synergy)

Business relationships (participation). The degree to which external resources participate in product development as team members is critical. External suppliers and customers need to participate fully in product development.

A. Transaction-based
B. Contractual
C. Joint venture
D. "Partnership" (total involvement)

TABLE 8.2B   Concurrent Engineering Self-Assessment Criteria: Requirements

	CE Environment				
	A	**B**	**C**	**D**	**N**
Definition	\<———————————————— Completeness ————————————————\>				D y n a m i c c o n t i n u o u s i m p r o v e m e n t
	Itemized requirements definition	Requirements traceability	Requirements weighting	Unambiguous specifications	
Schedule types	\<——————————————————— Parallel ———————————————————\>				
	Task duration based schedule	Calendar-based schedule	Event-based schedule	Continuous addition of value to the enterprise	
Planning/methodology	\<————————————————— Adaptability —————————————————\>				
	Bottom-up collation of task definitions	Top-down determination of task definitions	Synchronization of concurrent, interrelated tasks	Iteratively refined abstract plans	
Validation (specification to requirements)	\<—————————————————— Accuracy ——————————————————\>				
	Validation to itemized requirements	Validation of inter-related constraints	Validate to end user requirements	Validation to end use and product business strategy	

Training and education (awareness). The focus of training broadens beyond individual disciplines. For team members, this permits a great awareness of enterprise issues.

A. Team concepts
B. Multidiscipline understanding
C. Cooperative decision process
D. Synergistic knowledge discovery

Responsibility and authority (empowerment). The team is empowered to implement its decisions. With this authority comes the responsibility for the decisions, and motivation and rewards come to the team as a group rather than as individuals.

A. Member responsibility and rewards
B. Multidisciplinary group responsibility and rewards
C. Team decisions and responsibility
D. Team autonomy and rewards

Management decisions (perspective). Management perspective motivates the scope of decisions. Management's perspective broadens beyond short-term concerns to a stage that accounts for the total product life cycle.

A. Profit-based decisions and planning
B. Single-phase planning and investments
C. Multiphase planning and investments
D. Life-cycle-based decisions

TABLE 8.2C   Concurrent Engineering Self-Assessment Criteria: Communication

	CE Environment				
	A	**B**	**C**	**D**	**N**
Management of working data	*Control →* Local individual data management	Data structured for project-wide sharing	Program repository of working data	Enterprise repository of working data	D y n a m i c c o n t i n u o u s i m p r o v e m e n t
Data acquisition and sharing	*Accessibility →* As-needed data extraction	Data supplied by most knowledgeable source	Data available as generated.	Enterprise-wide availability of data	
Lessons learned feedback	*Experience →* Design guides with rationale and intent	Consolidated design guide with rationale	Rationale and weighting for each product development rule	Dynamic lessons learned feedback	
Decision traceability	*Legacy →* Individual decision rationale ownership	Project decision rationale ownership	Program decision rationale ownership	Enterprise decision rationale ownership	
Interpersonal	*Equality →* Member-specific terminology	"Common" terminology	Equal input/impact	Knowledge-based perspective	

Requirements (see Table 8.2B).   Requirements refer to external and internal constraints and assertions which affect the development of products. These external and internal drivers are categorized here as follows: product definition, scheduling capabilities, planning methodologies, validation plans, and documentation generation.

Definition (*completeness*). Product definition can be thought of as the process of capturing customer care-abouts and internal needs and translating them into specifications of product and process features to satisfy complete life-cycle needs. This includes all requirements, encompassing manufacturability, supportability, and upgradeability.

A. Itemized requirements definition
B. Requirements traceability
C. Requirements weighting
D. Unambiguous specifications

Schedule types (*parallel*). This category refers to the type of scheduling practices in place to support project planning and capability for simultaneous tasks.

A. Task duration–based schedule
B. Calendar-based schedule
C. Event-based schedule
D. Continuous addition of value to the enterprise

Planning/methodology (*adaptability*). This section refers to the methodology used to plan and monitor the program.

TABLE 8.2D   Concurrent Engineering Self-Assessment Criteria: Development Methodology

	CE Environment				
	A	B	C	D	N
Optimization	*Customer Satisfaction →*				D y n a m i c c o n t i n u o u s i m p r o v e m e n t
	Review-based optimization	Limited interrelated requirement optimization	Program-wide requirement optimization	Total weighted requirement optimization	
Data libraries (single master library source)	*Consistency →*				
	Control of preferred parts and process libraries	Controlled libraries if reusable modules and intent	Controlled technology-independent libraries	Controlled real-time library data from source	
Development process	*Controllability →*				
	Product-independent, repeatable, and consistent process	Measurement standards definition	Closed-loop control	Process improvement and optimization	
Reviews	*Noninterruptive →*				
	Schedule-driven product and process critiques	Event-driven review	Immediate issue resolution	Status reporting	
Measurements	*Information Content →*				
	Measurement using function-specific deterministic indices	Measurement using process-related deterministic indices	Measurement using heuristic predictive indices	Measurement using relevant, analytical, interrelated predictive indices	
Analysis architecture	*Hierarchical →*				
	Single-level modeling	Multi-level modeling	Mixed mode with multiple views	Mixed-signal, mixed-mode process modeling	
Verification	*Compliance →*				
	Member-dependent verification	Multi-discipline verification	Team verification	Correct by construction	

A. Bottom-up collation of task definitions
B. Top-down determination of task definitions
C. Synchronization of concurrent, interrelated tasks
D. Iteratively refined abstract plans

Validation (*accuracy*). Validation of the requirements is the process that determines if the specification meets all the requirements and if all specified processes will accomplish the intended result.

A. Validation to itemized requirements
B. Validation of interrelated constraints
C. Validation to end-user requirements
D. Validation to end use and product business strategy

Communication (see Table 8.2C).   Communication is the lifeblood of an enterprise. Strategies and common goals must flow out to every individual in order to mold the team into an efficient and productive unit. Feedback from knowledgeable individuals is essential in order to optimize design decisions and to improve the development, manufacturing, and support processes.

Concurrent engineering advocates the assembly of individuals knowledgeable about design, manufacturing, and support along with customers and suppliers into a team that has complete autonomy. Design decisions which affect product life-cycle cost, quality, and schedule are improved because the total enterprise is represented. This style of horizontal communication overcomes the hierarchical barriers to the exchange of timely and accurate information. The flattening of hierarchical organizations has been occurring for several years because communications have improved, vital information is accessible from easily usable data repositories, and individuals are empowered to make timely, informed decisions.

The timely exchange of accurate information is essential to rapid product cycles and cost minimization. However, it becomes increasingly difficult when team members are widely distributed, possibly throughout the world. Organization studies (Allen) have shown that the effectiveness of collaboration within an office building decreases by half for every 100 feet of separation between offices! Improved communication is essential to the success of concurrent engineering.

Management of working data (*control*). The early phases of a program (conception and feasibility evaluation) offer the greatest opportunity to improve product life-cycle cost and quality if a concurrent engineering methodology exists in the enterprise. The opportunity for benefits from concurrent engineering comes from its potential to eliminate the phases required to modify or optimize and redesign. Early input from downstream specialists and customer reviews of the embryonic design can result in a product which is optimally designed the first time. The forces driving design change will not be errors, but rather the injection of new technology or new lessons learned.

A. Local individual data management
B. Data structured for projectwide sharing
C. Program repository of working data
D. Enterprise repository of working data

Data acquisition and sharing (*accessibility*). For concurrent engineering to be successful, data must be available to be shared across the team. The basic concept is to create and enter data once and use them many times. The data consist of working data and released data that are directly applicable to the product under development and those associated with the product. This data sharing requirement is applicable to all program phases. Another factor is that data must be usable by the requester, and so data formats and user "views" are extremely important. For working data to be useful, certain levels of data management are necessary.

A. As-needed data extraction
B. Data supplied by most knowledgeable source
C. Data available as generated
D. Enterprisewide availability of data

Lessons learned feedback (*experience*). The concurrent engineering methodology states that designs should be influenced by downstream requirements. One of the most valuable sources of data is lessons learned from previous project/program experiences (those who ignore history are doomed to repeat it). Lessons learned are rarely used in current enterprises because they are not necessarily captured in usable form. If they are captured, designers are not necessarily aware of them, cannot readily access them, or do not know the rationale for them. The best source of this information is experienced individuals, but the turnover rate in many organizations results in a highly volatile corporate memory.

A. Design guides with rationale and intent
B. Consolidated design guide with rationale
C. Rationale and weighting for each product development rule
D. Dynamic lessons learned feedback

Design traceability (*legacy*). One purpose of lessons learned is to understand why design decisions were made. In any design process, the designer is faced with a bewildering array of trade-offs and decisions on a daily basis. Many times, a designer has a particularly important reason for specifying a particular component in a design. But the design intent is rarely understood by others in the enterprise.

A. Individual decision rationale ownership
B. Project decision rationale ownership

C. Program decision rationale ownership
D. Enterprise decision rationale ownership

Interpersonal (*equality*). This is the most important of all the dimensions of communication. The metric chosen to describe this dimension is equality. Communication can be impeded by personalities and egos. As the communication capability in the organization matures, there is a wider dissemination of relevant information to all members of the enterprise, and they receive equal treatment with regard to their need for information.

A. Member-specific terminology
B. "Common" terminology
C. Equal input/impact
D. Knowledge-based perspective

Product Development Methodology (see Table 8.2D).   The product development methodology must be understood by all and must encompass everyone affected by the process. The process must be predetermined, documented, and followed. The interaction of people, the interrelationships of tasks, and the timeliness of data must be comprehended. The capture of total requirements, the total product development process, the design of the manufacturing process, and the design of the product support processes are all included within the product development methodology.

Optimization (*customer satisfaction*). The primary goal of the organization is to deliver a product with the lowest cost, the shortest schedule, and the highest quality which results in customer satisfaction. Optimization of the product during the development cycle involves many factors. Of extreme importance is early tradeoffs among the functional disciplines so as to avoid optimization of one function at the expense of another functional area. The product should be viewed for producibility, testability, reliability, maintainability, and so on. Each of these areas must be examined to improve the robustness of the product when considering its manufacture and customer usage. As we move closer to the right side of the matrix, we increase the customer satisfaction, since a more complete enterprise team is involved in key product decisions.

A. Review-based optimization
B. Limited interrelated requirement optimization
C. Programwide requirement optimization
D. Total weighted requirement optimization

Data libraries (*consistency*). Data libraries consist of that set of data which is needed in order to design, analyze, produce, and test the product. It is assumed that there will be a single master library source coordinating all of the various discipline libraries.

A. Control of preferred parts and process libraries
B. Controlled libraries of reusable modules and intent
C. Controlled technology-independent libraries
D. Controlled real-time library data from source

Development process (*controllability*). Until product development is viewed as a process, it will be extremely difficult to begin the journey of continuous improvement. The development process must be controllable and measurable so as to be completely understood if process optimization is to occur. For this discussion, we will define the development process as all of the activities that occur during the program execution which, when properly performed, will efficiently place the desired product into production. This process includes all disciplines and begins at program conception and concludes at the release to manufacturing. It includes the development of the product support/operational support process and the design of the manufacturing processes. The key to process management is to identify the steps needed for improvement and the sequence by which to improve them. Areas of consideration should include standardization, commonality, modularity, adaptive manufacturing process, and variability reduction techniques, as examples.

A. Product-independent, repeatable, and consistent process
B. Measurement standards definition
C. Closed-loop control
D. Process improvement and optimization

Reviews (*noninterruptive*). One of the most glaring problems in current product development methodologies is the inadequacy of design reviews. There are many reasons for this, some of which are organizational, human behavior, market changes, and critical resource skills. Many product field failures can be attributed to lack of proper and complete design reviews. The goal of the highly effective organization is to standardize the review criteria and strive for a condition in which conflicts are resolved quickly and the design is correct by construction.

A. Schedule-driven product and process critiques
B. Event-driven review
C. Immediate issue resolution
D. Status reporting

Measurements (*information content*). The section on measurements refers to data collected in order to provide knowledge relevant to decision making. If results cannot be measured, then the process cannot be controlled, or performance improved.

A. Measurement using function-specific deterministic indices
B. Measurement using process-related deterministic indices
C. Measurement using heuristic predictive indices
D. Measurement using relevant, analytical, interrelated predictive indices

Analysis architecture (*hierarchical*). Analysis architecture involves the scope and range of applicability of analysis and simulation methods, plus their level of integration with supporting software tools. Existing tools and methods are discipline-specific, and each applies to a single product architectural level or product class. Also, data libraries to support existing tools are embedded within them and therefore are not available for transfer, expansion, or change. Capabilities must support concurrent engineering techniques.

A. Single-level modeling
B. Multilevel modeling
C. Mixed mode with multiple views
D. Mixed-signal, mixed-mode process modeling

Verification (*compliance*). Verification is a process to determine that the design is "correct." Correctness includes compliance with the "total" specifications, for a product that is producible and supportable, with major risks limited and managed. Verification is a continuous process that starts with the derivation of "total" requirements and continues through production and resolution of change orders. The verification process is adaptable to programmatic changes in requirements. Nonconcurrent engineering development processes rely heavily on prototype testing.

A. Member-dependent verification
B. Multidiscipline verification
C. Team verification
D. Correct by construction

Evaluation Technique—Improvement Road Map. The self-assessment matrix is a tool which provides each program in an enterprise with the means of measuring the status of its current concurrent engineering environment. As the reader might recall, the assessment matrix is in two parts: an influencing dimensions matrix (Table 8.1) and a concurrent engineering environment matrix (Table 8.2). The influencing dimensions matrix has nine categories:

Product complexity

Product technology

Program structure

Program "futures"

Competition

Business relationships

Team scope

Resource tightness

Schedule tightness

and the other matrix has four major categories which are attributes or characteristics of a program's concurrent engineering environment:

Organizational

Requirements

Communication

Development methodology

Each of the categories has several elements which together characterize that category. The concurrent engineering environment matrix provides a snapshot of the present capabilities within a program. The matrix is intended to be used as a self-measurement tool. It is not intended (nor should it be used) to compare one plant/staff/department with another. When it is used correctly, participants in each program can measure where the program is with the assessment matrix and where it should be with the influencing dimension matrix, and then plan on how to reach the appropriate (needed) concurrent engineering environment for the program.

Utilizing the influencing dimensions matrix, the first objective is to establish the appropriate concurrent engineering approach (A, B, C, D) based on the nature of the program as measured by the influencing dimensions. The influencing dimensions identified for a particular program will help guide the assessor to the required (should be) concurrent engineering implementation. The next objective is to use the environment matrix to determine where the program is (as is) with respect to the categories and key elements. When current capabilities are determined, the actions required to move the "as is" approach to the "should be" must be implemented. In almost all cases, movement to the "should be" implementation will be beneficial to a program. Such movement generally requires the investment of time and resources, and this should be weighed against the benefits to the program and the business unit. Long-term business plans, investment cost, implementation time, and the capacity to change the culture of the organization should be assessed against these benefits. However, the benefits of concurrent engineering should not be underestimated.

Concurrent Engineering Environment Evaluation.   By completing a self-assessment for all concurrent engineering attributes, the organization develops a current status of its concurrent engineering environment. In a strict interpretation of the matrix, the program's overall concurrent engineering capability is only as strong as its weakest concurrent engineering attribute, as coherency between the elements within a matrix column was built into the matrix. When a particular attribute is implemented with less capability, this conceptually acts as a bottleneck, reducing the capability for the whole system. Because the matrix was constructed with highly interrelated elements, an attribute with specific capability is feasible only when related attributes have similar capabilities. For example, immediate resolution of issues is not possible if issues are reviewed only on a periodic basis. For this reason, a cohesive and consistent solution is possible only when all attributes are implemented with consistent capabilities, represented by a single column. The matrix column represents a synthesis of the individual capabilities to provide a global view of an organization's overall ability to apply concurrent engineering methods.

After a critical self-examination is done, the self-assessment matrix can be used to generate an improvement road map based on the concurrent engineering implementation currently in place. By examining the descriptions for every attribute (matrix row) for each approach (matrix column), the "as is" environment is assessed. The improvement road map is generated by developing a plan to increase the concurrent engineering capability of attributes with characteristics to the left of the approach desired or needed to reach the "should be" environment.

Because all the approaches describe good concurrent engineering practice, the matrix is not intended as a rating tool [like the Software Engineering Institute (SEI) maturity rating]. Whether a particular approach is good enough for the needs of a program depends on the nature of the program. The environment has been designed to highlight weaknesses relative to a program's concurrent engineering needs. By comparing the current environment with the required environment, areas for improvement can be targeted and policies developed to overcome those identified weaknesses.

It is tempting to view the availability of automation technology as an enabler of concurrent engineering. A careful reading of the matrix, however, reveals that concurrent engineering is a new culture that must be instilled in team members. Automation of current, serial autonomous processes is a mistake and will only entrench current practices and stifle the emergence of a concurrent engineering culture. Additionally, automation by itself is not the answer. Automation should be viewed as an enabler or facilitator of concurrent engineering approaches.

The self-assessment criteria are focused on program requirements—what is necessary to develop a product. The matrix's self-assessment criteria are, therefore, applicable only to programs. The matrix is not meant to evaluate a company or a functional organization within a company. It is to be used by an organization to evaluate its concurrent engineering capability and determine its organizational needs relative to a specific program.

The matrix is a "snapshot" in time—a best view defining a concurrent engineering capability and what is needed. However, continuous improvement will, with time, cause new columns to be added to the right and eliminated from the left.

The matrix, its characteristics, and the influencing dimensions are provided as an attempt to describe specific characteristics which affect the successful execution of programs using concurrent engineering practices. The author expects that as more companies gain experience with concurrent engineering, additional insight will be gained into the key enablers and inhibitors of concurrent engineering implementation. This insight should then be used to refine the matrix.

Evaluation Technique.   A necessary step in establishing an effective concurrent engineering environment is to assess where the program is currently, relative to the concurrent engineering environment where it should be. This is the underlying principle behind application of the matrix.

1. *Assign the weighting factors for each of the influencing dimensions.* Weighting factors should reflect the relative importance of each factor to successful completion of the program. The weighting factor approach is not critical. What is important is for you to determine the relative importance of each of these dimensions to your long-term program success.

2. *Determine the appropriate level for each influencing dimension.* The level selected in Table 8.1 (A, B, C, or D) should reflect your assessment of how your program is influenced by each of the identified factors and therefore what is required for successful completion of this program. Your selection of levels is necessarily a subjective choice based on your evaluation of the impact or criticality of each influencing dimension on your program. You must resolve variations between levels. You can use majority rules or weighting techniques if there are differences in the relative importance of the influencing dimensions. Whatever subjective method is used, a single concurrent engineering environment needs to be established.

3. *Perform a self-assessment of the program.* For each concurrent engineering element within Table 8.2a through d, find the greatest capability for which all factors of that approach are currently in place. Define this as your current profile.

4. *Compare the current approach to the appropriate approach.* By comparing the appropriate approach, as determined in step 2, to the current approach, areas that need improvement are identified. All deficiencies should be brought up to the needed capability. The concurrent engi-

neering environment is no better than the weakest attribute. With the required attributes for each planned concurrent engineering approach established, you can assign responsibilities, set timetables, develop the necessary procedures, document plans, and measure progress toward implementing the identified changes.

Example Evaluation.   In order to aid the reader in understanding how to use the self-assessment technique, an example is utilized. First, you must put yourself in the shoes of the program manager and understand his assignment from his management. His charge is to develop a line of high-quality notebook-sized computers that will continually beat the competition to market with increased features and with highly competitive pricing, resulting in high market share. Some key product features, then, must be as follows:

Lightweight

Highly portable

Self-contained (keyboard, display, CPU, battery, data storage, etc.)

Sized to fit in a briefcase

Standard keyboard layout, feel

Compatible operating system

From the assignment, there are also a number of factors that drive product attributes/constraints and therefore product requirements:

Low-risk approach drives technology choices, design reuse, etc.

Price competitiveness drives material choice, design complexity, etc.

Market share drives the need for a reliable, testable, and producible product, etc.

There are aspects that affect the program, such as:

Multiple products under way at various stages (concept through production phasedown)

Short individual product life

Long-term production capacity

From the data provided thus far, you can already see the diversity of the requirements that affect the program and the product. It should also be obvious that a lot of planning (both short and long term) is required, that timely communication will be needed, that long-term capital investments will be requested, and that the individuals on the program must be focused on common goals if the program is to be successful. A concurrent engineering environment will greatly help in making this program successful. Let's evaluate this program by the criteria described in this section.

Using the influencing dimensions in Table 8.1, the program manager must use his program/product knowledge combined with company information (resources, constraints, policies, experience, etc.) to subjectively weight each dimension's importance to the program and then the level of impact/criticality of each dimension, from low impact (A) to high impact (D). This is illustrated in Table 8.3. Each influencing dimension will be discussed individually to provide insight into the thought process for marking the matrix.

Product complexity: The product design utilizes only common packaged devices that are auto-insertable into double-sided boards. The design is highly producible. *B.*

Product technology: The product design requires new applications of existing technology. Newer products' designs will require new capabilities from core technologies. *B now, but moving to C.*

TABLE 8.3   CE Influencing Dimensions Matrix Example

Influencing dimensions	CE environment level			
	A	B	C	D
Product complexity		X		
Product technology		X	X	
Program structure			X	
Program "futures"			X	
Competition				X
Business relationships		X		
Team scope		X		
Resource tightness	X	X		
Schedule tightness			X	X

Program structure: The program staff is moderately large, with the development staff in one location and the production staff in another. The development staff, although at one location, is distributed, since various products are at different stages of development. *C.*

Program "futures": Investments will be made in automation to facilitate manufacturing and test in production that spans product lines. Key suppliers have increased their capacity. *C.*

Competition: Competition is fierce in product features and price. Being first to market is critical. Perceived price-performance ratio is key. *D.*

Business relationships: Relationships are mostly commercial transactions; however, key suppliers are active contributors to the development staff. *Could be A or B—-selected B based on key supplier involvement.*

Team scope: Team works well together, but manufacturing is the dominant force. However, all requirements are considered. *B.*

Resource tightness: Resources are not tightly constrained in the beginning, since the strategy is to force into the existing market and become the leader. However, resources are expected to tighten quickly because of the strong price sensitivity of the product. *A going to B.*

Schedule tightness: Schedules are aggressive and will get more constrained, since you will want to increase market share and quickly introduce additional products. *C going to D.*

This completes the evaluation of the influencing dimensions on an individual basis. Next, consider the relative importance of the dimensions. If it is felt that they are all equally important, then the concurrent engineering environment for the program should be at least a B. Also, even though the program should be at a B now, the chart indicates a need to start the improvement process. As the effort becomes more of an established program with a product base, it will in several dimensions need to be at C, since it is moving to a more capable implementation of concurrent engineering. If the dimensions associated with C were weighted more heavily than the others, C may be the concurrent engineering approach needed now. Only the program manager can make those subjective decisions.

At this point, the "should be" concurrent engineering environment has been established for the program.

Now, it is appropriate to evaluate the program environment against the concurrent engineering self-assessment criteria of Table 8.2. This will help in evaluating where the concurrent engineering environment is relative to the criteria and where it should be, as shown in Table 8.4. The process to perform the self-assessment is much the same as the process used on the influencing dimensions. Each major category will be discussed.

TABLE 8.4   CE Environment Assessment Matrix Example

	A	B	C	D
Organization				
Team membership	X			
Team leadership	X			
Team member contributions		X		
Business relationships		X		
Training and education		X		
Responsibility and authority	X			
Management decisions			X	
Requirements				
Definition	X			
Schedule types			X	
Planning style			X	
Validation of specifications to requirements	X			
Communication				
Management of working data		X		
Data acquisition and sharing			X	
Lessons learned feedback		X		
Design traceability			X	
Interpersonal			X	
Development methodology				
Optimization		X		
Data libraries			X	
Development process		X		
Reviews		X		
Measurements	X			
Analysis architecture	X			
Verification		X		

Organization: The majority of the team members are product-oriented rather than program-oriented. Their product manager is appointed by management, as are the technical leaders. Manufacturing engineers are the dominant members, with advice coming from other design-influencing engineering disciplines. Their advice is considered and acted upon. The majority of supplier relationships are purchase order–based, but a few key suppliers provide advice during development. Training is primarily discipline-oriented, but any discipline can and is encouraged to attend. Performance awards are given to key individuals. Long-term investments are being made to penetrate the market.

Requirements: Primary requirements are documented. Periodic market surveys and competitive benchmarking will enhance traceability to user needs. Schedules are program event–driven—products will not be released until they are ready for production. This is because product reputation is critical. Planning and tasking acknowledges the interrelationships of tasks. Specifications are validated against requirements.

Communication: Product data are controlled within the program to allow data sharing upon demand. This allows lessons learned to be reviewed and their applicability determined. The program sharing of design and intention data across product projects encourages part commonality and design reuse. Design intent and major design approach and tradeoff decisions with their rationale are captured and stored at the program level. All the team members are focused on the product and their project goals.

Development methodology: The project tries to optimize its product across interrelated requirements. Data libraries have been established across the program to provide consistent application-independent data to all projects as well as complete product design data packages. The design methodology is documented and followed, providing a consistent approach and known analytical verification. The schedules are project-defined and are measured based on the established methodology. Single-level analyses are primarily conducted presently, but in light of the use of application-specific integrated circuits (ASICs), two-level analyses are being investigated. The verification process is very thorough and is conducted from various viewpoints to assure proper performance of the product in the user's hands.

Once this portion of your self-assessment has been completed, it is time to evaluate your current status as compared to the "should be" environment. As indicated, most checks fell under at least the B environment. Of the seven items that fell under A, four are being improved presently, so continue the good work. The other three areas need to be investigated and an improvement plan implemented.

If the "should be" environment was C, a larger improvement plan would need to be implemented. In either case, the self-assessment highlighted what needed to be done. It should be clear where you are and where you should be, and you have been provided with a road map to aid in improvement planning. Now make it happen!

A word of caution: This self-assessment is designed to help evaluate a particular situation. Readers are cautioned not to misapply or mislead themselves. The only thing that you will be hurting is yourself and your program. Concurrent engineering can provide you with closely knit teams, much broader than your directly reporting organization, that are focused on the team goals that will make their program successful, and you in the process.

8.4   IMPLEMENTING CONCURRENT ENGINEERING

As in all cross-functional efforts, it is critical that organizations start at the top. CEOs must revamp their corporations and set the example, day in and day out, through actions and words. The CEOs must obtain concurrent engineering buy-in from their total management staff and establish the right expectations. Up-front investment and long-term commitment are required to realize the success of concurrent engineering. These commitments are similar to the changes that we have experienced in quality programs (e.g., quality is free; inspecting out defects costs more than education programs to design in and build in quality).

An enterprise which has been infused with a zeal for concurrent engineering is constantly searching for new and better approaches through shared experiences and lessons learned. These lessons are rapidly translated into better products. Customer needs, technology, and production processes are advancing rapidly. The best products will be produced by a design methodology which can instantly release an improved product to an adaptable manufacturing process.

Most organizations are prevented from realizing these benefits because their manufacturing and support processes are rigid. Therefore, design iterations are normally viewed as bad and to be avoided. Clearly, iterations to correct design defects after design release must be avoided. However, iterations to incorporate a new technology, to incorporate an improved component with lower cost or better reliability, or to meet a new customer need must be performed rapidly. "Lessons learned" improvements come from experience with the product in the field. Rapid product introduction, rapid learning, and rapid product improvement are keys to improved competitiveness.

A fundamental product performance parameter for rapid product improvement is reliability. If a large inventory of spare components is required to support the product, changing and expanding this inventory with every design change will be prohibitively expensive. Products must be so reliable that spares are minimized. Trade studies are required to support improvement feasibility and direction.

Another expensive requirement that must be changed is that every product in the field should be upgraded to reflect the latest design change. Such a requirement stifles rapid product improvement. Every year, automobile designs provide greater safety, more reliability, and higher performance. Could you ever afford the cost of annual retrofits to your existing automobile? Can anyone afford the cost of this expectation in any product?

These are two significant barriers to rapid lessons learned, iterative product improvement, and rapidly improving competitiveness, which are benefits derived from concurrent engineering.

Concurrent engineering is not a new concept. Concurrent engineering–oriented companies have entrepreneurial team spirit, bringing together all of the best ideas from every member of the team, including exceptionally close relationships with key suppliers and customers. As organizations grow, they tend to move away from these fundamentals. Concurrent engineering is a culture which can also succeed in these larger organizations, especially with the aid of new technology which enables collaboration among larger, more dispersed teams.

8.4.1 Concurrent Engineering Assessment Caveats

The following caveats must be kept in mind throughout the assessment:

- Management must have established an atmosphere that allows concurrent engineering concepts to be realized and flourish.
- Cells within an environment are interrelated—baseline consistency within an environment is critical.
- Different approaches to concurrent engineering as described in the matrix do not imply differences in quality or achievement—they are appropriate to the size, complexity, and technology of different programs.
- Influencing dimensions, including business aspects, affect the required environment.
- Self-assessment provides a "snapshot" in time of the status of a concurrent engineering environment or the concurrent engineering needs of a particular program phase.
- Self-assessment is *not* a scorecard during proposal or program evaluations.
- Self-assessment is *not* a comparison between organizations.
- Self-assessment is *not* a strategic planning tool.
- The matrix is an evaluation tool, not a program to be implemented.
- Movement to greater concurrent engineering capability than is indicated will not necessarily operationally benefit this specific program's development, since the marginal operational benefit might not justify the amortized cost. However, continuous improvement needs to be a way of life in order to stay competitive.

8.4.2 Using Existing Concurrent Engineering Elements

Some aspects of CE have been in use for many years within industry. This has resulted in a sound foundation on which to build, as evidenced by utilization of a program-structured business approach, integrated development teams, and co-location of design-oriented specialty engineering disciplines as contributing members of the development team. Although informal benchmarking of CE implementation indicates pockets of excellence, industry often lacks across-the-board implementation.

When trying to determine key points on how concurrent engineering can be better applied within an industry, several implementation points seem to be critical:

- A clear (documented) understanding of engineering methodology and manufacturing and support processes is needed, followed by a decisive plan for implementation and improvement.

- "Top-down" systems engineering, in a broad sense, is required in order to succeed in understanding and documenting the complex requirements of customers, internal constraints, and knowledge of decision implications. CE principles are most effective if they are initiated with concept exploration and continue to be beneficial during the entire product development cycle [concept exploration, demonstration/validation, engineering and manufacturing development (EMD)].

- Each team member needs to have a broader understanding and awareness of every team member's functions, concerns, and value-added contributions to the overall team goal. The correlation and dependencies between design, manufacturing, and support must be understood. Broader methodology, tools, and training are needed to more effectively utilize incomplete and abstract data that are available early in the product development cycle.

- An overall cultural change is needed. This doesn't mean that it is necessary to revise a company's personality. Instead, a point-of-view change is needed—a change from a short-term/cost-driven orientation to a long-term orientation toward quality and continuous improvement. A sufficient point-of-view change combined with proper performance evaluation motivators (quality indices rather than first-quarter results) should encourage a concurrent engineering environment.

- Team members need to be at a compatible tool capability level so that each member has the opportunity to perform analyses from his or her perspective, provide timely inputs into the development process, and therefore influence the product development. If automation will be used as a CE enabler, all team members must be considered before compatibility can be reached. Even with team member automation compatibility, "in work" information must be accessible and shared within the team for success. A programwide information management system (system infrastructure versus tools) is needed to provide a common view of requirements, tradeoff decision rationale, working data (design, schedule, cost, etc.), lessons learned visibility, released data, etc. Common libraries are also needed for parts standardization efforts, design guide compliance, etc., to provide consistency of data across team members. "Design data" perspectives must evolve to "product data" perspectives that comprehend the entire product life cycle.

- Within each company, a concurrent engineering champion is needed to provide direction and coordinate the various discipline and automation champions who focus on concurrent engineering from their perspective. The CE champion would manage the definition (policy, procedure, methodology, automation, training, reward, etc.) development for CE under the cognizance of senior management and facilitate implementation across the company on a program-by-program basis.

- Involvement with one's total customer base (contracting/using/maintaining/monitoring) and with key suppliers during development must be improved. They must become part of the team, and viewed as "equal" members.

- Enablers for process monitoring, problem identification/resolution, and change implementation [product data management (PDM), statistical process control (SPC), etc.] need to evolve into an institutional culture.

- Program scheduling, funding, monitoring, and measurement need new perspectives. Key element completion criteria should be event-driven rather than calendar time–driven. Metrics should be based on pertinent process-related data, not necessarily items that are easy to measure. Metrics emphasis must be shifted from lagging (report card) indicators to CE process control and validation assessments.

- Design iterations and soft prototyping should be encouraged early and often (through software brass boarding, simulations, stereo lithography, etc.) prior to design release to refine a robust design.

- Most importantly, each of the above must be planned, measured, managed, and continuously improved in an effective manner.

Take an active role in utilizing as many of these CE implementation concepts as possible. Make CE the way of doing business.

8.5   RESOURCE INFORMATION FOR CONCURRENT ENGINEERING IMPLEMENTATION

Since the conception of the term *concurrent engineering,* industries from around the world have initiated changes to bring these concepts into their product development processes and to modify their organizational cultures. Each industry has developed its own approach to concurrent engineering and has implemented its plan of action within its organizations. Industries are slowly releasing information about their approaches to concurrent engineering/integrated product development/integrated product and process development (IPPD) in books (case studies), newspaper articles, conference papers, special features, etc. There is no single document that consolidates this information, nor is there a single approach that all industries are utilizing. Each industry has tailored the concurrent engineering concepts to fit its organizational style and is reaping the benefits that are consistent with customer-satisfying products and empowered employees on collaborative teams. Read trade journals, newspapers, and conference proceedings and be conscious of the concurrent engineering concepts. These articles usually don't directly state that this is an article about the benefits of concurrent engineering. They usually address success with a new product or describe a new approach to developing products. Don't expect the total answer to be provided, either. Industry holds approaches that lead to success as proprietary, and companies don't want to give too much of their competitive advantage away.

8.6   RELIABILITY ENGINEERING—A KEY CE PLAYER

Reliability engineering is key to concurrent engineering through product development teams. Each team comprises various disciplines that strive for a common goal. Each discipline can be then thought of as a link in a chain, as depicted in Fig. 8.6, with all links being essential to the

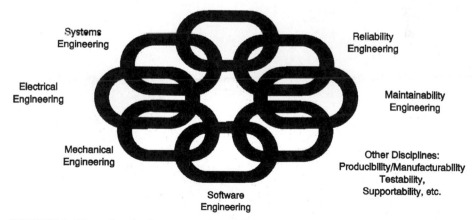

FIGURE 8.6   The product development team perspective.

chain. Reliability engineering is one of these essential links, since reliability, as a product attribute, is an essential ingredient to the success of any product. It is key to customer satisfaction. Customers expect their products to work on demand. To make product reliability a reality, reliability engineering must be involved throughout the complete product life cycle. To ensure that reliability is a key design requirement, it is essential that reliability engineering bring that perspective to the design teams developing products.

This chapter has provided an understanding of concurrent engineering and its concepts. With this understanding, reliability team involvement can be planned and reliability contributions can be defined to the team. Reliability needs to affect product proposals, aid product requirements definitions, perform analyses during product development, provide tradeoff inputs, generate documentation, etc., as has been defined in the other chapters of this book.

As a team member, reliability engineering must be aware of team dynamics and team synergy. Individuals cannot just be assigned to form a team with the expectation that the team will be successful. Team training is essential for teams performing in a concurrent engineering fashion. All members must be treated as equals, all contributions must be seen as critical, each individual must be recognized as a contributor, and each member must be a team player and a contributor. If any of these ingredients are not present, the team will flounder. (Refer back to Fig. 8.4.)

8.7 BIBLIOGRAPHY

1. Hutchison, Catherine, and Dennis Hoffman. "Implementing Concurrent Engineering." *High Performance Systems,* April 1990, pp. 40–43.
2. Winner, Robert, et al. "The Role of Concurrent Engineering in Weapon System Acquisition." *Institute of Defense Analysis Report R-338,* December 1988.
3. CALS/Concurrent Engineering Working Group—Electronic Systems, *First Principles of Concurrent Engineering: A Competitive Strategy for Electronic System Development,* May 12, 1992.
4. Hoffman, Dennis R. "An Overview of Concurrent Engineering." *1994 Annual Reliability and Maintainability Symposium Tutorial Proceedings Notes.*

CHAPTER 9
HUMAN-CENTERED DESIGN

Henry J. Kohoutek
MANAGEMENT AND ENGINEERING STRATEGIES
SSA INTERNATIONAL

9.1 INTRODUCTION

Successful accomplishment of a desired task using a piece of industrial machinery, a tool, or even a consumer electronic product depends not only on reliability of the equipment but also on the reliable performance of its user. Lately, this is becoming more and more problematic. The observable trends document that the ever-increasing functionality and overall capabilities of many products, reflecting the power of new technologies, are accompanied by growth in their operational complexity, causing higher probability of operator errors. Two basic responses to these trends are possible: improvement of users' skills, a challenge difficult to address successfully because of the simultaneous demand for new problem-solving skills in the users' application domain, or simplification of the products' design to make them more suitable for the user.

A similar situation is created by the need to continuously improve the productivity of low-skill workers. Here the favored solution is to introduce productivity-enhancing equipment, whose operation and control must be achievable with the level of skills available. A failure to meet this demand will result in undesirable productivity losses as a result of increased training time and expenses and an unnecessarily high level of operator errors.

In both of these situations, the best way to resolve the problem of assuring high reliability of the complete system, including its human element, is through reduction of its operational complexity. The progress in human factors–related engineering disciplines achieved during the last few decades makes it now possible to design products that are easy to use despite their internal functional or structural complexity. A good example is the automobile. When a car is rented at an airport, the rental agency realistically assumes that the driver will, in a few minutes, figure out how to operate the rented vehicle safely and effectively without any training or help. This is a significant accomplishment for the industry, given the wide range of cars and drivers. In contrast, the computer industry still has a long way to go to achieve comparable results. Its many products are still confusing, often annoying, and difficult to operate at their full performance potential.

The key to a strategy for development of products with the desired high levels of task reliability lies in three simple words: "Know the user." A model of a successful approach has already been presented by industries where the customer requirements define product quality. The reliability engineering and management communities should adopt a similar position, and let the equipment users define the interface characteristics required for high levels of task reliability. This will accomplish the first crucial and necessary step toward user-centered design.

9.1.1 Driving Forces behind Human-Centered Design

Leading companies and industries have already taken the initiative and accepted the challenge of reducing their products' operational complexity. Latecomers and laggards are forced to react to external driving forces that are much less controllable than the company's internal initiatives.

A driving force that is never enthusiastically embraced is *government regulations*. But the safety aspects of human-centered design are already regulated. Design specifications are encoded in standards published by international organizations (e.g., IEC and ISO), institutions of individual governments [such as the Canadian (CSA), Finnish (FEI), and Norwegian (NEMKO)], or product safety test houses (of which the best known are Underwriter's Laboratories, Inc. in the United States and VDE in Germany). Similar strict regulations and product licensing govern the design and qualification of communication equipment, with the goals of not only assuring operators' safety but preventing the degradation of the reliability of public communication networks as well.

In the past, *customer expectations* concentrated on issues of price-performance. Now, with the increased market maturity, the focus is on the product's overall value. The product's ease of installation, integration into existing environments, ease of learning, and ease of use are characteristics that protect the customer's already committed investments by assuring the reliability of the entire installed system. They represent a significant contribution to the product value, and to the company's ability to meet mature customers' expectations. In this way, the product's contribution to operators' reliability and productivity has become an aspect of competitive advantage.

The improved availability, quality, and effectiveness of *human factors expertise* are behind the developers motivation to acquire this expertise and apply it in order to achieve highest levels of customer satisfaction, expand to new markets through an ability to accommodate wider diversity of users, react to visible social trends and pressures, or benefit from the more effective systems approach to reliability engineering and assurance.

9.1.2 Design Management Issues

Human-centered design presents R&D management with a challenge: to integrate human engineering expertise and experts into existing environments. This is a pragmatic task, demanding monitoring of the contribution of this expertise to the overall goals of the R&D function, and to the company's financial results. Experience indicates that the most effective managers are achieving both

1. Short-term benefits, measurable by the reduction of the overall project cost and schedule through elimination of some of the redesign and retest activities or of a complete design iteration cycle
2. Long-term benefits in the form of lower total life-cycle cost for a product as a result of reduced after-sale support

New project cost elements include the cost of human factors resources, cost of interface with actual and potential users, and cost of development of special tools and tests. The practical key to the desired return on this investment, which should match or exceed the established project ROI goal, is concurrence of human factors activities with the general product design and development process phases and steps. In this aspect, cost-benefit evaluation of human factors work is identical to financial analyses of product reliability or of adding a new performance characteristic.

The inclusion of human factors experts in the existing R&D team creates some people management challenges, whose character depends on the underlying local cultural environment, which may or may not stimulate multidisciplinary teamwork. Integration of human factors experts with the established reliability engineering function will smooth out this inclusion.

9.1.3 Dominant Approaches to Human-Centered Design

The sources for human-centered design strategies and their implementation forms are many, some with high-tech and others with low-tech characteristics.

The most obvious and fundamental is, of course, *user involvement.* Users' feedback on their experience with existing products in the environment of their tasks includes information about their probable future expectation, but only indirectly. Market research methods may provide this information directly and in a quantified form. They all follow these five basic steps:

1. Identification of the design information that is important to the users
2. Measurement of the relevant profiles of typical users, and of the user community at large
3. Storage, retrieval, compression, and presentation of the collected data
4. Analysis, description, and interpretation of the accumulated user inputs
5. Evaluation of user inputs according to their usefulness for new product development

In practical terms, techniques such as focus groups, interviews, repertory grid, and choice behavior modeling are available for the stated purpose. It is prudent to use more than one technique to assure higher accuracy and credibility of answers to the formulated research questions.[1]

When the project has already been initiated, users can get involved also. This is usually done through a selected group, and may be formalized by a contractual agreement. Examples are:

- Participation in new product requirements formulation
- Participation in engineering testing and preliminary product evaluations, either at the company laboratory or at the customers' sites (so-called beta-site testing)
- Participation in preliminary product evaluation with the goal of suggesting improvements, as in the environment of repetitive prototyping

A routine customer and user involvement throughout the entire product life cycle has been adopted by *systems engineering* (SE), which is a discipline of developing and engineering complex systems. SE focuses on defining and documenting customers' needs and requirements early in the design cycle, and on subsequent design synthesis and its validation. It represents a serious attempt to address the complete scope of encountered problems through a systematic integration of specialty engineering groups into the project team, a structured process, and consideration of the needs and requirements of all customers, users, and suppliers. The SE process, which runs in parallel with the design core process, contributes unique steps such as problem definition, value system design, functional analysis, and requirements decomposition and deployment to assure that the full scope of identified needs is met in the most effective way. It introduces the application of special methodologies of risk management, methodologies of systems configuration management, and tools of cognitive ergonomics, all of which are suitable for systems reliability management and engineering as well. Because SE originated with the DoD need to manage major development programs, the main application challenge is in its proper scaling for smaller projects.

Human factors engineering is a profession fundamentally engaged in designing products that have a better fit with user needs and, through that, assuring higher user performance reliability and productivity. It focuses on the architecture, technology, and implementation details of the actual human–machine interface, within the constraints of the design project.[2] The ultimate goal is to maximize the performance of the entire system, including its human component. It is a problem-driven, tool-constrained, highly interactive discipline with close ties to implementation technologies, and motivated by a "user is never wrong" attitude. The current state of the art reflects both practical successes and lack of a solid general theory of human behavior.

The simplest strategy in human-centered design is assuring *compliance with existing interface standards.* They are well established for some product aspects, such as product safety and anthropometry. In contrast, standards for the cognitive aspects of the user interface design are only in an early stage of development. They are driven by political groups (trade unions, safety and health advocates), by end product users themselves (demanding transferability of training and stability of interface technologies), and by industries as well, when seeking productivity gains or business advantage. Besides the normal difficulties associated with standards' general acceptance,[3] user interface standards suffer also from their abstract nature and from "soft" descriptions of their topic of interest. The General User Interface, Windows, and VDT standards, all of which originated within the computer industry, are the currently most visible examples.

On the conceptual level, human-centered design strategies are challenging the long-established design principle "Form follows function." Ergonomists and human factors engineers are proclaiming that only a new principle, "Form follows the demands of the human–machine interface," leads to designs preferred by mature markets and users.

The internal success of all the approaches to human-centered design, from the traditional anthropometry-based ergonomics to the contemporary software-based user models, task animations, and virtual reality technology, is a strong function of their acceptance and support by the R&D management.

9.1.4   Summary

The ever-increasing complexity of contemporary products and the need for higher productivity and performance reliability of equipment operators at all levels of skill and all venues of human activity are the primary reasons for the growth in importance of human-centered design. The variety of design strategies developed in the past few decades provides a rich set of methodologies to cover all identified design needs, and documents the shift from the traditional narrow anthropometry to cognitive sciences as their base. The current developments point to the increasing influence of software-based models of human behavior, and of animation of behavior in defined work environment and task scenarios. But despite the highly visible demand for more user-centered designs and the improved availability of human factors expertise, the acceptance of human-centered design is still predominantly a function of R&D management attitude and support.

9.2   CONCEPTUAL FOUNDATIONS

In the practical sense, the success of human-centered design depends on R&D management's attitude and support. In the technical sense, the level of its contribution to the end-product value, to the improvement in users' performance and reliability, to the effectiveness of R&D processes, etc., depends on the available and applied specific human factors knowledge, both conceptual and practical. The conceptual part of this knowledge is essential because, in industrial environments, it provides a framework for selection of the most appropriate objectives, strategies, and implementation methodologies and tools and, in the field of research in human behavior, for experimentation. It also forms the basis for evaluation of both the professionalism and the effectiveness of the actual human-centered design work.

9.2.1   Device Paradigm

One of the fundamental explanations of the behavior patterns of a modern person surrounded by products and technologies is the device paradigm, which also indirectly describes the social impact of technology. It states that

> The presence of things has been replaced by the availability of commodities, which are procured through a device, and without personal involvement.

Examples actualizing the paradigm's meaning:

1. In the past, warmth in the house was procured through personal involvement in chopping and collecting wood, starting and maintaining a fire, cleaning up the fireplace, etc., by people surrounded by "things," such as wooden logs, kindlings, saws, axes, matches, and so on. Modern people procure this warmth as a commodity through a device called a "thermostat" without any personal involvement.

2. In the past, mapping of electric fields required involvement through knowledge of algorithms, actual numerical calculations, and drawing. Today, this map comes automatically out of a printer after a few parameter values and boundary conditions are entered and the "Execute" button for an appropriate computer program is pushed. Knowledge of the underlying algorithms and personal involvement are not needed any more; the whole task may be delegated to an administrative assistant.

The general meaning of this paradigm is that the current technologies minimize attention to means. The whole modern procurement process is so simple that it resembles magic, magic being an act of procuring ends without means. Frequent experiences with situations requiring neither attention to means nor personal involvement result in the establishment of particular expectations about the characteristics of new products and, by that, the stimulation of human-centered design. Finally, this paradigm leads to users' desire for products that are "invisible" during the work on their own task. This then leads to a controversial design strategy conclusion: that the product's internal structural complexity and technological sophistication are irrelevant from the customer's point of view. They are hidden, or present only cerebrally through users' mental model of the device.

Another consequence is a preference for products having the character of a commodity, i.e., products which can be discarded, are care-free, don't require maintenance, and are user-friendly. There are, of course, negative consequences also, such as the emerging dominance of products that are accessible to everyone but understandable by almost no one.

9.2.2 Cognitive Psychology

Studies in naive psychology indicate that "common-sense" approaches to defining users' behavior deserve a healthy dose of suspicion. Contemporary psychology requires serious experimentation in all four areas (or perspectives) of the human–machine interaction:

1. The *sensory-motor area,* which covers both perception and motor skills, involves a traditional view focused on anthropometric information about the user population.

2. The *affective* or *emotive area,* which considers users' feelings about the interaction with the machine, is concerned with topics such as hostility, frustration, intimidation, stress, friendliness, etc., but is plagued by experimental difficulties stemming from the dominance of introspective subjectivity.

3. The *temporal perspective* deals with the role of time in the interaction. Its concerns are often difficult to distinguish from those of the fourth area.

4. The *cognitive view* focuses on the ways in which users acquire and employ knowledge for the proper operation and control of a given tool or equipment. Primary notions of interest are human memory, information processing, and learning styles.

The importance of the cognitive perspective rests in the fact that two primary cognitive processes (learning and memory) are influenced by the selected interface technology, such as key words, function keys, menus, prompts, etc. Cognitive psychology, therefore, provides supplementary theories, concepts, and methods, as well as practical foundations for better human-centered design, especially for the actual human–machine interface. It also forms the foundation of *cogni-*

tive engineering, a discipline dealing with general issues of human problem solving with tools and human behavior in complex environments.[4]

9.2.3 Mental Models

The concept of mental models is gaining popularity in ergonomics and human factors literature, and is beginning to be perceived as one of the keys to good human-centered design. Mental models support a basic hypothesis, that understanding the user's cognition during work with the system assures sound design decisions and leads to better interfaces and products. When considering mental models, four topics must be addressed:

1. The target system, i.e., the system a person is learning or using
2. The conceptual model of that target system, invented to provide its appropriate (accurate, consistent, and complete) representation for use by trainers, designers, and specialty engineers
3. The user's mental model of the target system, which reflects the user's beliefs about it, guides its use, and enables the user to predict its operation
4. The designer's conceptualization of the user's mental model

The positions and relations of these topics in the user's and the designer's operation space are represented in Fig. 9.1.

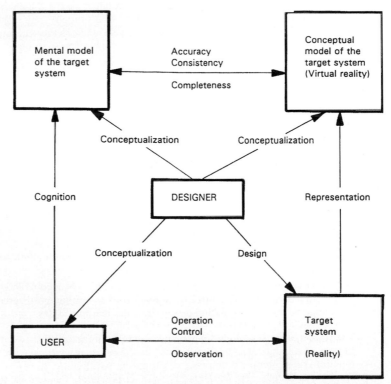

FIGURE 9.1   Positions and relations of models in user's and designer's operation space.

A few general observations have been deduced from the available data:

- Mental models are incomplete.
- People's ability to actualize them as behavior guides is severely limited.
- They are unstable, because people often forget the system's functional and structural details.
- They don't have firm boundaries.
- They are unscientific, frequently including elements of superstition.
- They are parsimonious by offering tradeoffs between physical action and mental complexity.

In general, people's mental models are often deficient in many ways, such as containing contradictions, errors, and unnecessary complicating concepts. Designers should, therefore, develop models which are more coherent and usable by choosing from proven approaches to model development, such as

- Strong analogy (e.g., a VDT is like a typewriter)
- Surrogate models, which provide a structural description of how the system works (e.g., flowcharts or state diagrams)
- Task-action maps, correlating commands with system activity
- Vocabulary models, listing terms in which the user encodes information about the system

As a practical example, it is reasonable to point to the GOMS model,[5] which postulates that cognitive structure consists of four components:

1. A set of hierarchical mental goals (G)
2. A set of cognitive, perceptual, and motor operations (O)
3. A set of methods (M) to achieve the goals
4. A set of selection rules (S) for characterization and choice from among competing methods

These components provide a framework for task analysis, and should give the system designer a rationally and empirically validated approach to the design of a reliable mental model of the interface.

9.2.4 The Human as a System Element

Every human-made system has functions that must be performed by humans. Even the most fully automated systems must be directed and maintained by specialists and experts. Therefore, appropriate consideration of the human element in all stages of the system development is necessary in order to achieve the best balance between the overall system reliability and its cost. Serious formal analyses may be required to optimally decompose the desired functionality and allocate it economically between automated equipment and the system's human component. This designer's interest is complemented by the user's need for the product's contribution to assurance of safety, minimal burden and discomfort during its operation, and elimination of factors causing degradation of the operator's performance and its reliability.[6]

Practical integration of responses to these requirements into the R&D process is done by allocating related unique steps throughout the product life cycle. Examples of those special steps are:

- Allocation of the planned functionality between equipment and the human element
- Assignment of particular functions to individuals and teams
- Analysis of user behavior
- Determination of requirements for skills, education, training, experience, and other personnel factors of the individuals and teams involved

- Development of operating procedures, time charts, etc.
- Human reliability estimation and apportionment
- Consideration of the consequences of undetected human errors
- Recommendation of system changes

Here is a partial list of human characteristics to be considered during allocation of the planned functionality between human and machine:
Favorable:

- Sensitivity to a wide variety of stimuli
- Ability to perceive patterns
- Ability to use judgment
- Ability to generalize
- Ability to adopt flexible procedures and to improvise
- Ability to discover new solutions to the stated problem
- Ability to handle unexpected events
- Ability to learn from experience

Unfavorable:

- Restricted range of stimuli
- Limited range of physical strength and mobility
- Sensitivity to extraneous factors
- Slow response to control signals
- Slow and inaccurate recall of data

Early consideration of the human element is essential for the optimality of the necessary trade-offs, for sound design decisions, for cost saving, and for avoidance of schedule difficulties related to late design changes. The best way to implement the necessary provisions is to assign responsibility for them to system engineers and/or human factors specialists.

9.2.5 Summary

Even though conceptual issues seldom command a high level of respect by practicing engineers and managers, familiarity with them provides a solid foundation for innovation, meaningful experimentation, better justified selection of strategies and methodologies, and ability to critically evaluate the actual human-centered design work.

9.3 HUMAN RELIABILITY

Inclusion of the human element in the system design has many project management and engineering implications, namely its impact on design for reliability, on reliability assurance during use, and on the overall system reliability management. Established hardware reliability practices, concepts, procedures, data banks, models, etc., are naturally being extended into the field of human reliability as well.

9.3.1 Concepts and Definitions

The need to deal with human reliability in a systematic way was identified in the 1950s.[7] Nowadays, a reasonable wealth of both research results and practical experience about this subject is available in the public domain. Human reliability as a specialty in the field of reliability engineering now has accepted definitions, taxonomy, analytical approaches, models, and so on, most of them very close to those of general reliability.[8] For example:

- *Human reliability* is the probability that humans will accomplish a given job or task successfully at any required stage in system operation within a specified time limit.
- *Human error* is the failure to carry out a specific task (or the performance of a forbidden action) that leads to disruption of the scheduled operation or results in damage to property or equipment.
- *Human performance reliability* is the probability that a human will fulfill all specified human functions subject to stated conditions.

The "ideal" situation for human reliability management and assurance would include an experimentally verified theory of behavior (an analogy to reliability physics) expressed in a variety of models, both causal and phenomenological, and data banks with identified quantitative patterns. This is not the current situation, and, considering the fundamental difficulties encountered by psychologists attempting to construct theories of human behavior, this ideal may never be reached. The best conceptual framework currently available are hardware–human analogies and metaphors, such as

Mission success	$\Leftrightarrow$	Task accomplishment
Failure	$\Leftrightarrow$	Error
Failure rate	$\Leftrightarrow$	Error rate
Environmental stress	$\Leftrightarrow$	Operator burden

Human performance reliability $R_{HP}(t)$ is given by a familiar formula:

$$R_{HP}(t) = \exp\left\{-\int_0^t h_{HP}(t)\,dt\right\}$$

where h_{HP} = time-dependent human error rate, which, in this case, may be constant or nonconstant
t = time

Mean time to human error (MTHE) is then expressed as

$$\text{MTHE} = \int_0^\infty \exp\left\{-\int_0^t h_{HP}(t)\,dt\right\}dt$$

The correctability function, concerned with the human capacity to compensate for self-generated errors, defines the probability $P_C(t)$ that an error will be corrected at time t, subject to operator's burden. Its definition has the form

$$P_C(t) = 1 - \exp\left\{-\int_0^t r_C(t)\,dt\right\}$$

where $r_C(t)$ = the time-dependent correction rate

Another useful concept in human reliability is the probability of occurrence of a function failure F_{ki} resulting from the kth error mode of the ith operation subtask. The following assumptions are given:

- A task function consists of a number of different subtasks.
- Each subtask may be carried out more than once.
- One or more error modes may be associated with each subtask.
- Error modes are independent.
- The entire task function may or may not fail as a result of an error.

When these hold, F_{ki} is given by

$$F_{ki} = m_{ki} M_{ki}$$

where m_{ki} = probability that an error of the kth mode occurs in the ith subtask
M_{ki} = conditional probability that if the mode k error of the ith subtask occurs, it will cause a total task failure

This model allows computation of the total probability of no task failure over the space of independent subtasks.

Human reliability in time-continuous operations under variable conditions, such as operating a motor vehicle in different weather, can be modeled using a Markov process.[9]

9.3.2   Human Errors

Selection of a desired error prevention strategy starts with classification of human errors in terms of human–machine incompatibility and with identification of the underlying mechanisms of behavior control from the three fundamental sets: knowledge-based, rule-based, and skill-based controls.

A variety of human error classification schemes have been proposed for different purposes and objectives. Practical approaches usually distinguish among operating errors, assembly errors, design errors, inspection errors, installation errors, and maintenance errors. The basic underlying causes of all of these errors can be classified into five basic groups:

1. Failure to perform a necessary task or subtask, or performing it out of sequence or incorrectly
2. Carrying out a forbidden task
3. Insufficient or poorly timed response to a contingency
4. Incorrect response to a problem situation
5. Unrecognized hazard

Besides the obvious importance of knowledge about the consequences of human errors, the probability P_{HE} that a human error will occur during a specific task is important as well:

$$P_{HE} = \frac{E_n}{E_{OP}}$$

where E_n = total number of known errors of a given type
E_{OP} = total number of opportunities for error

Actual or estimated values of this probability (or of other human failure–related parameters) for selected tasks are known, and may be found in reference materials such as Ref. 10, where associated assumptions are usually stated as well.

Examples:

1. Control operations		
• Reading gauge incorrectly	Error rate	0.005
• Closing valve incorrectly	Error rate	0.0018
• Actuating switch incorrectly	Error rate	0.0011
• Manually overriding automatic system	Response time	3.8 min
• Manually overriding false control signal	Response time	10.5 min
2. Workmanship		
• Component omitted	Error rate	0.00003
• Wrong component used	Error rate	0.0002
• Insufficient manual solder joint	Error rate	0.002
3. Inspection		
• Functional test of components	Error rate	0.00001
• Component functionality surveillance	Error rate	0.0001
4. Learning		
• Erroneous reading of instructions	Error rate	0.0645
• Misunderstanding of requirements	Error rate	0.0912
5. Design of controls		
• Joystick length 6 to 9 in.	Reliability	0.9963
12 to 18 in.	Reliability	0.9967
21 to 27 in.	Reliability	0.9963

Numerical values of reliability or error rate found in such reference tables are the starting point for human reliability predictions. Calculations then consider additional assumptions, such as:

• Independence of individual errors
• Gaussian or extreme value distribution of error rate of individual operators
• Gaussian or log-normal distribution of error rates across a population of operators

Given these assumptions, the calculations follow algorithms for failure rate compounding from hardware reliability.

Operator error rate or reliability estimates are helpful in selection of design alternatives and error prevention methods, in risk assessment, in planning of operator training or certification, and in prediction and management of the entire system reliability.

The following is an example of guidelines for dealing with human–computer interface errors. Objectives:

• Eliminate fear of failure
• Improve operators' productivity

Error prevention guidelines:

• Effective location and arrangement of keys and control buttons
• Consistent identification of commands by light, color, names, symbols, icons, etc.
• Confirmation request for all major actions

Guidelines for error correction:

• Automatic correction of typing errors

- Opportunity to respecify command or restart the activity from the beginning
- Ability to abort an operation that is under way
- Ability to undo previous command(s)
- Implicit acceptance or explicit rejection of commands

A more general approach to dealing with human errors assumes an analogy with adaptive industrial process controllers; it tries to describe an operator-related transfer function and studies its variations as a function of changing conditions using the mathematical apparatus of control theory. Other approaches which employ models of humans as information processors used in psychology, attempt to utilize information theory. The results of such studies are not yet significant enough to be of practical value to system and interface designers and educators.

9.3.3 Human Reliability Analysis Methods

Human reliability analysis is a branch of the analysis of overall human performance in industrial environments, and also a segment of the analysis of the reliability of complex systems. It benefits from the results and progress achieved in hardware reliability engineering. Currently, human reliability analysis is still dominated by research and development activity, but a variety of methods are already in practical use, especially in industries where human errors may cause system failures with serious consequences.

THERP (Technique for Human Error Rate Prediction) is the currently dominant approach. It answers the question, "What is the probability that a human implementer of a well-specified task will err seriously enough to cause task failure?"[11] THERP is an iterative deterministic method and therefore does not represent the complexity and variability of human behavior well. It follows this general outline:

1. Describe the goal of the task.
2. Develop a list of subtasks (steps) and of means for accomplishing them.
3. To capture their sequence and logic, translate the list of subtasks into a tree representation of the events.
4. Assign an error probability value to each subtask.
5. Compound these probabilities according to the event tree logic.
6. Evaluate the effect of individual errors on the task success.
7. Make appropriate recommendations.

THERP methodology suggests modifying tabulated generic error rate and probability values by so-called *performance shaping factors* reflecting the impact of high-stress situations or environmental burden. It also considers redundancy and mutual dependence of errors by including conditional probabilities in calculations. These conditional probabilities may be approximated by a simple formula:

$$\Pr\langle X \,|\, Y \rangle = a + b \, \Pr\langle X \rangle$$

where a and b are positive numbers between 0 and 1, obtained by expert judgment.

The final holistic answers that THERP methodology can deliver are

1. The probability that an error, or group of errors, will result in task (or system) failure
2. The probability that a subtask will result in a defined error situation

Current attempts to improve the THERP methodology aim at

- Elimination of the bias introduced by subjectively set values of the performance shaping factors
- Integration of THERP with risk assessment methodologies
- Addressing explicitly negative influences, such as conflict, hesitation, confusion, etc.
- Adapting THERP to dealing with symptom-oriented emergency procedures

Pontecorvo's method may be considered a variant of THERP, where error rates are modified by a ten-point scale rating of each subtask's difficulty or of the potential for error. This method fits best in situations where there are separate and discrete subtasks; it may be used for a single person acting alone or a person supported by backup.

An *event probability tree,* a tool which is a part of the THERP methodology, may be used independently also. It represents the critical human actions associated with the task or the system analyzed. This simple tool improves visibility of the logic of errors causing failures, and simplifies the traditional probability compounding.

EXAMPLE   From the event tree, calculate the probability of success S_T and of failure F_T for task T consisting of two subtasks a and b with success probabilities $S_a = 0.8$ and $S_b = 0.9$, respectively.

Event tree branches				
Subtask a	Subtask b	Branch probability	Task result	Task total probability
$S_a = 0.8$	$S_b = 0.9$	$0.8 \times 0.9 = 0.72$	Success	$S_T = 0.72$
$S_a = 0.8$	$F_b = 0.1$	$0.8 \times 0.1 = 0.08$	Failure	
$F_a = 0.2$	$S_b = 0.9$	$0.2 \times 0.9 = 0.18$	Failure	
$F_a = 0.2$	$F_b = 0.1$	$0.2 \times 0.1 = 0.02$	Failure	$F_T = 0.28$

The *personnel reliability index* expresses the reliability R_{PF} of personnel performance in a given task by the ratio

$$R_{PF} = \frac{n}{n + m}$$

where n is the number of instances of specially effective behaviors, and m is the number of instances of specially ineffective behaviors, both over equal periods of time.

The total reliability index R_{TOTAL} over i tasks is then the product of individual indices:

$$R_{TOTAL} = \prod_i R_{PF(i)}$$

In practical use, for example by the U.S. Navy, the value of individual indices $R_{PF(i)}$ is based on observation of well-defined jobs, such as equipment operation, equipment inspection, usage of reference materials, and compliance with safety procedures.

The *block diagram method* assesses the impact of critical human errors, which are assumed to be the "common cause" of system failure in redundant or parallel system configurations, and non-critical errors, which cause a failure of a single subsystem. Computations are done using traditional combinatorial reliability formulas.

Other methods of human reliability analysis consider factors such as time, the probability that erroneous task results will go undetected, the reliability of the operator's decision making, etc., using the computational apparatus of hardware reliability analysis.

In every application it is necessary to remember that all of these models and methodologies have a significant speculative component which degrades their utility. Models of human reliability and methods of its analysis are, currently, mostly reliability management tools, in contrast to hardware models that provide valuable contributions to engineering aspects of the system design.[12]

9.3.4   Human Reliability Data

The conceptual credibility of methods of human reliability assessment, analysis, prediction, etc., rests in their realism, logical accuracy, ease of understanding, and, to a lesser degree, computational simplicity. But their practical value is strongly influenced by the availability of credible data. At present, there still are not enough data about human reliability, especially in comparison to the volumes of data about hardware parts and components. This lack is impeding the contribution that reliability engineering can make to human-centered design.

Neither are there established data collection systems of the same magnitude and completeness as warranty reporting systems, repair information reporting systems, customer satisfaction surveys, or company internal product qualification systems. Even though the methodology for the development of data collection systems is well known, the difficulties in establishing a human reliability data system are substantial:

- Lack of sound and widely accepted theoretical foundations makes it difficult to identify which parameters are of significant importance and should be systematically monitored.
- There is a cultural sensitivity of equipment operators and users to the visibility of detailed information about their performance, especially information about individuals' mistakes and errors, with its potential for misuse and correlation to job security.
- Legal implications, in terms of liability or potential negligence, may be drawn from causal evidence of system failures.

In particular product situations, the main source of real use data is laboratory experiments employing anonymous subjects. Valid general data are available mostly from a variety of government agencies and independent test institutions, such as

- United States Nuclear Regulatory Commission
- United Kingdom Atomic Energy Authority
- U.S. Air Force Systems Command
- U.S. NAVY Electronic Laboratory Center
- Bureau of Naval Personnel
- U.S. Army Research Institute
- NASA
- Oak Ridge National Laboratory
- American Society of Safety Engineers
- IEEE
- Electric Power Research Institute

Ideal human reliability data provide information about types of tasks, environmental conditions, systems components and their characteristics, personnel motivation, skills, training experience, psychological stresses, and quality of systems documentation and operating instructions. The major source of data should also be indicated because of these sources' different characteristics.

Source	Fidelity	Flexibility
Real-world observations	High	Low
Field studies	Very good	Limited
Simulations	Good	Good
Laboratory experiments	Moderate	Good
Experts' judgment	Moderate	Limited
Analytical models	Low	High

Structuring the information from data banks leads to empirical formulas (models) supporting human-centered design and reliability decisions. Some of the relationships are already known; for example,

- Required minimum size of a VDT image for reliable shape recognition
- Recognition distance of emergency light, as a function of environmental light density
- Calculation of the total amount of rest required for a specific activity
- Decision time for visual identification of difference between items
- Quality inspector's error rate as a function of vigilance
- Impact of magnifying lens on determination of eye-hand distance

Understanding those data-based relationships allows identification and elimination of error-causing factors and conditions and, by that, improvement of human performance and its reliability, the overall goals of human-centered design.

9.3.5 Other Design-Related Issues

Even though the main technical system design concepts concentrate on human and equipment reliability, the quality of the human–machine interface, and available implementation technologies, other topics related to such factors as operating environment or system failure consequences must be considered also.

Operator burden, imposed on human capabilities by external conditions, can significantly influence human reliability, and may actually force errors. Sources of this burden are related to

- Conditions of the physical environment, such as noise, inadequate space, or improper lighting
- Time constraints imposed on the task
- Multiplicity of concurrent activities
- Remoteness of the operator from the equipment or real action
- Complexity of events observed
- Confusing diagnostic requirements
- Forced operation decisions in conflict with available behavior options and normal task intent
- Hostility in the operator's cultural and social environment

Failure recovery procedures and events are usually left out during the initial reliability analysis and modeling for two main reasons:

1. They are highly dependent on details that are difficult to anticipate during early design phases.
2. They increase the number of design alternatives, complicate selection from among them, and adversely affect project cost and schedule.

This is an agonizing strategy of deferring detailed design analysis and decisions until the main body of the design has already been fixed.

Accidents, despite their unexpected character, follow a definite sequence of events, i.e., a path through the branches of the event tree. They consist of a trigger event, various task successes and failures, and the final failure state. Safety engineering provides analytical tools for identification of the event sequence. Human behavior may trigger accidents and result in losses from them.

9.3.6   Summary

Understanding human reliability at the level of quantitative models and results of formal analyses is a major factor in the success of human-centered design. Conceptually, the strategies for development of this understanding are known, because of the problem's similarity to hardware reliability. The incompleteness of the understanding, characterizing the current state of the art, reflects lack of both solid scientific foundations and real-life data.

The methodology of general reliability engineering, especially its tools for analysis, prediction, and empirical modeling, is the main body of expertise suitable for adaptation to human reliability–related problem solving, inviting additional creative applications.

9.4   ERGONOMICS

Ergonomics is the traditional approach to assuring human performance reliability through application of anthropometric data to the design of products and work environments.

9.4.1   Introduction

Human physical limitations and fatigue, and related performance failures or injuries, have been with the working part of the population for its entire history. Modern technologies have eliminated much of the need for hard toil, but they have introduced new stresses related to mental functions, highly repetitive actions, or intimidating environment.

The efficiency consultants of the 1920s tried to adapt people to machines. The industrial engineers of the 1930s were the first to attempt to make industrial machinery easier to operate, but their approaches were shallow and nontechnical, mostly responding to social pressures by theatrics or fashion. Only after the alarming number of World War II aviator casualties, caused by the impossibility to attend simultaneously to the complicated controls of the airplane and the tactical requirements of air battle, were the ergonomists charged with support, or even leadership, in the necessary redesigns. Designing weapon systems remained a main focus of these professionals until recent years, when their expertise has been applied to a wide variety of products. Currently, they cover the field from shaping bathtubs to designing trading rooms for stock exchanges, with the highest concentration of effort involving computers and consumer electronic products.

Ergonomics is now perceived as a soft multidisciplinary technology, founded on knowledge of the anatomical, physiological, and psychological problems of people at work. Its overall goal is to enhance human performance and its reliability, which, in industrial environments, are usually measured in terms of output and outcome, and are a result of interaction among operator, equipment, and work procedures.[13] Reliability engineers' and managers' focus is, of course, on the reliability aspect of this performance.

Even though application of ergonomics, like that of other technologies, is governed by cost-benefit tradeoffs, the primary considerations involve issues of health and safety. But the cost of effective ergonomic solutions may be prohibitive (e.g., in underground mining). In such situations, human adaptability remains the only fundamental option, and it must be supported by effective motivation and training, avoidance of overload, and clear job instructions and aids.

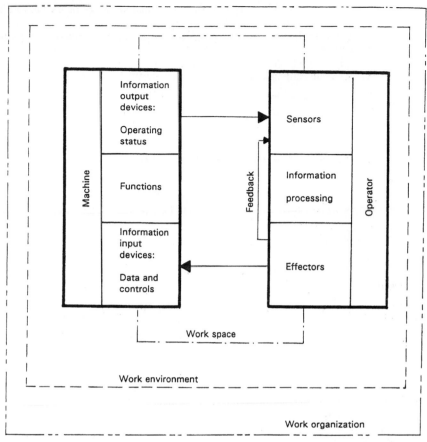

FIGURE 9.2 Human–machine system.

The majority of ergonomic solutions are based on a simple model of the human–machine system (see Fig. 9.2). Besides depicting the three types of basic functions of both the operator and equipment, this model distinguishes among:

- Work space, encompassing equipment, input/output devices, and the operator's sensors and effectors
- Work environment, which may include multiple work spaces, and is characterized by physical parameters, such as light and noise levels, temperature, humidity, vibrations, and radiation
- Work organization, which is defined by operator training, fatigue, diagnostic and decision load, etc., as well as by imposed performance goals, activity rates, supervisory control levels, and so on

They all should be considered components of the design. Because of their influence on the transfer of information between the operator and the machine, they affect the operator's information processing capabilities.

Ergonomics is a technical approach to the solution of the human–machine interface problem; therefore, the sophistication of the system model and its level of detail, accuracy, and precision have a strong influence on the effectiveness and quality of those solutions. Design guidelines, especially for considerations beyond traditional anthropometry, are the most practical tool in assuring this effectiveness and quality.

For example, here are ergonomic guidelines for the design of a user-computer interface.[14] Objectives:

1. Assurance that the operator-computer interface will be easy to use
2. Generality, allowing application to a majority of systems and interface technologies
3. Practicality, allowing measurement and testing

Guidelines:

Prompts and acknowledgments

- Prompt all inputs.
- Acknowledge all completed inputs.
- Responses to all inputs should indicate success or failure.

Making selections

- Provide visibility of choices to select from.
- Avoid making the user select from more than seven choices at a time.
- Never suggest a wrong choice.

Human culture

- Use accepted conventions.
- Use audio and visual symbols.
- Avoid computer designers' lingo.

Undo

- Assure the possibility of correcting anything the user can do.
- Integrate undo with error processing.

Help

- Use examples.
- Don't lose the context of the help.
- Allow the user to browse through help information.

Consistency/adaptability

- Use identical interfaces as often as possible.
- Consider differences between beginners and expert users.

9.4.2 Human Models

Development of conceptual and mental models of the system is a routine activity, sometimes conscious, sometimes unconscious, of both system designers and operators. Development of human models, accurate models of the system users and their behavior, is usually a task for human factors specialists.

The easiest models to develop are anthropometric models, which express the accumulated data about the dimensions of the human body both graphically and in tabular form, usually presenting the mean and the 5th and 95th percentiles. Many anthropometric standards, which often include information about human physical strength, also have been published. The book *Human Scale*[15] is the most widely used reference at the time of this writing. This information is being used for design of instrumentation consoles and panels, location of components for accessibility, etc.

In traditional design strategies the proposed product dimensions, even when in compliance with anthropometric standards, need to be verified by physical prototypes or mock-ups. Current CAD/CAE software opened a new way to eliminate physical prototypes through the use of graphic simulations and animation of the human body in motion. Software human models, combined with software models of the equipment in the work environment, can determine additional mechanical factors, such as the load on individual parts of the human body, that the equipment operator will see, determine precise motions during a task execution, and even analyze a multi-operator system.[16] These models form two major classes (with some overlap):

1. Dimensional models for human factors analysis
2. Mechanical models for kinematic and dynamic analysis

Even though this modeling field is quite young, some software packages are already commercially available on the open market.

Models of user cognition and behavior are needed in complex situations, namely for systems that

- Will be operated by a diverse group of users (e.g., a cellular phone system)
- Will significantly participate in decisions about actions leading to task success (e.g., automated all-weather landing systems for airplanes)
- Will be required to automatically adapt their functionality, or interface characteristics, to different individual users (e.g., multilingual interfaces)

The state of the art of these models is characterized by a focus on explaining only a narrow set of user behavior in preselected simple scenarios.

Human models are useful tools for the elimination of ineffective proposals for the human–machine interface. Currently, they cannot lead systematically to the optimum design. Just as behavior learned on simulators must be tested in real-life situations, model-based solutions must be verified by real users operating real equipment and performing real tasks.

From the viewpoint of human performance reliability, human models may eliminate catastrophic situations and systemic design errors, but their impact on actual human reliability remains uncertain.

9.4.3 Issues of Recent Interest

Every new technology development, and its application through new products, creates its own set of ergonomic issues and concerns. Currently, the dominant focus is on computers and on products of the consumer electronics industries.

The *video display terminal* (VDT) has become a very common feature of the work space. Ergonomic concerns have led to improvements in screen resolution, software production of characters extending above and below the line as in print, and increased refresh rates. But still there are many obstacles to reliable transmission of information from the display to the human sensors:

- Intraocular muscles, those responsible for focusing the eye, are much slower than external muscles. This leads to the requirement of direct vision from the screen to the source document.
- There can be image washout due to bright illumination of the screen, causing reduced contrast.
- Inflexible software can force operators to perform certain operations, such as indexing manually, instead of allowing rearrangement of the data on the screen.

To involve the users in helping to define requirements for VDT design, surveys have been conducted by regulatory and standards agencies, as well as by individual computer companies. Ignoring these requests, even those justified by users' emotions only, leads to lost sales.

For example, here is a minimum set of VDT users' common demands, identified by surveys, in the order of their importance:

- Keyboard supporting the national language
- Antireflection screen
- Brightness control
- Tilt screen
- Focus control
- Swivel screen
- Contrast control
- Other, such as flicker-free image and palm rest

With the operation of VDTs and their associated keyboards has emerged the issue of the consequences of *repetitive stress,* reflected in complaints about wrist and other pains, or even in repetitive stress injury (RSI). Even though injuries have been a problem in manufacturing factories for years, dealing with VDT-related RSIs is a new situation:

- In an office environment they are much more visible.
- There is a changed perception by the physicians dealing with them.
- Lawyers are now more frequently involved by filing suits against employers.
- There is a strong interest by social, governmental, and regulatory institutions.

The resolution of this problem is hindered by uncertainties about the root cause, lack of definite data, and lack of ergonomic design standards. Given this situation, companies are responding by new empirical designs of the complete workstation, specialized training, exercises for relief of stressed muscles, and support of recuperation programs for those with RSIs.

9.4.4 Summary

Ergonomics, a multidisciplinary soft technology founded on anatomical, physiological, and psychological knowledge of the problems of people at work, is contributing to the enhancement of human performance and its reliability in many fields, as well as to the elimination of risks of potential operator injuries caused by some aspects of the human–machine interface. The original anthropometric focus has now been extended to body kinetics and dynamics, especially through software-based human models that allow animation according to expected scenarios. Together with human factors engineering, ergonomics is beginning to address issues related to human cognition and information processing as well. Despite the significant progress that has been made, contemporary ergonomics still remains an empirically based design and human reliability assurance technology.

9.5 *DESIGN FOR USABILITY*

Attributes of a complex product that are related to its ease of use carry the common technical identification of usability. Usability belongs to the set of product quality attributes. The term *usability* is linguistically equivalent to the term *reliability.* An acceptable statement, "Low-reliability products are not very usable" may suggest that reliability is a subattribute of usability. The statement "The reliability of results from a difficult-to-use system is questionable" may lead to an opposite conclusion. This potential conflict is usually resolved in favor of dominance of reliability for practical reasons: the historical priority of reliability engineering (at least in industrial environment) and its more mature set of concepts, models, and methodologies and more extensive data banks. The familiarity of R&D management and engineering with issues of reliability confirms this solution's practical value also.

9.5.1 Usability Definitions

In the most fundamental sense, usability is a universal secondary quality attributable to classes of human-made objects. Recognizing usability as a secondary quality indicates that the difficulties encountered in its measurement are fundamental, not just a reflection of limited knowledge.

The linguistic equivalence of the concepts of usability and reliability indicates a strong potential for benefits stemming from reliability-usability analogies (Table 9.1). These analogies allow reinterpretation of the stress-strength model of reliability into the language of usability, using the

TABLE 9.1 Key Aspects of the Reliability-Usability Analogy

Aspect	Reliability	Usability
Concept	Universal secondary quality	Universal secondary quality
Characteristic	Probabilistic	Probabilistic
Deals with	Failure to accomplish a mission	Failure to accomplish a set of tasks
Goals	Reduce probability of failure Increase mission duration Improve strength/stress ratio Success in many environments	Reduce probability of failure Increase task complexity Improve skill/complexity ratio Success by many users
Object	Machine	Human–machine system
Success parameters	Mission stresses System's strength Failure-free mission duration Physical environment Product structural simplicity Mission criticality Reparability Fault tolerance	Task's complexity System's learnability Number of error-free subtasks Task environment Product operational simplicity Task criticality Ease of recovery Operator error tolerance
Improvement strategies	Design ruggedness Component technology Designer's competence Product simplification Expert support	User task-centered design Interface technology Designer's competence Task simplification Expert support

Note: Aspects in the "Reliability" column correspond horizontally to those in the "Usability" column on a one-to-one basis.

formalisms of hardware reliability.[17] The graphical representation of this analogy (see Fig. 9.3) is self-explanatory. This adaptation of the hardware stress (x)–strength (y) model will result in an *operational definition* of usability as the difference (or ratio) between the user's capability y to operate a given system and the complexity x of the benchmark set of tasks (BSOT). Like the reliability model, the operational definition of usability directly suggests key strategies for improvement:

- Reduction of the operational complexity x of the product will result in a virtual improvement in the user's capabilities y relative to a given BSOT domain. This is illustrated by the shift of $f(y)$ to higher values or by the shift of $g(x)$ toward lower complexity.

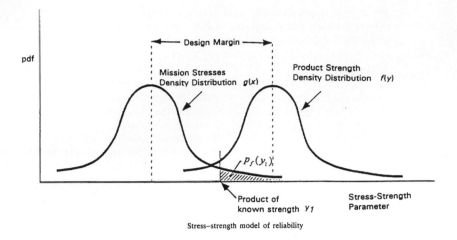

Stress–strength model of reliability

Complexity–user capability model of usability

FIGURE 9.3   Comparison of reliability and usability in models.

- User skill training and skill standardization will cause a favorable shift and narrowing of the $f(y)$ distribution.
- Elimination of the lowest-skill operators from the user population will result in truncation of $f(y)$ at the low end.

Each of these strategies has an analogy in the world of hardware reliability.

From the practitioner's viewpoint, the most useful seems to be the *contextual definition:*

> Usability is a balanced set of multiattribute product characteristics; it can be expressed by an overall generalized rating based on multiparametric measurements of individual attributes, which are rank-ordered according to users' preference, and weighted as per their importance to the success of the users' task.

An example of a contextual definition of usability of a personal computer written from the user's viewpoint is presented in Table 9.2. Definitions of identical form, but with different attributes, subattributes, rank orders, and weighted importance, may be developed to represent other viewpoints, such as those of systems administrators, service and support personnel, etc., or for different products. It is beneficial to define the subattributes in such a way that their decomposition into quantifiable design specifications and development of related metrics will not be difficult.

TABLE 9.2 Example of a Contextual Definition of Computer Usability from the User Viewpoint

Product attributes and subattributes	Metrics and goal	Rank	Weighted importance
Interface quality		1	Must
Hardware reliability	Failure rate $<x\%$/yr		
Safety	U.L. compliance		
Ergonomics	ZH 1/618 compliance		
Acoustical noise emission	50 dB maximum		
Display refresh rate	70-Hz minimum		
Glare	Compliance with internal standard		
VDT standards	Compliance required		
Screen–hardcopy match	Required		
Operator overall satisfaction		2	0.20
Initial product acceptance	7 out of 10		
Satisfaction with functionality	9 out of 10		
Satisfaction with performance	8 out of 10		
Overall satisfaction	8 out of 10		
Ease of use		3	0.25
Time to carry out BSOT	4 hours		
User error rate in carrying out BSOT	0.005		
Type of errors in carrying out BSOT	Recoverable		
Number of subtasks	20 max.		
Availability of action reversal	Required		
Task success rate	99.5%		
Operator burden level		4	0.10
Accuracy of messages	95%		
Short memory load	15 new mnemonics max.		
Availability of shortcuts	Required		
Time spent in waiting	<1 minute		
Ease of learning		5	0.40
Time to learn BSOT	4 weeks		
Retention over time	6 months		
Time to locate help information	5 min max.		
Number of references	2 max.		
Ease of installation		6	0.05
Time to install	<1 h		
Installation success rate	99%		
Level of help needed	No help		
Generalized overall rating.	Excellent		

Notes: BSOT: Benchmark set of tasks, relevant to the application domain.
      U.L.: Underwriter's Laboratories, Inc.
      ZH 1/618: Ergonomic requirements cited in the *German Safety in the Workplace Act.*

In traditional hardware reliability engineering, equipment probabilistic reliability is also formulated in terms of failure rate. An error-rate-based definition of human reliability may be developed in a similar fashion, with usability as one of its parameters. In general, the instantaneous human error rate $\epsilon(x)$ reflects three fundamental components:

1. Intrinsic error rate r, depending on the overall limitations of human motor sensory, physiological, and cognitive faculties.
2. Task complexity x, reflecting the number of subtasks n, and mental workload w. This task complexity can be mediated by usability U, which may be defined, for example, via the operational definition introduced earlier.
3. The external burden B.

In general, we can write

$$\epsilon(x) \equiv f(r;n;w;B;U)$$

Practical observations suggest that an individual's error rate is almost constant in the area of simple tasks, and grows exponentially with increasing task complexity. This characteristic can be expressed by an extreme value distribution:

$$\epsilon(x) = Be^{rx} = Be^{rg(n;w;U)}$$

where the complexity x is a function g of n, w, and U.

This reasoning also leads to the expression of human reliability R_H via the classic formula:

$$R_H(x) = \exp \left\{ -\int_0^x \epsilon(x)\, dx \right\}$$

with usability included in the expression for the task complexity x.

All these definitions of usability provide an opportunity for strengthening the link between reliability engineering and human factors engineering, and an opportunity for standardization of common concepts, language, and formats for presentation of the results of observations and experimentation.

9.5.2 Elements of Human Factors Engineering

Human factors engineering (HFE) is a profession engaged in designing products for a better fit with user needs, especially by improving their usability. HFE takes problems identified by the user, translates them into the technical language of usability concepts, and attempts to solve them by employing its unique knowledge and tools. In a narrow sense, it explores and solves issues related to the human–machine dialogue. Considering its practical aims, it benefits from the developmental approach to human psychology, reflected in this quite reasonable set of assumptions:

- Human behavior follows a set of meaningful rules.
- Knowledge is best acquired through action.
- Assimilation and accommodation of events drive mental development.
- Development occurs as a transition between qualitatively different mental stages.
- Regression is a natural part of mental development.
- Mental development appears to be holistic, but actually it progresses simultaneously along identifiable dimensions.

These assumptions reflect the belief that behavior can be best understood through a simple set of rules or laws, and that the complexity of adult human behavior can be explained by additive application of those simple rules. Such a belief is fundamental for a successful decomposition strategy to solve human–machine interface problems, especially when the equipment and the desired task to be accomplished with its help are both complex.

The tool set of the human factors engineers is similar in form to the tools of other designers, but with different technical contents.[18] It consists of

> *User-centered design specifications,* which are the starting point. They should be based on thoroughly researched user needs and requirements, be verifiable, and have all the desired attributes of clarity, consistency, completeness, and acceptability.
>
> *Industrial standards* complementing the user requirements.
>
> *First design principle* in designing the interface, reflected in the simple command "Know the user." This knowledge is fundamental for requirements and specifications realism, and must cover user characteristics, task profiles, and application environment.
>
> *Design framework,* which must go beyond traditional ergonomics and stress relentless pursuit of operational simplicity, while accommodating human diversity and technological complexity. It must support HFE work at all three levels: system architecture, design principles and guidelines, and particular solutions.
>
> *Models* of humans as system components and of human information processing. These are available with many different levels of detail. But analytical models like those that provide engineers in other fields with tools for prediction of design effectiveness and opportunities for improvement, are not yet readily available to HFEs because of insufficient engineering experience.
>
> *Design principles,* which are emerging from practical experimentation and empirical studies. Currently the most advanced areas are data displays, prototyping, acceptance testing, prevention of operator errors, and the balancing of human control with automation.
>
> *Design guidelines,* which provide practical refinements of design principles. In some areas, e.g., use of colors, such guidelines have a high level of detail. In others, such as use of natural language in interactions, HFE has barely scratched the problem's surface.

As an example, here are some design guidelines for VDT screens:

- Individual items must be legible and discriminable.
- The screen must not appear cluttered.
- Long numbers should be broken into groups of three digits.
- All screens and windows must have titles.
- Particular screen elements, such as titles and prompts, must have a unique and consistent location.
- Columns of data must reflect local cultural conventions, especially those concerning leading zeros, justification, and alignment.
- Unavailable options should be removed from the screen.
- The screen must accurately display objects and closely resemble hard copy.
- Screen feedback must follow the user's action in less than 2 s.

9.5.3 Design for Usability Process

The implementation of the adopted design strategy cannot rely solely on the intuition and experience of the designers, but requires a formally managed process which includes management documents, progress checks, and evidence of the design acceptance. Even though every design process is inherently creative, and therefore not completely predictable, design processes follow clearly identifiable stages, phases, and steps. The HFE design process is nonhierarchical (it is neither strictly top-down nor strictly bottom-up), is radically transformational, and often results in discovery of new goals, new requirements, specification inadequacies, and new design implementation alternatives.

To manage the total system quality and reliability during the design, it is necessary to identify contribution opportunities for the HFE team associated with each major phase of the product life cycle. Bringing the usability professionals directly into the design process as partners enriches the process's cultural environment, and permits a sharing of different areas of expertise for the benefits of the final design quality.

The first step in the design for usability process is always related to finding out relevant information about the users and their tasks. The key questions are:

- Who are the users?
- What tasks do the users perform?
- How often are the tasks performed?
- What are the tasks' time constraints?
- How are the tasks learned?
- Where are the tasks performed?
- How do the users access task-related information?
- What additional tools do users have?
- How do users communicate among themselves?
- What happens when a failure or error occurs?

These questions should be refined or decomposed to additional inquiries to assure the maximum possible information gain from the task analysis process step.

Virtually every design is evaluated before its release to production or customers. Because the design of the human–machine interface is difficult, the importance of frequent evaluations during the design process cannot be overstated. Usability must be repetitively assessed by systematic testing, which must reflect both basic perspectives: behavioral and subjective.

In a research environment, testing mainly takes the form of controlled experimentation. In industrial situations, testing is performed with the goal of obtaining an unbiased assessment of the design from the user's point of view. Test management issues are obvious: cost, impact on schedule, identified levels of potential improvement, and overall ROI. Testing effectiveness is assured only when the test complies with these three fundamentals: real users, real task, and actual product. Other issues, such as iterative scheduling, statistical controls, objective measurements, etc., should be addressed as well. A high level of application realism can be achieved in state-of-the-art human factors laboratories, which provide the usability-related content of product R&D reports.

For dealing with behavior in critical situations, many human factors laboratories employ simulators assuring full mission scenarios. A state-of-the-art facility at NASA Ames, developed jointly with NASA Langley and Lockheed Aeronautic Systems Co., contains three (cockpit, flight, and traffic control) simulators to explore automation concepts related to the long-term conflict between completely automated transport and the culturally desired continuation of the pilot's responsibility for flying the aircraft. Short-term goals relate to reduction and redistribution of the pilots' workload.

The users' subjective evaluations may be the key to the predictions of the overall success of a given interface, or product, in customers' applications. Sample surveys are the primary tool for this activity, and are widely accepted as means for providing statistically valid data. One of the preferred tools is QUIS methodology,[19] which expresses the acceptability of the interface through a ten-point score evaluation on the following six scales:

Terrible	$\Leftrightarrow$	Wonderful
Frustrating	$\Leftrightarrow$	Satisfying
Dull	$\Leftrightarrow$	Stimulating
Difficult	$\Leftrightarrow$	Easy
Inadequate power	$\Leftrightarrow$	Adequate power
Rigid	$\Leftrightarrow$	Flexible

More detailed questionnaires then investigate issues related directly to the display screen, interface terminology, learning, etc.

9.5.4 Summary

Human factors engineering emerged as an important contributor to the design of human–machine interfaces when their usability was identified as an attribute influencing many aspects of the user's performance effectiveness and reliability. The impact of usability on human reliability stimulated a search for credible analogies that would rationalize the transfer of the widely accepted reliability engineering methodologies into the field of human factors engineering. Many have been found, and so the close link between these two professions has been verified and found to be beneficial to both parties involved, as well as to the overall system design team. Similarity of concepts and methodologies points to similarity in tactics of participation in the design team also. The main difference is reflected in the amount of accumulated data and validated tools, which explains the still highly experimental character of human factors engineering.

9.6 RELATED SPECIAL TOPICS

Acceptance of the operator and other humans as system elements, to be considered during the system design, created a need for contributions from a wide variety of disciplines. Even though the ergonomists and human factors engineers are the two most visible professions associated with human-centered design, there are other aspects of the human presence which must be addressed through application of additional bodies of knowledge.

9.6.1 Safety

Issues related to safety have been receiving systematic attention since the early 1950s. Safety is usually perceived as freedom from those conditions that can cause damage or loss of equipment, or injury or death to human beings. System safety is often defined as the optimum level of safety, subject to resource and operational effectiveness constraints, attained by applying engineering and system safety management principles through the life cycle of a system. Typical tasks assuring successful design for safety are system safety planning and a variety of hazard analyses, the results of which impose design requirements and constraints. The overall goal is to prevent mishaps, eliminate hazards, and reduce risks. Because equipment damage and injury are extreme cases of system failure, the reliability engineering concepts and methods are applicable to deal with them. In a similar way, human factors engineering understands that human error may be the trigger event leading to system failure, and therefore has a vested interest in the elimination of hazards as well. Development of the safety program may benefit from the application of FMECA, by identification of failure modes that affect the safety of system personnel or may cause damage to the system itself.

Along with the application of many well-established engineering concepts, safety engineers have developed and perfected some concepts of their own. The most important is, of course, the concept of risk, encompassing the probability of failure, the severity of its consequences, and the overall exposure to it. In the extreme case, risk is perceived as the chance of death before one's "allotted" time. The most controversial methods are those that assign a value to harm done to the system personnel, especially assessing the value of lost life. In this situation the power of perception, social views, legal skills, and situation politics are the dominant factors, with safety engineering playing only a secondary role. Because of the social and political aspects of safety, most governments have established strict prescriptive regulatory requirements, which form the basis for both system design and system use. These issues only magnify the importance of human-centered design, because all aspects of systems personnel well-being are within its scope.

9.6.2   Design for the Handicapped

The overall general objective of human-centered design, accommodation of human diversity, points to the discipline's desire to provide human–machine interfaces that are suitable to people with disabilities as well. This intent involves two basic objectives:

1. Make machinery designed for the general user accessible to the handicapped also.
2. Promote the development of special equipment and tools aimed at betterment of handicapped people's quality of life.

Advances in computing technology seem to be fundamental for dealing successfully with both objectives.

Human-centered design for the handicapped is based on a set of general requirements, covering their needs from both a technical and a social point of view. These requirements must be projected into design specifications early in the design phase.[20] The principal requirements are

- *Practicality,* both technological and psychological
- *Adaptability* to a variety of similar handicaps
- *Portability,* to support handicapped persons' independence and self-sufficiency
- *Affordability,* considering the employment and income limitations of the users
- *Availability* through a wide variety of distribution channels
- *Serviceability,* in conjunction with emphasis on ease of maintenance by the users themselves
- *Operability,* and operation flexibility, to avoid confusion and delays in response

A more complex set of design requirements emerges when contemplating, for example, the design of intelligent prosthetic devices, which require an extremely sophisticated human–machine interface as well as physiological or biological compatibility with the user, or the design of complex functionality robotic systems for quadriplegics. These areas unquestionably still have a research character.

9.6.3   User-Driven Approach to User Manuals

The characteristics of user manuals significantly influence users' rate of learning of the interface-related knowledge and acquisition of operating skills. They are also fundamental to the quality and accuracy of the mental model of the system that users are unconsciously developing. A document usability evaluation concentrates on three basic attributes: accuracy, clarity, and information transfer effectiveness. A well-written user's manual leaves nothing to imagination.

The most frequent criticism encountered is that manuals appear to be written by and for system designers. This criticism, which is not surprising, reflects the dominance of system specifications over user requirements and needs. On the other hand, it has also been recognized that even the best manuals cannot make a difficult-to-use system more usable.

The answer proposed for reorganizing manuals may be summarized in these two strategies:

1. Manuals should be aimed at users according to their function in the system.
2. Every manual should contain two separate sections: a learning section and a reference section.

The practical aspects of writing a usable document require the skills of successful writers:

- Task analysis, or, in other words, finding out how the document will be used
- Language control, which means adapting the language of the document to the language of its users
- Graphic and typographic presentation control to enhance user comprehension
- Management of the document production, if needed

Design for usability experience also points to the fact that focusing on the technical aspects of writing at the level of words, sentences, paragraphs, etc., does not cover the whole set of encountered difficulties. It needs to be recognized that writing is a process and, therefore, should be treated and managed as a process. Successful document centers follow a model which identifies three process stages:

1. Predesign stage, in which the purpose, audience, scope, role, and constraints of the document are determined
2. Design stage, comprising selection of the content, its organization, and its expression in clear language with supporting graphics
3. Postdesign stage, which includes reviews, revisions, editing, and final document evaluation

It is generally understood that this process is recursive, with potentially many rewrites. This iterativeness should be perceived as a positive characteristic of the process, because it assures a better and more usable final version of the document.

9.6.4 Human Reliability in Manufacturing

The most visible activities of human factors and human reliability engineering are in the area of new products and systems development. But employees in manufacturing processes make mistakes too. Because the frequency and repetitiveness of these errors is quite low, taking corrective actions individually at the time of their occurrence is not effective. The practical approach, applied recently to those problems, focuses on foolproofing the operations and equipment.

Prevention again seems to be the most effective strategy. To successfully implement effective and efficient preventive actions, they must be based on plans reflecting credible analysis of potential and, especially, anticipated errors. A variant of the FMECA methodology, the error mode and effect analysis (EMEA), has been proposed and practiced in some manufacturing environments.[21] EMEA requires complementary risk analysis as well, to account for the value and extent of error consequences.

9.7 REFERENCES AND SUGGESTED READING

9.7.1 References

1. Vincent P. Barabba, "The Market Research Encyclopedia," *Harvard Business Review,* January–February 1990.
2. R. Rubenstein and H. Hersh, *The Human Factor,* Digital Press, Maynard, Mass., 1984.

3. B. L. Meek, "Changing People's Attitude: Personal Views," *Computer Standards and Interfaces,* Vol. 9, No. 1, 1989.

4. D. D. Woods and E. M. Roth, "Cognitive Engineering: Human Problem Solving with Tools," *Human Factors,* Vol. 30, No. 4, 1988, pp. 415–430.

5. D. E. Kieras, "Toward Practical GOMS Model Methodology for User Interface Design," in Helander (ed.), *Handbook of Human-Computer Interaction,* Elsevier, New York, 1988.

6. MIL-STD-1477, Department of Defense, Washington, D.C.

7. H. L. Williams, "Reliability Evaluation of the Human Component in Man-Machine Systems," *Electrical Manufacturing,* April 1988, pp. 78–82.

8. Balbir S. Dhillon, *Human Reliability,* Pergamon Press, New York, 1987.

9. Balbir S. Dhillon, "Stochastic Models for Predicting Human Reliability," *Microelectronics and Reliability,* Vol. 22, No. 3, 1982, pp. 491–496.

10. A. D. Swain and H. E. Guttman, *Handbook of Human Reliability Analysis with Emphasis on Nuclear Power,* U.S. Nuclear Regulatory Commission, Washington, D.C., 1983.

11. E. M. Dougherty, Jr., and J. R. Fragola, *Human Reliability Analysis,* John Wiley & Sons, New York, 1988.

12. Lars Bodsberg, "Comparative Study of Quantitative Models for Hardware, Software, and Human Reliability Assessment," *Quality and Reliability Engineering International,* Vol. 9, 1993, pp. 501–518.

13. Tom B. Leamon, "Ergonomics: A Technical Approach to Human Productivity," *National Productivity Review,* Autumn 1987, pp. 331–339.

14. Thom Foote-Lennox, "Ergonomic Guidelines for Computerized User Interfaces," *Proc. Comp. Standards Conf., IEEE,* 1986, pp. 38–41.

15. Niels Diffrient, Alvin Tilley, and David Harman, *Humanscale,* MIT Press, Cambridge, Mass., 1974, 1981.

16. Caren Potter, "The Human Factor," *Computer Graphics World,* March 1991, pp. 61–68.

17. Henry J. Kohoutek, "Reliability Engineering Concepts in Design for Usability," *Quality and Reliability Engineering International,* Vol. 10, 1994, pp. 133–141.

18. Henry J. Kohoutek, "Assuring Quality of the Human-Computer Interface," *Quality and Reliability Engineering International,* Vol. 8, 1992, pp. 427–440.

19. J. P. Chin et al., "Development of an Instrument Measuring User Satisfaction of the Human-Computer Interface," *Proc. CHI,* May 1989, pp. 213–218.

20. Henry J. Blaszcyk, "Computing and the Handicapped: The Challenge in Education," *Computer,* January 1981, pp. 15–17.

21. Takeshi Nakajo and Hitoshi Kume, "Assessment of Prevention of Human Errors in Manufacturing," *Proc. ICQC* (International Conference on Quality Control), JUSE (Japanese Union of Scientists and Engineers), Tokyo, B-5-02, 1987, pp. 291–296.

9.7.2 Suggestions for Further Reading

1. Rouse, William B. *A Human Centered Approach to Designing Successful Products and Systems.* John Wiley & Sons, New York, 1991.

2. Dhillon, Balbir S. *Human Reliability with Human Factors.* Pergamon Press, New York, 1987.

3. Dougherty, M., Jr. and J. R. Fragola. *Human Reliability Analysis.* John Wiley & Sons, New York, 1988.

4. Woodson, E. *Human Factors Handbook,* McGraw-Hill Book Company, New York, 1981.

CHAPTER 10
RELIABILITY INFORMATION COLLECTION AND ANALYSIS

W. Grant Ireson
PROFESSOR EMERITUS, STANFORD UNIVERSITY

An effective reliability program would be impossible without the collection, recording, analysis, and use of information obtained through the testing and operation of industrial, military, and consumer products. Without the information gained through experience with all manner of components, systems, machines, environments, user stresses, and human errors, the production of a reliable product would be like reinventing the wheel each time you wanted to build a wheelbarrow. The real purpose of developing a formal system of recording, analyzing, and retrieving information gained from our experiences is to enable us to design and build a better, more reliable device without having to repeat all the research, experimentation, design, development, and testing that has previously been carried out to achieve the current level of product reliability.

The goal of this chapter is to provide help regarding

1. What information to collect
2. How to record the information
3. How to analyze and summarize the information in order to condense great masses of data into simple tables, charts, or diagrams that are easily understood and used by all personnel concerned with quality and reliability
4. How to file or store the data for easy retrieval
5. How to use the accumulated knowledge contained in the files
6. Who should collect what kinds of information
7. How the data from many different originators should be transmitted to a central analysis and depository unit, and how they should be summarized and indexed for ready reference by all the different users

Unfortunately, there is no standard method or system for accumulating and recording reliability data. The basic principles are generally applied by most companies and organizations, but usually each designs its own system. The rapid development of computers (and especially the great advances that have been made in micro- and minicomputers) has led most companies and organizations to design a computerized system into which the data that were contained in manually manipulated files can be fed.

Each company should examine its sources of information, internal and external, and carefully plan a system that will take full advantage of the software programs that are currently available. The cost of developing a complete set of software programs is great, and if external sources of reliability data are not considered, the company-devised plan may increase the cost of operations for years to come. The list of goals for this chapter is the starting point for the development of a computerized system. Commercially available software can be adapted to the specific needs of the company.

Obviously the needs of a company manufacturing small home appliances will be substantially different from the needs of a company producing microcomputers or automobiles. However, the fundamental principles to be followed in the development of the data system are the same: "Make the shoe to fit the foot." In other words, tailor the system; start with commercial software for file management, input devices, retrieval methods, and data processing routines, and tailor them to the specific needs and conditions of the company. Some questions that should be answered before selecting either the hardware or the software are:

1. Are specific sources of reliability information related to my company and its products available? If so, in what form: printed matter, computer disks, computer tapes, microfiche, etc.?

2. What are the formats of the available sources? What problems will I encounter in translating the data into a format that I can use on my computer?

3. How important are the external data compared with my internally generated data? Will the external data be the major source of information for designers, or will they only supplement or verify internally generated information?

4. What is the most economical means of inputting the internal data into the files? Are there standard reporting forms in use, prescribed by customers, standardized for the industry, or individually designed by the company? Are electronic input devices available that are compatible with the computer?

5. Have each of the potential users of reliability data been polled to determine the specific data that should be recorded?

Figure 10.1 provides a comprehensive list of data sources and their uses.

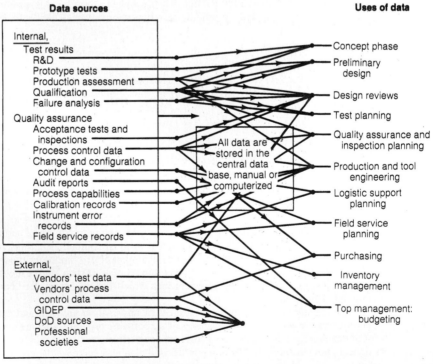

FIGURE 10.1   Data sources and uses.

10.1   *INFORMATION NEEDS TO BE MET*

The design, development, production, and maintenance of an acceptably reliable product should *not* be a new adventure each time a new product is developed or each time a modified product is developed from an older product. Each project should be a learning experience, and there should be complete documentation of all the relevant information so that the learning experience will not be repeated unnecessarily. The information gained in each previous project can be used in each of the phases, conceptualization, design, production, and support, to help assure that the reliability goals will be attained in a minimum time and with minimum expense. Each of the phases requires certain kinds of information, and the system for recording, storing, analyzing, and summarizing the information should be designed with the end uses clearly in mind.

The need to record data gained in each project cannot be overemphasized. Even though reinvention is terribly costly and time-consuming, too many companies depend on the memories of the employees to provide the data needed to speed up the development process for a new product, when it is well known that memory is not perfect, details are forgotten, persons with the required knowledge move to other organizations, and the time required to extract the desired information from available persons discourages the responsible persons at every phase from a complete review of previously learned facts.

10.1.1   Conceptualization Phase

The concept for a new or improved product usually is based on scientific facts which the designers believe can be applied through good engineering to accomplish a product goal. Usually many different basic science facts are involved in any product; most of them have probably been applied to similar products in the past, but one or more may never have been tried. The first question, then, is, "Do we have any experience in the application of this principle or fact to a successful product?" That information should be readily available.

Most companies develop a line of products and continue over the years to improve them, drop some, and add new ones as the science, technology, and market demands change. Each product and each improvement involves preliminary designs, prototype testing, and design reviews. The design reviews should provide permanent records of the experience gained in product development. They certainly will contain information about the concepts underlying the product's design, and the records regarding successes, failures, corrective actions, and design changes. Those histories can be used by the product designer to guide the development of the product concept by preventing the use of basic science ideas which have failed in the past and by identifying those that have resulted in successes. Thus the designer can concentrate on the ideas or concepts that have *not* been proved. This type of information is especially relevant because it represents not only the results of conceptualization and design, but the skills and knowledge of the personnel. It is better, in that regard, than any similar information collected by an agency or another company.

Similar information about the concepts can be obtained by reviewing the patents on similar products in the U.S. Patent Office. This method can be effective, but it is usually very expensive, requiring the services of highly qualified engineers and scientists. Also, it does not provide any information about the problems encountered in putting the product into production.

Some companies buy one or more units of all competing products, perform an autopsy, and assess the fine points of the competitors' designs. This method is helpful in developing ideas for future development, but it, like the patent review, is too late. The competitor's product is already on the market.

10.1.2   Design Phase

In the design phase the product concept is reduced to specific details regarding all the components, subassemblies, circuits, mechanical devices, etc. In this phase the information needs are much more specific and extend well into the information gained through production and maintenance. The basic principle for good, reliable designs is to use only proven components and production techniques.

The reliability information required can be summarized briefly as

1. The reliability of each component under specific uses, environments, and stresses
2. The effects of changes in stress, temperature, humidity, pressure, radiation, vibration, etc., on the reliability (derating effects and surge effects)
3. The ease or difficulty of diagnosing and repairing failed components
4. Special problems encountered in assembling the components into a complete system (danger or damaging components in the assembly operations)
5. Difficulties encountered in the manufacture or fabrication of similar design features (how to avoid degradation of the inherent reliability in the production processes)
6. Field use problems (kinds of failures or malfunctions resulting from use, misuse, and abuses by operators and/or maintenance personnel)

Even with the best of information systems, the design engineers will nearly always pose questions for which there are no data and no prior experience. However, an effective information collection, storage, and retrieval system will tend to minimize the problems requiring special testing or qualification during the design stage.

When the designs are extending the state of the art into new frontiers, the number and difficulty of the problems not covered by prior experience will increase. That is what puts the "research" into R&D. In other words, lack of information regarding some aspects of the design concept requires that research be performed to supply the needed data.

10.1.3 Production Phase

Production engineering and processing are always involved in the design reviews at all levels of development, and they need to have extensive records of prior production projects in order to be able to advise the designers about production problems that can be anticipated from the design features. The purpose is, of course, to prevent these problems at the design stage, before the release of the design to production.

Additionally, production engineering needs to have records that will help it solve production problems during the planning phases for production tooling, inspection, testing, etc. The specific kinds of information production engineering and processing needs to have include

1. Process capabilities for every machine tool and process in the plant
2. Accuracy and repeatability of the measuring devices issued to the setup and operating personnel
3. Specific machine operations and processes that have previously been difficult to control (in the statistical quality control sense)
4. Qualification of operating personnel to perform acceptably and to maintain the process standards
5. Control and identification of special materials, components, parts, etc., especially when a visual identification is impossible
6. Accuracy and repeatability of the incoming inspection and acceptance procedures and practices to assure that only "good" components, parts, materials, and subassemblies are submitted to the production facility

7. A record system that assures the identification of the causes of production-induced degradation of reliability

10.1.4 Support and Maintenance Phase

The support and maintenance phase begins when the product is delivered to the customer and is an extremely important aspect of the maintenance of customer loyalty and repeat sales. It is almost impossible to anticipate all of the stresses that the customer will apply to a product or to anticipate all the different environments to which the product will be subjected, but it is *very* important to try to anticipate these unstated requirements. The founder of one very prominent company in the high-technology field is frequently quoted as saying, "Make it idiotproof!" Data on prior experience will help the designers, production engineers, and maintenance engineers avoid the kinds of difficulties that will result in high frequencies of product failures, high warranty costs, loss of reliability and quality reputation, and loss of customers.

A chapter of this handbook is devoted to the problems of maintenance engineering, but these are some of the items of information that need to be considered in order to make the product more reliable after it is delivered to the customer:

1. Complete and accurate "change control records" to provide a record of engineering changes, models, components, subassemblies, etc., so that at any time the proper replacement part for equipment in the field can be identified and supplied.

2. Identification of the parts, components, and subassemblies that are most likely to fail, wear out, or degrade the performance, and how to make their identification and replacement easy and fast.

3. Typical abuses that users apply, and how to derate the components so that they can withstand the extreme stresses applied by these abuses.

4. Records by component, part, and subassembly of failures and malfunctions, along with the causes of the failures. The frequencies of such failures indicate the relative importance of solving problems of design, materials, production, maintenance, etc.

5. Records of failure analyses that indicate the necessity to change or upgrade processes or tools. These records are extremely important to the production engineers during the design reviews because they help avoid designs that approach the limits of capability of the tools and processes.

6. Records of the effectiveness of the field service personnel in identifying, correcting, and returning to service the malfunctioning equipment. These records will help to identify the personnel who need additional training, the need for better diagnostic tools, and the effectiveness of the maintenance manuals supplied to the personnel.

7. Records of failures or malfunctions resulting from inadequate, incorrect, or poorly prepared operation and maintenance manuals.

This lengthy listing of the needs for information emphasizes the importance of having a data system that will serve the needs of all the different phases of a product life cycle. These needs, obviously, overlap and are intertwined, so that the same information is frequently used by many different persons in the drive to attain high reliability. Therefore, it is imperative that the data collection and storage system be designed so that each responsible person has ready access to the desired information, is not overloaded with unnecessary or trivial data, and can interpret the data to solve his or her immediate problem. Economy dictates that the system be designed to prevent duplication of effort, storage facilities, retrieval facilities, and computer time, and that tasks that can be performed by the computer be programmed so that the personnel can spend time in using the information rather than in trivial tasks such as tabulating, counting, and reading unnecessary data.

10.2   *SOURCES OF DATA*

10.2.1   Internally Produced Information

Although a great amount of valuable reliability data is available from many external sources, most companies place more importance on their own internally generated information. This stems from the fact that the internally developed information reflects all of the "personality traits" of that organization: personnel capabilities in science, research, development, design, engineering, and production. It also reflects the capabilities of the machines, processes, tools, and management to produce products that comply in all respects to the design specifications. For these reasons, the internally generated data can be used to guide the reliability work with confidence that similar results can be obtained on new products.

Test Results.   Because a great many tests are conducted in every quality and reliability program, a test plan should be drawn up for every test to specify its purpose and to assure that there is a record of the environmental conditions, procedures, test equipment, personnel, model and serial number of items tested, and time and date. The test plan must provide a report form to be completed during the test to show all of the results of the various steps in the test procedure. The test plans and the test results should be in "hard copy" to be filed in the conventional way, but the results should be entered into the computer system as part of the overall test database for easy future retrieval.

Test results make up a great portion of the total data system supporting reliability programs. These results should include

Research tests

Prototype tests (see Chaps. 5, 7, and 12)

Environmental tests (see Chap. 12)

Development and reliability growth tests (see Chaps. 24, 25, 26, and 27)

Qualification tests (see Chaps. 5, 12, and 27)

Tests on purchased items (see Chap. 23)

Production assessment and production acceptance tests (see Chaps. 6 and 12)

Tests of failed or malfunctioning items (see Chaps. 13 and 14)

Each company must establish its own policy on how much and what kind of testing to do and how much and what kind of data to record in its data system. The preceding list of types of tests also refers the reader to various chapters of this handbook in which the authors describe the types of tests needed for more complex products and the kinds of data generated and their uses. Those chapters will be very helpful to the decision maker in determining how extensive the company's testing and data program should be.

Another source of information on test plans and requirements is the several Military Standards and Specifications which deal with many aspects of reliability programs. Appendix C provides an annotated bibliography of many of these references.

Quality Control Data.   The quality control group of every company routinely performs inspections and tests on products and equipment in order to maintain the quality standards necessary for reliable products. The results of those inspections and tests are very valuable in design reviews; selection of vendors for components, subassemblies, and supplies; and advising designers on the problems that may arise in production. Again, it is highly desirable to enter all of those data into a database so that they can be retrieved quickly and easily. The specific types of information needed are

1. Incoming inspection and test results, identified by component, part number, vendor, date code, and specification

2. In-process quality control data by part number, product, process, and machine
3. Calibration records of all measuring instruments used in research, testing, and production, including indicators of instrument error, repeatability, accuracy, and frequency of recalibration
4. Results of machine and process capability studies
5. Quality audit records and final test results

Field Support Data.   Practically every manufacturer must maintain some kind of field service facilities to provide repair and maintenance service to its customers. This is one of the most valuable sources of reliability information because it represents the customer's view of the reliability of the product. Every field service or maintenance activity should result in a report to the marketing department and to the reliability organization. The report should provide the following kinds of information:

1. Product, model, and serial number.
2. Part that failed or malfunctioned, part number, serial number (if serialized), vendor, and date code.
3. Nature of failure. What happened?
4. Cause of failure if determinable on site.
5. Action taken by service personnel.
6. Date and name of service personnel.

It is common to have a standard report form with many of the desired data preprinted in such a way that they can be checked off, so that there is no need to write out a lot of words. Even on such a form, however, there must be space for special remarks by the service person if the preprinted information is not adequate.

For example, many forms have a list of causes of failure ending with "other." Unfortunately, the other block usually gets checked more frequently than the specific blocks because the service person does not take the time to determine the cause and the other block is very convenient. Instead of other, there should be a space identified, "If none of the above, explain the cause." When such a preprinted form is used, a small notebook should be provided, with quick references to the various items printed on the form, to help assure that the service person makes a correct diagnosis and knows which items to check under each category.

It is also common practice, especially with high-reliability, expensive items, to have the failed part returned to the failure analysis group at the home plant in order to answer the following questions: (1) Can the part be repaired and put back into service? (2) Was the service person's diagnosis correct, and/or is there an underlying cause that should be corrected by a design change, change in production technology, change in supplier, or other corrective action? The returned parts can be sent to the failure analysis laboratory for a detailed autopsy, which may lead to important improvements in the design or production (see Chap. 13).

10.2.2 External Information Sources

There is a great amount of important reliability information available to all companies from external sources, and obtaining and using the appropriate information will save a lot of time, money, and errors. A brief discussion of the major external sources follows.

Vendors' Inspection and Test Data.   Every company ought to include quality and reliability requirements in every purchase contract for components, supplies, and subassemblies. These contracts should include requirements that the vendor supply, along with the product, the following information:

1. The vendor's written quality control and reliability program plans, which should show how the vendor assures a satisfactory product conforming to the buyer's specifications

2. Quality control records (Shewhart control charts and other process control reports) covering the specific run of the product

3. Inspection reports of the final product, by lot or batch, showing the lot number or identification, the number inspected, and the number of defects or nonconforming units found (and removed)

4. Reliability test documents showing the test procedure, environmental conditions, stresses applied, type of test equipment, length of test, number tested, number failed, and causes of failure

5. A computation of the estimated reliability for the buyer's specific use, and whether or not the part or item has been "qualified" under one of the Military Standard plans

In high-value items, the contract may specify that the buyer's quality and reliability engineers can witness the reliability tests and visit the plant to observe the workings of the quality and reliability program plans. This is especially important when the tests are long and expensive, and require very special test facilities. The Department of Defense usually has "resident representatives" stationed in the plants of major contractors to help assure that the products are properly manufactured, inspected, and tested, and to prevent duplication of the inspections and tests by the military agency.

Obviously the demands on the vendors should be tailored to the needs of the customer and the complexity of the product. It is true in most cases that appropriate contractual requirements can prevent duplication of inspections, tests, and failure analyses by the customer. This not only saves money but also helps to assure that there is a clear and complete understanding between the buyer and the seller as to just what the buyer wants.

The uses of the information obtained from vendors are

1. Evaluation of the vendors' abilities, quality, and reliability in order to select vendors for future orders

2. In the conceptualization and design phases, the estimation of the inherent reliability of the proposed product design

3. Decisions on which components to use in the design. Proven components reduce the time and cost of designing new products

4. For the reliability group, reduction in the investment in test equipment and the cost of running the tests

5. For the quality control group, planning of the incoming inspection and where and when to inspect in the production line

Government-Industry Data Exchange Program.   The most comprehensive external data source for all aspects of the design, production, and field support of highly reliable products is the Government-Industry Data Exchange Program (GIDEP). The program is authorized and funded by the U.S. government, but it is open to participation by all companies engaged in producing commercial, off-the-shelf items if the company uses and generates the type of data compiled and distributed through the exchange program. Some government agencies may contractually require participation by companies that produce items for the agencies. The program maintains four data "interchanges":

1. *Engineering data interchange:*   Engineering and qualification test reports, nonstandard parts justification data, parts and materials specifications, manufacturing processes, and engineering methodologies and techniques.

2. *Reliability-maintainability data interchange:*   Failure rate, failure modes, and replacement

rates on parts, components, and materials based on demonstration tests and field performance information.

3. *Metrology data interchange:*   Technical data on test systems, calibration systems, test equipment calibration procedures, measurement technologies; designated as a data repository for the National Bureau of Standards.

4. *Failure experience data interchange:*   When significant problems are identified regarding parts, components, processes, fluids, materials, safety, and fire hazards, the objective failure information is reported and added to the database.

In order to make use of GIDEP, a company must apply formally to become either a full participant or a partial participant for any one or all four of the interchanges. This means that the company must really participate by providing copies of all relevant reports, test results, etc., to the program in order to receive the benefits. A specific GIDEP representative must be appointed as the contact, and periodic progress reports on GIDEP activity and benefits must be submitted. The reports must not contain "classified" (security) information or proprietary information.

A fully participating company receives the microfilmed data banks, indexes, and all associated documentation. A partial participant receives all program materials and the indexes but not the microfilmed data banks. That company may request a loan of the microfilms of desired reports. Hard copy can be made from the microfilm, or, if the company has remote terminal equipment compatible with the GIDEP Operation Center's computer, it can make direct inquiry and print out the information if desired.

With hundreds of companies involved in all types of products participating, access to the GIDEP interchanges can save any organization great amounts of money and time, as well as prevent serious errors, in the development of high-reliability products. The information is applicable in all phases of the life cycle: conception, design, demonstration, production, and field support. The experience of a great many companies and government agencies has proved over and over that the GIDEP program saves millions of dollars every year.

To obtain complete information about GIDEP and how to become a participant, write:

Director, GIDEP Operations Center
Corona, CA 91720

Other Government Data Sources.   The Department of Defense has been the primary supporter of research in reliability since the early 1950s and has supported various activities to collect and disseminate data to contractors in order to improve the reliability of military hardware and to reduce costs. The GIDEP program is one of those activities. Two other agencies under the Department of Defense are

Technical Information Center   Reliability Analysis Center
Cameron Station   Rome Air Development Center (RBRAC)
Alexandria, VA 22304   Griffis AFB, NY 13441

The Department of Commerce maintains the

National Technical Information Service
5285 Port Royal Road
Springfield, VA 22161

Professional Organizations.   The Institute of Electrical and Electronics Engineers, 345 East 47th Street, New York, NY 10017, has developed a great amount of data regarding the reliability of electrical hardware of all types.

When using data obtained from external sources, there is always the question of its applicability to the total company environment. A simple example may help clarify this statement.

Environmental stress-test data, including derating factors or charts, can be obtained for many components. If you are using those data in designing a product, you must question several points:

1. How much of the variation indicated is the result of measuring instrument error compared with the instrument error in your plant?

2. How precise were the measuring instruments used in the tests compared with the precision of the instrumentation in your plant? For data to be truly trustworthy, the instruments must be capable of repeatedly measuring 0.5 standard deviation of the characteristic being measured or less.

3. Did the range of variables in the reported test data cover the range of the same variables that your product will encounter?

4. How many tests, cycles, or operations were carried out to produce the results reported?

5. Do the reported data compare favorably with your internal data?

10.3 DATA STORAGE AND RETRIEVAL METHODS

10.3.1 Manual

In spite of the tremendous advances that have been made in computers, manual methods are still important in the collection, retrieval, analysis, and application of reliability data. Most of the results of tests are first recorded manually on paper forms, then the appropriate portions of the data are transferred to the computerized database. Reliability testing programs usually require that specific conditions be observed and recorded in order to validate the test. The procedures must be followed precisely. The results or measurements of each step are then recorded. The plan and procedures are usually printed forms with provisions for the observations to be recorded after each step. All of this is necessary to assure that there is a complete, accurate, and permanent record of the test that can be used for many purposes.

Failures, malfunctions, or unscheduled maintenance actions require a written report. This report provides important information to the reliability and quality assurance engineers, designers, logistics support, and inventory management as the basis for corrective actions, redesign, production process improvement, and future product design. Most of these failures or unscheduled maintenance actions occur at remote locations and are handled by field service personnel who do not have immediate access to a computer terminal, but often carry portable computers for this purpose.

A copy of the written report normally accompanies the part, subassembly, or component as it is returned to the home facility for diagnosis, repair, or failure analysis. This helps to reduce the diagnostic time and assists the repair operation. Another copy of the report is sent to the data center, where the failure or malfunction is recorded in a specialized format in the computer data bank.

The design of the report forms is not standard, although some Department of Defense agencies have special forms to facilitate the transfer of the data to their computer database. Most companies design their own forms to provide just the desired data and avoid excess writing. Some examples of report forms are given here.

Figure 10.2 shows a form that was used by the Bureau of Naval Weapons for aeronautical-material deficiencies. Figure 10.3 shows a form used for electronic equipment failure or replacement by the Navy's Bureau of Ships. Figure 10.4 shows the codes used in Fig. 10.3 to indicate the nature of the failure. The list of codes and types of failures helps the service personnel to describe the failure more accurately, and the codes can be used in computerizing the report. Note that other information relative to maintainability, repair time, downtime, etc., is also recorded. A similar Air Force form is shown in Fig. 10.5.

Manually prepared reports are normally filed for future use or reference. Frequently only the most significant information is stored in the computer database, and at some future time it may be

Report Symbol BUWEPS 13070-3

| 1. Reporting Activity | | 2. Report Ser. No. | 3. Date Of Trouble | 4. Installed In Aircraft/Arrest. Gear/Catapult/Support Equipment Model Bung or Ser. No. | | 5. Aircraft Logbook Time |

| System, Set Equipment Or Engine | 6. Model Designation And Model No. | 7. Nomenclature | 8. Serial No. | 9. Time Meter Read./Logbook Time or Events (if applicable) Hour meter / Logbook hrs. / Starts / Landings |

| Unit, Component Accessory, Assembly Or Equipage | 10. Manufacturer's Part No. | 11. Nomenclature | 12. Serial No. | 13. Mfr's Code No. | 14. Contract No. | 15. Time Or Events Hrs. / Starts / Ldg's |

| Subassembly (Electronic) Or Primary Part Failure (Non-electronic) | 16. Manufacturer's Part No. | 17. Nomenclature | 18. Serial No. | 19. Mfr's Code No. | 20. Location (if applicable) |

| Supply Identification Item(s) Returned | 21. Federal Stock Number | 22. (RM, MR copies only) | 23. Quantity | 24. (RM, MR copies only) | 25. (RM, MR copies only) |

| Reason For Report (Check one) | 26. Removal Or Maintenance Action Required As A Result Of: | | | | | 27. Item overhauled by |

1 ☐ Failure/ Suspected Failure Or malfunction
2 ☐ Damaged due To Improper Maintenance/ Operation/Test
3 ☐ Damaged or Defective On receipt
4 ☐ Damaged Accidentally
5 ☐ Scheduled/Directed Removal, high time Overage, excess To requirements

DESCRIPTION OF TROUBLE

(If box 1, 2, 3, or 4 was checked in space 26, complete spaces 28 through 31. If box 5 was checked in space 26, leave spaces 28 through 31 blank.)

28. First Observed/Occurred During

| 1 ☐ Flight operations—Land based | 3 ☐ Pre-flight | 5 ☐ Conditional | 7 ☐ Overhaul/PAR | 9 ☐ Special directed inspection |
| 2 ☐ Flight operations carrier based | 4 ☐ Daily | 6 ☐ Calendar | 8 ☐ Shop maintenance bench test | 10 ☐ Normal operation of support equip., catapults, arresting gear, mirror landing sys. only. |

29. Symptoms— How Discovered Item

	D ☐ Incorrect display	I ☐ Low performance	N ☐ Overheating	
	E ☐ Inoperative	J ☐ Metal in oil	O ☐ Pressure out-of-limits	S ☐ Torque out-of-limits
A ☐ Excessive vibration	F ☐ Interference/Binding	K ☐ Noisy	P ☐ RPM out-of-limits	T ☐ Unstable operation
B ☐ High fuel consumption	G ☐ Intermittent operation	L ☐ None noticed	Q ☐ Surging/Fluctuates	U ☐ Visible defect
C ☐ High oil consumption	H ☐ Leakage	M ☐ Out-of-balance	R ☐ Temperature out-of-limits	V ☐ Other (Amplify)

30. Part Condition

								585 ☐ Sheared
007 ☐ Arced	130 ☐ Changed value	201 ☐ Distorted/Stretched	750 ☐ Missing	196 ☐ Shorted/Grounded
780 ☐ Bent	910 ☐ Chipped/Nicked	148 ☐ Eroded	008 ☐ Noisy	422 ☐ Soldering defect
135 ☐ Binding	999 ☐ Circuit defective	250 ☐ Frayed/Torn	450 ☐ Open	660 ☐ Stripped
429 ☐ Blistered/Peeled	160 ☐ Connections defective	001 ☐ Gassy	790 ☐ Out-of-adjustment	018 ☐ Tested OK—Did not work
070 ☐ Broken/Cracked	818 ☐ Contacts Burned/Pitted	381 ☐ Leaking	439 ☐ Plugged/Clogged	389 ☐ Unknown (Cannot disassemble)
900 ☐ Burned/Burned out	170 ☐ Corroded	730 ☐ Loose	576 ☐ Ruptured/Split/Blown	020 ☐ Worn—Excessively
120 ☐ Chafed/Galled	200 ☐ Dented	004 ☐ Low GM or emission	935 ☐ Scored	099 ☐ Other (Amplify)

31. Cause Of Trouble

A ☐ Design deficiency	D ☐ Faulty overhaul (Quality control)	G ☐ Fluid contamination	J ☐ Operator technique/ Adjustment	M ☐ Weather conditions
B ☐ Faulty maintenance (Quality Control)	E ☐ Faulty preservation/ Packaging	H ☐ Installation environment (Location in weapons sys.)	N ☐ Other parts primary cause	N ☐ Wrong part installation
C ☐ Faulty manufacturing (Quality Control)	F ☐ Foreign object	I ☐ No failure-replaced to improve sys. performance	L ☐ Undetermined (Cannot disassemble)	O ☐ Other (Amplify)

32. DISPOSITION OR CORRECTIVE ACTION: Select appropriate code(s) from list below and enter in boxes at left to indicate disposition or corrective action taken with respect to each of the items entered in spaces 6, 10, and 16.

Replaced And Returned To Supply
Code / Reason
A Hold 90 days
B Lack of repair facilities
C Lack of repair parts
D Lack of Tech. Pubs.
E Lack of personnel
F Beyond assigned maintenance level
G Other—(Defective on receipt, high time, directed removal, excess to requirements, etc.)

Code / Corrective Action
H Used as is
I Adj./Realign./Serv./Repaired in place
J Removed-Adj./Realign./Serv./Repaired-reinstated
K Removed-repaired-made RFI
L Removed-tested Ok-made RFI
M Removed-scrapped
N Surveyed
O Released for investigation and replaced (indicate custody in space 35)

Space 6 ☐
Space 10 ☐
Space 16 ☐

33. Maintainability Information | Hours | Tenths

Man-hours to locate trouble Space 10
Man-Hours to locate trouble Space 16
Man-hours to repair/replace/adjust
Actual time A/C was undergoing repair
Total time aircraft not flyable due to this malfunction

ACCESSIBILITY
S ☐ Satisfactory
U ☐ Unsatisfactory (Amplify)

SPECIAL
1 ☐ Frequent trouble item
2 ☐ Can be installed wrong (Amplify)

34. Component/Assembly, Subassembly Replaced With:
Mfr's Part No.
Serial No.
Mfr's Code No.

35. AMPLIFYING REMARKS (Furnish additional information concerning failure or corrective action not covered above. Do not merely repeat information checked above. Specify any severe operating conditions, such as hard landings, wheels-up landings, severe maneuvers, etc.)

| 36. Report Is: | | Signature | Rank/Rate | Date |
| 0 ☐ FUR | 1 ☐ AMPFUR | 2 ☐ Urgent AMPFUR | 3 ☐ Flight Safety AMPFUR | 4 ☐ Follow up report | | |

Associated Parts Repaired Or Replaced (Do not list any item reported above)	37. Part No. (Non-electronic parts) Or Part Ref. Designator (Electronic parts)	38. Part Name, Tube Type, Semi-Conductor Type Or Description	39. Mfr's Code No.	40. Failure Code (From space 30)	41. Disposition (Code from space 32)	42. Activity Repaired By
						Signature
						Rank/Rate / Date

FAILURE, UNSATISFACTORY OR REMOVAL REPORT
NAVWEPS FORM 13070/3 (10-62)   (Mail this copy to NATSF)   **F U R**

FIGURE 10.2 Example of failure, malfunction, and parts replacement report.

ELECTRONIC EQUIPMENT FAILURE/REPLACEMENT REPORT DD—787 (PROPOSED)   REPORT BUSHIPS 10550—1

1. DESIGNATION OF SHIP OR STATION

2. REPAIRED OR REPORTED BY

NAME	RATE

AFFILIATION
1. ☐ U.S. NAVY 2. ☐ CONTRACTOR 3. ☐ CIVIL SERVICE

3. TYPE OF REPORT (CHECK ONE)
1. ☐ OPERATIONAL FAILURE
2. ☐ PREVENTIVE MAINTENANCE (POMSEE)
3. ☐ PREVENTIVE MAINTENANCE (NOT POMSEE)
4. ☐ STOCK DEFECTIVE
5. ☐ REPAIR OF REPLACEABLE UNIT OR PLUG-IN ASSEMBLY
6. ☐ OTHER

4. TIME FAIL. OCCURRED OR MAINT. BEGAN
MONTH	DAY	YEAR	TIME

5. TIME FAIL. CLEARED OR MAINT. COMPL.
MONTH	DAY	YEAR	TIME

EQUIPMENT

6. MODEL TYPE DESIGNATION

7. EQUIP. SERIAL NO.

8. CONTRACTOR (NAVY CODE OR COMPLETE NAME)

9. FIRST INDICATION OF TROUBLE (CHECK ONE)
1. ☐ INOPERATIVE
2. ☐ OUT OF TOLERANCE, LOW
3. ☐ OUT OF TOLERANCE, HIGH
4. ☐ INTERMITTENT OPERATION
5. ☐ UNSTABLE OPERATION
6. ☐ NOISE OR VIBRATION
7. ☐ OVERHEATING
8. ☐ VISUAL DEFECT
9. ☐ OTHER, EXPLAIN

10. OPERATIONAL CONDITION (CHECK ONE)
1. ☐ OUT OF SERVICE
2. ☐ OPERATING AT REDUCED CAPABILITY
3. ☐ UNAFFECTED

11. TIME METER READING
A. HIGH VOLTAGE
B. FILAMENT /ELAPSED

12. REPAIR TIME
MAN-HOURS	TENTHS

REPLACEMENT DATA

13. LOWEST DESIGNATED UNIT (U) OR SUB-ASSEMBLY (SA)	14. LOWEST DES. U/SA SERIAL NO.	15. REFERENCE DESIGNATION (V-101, C-14, R11, ETC.)	16. FEDERAL STOCK NUMBER	17. MFR. OF REMOVED ITEM	18. TYPE OF FAILURE	19. PRIMARY OR SECOND-ARY FAIL?	20. CAUSE OF FAILURE	21. DISPOSITION OF REMOVED ITEM	22. REPL. AVAILABLE LOCALLY?
						P ☐ S ☐			Y ☐ N ☐
						P ☐ S ☐			Y ☐ N ☐
						P ☐ S ☐			Y ☐ N ☐
						P ☐ S ☐			Y ☐ N ☐
						P ☐ S ☐			Y ☐ N ☐

23. REPAIR TIME FACTORS

CODE	DAYS	HOURS	TENTHS	CODE	DAYS	HOURS	TENTHS

24. REMARKS

(CONTINUE ON REVERSE SIDE IF NECESSARY)

SRL-1

FIGURE 10.3 Failure and replacement report for data file input.

BLOCK 18 - TYPE OF FAILURE

QUICK REFERENCE LISTING OF MOST OFTEN USED FAILURE CODES
(IF PROPER CODE CANNOT BE FOUND, REFER TO ALPHABETICAL LISTING BELOW)

ELECTRON TUBES

CODE	TYPE OF FAILURE
002	AIR LEAK
007	ARCING, ARCED
960	BROKEN ENVELOPE
001	GASSY
380	LEAKAGE
004	LOW GM OR EMISSION
131	MARGINAL PART REPLACEMENT
009	MICROPHONIC
053	MISFIRES (THYRATRONS)
008	NOISY
003	OPEN FILAMENT
560	POOR REGULATION
011	SCREEN DEFECTS (CATHODE RAY)
005	SHORTED, INTERMITTENT
006	SHORTED, PERMANENT
018	TESTED OK, DID NOT WORK

TRANSISTORS AND SEMICONDUCTOR DIODES

CODE	TYPE OF FAILURE
741	ALPHA CUT-OFF LOW
744	BACK RESISTANCE LOW
739	BETA LOW
743	FALL TIME, EXCESSIVE
745	FORWARD RESISTANCE HIGH
742	Ico HIGH
737	OPEN, BASE-TO-COLLECTOR
735	OPEN, BASE-TO-EMITTER
156	RISE TIME, EXCESSIVE
734	SATURATION RESISTANCE HIGH
738	SHORTED, BASE-TO-COLLECTOR
736	SHORTED, BASE-TO-EMITTER
731	SHORTED, COLLECTOR-TO-EMITTER
749	STORAGE TIME, EXCESSIVE

PLUG-IN ASSEMBLIES

CODE	TYPE OF FAILURE
033	DRIFTS
088	GAIN, LOW
094	GAIN, NONE
360	INTERMITTENT OPERATION
387	LOW PERFORMANCE
099	MODULATION, LOW
096	MODULATION, NONE
022	NO OSCILLATION
442	OUTPUT, LOW
235	OUTPUT, NONE
258	OVERHEATS
560	POOR REGULATION
097	RESPONSE, POOR
091	SENSITIVITY, LOW
480	UNSTABLE

ELECTRICAL, ELECTRONIC

CODE	TYPE OF FAILURE
007	ARCING, ARCED
080	BURNED OUT
170	CHANGE OF VALUE
220	HIGH VOLTAGE BREAKDOWN
330	INSULATION BREAKDOWN
380	LEAKAGE
008	NOISY
430	OPEN
082	OPEN, INTERMITTENT
440	OPEN PRIMARY
431	OPEN ROTOR
470	OPEN SECONDARY
432	OPEN STATOR
433	OPEN WINDING
520	PITTED
005	SHORTED INTERMITTENT
006	SHORTED PERMANENT
612	SHORTED PRIMARY
630	SHORTED SECONDARY
613	SHORTED STATOR
540	PUNCTURED
097	RESPONSE, NONE
734	RISE TIME, EXCESSIVE
570	RUSTY
740	SATURATION RESISTANCE HIGH
935	SCORED
011	SCREEN DEFECTS (CATHODE RAY)
091	SENSITIVITY, LOW
738	SHORTED, BASE-TO-COLLECTOR
736	SHORTED, BASE-TO-EMITTER
731	SHORTED, COLLECTOR-TO-EMITTER
005	SHORTED, INTERMITTENT
006	SHORTED, PERMANENT
620	SHORTED PRIMARY
630	SHORTED SECONDARY
613	SHORTED STATOR
600	SHORTED TO CASE
610	SHORTED TO FRAME

OTHER COMMON TYPE OF FAILURE CODES — ELECTRO-MECHANICAL, MECHANICAL, CHEMICAL

CODE	TYPE OF FAILURE
710	BEARING FAILURE
780	BENT
040	BINDING, MECHANICAL
070	BROKEN
720	BRUSHES, IMPROPER TENSION
120	BRUSH FAILURE
160	CHATTERING
140	CONTACTS, CONNECTION DEFECTIVE
170	CORRODED
210	DETENT ACTION POOR
226	EXCESSIVE PLAY
567	HIGH CONTACT RESISTANCE
730	LOOSE
790	OUT OF ADJUSTMENT
570	RUSTY
770	SLIP RING OR COMMUTATOR FAILURE
144	SPEED INCORRECT
630	STICKY
943	STRUCTURAL FAILURE
020	WORN EXCESSIVELY
615	SHORTED TO GROUND
440	SLIPPAGE
770	SLIP RING OR COMMUTATOR FAILURE
026	SOLDER JOINT DEFECTIVE
144	SPEED INCORRECT
630	STICKY
749	STORAGE TIME, EXCESSIVE
660	STRIPPED
943	STRUCTURAL FAILURE
018	TESTED OK, DID NOT WORK
947	TORN
965	TUNING DRIVE DEFECTIVE
670	UNBALANCED
480	UNSTABLE
690	VIBRATION EXCESSIVE
700	WEAK ELECTRICALLY
964	WINDOW SUCK-IN (MAGNETRON)
020	WORN EXCESSIVELY

ALPHABETICAL LISTING

CODE	TYPE OF FAILURE
002	AIR LEAK
741	ALPHA CUT-OFF LOW
007	ARCING, ARCED
744	BACK RESISTANCE LOW
710	BEARING FAILURE
780	BENT
739	BETA LOW
226	BRITTLE
060	BROKEN
960	BROKEN ENVELOPE
015	BROKEN GLASS
720	BRUSHES, IMPROPER TENSION
120	BRUSH FAILURE
160	CHATTERING
910	CHIPPED
180	CLOGGED
140	CONTACTS, CONNECTION DEFECTIVE
170	CORRODED
190	CRACKED
200	DENTED
210	DETENT ACTION POOR
230	DIRTY
035	DRIFTS
226	EXCESSIVE PLAY
743	FALL TIME, EXCESSIVE
240	FLAKING
745	FORWARD RESISTANCE HIGH
250	FRAYED
270	FROZEN
280	FUNGUS EFFECT
088	GAIN, LOW
094	GAIN, NONE
001	GASSY
080	BURNED OUT
290	GROOVED
300	GROUNDED
567	HIGH CONTACT RESISTANCE
320	HIGH VOLTAGE BREAKDOWN
742	Ico HIGH
	INSTALLED IMPROPERLY
330	INSULATION BREAKDOWN
360	INTERMITTENT OPERATION
370	JAMMED
380	LEAKAGE
730	LOOSE
013	LOOSE BASE
012	LOOSE ELEMENTS
400	LOSS OF RESIDUAL MAGNETISM
004	LOW GM OR EMISSION
387	LOW PERFORMANCE
225	MANUFACTURER'S DEFECT (EXPLAIN)
131	MARGINAL PART REPLACEMENT
040	MECHANICAL BINDING
009	MICROPHONIC
053	MISFIRES (THYRATRONS)
730	MISSING
099	MODULATION, LOW
096	MODULATION, NONE
008	NOISY
022	NO OSCILLATION
	NOT DETERMINED
737	OPEN, BASE-TO-COLLECTOR
735	OPEN, BASE-TO-EMITTER
430	OPEN
003	OPEN FILAMENT
082	OPEN, INTERMITTENT
460	OPEN PRIMARY
451	OPEN ROTOR
470	OPEN SECONDARY
432	OPEN STATOR
433	OPEN WINDING
099	OTHER, EXPLAIN
161	OUTPUT INCORRECT
462	OUTPUT, LOW
255	OUTPUT, NONE
790	OUT OF ADJUSTMENT
258	OVERHEATS
727	PEELING
	PINCHED
520	PITTED
010	POOR FOCUS
156	POOR RECOVERY TIME
560	POOR REGULATION
944	POOR SPECTRUM (MAGNETRON)

FIGURE 10.4 Examples of codes to identify failed components and types of failures.

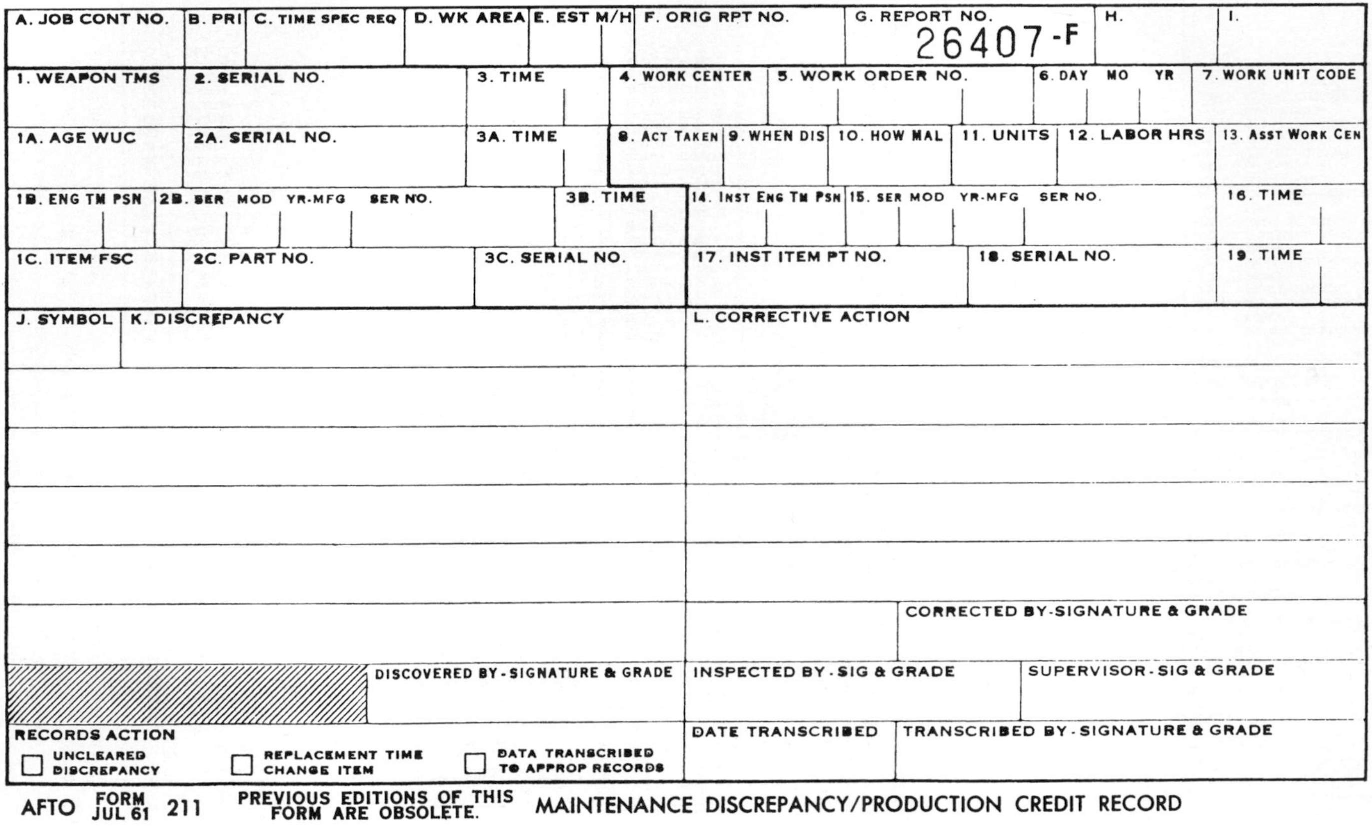

AFTO FORM 211 JUL 61   PREVIOUS EDITIONS OF THIS FORM ARE OBSOLETE.   MAINTENANCE DISCREPANCY/PRODUCTION CREDIT RECORD

A. JOB CONT NO. | B. PRI | C. TIME SPEC REQ | D. WK AREA | E. EST M/H | F. ORIG RPT NO. | G. REPORT NO. 26407-F | H. | I.

1. WEAPON TMS | 2. SERIAL NO. | 3. TIME | 4. WORK CENTER | 5. WORK ORDER NO. | 6. DAY MO YR | 7. WORK UNIT CODE

1A. AGE WUC | 2A. SERIAL NO. | 3A. TIME | 8. ACT TAKEN | 9. WHEN DIS | 10. HOW MAL | 11. UNITS | 12. LABOR HRS | 13. ASST WORK CEN

1B. ENG TM PSN | 2B. SER MOD YR-MFG SER NO. | 3B. TIME | 14. INST ENG TM PSN | 15. SER MOD YR-MFG SER NO. | 16. TIME

1C. ITEM FSC | 2C. PART NO. | 3C. SERIAL NO. | 17. INST ITEM PT NO. | 18. SERIAL NO. | 19. TIME

J. SYMBOL | K. DISCREPANCY | L. CORRECTIVE ACTION

CORRECTED BY-SIGNATURE & GRADE | SUPERVISOR-SIG & GRADE

DISCOVERED BY-SIGNATURE & GRADE | INSPECTED BY-SIG & GRADE

DATE TRANSCRIBED | TRANSCRIBED BY-SIGNATURE & GRADE

RECORDS ACTION
☐ UNCLEARED DISCREPANCY
☐ REPLACEMENT TIME CHANGE ITEM
☐ DATA TRANSCRIBED TO APPROP RECORDS

FIGURE 10.5   Example of maintenance report for data input to database.

helpful to go back over these reports to extract some other information. For example, it may be desirable to study the differences in the detection and repair time for some equipment at different field stations or repair centers. The information might be necessary in order to determine if additional training, new test equipment, or better instructions are needed for some service facilities.

10.3.2   Computerized Data System

The design of a computerized data system is a complex task requiring a lot of time of highly qualified programming specialists. At the beginning of this chapter attention was drawn to the needs for information, especially reliability information. A computerized data system should satisfy the needs enumerated in Sec. 10.1 and the needs of management. For example, the analysis of quality and reliability costs (see Chap. 4) requires that defects, failures, and unscheduled repairs be reported not only for reliability analysis purposes but for the accumulation of failure costs or costs of "unquality."

The same input information frequently is needed by several groups, but their uses of the information differ, and thus they desire the outputs to be different. Some persons will want only summaries of data, while others will want a more detailed output. This requires that the coding of the information be set up in such a way that the information can be retrieved and even analyzed in different ways. To satisfy all of these needs with one central database becomes a very difficult programming problem that could require several person-years of programming effort.

Many software companies have developed very effective database management programs that can be used "as is" or adapted to the specific reliability needs. These programs may appear to be very expensive, but they are usually more economical than trying to develop your own data management programs.

Many companies find it beneficial to set up several databases. One system could be just for test data to handle the results of all tests during development and production. One system might be used just for field service failure and malfunction data and analysis. A third system could be set up just to handle the quality and reliability records of the vendors. Another system could be used for in-plant quality control, process control, and process capability studies. Computer storage capacity is very inexpensive now, and the additional storage space required by separate data systems will usually be more than recovered in programming and operating time for single systems.

The inputting of data into the computer(s) can be accomplished in several ways. Computer terminals can be located throughout the plant facility and even in facilities hundreds or thousands of miles away and the data transmitted to the mainframe. The actual input may be done by a person typing in the data according to a specific format, or it may be automatically fed into the computer from sensors or detectors on the test equipment, measuring devices, or machines. Facsimiles of handwritten reports can be transmitted to the data center and the data entered manually. In all cases the data must be coded or indexed so that it will be stored in the proper place in the computer.

The rapid introduction of relatively inexpensive powerful personal computers and high-density floppy disks has resulted in substantially changed thinking about computerization of reliability and quality databases. Many companies are providing their engineers with desktop computers that can access the mainframe to either extract data, make complex computations, or store data, but with most of the engineer's work being performed on his or her own computer. A large number of software companies are providing a wide range of program packages that will perform most of the manipulative and analysis activities for reliability and quality work. Anyone interested in such programs for personal computers should examine the magazines *Quality,* published by Hitchcock Publishing Company, and *Quality Progress,* published by the ASQC, for the names and addresses of these software companies.

The whole field of computers and software is changing so fast that it is not advisable to try to list computers and software by name. Probably the available hardware and programs will be entirely different within a year. Persons planning new systems should make a thorough investigation into the available systems before contracting for either hardware or software.

A warning is in order. Where several different people need to have access to the same basic data, the main data files should be in the mainframe or network server. Otherwise, there will always be some person(s) who has (have) not received and entered the latest data into his or her (their) computer files. Unknowingly, the analysis will be incomplete.

10.4   GOALS OF DATA COLLECTION AND ANALYSIS

The goal of a data system is to convert the massive amount of information that is accumulated in all of the different divisions of an organization into an organized form that will enable all of the people to obtain the information they need rapidly and in a form that they can use with confidence in performing their assigned duties. There are many ways in which this organized information can be stored and presented to the users. A computer data bank is one way, but it has some disadvantages. In many companies the most commonly used data are printed for inclusion in loose-leaf binders, which become the source documents for action.

Some common types of information contained in these binders are

Derating charts for various standard electronic components

Mean time to failure (MTTF) for various components or parts under different stresses and environments

Process capabilities for the different production equipment and processes (by individual units)

Qualified parts lists with operating characteristics and design criteria

"Design rules"—standard design practices

Standard operating procedures, test procedures, etc.

The basic idea is that once something has been proved acceptable or effective, that information should be made available to every potential user in an easily accessible form, whether through a computer or in information binders. This will have many good effects:

It reduces the number of decisions a person has to make and saves time.

It helps to assure that proven parts, processes, and designs will be specified unless the requirements cannot be met with the proven items.

It helps quality assurance and purchasing to obtain acceptable parts and components from vendors.

It helps to assure that work will be assigned to the machines and processes that are capable of meeting the specifications.

It reduces the time and expense of prototypes, prototype testing, and design reviews, and speeds up the release to production.

It improves the capability to estimate the reliability of a product at the design stage.

Obviously the nature of the products and their intended uses will be the major factor in deciding how extensive the data collection and analysis system should be. A company stamping out stainless steel flatware will need a very simple data system, but one producing television sets, videocassette recorders, and video cameras must have a much more complex system to serve its needs. Determine who needs information, what they need, how they will use it, frequency of need, and the desired form of the information before trying to decide what kind of system to install and how extensive it should be. Do not let a salesperson sell you a system that is not appropriate for your immediate and foreseeable needs!

CHAPTER 11
DESIGNING EXPERIMENTS TO MEASURE AND IMPROVE RELIABILITY

Spencer Graves
*CENTER FOR QUALITY AND PRODUCTIVITY
IMPROVEMENT
UNIVERSITY OF WISCONSIN
MADISON, WI
AND SPENCER GRAVES ASSOCIATES
SAN JOSE, CA*

Tom Menten
*HEWLETT-PACKARD CO.
BOISE, ID*

The extensive reliability testing programs of many companies can generally provide much more information than traditional reliability engineers realize; conversely, problems sometimes arise later because reliability testing provides *less* information than many experimenters believe (Secs. 11.1.3 and 11.2.3). The opportunities are illustrated by the ball bearing experiment described in Sec. 11.1, which identified a product design that was much more reliable than what the engineers had previously thought possible, with a production cost less than anticipated.

Two potential problems illustrate the threats hidden in traditional testing:

1. The prototypes and test conditions employed in reliability testing are not always representative of standard usage of production units; see Sec. 11.5.2 for an example of this.

2. The acceleration formulas often used to convert the results from stress tests to warranty cost estimates are not always as accurate as people think and sometimes lose validity as technology progresses.

In today's increasingly competitive markets, leading engineering organizations need to capitalize on every available opportunity to increase the quality of their products while reducing development cycle time[1,2] and manufacturing cost. This chapter describes how an organization can meet these challenges while producing product designs that are more robust with regard to variations in environment, usage patterns, components, and subassemblies (Sec. 11.3.5). This approach includes being as aggressive as feasible[3] with accelerated testing.[4] Any use of accelerated testing, whether mildly or severely aggressive, will benefit from designed experiments to evaluate and improve the acceleration formulas employed.[5] The roughly $200,000,000 loss suffered by a large U.S. company[6] because of the unexpectedly short life of a lubricating oil illustrates the risks of using an acceleration formula without ongoing empirical validation: The oil broke down at rates much higher than predicted by the acceleration formula that had been used for years in that company. This case illustrates the general principle that as technology evolves, the formulas that have been used in the past sometimes lose their validity (Sec. 11.2.3). Similarly, it is wise to compare reliability predictions with actual performance in the hands of customers, as discussed in Sec. 11.7. Additional examples and discussion can be found in the recent book *Reliability Improvement with Design of Experiments.*[7]

This chapter begins by discussing, in Sec. 11.1, an experiment performed on the life of ball bearings. This example shows what we mean by a statistically designed experiment, illustrates the appreciable economic advantages of using such experiments to measure and improve reliability, and can help the reader understand how traditional reliability testing provides less information than many experimenters realize. Section 11.2 discusses examples of engineering experiments related to reliability. Section 11.3 introduces some basic ideas of experimental design strategies. Some of the simplest designs and their analysis are described in Sec. 11.4. Other designs and analysis techniques are discussed in Sec. 11.5. When the testing budget is substantial or the time pressures warrant, more sophisticated techniques can be employed. Some of these are briefly introduced in Sec. 11.6; after reading this section, an engineering manager should have a better idea about when the budget and objectives for a project might best be served by obtaining expert help in experimental design and analysis.

It is now standard in the literature on quality and reliability to visualize product development as cyclical,[8,9] following the Shewhart/Deming "Plan–Do–Check–Act" improvement cycle; Sec. 11.7 discusses the use of warranty data in this context. Companies that don't make effective use of warranty data often perpetuate quality and reliability problems from one model to the next in a product line, with dire consequences for customer satisfaction, market share, and profitability.

11.1   EXPERIMENTAL DEVELOPMENT OF COMPETITIVE ADVANTAGE

This section presents an example of a designed experiment for reliability improvement. It illustrates both the potential gains from experiments of this nature and the losses sometimes suffered by less systematic experimenters who overestimate the generalizability of the results.

By design of experiments (DoE), we refer to the development of procedures for systematically manipulating product (or process) design or environmental factors in order to evaluate their impact on some measure of performance. The Swedish company SKF conducted a ball bearing experiment described by Hellstrand[10] and summarized in Fig. 11.1 that illustrates many valuable lessons about experimental design.

The ball bearings used in this experiment consisted of several balls, inner and outer rings, and a cage that maintained a constant spacing between the balls. The design engineers thought that the product life might be most sensitive to the cage design, the heat treatment of the inner ring, and the osculation of the outer ring. (Osculation, in this context, is the ratio of the ball diameter to the radius of the arc of the raceway.) The engineers knew a lot about the design of ball bearings but were uncertain about the effects of these three factors. They therefore made inner rings with two different heat treatments (standard and modified), outer rings with two osculations, and cages of two designs and assembled them into bearings with the eight combinations shown in Fig. 11.1. They then ran the test bearings with a load much higher than the design load and measured the time to failure to get the numbers in the circles at the corners of the cube in Fig. 11.1.

11.1.1   A Surprising Interaction

The first thing to note from Fig. 11.1 is that the longest life came from combining the modified inner-ring heat treatment with the modified outer-ring osculation. The two points of the cube with this combination of factor levels lasted approximately 100 h; the other six lasted only 20. When the combined effect of two (or more) factors is different from the sum of the individual effects, the result is called *interaction*. Hellstrand said this "interaction was not previously known in the existing life theory for bearings, and it could only have been found by applying the factorial design technique of studying all variables at the same time."

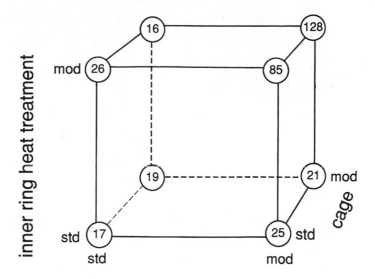

outer ring osculation

(= ratio of ball diameter to the radius of the arc of the raceway)

FIGURE 11.1   SKF ball bearing experiment.

This is not an isolated example; Zangwill noted that interaction effects between experimental factors "seem almost magical or paradoxical, yet are often rampant in engineering design."[11] Interactions involving more than two factors can also be defined and are occasionally important, although they are not as common as two-factor interactions such as the one in this example. Not all interactions are as dramatic as the example in Fig. 11.1, but economically important interactions are relatively common.

11.1.2   Value of Noncontributory Factors

A second lesson from Fig. 11.1 is that there were no appreciable differences between the two cage designs. Hellstrand said, "This is very important information as use of one of these cage designs results in a reduction of the cost of the bearing. In general, knowledge of non-contributory factors is very important. Such information can often be used to reduce the final product cost." The economic importance of such results can be enormous because any money saved from using design options that reduce cost without decreasing quality can be used in three different ways:

1. A cost reduction of this nature with no change in sales price converts dollar for dollar into profit.

2. Alternatively, such a change allows the organization to reduce its price by the amount saved while maintaining its previously planned profit target.

3. If the engineers are having trouble meeting the overall product cost and performance targets for the design, a saving in one place allows greater flexibility in other aspects of the design.

11.1.3   The Limited Generalizability of Results from Testing Identical Units

A more subtle problem occasionally appears when identical prototypes are tested: Traditional statistical inference applies, strictly speaking, only to other units produced by the same process as the prototypes tested. Deming called this an "enumerative" study.[12]

Unfortunately, the process during production will be different, and early production will be different from late production. Deming called any study whose purpose is to predict the future an "analytic" study.[13] The experimenter hopes that prototype test results will help to predict the performance of production units. However, the step from prototypes to production requires a "leap of faith" that is separate from statistical theory. By building more variability into the prototypes with a designed experiment such as Fig. 11.1, we can increase our (nonstatistical) confidence in using test results from prototypes to predict production.[14] This, in turn, reduces the chances of unpleasant surprises due to changes in production.

11.1.4   The Cost of a Designed Experiment versus Testing Identical Units

A third observation regarding Fig. 11.1 is that SKF's test plan might have called for testing eight units without the experiment reported in Fig. 11.1. If it did, the experiment reported in Fig. 11.1 employed the same number of test runs, and so it did not appreciably increase SKF's testing budget. There was only a modest increase in expense for planning, managing the experiment, and evaluating the results. The experimenters (1) made two sets of four inner rings, with each set having a different heat treatment, (2) prepared outer rings with two different osculations, and (3) made four each of two different cage designs. Making two sets of four inner rings, outer rings, and cages was somewhat more expensive than making eight of one design. There also was an increased expense in assembly to ensure that all combinations were properly assembled so that each point of the cube was represented only once. With more than three factors, the conduct of the experiment can sometimes require a substantial effort just to ensure that each set of experimental conditions is tested as planned. However, with between one and three factors, the administrative cost is typically noticeable but not excessive.

The modest increase in administrative expense for the experiment in Fig. 11.1 was amply repaid: SKF was able to increase the lifetime by a factor of five and use a cheaper cage design, confident that it would not jeopardize product quality.

In addition, the data in Fig. 11.1 can be used to help set specifications on how tightly process control must be maintained on the heat treatment and the osculation to ensure quality (Sec. 11.4.4). Conversely, if only the standard level of all three factors had been tested, the engineers not only would have missed the surprising increase in lifetime resulting from interaction, they would not have known how tightly to maintain process control.

Without this experiment, they might also have used the more expensive cage design in the mistaken belief that it contributed to product life when it did not. In such cases, the company may never know that a cheaper alternative exists unless a competitor starts to sell a product at a price below the company's costs. Especially with foreign competitors, people often attribute a competitor's lower price to lower wages or governmental subsidies that the competitor may get but that are not the key to its ability to sell below your costs.[15] By the time one becomes aware of a threat of this nature, it can be extremely difficult to recover.[16] The best defense is a good offense, with an aggressive program of experimentation that tests various design alternatives with little or no incremental cost during reliability testing.

If this experiment had show *nothing*—e.g., if all eight ball bearings had failed in roughly 20 hours—the data could have been analyzed and used as though all eight ball bearings had been identical (as mentioned in Sec. 11.3.3, with exceptions as discussed in Sec. 11.5.2). This fact allows companies to convert their traditional confirmation testing into designed experiments, with substantial increase in the value of the results obtained, almost regardless of the outcome.

11.1.5   Contributions to the Long-Term Competitiveness of the Organization

A final lesson from Fig. 11.1, and in some cases the most important, is that the previously unsuspected interaction suggests that there is an opportunity to develop improved theory and design rules for this type of product that can be used in the design of similar products in the future. The easiest and cheapest way to capitalize on such opportunities is to look for similar interactions in the design of future products until patterns begin to emerge. In other cases, it may be best to spin off the development of technology into separate projects, e.g., to develop sufficient understanding of the important interaction in Fig. 11.1 to be able to use it in future products. Clark and Fujimoto said, "Firms searching for competitive advantage in the 1990s may well find it in the ability to solve" the problems of managing technology development to optimally support product development.[17]

11.1.6 Ease of Documentation

Another benefit of experiments like those described in this chapter is that they are easily documented in such a way that their value endures far longer than the typical lab notebook entry. Such documentation is becoming increasingly familiar in engineering and management meetings in some organizations and provides a convenient structure for discussing results and contributing to design rules used in future products.

11.1.7 Summary

The ball bearing example has illustrated a relatively simple experimental design that produced several economically important results:

1. An *important interaction* between two factors (inner-ring heat treatment and outer-ring osculation) that substantially increased the value (in this case the life) of the product
2. An *unimportant factor* (cage design) that allowed the use of a cheaper design option
3. *Information on* the relative importance of *process control* in two factors (inner-ring heat treatment and outer-ring osculation)
4. A possible source of *long-term competitive advantage* based on using the interaction in other designs

This example illustrates another point: Often, the analysis required to interpret the results of experiments is no more sophisticated than that shown in Fig. 11.1. Conversely, if the experimental design does not allow a certain question to be addressed, no level of mathematical sophistication can compensate.[18] In between, there are situations where sophisticated mathematics can be used to reduce the time or the number of tests; see Sec. 11.6.

In the remainder of this chapter, we introduce a number of options for designing and analyzing engineering experiments.

11.2 ALTERNATIVE RESPONSE VARIABLES AND ACCELERATED TESTING

The ball bearing experiment illustrated the employment of a factorial experiment to directly study and improve product life. In this example the "response" variable was "time to failure," and the connection to product reliability was immediate and clear. This section describes other experiments relating to reliability. In some of these examples, the response measured was not product life. An understanding of these applications can help managers and engineers plan product testing.

11.2.1 Product Life

As in the ball bearing experiment, the simplest applications of factorial experiments to reliability use "time to fail" as a response variable, although "time" may be measured in cycles or repetitions rather than hours. One problem with such experiments is the substantial project time required to get all units to fail. Pressures on time and cost often limit the quantity of data collected in such experiments.

Data scarcity requires us to make the best possible use of our data. Project time constraints may require us to conduct experimental runs simultaneously. Both influences strongly indicate the use of planned multifactor experiments.

"Time to fail" data are often "censored"—that is, some tested units may not have failed during the experiment. For these units, we know only that the lifetime was longer than the recorded running time, or "censoring time." The design and analysis of lifetime experiments involving censoring is an important subject with a vast literature, briefly introduced in Sec. 11.6.2.

11.2.2 Estimating Reliability from a Strength Distribution

In many cases, the probability of failure of a device can be estimated as the probability that a critical parameter is below a specification limit. For example, a volume customer for a connector was having a major problem with breakage during assembly. Specifically, the torque applied to mounting screws would sometimes exceed the strength of the connector. Negotiations between supplier and customer revealed that the problem was primarily with connectors that could not withstand about 8 ft–lb of torque. The vendor suggested that a stronger but more expensive connector should solve the problem. An engineer with the customer suggested that a manually installed washer might be needed, at a still higher cost.

A two-factor experiment was planned to evaluate the relative efficacy of the four possibilities: the current and the alternative connector, with and without the washer. At each of the four sets of conditions, the torque required to break each of 32 connectors was measured. The data from one of the four sets of conditions are shown in Table 11.1.

TABLE 11.1 Connector Strength Measurements: Redesign; No Washer

35.9	28.4	25.2	17.6
32.7	18.4	34.7	22.9
28.8	34.2	23.2	36.9
22.2	19.0	22.5	15.2
22.0	36.9	33.8	19.7
29.5	17.4	22.4	25.9
32.6	12.8	25.5	21.9
39.8	25.1	35.4	28.3

We now want to estimate the portion of the product distribution that will fall below 8 ft–lb. This estimate is quite sensitive to what we assume regarding the probability distribution of the strengths. Two standard candidates are the normal and the lognormal distributions. Physical measurements are often lognormally distributed; this is consistent with engineering intuition that 0.5 ft–lb sounds like a broad range for a 1 ft–lb connector but incredibly precise for a 10 ft–lb connector, for example, whereas ± 30 percent sounds like it has roughly the same meaning for both connectors.

With this in mind, we made normal probability plots of the numbers in Table 11.1 and of their logarithms. Both lines were quite straight, suggesting that either assumption was reasonable. The average and standard deviation of the numbers in Table 11.1 are 26.8 and 7.5. Therefore, the mean is 2.5 standard deviations above the specification limit of 8 [computed as $(26.8 - 8)/7.5$]. A table of the normal probability distribution suggests that 0.6 percent of the connectors with the new design may be below spec. The comparable analysis of the logarithms produced a much smaller estimate for the probability of connectors that break with less than 8 ft–lb of torque. The same computations based on the tests with washers indicated higher strengths, but not enough to justify the extra expense. More precise computations could be made based on the theory of statistical tolerance intervals.[19] However, even this refined theory does not consider the fact that production units are different from prototype units and may vary over time. Experience in subsequent production indicated that this calculation gave a reasonable answer, and the breakage problem essentially disappeared with the alternative connector.

The point here is that a reliability problem was resolved through strength testing in a two-factor experiment.

11.2.3 Experimental Determination of Accelerated and Surrogate Testing Protocols

In many cases, reliability can be estimated from previous knowledge about the relationship between "surrogate" characteristics that are difficult to measure and characteristics that are easier to measure. A special case of this is accelerated life testing, where units are tested under extreme stress and an *acceleration formula* is used to estimate the lifetime under normal usage from the lifetimes observed under stress.

For example, a reliability engineer for a company that makes computer printers suggested that the rate of paper jams or "misfeeds" might be related to the "pull strength," i.e., the force required to dislodge the paper from the rollers. An experiment was designed to test this theory. If correct, it could substantially reduce the testing required because the pull strength could be measured in seconds, whereas the misfeed rate required months.

In this experiment (see Table 11.2), two rollers were used, a new roller ("+" in Table 11.2) and a worn roller ($-$), with two media types, rough ($-$) and smooth (+). The experimental objective was to determine the relationship of pull strength to misfeed rate over the space bounded by these four conditions.

TABLE 11.2 Pull Strength and Misfeeds for Paper Feed Mechanism

Random order	Standard order	A Roller	B Media type	AB	Pull strength	Misfeed rate (coded units)
	1	$-$	$-$	+	8.4	5
	2	+	$-$	$-$	6.1	20
	3	$-$	+	$-$	7.9	5
	4	+	+	+	5.9	26

(hypothetical)

A simple plot of the pull strength and misfeed rate numbers in Table 11.2 suggested that a linear relation might be plausible, which least-squares regression estimated as follows:

$$\text{Misfeed rate} = 37 - 24 \times \text{pull strength}$$

The numbers in Table 11.2 are hypothetical results, used to convince managers and others of the need to perform an experiment of this nature. The actual results turned out to be quite interesting. Related experiments are currently being planned that are expected to result in a proprietary testing procedure that will allow the company to substantially reduce the testing required of new product designs.

Either surrogate or accelerated testing can fit naturally into a concurrent engineering program:

1. Tooling and initial production can begin based on surrogate or accelerated tests that suggest that this is a good gamble. Meanwhile, testing continues in order to confirm (or refute) the predictions and to evaluate the formulas used to make the predictions.

2. When tooling is completed and products are ready to ship, the additional experimental results can be reviewed to confirm that the new data support the previous decision. The $200,000,000

disaster described in the introduction to this chapter could have been avoided if this procedure had been followed.

3. Even if subsequent testing supports final release of a particular design, the data may still uncover deficiencies in the accelerated or surrogate testing procedure. This may require modifications of test procedures for future products.

11.2.4   Problem Resolution Using Disassembly-Reassembly

A print quality problem was identified in a prototype of a new printer. Ordinarily, a print quality problem might not be considered a reliability issue. However, this problem was noticeable only under relatively rare circumstances that would be likely to generate a service call, and a reliability engineer was working on the problem.

The design team had two prototypes, only one of which exhibited the problem. In a brainstorming session, several subassemblies were identified as potential sources of the problem. After some discussion, the sources were classed as either electronic or mechanical. The two printers were disassembled and the electronics switched, and a resulting test instantly identified the problem as mechanical. The mechanical portion of the printer was further divided into major subassemblies and the process repeated until the source of the problem was isolated, understood, and fixed.

This kind of situation occurs often in industry, and this "disassembly-reassembly" technique is widely used to good effect.[20]

11.3   EXPERIMENT DESIGN STRATEGIES IN RELIABILITY

This section relates some time-tested principles of experimentation to reliability.

11.3.1   Iterative Experimentation

There are two points to be made here:

1. When experiments are planned and conducted in a sequence, each experimental design is improved by information from previous experiments.
2. This fact has implications for budgeting and allocation of experimental resources. The standard advice is, wherever feasible, to spend no more than 25 percent of the available experimental budget on any one experiment.

Box, Hunter, and Hunter[21] noted that scientific investigations are inherently iterative in nature: What is learned from one stage of experimentation is often valuable in generating more elaborate hypotheses and theories regarding sources of problems and potential solutions, which improves the effective use of precious experimental resources at the next stage. When the time to perform an experiment is relatively short, detailed planning is often not justified; relatively simple experiments like that in Fig. 11.1 or even simpler 2×2 designs (with only two factors rather than three at two levels each) often suffice. These are simple to plan and analyze and are much more powerful than the traditional one-factor-at-a-time approach. When nonlinearities are a major issue, more sophisticated experiments, such as those outlined in Sec. 11.5.2, should be employed. More sophisticated experimental designs might be called for when the time required to perform the experiment is long or the experimental budget is substantial; see Sec. 11.6.

In most cases a sequential experimental strategy will be more effective and easier to implement than a single comprehensive design. In addition, the realities of work in real organizations often interrupt large experiments. Process instabilities during such interruptions sometimes destroy the effectiveness of large experiments unless that possibility was explicitly considered, e.g., by blocking, as described in Sec. 11.3.6.

Many reliability applications are severely constrained in chronological (project) time. This creates a strong preference for conducting experiments in parallel rather than sequentially. Even here, however, there will be an iteration of product designs and prototypes, and so a sequential strategy may still be preferred. For example, early experiments may be designed to identify trouble spots and important environmental factors that can be studied in more detail in later experiments. The alternative of a single, large, "comprehensive" experiment may provide more detail than is needed about many issues that are irrelevant to current design questions but lack important detail about critical engineering design questions.

An initial foray into a subject might start with a "screening" or "main effects" design,[22] which may test seven factors in eight runs, for example. This might be followed by a full factorial design, such as the experiment in Fig. 11.1 (called "full factorial" because it consists of all possible combinations of all the levels of all the factors), and may extend to explore a "response surface" as outlined in Sec. 11.5.2. Each experiment may yield new insights into what factors might be most important, which factor levels are of most interest, and even what response variables to use and how to measure phenomena of interest.[23] Each subsequent design will benefit from what was learned at earlier stages. Furthermore, the use of more than one experiment provides a confirmation of results, increasing our confidence in the applicability of the tests on prototypes to production (as discussed in Sec. 11.1.3).

In sum, virtually all good scientific and engineering investigations proceed sequentially. It is wise to consider this in planning product development testing.

11.3.2 Number of Runs in an Experiment

Many of the techniques used to analyze and interpret the results of an experiment work better for larger experiments. In particular, randomization (Sec. 11.3.6) and certain data analysis techniques work better for experiments with 16 conditions and less well for smaller experiments. Thus, when there is considerable flexibility in the number of runs that can be conducted, designs having 16 runs should often be considered.

In some contexts, 16 runs is considered a large experiment. When an experiment of this size is considered, if total time for the project is not an issue, a 16-run experiment can often be "fractionated" into two "halves" of eight runs each, to be run sequentially.[24] The first half can be run, analyzed, and interpreted before it is decided whether the other eight runs are needed.

In reliability testing of things like light bulbs, it is common to place a large number of items on test and then count the number that failed after a certain period of time. A manufacturer of light bulbs might want to know, for example, that the bulbs performed well between 95 and 125 V, at 0 and 120°F, and at 0 and 99 percent humidity. Test chambers could be built to support an experiment of this nature following the basic experimental layout discussed with Fig. 11.1. The question of how many bulbs to test in each chamber is discussed in papers by Bisgaard and Fuller.[25,26] This same methodology can be used to obtain a conservative estimate of the sample size required when testing involves recording actual lifetimes subject to censoring or when lifetime data are recorded only to the nearest day, for example, as outlined in Sec. 11.6.2.

11.3.3 Hidden Replication

The analysis of two-level factorial experiments (Sec. 11.4.3) proceeds by partitioning the runs into two equal groups in many different ways: One group will contain all observations with a given factor or interaction at the high level, while the other group will contain all the observations at the low level. The estimated effect of a factor or interaction is then computed as the difference between the averages of the observations in each of the pairs of two groups.

In this way, the estimated effect of each factor is obtained as though it were the only factor in the experiment. In the experiment of Fig. 11.1, for example, the eight runs provide four "hidden replicates" of the comparison of the modified to the standard heat treatment, and two hidden replicates of each combination of levels of heat treatment and osculation. This effect is known as *hidden replication* and offers us a reason to include another factor in any circumstance when we would consider replicating the experiment (in addition to increasing our confidence in applying the results of prototype tests to production, as discussed in Sec. 11.1.3).

11.3.4 Choice of Response Variables

There are several aspects of the response variable that are relevant to reliability experiments: (1) how many are there, (2) of what data type (e.g., continuous variable such as voltage or lifetime, or discrete such as counts of defects, or 0 or 1 for whether a unit survived to a certain time), and (3) should the data be transformed for analysis?

Multiple Response Variables and Covariates.   While adding an additional factor may double the number of runs in the experiment for a full factorial, adding a response variable does not affect the number of runs. For example, print quality might be measured as well as the lifetime of a printer; candlepower might be measured in addition to the lifetime of light bulbs.

There are many possible reasons to include more than one response variable. Multiple response variables can sometimes help us to address additional experimental objectives. For example, we may estimate both "time to jam" and "time to fail" for a paper-feeding mechanism. If we wish to develop a quicker measure of ball bearing reliability, we might introduce a wear measurement while we continue to test to failure. The experimental results may enable us later to estimate "time to fail" by extrapolating wear in a shortened test, as in the example in Sec. 11.2.3.

We can also measure *covariates,* which are other factors such as humidity and temperature that may affect the response variable(s) but are not controlled in the experiment. When we measure an uncontrolled variable, we know how much it changed over the course of the experiment. If it changed very much, we can analyze the experiment using regression, including all the controlled factors in coded units, e.g., ± 1 for "std" and "mod" as in Table 11.3, and including the numerical values of the covariates directly in the regression equation.

TABLE 11.3   Estimated Effects for SKF Ball Bearing Experiment

Random order	Standard order	*A* Osculation	*B* Heat treatment	*C* Cage	*AB*	*AC*	*BC*	*ABC*	Hours
	1	−	−	−	+	+	+	−	17
	2	+	−	−	−	−	+	+	25
	3	−	+	−	−	+	−	+	26
	4	+	+	−	+	−	−	−	85
	5	−	−	+	+	−	−	+	19
	6	+	−	+	−	+	−	−	21
	7	−	+	+	−	−	+	−	16
	8	+	+	+	+	+	+	+	128
Estimate of effect		45.25	43.25	7.75	40.25	11.75	8.75	14.75	42

Covariates, like multiple response variables, do not directly affect the number of runs in the experiment. If the covariates explain a significant portion of the variability in the data, they can make the impact of the factors manipulated in the design much easier to detect. If temperature, for example, has a major impact on reliability but is neither measured nor controlled, its effect becomes part of measurement and replication error. The resulting increase in error increases the difficulty of seeing the effect of the experimental factors. The worst case is when a covariate, e.g., temperature, is high when one of the experimental factors, e.g., line voltage, is high and low when that factor is low. In that case, the results may indicate that the line voltage had an important impact on reliability when the effect was really due to the unmeasured covariate, temperature. This is called *confounding* and can be managed using randomization and blocking; see Sec. 11.3.6.

Failure Time versus Strength/Stress Measurements or Wear and Performance Degradation. Failure times (or lifetimes) are a very frequently used but often troublesome response measurement for reliability experiments. Considerable testing and project time are often required in order to observe *any* failures, and it is often impractical to test until all units have failed. This latter circumstance leads us to the subject of censored data, discussed in Sec. 11.6.2.

An alternative to testing to failure is to develop a strength measurement for which the corresponding stresses are understood, as suggested by the example of Sec. 11.2.3. If such relationships can be found, tests that previously required months of project time can often be replaced by tests performed in minutes. As noted in the introduction to this chapter and in Sec. 11.2.3, such relationships sometimes lose their validity as failure modes change with the evolution of technology. In some cases, one way to reduce time to market is to release the design to production based on stress-testing relationships developed with earlier products while continuing testing to ensure that the relationships still hold. If there is a problem, it is likely to be identified before it becomes a serious issue with customers. Money spent on manufacturing preparation will have been wasted. However, that should happen relatively infrequently and so should be a fairly good risk. Moreover, the formulas can be improved for use in future product development efforts.

Transforming the Response Variable.   Engineers are quite familiar with the advantages of plotting functions on log or semilog graph paper, in which one or more of the axes is constructed on a logarithmic scale. A simplicity of description can often result. Transformations can similarly be used to simplify the description and analysis of results of a multifactor experiment, as noted in Sec. 11.6.1.

11.3.5 Location, Dispersion, Robustness

Many experiments are conducted to estimate the "typical" or average life at each set of experimental conditions. In some cases, the standard deviation (also called dispersion) of results at each set of experimental conditions is of interest. For example, the connector strength experiment (Sec. 11.2.2) concerned components whose average strengths were all above the typical applied stress. In this case, the standard deviation of strengths could be the primary determinant of the percentage that fail. For conditions where the standard deviation is negligible, virtually no units will have less than the minimum strength. If, under a different set of conditions, the standard deviation is huge, a substantial portion of the units could fail. If some product design factors can be found that influence dispersion and not the average response, whereas others affect the average but not the dispersion, the dispersion can be minimized while the average response is set at the desired target. The simplest way to search for such opportunities is to perform multiple tests at each set of experimental conditions so that a standard deviation can be estimated for each,[27] though other techniques exist.[28]

These considerations are related to the issue of designing products that are robust to variations in environmental factors and to variability and degradation of components. A substantial literature on this subject has recently been developed.[29]

11.3.6 Randomization and Blocking

The purpose of an experiment is to understand the sources of variability in the response variable, e.g., how long it takes a unit to fail and what factors influence the lifetime. In the design of an experiment, we choose to control some variables (the experimental factors) and hold other variables constant. However, there may be yet other factors that influence the response variable of which we are as yet unaware. Suppose, for example, that we are conducting an experiment to eliminate cracks in welded material. One experimental variable is the preheat temperature, and all the tests at the high temperature are run on one day while all the tests at the low temperature are run on the next day. However, unbeknownst to the experimenters, the welding rod used on the second day came from a different batch. Thus, a significant effect due to batch of welding rods would be incorrectly attributed to or "confounded with" preheat temperature.

There are two ways to avoid this kind of misleading result: randomization and blocking. By randomization, we mean using random numbers to decide (1) the order in which to perform the runs and/or (2) which sample of material, which device, or which test station to use with each run. Blocking can be thought of as including an additional experimental design factor to account for differing batches of material or devices, or possible changes from one day or shift of operations to the next.

In the welding example, if the run order had been randomized, the effect of the unsuspected effect due to the batch of welding rods would appear in measurement and replication error—noise—rather than giving us the false impression that the preheat temperature was important. Alternatively, if the experiment had been blocked by day, then the effect of the batch of welding rods would appear as a significant effect due to blocking (i.e., the day) but would not disturb our ability to evaluate the effect of preheat temperature and other variables. For more on randomization and blocking, see Box, Hunter, and Hunter.[30]

Many industrial experiments involve factors that are difficult to change, such as the temperature of a large oven that requires a long time to stabilize or the chemical composition of a large reacting vessel. As a practical matter, such experiments are rarely randomized, leaving main effects confounded with blocks. The result is called *split plotting*.[31] The analysis of data from such experiments requires techniques that are beyond those described in this chapter. Experimenters who are unfamiliar with those techniques often ignore the split plotting in their analysis. This increases the chances of misinterpreting the results of the experiment. As long as the conclusions are validated with subsequent tests, however, experimenters usually will not be led too far astray by ignoring split plotting.

11.4   TWO-LEVEL FACTORIAL DESIGNS AND ANALYSIS

The design of the experiment (DoE) employed in the ball bearing study (Sec. 11.1 and Fig. 11.1) is called a *two-level factorial* design because each *factor* (heat treatment, osculation, cage design) was studied at two *levels* (standard and modified). Such experiments are the workhorse of industrial DoE application. These designs are easy to apply, analyze, and interpret. In addition, these designs can be run in "fractions"[32,33] that allow, for example, 15 factors to be evaluated simultaneously in only 16 runs. So-called fractional factorial designs include designs whose purpose is to screen a large number of factors to find the most important ones in relatively few runs.

Since two-level designs are relatively simple and are so pervasively used, we devote this section to them.

11.4.1   The Language of DoE

The ball bearing experiment of Fig. 11.1 consists of exactly one replication of every combination of three factors studied at exactly two levels each. The result is denoted as a $2 \times 2 \times 2 = 2^3$ factorial experiment and has $2^3 = 8$ runs. Another view of Fig. 11.1 is provided in Table 11.3. The experimental design factors—those aspects of the experiment, such as the outer-ring osculation, that are systematically manipulated in the experiment—have been coded as "Factor A," "Factor B," and "Factor C," while the levels "std" and "mod" have been coded as "$-$" and "$+$," respectively.

By extension, an experiment testing every combination of k factors studied at two levels each is called a 2^k. Many analysis techniques can be applied to the results of a 2^k experiment, depending on the objectives and design. In this section we illustrate a few of the simplest. Another technique, analysis of means, is discussed in Sec. 11.5.1.

11.4.2 Graphical Analyses

The *"cube plot"* of Fig. 11.1 illustrates a first and important way of analyzing the results of a 2^3 design. This figure quickly conveys a graphical image of the design and communicates the results in a format that is easy to read, present, and discuss. With $k = 4$, a 2^k design can be portrayed in two cubes side by side, with the first cube being the runs with the fourth factor at "$-$" and the second being the runs with the fourth factor at "+." A 2^6 design can similarly be portrayed as a cube of cubes, with three of the factors indicating which of the smaller cubes to access.

With a 2^k design with $k>3$, if only two or three factors seem important, it is common to report the results in a square or cube and suppress the unimportant factors in a visual display whose purpose is to communicate the main results quickly to a large audience.

Having determined that the cage design in Fig. 11.1 is unimportant, we next prepare an *interaction plot*. This portrays the response (lifetime) vs. one important factor (outer-ring osculation), with separate lines for the levels of other important factors. Four lines are plotted in Fig. 11.2, one for each combination of inner-ring heat treatment and cage design. We could create a similar plot with only two lines, averaging the lifetimes for the two different cage designs for each combination of heat treatment and osculation. A plot with only two lines would show the interaction better, but Fig. 11.2 also gives us an idea of how much the lifetime might vary from one ball bearing to the next, assuming that the cage design has no effect. (We could also plot lifetime vs. inner-ring heat treatment with separate lines for the levels of outer-ring osculation.)

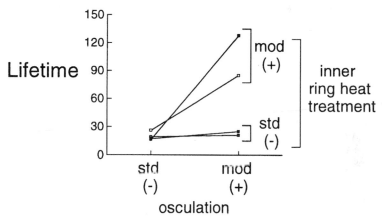

FIGURE 11.2   Interaction plot for the ball bearing experiment.

Whereas a cube plot like Fig. 11.1 helps an audience understand quickly both the experimental design and the results, the primary message in the results can often be understood more quickly from an interaction plot. The two are often presented in this order to convey quickly and succinctly the experiment performed and the results obtained.

In addition to the cube and interaction plots, it is often useful to have numerical estimates of the magnitude of effect. We now describe how this is done for a 2^3 design like Fig. 11.1.

11.4.3 Numerical Estimation of Effects

Each leg or edge of the cube plot of Fig. 11.1 can be thought of as an experiment involving just two runs. Viewed this way, there are four separate experiments on the effect of heat treatment, with each experiment conducted under different conditions of cage design and outer-ring osculation. The *estimated effect* of each design factor is the difference between the responses at "+" and "−" levels of the factor, averaged over all four of these two-run experiments:

$$A \text{ effect (osculation)} = [(25 - 17) + (85 - 26) + (21 - 19) + (128 - 16)]/4 = 45.3$$

$$B \text{ effect (heat treatment)} = [(26 - 17) + (85 - 25) + (16 - 19) + (128 - 21)]/4 = 43.3$$

$$C \text{ effect (cage design)} = [(19 - 17) + (21 - 25) + (16 - 26) + (128 - 85)]/4 = 7.75$$

Here, *A*, *B*, and *C* are the three factors in Fig. 11.1, as identified in Table 11.3. Note that the numbers with a "+" sign in the above expression for *A* are those in rows with "+" in the *A* column of Table 11.3, while those with a "−" sign are those in rows with "−" in the *A* column. This is also true for *B* and *C*.

These four experiments afford the opportunity to assess the stability of the effect of heat treatment under varying conditions of the other important factors. In this case, it appears that with the standard outer-ring osculation, the effect of inner-ring heat treatment is small ($9 = 26 - 17$ and $-3 = 16 - 19$); with the modified outer-ring osculation, the effect of inner-ring heat treatment is huge ($60 = 85 - 25$ and $107 = 128 - 21$). This phenomenon—that the effect of one factor depends on the level of another—is called *interaction*. When interactions can be ignored, the effect of *A* is essentially the same for all levels of *B* and *C*; when important interaction effects are present, the main effects of the primary factors included in the experimental design cannot be interpreted without considering the interactions. Thus, in Fig. 11.1, the effect of osculation cannot be described without simultaneous reference to heat treatment, and vice versa.

To understand osculation, we compare the design points on the right face of the cube to those on the left. To understand heat treatment, we compare the points on the top face to those on the bottom, and to understand the effect of cage design, we compare the back face to the front. In this way, we reuse the data for three analyses.

By rearranging the terms in the above expression, we see that the effects can also be described as the average of the responses at the high level of the factor minus the average at the low level. Thus, for *A*, we can write

$$A \text{ effect (osculation)} = (25 + 85 + 21 + 128)/4 - (17 + 26 + 19 + 16)/4 = 45.3$$

By analogy, the estimated effect for the *AB* interactions can be obtained by subtracting the average response with "−" signs in the *AB* column of Table 11.3 from the average response with "+" in that column:

$$AB \text{ effect} = (17 + 85 + 19 + 128)/4 - (25 + 26 + 21 + 16)/4 = 40.25$$

The other interaction effects are similarly estimated:

$$AC \text{ effect} = (17 + 26 + 21 + 128)/4 - (25 + 85 + 19 + 16)/4 = 11.75$$

$$BC \text{ effect} = (17 + 25 + 16 + 128)/4 - (26 + 85 + 19 + 21)/4 = 8.75$$

$$ABC \text{ effect} = (25 + 26 + 19 + 128)/4 - (17 + 85 + 21 + 16)/4 = 14.75$$

One more point: The signs in the *AB* column of Table 11.3 (and in the other interaction columns) are the products of the corresponding signs in the columns for *A* and *B*. The columns of Table 11.3 labeled *A*, *B*, and *C* serve two purposes. During the implementation of the experiment, these columns specify the different levels of the experimental factors for each run. During the analysis phase, these columns describe the calculations needed to estimate the effects, as in this illustration.

The next section discusses one use of these estimated effects.

11.4.4 Regression Line

If a regression equation is fit to Table 11.3, the resulting parameter estimates will be one-half of the effects just estimated. Thus, ignoring cage design, we get the following:

$$y = \text{lifetime} = b_0 + b_A A + b_B B + b_{AB} AB + (\text{error})$$

where b_0 = grand mean = 42.12
b_A = 22.62
b_B = 21.62
b_{AB} = 20.12

and A and B are the heat treatment and osculation in coded units (± 1). For example, with $A = B = -1$, this equation gives us $y = 42.12 - 22.62 - 21.62 + 20.12 = 18$, which is halfway in between 17 and 19, the lifetimes of the two ball bearings tested at $A = B = -1$.

While we don't have information on lifetime for other combinations of A and B, in many cases, this equation won't be grossly inaccurate for A and B between, say, -1.2 and -1.2. It was suggested in Sec. 11.1.3 that this equation could be used to evaluate how tightly process control should be maintained on heat treatment and osculation. To see how this can be done, suppose that control charts from previous production indicate that heat treatment and osculation can be held within ± 20 percent. Then we estimate

$$y = 90 \text{ for } A = B = 0.8 \ (= 1 - 20\%)$$

and

$$y = 124 \text{ for } A = B = 1.2 \ (= 1 + 20\%)$$

For this example, the reader can verify that in all other cases with A and B between 0.8 and 1.2, estimated lifetime will be between 90 and 124. If the company wanted to offer a warranty figure on lifetime, more tests in this range and a more careful statistical analysis[34] should be performed. However, this kind of calculation could be applied to any experiment of the nature described here to obtain a rough idea of whether existing process controls are adequate.

11.4.5 Dot Diagram of Estimated Effects

One question often asked of effects calculated as in the previous section is whether any might be larger than would be expected from experimental error. One approach to this question is simply to plot the estimated effects (or regression parameters) on an axis as in Fig. 11.3. From this, it is instantly obvious that the estimated effects for A, B, and AB are all larger than 40, while all other

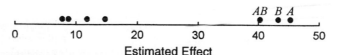

FIGURE 11.3 Dot diagram of estimated effects for SKF ball bearing experiment.

estimated effects are within ± 15. The large separation of these effects from the others supports the tentative conclusions already reached in discussing Figs. 11.1 and 11.2.

There are many other methods for analyzing data from designed experiments.[35–39] If multiple observations are taken under some of the test conditions, an analysis of means could be used; see Sec. 11.5.1.

11.4.6 Interpreting the Results

The analysis of the SKF ball bearing experiment revealed three effects (*A, B,* and *AB*) of quite different magnitude from the remaining effects. It will not always happen that the results are so evident, but neither is this unusual for a designed experiment. Heat treatment and osculation both had large main effects, and the interaction of these factors was equally large. Because the interaction is large, we know that the main effects cannot be interpreted by themselves, and indeed, when we examine the simple main effects, we see that long-lasting bearing assemblies are achieved only when both the heat treatment and the osculation are modified, and that they are achieved with either cage design.

11.4.7 Checking Assumptions and the Model

Our analysis and interpretation of the results of two-level factorial experiments relies on the accuracy of a few underlying assumptions concerning the behavior of the experimental error. In particular, we usually assume that the deviations from the average lifetimes are uncorrelated and have a constant standard deviation, not influenced by the experimental design factors. This assumption could be violated if, for example, the heat treatment used for the ball bearing experiment had drifted significantly during the course of preparing the inner rings and if the ball bearings were assembled in the standard order listed in Table 11.3. As another example, if the balls were made in four batches, with one of the four batches having an appreciably longer life than the other, and those balls were placed in the bearings with modified heat treatment and modified osculation, the lifetime tests might appear as in Fig. 11.1.

A standard technique for reducing the chances of being bothered by such problems is to randomize the assignment of materials (e.g., balls, cages, inner and outer rings) to bearings; the assignment of bearings to test fixtures, conditions, and measurement equipment; and/or the order in which equipment is tested. (By randomization, we mean assignment by use of random numbers.) The ultimate answer with any experiment is obtained from follow-on testing of the results of the experiment to ensure that you get substantially what was predicted based on the results of the experiment.

11.5 OTHER DESIGNS USEFUL IN RELIABILITY

Up to this point, we have primarily discussed the design and analysis of simple two-level experiments. In some cases, it is common to experiment with factors at more than two levels. This can occur when trying to decide between, for example, alternative vendors or materials, or when nonlinearities are suspected. These two cases are discussed in this section.

11.5.1 Analysis of Means (AOM)

Ellis Ott[40] described an experiment to evaluate the effect of copper on the corrosion of metal containers storing a certain substance. Reagents with 5, 10, and 15 PPM (parts per million) copper were stored in 240 metal containers (80 containers at each of the three levels of copper). Later, the containers were examined for the presence or absence of corrosion. The results are summarized in Table 11.4. If the numbers of corroded containers had been at three different points in time

TABLE 11.4 Results of Corrosion Experiment

Level of copper (PPM)	Number of failed containers (out of 80)
5	14
10	36
15	47

rather than three different concentrations of copper, it would be natural to make a *p* chart of these data. The use of statistical control chart calculations when the horizontal axis is not time is called an *analysis of means.*[41]

Figure 11.4 is the analysis of means suggested by Table 11.4. *UCL* and *LCL,* the upper and lower control limits, are 3-sigma limits from the standard control chart calculations. If the horizontal axis were time, we would say that a nonrandom appearance such as a point outside the control limits would suggest that the process changed from one time period to the next.

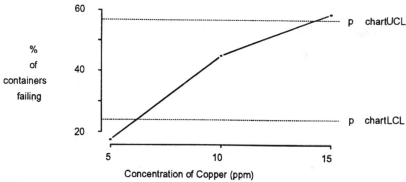

FIGURE 11.4 *p* chart analysis of means of corrosion.

In Fig. 11.4, since two of the three points are outside the limits, it suggests that increasing the copper concentration increases the corrosion. Moreover, a comparison of the first point with the last point provides an estimate of how much the corrosion increases with a given increase in copper concentration.

If all the points had been inside the control limits, we would want to know how small a difference could be detected by this experiment. One simple answer to this question is *UCL* − *LCL:* Differences appreciably less than *UCL* − *LCL* are so small they would not be likely to be detected by this experiment. Differences appreciably greater than *UCL* − *LCL* would be likely to be detected. We say "likely" in discussing the results of experiments because any engineering evaluation is subject to random variation. Therefore, the experimental method occasionally suggests a difference when none exists (a Type I error), and at other times suggests no difference when there is one (a Type II error).

There are three ways to reduce *UCL* − *LCL* if it is too large:

1. Increase the number of observations for each point (in this case, the number of containers at each concentration).
2. Reduce the random variability in individual observations, e.g., by getting a more precise measurement of the amount of corrosion.
3. Make the alternatives more distinct, e.g., by testing 5, 25, and 100 PPM, rather than 5, 10, and 15.

Interpretation of Analysis of Means (AOM).   The information from Fig. 11.4 might be used in two ways by the company that makes the containers. First, the company might include with its information for customers a warning that the container is not rated for storing reagents like those in this experiment unless the reagent has less than, say, 0.1 PPM copper. Before doing that, it would be wise to repeat the test at lower concentrations of copper to make sure that the apparent linearity in Fig. 11.4 continues to those lower concentrations. Second, the company might use this information to justify development of a container that was insensitive to copper-mediated corrosion.

Analysis of Means with $\overline{X}$–S Charts.   If corrosion could have been measured on a scale from 1 to 5, say, rather than simply as yes vs. no, it might have been possible to perform this experiment with appreciably fewer than 240 containers. Data of that nature could be analyzed using an $\overline{X}$–R or $\overline{X}$–S chart. Of course, if precision metrology were available but very expensive, while the containers were cheap, it might still make sense to do precisely this experiment.

Control chart techniques can be used to analyze data from many experiments. For example, an engineer may get samples of material or components from several vendors. A control chart can be made with one point for each vendor. If tests are made on only a few samples from one vendor and on many from another, the data can still be analyzed using an $\overline{X}$–S chart that lets the control limits vary with the sample size. Some statistical packages (e.g., Minitab) make control charts for $\overline{X}$ and S that adjust the control limits for the number of observations in each subgroup.[42] An analysis of means can be performed with data from any experiment, including a 2^k, if multiple observations are performed under some of the sets of experimental conditions.

In industry, one should usually consider both average and standard deviation. For example, suppose an engineer obtained samples of fasteners from two vendors and tested the tensile strength. Vendor A averaged 100 lb with a standard deviation of 10 lb. Vendor B averaged 120 lb with a standard deviation of 20 lb. Assuming that these numbers are representative of future production (and the numbers are normally distributed), the weakest fastener in a batch of 1000 will be roughly three standard deviations below the average. This is 70 lb for vendor A and 60 lb for vendor B. If fasteners weaker than 60 lb are likely to cause problems, one could expect fewer problems with vendor A than with vendor B, even though B has a larger average strength. In general, it is wise to look at the S or R chart as well as the $\overline{X}$ chart.

11.5.2 Multifactor Experiments with More Than Two Levels

Small two-level experiments are powerful and relatively inexpensive ways to address many engineering questions. They provide much more information than the same number of tests replicating only one set of conditions. Occasionally, however, two-level experiments run into trouble because

1. Materials or components used in production are different from the prototypes tested.

2. The response measured is highly nonlinear in a quantitative variable in the test.

In this section, we describe a disaster story that resulted from testing only two levels and how to avoid such situations easily and cheaply.

A Disaster Showing the Weakness of Two-Level Experiments.   A leading producer of biochemical assay equipment designed a specialized piece of equipment that generated a potentially corrosive aerosol that sometimes coated the inside of the device during use. The company tested the unit at high and low usage and found no problems. It started making units and shipping them to customers. Six months later, a reliability problem emerged: Parts inside the unit were being eaten by the corrosive aerosol.

What went wrong? Were production units so different from the prototypes tested? After some investigation, the company found that the corrosion process occurred where air contacted the edge

of an aerosol droplet on metallic surfaces. At low volumes, there were not enough aerosol particles to worry about. At high volumes, the inside of the unit became so coated as to virtually eliminate the air-aerosol-metal interface and thereby the corrosion. Only at medium usage were there
problems.

Cheap Insurance: Add a Center Point. This corrosion disaster is important because it illustrates a broader phenomenon: Two-level designs are great because most situations are approximately linear. Occasionally, however, the situation is highly nonlinear, as in this case. Without
getting into the details of the original test plan, we assume here that it was a 2^3, which has 8 runs.
(The ball bearing example of Fig. 11.1 is a 2^3.) We assume that one of the three factors was usage
rate. If the project team suspects that there might be nonlinearities, an obvious alternative might
be to test each of the three factors at three levels, giving a 3^3 design. However, this experimental
design requires 27 runs. This is more than three times as many runs as a 2^3, and in many cases this
design does not provide much more information than is available from a 2^3.

Instead, cheap insurance against nonlinearities can be obtained by starting with a design with
a center point (cp), as illustrated in Fig. 11.5, and adding other runs later as needed.

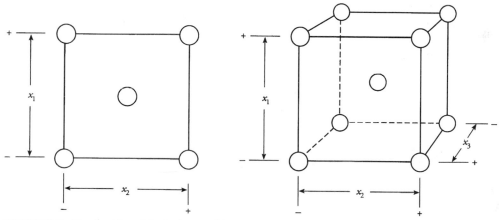

FIGURE 11.5   Two-level factorial experiments with center points.

Designs with center points will detect nonlinearity, but they won't tell which experimental factor was the source of the nonlinearity. However, appreciable nonlinearities are sufficiently rare
that it is usually good management to follow a sequential approach. First run a two-level design
with a center point. Then, if nonlinearity is detected, other tests can be performed to determine
which factor(s) are responsible for the nonlinearity. In most cases, these other tests are not needed. A sequential strategy allows the organization to spend money on these extra tests only when
they are needed.

The search for nonlinearities is often sharpened by performing multiple runs at the center
point. The replicated runs also provide an independent estimate of the error or random variability in the experiment.

Central Composite Designs. A 2^3 + cp includes 9 of the 27 runs in a 3^3. When nonlinearities
are detected, one might be inclined to perform the other 18 of the 27, thereby obtaining a full 3^3.
Much can be learned from a 3^3 experiment. In many engineering applications, however, one can
learn almost as much more quickly and cheaply by employing a sequential strategy,[43] as just
described. In the rare cases in which appreciable nonlinearity is indicated in an initial experiment,
many seasoned experimenters will then run only the points on the faces of the cube or "star

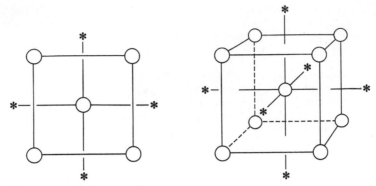

FIGURE 11.6   Central composite designs.

points" that lie just outside the cube, as illustrated in Fig. 11.6, while replicating the center point to guard against a shift in materials or metrology that might have occurred between the running of the 2^3 + cp and the face or star points. These are called *central composite designs* because they are constructed as a composite of a two-level design with a center point and a separate experiment with only the face or star points.

With three factors, the central composite design includes 16 runs, 9 with the 2^3 + cp, 6 face or star points, and (at least) one replication of the center point, as just mentioned. This is still 11 fewer runs than a full 3^3. Not only is the sequential strategy cheaper than the full 3^3, but the runs on the face or star points are rarely performed in practice because other ideas occur to the experimenters in the course of the initial 2^3 + cp that convince them either to stop there or to spend their remaining experimental budget in other ways.

A central composite design provides enough data to estimate a full quadratic in all factors. Many software packages include regression analysis capabilities that can analyze data from a central composite design fairly easily. For more information on the design and analysis of such experiments, see the books by Box, Hunter, and Hunter[44] and by Box and Draper.[45]

2 × 2 + cp with Five to Nine Points.   The simplest design that allows for interaction and a check for nonlinearity is a 2 × 2 + cp (where cp = center point, as in Fig. 11.5). Designs of this type require five or more runs; various alternatives with five to nine runs are outlined in Table 11.5. Whenever feasible, the order of production and assembly should be randomized; similarly, if the testing is not concurrent, then the order of testing should also be randomized. If randomization is not employed, then any tool wear or drift in calibration of test instruments can be confounded with the effect of one of the factors, making it look as if there is an effect when there is none or vice versa. When the experiment is not randomized, it should at least be blocked,[46] as discussed in Sec. 11.3.6.

TABLE 11.5a   2×2 + cp with 5 to 9 Points

Number of runs	Design
5	2×2 + cp
6	2×2 + 2cp
7	2×2 + 3cp or 2×2 + cp + 2^{2-1}
8	2×2 + 4cp or 2×2 + 2cp + 2^{2-1}
9	2(2×2) + cp

TABLE 11.5b $2 \times 2 + k(\text{cp})$

Run number (standard order)	k	Factor 1	Factor 2	
1		−	−	⎫
2		−	+	⎬ 2×2
3		+	−	⎭
4		+	+	
5	1	0	0	⎫
6	2	0	0	⎬ cp
7	3	0	0	⎭
8	4	0	0	

TABLE 11.5c $2 \times 2 + k(\text{cp}) + 2^{2-1}$

Run number (standard order)	k	Factor 1	Factor 2	
1		−	−	⎫
2		−	+	⎬ 2×2
3		+	−	⎭
4		+	+	
5		−	−	⎫ 2^{2-1}
6		+	+	⎭
7	1	0	0	⎫ cp
8	2	0	0	⎭

TABLE 11.5d $2(2 \times 2) + \text{cp}$

Run number (standard order)	Factor 1	Factor 2	
1	−	−	⎫
2	−	−	
3	−	+	
4	−	+	⎬ 2(2×2)
5	+	−	
6	+	−	
7	+	+	
8	+	+	⎭
9	0	0	cp

11.6 OTHER ANALYSIS TECHNIQUES

In this section, we alert the reader to other approaches to analyzing the data from a reliability experiment which can sometimes squeeze an extra 10 to 60 percent of information from a given experimental budget. Some of the more sophisticated techniques for reliability data analysis require extensive study to understand how to use them. They are appropriate when a reduction of 10 to 40 percent in the cost or duration of the experiment could pay the consulting fee of a good industrial statistician who is knowledgeable in these specialized techniques. We begin this section with some simpler things that an engineer can do with lifetime data.

11.6.1 Time to Failure versus Failures per Unit Time

With lifetime experiments, it often makes sense to analyze some transformation of the lifetimes. In many cases, lifetimes seem to be lognormally distributed, i.e., the logarithm of the lifetime seems to follow a normal distribution.

In other cases, people find it more sensible to analyze the reciprocal of lifetime, failures per unit time. In many cases, an analysis of time to failure (lifetime) identified significant interactions, while analyzing failures per unit time (1/lifetime) indicated only that the main effects were important. The logic of this can be seen as follows:

$$\frac{1}{b_0 + b_1 x_1 + b_2 x_2} = a_0 + a_1 x_1 + a_2 x_2 + a_{12} x_1 x_2 + \text{remainder}$$

The reciprocal of a simple model with only main effects ($b_1 x_1$ and $b_2 x_2$) can be expanded in a Taylor series, which will generally have a nonzero cross-product or interaction term ($a_{12} x_1 x_2$).

In addition, the deviations from the model of failures per unit time often have more nearly constant variance than deviations of time to failure. Since constant variance of deviations is one of the assumptions of many analysis techniques, this is another reason to analyze failures per unit time rather than time to failure.

Occasionally, people find justifications for analyzing other transformations of the lifetime, like the reciprocal square root of lifetime.[47] Various transformations of the ball bearing data of Fig. 11.1 were analyzed, including logarithm, reciprocal, and reciprocal square root of lifetime. For this particular data set, all analyses gave essentially the same answer. It is often wise to perform multiple analyses of the same data. If one gets the same answer from the different analyses, one can feel more confident in the results and select the particular analysis that communicates those results in the simplest and most direct way. If one gets different answers, one must then think about which method seems to be most appropriate. If, for example, one model does not have an interaction while others do, the simpler model is often preferred.

11.6.2 Censoring

Censoring means that the failure time is known if it is small; however, if the unit is fairly reliable, we know only that it outlived the test period. Actual failure times are recorded for all units failing on test, while the fact of censoring and the censoring time are recorded for all other units. In other, so-called quantal response situations, test units are inspected every day or week and failure times are known only to the interval in which they occurred.

Censored and quantal response data present complications for both planning and analysis of experiments. For example, the relative widths of confidence intervals for parameter estimates will depend on the number of failures observed. For selecting sample sizes, a conservative approach is to pretend the response is yes or no: did or did not fail within the experimental period; see Sec. 11.3.2.

Many reliability engineers and managers will not have the time to master these techniques. In many reliability experiments, it suffices to increase the sample size and test time by 50 to 100 percent and use simpler analysis techniques. When the cost of the experiment is high or the value of early completion is sufficient, a logical option is to increase the budget for analysis in order to reduce the budget for testing. This could entail a variety of analysis techniques,[48] including a variety of graphical techniques and so-called maximum likelihood estimation with log-linear regression models.[49] In such cases, it is often appropriate to hire a consultant with demonstrated competency in the design and analysis of reliability experiments to help design the experiment and analyze the results within your time and budget constraints, as discussed in the next section.

Of course, in the rare industrial situation where one is willing to test to determine the entire lifetime distribution, not just the failures anticipated under warranty, the drastic (e.g., 60 to 80 percent) censoring suggested here is not appropriate.

11.6.3 When to Hire a Consultant

In general, there are two approaches to dealing with technical questions beyond one's knowledge:

1. Try different options, e.g., different analysis techniques or different data transformations. Often, one analysis will give a clearer answer than the others, but the others will still suggest the same conclusion, or at least won't contradict it severely. In such cases, it hardly matters which analysis method (e.g., using lifetimes or failures per unit time) is "correct."

2. Hire an expert. This is preferred if a 20 to 40 percent savings in the testing budget (in time or money) could pay the consultant's fee.

A third option, elaborating on (1) when (1) is equivocal, is to run more experiments.

11.7 WARRANTY DATA

It is always wise to compare reliability predictions with the actual performance of the product. This can help engineers evaluate the procedures they use to estimate reliability and can be of great value in designing future products. This is especially critical with accelerated testing. A comparison of the results of accelerated tests with field failures can help engineers and managers decide how aggressively they can pursue accelerated testing in future product development projects.

Engineering, including reliability engineering, involves assumptions. Critical assumptions may be wrong, either initially or through changes during production. Sometimes an unadvertised change by a vendor, for example, is identified via reliability problems.

Many companies have systems for obtaining information about the reliability of their products in actual use. Companies that do not have extensive warranty tracking facilities often use some of their own products internally and can track those units carefully to help confirm or improve their reliability evaluation practices.

The importance of this is illustrated by a recent report by Joe Juran[50] of a major U.S. company that invented an important product and dominated that market for years. New competitors entered the market, defeated the inventor's patents, and came to dominate the market with products of lower cost and higher quality and reliability. With the company facing possible bankruptcy, the executive team for the early leader called Juran. He asked to see the warranty data. They found that year after year, the new models added new features while perpetuating the dominant failure modes of previous models. Meanwhile, competitors had designed many of the previous failure modes out of subsequent models. The service department of the early leader was a profit center, whose "profits" had helped to blind senior executives to the need to design more reliable products. The early leader finally understood the problem, began designing more reliable products, and has since regained profitability and market share.

11.8 SUMMARY

The main point of this chapter is that virtually any reliability experiment that does not systematically vary design or environmental factors is wasting resources: The experimenters often learn less than they could for the same budget and may not get as much information as they think (Sec. 11.1).

In this chapter, we discussed reliability improvement through experimentation (Sec. 11.2), various experimental designs (Secs. 11.3 and 11.5), and analysis strategies (Secs. 11.4 and 11.6). It was our goal that a typical reliability engineer could, with this chapter,

1. Design and analyze simple reliability experiments using the techniques discussed in Secs. 11.1 to 11.5.

2. Know when the situation would justify seeking outside help (Sec. 11.6).

11.9 REFERENCES

1. G. Stalk and T. M. Hout, *Competing Against Time,* Free Press, New York, 1990.

2. W. I. Zangwill, *Lightning Strategies for Innovation,* Lexington Books, New York, 1993.

3. G. J. Hahn, "The Design of Experiments: Broadening the Perspective," presentation at the Annual Meeting of the American Statistical Association, General Electric, Schenectady, N.Y. (Aug. 17, 1994).

4. W. Nelson, *Accelerated Life Testing,* Wiley, New York, 1990.

5. L. W. Condra, *Reliability Improvement with Design of Experiments,* Marcel Dekker, New York, 1993.

6. "GE Refrigerator Woes Illustrate the Hazards in Changing a Product," *Wall Street Journal,* May 7, 1990, pp. A1, A5.

7. L. W. Condra, *Reliability Improvement with Design of Experiments,* Marcel Dekker, New York, 1993. This book is an excellent follow-on to this chapter, with much good practical advice. It contains a minor flaw in that it asserts that "until recently, the only effective application of [U.S.- and British-style design of experiments, as opposed to "experiments"] was in agriculture." In fact, experimental design was used effectively in the Guiness Brewery in Dublin, Ireland, in the 1920s, in the British textile industry in the 1930s, and in Imperial Chemical Industries in the 1940s. DuPont, Eastman Kodak, General Electric, the Bell Telephone System, and various food companies have conducted numerous U.S.- and British-style designed experiments over the past 40 years, as have the numerous clients of Dorian Shainin (K. R. Bhote, *World Class Quality,* American Management Association, New York, 1991). See S. Bisgaard, "The Early Years of Designed Experiments in Industry," *Quality Engineering,* **4**(4):547–562 (1992). For early applications of designed experiments to reliability and life testing, see G. E. P. Box, "Problems in the Analysis of Growth and Wear Curves," *Biometrics,* **6**(4):362–389 (1950); M. Zelen, "Factorial Experiments in Life-Testing," *Technometrics,* **1**(3):269–288 (1959); A. J. Hitzelberger, "Improve Your Reliability," *Industrial Quality Control,* **24**(6):313–316 (1967); G. E. P. Box and D. R. Cox, "An Analysis of Transformations," *Journal of the Royal Statistical Society,* ser. B, **26**:211–252 (1964). The only deficiency in Condra's book of any practical importance known to the present authors is his advocacy of three-level orthogonal arrays such as L_{18}. Since G. E. P. Box and K. B. Wilson, "On the Experimental Attainment of Optimal Conditions," *Journal of the Royal Statistical Society,* ser. B, **13**, 1–45 (1951), it has been known that these three-level designs make generally less efficient use of data than the sequential strategies and central composite designs discussed in Sec. 11.5. When three-level designs are compared with the sequential strategy advocated by Box and Wilson and discussed in Sec. 11.5, the difference in efficiency can be substantial. On the other hand, this loss of efficiency is negligible in comparison to the gains typically achieved from using a three-level design over the traditional one-factor-at-a-time approach too often used in engineering experiments.

8. S. Bisgaard, "A Conceptual Framework for the Use of Quality Concepts and Statistical Methods in Product Design," *J. Eng. Design,* **3**(1):31–47 (1992).

9. S. B. Graves, W. P. Carmichael, D. Daetz, and E. Wilson, "Improving the Product Development Process," *Hewlett-Packard Journal,* 71–76 (June 1991).

10. C. Hellstrand, "The Necessity of Modern Quality Improvement and Some Experience with Its Implementation in the Manufacture of Roller Bearings," *Phil. Trans. of the Roy. Soc.,* **A 237**:529–537 (1989).

11. W. I. Zangwill, *Lightning Strategies for Innovation,* Lexington, Macmillan, New York, 1993, p. 292.

12. W. E. Deming, *Out of the Crisis,* MIT CAES, Cambridge, Mass., 1986, p. 132.

13. Ibid.

14. R. D. Moen, T. W. Nolan, and L. P. Provost, *Improving Quality through Planned Experimentation,* McGraw-Hill, New York, 1991; W. E. Deming, *Out of the Crisis,* MIT CAES, Cambridge, Mass., 1986, p. 132.

15. M. E. Porter, *The Competitive Advantage of Nations,* Free Press, New York, 1990.

16. P. F. Drucker, *Innovation and Entrepreneurship,* Harper & Row, New York, 1985.

17. K. B. Clark and T. Fujimoto, *Product Development Performance,* Harvard Business School Press, Boston, 1991, pp. 319–324.

18. G. J. Hahn, "The Design of Experiments: Broadening the Perspective," presentation at the Annual Meeting of the American Statistical Association, General Electric, Schenectady, N.Y. (Aug. 17, 1994).

19. R. E. Odeh and D. B. Owen, *Tables for Normal Tolerance Limits, Sampling Plans and Screening,* Marcel Dekker, New York, 1980.

20. E. R. Ott, *Process Quality Control,* McGraw-Hill, New York, 1975, pp. 106–107.

21. G. E. P. Box, W. G. Hunter, and J. S. Hunter, *Statistics for Experimenters,* Wiley, New York, 1978.

22. Ibid.

23. G. E. P. Box, "Sequential Experimentation and Sequential Assembly of Designs," *Quality Engineering,* **5**(2):321–330 (1993).

24. G. E. P. Box, W. G. Hunter, and J. S. Hunter, *Statistics for Experimenters,* Wiley, New York, 1978.

25. S. Bisgaard and H. T. Fuller, "Sample Size Estimates for Two-Level Factorial Experiments with Binary Responses," *Center for Quality and Productivity Improvement Report No. 91,* University of Wisconsin, Madison, 1992; to appear in *Journal of Quality Technology.*

26. S. Bisgaard and H. T. Fuller, "Analysis of Factorial Experiments with Defects and Defectives as the Response," *Center for Quality and Productivity Improvement Report No. 119,* University of Wisconsin, Madison, 1994; to appear in *Quality Engineering.*

27. V. N. Nair and D. Pregibon, "Analyzing Dispersion Effects from Replicated Factorial Experiments," *Technometrics,* **30**(3):247–257 (1988).

28. G. E. P. Box and D. Meyer, "An Analysis for Unreplicated Fractional Factorials," *Technometrics,* **28**(1):11–28 (1986).

29. G. Taguchi (G. Taguchi and Y. Wu, *Introduction to Off-Line Quality Control,* Central Japan Quality Control Association, Nagoya, Japan, 1985) has recently popularized the idea of selecting the value of product design parameters to make the product robust to variations in environmental factors. As S. Bisgaard noted ["The Early Years of Designed Experiments in Industry," *Quality Engineering,* **4**(4):547–562 (1992)], the use of designed experiments to design products robust to environmental variation was advocated as early as 1964 by S. E. Michaels, "The Usefulness of Experimental Design," *Applied Statistics,* **13**(3):221–235 (1964). The concept has been used for some time in metrology, where it is known as "ruggedness"; see G. T. Wernimont, *Use of Statistics to Develop and Evaluate Analytical Methods,* Association of Official Analytical Chemists, Arlington, Va., 1985. See also G. E. P. Box and C. Fung, "Studies in Quality Improvement: Minimizing Transited Variation by Parameter Design," *Center for Quality and Productivity Improvement Report No. 8,* University of Wisconsin, Madison, 1986; G. E. P. Box and C. Fung, "Is Your Robust Design Robust?" *Center for Quality and Productivity Improvement Report No. 101,* University of Wisconsin, Madison, 1993; S. Bisgaard, "Designing Experiments for Tolerancing Assembled Products," *Center for Quality and Productivity Improvement Report No. 99,* University of Wisconsin, Madison, 1993; G. Box and S. Jones, "Designing Products that Are Robust to the Environment," *Center for Quality and Productivity Improvement Report No. 56,* University of Wisconsin, Madison, 1990; D. M. Steinberg and D. Burnsztyn, "Noise Factors, Dispersion Effects and Robust Design," *Center for Quality and Productivity Improvement Report No. 107,* University of Wisconsin, Madison, 1993; V. N. Nair, "Taguchi's Parameter Design: A Panel Discussion," *Technometrics,* **34**(2):127–161 (1992).

30. G. E. P. Box, W. G. Hunter, and J. S. Hunter, *Statistics for Experimenters,* Wiley, New York, 1978.

31. G. E. P. Box and S. Jones, "Split-Plot Designs for Robust Product Experimentation," *Journal of Applied Statistics,* **19**:3–26 (1992).

32. G. E. P. Box, W. G. Hunter, and J. S. Hunter, *Statistics for Experimenters,* Wiley, New York, 1978.

33. S. Bisgaard, *A Practical Aid for Experimenters,* Starlight Press, Madison, Wis., 1988.

34. S. Bisgaard, "Designing Experiments for Tolerancing Assembled Products," also *Center for Quality and Productivity Improvement Report No. 99,* University of Wisconsin, Madison, 1993; to appear in *Technometrics.*

35. C. Daniel, *Application of Statistics to Industrial Experimentation,* Wiley, New York, 1976.

36. Box, Hunter, and Hunter, *Statistics for Experimenters,* Wiley, New York, 1978.

37. G. Box, "Finding Bad Values in Factorial Designs," *Quality Engineering,* **3**(3):405–410 (1990–91).

38. S. Bisgaard, *A Practical Aid to Experimenters,* Starlight Press, Madison, Wis., 1988.

39. D. J. Wheeler, *Tables of Screening Designs,* 2d ed., SPC Press, Knoxville, Tenn., 1989.

40. E. R. Ott, *Process Quality Control,* McGraw-Hill, New York, 1975, pp. 106–107.

41. Ibid.

42. D. C. Montgomery, *Introduction to Statistical Quality Control,* 2d ed., Wiley, New York, 1992, pp. 235–241.

43. G. E. P. Box, "Sequential Experimentation and Sequential Assembly of Designs," *Quality Engineering,* **5**(2):321–330 (1993).

44. Box, Hunter, and Hunter, *Statistics for Experimenters,* Wiley, New York, 1978.

45. G. E. P. Box and N. R. Draper, *Empirical Model Building and Response Surfaces,* Wiley, New York, 1987.

46. Box, Hunter, and Hunter, *Statistics for Experimenters,* Wiley, New York, 1978.

47. G. E. P. Box and C. A. Fung, "The Importance of Data Transformation in Designed Experiments for Life Testing," *Center for Quality and Productivity Improvement Report No. 121,* University of Wisconsin, Madison, 1994; to appear in *Quality Engineering.* See also G. E. P. Box and D. R. Cox, "Analysis of Transformations," *Journal of the Royal Statistical Society,* ser. B, **26**(2):211–252 (1964).

48. W. Nelson, *How to Analyze Reliability Data,* American Society for Quality Control, Milwaukee, Wis., 1983, or W. Nelson, *Applied Life Data Analysis,* Wiley, New York, 1982.

49. J. D. Kalbfleisch and R. I. Prentice, *The Statistical Analysis of Failure Time Data,* Wiley, New York, 1980; M. Hamada and S. K. Tse, "On Estimability Problems in Industrial Experiments with Censored Data," *Statistica Sinica,* **2:**381–392 (1992); M. Hamada and C. F. J. Wu, "Analysis of Censored Data from Highly Fractionated Experiments," *Technometrics,* **33:**25–38; I. Gertsbakh, "Orthogonal Design of Life Testing with Replacement: Exponential Parametric Regression Model," *Center for Quality and Productivity Improvement Report No. 112,* University of Wisconsin, Madison, 1994.

50. J. M. Juran, "Made in USA: A Renaissance in Quality," *Harvard Business Review,* pp. 42–48 (July–August 1993).

CHAPTER 12
ACCELERATED TESTING

Charles Schinner
HEWLETT-PACKARD COMPANY

12.1 DEFINITION OF THE ACCELERATED TESTING PROCESS

Throughout our school years, we have been conditioned to associate the word *test* with passing or failing. One of the dictionary definitions of *test* is "to subject to trial and examination." When applied to the product development process, *test* could mean "to achieve certain measurable criteria." If the criteria are achieved, the unit *passes*; if not, the unit *fails.* This type of test results in a binary decision.

Accelerated testing is not something a product passes or fails in this sense. Instead, it is a process which attempts to transform hidden (latent) defects into detectable failures by the use of stresses more severe than those expected in the user environment. Accelerated testing during product development is *intended* to cause failures. When this is done in conjunction with failure analysis and corrective action, the product's reliability can be increased. In the next few pages, we'll take a closer look at the implications of this strategy, and how to implement it.

12.1.1 Simulation versus Stimulation

The intent of many traditional test plans is to *simulate* (reproduce) the product's use environment. This is accomplished by measuring or estimating the stress levels experienced by the product in normal use, and then duplicating them. For example, nonoperating temperature test limits may be determined by shipping temperature recording instrumentation along with a product, in the same container. As long as the recording device is small in comparison to the product and is in a container of the same or similar physical size and packaging materials, results will be reasonably accurate. After shipping the container to several different locations (to increase sample size), a composite temperature profile may be compiled. During product qualification, a new product can be subjected to this profile to verify its performance when exposed to the expected shipping stresses.

This type of testing can be very time-consuming and does not yield quantifiable information about a product's design margin. (In this context, *margin* is defined as the difference between the most extreme stress the product experiences in use and the least extreme stress which will cause the product to fail immediately.) Variability of manufacturing processes, at both the component and product levels, can decrease design margin to the point where reliability is degraded. On the other hand, an overdesigned product (with excessive margin) may result in higher costs (and prices), allowing the competition an opportunity to gain market share at your expense.

One of the objectives of accelerated testing is to determine the product's strength limits by applying stresses high enough to *stimulate* (excite) failures. In recognition of the fact that not all use environments subject a product to the same (much less constant) stress levels, and that there are always unit-to-unit variations in the robustness of products, stress and strength are usually graphed as probability distributions. Each component's parameters will vary over some range of values, resulting in variations in robustness and performance from product to product. The stresses experienced by these components will also vary, depending on their application in each product and the end use environment.

In the ideal case, there would be no overlap in the product strength and use stress distributions—the weakest product would be able to withstand the most severe use environment. Failures would occur only when the applied stress exceeded the product's strength. (One can argue about whether failure occurs instantaneously or gradually. Whichever is the case, more failures occur as stress increases.)

As a product ages, various wearout failure mechanisms (mechanical, electrical, and chemical) are at work, and a product's strength distribution gradually changes, usually toward less robustness—products become weaker. Eventually, the two distributions overlap significantly, and within the overlapping region, failures are very likely (Fig. 12.1). This deterioration may require many hours to occur under normal user environmental conditions, and can be detected by the traditional reliability test methods—long-duration, constant-stress operation.

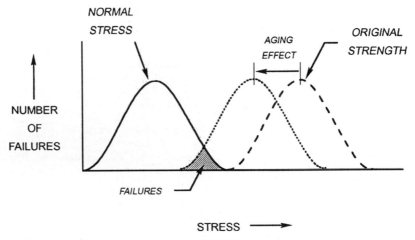

FIGURE 12.1   Failures increase as strength decreases.

To force an overlap more quickly, the applied stress distribution may be shifted toward the right, so that an overlap in the two distributions occurs immediately (Fig. 12.2). If this is done correctly, with knowledge of the technologies involved, it is possible to discover the location and nature of the lower edge of the product strength distribution very quickly. The test time to verify adequate design margin can be reduced by orders of magnitude, and fewer test samples are required than in a normal stress life test.

12.1.2 Acceleration

The primary purpose of the *accelerated* testing process is to achieve reliability improvement as quickly as possible. Because we don't know the precise nature of a product's weaknesses ahead of time, we must resort to the application of an assortment of stresses. Our basic assumption is that subjecting a product to elevated stress will cause failures to occur more quickly. A classic example of this is chemical reactions, where reaction rate has been found to increase exponentially with temperature (the Arrhenius relationship). Therefore, accelerated testing can be called a productivity tool, because when a small sample of a product population is subjected to multiple, elevated stresses, a larger number of failures will occur in a shorter time.

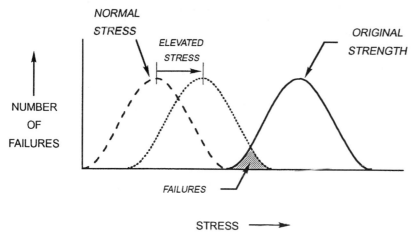

FIGURE 12.2 Failures increase as stress increases.

12.1.3 Accelerated Testing Process

The purpose of accelerated testing is to discover and eliminate the *root causes* of the failures which occur during the evaluation process. (Use of the word *evaluation,* instead of *test,* is intended to emphasize this. The accelerated evaluation testing process is not something you pass or fail—it is intended to *discover* and *evaluate* a product's weaknesses.) Experience shows that every new product contains at least one unwanted failure mechanism, with many products containing more than one. It is not desirable to wait for the customer to discover these failure mechanisms in use, because of the high cost to both producer and user. Perhaps a perfect product has been designed and produced sometime in the past, but that is becoming increasingly unlikely with today's rapidly changing, complex technologies and manufacturing processes, and especially with the increasing pressure to get new products to market quickly, reducing the time available for testing.

The adage that "a chain is only as strong as its weakest link" really applies in the field of reliability engineering. It is particularly applicable to products that don't use redundancy to ensure product performance after a single component or subsystem failure (known as "reliabilitywise-in-series" systems). Such systems require *all* components and subsystems to operate properly if the system is to perform correctly. Commercial products are usually of this type, avoiding redundancy for cost reasons. Accelerated stress testing exposes their weak links, which are usually the causes of failures. The goal is to remove these weak links through the application of stress testing to failure, root cause analysis, corrective design changes, and reevaluation. With each iteration, the product's reliability will increase.

This highlights an interesting dilemma. Contrary to our culture and education, in which failure is to be avoided, failures during accelerated testing are *desired!* Failures are data, and so no failures mean no data. During the evaluation testing process, more can definitely be learned from failures than from success!

12.2 PRODUCT DEVELOPMENT PROCESS

Accelerated testing should be employed *during* the development phases of a new product, so that changes in design and manufacturing processes can be effected with minimum cost and delay. It is important to discover and eliminate the product's failure modes *as early as possible* in design. During the early design phases, the freedom to change the design (to eliminate a failure mecha-

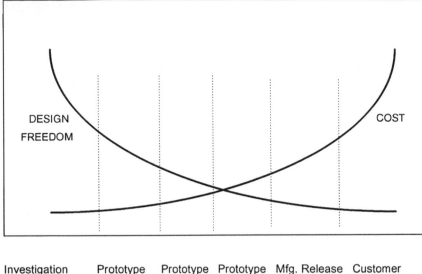

FIGURE 12.3 Design freedom versus cost tradeoff.

nism) still exists (Fig. 12.3). As the start of production approaches, fewer design, component, or manufacturing process changes are possible because they cost too much (retooling) or delay sales introduction. In the worst case (after manufacturing release), it may be impossible to eliminate the failure mechanism by design, and 100 percent screening (inspection test) in manufacturing may be the only alternative. Any screening process, whether at the component level or at the finished product level, adds directly to the cost of the product because it is an additional manufacturing process step. Total product cost includes field returns and warranty repairs, and improved reliability will reduce those costs as well. Therefore, it is *strongly* recommended that prototypes of all new products be subjected to the accelerated testing process during the earliest possible design phase.

12.3 EVALUATE–ANALYZE–CORRECT–VERIFY CYCLE

Reliability growth (improvement) is accomplished by repetitive application of an evaluate–analyze–correct–verify (E–A–C–V) cycle (Fig. 12.4). Unless the design, components, and manufacturing processes are very mature, this cycle will uncover significant reliability improvement opportunities. The first step in the cycle is to decide which stresses to use and at what levels. To do this, consider what stressful environments (if any) the product will have to survive, and what the possible failure modes of the product are. For example, an electronic product intended for use in vehicles must be able to withstand more extreme and rapidly changing temperature, humidity, contaminants (dust and oil), vibration, mechanical shock, and radio-frequency interference than a similar product intended for home or office use. There is no industry standard stress test plan. Each test plan will be different, because of the diversity of products and use environments.

Once the appropriate stresses have been selected, their severity levels must be determined, either by using stress levels commonly discussed in the literature or by developing tailored pro-

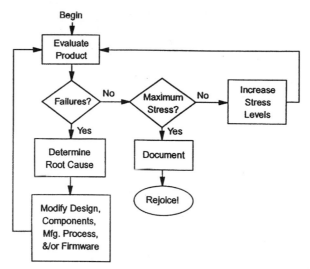

FIGURE 12.4   Evaluate–analyze–correct–verify process flowchart.

files based on measurements or customer requirements. The advantage of starting with commonly used stress levels is that other equipment has successfully survived them, which lends credibility. The profiles presented in this chapter were chosen in this manner.

12.3.1   Process

The intent of this particular accelerated testing philosophy is to eliminate hidden (latent) failure modes. The idea is to avoid sorting (screening) the good from the bad, and instead to improve the overall product reliability *by design*. With any manufacturing screen, additional cost is incurred for every unit tested. This overhead can never be eliminated unless the cause of the defects is eliminated (corrective action), which can be done if the E–A–C–V cycle is used during the product development.

The evaluate–analyze–correct–verify cycle should also be used during the development of the manufacturing processes, to uncover problems caused by them. (Manufacturing process damage is a leading cause of both quality and reliability problems. For example, when surface mount construction was first introduced, many components were overstressed by the rapid temperature changes inherent in reflow soldering, and these damaged components caused reliability problems in the field.) Using E–A–C–V, manufacturing processes can be improved from a reliability and quality standpoint. Once the reliability goals have been achieved, some parts of the testing process may be reduced to sampling plans rather than 100 percent test. Of course, the sample sizes for such plans must be determined using proper statistical methods.

Repeated application of E–A–C–V will facilitate the achievement of acceptable reliability in a short time. Unfortunately, it is very difficult to predict how many iterations will be necessary. To minimize the number of cycles, the maturity of the failure analysis, design, manufacturing, and component processes must be high. The number of evaluation cycles will be determined by your ability to resolve failures to root cause, corrective action implementation time, and the credibility of the evaluation data. To minimize the impact of an unknown number of cycles on the development schedule, E–A–C–V must be started as early in product development as possible.

12.4   DEVELOPING THE ROAD MAP—PHASE ONE

1. Form a multidiscipline team
2. Determine appropriateness of accelerated testing
3. Identify and prioritize reliability risks
4. Determine assembly level for evaluation: component, subassembly, or complete product
5. Decide sample size
6. Select stress stimulus portfolio
7. Select a flaw detection methodology
8. Validate the test system
9. Use a defect tracking system

12.4.1   Form a Multidiscipline Team

A multidiscipline team will be needed to engineer and manage the application of accelerated evaluation testing to a particular product. The team should include, or have access to, the following:

- A program manager to act as a guide through the process and remove any barriers that prevent the team from succeeding
- Engineering resources to perform unit filtering (selection and elimination), failure analysis, diagnostic software development, and implementation of corrective action for both hardware (electrical and mechanical) and software

12.4.2   Determine Appropriateness of Accelerated Testing

First, the team must determine if accelerated testing is appropriate for this product. Accelerated testing will not expose all reliability or product quality problems. For example, most firmware and software (SW) bugs are not detected using physical stresses on the electronic components. However, firmware and software are used during product evaluation, and sometimes bugs will be discovered simply because of that use. Consequently, it is important to make sure that any suspected software failures during hardware evaluation are analyzed and corrected, and also to perform other tests specifically designed to evaluate software and firmware.

Another example of a situation in which accelerated testing may or may not be appropriate is when determining mean time between failures (MTBF). The accelerated testing philosophy presented here utilizes stress levels *far* above those of normal use, and so any calculation of MTBF made from these data will be valid only for those extremely stressful conditions. If the failure rate acceleration factor for the elevated stress is not known, MTBF in actual use conditions cannot be estimated. A related problem is that there is never a single acceleration factor. Each failure mechanism responds to stress differently, and each component of the product has several different inherent failure mechanisms. Using accelerated test data plus a single acceleration factor can result in an MTBF estimate which is erroneous and misleading.

12.4.3   Identify and Prioritize the Major Reliability Risks

Every product should be evaluated for reliability risk. Examples of reliability risks are circuits which switch high power (power supplies and electromechanical component drivers), new design architectures, and new technologies. For each risk type, the appropriate level for evaluation should be determined: component, subassembly, or complete product. Formal techniques such as FMEA or FTA (see Chap. 6) are good tools for doing this.

12.4.4 Determine Appropriate Assembly Level for Evaluation

The level of assembly to be evaluated is usually chosen to reduce the total cost. It may be best to evaluate individual components, subassemblies (power supplies or single printed circuit assemblies), or even the complete product. No one level of assembly will be appropriate every time, because of the diversity of product types and applications. The following guidelines may help in determining the level of assembly for your product, considering engineering time, equipment, and cost of evaluation.

The first consideration is custom test plan design. This includes mechanical, electrical, and software considerations. Depending on what assembly level is selected for evaluation, the cost (in time and money) can be optimized.

For example, consider testing at the component level. To be sure of detecting all major failure mechanisms, a different test setup would be required for each component in the system. Even maximizing leverage by having one test for all resistors, one for all capacitors, and so on, many test setups would be needed in evaluating all the types of components in a product or system. A better solution would be to let the component manufacturers perform component-level testing. They almost certainly have more advanced knowledge of the appropriate test methods, and they have developed cost-effective methods of testing their components.

At the subsystem level, test fixture design can be easy or difficult. Power supplies are usually easy to fixture, because they require only input power, a passive load, and voltage monitoring, and most are not affected by the length of the input or output cable. By contrast, disk drive electronics are generally difficult because they cannot be separated from the disk mechanism by more than a few inches. Unfortunately, the minimum distance possible in a test setup is a function of the equipment being used for imposing stress on the equipment under test. An environmental test chamber (capable of simultaneous temperature, humidity, and sometimes altitude and vibration stresses) typically has a wall thickness of more than 4 in. For a product that is sensitive to cable length, this distance alone could make fixturing impossible.

The physical size of the product may also increase the cost of evaluation. A large product requires larger equipment to impart stress. Consider the differences in size between a hand-held calculator, a television set, and a complete aircraft electronic control system. To be able to apply stresses such as temperature and humidity to each of these products, the environmental facility must grow in size as the physical size and power dissipation of the product increases, and the associated cost of this equipment would also grow. For very large products, it is usually more practical to split the system into smaller subsystems for separate evaluation (see Table 12.1).

TABLE 12.1   Product Assembly Level Versus Evaluation Requirements

Integration level	Cost	Time required	Expertise required	Evaluation recommendation
Component	High	Long	High	Cost-effective when the component manufacturers perform evaluation
Subassembly	Medium	Medium	Medium	Appropriate when system can be partitioned without major interaction, when system's physical size is too large, or when cable length is not a concern
System	Low	Short	Low	Appropriate when system physical size is manageable or there is too much subsystem interaction

12.4.5 Decide Sample Size

If every failure mechanism were present in every product, testing only one unit would be suffi-cient to detect and correct all of them. Unfortunately, because of component and manufacturing process variability, every unit is *not* exactly like every other unit, and more than one must be test-ed to prove the absence of serious failure mechanisms. In real-life tests, some units will fail and others won't. Assuming that the failure mechanism is present in only a small percentage of the units, testing with small sample sizes just doesn't cut it! With a sample size of 2, you have only a 20 percent probability of detecting a defect that occurs in 10 percent of the products. Even test-ing 10 units raises the probability only to slightly over 60 percent. Figure 12.5, based on the bino-mial distribution, shows that a sample size of 22 is necessary to achieve 90 percent probability of detecting a defect that occurs in 10 percent of the population. If you are concerned with achiev-ing really low defect densities, say 1 percent or less, the sample sizes become really large.

Probability of Observing

One or More Defects

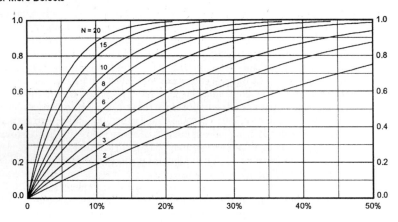

Percent of Products with Defect

FIGURE 12.5   Probability of observing defect, with a given sample size.

Complications.   Barring interaction, the failure rate of a complete product is the sum of the fail-ure rates of its parts. A product with a 5 percent failure rate may be made up of five subassem-blies, each with a 1 percent failure rate. Failure rate is usually not constant with age; a newly manufactured product may exhibit "infant mortality," a higher failure rate than that of a product that has been in use for a year or more, while an old product may also have a higher failure rate because of wearout. Another characteristic of a newly designed product is multiple failure mech-anisms, with 20 percent of the mechanisms accounting for 80 percent of the total failures (the Pareto principle). A mature design will usually have a more equal distribution of number of fail-ures versus individual failure mechanisms.

The obvious consequence of both of both of these characteristics is that you should expect a newly designed, newly manufactured product to have a high initial failure rate with a variety of failure modes. To uncover and eliminate these will probably require several E–A–C–V cycles and a variety of stresses.

12.4.6 Select Stress Stimuli Portfolio

The next step is to decide which stresses to employ, at what intensity levels. Gather as much infor-mation as possible about the potential failure modes of the product and what stressful environments (if any) the product is expected to endure. There is no standard stress stimulus portfolio—each product program will have differences because of the vast diversity of products.

To illustrate the stress selection process, consider an evaluation plan for a typical electronic product which specifies temperature cycling without product power being applied. Since the unit is not operating, no functional failure modes will be discovered. Such failure modes would include firmware bugs, race or other timing problems, power dissipation effects, and temporary component parameter changes with temperature. The only failure modes that could be detected are permanent, catastrophic ones such as fractures due to thermal expansion mismatch or permanent component parameter shifts. If the problems that occur in the field or in other related testing are related to functional failures, this evaluation plan would *not* detect or prevent them. It would then be possible for these problems to occur after the product is delivered to the customer. (The author's experience has repeatedly confirmed this possibility.)

Broad-Spectrum Stress Portfolio.   Utilization of a broad-spectrum stress portfolio will solve the problem of selecting the right stresses. This approach uses a wide variety of stresses that have demonstrated success in failure acceleration. These stresses may be used sequentially or in simultaneous combinations. Utilizing this approach maximizes the probability of finding the largest number of failure modes in the shortest time. The portfolio of stresses should be flexible; modifications are justified when the data trends indicate changes.

Each stress has the ability to expose certain failure modes (Table 12.2). When multiple stresses are used (simultaneously or sequentially), the failures will consist of some duplicated modes and many unique modes. An example of this is "cold" (incomplete or crystallized) solder joints. Applying either temperature cycling or random vibration stress may cause these weak solder joints to fail, and hence be discovered. This is an example of a failure mode that can be accelerated by either of two stresses. Other failure modes that are most easily discovered by temperature cycling are fracture or fatigue due to mismatch of coefficient of thermal expansion, parametric drift, and malfunction due to lack of thermal design margin. Failure modes that are best discovered by vibration include lead fatigue due to inadequate component mounting, poor screw retention methods, and lack of mechanical design margin.

Evaluation of the product using *simultaneous* multiple stresses will maximize the number of failure modes that can be discovered in a short time. Also, some failure modes may go undetected if only a single stress evaluation plan is used. A suggested broad-spectrum stress portfolio contains temperature (with rapid changes between temperatures as well as dwell time at each extreme), random vibration (single- or multiple-axis), power-line variation (frequency and voltage), internal power supply output variation (to test other circuits powered by it), clock frequency variation, worst-case software exercising algorithms, and maximum load or power dissipation conditions.

Application of alternative (nontraditional) stresses should also be considered, because of diverse user environments and product functions. Examples include user abuse (spilled liquids, intentionally erroneous data sequences, rough handling) and chemical or air pollutant susceptibility.

Despite its power, this evaluation approach will probably *not* uncover *all* failure modes in a product. Failure mode detection will be limited primarily by sample sizes, differences between prototype fabrication and actual manufacturing environments, and the possibility of unrepresentative components. Process variations that did not occur in the development phase may cause additional failure modes once in manufacturing. During a product evaluation, it is recommended that actual manufacturing processes and components be utilized if possible. Evaluate as many products as is economically possible, to look for process variations.

Different types of products will have different dominant failure modes. For example, mechanical products which involve bearings will have failure modes not found in all-electronic products.

Don't limit the stress portfolio when evaluating a newly designed product, because it is nearly impossible to predict the response of a new design to the whole portfolio of stresses. In subsequent E–A–C–V cycles, the portfolio may be gradually narrowed, to zero in on the mechanisms that have been discovered and need to be eliminated. This contrasts with other evaluation strategies, where the stress levels are dictated by specifications.

TABLE 12.2 Stresses and Their Primary Failure Mechanism

Stress	Traditional failure mechanisms
Moisture	Hermetic seal failure Corrosion Dendrite growth Ionic migration High-impedance circuit stability
Elevated temperature	Chemical reaction rates increased Ionic migration Digital timing margins changed Silicon diffusion processes increased
Temperature cycling	Circuit thermal stability Differential expansion rates Chemical contamination Soldering quality Heat transfer design Cracked die Open/shorted metallization Wirebonds: broken, lifted, or overbonding Die attach defects Passivation defects Bimetallic micromotion corrosion Crystal imperfections
Random vibration	Large component mounting Fastener selection and mounting Fatigue Heat sink mounting Soldering quality
Power-line variation	Circuit design margin Power dissipation Output regulation
Power cycling	Protection circuit transient Power dissipation In-rush current response
Maximum output loading	Maximizes internal component temperature rise and power dissipation levels
Product monitoring	Operation verification Firmware operation Intermittent failures

Caution must be exercised when a stress apparently does not contribute to failure discovery for a particular product. You should not arbitrarily or prematurely eliminate this stress from the portfolio solely on the basis of one failure-free cycle. It is likely that there are some failure modes present that will not be uncovered during the evaluation, but will arise later as a result of an unexpected process variation, component deviations, or latent design problems. These modes are just as important as the ones discovered during evaluation, and their root causes must be determined. Once that is done, appropriate new stress tests may be added to the stress portfolio to help uncover similar weaknesses in future products. This constant addition of knowledge to the portfolio is a necessary part of the accelerated testing process, and tends to result in expansion of the portfolio with time.

12.4.7 Select Defect Detection Methodology

While the unit under test (UUT) is being stressed, *continuous* measurements could be made to determine if the unit is still functioning correctly. Alternatively, *intermittent* measurements could be made prior to, during, and/or after the application of stress. There are advantages and disadvantages to each alternative.

The most common practice is to perform a performance test prior to and continuously during the evaluation. This allows determination of exactly *when* a failure occurs, and hence what stress apparently caused it. Also, running a continuous performance test throughout the evaluation may detect intermittent failures that would otherwise go undetected. An example might be a component whose parameters change more than the surrounding circuit can tolerate under stress conditions, but return to normal when the stress is removed. To detect this phenomenon, the unit under evaluation must be continuously performance tested. The main disadvantage of continuous performance testing is the associated cost, which can range from as little as the cost of a standard I/O cable to as much as the cost of a custom test fixture and custom software. A good reference that describes the benefits of continuous testing during evaluation is Paul Parker's "ESS Case Study."[21]

If a performance test is used only prior to and after stress, there are no data on how the unit under evaluation performed *during* the application of stress. Worse yet, all that can be detected are the "hard" (permanent) failures. The "soft" (intermittent or temporary) failures are likely to be undetected, and will surely cause problems later. The author has observed that only a small percentage of all failures are of the permanent variety, and that continuous testing during the stress application yields a high return on the investment.

12.4.8 Validate the Test System

Evaluation failures may be divided into three major categories: those related to the unit under evaluation, those related to the test system, and those related to overstress. This section will address test system–related failures.

The performance test system must be validated before any evaluation of units under stress conditions is begun. This may be done with a small number of units at nominal stress. An example would be the user's normal ambient temperature, humidity, and altitude; nominal power-supply voltages (ac mains or dc); and any other similar user conditions. A product destined for portable or mobile use would obviously be expected to survive a wider range of stresses than one for laboratory or office use, and so the nominal stress might be different.

One method of validation is to run the performance test system (with a sample unit) through all available test procedures, using multiple permutations of the final product configuration. This means employing or simulating various cable lengths, software or firmware revisions, and test equipment, and even simulating possible failure modes to see if they are detected. While not all possible failure modes of the unit under evaluation can be simulated, the system should tolerate the most probable error messages. Special software (or firmware) to decode failure messages which are "catch-all" categories, combining inputs from several subsystems or components, should be considered. It may be worth the effort to create separate error messages for each subsystem, to help focus troubleshooting and find the root cause more quickly. After coding, it is also important to test these error messages by simulating a fault condition in each subsystem. Typically, these error messages are not tested as thoroughly as the main code, and you may find a bug or a misleading message.

For product evaluation, special test hardware, software, and fixtures will be required. It is a good idea to borrow as much as possible from existing or soon-to-be-existing tools. An example might be using software intended for service and repair applications. Development of this software is usually done in parallel with product development, and one may be able to use it to stimulate the equipment under test. An added advantage is that evaluation of both the product and service software will have then been achieved. Be cautious—field service software is also likely to contain bugs!

12.4.9 Use a Defect Tracking System

When failures occur during the qualification process, some means must be established to capture and record this information, and especially to track resolution of the causes of failure. This process is commonly known as *design defect tracking* (DDT). Using DDT ensures that quality and reliability receive the same visibility as the project schedule and cost. The goal of DDT is not just to prevent known failures from slipping through the cracks, but also to facilitate understanding the root causes of all failures. The DDT process tracks the progress of all observed failure mechanisms uncovered during all phases of the design qualification process. It may be used to track both hardware and software defects. A simple system such as a spreadsheet program may be used, but it should be available on a network so that it is accessible and easily viewed by all. Tables 12.3 and 12.4 illustrate what is needed.

TABLE 12.3 Contents of a DDT Report

Column heading	Description
Defect number (DT #)	An ID number assigned as each failure mechanism is logged into the defect process
Test	The test activity where the failure occurred
Priority	Importance to the project before advancing to the next phase: 1 = Must be fixed prior to moving to the next phase 2 = May need to be fixed before moving to the next phase (judgment call) 3 = Does not hinder project progress into the next phase
Failure description	A short description of the failure mechanism, what happened under which conditions, etc.
Owner	The name of the person responsible for investigating the failure mechanism—primary contact for updating the information that has been learned and what corrective action is pending
State	Where the failure mechanism lies in the resolution process: 4 = Unknown, failure observed, no analysis yet 3 = Understood, the root cause of the failure mechanism has been determined (by analysis) 2 = A fix has been designed, to eliminate the failure mechanism 1 = The fix has been verified (by repeating the test process which found the problem) 0 = The solution has been implemented through either a design, component, or manufacturing process change

12.4.10 Definitions of States

State 4—Unknown.   A failure occurred during the qualification process, and the root cause is unknown. The symptoms of the failure must be documented, along with the owner, state, and details of the unit number and test step where the failure was detected. This will aid in finding the root cause of the failure. The date of the failure may also be recorded, to enable better estimation of the time needed to resolve different failure mechanisms during a project retrospective. Priority tends to be subjective, and should be decided by the project manager in collaboration with the design team.

TABLE 12.4 Example of a DDT Log (Priority Order)

DT #	Test	Priority	Failure description	Owner	State	Serial number
6	STRIFE	1	Continuous resets	CS	4	101, 75, 102
12	BEST	1	Cold solder joints on the transformer	DW	4	75
16	ESD	1	Processor does not respond to keyboard input after discharge	SS	3	88
20	Vibration	1	Capacitor fatigued leads—fell off after 3 min	HF	3	75
21	Vibration	1	Loose screws on power supply board	DW	2	102, 103
23	RFI	1	Emissions worst than rev A unit by 5 dB	ED	2	103
5	BEST	1	Ripple on +5 V output above specification by 200 mV	HF	2	98
8	System	2	"Busy" LED does not light up when unit is operating—LED OK	IW	4	75, 88, 99
10	STRIFE	2	Blown fuse at 120 V ac/60 Hz	HF	3	98, 99
3	Service	2	Access cover difficult to remove	TD	2	99
7	Audit	3	Ink smudge on top case	DW	4	88
24	System	3	Weld line on rear bottom cover	DW	4	All
17	Humidity	1	Leakage between mains and ground of 10 mA	HF	0	100
18	Power line	1	Shorted rectifier when operating at 240 V ac/60 Hz	HF	0	90
19	Package drop	1	Unit right plastic housing cracked after drop on right side	DW	0	101
15	STRIFE	2	Ball cam does not rotate	TD	0	98
13	Software	2	Incorrect output with test plot	IW	0	99
22	Package drop	3	Loose screws on front panel	DW	0	101

State 3—Understood. To enter this state, the root cause of the failure must have been determined. Asking why five times* usually gets to the root cause of the failure. For example, suppose that during accelerated testing, unit 99 failed. Why did unit 99 fail? There is no output from the power supply. Why did the power supply fail? The input rectifier shorted and the main fuse opened. Why did the rectifier short and the fuse open? The rectifier shorted because the steady-state peak current exceeded the component specification, and the fuse opened because the rectifier shorted. Why was the rectifier current specification exceeded? The output load was increased because additional features had been added to the product.

State 2—Fix Designed. Once the failure root cause has been determined, action plans are needed to prevent this failure mode from recurring. Designing out the failure mechanism will ensure that this failure will not return. In state 2, modifications to either the design or processes are proposed and hypotheses tested. The criterion for moving on to the next state is that the proposed changes *prevent* this failure from recurring. Using the above example, some action items would be to recalculate the worst-case maximum load current and select components which can tolerate this level of stress.

State 1—Fix Verified. Once the changes have been proposed, they must be tested to determine whether they really solve the problem without introducing new failure modes. When a design change is made, the possibility of introducing new failure modes is very high, particularly if the

---

*The number five is symbolic; ask why until a stage is reached that allows the root cause to be understood and a permanent fix proposed.

process is hurried. Thorough testing is needed to verify the solution. Suppose, in our example, new rectifiers with increased current capacity are installed in the power supply. Current, voltage, and temperature measurements must be made to verify that these devices are operating within specifications and stress ratings, and to assure that some other component (such as the ripple filtering capacitor) has not been overstressed by the changes.

State 0—Solution Implemented.   Once a fix has been verified, an implementation plan must be formulated. Are the new parts available by the date needed and in the quantity required? Can changes to the printed circuit layout be accomplished? Has a component been selected that is too large for the space available for it? Questions must be asked to determine when and how the solution can be implemented in the final product. Once the original failure mechanism has been successfully eliminated with no new failures introduced, the defect may be deleted from the active tracking list. (Even then, a permanent record should be kept, to show progress and status to management.)

12.5   PERFORMING THE EVALUATION—PHASE TWO

1. Characterize the unit under evaluation.
2. When failures occur, perform failure analysis to determine the root cause.
3. Take corrective action, based upon the results from failure analysis.

12.5.1   Characterize the Unit under Evaluation

To properly evaluate design margin, each stress factor should be divided into small increments. Instead of just testing with high and low extremes of stress (two conditions), stepping between those limits in small increments may reveal pockets of product nonperformance. When displayed in a graphical form, the resulting boundary between acceptable and unacceptable operation (at various stress levels) may take on the shape of a Schmoo. (A Schmoo was a furry, pear-shaped creature in the L'il Abner cartoon strip of the 1950s. The characteristics of a "Schmoo plot" are irregular, somewhat circular boundaries, not unlike the lines on a contour map.) The following process will help define the area of operability when one or more stress factors vary.

For our example, an ac line-operated dc power supply was being characterized. The turn-on performance was intended to be characterized over a range of ambient temperatures and ac input voltages. The performance criterion is whether or not the (switching) power supply starts up correctly every time ac power is applied. If any one of the five output voltages was outside the specified range, the supply was deemed to fail. When the results were evaluated, the design was modified in ways that yielded a wider range of operation with minimal cost impact on the design. Table 12.5 shows the evaluation data of the original design. The shaded portions of the table represent conditions outside of the published specifications. (NT is the designation for not tested.) The number in each cell represents the number of supplies which functioned correctly under those conditions.

12.5.2   Perform Failure Analysis to Root Cause

Determine, through failure analysis, the root cause for all failures. Most failures will be due to flaws in the design, components, or manufacturing processes. A few failures will be caused by the test system or the overstress conditions during the test. Correct any deficiencies in the test system and repeat the evaluation. Blaming a failure on overstress is not a simple matter. For example, if an IC fails at a temperature just above its data sheet specification, one should not automatically assume that overstress is the cause, and do nothing. Typical commercial ICs are rated for 70°C operation. If the part malfunctions at 75°C, this may be an indication that a generic problem exists within the part or

TABLE 12.5a Supplies without Brownout Modifications (6 Units)

Input V ac, 60 Hz	Ambient temperature					
	−20°C	−10°C	0°C	10°C	20°C	30°C
80	0	3	3	3	3	4
88	3	3	3	4	4	4
100	4	4	4	4	5	6
110	4	4	4	5	6	6
120	4	5	5	6	6	6
130	5	5	6	6	6	6
134	5	5	6	6	6	6
140	5	5	6	6	6	6
150	NT	NT	NT	NT	NT	NT

TABLE 12.5b Supplies with First-Pass Modifications (15 Units)

Input V ac, 60 Hz	Ambient temperature					
	−20°C	−10°C	0°C	10°C	20°C	30°C
80	2	3	8	11	11	11
88	11	11	11	11	11	14
100	11	11	13	15	15	15
110	13	15	15	15	15	15
120	15	15	15	15	15	15
130	15	15	15	15	15	15
140	15	15	15	15	15	15
144	15	15	15	15	15	15
150	15	15	15	15	15	15

TABLE 12.5c Supplies with Final Modifications (15 Units)

Input V ac, 60 Hz	Ambient temperature					
	−20°C	−10°C	0°C	10°C	20°C	30°C
80	NT	10	11	11	11	12
88	NT	11	11	11	13	15
100	NT	12	14	15	15	15
110	NT	15	15	15	15	15
120	NT	15	15	15	15	15
130	NT	15	15	15	15	15
140	NT	15	15	15	15	15
144	NT	15	15	15	15	15
150	NT	15	15	15	15	15

design. If this part failed at 75°C, others may fail at 70°C or even lower, depending on the failure distribution of the population. A larger sample may have to be tested to determine the true root cause of the failure or the existence of adequate margin. When you can verify your hypothesis by causing the same failure at will, you can be confident that the root cause has been found.

Often, questions are raised about the "validity" of failures detected by accelerated stress evaluation. The argument is that if one goes far beyond the component's or design's stress limits, then failure is inevitable and bears no relation to "real life." In response to this general fear, experience has shown that only through forcing failures and eliminating the weaknesses that allowed them can improvement be made. In most situations, failure analysis data will answer the overstress question. From the author's experience, the majority of failures will be due to design, component, or process deviations and not simply to overstress caused by the evaluation process. Accelerated stress testing shortens the time to failure, but it doesn't necessarily *cause* failures that would not otherwise have happened.

12.5.3   Perform Corrective Action

This step generates the payoff for all the work of the preceding two. Once the suspected root cause has been discovered, you will be able to generate many ideas on how to eliminate it. Choose the most attractive one, and try it. This may be something as simple as replacing one or more components with others that are more robust or have different specifications. Your freedom to make a more significant change is usually circumscribed by cost and schedule considerations. The most important point is to make sure that, whatever the change, it is thoroughly tested, to ensure that the original problem is fixed and no new ones are introduced. That means repeating at least the stress test that propagated the failure, and possibly others as well. The more significant the design change, the more thorough the retesting should be.

12.6   FORMULATE THE EVALUATION REGIMEN—PHASE THREE

An alternative to the broad-spectrum stress portfolio is to base the stress profiles upon either corporate or industry specification levels. This is risky, and should only be used as a starting point. It is important to expand beyond specification limits, because of the limited sample size. If you test only to the specification limits on a small sample, you have very little confidence that there is any design margin, and you may not discover infrequent but serious failure modes that affect a fraction of the population. By expanding the profiles, one can maximize the failure modes without creating unrealistic failure modes.

12.6.1   Using Multiple Stresses

Typically, testing of products for conformance to specifications uses in test protocols that evaluate the effects of individual stress conditions upon product performance, one at a time. Product specifications are often written that way, too—apparently assuming that a product only sees one stress at any one time.

Suppose, for example, that a product is to be evaluated over a range of temperatures and ac line voltages. Typical test protocols will evaluate the influence of temperature while ac line voltage is held constant at its nominal value. At a different time, ac line voltage is varied, while temperature is held constant at nominal ambient. This results in the *implied* area of performance shown in Fig. 12.6. The reason this is an implied performance area is that only the corner points have been tested, and it is assumed that the product will operate successfully at all other points contained within the area (combinations of the two stresses). If no other testing has been performed, this may be a dangerous assumption!

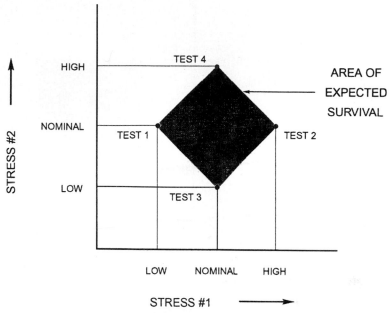

FIGURE 12.6   Varying one stress at a time.

In the real world, a product is likely to experience two or more stresses at once. This would suggest a new test paradigm, one in which multiple stresses are applied to a product simultaneously. At the very minimum, all combinations of extreme values for each stress condition should be tried. In the example of temperature and ac line voltage, this would result in the protocol summarized in Table 12.6.

TABLE 12.6   Four Corners Test Protocol

Temperature	Ac line voltage
Low	Low
Low	High
High	Low
High	High

The resulting area of implied performance is shown in Fig. 12.7; note that with no more tests than were performed before, the included area is twice as large. If a few intermediate points are tested as well, confidence that the implied performance area is a real performance area is greatly increased. Multiple, simultaneous stresses also tend to discover failure mechanisms that cannot be activated by one-at-a-time stresses. For example, moisture and voltage together cause corrosion and dendritic growth that simply does not happen with either stress alone.

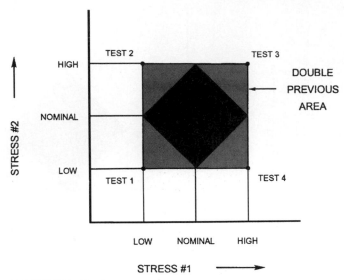

FIGURE 12.7 Varying two stresses at once.

12.6.2 Determining Acceleration Factors

When elevated stresses are used to evaluate product reliability, the failure rate is being accelerated (or the time compressed). Unfortunately, not every failure mechanism is accelerated by the same factor for a given stress. If sufficient data for each failure mechanism exist (a minimum of five failures at each stress level), then each mechanism's acceleration factor may be estimated. Otherwise, about all you can do is combine all the failures for the test process into one group and calculate an average acceleration factor. Because of the small sample sizes, you will be lucky to have two occurrences of the same failure mechanism occurring, and so an average factor for the whole product is generally calculated.

In order to make this calculation, tests must be performed at two or more different stress levels, the first under normal product use conditions and the others at elevated stress conditions. The test at normal stress may require a larger sample size or longer duration, so that there is a chance for failures to occur at the lower (nonaccelerated) rate.

Analyze the data at each stress separately, using an appropriate life distribution (the Weibull will work satisfactorily in most cases). Use the same statistical model to analyze both testing results (normal and elevated), and determine the time at a selected percent failing. The acceleration factor is determined by dividing the Xth percentile life at normal stress by the same Xth percentile life at elevated stress. If you have enough data, X should be the 50th percentile or higher, but using a lower number (like the 10th percentile) is OK so long as it is based on five or more failures. Since the life is expected to be shorter at elevated stress, this factor should be greater than 1. If it is not, then the elevated stress conditions were not accelerating any failure mechanisms.

$$\text{Acceleration factor} = \frac{X\text{th percentile life (normal stress)}}{X\text{th percentile life (elevated stress)}} \qquad (12.1)$$

There are economic consequences of performing the testing necessary to make this calculation. First, the number of units needed for testing has doubled, at least. Typically, the number of working prototypes of a new product is very limited, and this larger number of units for test may simply not be available. Second, the facilities and resources needed to perform the testing have also been increased. Third, the time involved increases. The portion of the test conducted at normal use conditions may take many months if the expected life is long. Given the pressure on new product development to shorten time to market, a possible solution is to put more units on test in order to reduce the time to five or more failures. This may not be feasible, either. The net result is that, while estimating an acceleration factor is analytically seductive, it is very costly, and may be misleading. The philosophy of forcing failures, then finding and eliminating the root causes, will improve reliability by a large factor, without your ever knowing the acceleration factor.

12.6.3 Arrhenius Reaction Rate Model

In 1889, Swedish physicist and chemist Svante August Arrhenius empirically demonstrated chemical reaction rates, using sucrose inversion, that increased exponentially with temperature. More recently, electronic components have been observed to exhibit this same exponential behavior of failure rate with temperature. The expression which is most commonly used for failure rate λ as a function of temperature is

$$\lambda(T) = e^{-qE/KT} \tag{12.2}$$

where q = electron charge (1.60206×10^{-19} C)
  E = activation energy (eV)
  K = Boltzman's constant (1.38044×10^{-23} J/K)
  T = temperature (K = °C + 273.18)

Using this expression, an acceleration factor S relating the failure rates at two temperatures may be determined:

$$S = \frac{\lambda_{T2}}{\lambda_{T1}} \tag{12.3}$$

Substituting from Eq. (12.2),

$$S = e^{-(qE/K)(1/T2 - 1/T1)} = e^{11,605\,E(1/T1 - 1/T2)} \tag{12.4}$$

The activation energy term E determines the slope of the acceleration factor with temperature (Fig. 12.8). A small activation energy characterizes a failure rate which is not strongly temperature-dependent. Activation energies of 1.0 eV or greater signify strong dependence on temperature. An activation energy of approximately 0.53 eV results in the failure rate doubling for every 10°C increase, which is common for ordinary bipolar semiconductor components. Each and every failure mechanism has its own characteristic activation energy, and so a component containing a mix of failure mechanisms will have a failure rate versus temperature which does not have a constant slope, but varies depending on the actual temperature.

The activation energy E of a failure mechanism may be determined experimentally by the following procedure:

1. Perform a stress test at three or more different temperatures.
2. Plot the resulting data (special graph paper is available).

   X axis; reciprocal of temperature ($1/T$)

   y axis: natural log of life (L)

3. Draw the best fit line.
4. The slope is equal to $B = E/K$.

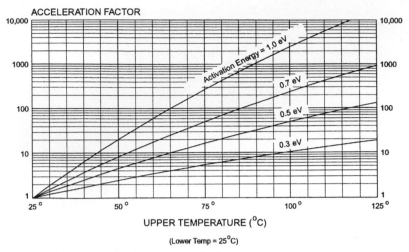

FIGURE 12.8   Acceleration factor versus activation energy.

Estimation.   If the identity of the failure mechanism is known and the activation energy has been previously determined by the manufacturer, use that value. A word of caution, however: The same mechanism in components from different manufacturers may have different activation energies because of differences in process and design. Also note that the activation energies listed are often determined at very high temperatures (200°C), which may not be applicable to your situation. The activation energy values may further be affected by electrical stresses on components.

12.7   *HEAT TRANSFER VS. AIR SPEED IN CHAMBERS*

A printed circuit assembly (PCA) containing a typical mix of integrated circuits and discrete components (Table 12.7) was the subject of a thermal response experiment. Fifteen components were monitored with a multichannel thermocouple temperature measurement system. The PCA was then tested in two different environmental test chambers. In both chambers, the air velocity across the PCA was varied between two values. For all cases, the air speed was as uniform as possible across the PCA. The same thermal profile was used in both chambers (Fig. 12.9). The ambient air

TABLE 12.7   Monitored Components on PCA

Quantity	Description
1	40-pin plastic DIP
2	14-pin plastic DIP
2	28-pin plastic DIP
2	Small-signal diodes
1	TO-5 transistor with heat sink
3	TO-220 transistors without heat sinks
2	TO-220 transistors sharing same heat sink
2	Power rectifiers

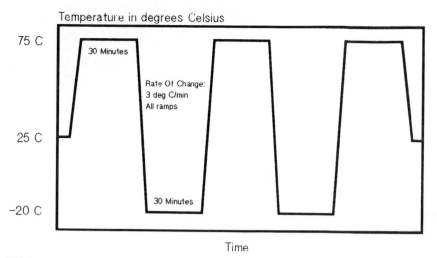

FIGURE 12.9   Test chamber temperature profile.

temperature was measured with a thermocouple mounted adjacent to the PCB. The resulting air speed was measured with an ALNOR 6000BP velometer. Data were acquired for each component with a delay time of 6 s between samples.

The temperature measurement data were analyzed using a personal computer statistics package. The data for the 15 components were averaged and plotted for each air speed, as shown in Fig. 12.10a to d. The recorded ambient air temperature is shown as a solid line. The plus symbols indicate the resulting averaged component data. The thermal time constant τ was also calculated. τ is a measure of how long the components' temperatures lag behind the ambient chamber tem-

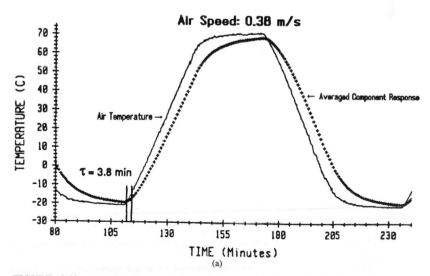

FIGURE 12.10   Averaged component thermal response. (*a*) Air speed 0.38 m/s.

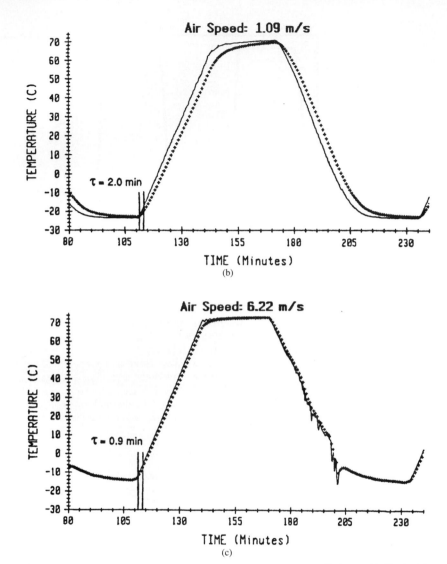

FIGURE 12.10 (*Continued*) Averaged component thermal response. (*b*) air speed 1.09 m/s; (*c*) air speed 6.22 m/s.

perature. Figure 12.11*a* displays the relationship of air speed to the calculated τ. The air speed axis was then transformed using the inverse square root of the air speed. An approximately linear relationship results, as shown in Fig. 12.11*b*. This type of relationship is also predicted by theory.[18]

12.7.1   Conclusions about Air Speed

The rate of thermal component response on a PCA is a function of the air speed inside the chamber. From Fig. 12.11, the reduction in the thermal time constant τ is negligible above 6.22 m/s (1225 ft/min). τ is approximately doubled as the air speed drops to 1.09 m/s (215 ft/min). A minimum air speed of 3.66 m/s (720 ft/min) or greater is recommended, as a starting point. This is the

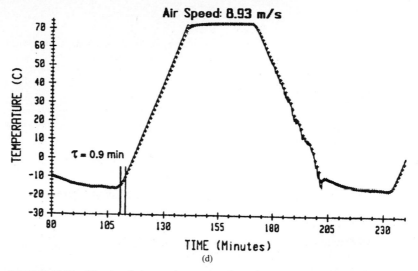

FIGURE 12.10 (*Continued*) Averaged component thermal response.(*d*) air speed 8.93 m/s.

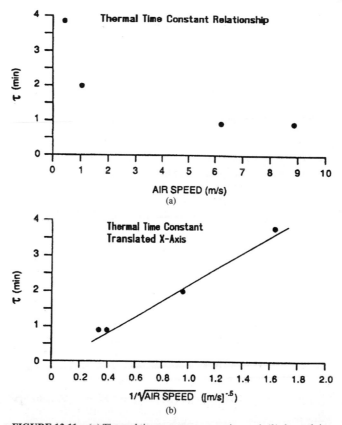

FIGURE 12.11 (*a*) Thermal time constant versus air speed; (*b*) thermal time constant—translated *x* axis.

midpoint between 6.22 m/s and 1.09 m/s. If components with large thermal mass or product air flow restrictions exist, the minimum air velocity may need to be increased.

12.8 PROFILE TAILORING PROCEDURE

The following is an outline of a procedure for tailoring a thermal cycle profile to a generic product and chamber.

12.8.1 Temperature Extreme Selection

This may be the most difficult number to select, from a political point of view. The reason, of course, is the conflict between the need to find and eliminate weaknesses in the product by forcing failure, and the pressure to pass the product as quickly as possible, so that it can be released to manufacturing and sales. At an absolute minimum, the test temperature extremes should be equal to or 5°C beyond the product's specified operating temperature limits, at each extreme. If you have the freedom to pursue an aggressive regimen, a test temperature range 15 to 25°C beyond specs at each extreme is suggested.

12.8.2 Temperature Change Rate Selection

Very little has been published on this subject. Rates of change of less than 1°C/min don't stress the product enough to accomplish the goal of accelerating failures, and will also make the test rather lengthy. The author uses nonlinear temperature change rates that are in excess of 10°C/min, and has found these effective in uncovering latent product faults without causing inappropriate failures or product damage. Some practitioners advocate linear temperature ramps with equal heating and cooling slopes, but this is hard to do and has not been proven superior.

The most serious implication of demanding a linear temperature change rate is capital equipment cost. To maintain a constant change rate, additional cooling and heating capacity are required, far beyond what is necessary for the same temperature change without the linearity requirement. In addition, cooling (for the same number of BTUs) is much more expensive than heating. Recent experiments replacing traditional CFC refrigeration with liquid nitrogen (LN_2) have shown that this makes linear cooling ramps somewhat easier to generate. However, even if temperature were changed instantaneously (thermal shock), the thermal response of a component is nonlinear (exponential), and so the argument is somewhat moot.

12.8.3 Component Prioritization

Because each component has a different thermal mass and coupling to the cooling medium (air, usually), the rate of change of temperature of each will be different. To be sure of maximizing the reliability improvement, you should decide which components should experience the most rapid rates of change, and design your test accordingly. The author's recommendation is to give priority to the semiconductor devices and other active, complex components when considering temperature change rates. The larger, simpler passive components (coils, transformers, etc.) need only experience the complete thermal cycle, from one extreme to the other, and generally cannot be forced to change at rates as high as 10°C/min. These components may determine the dwell time at the extremes, but not the rate of change.

12.8.4 Understand the Thermal Response

Without careful planning or long experience, it is easy to make mistakes which reduce the effectiveness of temperature cycling—for example, failing to remove a product's covers, exceeding the thermal mass limit of the test chamber, or having insufficient air flow in the chamber because of overfilling the chamber with product. If the component thermal response had been measured in the above situations, the results would have shown that the test was not effective. The assumption that all products will have similar thermal responses is not valid. Also, not all chambers are created equal with respect to their heat transfer ability. The thermal interactions of the chamber and product *must* be understood before an effective program can be achieved.

12.8.5 Options for Achieving Thermal Cycling Rates

If, upon analysis, the priority components do not experience the desired temperature change rates or ranges, there are several options for improving their performance. If the empty chamber is not capable of achieving the desired change rates, then additional heating and/or cooling is required. If the chamber has the basic ability to change air temperature at the desired change rate, the following ideas are suggested:

1. Minimize the thermal mass placed inside the chamber. This includes product, holding and interfacing fixtures, and perhaps even the chamber interior walls. The larger the thermal mass, the more energy is required to change the temperature quickly. Examples of ways to reduce thermal mass are to minimize the product enclosures that are not required for fixturing and to use aluminum instead of steel for product fixtures.

2. Increase the air speed around the components. This is accomplished by removing product enclosures, channeling the air stream past the components with baffles and deflectors, and adding additional blowers to boost the air speed. [Recall the author's recommendation of a minimum of 3.66 m/s (720 ft/min) air speed.]

Modify the temperature profile and/or temperature controller. Add temperature overshoots beyond the intended profile extremes to maximize the component change rates. If the air temperature is correct, but the components do not achieve the intended temperature extremes or change rates, increase the dwell time at the extremes. In Fig. 12.12, a temperature overshoot was intentionally added, to achieve the desired $-20°C$ temperature limit faster.

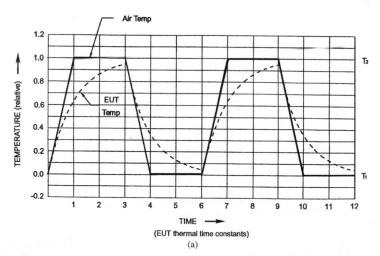

FIGURE 12.12 Tailored test chamber temperature profiles. (*a*) Thermal response of EUT to temperature cycling.

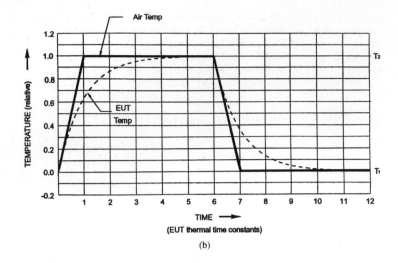

(b)

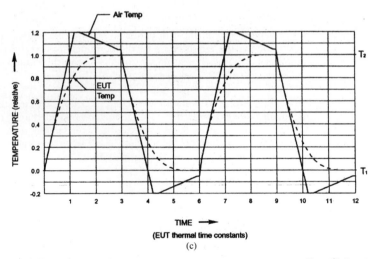

(c)

FIGURE 12.12 (*Continued*) Tailored test chamber temperature profiles. (*b*) longer dwell time compensates for slow response; (*c*) internal overshoot compensates for slow response.

Another method of achieving a similar result is to mount an equivalent component thermal mass on the chamber-controlling thermocouple. This will simulate the response of the components in the chamber. When these methods are used, it is a good idea to make extra measurements to assure temperature consistency throughout the test chamber.

12.8.6 Conclusions

The effectiveness of thermal cycling, as a technique for discovering product defects, is dependent entirely on the component's rate of thermal response. Without an understanding of the component's thermal response, reliability improvement programs using accelerated testing or environmental stress screening (ESS) may not be effective. In the author's experience, the *majority* of thermal cycling programs are flawed. Additional product development expense is incurred with these temperature cycling programs. This leads, in turn, to an increased product price. If these programs are not effective at finding product defects and improving reliability, the investment may be squandered and future reliability improvement programs jeopardized. Using these analysis techniques will contribute to the goal of improving product quality and increasing customer satisfaction.

12.9 CASE HISTORIES

12.9.1 Computer Circuit Cards

This example is taken from Dennis Pachucki, "Analysis of Concurrent Tri-Axial Random Vibration and Thermal Cycling Applied to Computer Circuit Cards.[20]

- Three groups of 200 computer PCAs
- Group 1: Vibrated at ambient
- Group 2: Four thermal cycles (0°C to 55°C) with vibration at dwells
- Group 3: One thermal cycle (vibration with power off, −35 and 85°C), three thermal cycles (0°C to 55°C) with vibration
- 25 G_{rms} triaxial quasi-random vibration
- 20°C/min component temperature changes
- Doubling of failures from Group 1 to 2 and from Group 2 to 3

12.9.2 High-Density Surface-Mount Circuit Card

This example is taken from Paul Parker, "ESS Case Study of a High Density Surface Mount Circuit Card."[21]

- 17,000 computer PCAs tested.
- Thermal cycling and product monitoring.
- Results showed an infant mortality failure rate.
- Product average temperature rate of change: 6 to 18°C/min (Table 12.8).

TABLE 12.8 Thermal Cycle Profile

	Start, °C	End, °C	°C/min
Pull down	70	0	9
Pull down	70	−20	6
Pull up	−20	70	18

12.9.3 Summary of Case History Results

- Dynamic testing and monitoring during thermal cycling can more than double the effectiveness of the process in detecting potential defects.
- Rapid root cause analysis and corrective action continue to be essential to a successful ESS program.

- Component supplier quality is not constant and must be continually monitored. The same applies to the board assembly process. As a result, to assure total process control, some form of ESS will always be required.
- Accurate collection and analysis of ESS results are required to fully optimize the process, and to understand the impact on product reliability.

12.9.4 Original Equipment Manufacturer (OEM) Power Supply Example

The accelerated stress evaluation process was applied to a "mature" OEM power supply. Changes in the standard design had been made to include additional overvoltage and overcurrent circuit protection. The supply was a high-efficiency switching design with a maximum dc power output of 200 W. There was no vendor history available, nor could the vendor supply sufficient reliability information. Past experience with other switching supplies had yielded unacceptable field reliability.

Since there was no product history to base stress selection upon, the broad-spectrum stress portfolio was utilized. The stresses selected were temperature cycling, random vibration, power-line voltage cycling, high humidity, and maximum output loading. Refer to Table 12.2 for the traditional failure modes for each stress.

The evaluation process consisted of three sequentially conducted evaluations. The order of evaluation was a high-humidity soak, temperature cycling, and then vibration. The humidity soak was placed first to maximize the observation time for corrosion and crystal growth problems. Either the vibration or the temperature cycling could have been performed next. Because of the number of combined stresses involved, temperature cycling was chosen. The series concluded with the random vibration profile.

The product outputs were loaded to maximum rated current. This resulted in maximum internal power dissipation and highest component temperature rise. Our test system measured output voltages sequentially. The measurement cycle time was roughly 5 s to measure all outputs of ten supplies.

The evaluation process began as soon as engineering prototypes were available. Evaluation of five different design generations occurred during the product development phase. Each subsequent design generation contained design, component, and process modifications in response to failure modes uncovered. The complete evaluation process took 40 h of testing to evaluate 10 units. Table 12.9 contains a summary of the failure modes discovered.

TABLE 12.9 Failure Modes Uncovered

Area of change	Failure description and fix
Design	Components reversed due to incorrect documentation
	Upon power cycling, main fuse blows. Change to slow-blow fuse
	Fatigued component leads due to the device supporting the heat sink. Heat sink design changed to mount onto PCA
	Cores of coils were loose. Epoxy cores in place
	Add locking hardware on loose screws
	Coil and electrolytic capacitor leads fatigued. Mounting changed to remove stress
Component	Shorted primary input rectifiers. Use a higher-quality vendor for the part
	Leakage current between mains and ground exceeded specifications. Changed filter capacitor part type due to poor process control
Manufacturing	Units were dead on arrival. Added monitoring of manufacturing burn-in process
	Loose screws. Added manufacturing process control
	Cold or broken solder joints. Train operators to solder
	Schottky diode shorted due to poor thermal contact with heat sink. Change manufacturing process to keep the heat sink flat
	Misinsertion of DIP packages, tucked lead. Modified insertion tooling

12.9.5 Power Supply Evaluation Conclusions

Temperature cycling, power-line variation, and random vibration failure modes accounted for 70 percent of the total observed (Fig. 12.13). While that may seem impressive, remember that if these were the only stresses used, 30 percent of the failure modes would have gone undetected. Also note that 26 percent of the failure modes were discovered in areas unrelated to the accelerated evaluation—incoming inspection and final assembly. This demonstrates the inability of this evaluation plan to detect *all* failure modes.

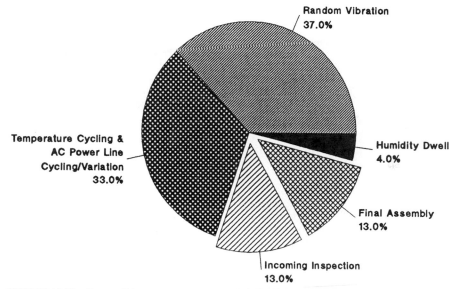

FIGURE 12.13   Test conditions versus failure mechanisms discovered.

"If It Works—Don't Fix It!"   Modifications of the design, components, or manufacturing process can and do frequently produce new failure modes. Always verify major changes. Figure 12.14 shows this phenomenon. In the later phases of the project, new failure modes suddenly

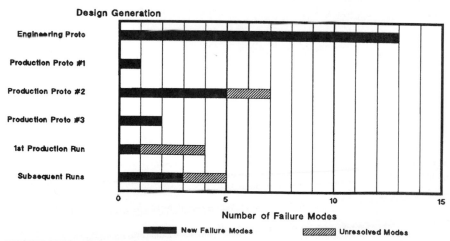

FIGURE 12.14   Design generation when failure modes occurred.

appeared. These were primarily due to changing the manufacturing line location from the prototype to the standard production environment.

Failures that required either design, component, or manufacturing changes are displayed in Fig. 12.15. Nearly half the modes required manufacturing changes. A large percentage of the remaining failures were design deficiencies. If the product design verification process was mature, then the number of design-related failures should have been minimized.

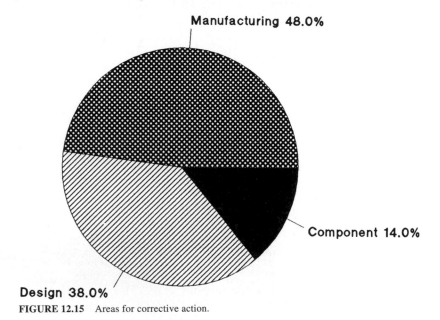

FIGURE 12.15 Areas for corrective action.

Obtain an up-front commitment to follow the evaluate–analyze–correct–verify process. This type of program is a radical departure from traditional reliability techniques, and may be viewed with fear or suspicion. Only the resulting root cause failure analysis data will indicate the required corrective action. Don't jump to corrective action without finding the *true* failure root cause, or the likely result will be to induce new failure modes.

12.10 BIBLIOGRAPHY

1. Bailey, R. A., and R. A. Gilbert. "Strife Test for Reliability Improvement." *Proceedings of the Institute of Environmental Sciences,* 1981.
2. Beaton, Bradford P. "TARGET: Thermal Accelerated Reliability Go–No-Go Environmental Testing Dynamic Board Thermal Shock Using a Single Liquid Fluorocarbon Bath." *Proceedings of the IEEE CHMT IEMT Symposium,* San Francisco, September 1991.
3. Burgaud, Georges G., and Gilles Brullon. "Use of Built-in Self Test to Control Electronic Board ESS." *1994 Institute of Environmental Sciences Annual Technical Meeting,* pp. 169–173.
4. Davis, Don. "Understanding the Economic Benefits of ESS." *Evaluation Engineering,* August 1987, pp. 42–45.
5. Diekema, Jim, ed. *The Environmental Stress Screening Handbook.* Thermotron Industries, Holland, Mich., 1987.
6. Englert, Paul, Bharat Thakkar, and Ted Bonds. "Template for an Environmental Stress Testing Program." AT&T Bell Laboratories, Whippany, N.J.

7. Fedraw, Ken, and John Becker. "Impact of Thermal Cycling on Computer Reliability." *1983 Proceedings of the Annual Reliability and Maintainability Symposium,* IEEE, #0149-144X.

8. Fuqua, Norman. *Reliability Engineering for Electronic Design,* Marcel Dekker, New York, 1987.

9. Hobbs, Greg. "Stress Screening—Some Misconceptions Clarified." IES San Diego Chapter Meeting, May 1984.

10. Hobbs, Greg. "Development of Stress Screens." *1987 Proceedings of the Annual IEEE Reliability and Maintainability Symposium,* Philadelphia, Pa., Jan. 27–29, 1987, pp. 115–119.

11. Institute of Environmental Sciences. *Proceedings of the Third National Conference and Workshop on Environmental Stress Screening of Electronic Hardware,* IES, Mount Prospect, Ill., Sept. 10–13, 1984.

12. Jawaid, Shams, and Kevin Crook. "Linear Ramp Chambers and Thermal ESS." *Evaluation Engineering,* June 1992, pp. 70–79.

13. Kececioglu, Dimitri, *Reliability and Life Testing Handbook,* vols. 1 and 2. Prentice-Hall, Englewood Cliffs, N.J., 1993.

14. LoVasco, F., and K. Lo. "Relative Effectiveness of Thermal Cycling vs. Burn-in." *Proceedings of the 42nd Electronic Components and Technology Conference,* San Diego, May 1992, pp. 185–189.

15. Molenarr, George. "Chamber Design for Environmental Stress Screening." George Molenarr, 10877 Lake Thames Drive, Cincinnati, OH 45242, 1987.

16. Morgen, Richard J. "Random Vibration 'Out Shakes' Swept-sine." *Machine Design,* Feb. 6, 1992, pp. 58, 60.

17. Nelson, W. *Accelerated Testing,* Wiley, New York, 1980.

18. Reynolds, William C., and Henry C. Perkins, *Engineering Thermodynamics,* McGraw-Hill, New York, 1970, p. 563.

19. Rondeau, Herbert F. "Predicting IC Failures." *Machine Design,* Mar. 20, 1980, pp. 78–82.

20. Pachucki, Dennis. "Analysis of Concurrent Tri-Axial Random Vibration and Thermal Cycling Applied to Computer Circuit Cards." *Proceedings of the Institute of Environmental Sciences Annual Technical Meeting,* May 1994, pp. 143–149.

21. Parker, T. Paul. "ESS Case Study of a High Density Surface Mount Circuit Card." *Proceedings of the Institute of Environmental Sciences Annual Technical Meeting,* May 1991, pp. 393–402.

22. Schinner, Charles. "Optimization of Thermal Cycling Profiles." *Institute of Environmental Sciences Annual Technical Meeting,* May 1991.

23. Schinner, Charles. "The Board Electronic Strife Test (B.E.S.T.) Program." *Reliability Review,* vol. 8, June 1988, pp. 310–317.

24. Schinner, Charles. "Reliability Growth through Application of Accelerated Reliability Techniques and Continual Improvement Processes." *Institute of Environmental Sciences Annual Technical Meeting,* May 1991.

25. Smithson, Stephen A. "Shock Response Spectrum Analysis for ESS and STRIFE/HALT Measurement." *Test Engineering and Management,* December/January 1991–92, pp. 10–14.

26. Smithson, Stephen A. "Effectiveness and Economics—Yardsticks for ESS Decisions." *Proceedings of the Institute of Environmental Sciences,* 1990, pp. 737–742.

27. Striberg, Colleen E. "Process Automation as Applied to Margin and Abuse Testing." Master's Thesis, California State University, Chino, Spring 1993.

28. Suesy, Cliff. "Achieving Phenomenal Reliability Growth." *Proceedings of ASM Reliability Conference,* Los Angeles, March 1987, pp. 79–88.

CHAPTER 13
FAILURE ANALYSIS SYSTEM— ROOT CAUSE AND CORRECTIVE ACTION

John R. Adams
VALLEYLAB, INC.
BOULDER, COLORADO

13.1 INTRODUCTION

Failures have a root cause (the fundamental source of the failure), and most are understandable if an organized approach to analysis is used. This chapter will explain how to analyze failures and determine their root cause. With true root cause determination, costly failures can be eliminated by developing effective corrective action. This is an essential part of any continuous reliability improvement process.

Failures cannot be eliminated without an understanding of the root cause. Curiously, some people jump to the conclusion that the symptom of the failure is the cause. This is rarely true. For example, the loss of short-term memory may be blamed on a headache, the symptom. But only an investigation to find the root cause (perhaps a brain tumor) will eliminate the failure.

This points out another important fact about failure analysis. These principles can be applied to all areas of life in which failures occur. This chapter describes primarily engineering designs with an emphasis on electronic equipment, but the methods described will work well on the failure analysis of almost anything.

Knowledge and an understanding of the common failure mechanisms is valuable in performing failure analysis. Chapter 14 of this handbook, "Physics of Failure," explains many of these failure mechanisms. Knowledge of the components involved is also very important. See Chap. 16, "Electronic Component Reliability." There are also many good handbooks* on electronic components.

13.1.1 Definitions

Failure: Nonperformance in terms of meeting the normal and/or specified operating characteristics of the component or system.

Failure analysis: A closed-loop system for investigation of failures to discover the root cause and develop a corrective action to eliminate that cause.

Root cause: The fundamental source of the failure. Root cause means the underlying basic reason for the failure. Any cause higher up is a symptom or a result of the root cause. For exam-

---

*One fine collection is the three-volume handbook by General Electric, *Component Technology and Standardization.*

ple, a burned-up resistor in a circuit might seem to be the logical cause for the failure of that circuit. However, careful analysis may find that the resistor was perfect in every way and that the designer had failed to use the correct size (wattage) for this application. In that case, the root cause was an error in design. The perfectly good resistor was overstressed (beyond its specifications) by the design. This distinction is the essence of good failure analysis: determine the *root cause* of failures.

Corrective action: The *proven* elimination of a root cause and therefore of the failure mechanism. Proof of proper corrective action is best accomplished by demonstration.

13.1.2   Purpose and Benefits of Formal Failure Analysis

There are many good reasons for formal failure analysis. A few of the best reasons are listed below:

- Improve reliability
- Increase customer satisfaction
- Lower product cost

Improve Reliability.   Reliability is best achieved by designing it into the product at the beginning of the design process. This includes specifying reliable parts from quality manufacturers. It also includes defining the manufacturing or assembly process so that a reliable design is not compromised.

When this has not been done or has been done only partially, then the next best way to improve reliability of a new or an existing product is to perform failure analysis (down to the root cause) and take appropriate and effective corrective action.

Increase Customer Satisfaction.   There has been a lot written and said about TQM (total quality management). The most basic concept in all of this has been to strive for customer satisfaction. Reliability (the absence of failures) is certainly one major characteristic that increases customer satisfaction. *Failure analysis* down to the *root cause* combined with effective *corrective action* eliminates failures and improves reliability and therefore customer satisfaction.

Lower Product Cost.   The cost of failures is much more than the cost of the repairs, which in itself can be a significant expense. Repeated failures of a product will soon discourage customers from buying that product. Failures that may cause harm or a hazard to the customer may result in a major recall of the product, which can be extremely expensive. Table 13.1 shows the estimated costs of failures in various operations in a company. Elimination of these failures will lower product cost and increase company profits.

13.1.3   Failure Analysis Is Performed by the Entire Organization

Failure analysis is a team task. It requires the involvement of many employees, from top management down to the test technician in production. A successful FA team needs a dedicated leader who has the desire and training necessary to get the whole organization interested in and committed to the elimination of failures. More on this later in the chapter.

13.1.4   Failure Analysis Process Overview

There are three major aspects to a good failure analysis system:

1. Organization
2. Process of failure analysis
3. Corrective action

First, a defined *organization* with overall responsibility is needed for implementing a failure analysis system in any company. It needs to have a charter authority and a clear reporting path.

Second, there must be a defined *failure analysis process* via a procedure for discovering the root causes of failures.

Third, there needs to be a defined process for taking effective *corrective action* to eliminate the root cause of each failure. Good documentation is part of the corrective action.

TABLE 13.1   Cost of Failures

Location	Description of failure cost	Estimated cost
Field	Repeated failures: Loss of customer Recall if hazardous	High because it can grow
Field	Each time: Shipping (two ways) Cost of repair Cost of salesperson's time	$200.00 $100.00 $100+
Production	Cost of repair/rework Cost of overtime (unplanned) Cost of scrap	$40.00/h $40.00/h Actual price of parts
Incoming	Loss of time to replace the parts Extra, higher inspection levels—labor hours	Depends; if it delays production, $10,000/day? $40/h

Note: These estimates are based on 1994 labor rates with moderate overhead.

13.2   ESTABLISHING A FAILURE ANALYSIS SYSTEM

A failure analysis system is a companywide activity. It must involve design, manufacturing, purchasing, quality (via incoming inspection and component engineers), reliability, and management. The last is not least. Without management support, the efforts by others are likely to go unrewarded, and therefore eventually fail.

13.2.1   Getting Management Support

Management will support a failure analysis system when it understands the value and is shown the return on investment (see Table 13.2). Failures are very expensive, as mentioned above. This expense can be much more than the cost of returns and repairs. Loss of large customers because of failure-prone products can wipe out a previously successful company. Recall of products because of failures can also be extremely expensive.

The information provided in this chapter, and some of the references, will prepare the reader to show management that a failure analysis system is an important investment.

Start with a small lab with one or two dedicated engineers and/or technicians. Then find the expensive failures and determine the root causes of those failures. Finally, help the failure analysis team to develop corrective action that will eliminate this failure forever. Write the failure report and present it to management with the cost savings clearly indicated. After this achievement, management should provide solid support.

13.2.2   Defining the System

After management approval, or perhaps as a step in getting that approval, define the failure analysis system. This is best done as a *business plan* with an *operating procedure*. These are presented

TABLE 13.2   Cost of Failures for Calculation of ROI for Failure Analysis Systems

Field failure	Repair time, $	Sales lost	Salesperson's time	Shipping, etc.	Miscellaneous: phones, etc.
Total $ = $	1out of every 3 sales	1 for every 3 ($XXXX)	$100	$200	$50 to $1000

Multiply failure cost times the yearly number of failures.

Failure analysis system yearly cost	Laboratory*	Equipment†	Technician	Leader
Total = $85,000/yr‡	$4000	$1300	$30,000	$50,000

*The laboratory cost is figured on cost to lease 400 sq ft per year at $10/ft². This varies from one location to another. Ask facilities for this value.

†The cost used for the equipment is usually one-tenth the initial cost because of ten-year depreciation.

‡If the cost of the failure analysis system is greater than the field failure costs by very much, add in the production costs and the cost of buying spare parts. If field failure and production costs combined are not more than the cost of the FA system, it will not be a good investment.

below. Keep the plan and procedure as simple and straightforward as possible. They can be modified and expanded as needed.

13.2.3 Elements of a Failure Analysis Business Plan

Note: This is an outline of a basic business plan. An example of a business plan is presented in the appendix to this chapter. Any plan for a specific company must be based on the actual facts and operating policies and procedures for that company.

Purpose.   State that the purpose of a failure analysis system is to

- Eliminate costly failures
- Increase reliability
- Increase profits

Statement of the Problem to Be Solved.   The problem is too many expensive failures that are hurting the reputation and the profits of the company.

Return on Investment.   Prove that the cost of failures now is very high, and show that a failure analysis system is not expensive and will pay for itself in a short time.

Action Plan

- Set up a failure analysis laboratory.
- Train the failure analysis team.
- Train all the employees that are involved.
- Perform failure analysis and corrective action until the critical and costly failures are eliminated.

Failure Analysis Process and Procedure.   This section of the business plan is outlined in Sec. 13.2.4, "Failure Analysis Process (Outline)," and should be included here in the business plan.

Results.   Monthly reports will show progress and a list of the failures that have been eliminated. The cost savings associated with the results of the corrective action will be reported as the difference between the cost of failure analysis and the cost savings resulting from elimination of the failures.

13.2.4   Failure Analysis Process (Outline)

See Fig. 13.1.

1. Identify, report, and collect data on all major failures in production and in the field.
2. Determine the failure rate and the cost of the failures.
3. Calculate the MTBF (1/failure rate) for all field failures.

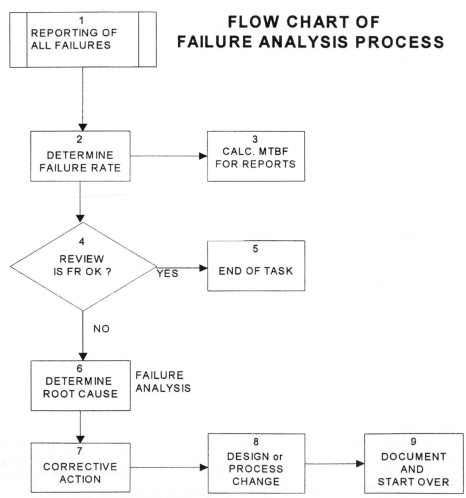

FIGURE 13.1   Flowchart of the failure analysis process.

4. Select the top three failures (selected by expense to the company) for analysis, using Pareto analysis methods. These can be easily determined when the costs involved with each failure are reviewed.

   The cost of a single failure may be small. But if this failure is occurring at a high rate (>10 percent), then the overall cost of that failure will be high. Conversely, a low level of failures of a very expensive assembly in the field could also cause a very high cost in terms of return, repair, and loss of customers. Assign values to each failure and plot them as shown in Fig. 13.2a.

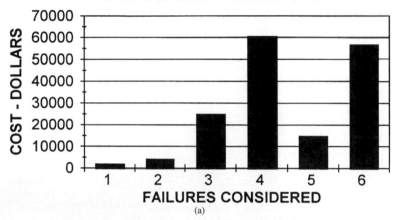

FAILURE COST
GRAPHICAL EVALUATION

	A	B	C	D	E	F	G	H	I
2				COST S				TOTAL	Failure
3	Failure Description	Repair	Ship'g	Spares	Lost	Recall	Qty.	COST	No.
4					Sales	Scrap			
5	1.Production Assy 1	10		35			40	1800	1
6	2.Mother board	75		750			5	4125	2
7	3.DOA in Field	120	100	2000	4000		4	24880	3
8	4.Production Assy 2	55		550			100	60500	4
9	5.Field Unit XYZ	120	200	100		not yet	35	14700	5
10	6.Accessory(Fld.)		25	25		25	755	56625	6

(b)

FIGURE 13.2 (*a*) Failure cost evaluation; (*b*) data for failure cost evaluation graph.

5. If the cost of failures is acceptable or the rate of failures is below the predetermined acceptable level, end this review and start again at step one in the next reporting period.

6. Perform failure analysis to determine the root cause of each failure selected. Section 13.3 explains this in great detail.

7. Determine the most effective corrective action to eliminate the cause of each failure.

8. Make the design, production, or vendor design/production changes to eliminate the root cause and thereby the failures.

9. Document all of the above and start all over again.

13.2.5 The Failure Analysis (FA) Team

Failure analysis is a team task. Suggested members of the primary team are

- **Reliability engineer,** for the failure analysis
- **Design engineer,** for understanding the design
- **Production engineer,** for understanding the assembly processes
- **Quality assurance engineer,** for the failure history data

The team alone is not sufficient to develop a first-class failure analysis program in the company. Many employees are involved, from top management down to the production workers.

Each employee involved in testing and evaluation of the product must be trained to understand the importance of recording every failure accurately and notifying one of the FA team members of the problem. Without an FA system, most failures go unreported and therefore continue until the company's customers get angry. Management must recognize the value of this important tool and support the failure analysis system completely for continued success.

13.2.6 The Failure Analysis Laboratory

A failure analysis laboratory does not have to be expensive or elaborate. Analysis of failures can be accomplished with simple test equipment that is probably available now and a couple of good microscopes. Table 13.3a and b provides a list of the equipment required for a small failure analysis facility for a medium-sized company ($100 million in sales). Failure analysis can be done with less equipment than is shown by using outside labs and renting the more expensive equipment when it is needed. Having the equipment listed will speed up the process and provide better results for a good start.

TABLE 13.3a Failure Analysis Requirements List

1. Laboratory facility with approximately 400 sq ft
2. Four workbenches with 240-V and 120-V outlets
3. Shelves and cupboards on at least two walls
4. Test equipment listed in Table 13.2b
5. A Certified Senior Electronics technician with at least four years' experience in troubleshooting complex electronic equipment

13.2.7 Training

Training the FA Team.   The FA team must be trained to be familiar with what it takes to build a failure analysis system (understanding this chapter). The FA team must become leaders to train the rest of the company.

Training the Rest of the Company.   Train the personnel in the key areas first. Look for the areas with the most failures, and pick out the most expensive failures to work on first. Production is a likely area because of the volume. Field failures may be a close second or first because of the higher cost of failures in the field. Solving one $20,000 failure is better than solving ten $1000 failures. The cost of failures to the company is the best way to set priorities. See the chart in Fig.

TABLE 13.3*b*   FA Laboratory Equipment (Minimum List)

Failure analysis laboratory equipment list (minimum) for startup phase	Estimated cost
1. Zoom stereo microscope, 10× to 200× power, with camera mount (third lens tube)	$3000
2. Metallurgical microscope, 100× to 600×, with camera mount (matching above)	$3000
	$3000
3. Polaroid camera for above	$1500
4. Oscilloscope, dual channel, 100 MHz	$3000
5. Power supplies (2), tri-outputs at 5 V through 30 V	$1000
6. DVM—$4\frac{1}{2}$-digit, high quality	$500
7. Electric drill ($\frac{3}{8}$ in.) and large set of bits; Dermal™ rotary type tool with circular saws, carbide disks, reamers, routers, etc.	$400
8. Tools: Set of files, hacksaw, cutters, clippers, pliers, wrenches, screwdrivers	$200
9. Tabletop vise (4-in jaws) and small hobby vise, multiposition	$150
10. Tabletop grinder (fine and coarse wheels)	$100
Total estimate for the above equipment	$12,850

Depreciated over ten years = $1300 per yr

13.2*a.* Monitor all failures, and get these data into a list (a database program, if computers are available). Then determine the priority list of failures by evaluating the cost to the company.

Comment.   Failure analysis is a companywide task that requires support from the people who find the failures and cooperation from the people who are causing the failures. That includes nearly everybody. *Always address the problem* and not individuals or groups of people. This is good advice for every activity, but it is especially true in failure analysis, where the focus is on serious problems. The causes of failures are many and varied. Nobody wanted to create them, and everybody should want to eliminate them. The FA team has the ability and knowledge to accomplish this.

13.2.8   Selection of the Important Failures

Prioritization of Failures.   Ever hear of finding treasure in the backyard? Well, every company without a failure analysis system has one.

Eliminating the most expensive failures in the company is worth a lot of money. Finding these failures and eliminating them by correcting the root cause will save enough to start a successful failure analysis lab. This has happened in several successful companies. The cost of failures is estimated in Table 13.1. A graphical analysis of the actual cost of failures will help in determining which failures to fix first. Figure 13.2*a* and *b* shows examples of various failures and the cost to the company of those failures. From the graph, the three failures to be analyzed clearly stand out.

Reporting System.   Start a reporting system. Have the field, production, and incoming/receiving managers send reports on the number and type of failures they have each month; they probably are doing this now for their managers. Get accounting to estimate the actual cost of each of the parts that are replaced. Then get an estimate of the labor required to repair or replace these parts. Also, have these areas save and collect the failed parts and assemblies. These will be needed later for analysis.

Determine the cost of delays in production (usually thousands of dollars per day) and/or the recall of products from the field because of repetitive failures in the field.

13.2.9 Results and Progress Reports

Start with the fewest employees that can handle the work. This depends on the size of the company and the volume of products.

Using the failure information gathered, select the most expensive failures. Perform the failure analysis (using the methods described in Secs. 13.3 through 13.5). Then determine and implement effective corrective action as described in Sec. 13.7. Publish the results in a report that clearly outlines the cost savings now and in the future. With this type of information, management will want to support a failure analysis system.

13.3 ROOT CAUSE FAILURE ANALYSIS

See the flowchart of failure analysis in Fig. 13.3.

13.3.1 Detection of the Failure

When a failure is detected, the failure analysis must start immediately. This is why it is necessary to train most of the company in failure analysis. Whoever finds the failure must record all the conditions of the equipment and the environment of the operation at that time. Collecting this failure evidence can make the difference between whether the root cause is determined in a reasonable time or not. The use of a formal failure reporting form may help. A basic (generic) failure report form is shown in Fig. 13.4.

Any form will work as long as it records the same essential information.

13.3.2 Collection of Failure Data

For each failure, start a formal failure analysis file (folder) with a log number so that it can be tracked. Include the initial findings using a failure report form similar to Fig. 13.4.

Look up the history of this part. If there isn't one, create a database of previous failures that is sorted by assembly or component part number or description. A history file of failures is essential to any good failure analysis system.

13.3.3 Visual and Physical Examination

Make an external visual and physical examination to see if the component has been damaged or modified in any obvious way. Look for shipping damage or evidence that the component has been dropped or abused in any manner.

13.3.4 Verification of the Failure

Verify the failure symptom, if possible. Test the assembly for its normal functioning under the same conditions as reported on the initial failure report and record the symptoms.

If the unit fails as before, record this and move on to the next step.

However, sometimes the assembly will operate per specifications when it is moved into the failure analysis lab. This can be true for many reasons that will be explained later as we discuss the basic causes of failures.

If the failure cannot be duplicated, take the component (assembly) back to its original setting, where the failure was first detected. Reinstall it exactly as it was. (This is another reason why the initial failure data in step one are so important.) If the failure has disappeared, then look for the logical reason. Usually it will be a connection that was or is intermittent, or the component itself will have an intermittent failure mechanism. Stress test the assembly through temperature limits

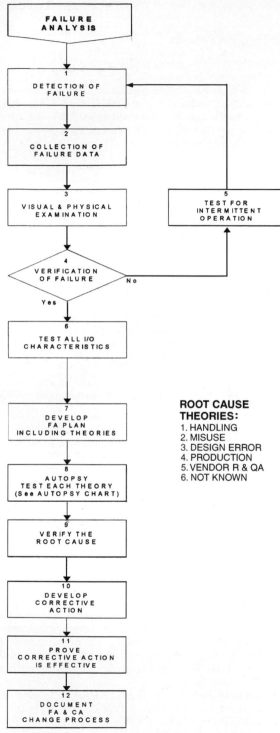

FIGURE 13.3 Failure analysis flowchart.

Report #	Assy P/N	Name	Dept.	Date: Time:
Symptoms of Failure:			Environmental Conditions: Temp._____ Hum._____ Other_____	

Vendor:	Date Code:	S/N:	Operation at time of Failure:

Trouble shooting or checkout that was done to evaluate the mis-operation and determine that the equipment had failed.   Include an estimate of Time spent.

Was the Failure symptom consistent during Trouble-shooting? Yes[ ] No [ ]	If [No] , how did the symptoms change ?			
Catagory of Failure Mode.	Injury ? Yes[ ] No [ ]	Smoke / Fire ? Yes[ ] No [ ]	Elect. Shock? Yes[ ] No [ ]	Leaks ? Yes[ ] No [ ]
Criticality of Failure : Need FA ; 1 Day [ ] 1 Wk. [ ] 1 Mon. [ ]	Cost = _____	Time Lost ? _____ Hours		

Other Misc. Information pertinent to the Failure:

FIGURE 13.4   Failure report form.

and with low levels of vibration and shock to stimulate the unit to fail again. If it does not fail again, just wait. If it never fails again, assume that the operator made an error or that a dirty connection was cleaned in the process of disassembly and reinstallation.

13.3.5   Test All Characteristics

Perform nondestructive tests for normal and expected characteristics on all the inputs and outputs.

   Note: There may be a need to have good assemblies and parts for comparison of the I/O characteristics. It depends on how well this assembly is known. A good assembly is very useful for practicing the disassembly and testing. Record all the measurements. These practice assemblies will be valuable as a reference as the analysis continues. Things may change as the unit is disassembled.

13.3.6 Failure Analysis Plan

Planning is difficult and takes extra time; however, in failure analysis, *planning is essential.*

Review all the data from the original input along with the data and observations just recorded. Then review the schematics and the design specifications. And then, brainstorm with the FA team on what the likely root causes that would result in this particular failure may be. (Basic root causes are described in Sec. 13.4.) This is where the training and experience are really put to use. If the team identifies likely theories as to the root causes, it is highly probable that the root cause will be found.

Use fault-tree analysis (FTA) to determine the components that are likely to be causing the failure symptoms. FTA is described in Chap. 6 of this handbook. FTA is a logical way of determining what subassemblies can cause the output failure that was observed. But don't stop there. Get a good unit from production and test it. Study it. Also study the schematics and the assembly drawings. Know as much about this assembly as is possible before starting the disassembly. Each theory must be tested in the autopsy analysis, as explained in Sec. 13.5.

The failure analysis plan is important, but it is a guide, not a law. It must be a "living" document; follow it until it doesn't make sense. Then regroup and revise the plan.

13.4 BASIC ROOT CAUSES

The following should be the basis of the team's theories as to the root causes in planning the failure analysis.

There are *six* primary root causes of failures that are commonly found through failure analysis:

1. Shipping or handling damage
2. Misuse of the product
3. Design error
4. Production process error
5. Vendor error
6. Unknown

1. Shipping or Handling Damage. This is usually found during the first visual inspection of the shipping package and/or the outside case of the unit. It is then confirmed during autopsy by finding a broken part inside. Another clue is that many parts are bent in one direction (and some broken) because of the shock of dropping the case.

2. Misuse of the Product. Misuse occurs when the product is used in a way that is not covered by or is well beyond the specified conditions of operation. An example would be a computer being operated in the very hot environment of a mobile home while traveling over a very bumpy road. To make things even worse, the power supply might vary well below and above the specified range. Failures in this case are likely to be caused by this misuse. Misuse can be a difficult root cause to prove. Basically, all other causes of an overstressed part have to be eliminated before the customer is suspected. Then, some very tactful and skillful conversation must be held with the customer. Involve the salesperson directly with the customer, if at all possible, to find out how the unit was used. Another clue to this root cause is that product is failing only with one customer and not with other customers. Be careful on this clue. For example, the use of walkie-talkies by maintenance caused equipment to fail at only one customer's location. Good root cause failure analysis revealed that radio-frequency interference (RFI) caused these units to fail. This was a design error in not designing for, or testing for, RFI.

3. Design Error. If a part has failed and has the appearance of overstress, then a design error is likely. An analysis of the application of that part is necessary. This is called stress analysis by most reliability engineers. Calculate the maximum voltages, currents, power, and internal temperatures for the worst-case conditions specified for operation. Then compare those values to the specified maximum operating parameters recommended by the vendor of that part. A good design will always have application stress with a safety margin of 10 to 50 percent below the vendor's maximum values.

If the application stress causes the part to have one of these parameters (volts, amperes, watts, or degrees) above the vendor's ratings, the root cause of the failure is very likely a design error. This root cause can be easily proven by testing some known good parts in this overstressed condition.

For example, a transistor that has a maximum voltage rating (V_{CEO}) of 45 V and is used in a switching application for 28 V fails. What is wrong? A switching application into an inductor often produces two times the voltage being switched across the transistor. In this case, this would result in spikes of 56 V. The transistor may work for a while, but eventually it breaks down and shorts out under repeated spikes of voltage above its rating.

4. Production Process Error. Production errors are usually obvious and therefore easy to find. Examples are nicked traces on PC boards, cold solder connections, broken leads, loose parts, and loose connectors. But sometimes these errors are not so obvious. For example, one company had a high failure rate of a large power transistor. The application was not derated as much as it might have been because it was more expensive to buy four better transistors than to fix the 1 percent field failure rate that was expected. However, the failure rate was over 10 percent and rising. Something was wrong.

Examination of the failed transistors showed that they had failed with collector-to-emitter shorts. This is usually a symptom of high-voltage overstress. Testing the machines for higher than normal voltage proved that high voltage was not the reason. Therefore, it must have been something in the production process. A review of the production processes exposed the problem within a couple of hours. The large 0.04-in. leads on these germanium transistors were being bent over to hold the transistor onto the board prior to wave soldering. This broke the glass-to-metal seals, which were now hidden from view. This production error was quickly and easily corrected. The failure rate returned to its usual 1 percent. Later, reliability and component engineers developed a screening specification to be applied to the transistors at a nominal extra cost and the failure rate went down to a normal 0.05 percent, where it should have been from the start.

When you suspect production process failure, walk around in the production area and ask the supervisor to tell you how parts are handled and assembled into the product. In most cases, this is an eye-opening experience.

5. Vendor Error. Vendor errors can include any or all of the above basic causes. For example, a supplier of power supplies could have failures because of poor quality control on the part of a supplier. Sometimes the chain of investigations leads back to a supplier of the basic materials, as shown in Fig. 13.5.

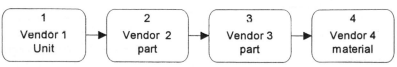

FIGURE 13.5 Vendor chain to find root cause.

The best clue to a vendor defect as the root cause is the failure of one specific vendor's parts, especially if all defective parts come from the same data code and there are no failures from a second source. Vendor problems are discovered after the design and production theories have been proven invalid (not the source of the problem). Vendor problems usually happen suddenly, without any previous history. And they may disappear just as quickly. Unfortunately, if they are not found and corrected, they will probably show up again in the future.

Comment: Do not eliminate (change) vendors because they have a problem now and then. Experience with vendors has shown that all vendors make mistakes. The best vendors admit it quickly and solve the problems on their own through effective failure analysis and corrective actions. Poor vendors usually put themselves out of business. What is true for vendors is also true for any company. Work with vendors to trace the failure mechanisms to the root cause. The root cause may be in the materials supplied to the parts vendor by a primary vendor.

If some of your vendors are small companies without the advantage of failure analysis systems, lend (or buy) them a copy of this reliability handbook. It will pay good dividends.

6. Unknown.   Some small percentage of failures may not be resolved to a known root cause. The experience, knowledge, and techniques of the investigator will determine the percentage in this category. The more analysis done, the better the success ratio. Practice does make perfect in this activity. Remember to use the FA team for brainstorming the problem, and review the FA plan and all the data taken so far. Going much further may be difficult and prohibitively expensive. If this is true, move on to another failure. If the original failure is persistent, there will be another chance (more failures), and analysis will resolve it later.

Caution: A critical failure that causes a hazard for customers should be analyzed until a root cause is found. If there are only one or two failures, recreate the conditions and try to cause additional failures on units from the same lot. Stress some good parts to accelerate the failures. Get outside help if necessary.*

General Comment: Look for the root cause as described in the six primary areas described in Sec. 13.4. Do not stop the analysis at just a symptom.

For example, a burned-up resistor is found. One could *assume* that it was a bad vendor part and close the FA file. This would not be proper. Stress analysis needs to be performed to determine that the design was derated properly. There needs to be a review to see that the production process was being done according to specifications, and a search to prove there were no bad or mismarked parts in that lot of resistors. Then maybe the root cause will really be found.

13.5   *AUTOPSY PROCESS*

See the flowchart in Fig. 13.6.

CAUTION: This is a very critical phase of the analysis.

Some of the information in this section is repetitive. This was done on purpose because this is an important and difficult part of failure analysis. (Also see Chap. 14, "Physics of Failure.")

A. Carefully disassemble the failed unit to find the part or parts inside that failed. Practice on a good, disposable component first if it is not *absolutely* certain that how to open the component without damage is known.

---

*There is a wealth of help and assistance in industry, probably at a company close by.
1. Large electronics companies, if they have a failure analysis laboratory, will usually accept outside jobs at a high, but reasonable, fee; $500 minimum was typical in 1993.
2. Professional failure analysis laboratories advertise in many IEEE magazines, especially the *IEEE Transactions on Reliability.*

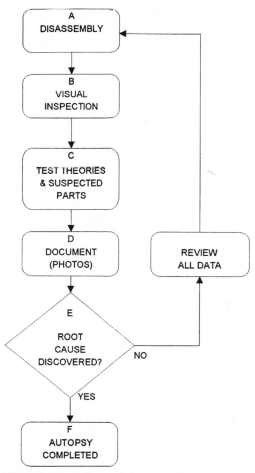

FIGURE 13.6   Failure analysis autopsy flowchart.

B. Make a visual inspection before doing anything else. Record anything unusual. Photographs are worth the cost and effort at this point.

C. Test the suspect parts. Know which parts are suspect from the I/O characteristics recorded before. Trace back from the incorrect input or output to the parts that connect to it. Keep testing each part, going away from the nonfunctional input or output. The part that caused the abnormal characteristic will be there. Parts in this investigation include solder connections, traces, PCBs, etc. Anything in the assembly can be the site of the failure.

If the failed assembly is a discrete part (like an IC), this will require additional training and equipment that is beyond the scope of this chapter. Most small- to medium-size companies do not need this capability and can usually obtain failure analysis on discrete parts from vendors or from failure analysis laboratories that specialize in this very difficult craft. The technique is the same, testing the inputs and outputs and then a careful disassembly. The disassembly is now a matter of art and experience with some very sophisticated and dangerous tools. Boiling nitric acid is one of those tools. This is the reason this is not being covered in a basic description of a failure analysis autopsy.

If several parts have failed, a review of the schematic will be necessary to determine which one of these parts can cause the others to fail when it fails in a certain way. Then prove that this is what happened by recreating the failure on a breadboard or another unit.

D. Document everything. Document (with notes, photographs, drawings, etc.) the evidence of the failure immediately as it is observed. Many times something important is lost as the disassembly takes place. For example, if a PCB connection is unsoldered to remove a suspected bad part, the evidence of a cold or cracked solder joint may be lost. When the part tests good and the unit works when it is replaced, it will then be necessary to have a photograph of that solder connection to determine the cause. Take the necessary time and take lots of data, look at all connections, and record the state before breaking the connection. Test before and after each part is removed.

E. If the root cause has not been discovered at this point, get the FA team together and review the progress to this point. Develop some new theories and then start all over again with more disassembly, more visual examinations, more testings and analysis. Get help from consultants or FA laboratories. If the failure is still there, someone can usually find it.

This is also the time to decide if resolving this failure is worth a lot more expense and time. If it is not critical, it may be prudent to move on to the next most important failure.

F. Case closed: The autopsy is done. Write the report and do one of two things:

If the root cause was found, verify that it really was the root cause. That is the next step in the failure analysis flowchart.

If the root cause was not found and it isn't worth it to go any further, file the report and wait until the next time. If there really is a root cause, chances are that it will cause more failures. If a root cause was not found, a random unexplained failure may have happened and been lost in the disassembly. It is likely that it will be a long time before another one just like it happens again.

13.6 DESTRUCTIVE PHYSICAL ANALYSIS

Destructive physical analysis (DPA) uses the same analysis techniques described above. The purpose of this procedure is to examine new or competitive devices to determine the internal structures and therefore evaluate the inherent reliability and quality of those devices and the manufacturing processes.

In critical components, such as spacecraft parts, DPA is used to ensure the quality and reliability of each lot of devices to be used in the critical assemblies. It is required for all critical parts in most manned spacecraft projects. DPA is good for learning about the destructive disassembly aspect of failure analysis.

DPA provides an understanding of how the component is manufactured and the processes of the manufacturing. This is extremely helpful in analysis of failures.

Also, an understanding of how the component functions under various conditions is necessary in order to understand the modes of failure and how these are associated with the design and manufacturing processes.

For example, critical parts for a pacemaker are being purchased from a sole-source vendor at very high cost. Testing the parts is expensive and can use up valuable life of the part.

Let's assume that the part in this case is a low-power microprocessor. One hundred percent inspection is not possible. This is a good reason for performing DPA. When the vendor supplies a lot (a group) of devices, take a 10 percent random sample and perform DPA on these parts. The parts will be tested for characteristics and performance and carefully disassembled (just as in a failure analysis autopsy). Then all of the assembly processes will be tested. Lead bond strength will be measured. The surfaces will be tested for contaminants. Microscopic or SEM photographs will be taken and studied to find any flaws in the manufacturing processes. The trace widths and thickness can be measured and compared with design requirements. The purity of the materials can be evaluated. This can be as thorough or as quick as desired. But it will provide the final answer as to the quality of the remaining parts in that lot. Laboratories with professionals trained in this science charge handsomely for their abilities and expertise, but they usually are worth every penny if the need for quality and reliability is high enough. If the pacemaker in this example was helping a loved one's heart function, it would be worth the time, effort, and cost.

13.7   CORRECTIVE ACTION

13.7.1   Short Term

Short-term corrective action is usually necessary while long-term changes in design or significant changes in production are being implemented. These long-term changes usually must be done in accordance with a schedule and fully documented. Thus, short-term corrective action needs to be taken to limit the effect of the failure mode (root cause) while the long-term corrections are being made. A team effort is highly recommended to devise a cost-effective and immediate action that will provide sufficient correction of the failure. As an example of this, let's suppose that the output stage of a power supply was poorly designed and cannot supply the rated maximum power (current and voltage) at 45°C. The long-term corrective action is to redesign the output stage with a larger heat sink and more efficient power transistors. The short-term solution might be to increase the heat sink surface area by placing small additional heat sinks (rather expensive) between the power transistors and the existing heat sink. This will increase the heat dissipation sufficiently to just meet the specification. The long-term solution of a redesign will be much less expensive and provide a design margin, so that the unit will operate at least 10°C below the specified maximum operating temperature.

13.7.2   Long Term

As stated in the example above, the long-term solution must be a complete and well-thought-out one that absolutely eliminates the root cause of the failure and provides ample safety margin so that the specified operation conditions are *easily* met. If long-term solutions can be made in adequate time, don't take any short-term action.

13.7.3   Implement and Verify the Corrective Action

It is necessary, and desirable, to prove that the corrective action is effective, not only in eliminating the cause of failure, but also in providing some degree of safety margin. The following steps should be performed on a prototype or the first production units after the corrective action has been implemented:

1. Test the unit to determine that it operates normally through the minimum and maximum ranges of operating conditions. These should include temperature, vibration and shock, line voltages, etc. This testing is done to ensure that the changes, even if they fix the root cause of the failure, have not changed the other operating characteristics of the unit.

2. Stress-test the unit at 10 percent above the specified minimum and maximum ranges (as above) to determine that the unit has adequate design margin. If this unit can cause a harmful or dangerous output, then test the unit for all known single-fault failures that can cause a dangerous output. It might be wise to also test for a higher margin of safety, such as 20 percent.

3. Finally, simulate the conditions that caused the original failure or failures, and make sure that these conditions do not cause a failure now. A minimum of 48 h of power on/off life cycles during simulation is highly recommended to determine the reliability of the corrective action

implementation. A longer life test may be necessary if the previous root cause was a long-term degradation of some part to the point of failure. If the basic root cause of the failure has been determined and an effective corrective action has been designed, it should be possible to turn the failure on and off.

13.8   CLOSING THE LOOP

13.8.1   The Failure Analysis File

The failure analysis file should contain the following information:

- The initial failure report with all the information about the conditions which preceded the failure.
- The initial failure analysis log, which identifies the unit, its serial number, date code, etc.
- The total failure analysis procedure plans, notes, and findings.
- A complete description of the root cause of the failure and the corrective action taken to eliminate this failure mode.
- A test report that shows verification of the root cause and that the corrective action was effective.
- Documentation to show that corrective action was implemented on a specific date. This will be helpful in answering various authorities' questions about additional failures in units prior to the corrective action.

13.8.2   Product Development Procedures

To ensure that the lessons learned do not have to be learned again (the hard way), change the company's product development procedures. Find the appropriate section that covers the appropriate part of the process and change it so that this root cause will not occur in the next product. If there isn't such a working document, now is the time to initiate one. The stockholders of the company will be forever in the FA team's debt.

13.8.3   Closing the Loop (The Final Word)

1. Send the failure analysis and corrective action information to all design and production managers, for distribution to all engineers. This should help prevent making the same mistakes in the near future.
2. Put the information in the design guidelines, manufacturing standards, and/or component evaluation procedures. This should help prevent making these common mistakes in the long-term future.

13.9   BIBLIOGRAPHY

1. MIL-STD-785, Task 104, "Failure Reporting, Analysis, and Corrective Action System (FRACAS)."
2. MIL-STD-781, Sec. 4.6, "Failure Reporting, Analysis, and Corrective Action System (FRACAS)."
3. Rome Airforce Development Center (RADC), *Reliability Engineers Toolkit*, Gritfiss Airforce Base, New York.
4. MIL-STD-883, *Test Methods for Microcircuits.*
5. ICE, *Visual Inspection Criteria for IC's and Hybrids.*
6. General Electric, *Component Technology and Standardization,* 3 vols.

7. IEEE Reliability Physics Symposium Tutorials (1990–92).

8. Dicken, Howard K. *How to Analyze Failures of Semiconductor Parts.*

9. Devaney, John R. *Failure Analysis Mechanisms, Techniques and Photo Atlas,* 1983.

Three tutorials published in *International Reliability Physics Symposium—Tutorial Notes,* 1990.

10. Mann, J. E. "The Start-up of a Failure Analysis Laboratory."

11. Murphy, Charles R. "A Review of the Failure Analysis Process for Parts."

12. Beall, James. "Failure Analysis Outline Guide for Semiconductors/Integrated Circuits."

APPENDIX: EXAMPLE OF A FAILURE ANALYSIS BUSINESS PLAN

Purpose

The purpose of this failure analysis system business plan is to define the requirements and the benefits of a failure analysis system. A failure analysis system can eliminate critical and costly failures that hurt the company's reputation for reliability and eat away at profits.

The Problem to Be Solved

Repetitive failures hurt business. The company may be losing customers because of poor reliability. Return and repair costs are higher than forecasted budget. A recall of some products is likely if the failure rate grows.

Return on Investment

The cost of failures, especially in our customers' use, is far greater than the cost of setting up and maintaining a failure analysis system. The failures that are eliminated in production will be extra returns on the investment.

Table 13.1 shows the estimated costs of failures of equipment in the customers' location. And Table 13.2*a* and *b* shows the facilities needed and the estimated cost of setting up a failure analysis system.

Multiply the number of failures over the last year and compare that with the cost of the failure analysis system.

Action Plan

(Include the flow diagrams for failure analysis shown elsewhere in this chapter.)

1. Locate and modify a laboratory area of 500 to 1000 ft^2.

2. Acquire or lease the equipment shown in Table 13.2*a* and *b*.

3. Assign or hire a Certified Electronic Technician with about 4 years' experience in troubleshooting electronic equipment.

4. Train the technician and all the production test personnel in failure analysis detection and reporting. Use this chapter and its references to learn how to do this.

5. Begin failure analysis in accordance with the failure analysis process described below.

Failure Analysis Process and Procedure

This part of the plan should be similar to the failure analysis process outline in Sec. 13.2.4.

Results

All failure analysis activity shall be summarized in monthly reports. These reports will list the failures that have been identified, analyzed, and corrected. The cost of the failures will be estimated, and the cost of the analysis and the correction action will also be calculated. The difference between these will be listed as a return on investment.

CHAPTER 14
PHYSICS OF FAILURE

Jack S. Smith
LOCKHEED PALO ALTO RESEARCH LABS

14.1 INTRODUCTION

This chapter's objective is to provide the reader with an understanding of the underlying causes of failure of any product, whether it be a transistor or a bridge. Since the author's experience is with electronics, the examples posed will deal with electrical devices, but the principles are universal. The chapter discusses the fundamental failure mechanisms, one or more of which operate on virtually any product. The extension of those mechanisms to a particular situation will permit the selection of the right product for the application. In addition, a knowledge of those mechanisms should give insight into how to redesign or test for high reliability.

It is imperative to have a firm grasp of how the ideal product works before one can begin to enumerate the many ways in which it can fail. That is, upon what principles does the flawless device rely for operation? With those principles understood, the product can be examined to determine the physical construction details that permit those principles to function. Once construction details are known, one can assess whether the margins in physical construction are large enough to accommodate the myriad ways in which the product can fail. It is unrealistic to think that failure modes and mechanisms can be anticipated without a knowledge of both construction and principles of operation.

In this chapter, the principles of operation for a semiconductor PN junction (diode) are discussed. We choose the more complex PN junction primarily to provide the reader with a feel for the level of detail one must delve into to obtain a high-quality product. The operation of other electronic devices, such as resistors and capacitors, relies more heavily on basic material properties. The operational simplicity of passive devices has caused them to be given only brief consideration in the search for high reliability. Since all the parts in an electronic system must function flawlessly, each part must be given the thorough review often accorded only complex integrated circuits. Because the operational principles of simpler devices are considerably easier to understand, that review should take less time, but it cannot be ignored.

The mechanisms cited in this chapter operate broadly against virtually any man-made device, ultimately preparing the device for the scrap heap. These failure processes can be slowed down but never stopped. Construction details determine whether these processes have been slowed to the point where obsolescence wins the race. The part designer's goal is to have obsolescence win that race at the lowest possible cost.

While a complete catalog of failure mechanisms can be reviewed, it is difficult to anticipate the variety of ways in which a device might be constructed. There is neither room nor a sensible method of presentation that will help the user connect details of construction with part reliability.

This chapter provides only the *functional form* for the rate at which events occur. It should become clear that space, i.e., distance, and the time to fail are intimately connected. Once the form is known, it is up to the reader to insert the *parameter values* specific to the materials involved and determine the *limits* of integration. The parameters and limits are determined from material properties and dimensions, which are obtainable only from the details of construction.

14.2   DEVICE PHYSICS AND RELIABILITY

The electrical characteristics of any device depend upon the device's material properties and design. A resistor, for example, relies on the resistivity of the material chosen to form the resistor and the shape of the resistor to determine the specific resistance level. If the value of the resistor meets design, this does not mean that the device is of high quality and reliability. The failure of a device's performance to meet design should be a signal for further investigation. Only after the cause of the discrepancy is identified can a judgment as to quality and/or reliability be made. To be aware of the signal, one must know the connection between design and electrical performance.

The following section will draw upon semiconductor theory in order to make the connection between construction and electrical characteristics. There are many fine books[1-5] on the subject. Perhaps the best for beginners is Grove's *Physics and Technology of Semiconductor Devices.*[5] For those wishing to explore the subject more deeply, Sze's *Physics of Semiconductor Devices*[3] is recommended. Describing the development of the theories is not within the scope of this handbook. We shall only give the results of many years of hard work and illustrate how those results can be used to advantage.

14.3   SEMICONDUCTOR JUNCTIONS

The semiconductor junction is the heart and soul of virtually every modern active circuit, integrated or otherwise. An intimate knowledge of PN junction theory is essential for those responsible for obtaining and assessing microcircuit products. The dc current-voltage characteristic of a junction is given by[6]

$$J = J_s \, (e^{qV/kT} - 1) \tag{14.1}$$

where J = current density, A/cm^2
V = potential across the PN junction, V
J_s = saturation current density, A/cm^2

Notice that the equation is given in current density J, i.e., current per junction area. Here is the first hint of construction details determining the final measurable result; namely, you apply a voltage across a junction and you get a current the magnitude of which is directly proportional to the junction area.

The diode equation is deceptively simple. For it to be useful we need to know more about V and J_s.

$$V = \phi_{Bi} + V_a \tag{14.2}$$

where ϕ_{Bi} = built-in potential, V
V_a = applied voltage

ϕ_{Bi} results from the formation of the PN junction. A junction with no externally applied potential is in a state of dynamic balance. A small voltage in the forward direction immediately upsets that balance. ϕ_{Bi} is responsible for the forward voltage drop observed in diodes (approximately 0.5 V for silicon junctions). The reader should consult device physics texts for the complete story. The built-in potential ϕ_{Bi}, in turn, is given by[7]

$$\phi_{Bi} = \frac{kT}{q} \ln \frac{N_A N_D}{n_i^2} \qquad (14.3)$$

where k = Boltzmann's constant
       T = absolute temperature
       q = electronic charge
      N_A = acceptor concentration
      N_D = donor concentration
      n_i = free carrier concentration

The term kT/q is merely a constant (approximately 26 mV). N_A and N_D are the concentrations of dopants on either side of the junction, and n_i is known as the free carrier concentration.[8] You will see n_i in many semiconductor equations. This term is a powerful function of temperature, namely, $n_i^2 = N_C N_V e^{-Eg/kT}$, and is largely responsible for the temperature sensitivity of many semiconductor properties. E_g, N_C, and N_V are material properties describing the band gap and concentration of available energy states in the conduction and valence bands, respectively. In the case of Eq. (14.3) we are taking the natural log of the exponential function in n_i. This action removes the exponential, making ϕ_{Bi} a linear function of temperature. It is so linear, in fact, that ϕ_{Bi} can be used as a good thermometer. Notice that Eq. (14.3) has reduced us to material properties, i.e., n_i, and construction details, N_A and N_D. The spatial distribution of the dopants within the semiconductor will determine the built-in potential and hence the forward voltage drop of a PN junction diode. The forward voltage drop is seen in Fig. 14.1 to be approximately 0.6 V.

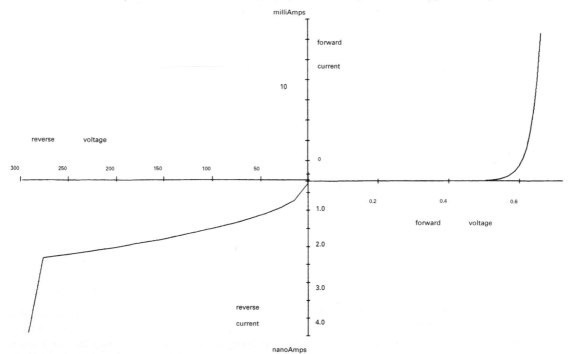

FIGURE 14.1   Current-voltage characteristics of a typical silicon diode. These characteristics were calculated using the diode equations in this section. The results are for a planar one-sided step junction with the more lightly doped side at 1×10^{15} atoms/cm³, a temperature of 300 K, and an area of 100 mil². Mobility and critical electric field data were taken from Refs. 2 and 3, respectively.

Having arrived at an expression [Eq. (14.3)] which depends only on geometry and material properties, we take this opportunity to show how this knowledge can be used to assess reliability. Suppose that for one reason or another you are concerned that the forward voltage drop in a selected silicon diode remain constant in time at a particular temperature. In other words, for your application, a failure would result if ϕ_{Bi} changed. A look at the relationship should give you a great deal of satisfaction. In silicon the dopants (N_A, N_D) do not move around at room temperature. They are locked in the silicon lattice. For the same reason, N_V and N_C will never change. And KT/q is a constant at a given temperature. One could expect the forward voltage drop of a diode to remain constant over the foreseeable life of the product. One remaining reason to be content is that the local electric fields in the forward voltage direction are small. More on this subject later.

The saturation current density J_s is given[9] by Eq. (14.4) for an abrupt junction biased in the reverse direction with the P dopant concentration far higher than the N dopant concentration. An abrupt junction occurs when one side of the PN junction has far fewer dopants than the other side, leading to an abrupt dopant-level transition.

$$ J_s = q \sqrt{\frac{D_p}{\tau_p}} \; \frac{n_i^2}{N_D} + \frac{q n_i W}{\tau_e} \tag{14.4} $$

where D_p = diffusion coefficient
τ_p = minority carrier lifetime (holes in N-type material)
τ_e = effective lifetime of carriers in the space charge region
W = space charge width

The first term on the right is called the diffusion current density, while the second is the recombination-generation current density. The original Shockley equation contained only the diffusion component. The recombination-generation current was not discovered until silicon PN junctions were made and the expected dramatic reduction in reverse current failed to materialize. The explanation was given by the now famous Sah, Noyce, Shockley paper[10] on midgap recombination-generation centers in silicon. Every serious student of semiconductor physics should study this paper.

For wide-bandgap materials such as silicon, the diffusion component is ignorable at room temperature. Germanium, with a bandgap half that of silicon, has a diffusion term which dominates even at room temperature. D_p is given by $(kT/q)\mu_p$, where μ_p is the mobility of holes in N-type material. τ_p is the lifetime of holes in N-type material, and N_D is the density of donors in the N-type material. Since the vast majority of the electronics world is based on silicon, we concentrate on the second term in Eq. (14.4), the recombination-generation term. Equation (14.5) describes the space charge width W.[11] Notice that it is dependent on the applied voltage.

$$ W = \sqrt{\frac{2K_s \varepsilon_0}{q} \left(\frac{N_A + N_D}{N_A N_D} \right) (\phi_{Bi} + V_a)} \tag{14.5} $$

where K_s = dielectric constant of silicon
ε_0 = permittivity of free space

Examining Eqs. (14.4) and (14.5), we are back to material properties, i.e., dielectric constant ($K_s \varepsilon_0$) and n_i, and geometry, i.e., the doping profiles N_A and N_D. The remaining undefined term τ_e is the effective minority carrier lifetime. It is determined by[12] $\tau_e = 1/\sigma v_{th} N_t$, where N_t is the trap density, σ is the capture cross section, and v_{th} is the thermal velocity. Any unwanted impurities or crystal defects which sit at midgap in the space charge caused by the reverse bias are potential sources of hole-electron pairs by serving as stepping stones for carriers between the valence and conduction bands. Once a hole-electron pair is present in the space charge, the electric field sweeps the carriers to the contacts, giving rise to current. All such impurities present are lumped together to become the trap density. The process of forming a PN junction (for example, by high-temperature diffusion) tends to drive in unwanted impurities, creating a certain number of dislocations, impurities, etc., all of which form trap centers N_t in the critical high field region, i.e., the space charge of the junction.

As the applied voltage continues to increase in the reverse direction, there comes a point when the current no longer behaves according to Eq. (14.4) (see Fig. 14.1 at -280 V). The junction has been sent into breakdown. "Breakdown" is an unfortunate term, since it implies damage being done to the junction. It is certainly easy to cause permanent damage while in the breakdown conduction mode, but merely entering this conduction mode does not mean that permanent damage has been done. The conditions required to obtain permanent damage will be discussed later [see Sec. 14.5.5, "Electrical Overstress/Electrostatic Discharge (EOS/ESD)"]. Avalanche breakdown is the prevalent form in most devices. Functionally, the voltage at which avalanche breakdown occurs for a one-sided abrupt junction is described in Eq. (14.6).[13]

$$V_{Br} = \frac{K_s \varepsilon_0 E_{cr}^2}{2q N_B} \tag{14.6}$$

where N_B = dopant concentration (more lightly doped side)
E_{cr} = critical field strength

N_B is the dopant concentration on the more lightly doped side, whether N-type or P-type. E_{cr} is the critical field at which avalanche begins to take place. E_{cr} is a slowly varying function of the dopant type and concentration, being approximately 5×10^5 V/cm.[14] An extremely important consideration for the correct use of Eq. (14.6) is to include field enhancement effects that occur when the doping profile is not planar. Unfortunately, almost all modern microcircuit junctions form a sharp curve as they come to the surface. This is the natural consequence of the diffusion process. The curvature will cause field enhancement. It is this enhanced field which needs to be used when calculating breakdown voltage.

The equations describing the current-voltage characteristics of a diode must not be treated as if they were simply theoretical expressions. Many years of effort went into their development, and certainly for silicon they are remarkably accurate. It would be a mistake not to use them to advantage. As with the resistor analogy, one can calculate the current which will be observed when a known voltage is applied, provided the geometry (doping profile and junction area) is known. Visual examination of the chip will reveal the junction area, and the doping profile can often be measured using capacitance-voltage techniques.[15–18] The radius of curvature can be measured using metallographic cross sections accompanied by stains to bring out the junction. Often such details as radius of curvature are not required. They can be estimated with sufficient accuracy to allow the user to form a judgment as to quality and reliability. The reverse bias current is by far the most important when attempting to assess the quality of a product. The high fields which are afforded in this direction permit serious faults to show up as extra current beyond what would be calculated.

We have spent a great deal of time on the diode equations. The justification is their use in assessing the quality and reliability of microcircuits (normally input and output junction characteristics can be conveniently measured). No less useful are the equations which govern multijunction devices such as bipolar and MOS transistors. There is not enough space to deal with them all here. Let us explore how the knowledge of a junction's current-voltage characteristics can be used for the purpose of achieving high quality and reliability.

14.4 PARTS SPECIFICATIONS

Manufacturing any semiconductor device involves a myriad of technical and economic tradeoffs. Perhaps the most difficult technical challenge is to maintain functionality over a wide temperature range. As we observed, the term n_i appears in many of the fundamental equations. To produce large quantities of a finished product, the chip maker would like to have the loosest possible electrical specifications. Working against this objective is the requirement to have a useful device. As a consequence, the manufacturer often exceeds, by orders of magnitude, a number of the specifications the chip must meet at room temperature. One prominent example is the input leakage current of a high-speed logic chip. Usually, this particular specification is 1 μA at a reverse bias of 5 V. A typical input junction (area, doping profile, and minority carrier lifetime) will lead to calculated reverse currents of less than 10 pA (five orders of magnitude less than the specification!). Under these circumstances, there is no connection between part quality and specification. In fact, if you were to measure 0.5 μA of current for a 10-pA device, you should reject the device for use even though specifications are met. The 0.5-μA device might be perfectly reliable despite its anomalously high current. The difficulty is that one must know the reason for the extra current before a judgment on reliability can be made. Normally, such an investigation is not justified. It is simpler to reject the device.

The mere fact that a part meets its electrical specifications does not mean that the part is high quality or highly reliable. It only means that the part is capable of meeting certain performance criteria. Quality and reliability are another matter. Inferring more is often a major error.

Quality and reliability assessment must be measured against the design, not the electrical specification. This is where a knowledge of construction details enters the equation. Even if the part's electrical characteristics meet its intended design features, we are assessing only the quality of manufacture. It has been observed that quality and reliability go hand in hand. This is largely true, since one usually designs with reliability in mind. Deviations from design often affect reliability as well as quality. Therefore, though the material discussed above is used mainly to achieve quality, one often gains reliability in the process. The subject of the rest of this chapter is those mechanisms which operate to cause the failure of an electronic part and thus directly alter reliability.

14.5 FAILURE MECHANISMS

After an electronic part (or bridge, for that matter) has been constructed, the only way in which it can fail is by having some portion of it move. In other words, matter must move to change (electrical) performance. This trivially obvious statement is often forgotten in the search for reliability, yet it is the essential basis for any scheme which is supposed to provide assurance of long life. The only way matter can move is if it is provided with energy. Energy takes many forms. A discussion of the various energy forms will give the reader the insight necessary to spot design weaknesses and construct meaningful reliability tests for the product of interest.

14.5.1 Thermal Energy

While we don't perceive it, all matter is in a state of motion as the result of being at a temperature above absolute zero. The potential for the displacement of atoms is always present. Whether the atoms can actually be displaced depends upon the local forces binding one atom or molecule to another. In the case of a liquid or a gas, the binding forces are small compared to the kinetic energy each molecule possesses, and hence the atoms or molecules are free to wander with little hindrance throughout the bulk of the material, and throughout space as well. For this reason, it is necessary to offer our own binding force in the guise of a container to hold the liquid or gas. For solids, the binding forces are higher than the kinetic energy of the molecules. The atoms or molecules are more or less confined to a fixed location. But even in this situation there is a finite chance that one molecule can gain sufficient energy to escape its local restraining forces and move to a new location. Physicists view this procedure as a particle in a box (see Fig. 14.2).

The walls of the box represent the restraining forces on the particle. The particle is assumed to be in a constant state of motion, hitting the box walls and being *perfectly* reflected with no loss or gain of kinetic energy. In reality, the particle imparts energy to the walls and/or the walls impart energy to the particle, since they are both in a state of motion. In thermal equilibrium, the two energies are equal, permitting the assumption of perfect reflection to hold on average. Occasionally, the particle hits the wall at a precise moment in the wall vibration such that more

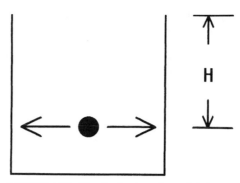

FIGURE 14.2 The simple view of a particle bouncing back and forth between the walls of an energy box. To escape, the particle must gain an energy of H.

energy is received than is lost. When this happens, the particle is viewed as moving a little higher in the box. Should this sequence happen several times in succession, the particle can gain enough energy to climb out of the box. Once out of the box, it is free to move until it falls into another box. The particle gives up its extra energy, as the result of the fall, to the walls of the box. Then the statistical process is repeated.

The difference between box rim and particle energy is the amount of energy the particle must have in order to escape. The height of the particle in the box is the amount of energy the particle possesses. The probability that the particle has energy exceeding the box height is exp $(-H/kT)$. This term, as suggested by Boltzmann,[19] arises from the second law of thermodynamics, which expresses entropy as a measure of the disorder among atoms. The probability that a particle will jump out of the box and land to the right [see Eq. (14.7)] is simply the product of the number of times n is heading in the right direction per unit time and the probability that it has enough energy [the exponential term in Eq. (14.7)]. v is called the frequency factor since its units are occurrences/s, while the exponential is a dimensionless evaluation of probability.

$$R = v e^{-H/kT}$$

(14.7)

where R = the rate at which a particle jumps to the right of the box in Fig. 14.2
 v = frequency factor
 k = Boltzmann's constant
 T = temperature, K
 H = escape energy

Equation (14.7) should be recognized as having the same form as the Arrhenius equation for the reaction rate constant so often employed in discussions on reliability in this book and elsewhere. This is no accident! The theoretical basis for the Arrhenius equation is an analogy with the thermodynamic equilibrium constant.

Suppose we have a number of boxes filled with particles adjacent to a number of empty boxes (see Fig. 14.3). At the center of the filled boxes, as particles surmount their barrier, whether they head right or left, they are not likely to find an empty box to fall in or they are likely to collide with other particles out of their boxes. Most of the particles in the middle, finding nowhere to go, simply fall back to their original home. Some will wander until they find an empty box. Because of the symmetry in the center, it is easy to see that on average all the boxes remain full. But the filled boxes next to the empty ones (on the right or left edge) do not enjoy this symmetry. The particle in the end box has an equal probability of exiting the box to the left or to the right. Should it go left, it finds no empty boxes or collides with another particle so liberated. On the other hand, should it go right, there are plenty of empty boxes into which it can fall. As time goes on, the original filled boxes lose their particles and the empty boxes gain particles. Over a sufficient period

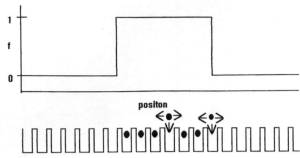

FIGURE 14.3   The statistical picture of particles in a series of boxes. Those particles in the center are likely to simply effect an exchange, causing no change in the arrangement. The particles on the edge will see nonfilled boxes, resulting in a distribution change.

of time, the particles become evenly distributed among all the available boxes. Each enjoys an occasional outing, only to fall into a box made vacant by another. At any instant in time, during the process of coming to equilibrium, the rate at which particles depopulate a region will be proportional to the *difference* between the concentrations of filled boxes on the left and on the right— that is, proportional to the *concentration gradient* of filled boxes. The process described is diffusion, shown graphically in Fig. 14.4.

14.5.2   Diffusion

Diffusion is as inevitable as life leading to death, and for the same reason. If there is a concentration of one type of matter in intimate contact with a concentration of another type, the two will eventually mix, *given enough time*. This is a reflection of the second law of thermodynamics, that ordered states are destined to disorder. In the case of matter, the exchange process is called *diffusion*.

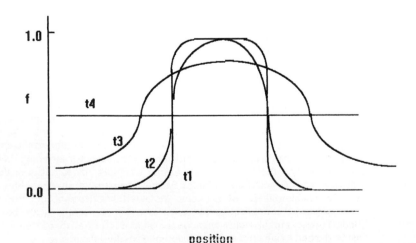

FIGURE 14.4   A depiction of the frequency of box occupation in space and time, where time $t_1 < t_2 < t_3 < t_4$. The equilibrium distribution is determined by the ratio of the number of particles to the number of states (boxes).

The rate at which matter moves across a boundary of unit cross-sectional area a is called *flux*. The flux at any instant in time will be determined by the gradient in the concentration as given by Eq. (14.8).

$$F(x) = -D\frac{\partial C}{\partial x}$$

(14.8)

where C = concentration
$\quad\quad D$ = diffusion coefficient

The diffusion coefficient is given[20] by

$$D = va^2 e^{-qH/kT}$$

(14.9)

where a = lattice spacing

The diffusion coefficient is a function of the species doing the diffusing and the particular medium. For a specific volume, the rate at which the concentration of a species will change is determined by the difference between the flux of atoms entering the volume and the flux leaving. The time dependence of concentration of a species in a volume is given by

$$\frac{\partial C}{\partial t} = \frac{dF}{dx} = D\frac{\partial^2 C}{\partial x^2}$$

(14.10)

The rate at which this exchange occurs can span many orders of magnitude in time, from microseconds to eons and beyond, depending upon the activation energy and frequency factor found in D and the divergence of the concentration in space. So while the process has been formalized, it has not been quantified. Equation (14.10) is not particularly useful unless one has specific values for the various terms. But once D is known, the structure of the body identified, and a criterion for failure established, the time it takes for one material to reach a certain concentration inside another (and perhaps failure) can be calculated with relative ease.

Impurities such as phosphorus and boron enter silicon substitutionally with diffusion constants measured in micrometers per hour at temperatures of 1200°C. At 900°C, the diffusion constant drops to approximately 1 μm/100 h. As can be seen, the manufacture of a silicon PN junction requires very high temperatures for relatively long periods of time. But once it is made, at use condition, i.e., room temperature, the atomic species are virtually locked in place. This brings us to a generalization for reliability. Usually, the higher the temperature required to make a device, the more reliable the device will be at use condition.

The diffusion of a species within a solid takes place in a number of ways. In a crystal lattice, where the atomic size of the species is small compared to the lattice spacing, diffusion often occurs interstitially. Otherwise, space must be made available by an atom's leaving its lattice site, permitting the diffusing species to occupy the vacancy. This is substitutional diffusion. If there are flaws in the crystal, space is made available for a species to move more easily. Generally speaking, flaws such as stacking faults, dislocations, and slip planes offer a highway for diffusing species. The surface of a solid offers the greatest space for any diffusing species. As a result, diffusion constants at surfaces or boundaries between layers of dissimilar material are generally much higher than in the bulk.

Surfaces and boundaries should always be examined with great care, since this is usually the conduit for a mobile species. In modern electronics, the oxide of silicon is a highly effective diffusion barrier, tying the surface down and thereby preventing atomic species from migrating. Lateral diffusion of species at the SiO_2–Si interface is virtually the same as that within the silicon itself. The effect is to push the high-mobility region away from the active silicon–silicon dioxide interface and place it at the top of the silicon dioxide, where electrical operation is passive.

Materials which do not enjoy a good passivating surface cannot hope to achieve the same stability that silicon offers. This is gallium arsenide's biggest weakness. Until a good passivating layer is found for gallium arsenide, its use will be limited to specialty electronics which trade unique electrical properties for long-term stability.

14.5.3   Field-Enhanced Diffusion

Obviously, an electronic part is subjected to electrical energy during its operation. The electric fields present during operation offer an additional way in which energy can be added to the system. Take, for example, the diffusion of an ionic species in the presence of an electric field. The field acts directly on a charged species and, at the same time, the medium containing the charged species.

The electric field affects the medium by tilting all the boxes, as shown in Fig. 14.5. The result is that the height of the box is lowered on one side compared to the other.

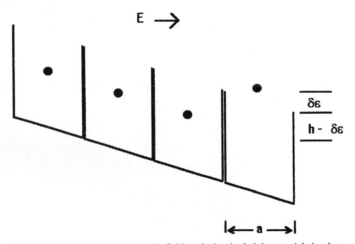

FIGURE 14.5   Effect of an electric field on the barrier height a particle in a box must overcome in order to escape.

The field alters the activation energy for the diffusion constant[21] by $E_a = (q/kT)\,(H - \frac{1}{2}a\varepsilon)$, where ε is the electric field and a is the lattice spacing (the distance over which the field can affect the barrier). Normally, fields are small, and so barrier lowering is negligible. However, since the field modifies the activation energy, such lowering can have profound effects at elevated temperatures and should not be forgotten.

The charged species (being directly influenced by the electric field) has a certain mobility μ under the action of an electric field. Thus the flux (number of species passing a unit area in unit time) will depend on the concentration of ionic species, the sign of the charge, the mobility, and the field. For a positively charged species of ionic concentration C, the flux[21] will be

$$F(x) = -D\frac{\partial C}{\partial x} + \mu\varepsilon C \qquad (14.11)$$

where μ = mobility of species in the medium
ε = electric field

It can be shown that the mobility and the diffusion constant are related by $D = (kT/q)\mu$, Einstein's relationship. To find the rate of change of concentration of charge within a specific volume, we need to differentiate Eq. (14.11). Assuming that the electric field is constant in the medium, its gradient is zero. The gradient in mobility is normally zero for an isotropic material. Under these circumstances, the effect of a field on the rate of change in concentration with time is

$$\frac{\partial C}{\partial t} = -D\,\frac{\partial^2 C}{\partial x^2} + \mu\varepsilon\,\frac{\partial C}{\partial x} \tag{14.12}$$

Equations (14.11) and (14.12) should be compared with Eqs. (14.8) and (14.10).

Charged species in semiconductor electronics can be particularly troublesome for high reliability. The operation of a typical semiconductor device depends upon extremely low concentrations of dopants within the silicon. For example, a PN junction with a reverse-bias breakdown voltage of 30 V relies on a concentration of dopants of approximately 100 dopant atoms for every billion silicon atoms. Clearly it takes very little migration to change that breakdown voltage. We have already pointed out that the surface of a semiconductor device permits relatively easy migration. Thermally the migration is in all directions. But electrically the migration is enhanced in a direction determined by the electric field. The result is to concentrate ionic species at specific locations. Those locations usually have the biggest impact on electrical performance. A good example of this effect is surface inversion.

Suppose there is a negatively charged species on the surface of the silicon dioxide as shown in Fig. 14.6. The action of a reverse bias on the junction is to create fringing fields, which penetrate the silicon dioxide as shown. These fields then cause the negatively charged species to accumulate over the surface of the N-type material. The negatively charged surface, in turn, attracts positive charges in the N-type silicon at the surface. If the concentration of surface charge is sufficiently large, the conductivity of the silicon changes type, i.e., it goes from N-type to P-type. This is called an *inversion layer.* Naturally, the inversion layer alters the geometry which the manufacturer was so careful to control. In the case shown, we can expect a reduction in the breakdown voltage and a small increase in the reverse leakage current. While this might not be a significant problem for a simple diode (unless, of course, one is relying on the breakdown voltage to control some other aspect of a circuit), for a transistor, the inversion could short the base region, rendering the device useless. The amount of charge required to cause the inversion layer is approximately 10^{11} atoms/cm^2 on the surface. A typical base width at the surface is 26 μm. If the inversion layer is 10 μm wide, the total number of charged atoms needed to create the inversion layer is 260×10^3 atoms. Since 1 cubic micrometer contains approximately 10^{10} atoms, very little contamination is needed to create this problem.

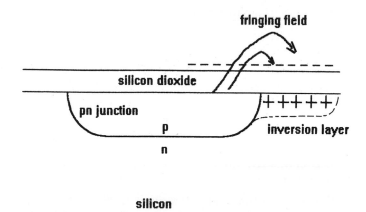

FIGURE 14.6 Fringing electric field causing the accumulation of negatively charged species on the surface, which in turn create an inversion layer in the silicon.

Since the mechanism just described depends on the fringing field at the surface, the manufacturer could reduce this mechanism by simply increasing the thickness of the oxide, thereby confining much of the fringing field to within the oxide, where, if the oxide is clean, there are negligibly small amounts of contaminants and those that are present have difficulty moving.

14.5.4   Electromigration

Referring to Eq. (14.12), suppose that in addition to a concentration of positively charged ion species C acting under the influence of an electric field E there are a quantity of negatively charged particles drifting in the opposite direction to the field. One consequence of the interaction between these oppositely charged particles is a momentum exchange with C. This momentum exchange would be such as to slow down the movement of C in the direction of the field. We would expect the interaction to be in direct proportion to the flux of negative charges, so that Eq. (14.12) becomes

$$ F(x) = -D\frac{\partial C}{\partial x} + (\mu \varepsilon - mJ_n)C \tag{14.13} $$

where J_n = conducting species opposite in sign to C
m = interaction proportionality constant

In metals where the conductivity is high, such as aluminum, local electric fields are difficult to support, being wiped out by a large electron flux (current). Under these circumstances, $\mu \varepsilon$ is small compared to mJ_n (the species made up of electrons) in Eq. (14.13), and ions can be transported in the same direction as the carriers of current, *opposite* the direction of the electric field (for N-type conductors)! Those ions (C) can be the conductor material itself. This effect is called electromigration, and it is a constant concern to manufacturers of microcircuits. As the density of active elements on a chip increases, the space available for the metal interconnects shrinks, forcing the interconnects to carry higher current densities and thereby bumping into the limits posed by electromigration.

Electromigration has been studied in detail by a number of workers.[22–25] Its effects on interconnects are expressed in terms of the mean time to fail (MTTF). General consensus shows the behavior to be given by[22]

$$ t_f = AJ^{-n}e^{-\Delta H/kT} \tag{14.14} $$

where A = material-dependent constant
J = current density
n = current-dependent exponent

The activation energy H is associated with the rate of self-diffusion for the conductor material. For low current densities, i.e., 10^3 to 10^5 A/cm^2, the value of n is typically 1. For current densities between 10^5 and 10^6 A/cm^2 where care has been taken to keep the conducting stripe from rising in temperature, the n value ranges between 1 and 3. Factors which influence the magnitude of H and A are the nature of the conducting film (i.e., the number and size of grain boundaries), the presence of impurities, the nature of the conductor surface (i.e., whether an insulating film has been deposited on the conductor surface), and of course the temperature and temperature gradients within the conducting film. The life of a part will be compromised if the current density exceeds 2×10^5 A/cm^2.

14.5.5   Electrical Overstress/Electrostatic Discharge (EOS/ESD)

Strictly speaking, EOS/ESD is not a reliability issue. However, because it occurs so frequently (approximately half of the failures encountered in a production facility are due to EOS/ESD) and has such a strong impact on the cost of production, we include a rather lengthy discussion of the subject in this handbook.

Electrical overstress is exactly what the term says. Electrical energy delivered to a part simply overwhelms the part and causes it to fail. It causes the failure by raising the temperature of some portion of the device to a point where material undergoes a phase change (melts or vaporizes). ESD is a subset of EOS in that the source of the electrical energy is an arc discharge, typically off a human being. We will discuss the fundamentals of EOS in the following section. Everything discussed also applies to ESD; it is simply a matter of employing the correct parameters.

The electrical energy in an EOS event is delivered to a part by the mechanism of Joule heating, i.e., the familiar I^2R relationship, where I and R are current and resistance, respectively. A more general form of Joule heating is IV, and when the deposited electrical energy is normalized to a unit volume, the term $J(x)E(x)$ or $P(x)$ is used. $J(x)$ and $E(x)$ are less familiar terms for the engineer but are far more powerful in their meaning. $J(x)$ is the local current density, measured (using the cgs system) in A/cm^2, whereas $E(x)$ is the local electric field, measured in V/cm. The product of these terms provides the local power density $P(x)$ in W/cm^3. We include the spatial term (x) to emphasize that electrical energy often is deposited in very localized regions of the material of interest.

When electrical energy (or any energy, for that matter) Q is absorbed in a volume of material, the temperature of the material rises according to the relationship $Q = c\rho V \Delta T$, where c is the heat capacity of the material, ρ is the material's density, V is the volume, and T is the temperature. When energy is delivered to material at a given rate (power, e.g., joules per second or watts), the temperature rises at a rate $P = dQ/dt = c\rho V\, dT/dt$. Since $c\rho$ appears frequently in heat equations, it is convenient to replace it with c_v. If the relationship is normalized to volume (bearing in mind that power can be a function of space), the expression becomes

$$c_v \frac{dT}{dt} = P(x) \tag{14.15}$$

Once a portion of a material is warmer (or cooler) than its surroundings, thermal energy flows into or out of the localized volume. It does this for exactly the same reasons a concentration of material diffuses from one region to another. In the case of thermal energy, the flux of heat crossing a unit area is proportional to the gradient in temperature, i.e., $F(x) = k\, \partial T/\partial x$. The proportionality constant k is the thermal conductivity of the material. As in the diffusion discussion, the rate at which a volume of material will gain thermal energy is proportional to the gradient in the flux. This gives the expression

$$c_v \frac{dT}{dt} = k \frac{\partial^2 T}{\partial x^2} \tag{14.16}$$

Combining Eqs. (14.15) and (14.16) and collecting the temperature terms gives the generalized EOS/ESD equation[26]

$$c_v \frac{dT(x, t)}{dt} - k \frac{\partial^2 T(x, t)}{\partial x^2} = P(x, t) = \sigma E^2(x, t) \tag{14.17}$$

where $E^2(x, t)$ is the electric field and σ is the electrical conductance. The expression (x, t) is included to emphasize that both temperature and power are spatial and temporal quantities. Equation (14.17) represents the only way in which electrical energy can alter the temperature for a given system of materials. It is very general, holding for all materials. It should be recognized that if the power density term were missing, we would have the diffusion Eq. (14.10). Or, from another perspective, the diffusion equation does not include an external way of adding mass, which would be equivalent to the $P(x, t)$ term. It is curious that a concentration of material behaves in the same way as a concentration of heat energy. But if thermal energy is thought of as taking on the behavior of particles (called phonons), that strangeness disappears. Phonons behave in the same manner as matter, migrating to regions of lower concentrations simply because "locations" (empty energy states) are available. One significant difference between the diffusion of matter and the diffusion of thermal energy is that the diffusion coefficient D is a strong function of temperature, whereas the thermal diffusion constant k is usually no more than linearly dependent on temperature. Exceptions to this statement occur when the radiation from the thermal energy (blackbody) is transparent in the conducting medium or when lighter, more mobile thermal energy carriers, such as electrons, come into play.

It should come as no surprise that Eq. (14.17) is normally solved by computer.[27,28] There are two situations in which its solution is relatively straightforward. The left-hand side of Eq. (14.17) includes two items. The first term describes the accumulation over time of thermal energy (or matter) within a region of space, while the second term describes the temporal behavior (flow) of thermal energy (or matter) already accumulated within a space. If one assumes no significant energy flow over a specific time period (the second term is zero), we have an adiabatic condition, i.e., no energy is lost to the surroundings. On the other hand, if the energy in a particular volume never changes (i.e., the first term is zero), but energy is flowing through the volume, we have the steady-state heat flow condition. Both situations, representing the extremes in the equation, yield surprisingly easy solutions.

EOS in Metals.   Take, for example, the adiabatic heating of a metal stripe, such as metallization in a microcircuit. Assuming no heat loss (i.e., the second term on the left is zero), we begin with $c_v \, dT/dt = P(x, t)$. But power is $J \times E$ and $J = \sigma E$, where σ is the electrical conductance of the metal. This leaves us with $c_v \, dT/dt = J^2/\sigma$. Separating the time- and temperature-dependent terms gives $\sigma c_v \int_{t_0}^{T_m} dT = \int_0^\tau J^2 \, dt$. Notice that we have assumed no temperature dependence in σ and c_v. c_v is normally only weakly dependent on temperature, but for many materials σ is highly temperature-dependent. σ should be brought under the temperature integral, but this complication is avoided in our example. If the current density J is a constant during the time t of its application, J^2 can be removed from under the integral. T_0 is the starting temperature of the metal, and T_m is the temperature which you decide will cause failure. We choose here the melting temperature of the metal being used. Performing the integration yields the algebraic expression $\sigma c_v (T_m - T_0) = J^2 \tau$. J^2 can be replaced with I/wd, i.e., the current divided by the cross-sectional area of the metal stripe (the product of the stripe's width w and thickness d). If we're interested in knowing the amount of current necessary to raise the stripe to the melting point, we simply solve for I. This becomes

$$I_m = \sqrt{\sigma C_v(T_m - T_0)(wd)^2/\tau} \qquad \text{for the adiabatic condition} \qquad (14.18)$$

Expression (14.18) applies to any material under the assumptions in its development. It is up to the user to be specific. For example, starting at room temperature ($T_0 = 22°C$), if the stripe is made of aluminum, we have the material properties $T_m = 660°C$, $\sigma = 0.5 \times 10^6$ mho/cm, and $c_v = 2.4$ J/cm^3 · °C; the geometry of the stripe might be 10 μm wide by 1 μm thick; and the concern might be a 1-μs-wide current pulse. In this case the stripe will fail, i.e., reach the melting temperature 660°C, at a current of 277 mA.

The expression contains the material properties T_m, σ, and c_v, which collectively are under the manufacturer's control by the selection of the material. Under the manufacturer's direct control are the *design* w and d. The user, on the other hand, can control only the ambient environment T_0, current amplitude, and pulse duration. The observed hierarchy of influence is encountered for virtually every item dealt with by the user. And as always, the details of construction are required for a specific answer.

An extremely important property for EOS is the temperature behavior of σ. Figure 14.7 depicts the current going through a structure as having to pass through a number of resistors in parallel.

Let us suppose that for thermal reasons the center resistor is at a higher temperature. If the temperature dependence of the resistor is such that conductance decreases (resistance increases), then current is spread to the other resistors. Because the power is being distributed among all the resis-

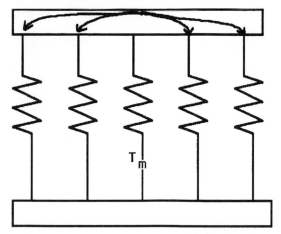

FIGURE 14.7 Depiction of the ballasting effect of resistors with a positive temperature coefficient of resistance. The resistance of the center resistor has increased because power dissipation has made the resistor hotter; i.e., it has a positive coefficient of resistance.

tors, there is a stabilizing effect. This is the ballasting principle (see Fig. 14.7). But suppose the warmer center resistor becomes more conductive (resistance decreases)! Now more current is invited to pass into the center resistor, which causes it to become even hotter. This is an unstable situation in which electrical energy is focused (see Fig. 14.8), leading to very high temperatures, melting, vaporization, and failure. This instability is called *thermal second breakdown* in silicon junctions.

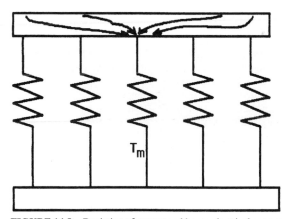

FIGURE 14.8 Depiction of current and hence electrical energy focusing, which occurs when the resistance of the center resistor decreases with increasing temperature; i.e., the resistor has a negative coefficient of resistance with temperature.

Figure 14.9 shows the temperature dependence of conductance for metals, semiconductors, and insulators using arbitrary scales. As can be seen, metals are stable over a wide temperature range (i.e., they have negative temperature dependence), whereas semiconductors possess negative dependence below some critical temperature (called the intrinsic temperature). Beyond the intrinsic temperature, the conductance increases with temperature, and hence they are unstable.

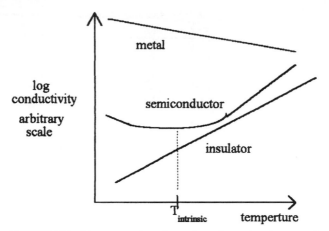

FIGURE 14.9   Temperature dependence of metals, semiconductors, and insulators. Notice that metals always ballast energy, whereas semiconductors, depending upon the temperature, either ballast or focus energy. Insulators always focus energy.

Finally, insulators, except, perhaps, at very low temperatures (not shown), have a positive temperature dependence at all temperatures and are always unstable. In EOS, once a material has a positive thermal coefficient of conductance, the focusing effect of electrical energy is so rapid and intense that it takes very little extra energy to get some portion of the material to melt or fail. Insulators and devices whose properties depend on insulators (such as MOS transistors) are often called voltage-sensitive devices, to emphasize how little energy is required to destroy them. Such devices are usually protected with devices which are energy sensitive, such as diodes or resistors.

EOS in Insulators.   The fact that insulators are unstable at use condition should be a source of concern for long-term reliability. Insulators really have no safe voltage, under dc conditions.[29,30] If even the slightest current is permitted to pass through the insulator, the warming effect and the subsequent influence on conductance (thermal runaway) can lead to thermal second breakdown. Insulators are saved from self-destruction by the heat flow from the warmer region to the surrounding material. Unfortunately, most good electrical insulators are poor thermal conductors. For dc use conditions, one should strive to use capacitors or MOS technology at electric potentials (i.e., voltages) less than 20 percent of the insulator breakdown voltage, preferably less than 1 percent. The lower the use potential, the better.

EOS in Junctions.   Early work[31] in studying EOS in semiconductor junctions showed that an estimate for the power density required to kill a junction depended to a great extent on the junction area (see Fig. 14.10).

This curve, called the Wunsch-Bell curve, was developed for the aerospace industry to assess EMP (electromagnetic pulse) vulnerability. While the curve has immediate and obvious uses, its most disturbing feature was the broad error range imposed on its use. These error bands caused other investigators to question whether the mechanism (i.e., Joule heating) was the only item operating. P. Mars[39] showed that the most significant temperature was the intrinsic temperature. This accounted for the factor of 10 lower than silicon melting point. Subsequent investigations showed that one could get a better correlation with junction perimeter.[32] It took the development of a two-dimensional model[27] for a junction being exposed to EOS to show that it is the con-

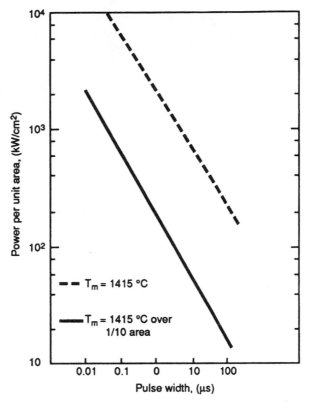

FIGURE 14.10 The Wunsch-Bell curve, showing the amount of
power per unit junction area required to cause failure. [D. C. Wunsch
and R. R. Bell, *IEEE Trans. Nuclear Science,* NS-15(6): 251 (1968).]

struction details which determine the energy required to cause failure. The most important results
of that later investigation are depicted in Fig. 14.11.

Shown is a mapping of the material which has reached the intrinsic temperature [the bottom of
the valley in the semiconductor conductance curve (Fig. 14.8)] as the result of a single EOS pulse.
Three cases are shown. The first is that of a junction pulsed by EOS in the forward direction. The
intrinsic temperature region (shown shaded) is determined largely by the area of metal contact to the
silicon and involves a large amount of material. In the second, for a reverse-bias EOS event, the
intrinsic temperature region has shrunk to that of the junction depletion. Because of the volume of
material involved, it becomes clear why the reverse-bias EOS requires much less energy to cause
failure. The third figure depicts the intrinsic temperature for a reverse-bias junction where the buried
layer has been removed far from the junction. Notice that the intrinsic region is the smallest of the
three cases, confined to the surface and largely determined by the junction perimeter. The large variations observed by Wunsch and Bell could be ascribed to details of construction and replacing the
melting temperature of silicon with the intrinsic temperature of the silicon material.

14.5.6 Mechanical Energy

The details of why mechanical energy causes materials to fail can be found in Chap. 18,
"Mechanical Stress and Analysis."

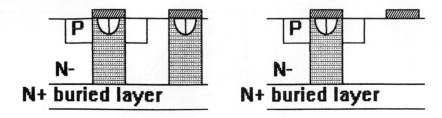

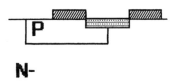

FIGURE 14.11   Intrinsic temperature maps of PN junction with buried layer. The shaded area roughly depicts the region which has reached the intrinsic temperature as the result of a single EOS pulse. The top figures show the effect of forward (left) and reverse (right) bias. The bottom figure is the reverse-bias situation with the buried layer far removed.

In electronics, mechanical energy comes from shock and vibration or from differences in thermal expansion in the various materials. Two prime areas of concern for mechanical effects are

Creating open circuits.   This can happen, for example, in microcircuits, where the weakest mechanical link is the flying lead wire bond to either lead frame or chip. In the case of PC boards, solder joints and plated through holes are the Achilles heel. Silicon is a brittle material, and it is subject to crack initiation and propagation in much the same manner as glass. As the silicon chips get larger, stress induced by differences in thermal expansion (and repeated thermal cycling) between chip and header leads to chip cracking. When a crack forms but does not open, we have the most difficult problem for an electronic system. The crack often manifests itself as an intermittent failure. Slight changes in temperature are enough to cause an open circuit, whereas a further change in temperature restores the contact.

Creating construction flaws.   Mechanical energy can create a flaw which by itself is harmless, but which permits other mechanisms to act. The classic example is package hermeticity. When we discuss corrosion, we point out the importance of preventing an electrolyte from being present inside the package. Mechanical forces can compromise that hermeticity, letting moisture penetrate the package and setting up the conditions required for corrosion.

14.5.7 Chemical Energy

It is not the energy derived from a chemical reaction which is of concern from a failure mechanism standpoint, but rather the effect of chemical reactions on reliability in the first place. Perhaps the most important reactions are those which involve the corrosion of metals. In all my years of exploring failure mechanisms and performing failure analysis, I have yet to run into a corrosion problem which did not involve a metal. The corrosion of metals is almost always electrochemical in nature.

Electrochemical corrosion usually involves an electrically conductive liquid (typically water) that serves as the essential medium through which all chemical reactions take place (electrolyte). Metal ions leave one electrode, go into solution (anodic reaction), migrate through the electrolyte under the influence of an electric field, and ultimately are deposited onto a second electrode (cathodic reaction). A metal ion enters the solution through oxidation, i.e., it loses electrons while being ionized. Thus an anodic reaction produces electrons. In order for the metal ion to be deposited on an electrode, it must receive electrons. Thus the cathodic reaction consumes electrons. If there is no electrical connection between the two electrodes, it is easy to see that a charge will build up on the two electrodes in such a way as to cancel the electric field driving the reaction, thus bringing the reaction to a halt.

If the two electrodes are electrically connected, then the electrons produced by the anodic reaction can provide the electrons consumed by the cathodic reaction. The current observed is called the *corrosion current*. The corrosion current and amount of metal corroded are related by Faraday's law:[33]

$$i_{cor}t = \frac{nFw}{M} \tag{14.19}$$

where i_{cor} = corrosion current, A
$\quad\quad t$ = time, s
$\quad\quad n$ = number of electrons involved in metal dissolution
$\quad\quad F$ = Faraday's constant (96,480 C/mol)
$\quad\quad M$ = molecular weight of the metal, g
$\quad\quad w$ = mass of corroded metal, g

Because the reaction is coupled, the anodic current must equal the cathodic current. But there is no such demand on the current density at either electrode. If the anode is small compared to the cathode, corrosion of the anode takes place quickly. For the opposite case, deposition on the cathode will be rapid. Grain boundaries, inclusions, residual stress, etc. have the effect of localizing the metal dissolution sites (anodic) to specific regions. The combination of small anodic sites and a large cathode results in severe pitting or cracking.

The details of corrosion are complicated and are affected by many elements. Among the most important are

1. Metal composition and properties
2. The chemical, physical, and electrical properties of the electrolyte
3. Surface films and their properties, particularly coherence, resistivity, and diffusion coefficients

The two reactions (anodic and cathodic) are coupled. The rate of dissolution of the metal electrode will be determined by which reaction rate is the slowest! Suppose the concentration of dissolved metal cations (oxides, hydroxides, metal salts) at the cathode is so high that they precipitate out. The resultant porous film on the cathode can interfere with the further transport of metal ions to the electrode, thereby slowing down the coupled reaction. Furthermore, there are times when the rate is determined by the electronic conductivity of the surface of a film.

The anodic and cathodic reactions can be studied individually by altering the potential applied to the electrodes. The currents observed take on the form[34]

$$i = i_0 \left\{ \exp\left(\beta \frac{nF}{RT} \eta \right) - \exp\left[-(1-\beta) \frac{nF}{RT} \eta \right] \right\} \tag{14.20}$$

where i = corrosion current density
 i_0 = exchange current density
 R = gas constant
 T = absolute temperature
 β = symmetry coefficient (approximately 0.5)
 n = number of electrons involved in the reaction
 F = Faraday's constant
 η = overpotential

The overpotential is a measure of how far the reaction is from equilibrium, i.e., $E-E_0$, with E being the applied potential. Notice that the relationship has a form similar to that of solid-state diffusion in that it is temperature-activated.

It is possible to compute from thermodynamic data the domain of stability of a metal, its oxides, and other species when in an aqueous solution. The overpotential versus pH of the solution is expressed graphically in what is known as a Pourboix diagram.[35] Such diagrams can show at a glance whether corrosion is possible.

14.5.8   Corrosion Control in Electronics

Compared to the other failure mechanisms described, corrosion is by far the fastest. This is not surprising, since ionic motion in a liquid is very rapid compared with solid-state diffusion either in the bulk, at interfaces, or at surfaces. From what has been described, it should be obvious that to avoid corrosion, one must avoid the presence of an electrolyte, i.e., moisture. Phipps and Rice,[36] as a result of studying atmospheric corrosion, put it succinctly when they said,

> We might summarize that the RH [relative humidity] is generally a dominant factor. For most metals, oxides are always present on the surface except at the bottom of growing pits, but no corrosion is detected below 15% RH and most metals corrode above 75% RH regardless of the particular gas concentrations or species in the accelerated atmospheric test.

In microelectronics, the quantity of metal is very small. Therefore, virtually any corrosion is likely to lead to immediate failure. Those devices which use thin-film resistors of materials such as Nichrome are especially vulnerable since such films are only 250 Å (approximately 50 atomic layers) thick.

The first line of defense, then, is to encapsulate electronics in a dry, hermetically sealed environment. The life of the part depends upon maintaining that hermeticity throughout further processes and use.

A second line of defense (which it is essential to employ in addition to the first line if thin films are involved) is to coat the part with a moisture barrier. A barrier that seems to work very well is silicon nitride. Care must be taken to ensure complete coverage of the coating, especially where there are sudden changes in topography, as is the case for the metal contact to the thin-film resistor. Cracks in transparent materials are virtually impossible to detect when viewed with a microscope's bright-field illumination. Even using a scanning electron microscope to spot cracks in glassivation is very difficult. Perhaps the best way to ensure a good coating is to perform the "water drop" test[37,38] on a sample basis. (We are letting electrolytic corrosion work for us.) The test involves applying 5 V across a thin-film resistor, applying a drop of water (which normally completely covers a chip), and waiting for 3 min with the voltage maintained. If the resistor has not changed value, the coating can be considered intact and free of cracks. If there are cracks in the passivation layer, the resistor will normally open within 30 s. For Nichrome resistors, visual inspection of a failed unit will reveal a hairline of missing resistor material.

The third line of defense, often not viable, is to choose noble metals such as gold or platinum. Noble metal finishes have been used to advantage in controlling corrosion. One has to be careful that the metal finish is complete, with no pores, cracks, or openings; otherwise a galvanic cell can be set up which will speed corrosion. Perhaps it is not common knowledge, but gold and even more so silver can migrate in the presence of moisture and grow dendritically to the point where two close electrodes become shorted.

For the PC board assemblies, an important source of corrosion is resin fluxes used in soldering. The lowest acid content fluxes consistent with obtaining good wetting should be used. Then careful cleaning is required to remove residue flux from the board and its components. When finished, the assembly should be protected with a moisture barrier coating such as silicone, since it is impractical to hermetically seal large enclosures.

In summary, if moisture can be eliminated or controlled, corrosion generally is not a problem for electronics. If it cannot, corrosion becomes a very important cause of electronic failure, second only to EOS/ESD.

14.6 PRACTICAL PROBLEMS ENCOUNTERED

The mechanisms described have been generalized, and therefore it is difficult to see how to use them to advantage. It is for this reason that I choose to end this chapter by recounting in grab-bag fashion specific failure mechanisms which through oversight in parts selection or for whatever reason have cost industry an enormous amount of money in returns and rework. Had attention been paid to the aforesaid mechanisms, these costs could have been largely averted.

A large number of diodes are encapsulated in glass. When they are placed on PC boards where no relief is allowed in the leads, stress through differences in thermal expansion cracks the glass package, thereby compromising the package hermeticity.

In integrated circuits, the glass-to-metal seal in the hermetic package is easily fractured if the right materials are not used. The loss of hermeticity allows the penetration of moisture. The presence of moisture on the surface of silicon dioxide enhances the mobility of ionic species, resulting in either corrosion or inversion layer formation.

Silicon junctions, when forced to conduct a large current in the forward direction or a small current beyond breakdown, have the nasty effect of injecting electrons into the silicon dioxide passivation layer. These charges, by means of induction, alter the junction surface doping, changing the junction's radius of curvature and thereby altering the breakdown voltage.

Unprotected MOS devices are easily damaged by EOS/ESD.

MOS technology has inherent in its design parasitic PNPN paths which under the correct conditions form regenerative circuits leading to latchup. A knowledge of the influence of temperature, geometry, and doping levels on transistor gain can reduce or eliminate latchup paths.

Thin-film resistors should certainly be avoided if possible, since they are very subject to corrosion. If they cannot be avoided, hermeticity and moisture barrier layers are a must.

Caution should be used when employing hermetic seals for moisture protection. The materials being sealed in may be a source of moisture, such as incompletely cured epoxy.

Insulators are inherently unstable from a thermal breakdown point of view. Subjecting an insulator to a steady-state electric field is always a dangerous business. When this must be done, the applied electric field should be as low as possible. The field should never be higher than 20 percent of the breakdown strength of the insulator. A better goal is 1 percent.

Metallic couples (such as gold–aluminum) are subject to Kirkendahl voiding, resulting in a weakened joint.

Large steady-state current densities in conductors will result in electromigration failure, whether in microcircuits or in fuses.

14.7 SUMMARY

Parts specifications and the electrical properties of a device are only weakly connected and should never be looked upon as assurance that a part is reliable. The motion of matter is the ultimate cause of failure, and such motion requires energy. We have reviewed the dominant forms of energy and shown that they mostly follow the pattern of the Arrhenius rate equation. Making the connection between these factors and the expected life of a part requires a knowledge of the principle of operation and the construction details for the part. The reliability and quality of a part are related to the construction details of the part and not simply to a count of the number of parts or active devices in the circuit. By paying close attention to these details in relation to the various failure mechanisms, great gains can be made in increasing yields, lowering costs, improving quality, and achieving high reliability.

14.8 REFERENCES

1. R. P. Nanavati, *An Introduction to Semiconductor Electronics,* McGraw-Hill, New York, 1963.

2. W. W. Gartner, *Transistors: Principles, Design, and Applications,* D. Van Nostrand, New York, 1960.

3. S. M. Sze, *Physics of Semiconductor Devices,* Wiley-Interscience, New York, 1969.

4. E. H. Nicollian and J. R. Brews, *MOS (Metal Oxide Semiconductor) Physics and Technology*, John Wiley & Sons, New York, 1982.

5. A. S. Grove, *Physics and Technology of Semiconductor Devices,* John Wiley & Sons, New York, 1967.

6. W. Shockley, *Bell Sys. Tech. J.,* 28: 4, 1949.

7. A. S. Grove, *Physics and Technology of Semiconductor Devices,* John Wiley & Sons, New York, 1967, p. 157.

8. E. H. Nicollian and J. R. Brews, *MOS (Metal Oxide Semiconductor) Physics and Technology*, John Wiley & Sons, New York, 1982, p. 35.

9. A. S. Grove, *Physics and Technology of Semiconductor Devices,* John Wiley & Sons, New York, 1967, p. 103.

10. C. T. Sah, R. N. Noyce, and W. Shockley, "Carrier Generation and Recombination in P-N Junction and P-N Junction Characteristics," *Proc. IRE,* 45: 1228 (1957).

11. A. S. Grove, *Physics and Technology of Semiconductor Devices,* John Wiley & Sons, New York, 1967, p. 159.

12. Ibid., p. 136.

13. S. M. Sze, *Physics of Semiconductor Devices,* Wiley-Interscience, New York, 1969, p. 114.

14. Ibid., p. 210.

15. M. Kuhn, *Solid-State Electronics,* 13: 873 (1970).

16. R. Castagne and A. Vapaille, *Surface Science,* 28 (1971).

17. A. Goetzberger and E. H. Nicollian, *J. Applied Physics,* 38: 4582 (1967).

18. E. H. Nicollian and J. R. Brews, *MOS (Metal Oxide Semiconductor) Physics and Technology*, John Wiley & Sons, New York, 1982, pp. 581–644.

19. F. W. Sears, *Thermodynamics, the Kinetic Theory of Gases and Statistical Mechanics,* Addison-Wesley, Reading, Mass., 1953, p. 286.

20. P. Shewmon, *Diffusion in Solids,* A Publication of The Minerals, Metals & Materials Society, Warrendale, Pa., 1989, p. 75.

21. A. S. Grove, *Physics and Technology of Semiconductor Devices,* John Wiley & Sons, New York, 1967, p. 37.

22. J. R. Black, *Proc. IEEE 6th Intl. Rel. Phys. Symp.,* 148 (1967).

23. C. J. Wu and M. J. McNutt, *IEEE 21st Intl. Rel. Phys. Symp.,* 24–31 (1983).

24. H. A. Shaft, T. C. Grant, A. N. Saxena, and C. Y. Kao, *IEEE 23rd Intl. Rel. Phys. Symp.,* 36–39 (1985).

25. B. K. Liew, N. W. Chueng, and C. Hu, *IEEE 27th Intl. Rel. Phys. Symp.,* 215 (1989).

26. H. S. Carslaw and J. C. Jaeger, *Conduction of Heat in Solids,* Clarendon Press, Oxford, 1959, p. 134.

27. N. Kusnezov and J. S. Smith, *Electrical Overstress/Electrostatic Discharge Symposium Proceedings,* 132 (1981).

28. J. S. Smith and N. Kusnezov, *Microwave Systems News,* 12(3): 67 (March 1982).

29. J. J. O'Dwyer, *The Theory of Dielectric Breakdown in Solids,* Oxford University Press, 1964, pp. 46–58.

30. J. S. Smith and F. Zitko, *Dielectric Breakdown in Propellants,* Chemical Propulsion Information Agency, Pub. 509, vol. 1, p. 327.

31. D. C. Wunsch and R. R. Bell, *IEEE Trans. Nuclear Science,* NS-15(6): 251 (1968).

32. W. D. Brown, *Semiconductor Degradation by High Amplitude Current Pulses,* Sandia Laboratories Document DC-DC-72-2288, Albuquerque, N.M., July 1972.

33. D. W. Shoesmith, "Kinetics of Aqueous Corrosion," *Metals Handbook,* 9th ed., ASM International, Metals Park, Ohio, 1985, vol. 13, pp. 29–36.

34. G. R. Brubecker and P. B. P. Phipps, *Corrosion Chemistry,* ACS Symposium Series, 1979.

35. D. L. Piron, "Potential vs. Ph (Pourboix) Diagrams," *Metals Handbook,* 9th ed., ASM International, 1985, vol. 13, pp. 17–28.

36. G. R. Brubecker and P. B. P. Phipps, *Corrosion Chemistry,* ACS Symposium Series, 1979, p. 243.

37. V. C. Kapfer and J. J. Bart, "Thin Film Nickel-Chromium Resistor Failures in Integrated Circuits," *IEEE 10th Intl. Rel. Phys. Symp.,* 176 (1972).

38. W. M. Paulson, "Further Studies on the Reliability of Thin-Film Nickel-Chromium Resistors," *IEEE 11th Intl. Rel. Phys. Symp.,* 61 (1973).

39. P. Mars, "Thermal Analysis of P–N Junction Second Breakdown Initiation," *International Journal of Electronics,* 1972, Vol. 32, No. 1, pp. 39–47.

CHAPTER 15
MAINTAINABILITY AND RELIABILITY

Richard Kowalski
ARINC
Annapolis, Maryland

15.1 INTRODUCTION

While reliability is concerned with reducing the frequency or severity of system failure, maintainability is concerned with reducing the duration of the system failure downtime and the effort required to restore a system to operation after a failure.

The origins of maintainability as an engineering discipline can be traced back to the 1950s.[1] In 1957, the Advisory Group on Reliability of Electronic Equipment (AGREE) published a report compiling recommendations of nine AGREE task groups, including a recommendation for improved maintainability design to reduce the burden of supporting fielded equipment. By 1960, the *Reliability Training Text*[2] advised that "Recently more emphasis has been placed on maintenance. Equipment is becoming more difficult to maintain due to complexity, high-level skills required, and the generally inadequate consideration of maintenance needs in the original design." These comments are not surprising since, at that time, maintenance was frequently considered to be a user (as opposed to designer) responsibility and the tradeoffs between acquisition and life-cycle cost were not well established.

Early efforts to address system maintenance generally took the form of design guidance material; for example, a series of articles on maintainability appeared in *Machine Design* magazine in 1956. In the late 1950s and early 1960s, maintainability research began to develop measures for quantifying maintainability attributes of equipment, identifying the different types of maintainability activity (i.e., preparation, securing repair parts, administrative delays), and identifying tasks that would contribute to a maintainability assurance program.

In 1966 the initial versions of MIL-STD-470, *Maintainability Program Requirements*; MIL-STD-471, *Maintainability Demonstration*; and MIL-HDBK-472, *Maintainability Prediction* were issued. During the next decade, tubes were replaced by transistors, then integrated circuits, and computers evolved from the ENIAC (electronic numerical integrator and calculator; the first digital computer, developed in the early 1940s) and MARK I to the IBM system 360.

By the end of the 1970s, a substantial collection of maintainability design analysis, prediction, and testing procedures and tools had been developed for electronic designs. Maintainability had evolved into a discipline that was working hard to adapt to new problems that were a part of the continuing evolution of electronic equipment. For example, testability, a word which first appeared in the literature in 1965, had become enough of a discipline so that research and study efforts were documented in the *RADC Testability Notebook* in 1982.[3]

The past decade has seen the digitization of many electronic functions, the increased use of complex microcircuits, extensive incorporation of software into electronic systems, and increased system functionality and complexity. The basic principles of how to design for the rapid diagnosis and removal of faults have not changed. However, the growth and accessibility of computing power and the development of new design tools [i.e., computer-aided design (CAD) packages] provide new capabilities for the analysis and improvement of maintainability characteristics of modern systems.

15.1.1 Maintainability and Reliability

Maintainability is now sometimes associated with such other disciplines as reliability, system safety, and human factors engineering under the names of effectiveness engineering, product effectiveness or, with the addition of quality assurance, product assurance. In some companies, maintainability is included as part of integrated logistics support (ILS). The organization is frequently a reflection of a company's prime customer organization or of the type of product the company makes. In any case, maintainability is still closely related to reliability.

Maintainability data are incorporated in the *failure modes and effects analysis* (FMEA), which is prepared by the reliability organization. System or equipment availability depends on parameters which relate to reliability and maintainability, and failure-rate data of various types provided by the reliability organization. These data are required to calculate many maintainability parameters, as will be seen later in this chapter.

15.1.2 Maintainability as a Program Element

The process of implementing maintainability in a program is represented in Fig. 15.1. The first block in Fig. 15.1 represents the operational requirements which include or determine the require-

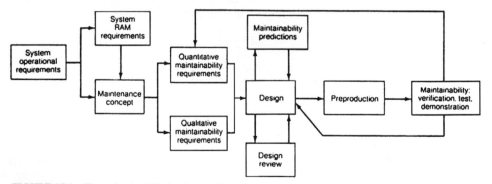

FIGURE 15.1   The maintainability implementation process.

ments for reliability, availability, and maintainability. A maintenance concept is developed to meet operational requirements and to enhance system or equipment maintainability. The maintenance concept is responsive to both the operational requirements and the reliability, availability, and maintainability requirements. The latter requirements are frequently established by a process of trade and sensitivity studies designed to maximize the system's reliability, availability, and maintainability characteristics while minimizing cost, generally life-cycle cost. The maintenance concept is concerned with the levels of maintenance (organizational, intermediate, and depot); the geographic locations of operational and support equipment; and other such concerns as the frequency of periodic tests and the acceptable amount, if any, of scheduled maintenance. The maintenance concept will be covered in greater detail in Sec. 15.4.4.

After the maintenance concept and the top-level maintainability requirements have been established, qualitative and quantitative maintainability requirements at the equipment level are developed and imposed on the designers, frequently as the maintainability section of a contract end item (CEI) or other form of product specification. During the design process, maintainability engineers keep an ongoing interface with the designers to help the latter meet the maintainability requirements. In addition to the everyday informal contacts, maintainability engineers prepare predictions to estimate the degree to which quantitative requirements are being met. They also participate in both internal and customer design reviews to discover any discrepancies in meeting either qualitative or quantitative requirements.

The final step in the maintainability implementation process is the performance of maintainability demonstration tests to measure, with actual hardware and software and under near-operational conditions, the success with which maintainability has been incorporated into the design. When differences exist between the requirements and the test results, redesign may be required or the quantitative maintainability requirements may have to be reallocated or revised, as shown by the feedback arrows in Fig. 15.1.

15.2   MAINTAINABILITY TERMS AND DEFINITIONS

15.2.1   Maintainability Defined

In MIL-STD-721C,[4] *maintainability* is defined as follows:

> The measure of the ability of an item to be retained in or restored to specified condition when maintenance is performed by personnel having specified skill levels, using prescribed procedures and resources, at each prescribed level of maintenance and repair.

From this definition, it is clear that maintainability measures are conditioned on assumptions on the skill levels of maintenance personnel, the specific procedures and resources associated with the maintenance process, and the levels of maintenance defined for the system. Thus, while the reliability engineer is concerned with many physical characteristics that affect system components, such as temperature, humidity, shock, and vibration, the maintainability engineer will be concerned with the physical partitioning of a system into repairable items; the accessibility, weight, and volume of these items; the skills and training of maintenance crew; and the availability of the appropriate tools and equipment for conducting maintenance activities.

The definition also points out the difference between the terms *maintainability* and *maintenance*. Maintainability is design-related and must be incorporated into the design of a system or equipment during the conceptual design, definition, and full-scale development phases of the life cycle. Maintainability expenditures are heaviest during these periods and taper off to sustaining engineering efforts once the system or equipment is in the field.

15.2.2   Maintenance Defined

In contrast with maintainability, *maintenance* is operations-related since it refers to those activities undertaken after a system is in the field to keep it operational or to restore it to operational condition after a failure has occurred. Expenditures for ILS, which includes maintenance, are primarily for planning during conceptual and definition phases and are relatively small, increasing during full-scale development as more details of the design become available, and are heaviest after the equipment is in use. These latter expenditures are often the user's responsibility rather than the designer's or manufacturer's.

15.2.3   Maintenance-Time Definitions

Two basic time elements are used to describe or measure the maintainability characteristics of a system: uptime and downtime.

Uptime is the time during which an item is operating or in a condition to perform its required functions. *Downtime* is the time during which an item is not in a condition to perform its required functions. Figure 15.2 shows the relationship between the various elements of downtime. The principal elements of downtime are maintenance time, delay time, and modification time.

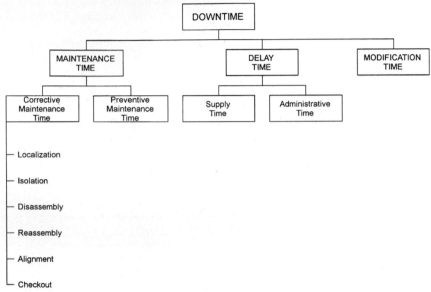

FIGURE 15.2   Maintenance downtime elements.

15.2.4   Maintenance-Time Composition

Maintenance time consists of

- *Corrective maintenance time,* which includes all the actions performed, as a result of failure, to restore an item to a specified condition
- *Preventive maintenance time,* which includes all actions performed to retain an item in a specified condition by providing systematic inspections, detection, and prevention of incipient failures.

15.2.5   Corrective-Maintenance-Time Composition

Corrective maintenance time consists of

- *Localization time:* determining the approximate location of a fault
- *Isolation time:* determining the location of a fault to the extent necessary to effect repair
- *Disassembly time:* opening the item to the extent necessary to make accessible the item that is to be replaced

- *Interchange time:* removing the item to be replaced and installing the replacement item
- *Reassembly time:* the inverse of the disassembly effort
- *Alignment time:* performing the adjustments necessary to return the item to specified operation
- *Checkout time:* tests or observations to confirm proper operation

Although these corrective-maintenance-time elements were initially defined for military equip-ments when tubes were the workhorses of electronic equipments, they can, with some interpreta-tion, be applied to maintenance activity on modern equipments to provide more detailed insight into the maintenance process.

15.2.6 Delay-Time Composition

Delay time is that element of downtime when no maintenance is being accomplished. It consists of

- *Supply time:* the time spent awaiting a needed replacement item.
- *Administrative time:* that part of delay time not attributable to supply delay. This can include the time required for a maintenance team to reach the equipment, time spent completing paper-work or searching for information, or other delays which depend on the operating and supply environments.

15.2.7 Modification Time Defined

Modification time is the time necessary to introduce change(s) to an item to improve its charac-teristics or to add new capabilities and characteristics.

15.3 *MAINTAINABILITY PARAMETERS*

Quantitative maintainability requirements are based on desired limitations on system or equip-ment downtime, maintenance labor-hours, or on system or equipment availability requirements. For example, *mean time to repair* (MTTR) is a basic maintainability measure. It is the mean cor-rective maintenance time for an item. MTTR and various percentiles of the repair-time distribu-tion may be limited to avoid the consequences of system failure (as when the maintenance time to repair a failed aircraft radio becomes so long as to impact flight schedules); maintenance labor-hours (MLH) is the total corrective and preventive labor-hours per year required to keep the sys-tem operating (not including supply or administrative delay times). For some systems, this parameter is divided by the expected annual system operating hours to produce the rate-based parameter maintenance labor-hours per operate-hour (MLH/OH). These parameters are signifi-cant contributors in determining a system's life-cycle costs (LCCs) and must be limited to control the LCC budget; availability is frequently used for systems that operate continuously and can be characterized as being in an "up" or "down" state. It measures the probability that a system is operating satisfactorily at a random point in time.

15.3.1 Duration-Based Parameters

Mean Time to Repair (MTTR).   MTTR is probably the most widely used maintainability pa-rameter. It is the mean of the times to repair weighted by the probability of occurrence. MTTR is calculated from the equation

$$\text{MTTR} = \frac{\sum\limits_{i=1}^{N} \lambda_i \cdot \text{MTTR}_i}{\sum\limits_{i=1}^{N} \lambda_i} \tag{15.1}$$

where λ_i = failure rate of the ith repairable unit
MTTR_i = mean time to repair the ith repairable unit
N = number of units in the system

When based on maintenance activity over a given period of time, MTTR is given by the total corrective maintenance time divided by the number of corrective-maintenance actions performed. The parameter $\overline{M}_{ct}$ (mean corrective-maintenance time) is often used interchangeably with MTTR.

Median Corrective Maintenance Time (Md$_{ct}$).   To determine the time within which half of the maintenance actions can be accomplished, the median time is used. Its calculation depends on the probability function that describes the repair-time distribution. For a normal distribution, the median and mean are collocated and the Md$_{ct}$ equals the MTTR.

For a lognormal distribution

$$\text{Md}_{ct} = \text{antilog} \ \frac{\sum\limits_{i=1}^{N} \lambda_i \cdot (\log M_{ct})_i}{\sum\limits_{i=1}^{N} \lambda_i} \tag{15.2}$$

For an exponential distribution, the fundamental maintainability parameter is the repair rate μ, which is the reciprocal of the MTTR. The median time to repair is given by

$$\text{Md}_{ct} = 0.69 \ \text{MTTR} = \frac{0.69}{\mu} \tag{15.3}$$

Maximum Corrective Maintenance Time (M$_{max\ ct}$). $M_{\text{max ct}}$ measures the time to perform a specified percentage of all potential repairs. The 90th and the 95th percentiles are most commonly specified. Again, the details of the calculation depend on the probability function that describes the repair-time distribution.

For the normal distribution

$$M_{\text{max ct}} = \text{MTTR} + z \cdot \sigma_{\text{MTTR}} \tag{15.4}$$

where z = 1.28 or 1.65 for the 90th and 95th percentiles and σ_{MTTR} is the standard deviation of the system's repair-time distribution.

For the lognormal distribution

$$M_{\text{max ct}} = \text{antilog}(T + z \cdot \sigma_T) \tag{15.5}$$

where z = 1.28 or 1.65 for the 90th and 95th percentiles
T = mean of the logarithms of the repair times
σ_T = standard deviation of the logarithms of repair times

For the exponential distribution

$$M_{\text{max ct}} = z_e \cdot \text{MTTR} \tag{15.6}$$

where z_e = 2.312 or 3.00 for the 90th and 95th percentiles.

Preventive Maintenance (PM).   Preventive-maintenance demands are often calculated on an annual basis. This is accomplished by establishing an annual frequency for each preventive-maintenance task and using these frequencies to weight the time to accomplish each task. When preventive-maintenance tasks are not calendar-driven (e.g., when they occur after so many operating cycles or flights or other pacing events), an operating scenario is used to convert operational time to calendar time. Then, the annual mean preventive-maintenance (PM) time is given by

$$PM = \sum_{i=1}^{N} f_i T_i \qquad (15.7)$$

where f_i is frequency of the *i*th task and T_i is duration of the *i*th task.

When a mean value is required, the mean preventive-maintenance time is given by

$$\overline{M}_{pt} = \frac{\sum_{i=1}^{N} f_i T_i}{\sum_{i=1}^{N} f_i} \qquad (15.8)$$

Mean Active Maintenance Time ($\overline{M}$). $\overline{M}$ is the mean time to perform a corrective or preventive-maintenance action. This time excludes supply delay and administrative delay times. It is given by

$$\overline{M} = \text{MTBM}(\lambda \cdot \overline{M}_{ct} + f_{pt} \cdot \overline{M}_{pt}) \qquad (15.9)$$

where MTBM is the mean time between corrective and preventive-maintenance actions, given by

$$\text{MTBM} = \frac{1}{\lambda + f_{pt}} \qquad (15.10)$$

where λ is the failure rate and f_{pt} is the preventive-maintenance rate.

Mean Downtime (MDT).   MDT is given by

$$\text{MDT} = \overline{M} + \text{SDT} + \text{ADT} \qquad (15.11)$$

where SDT is the average *supply delay time* and ADT is the average *administrative delay time* associated with each maintenance action.

15.3.2 Total-Labor Parameters

Although elapsed time measures are important indicators of maintainability, the total labor effort is also a concern for many systems. If elapsed time is reduced by increasing the human resources allocated to some of the maintenance tasks, then the overall labor effort may need to be estimated to assess the labor requirements for the system.

Mean Maintenance Labor-Hours Per Year (MLH/Y).   MLH/Y represents the sum of the corrective and preventive-maintenance labor-hours expended in a year. Once MTTR and PM times are estimated, MLH/Y is calculated as follows:

$$\text{MLH/Y} = \frac{\text{system operating hours per year}}{\text{MTBF}} \cdot (\text{MTTR}) \cdot (\text{average}$$

number of corrective-maintenance personnel per repair) + (PM) · (average   (15.12)

number of preventive-maintenance personnel per task)

An alternative equation for this parameter is

$$\text{MLH/Y} = 8760 \sum_{i=1}^{n} \frac{N_i d_i \, \text{MTTR}_i}{\text{MTBF}_i} + \sum_{j=1}^{m} (N_j f_j T_j) \qquad (15.13)$$

where N_i = average number of corrective-maintenance personnel to repair ith element
     d_i = duty cycle of ith element (fraction of calendar time ith element is active)
  MTTR_i = MTTR of ith element in hours
  MTBF_i = MTBF of ith element in hours
     N_j = average number of preventive-maintenance personnel to perform jth preventive-maintenance task
     f_j = frequency of jth preventive-maintenance task (times per year performed)
     T_j = duration of jth preventive-maintenance task in hours

Variations of MLH/Y.   By changing the interval over which the maintenance hours are evaluated, the parameter MLH/Y may be tailored to other needs. For example

1. Maintenance labor-hours per operate-hour (MLH/OH)
2. Maintenance labor-hours per operating cycle (MLH/cycle)
3. Maintenance labor-hours per month (MLH/month)
4. Maintenance labor-hours per maintenance action (MLH/MA)

15.3.3   Availability

Availability parameters generally have the form

$$1 - \frac{\text{downtime}}{\text{total time}} = \frac{\text{uptime}}{\text{total time}} \qquad (15.14)$$

Variations in calculating downtime and uptime provide for different availability parameters. Three frequently used measures of availability follow.

Inherent Availability.   *Inherent availability* is defined as follows:

$$A_i = \frac{\text{MTBF}}{\text{MTBF} + \text{MTTR}} \qquad (15.15)$$

This represents the probability that a system, when used under stated conditions, without consideration for any scheduled or preventive action, in an ideal support environment (available tools, spares, personnel, data, etc.) shall operate satisfactorily at a given point in time. It excludes preventive maintenance, supply delay, or administrative delay time.

Achieved Availability.   *Achieved availability* is defined as follows:

$$A_a = \frac{\text{MTBM}}{\text{MTBM} + \overline{M}} \qquad (15.16)$$

This represents the probability that a system or equipment, when used under stated conditions in an ideal support environment (available tools, spares, personnel, data, etc.) shall operate satisfactorily at a given point in time. This form of system availability includes both corrective- and preventive-maintenance tasks (but not supply delay and administrative delay times).

Operational Availability.   *Operational availability* is defined as follows:

$$A_o = \frac{\text{MTBM}}{\text{MTBM} + \text{MDT}}$$

(15.17)

This represents the probability that a system or equipment, when used under stated conditions in an actual operational environment, shall operate satisfactorily at a given point in time. In this case, the calculation for achieved availability has been modified to include supply delay and administrative delay times. Operational availability represents what is likely to be seen by the user since it includes the maintenance delays that can be expected to occur in actual operation.

15.4   THE MAINTAINABILITY PROGRAM

15.4.1   Maintainability Engineering

Maintainability engineering is performed for one or more of the following reasons:

1. To influence design to achieve ease of maintenance, thus reducing maintenance time and cost
2. To estimate downtime for maintenance which, when compared with allowable downtime, determines whether additional design or redundancy is required to provide acceptable continuity of a critical system function
3. To estimate system availability by combining maintainability data with reliability data
4. To estimate the labor-hours and other resources required for performing maintenance, which is useful for determining the costs of maintenance and for maintenance planning

This section describes how to identify maintainability program tasks, how to determine what maintainability parameters are appropriate for the project, and how to develop a maintenance concept for a system.

15.4.2   Establishing a Maintainability Program

A maintainability program includes the management and technical resources, plans, procedures, schedules, and controls for accomplishing the planned efforts.

Objective of a Maintainability Program.   The objective of the maintainability program is to ensure that the maintainability requirements of the project are met. Also, since the maintainability program is not accomplished in a vacuum, it includes efforts to interface with other engineering and related disciplines, such as reliability, safety engineering, and training. MIL-STD-470B, *Maintainability Program for Systems and Equipment,*[5] contains and describes a comprehensive list of tasks that could be included in a maintainability program. These tasks, together with guidance on task selection and the program phase(s) when each task is applicable, are listed in Fig. 15.3.

Program Phases.   The program phases listed in Fig. 15.3 are defined as follows:

1. *Conceptual phase:* the identification and exploration of alternative solutions or solution concepts
2. *Demonstration and validation phase:* the period when selected candidate solutions are refined through extensive study and analysis; hardware development, if appropriate; and test and evaluation

Task no.	Task title	Task type	Program phase			
			Concept	DEMVAL	FSD	Prod.
101	Maintainability program plan	MGT	N/A	G(3)	G	G(3)(1)
102	Monitor/control of subcontractors and suppliers	MGT	N/A	S	G	G
103	Program reviews	MGT	S	G(3)	G	G
104	Data collection, analysis, and corrective-action systems	ENG	N/A	S	G	G
201	Maintainability modeling	ENG	S	S(4)	G	C
202	Maintainability allocations	ACC	S	G	G	C
203	Maintainability predictions	ACC	N/A	S(2)	G	C
204	Failure modes and effects analysis (FMEA); maintainability information	ENG	N/A	S(2)(3)(4)	G(1)(2)	C(1)(2)
205	Maintainability analysis	ENG	S(3)	G(3)	G(1)	C(1)
206	Maintainability design criteria	ENG	N/A	S(3)	G	C
207	Preparation of inputs to detailed maintenance plan and logistics support analysis (LSA)	ACC	N/A	S(2)(3)	G(2)	C(2)
301	Maintainability demonstration (MD)	ACC	N/A	S(2)	G(2)	C(2)

Key: S—selectively applicable; G—generally applicable; C—generally applicable to design changes; N/A—not applicable; ACC—maintainability accounting; ENG—maintainability engineering; MGT—management; (1) Requires considerable interpretation of intent to be cost-effective; (2) MIL-STD-470B is not the primary implementation document (other MIL-STDs or statement-of-work requirements must be included to define or rescind the requirements; e.g., MIL-STD-471A must be imposed to describe maintainability demonstration details and methods); (3) appropriate for those task elements suitable to definition during phase; (4) depends on physical complexity, packaging, and overall maintenance policy of the system or unit being procured.

FIGURE 15.3   MIL-STD-470B maintainability task application matrix.

3. *Full-scale development phase:* the period when the system and the principal items necessary for its support are designed, fabricated, and tested

4. *Production phase:* the period from production approval until the last system is delivered and accepted

Maintainability Tasks.   Table 15.1 contains brief descriptions of each maintainability task that appears in Fig. 15.3. The MIL-STD discourages the indiscriminate application of these tasks and frequently emphasizes that "Task descriptions are intended to be tailored as required by their users as appropriate to particular systems or equipment program type, magnitude, and need." (Tailoring can include deletion of the task.) Extensive descriptions of each task are included in the MIL-STD together with additional guidelines for their selection and application. MIL-HDBK-338-1A, *Electronic Reliability Design Handbook,*[6] describes the relative importance of maintainability program elements during various system development life-cycle phases and provides detailed subtask listings and decision points for each system development phase.

    Together, these documents provide an excellent reference for defining the content and scope of a maintainability program, even when the subject equipment is not scheduled for DoD (Department of Defense) use.

TABLE 15.1   MIL-STD-470B Maintainability Tasks

Task number and name	Description
100 series: Program Surveillance and Control	
101: Maintainability Program Plan	Develop a maintainability program plan which identifies and ties together the maintainability tasks required to accomplish program requirements.
102: Monitor/Control of Subcontractors and Suppliers	Include appropriate quantitative maintainability requirements in specifications for subcontractor and vendor items. Provide for appropriate surveillance and management control of subcontractor and supplier maintainability programs so that appropriate management action can be taken, if required.
103: Program Reviews	Review and assess the maintainability program and maintainability at scheduled milestones.
104: Data Collection, Analysis, and Corrective Action Systems	Establish a maintainability and diagnostics effectiveness data collection system to augment predictions during design and for evaluation of demonstration results.
200: Series: Design and Analysis	
201: Maintainability Modeling	Develop appropriate maintainability models based on design characteristics which impact maintainability. The models should reflect the appropriate levels of maintenance for the system.
202: Maintainability Allocations	Allocate maintainability requirements to levels necessary to establish requirements for designers. Requirement imposed on subcontractors and suppliers shall be consistent with the allocations.
203: Maintainability Predictions	Develop predictions, in quantitative terms, of the maintainability parameters for the planned design configurations. If necessary, account for each maintenance, operational, and maintenance concept at each level of maintenance specified for the system. Compare the predictions with specified requirements to determine the adequacy of the design and the need for corrective action.
204: Failure Modes and Effects Analysis (FMEA)—Maintainability Information	Provide maintainability information to support any (FMEA) activity
205: Maintainability Analysis	Conduct analyses and tradeoffs to determine the relative advantages of alternative design approaches and maintenance concepts. Use tradeoffs to provide a cost-effective design that meets system objectives.
206: Maintainability Design Criteria	Develop design criteria to guide the design process. Typical goals shall include adequate accessibility; reduced maintenance activity, downtime, support costs, and maintenance personnel; and adequate built-in test capability.
207: Preparation of Inputs to the Detailed Maintenance Plan and Logistics Support Analysis (LSA)	Provide outputs and results of maintainability analysis that impact maintenance and support planning. Typical factors could include personnel skill levels and numbers, training requirements, and support equipment and tool needs.
300 Series: Evaluation and Test	
301: Maintainability Demonstration (MD)	Conduct formal tests to demonstrate that maintainability and testability requirements have been met.

15.4.3 Selecting Maintainability Requirements

Every system is developed in response to satisfying some set of needs or some operational functionality. The maintainability requirements for the system are one element in meeting these needs or functions. Although the best maintainability measures are determined by the unique circumstances associated with each system, Table 15.2 can be used as a guide to selecting maintainability measures for a particular system or for assessing the "reasonableness" of specified maintainability measures.

Table 15.2 identifies five conditions associated with a system or the distribution of maintenance times for the system. Although there are seven conditions listed, conditions 1 and 1*a* and 4 and 4*a* are complementary and mutually exclusive, thus reducing the overall list. The table also contains six columns that identify the major choices for maintainability requirements for a system. The entries in the table link the various system conditions to the requirement choices.

To select maintainability measures to be applied at the organizational level (i.e., where the physical replacement at the equipment will be made), the analyst should check each condition, determine whether it applies to the system, and, if so, circle any x's that appear in the row. Once the five conditions have been analyzed, the analyst can check each column to determine which requirements have the most "hits." These would be the primary candidates for system requirements. For example, if system availability is important (condition 1), limiting maintenance times is important (condition 2), and the maintenance distribution is known to be symmetric (condition 4*a*), then the third column of requirements (means and maxima) would appear to be the most appropriate since it contains an x for each of these conditions.

Table 15.2 is not intended to provide an exhaustive set of conditions that would influence measure selection, but it does contain some of the most commonly encountered conditions. Other rows of conditions could be added to reflect experience with a specific type of system being analyzed.

15.4.4 Developing a Maintenance Concept

The maintenance concept describes a maintenance process which will meet the system's operational requirements. It may be developed by either the customer or the contractor. It must be developed early in the life of a maintainability program to document the operational environment and constraints of the system being developed and the assumptions to be used in developing maintenance guidelines and policies.

TABLE 15.2 Maintainability Requirement Selection Matrix

System condition or requirement	M_{ct} and M_{pt}	$M_{max\ ct}$	M_{ct} and $M_{max\ ct}$; M_{pt} and $M_{max\ pt}$	Md_{ct}	Md_{ct} and $M_{max\ ct}$	MLH/OH
1. Availability is important.	x		x		x	
1*a*. Availability is not important.		x		x	x	x
2. Maintenance time limit is important.		x	x		x	
3. Limiting labor is important.						x
4. Maintenance time distribution is lognormal or skewed to the right.			x	x	x	
4*a*. Maintenance distribution is not right-hand-skewed.	x		x			
5. Maintenance time distribution is unknown.	x		x	x		

Ideally, the developer reviews the maintenance concept with the customer or user early in the life of the program to ensure that the developer has a proper understanding of the final use of the system. If the system is being developed for a general market (as opposed to an individual customer as in the case of military procurement), then focus groups that reflect the user population can be interviewed to confirm the realism and adequacy of the information in the maintenance concept. In any case, the maintenance concept becomes the basis for developing design guidelines, conducting equipment tradeoffs, and, in general, evolving the maintainability design of the system. Also, as the system design evolves, further amplification and clarification of the maintenance concept will be possible and should result in periodic updates to the document.

The following types of information should be included in the maintenance concept.

System usage profiles: Is the system used continuously (weather radar)? So many hours per day (office copier)? Or at irregular intervals (backup power generation equipment)? Are there different usage profiles? And if so, what are they? Are there degraded modes of operation possible? How many and what are they?

Physical or geographic deployment: How far apart are the different portions of the system? What are the maximum expected separations? Can personnel at the different locations communicate with each other if necessary?

Reliability, availability, and maintainability (RAM) requirements for the system: Are the customer's requirements imposed on the entire system, or are they initially allocated across portions of the system? Are there different values for different operating modes of the system?

Maintenance levels: Maintenance may be performed at three levels: organization, intermediate, and depot. *Organization* or *on-equipment* refers to the removals and replacements actually made on the system. At this level, the least-specialized maintenance skills, test equipment, and facilities are found. At the intermediate level, more complex tasks are accomplished often in a facility close to or centrally located with respect to several using activities. A depot will have the most specialized skills and equipment for repairs to accomplish any level of repair that removed items or their components may require. Other systems may have two levels of maintenance: organization and a combination of intermediate and depot. Which of these models (or what other model) applies to this system?

Scheduled and preventive maintenance: Are there restrictions on site availability for scheduled or preventive maintenance? Can the system operate in a degraded mode while these activities are performed? Are there operational limitations on the length of scheduled or preventive-maintenance times?

Maintenance workforce constraints: Are there physical, safety, union, or other constraints on the size of a maintenance crew? Are there limits on the length of time a single maintenance person can work (e.g., high noise levels)? Are there minimum times needed to get a maintenance crew to the site of the repair (e.g., unattended systems)?

Physical limitations: What are the practical limits on the weight, volume, size, or other characteristics of repairable items? Do these change with the size of the maintenance crew?

Test and support equipment availability: What test and support equipment is expected to be at the site? What will the maintenance crew be expected to bring?

Sparing and supply policies: What can be expected to be on site? To be local? How will other needs be met?

The maintenance concept frequently responds to high availability requirements by establishing that organization-level maintenance shall consist of removal and replacement of a faulty unit. Such a unit is called a *line replaceable unit* (LRU). This is normally the fastest way to return a system to full operational capability after a failure has occurred and therefore contributes to high availability.

15.5 MAINTAINABILITY ALLOCATION

System-level maintainability requirements are not generally useful as design requirements because they do not address the equipment levels at which design and maintenance will occur. Maintainability allocation is a continuing process of apportioning high-level requirements to lower levels of an equipment, as necessary, to establish figures that designers can work to. The maintainability allocation process is similar to that used for reliability except that, in the case of maintainability, both failure rates and MTTR estimates are used to perform the allocation. As a result, reallocation of maintainability requirements may be triggered by changes in either the reliability or maintainability estimates for the system.

Table 15.3 provides an example of the allocation process in the case for a system with a 60-min MTTR requirement which is composed of three subsystems, X, Y, and Z. In practice, allocations can be accomplished using a spreadsheet program or any of several reliability and maintainability software packages. The subsystems are identified in the first column. Working from the bottom up, failure-rate estimates (or allocations based on system reliability requirements) and MTTR estimates (based on previous experience or preliminary predictions) are entered in columns A and B. These data are used to calculate an initial MTTR allocation, which is 66 min. To meet the system requirement of 60 minutes, it is necessary to adjust the initial MTTR allocations by a proportionality constant k, which is given by

$$k = \frac{\text{required system MTTR}}{\text{initial MTTR allocation}} \tag{15.18}$$

In this case, $k = 60/66$.

The initial MTTR allocations are multiplied by the proportionality constant k, and the results are entered in column D (rounded to the minute). Then, MTTR is recalculated to confirm that the revised allocations are adequate to meet the system requirement (column E).

To the extent necessary to support the design and maintenance efforts, a subsystem allocation can now be used as the basis for lower-level allocations using the same technique.

The final allocated values, $k(\text{MTTR})$, now become the design requirements for these subsystems and the basis of allocations at lower levels of the equipment. If, as the design matures and more information becomes available, it appears that one or more of these values will not be met, it may be necessary to recompute the allocations using updated failure rates and maintenance times.

15.6 MAINTAINABILITY DURING THE DESIGN EFFORT

Maintainability and availability analyses are performed throughout a system's design and development phases to:

1. Establish the most cost-effective ways to minimize the need for maintenance and to achieve required maintainability performance (tradeoff analysis)

TABLE 15.3   Maintainability Allocation Example

	A Failure rate ($\times 10E{-}6$)	B MTTR, min	C A*B, min	D k*B, min	E A*D, min
Subsystem X	200	60	12,000	54	10,800
Subsystem Y	300	30	9,000	27	8,100
Subsystem Z	500	90	45,000	82	41,000
Sum	1000	—	66,000	—	59,900
		Initial MTTR	66	Revised MTTR	60

2. Quantify maintainability requirements at the design level (allocation)

3. Evaluate the design for its conformance with both qualitative and quantitative maintainability requirements (prediction)

4. Generate maintainability data for use in maintenance planning and logistics support analyses (LSAs)

15.6.1 Maintainability Design Guidelines

The maintainability engineer or organization will establish maintainability design guidelines for the design team to use during the development effort. Maintainability design guidelines are normally used to define the physical or operational maintainability characteristics desired in a system or equipment. They frequently reflect an organization's or industry's experience with similar equipments and provide an opportunity to introduce lessons learned on previous designs into new designs. These guidelines subsequently provide a basis for both qualitative and quantitative evaluations of the design as it develops.

Design guidelines often take the form of maintainability design ground rules or criteria. They describe such requirements as accessibility, ability to detect and isolate a failure, weight limitation of replaceable units, and dimensional limits to allow replaceable units to be transported from their installed locations to a repair facility or for shipment to a manufacturer's facility. Design guidelines frequently include human factors criteria such as those found in MIL-HDBK-1472D, *Human Engineering Design Criteria for Military Systems, Equipment, and Facilities*[7] or MIL-HDBK-761A, *Human Engineering Guidelines for Management Information Systems.*[8]

Design Ground Rules.   At the highest or least-detailed levels, maintainability design guidelines are often derived from the maintenance concept and may be referred to as maintainability ground rules. For example, (1) the design shall preclude the need for scheduled maintenance and (2) built-in test (BIT) and built-in test equipment (BITE) shall be the primary method of fault detection.

Maintainability Design Criteria.   Maintainability design criteria are often more specific than ground rules and are frequently oriented to specific types of equipment such as fluid systems, electronics, instrumentation, cables and connectors, and test equipment, regardless of where these items are located in the system. Maintainability design criteria may also be grouped by subsystem or specific equipment. In either case, these design criteria must be tailored to the specific types of equipment found in the system and to the requirements of the system. Examples of maintainability design criteria are as follows:

1. General design features:
   a. The design shall preclude the possibility of damage to the equipment during maintenance and servicing. Guards shall be provided to protest delicate parts exposed during servicing.
   b. Minimize the need for special tools.
   c. Part reference designations shall be located next to each part and shall be legible and permanent.
   d. Keying, size, or shape shall be used to ensure that removable parts are reassembled in the correct position.
   e. Guide pins shall be provided for alignment of modules or high-density connectors.
   f. Handles shall be provided for removable units weighing over 10 lb or whose shape makes them difficult to handle.
   g. Sharp edges, corners, or protrusions which could cause injury to personnel shall be avoided.
2. Mounting and location of units:
   a. Provide for the removal and replacement of LRUs without the removal of unfailed units.
   b. Provide for the removal and replacement of LRUs without interrupting critical functions.
   c. Provide clear access to all LRU locations. Mount units to chassis or structure rather than on other units.

     d. Mount heavy units as low as possible. Label each access for units which can be reached through it.
3. Test, checkout, and calibration:
     a. Fault-isolation test circuitry shall not cause failure of the circuit under test.
     b. Test points on printed-circuit boards (PCBs) shall be located to permit in-circuit testing.
     c. Calibration and adjustment controls which are intended to have limited motion shall be provided with adequate stops to prevent damage.
     d. All adjustments shall be designed to be common in their displacement response (i.e., clockwise, right, or up to increase).
4. Cables, wiring, leads, and connectors:
     a. Provide clearance around connectors for adequate viewing and hand access.
     b. Route cables to facilitate tracing, removal, and replacement.
     c. Provide service loops in cables and harnesses to facilitate installation, checkout, and maintenance.
     d. Code or label wires and cables throughout their length for easy identification.
5. Additional areas for design criteria include guidelines and policies for:
     a. Interchangeability of system units.
     b. Use of standard or otherwise approved parts.
     c. Limiting the number and skill levels of personnel required for maintenance.

15.6.2 Design Analysis

During the design and development phases, maintainability engineers will iteratively review the hardware development effort to ensure that existing maintainability allocations (Sec. 15.5) are reasonable, to determine how well the design conforms to maintainability guidelines and criteria, to determine how to avoid or reduce the need for maintenance, and to predict the relevant maintainability parameters (Sec. 15.8) to ensure that the design will meet its maintainability requirements. These tasks require both cooperation and coordination with the reliability and logistics organizations as well as the design group.

Reviews of emerging designs make use of conceptual sketches, preliminary design sketches, hardware designs, and the information available on the BIT and support equipment philosophies and their subsequent implementation in the system. The review may rely on paper-based sketches and drawings. When the design team is using computer-aided design and computer-aided engineering (CAD-CAE) tools, the designers will be using workstations or other types of computer terminals, the design information will be captured in electronic files and databases, and real-time communication between the designer and maintainability engineer and between the engineers and the databases is possible using a videodisplay terminal and input devices such as a keyboard, light pen, or mouse. As the design progresses, the maintainability engineer will also have access to either physical or, in the case of CAD-CAE tools, computer-based mockups of the equipment (which often permits more rapid analysis of the what-if questions).

Typical questions to be addressed during these reviews are as follows:

1. What are the principal failure modes for this item (replaceable unit, module, etc.) and how do they impact system operation?

Then, for each failure mode:

2. How does the operator or user know that this item has failed?
3. How does the BIT identify this failure?
4. What support equipment or tools are needed to localize or isolate these failures?
5. Can the system uniquely identify the failed item? If not, how many possible units are included in the "ambiguity group" associated with this failure?
6. What is the remove and replace sequence for this failure? How big a crew is needed and what, if any, special tools do they require?

7. What training or experience is required for this maintenance action?
8. What are the hazards to the crew or to the equipment in accomplishing this maintenance action?

And most importantly:

9. What can be done to reduce or eliminate the frequency of this failure mode?
10. What can be done to reduce the maintenance time required to accomplish this maintenance action?

Maintainability engineers will also participate in tradeoff studies. Typical topics to be addressed are

1. How will various equipment partitioning schemes affect system maintainability?
2. How will design alternatives impact maintenance skills, tools, or maintenance crew size?
3. How will part selection alternatives (that may be driven by reliability requirements) affect the system's maintainability characteristics?

In conducting or contributing to tradeoff analyses the maintainability engineer will, at the least, make use of allocation and prediction models to assess the impact of the tradeoff on maintainability parameters. At the other extreme, it may be necessary to assess the alternatives in terms of a system LCC model which includes the system's acquisition cost, repair costs for several years, spares or inventory costs, and other costs of ownership affected by the tradeoff.

15.6.3 Design Reviews

A series of formal design reviews are typically included as project milestones to allow the program manager, senior management, and the customer or user to assess the maturity of the design effort, the adequacy of program planning, and the technical and contractual adequacy of the program's efforts.

Two of these milestones are termed the *preliminary design review* and the *critical design review.*

Preliminary Design Review.   Preliminary design review (PDR) is a formal review of the basic design approach for a system or a functionally related group of lower-level elements. It is typically conducted after the hardware development specification is available. During the PDR, the maintainability portion of the review will address:

1. Quantitative maintainability requirements and a comparison with available preliminary predictions
2. Preventive-maintenance schedules
3. Repair-rate sources and prediction methods
4. Maintainability characteristics of the system
5. Provisions for operation of BIT and other support equipment used in maintenance
6. Conformance to the maintainability ground rules and design criteria
7. Plans for maintainability demonstrations
8. Actions to be taken if current predictions do not meet requirements.

Critical Design Review.   *Critical design review* (CDR) is a formal review for a system or functionally related group of lower-level elements prior to fabrication or production release to ensure that the detailed design solutions satisfy the requirements established by the hardware development specification. During the CDR, the maintainability portion of the review will address:

1. Quantitative maintainability requirements and a comparison with available preliminary predictions
2. Preventive-maintenance schedules for compatibility with system requirements
3. System conformance to the planned-maintenance concept and unique maintenance procedures (if any)
4. Methods for automatic, semiautomatic, and manual recovery from failure
5. Conformance to the maintainability ground rules and design criteria
6. Details of plans for maintainability demonstrations

15.7   TESTABILITY

15.7.1   Introduction

The maintainability performance of many systems is often influenced by the adequacy and efficiency of the system's test and diagnostic system (hereafter referred to as the *test system*). A test system serves two basic functions: to provide for fault detection, and to provide isolation to the defective item. At the same time, the test system must itself be reliable and its failures should not affect system performance.

The scope and complexity of the test system is related to the complexity and the operational and maintainability requirements for the host system. When it is not possible or feasible to detect all faults with built-in test (BIT) or to isolate all faults to a single or acceptably small number of replaceable units, the test system will require the use of external test equipment and manual procedures.

Test systems cannot be designed and implemented as an afterthought or add-on to the host system's development efforts. They must be developed as an integral part of the system design process. As a result, the maintainability engineer is responsible for including the test system development in the project's maintainability program. When the system is sufficiently complex, it may be necessary for the maintainability project leader to establish a separate testability program and assign and coordinate maintainability resources to that effort.

An integrated diagnostics approach is required to achieve a system's diagnostic capability. Such an approach is achieved by integrating all related pertinent diagnostic elements. Although a specific engineering organization is typically responsible for developing the test and diagnostic system, the process for developing an integrated diagnostics approach includes interfaces between design, engineering, testability, reliability, maintainability, human engineering, and logistic support analysis.

15.7.2   Testability Terms and Definitions

Testability.   In MIL-STD-2165A, App. C,[9] *testability* is defined as follows: "A design characteristic which allows the status (operable, inoperable, or degraded) of an item to be determined and the isolation of faults within the item to be performed in a timely manner." From this definition, it is clear that testability includes both the detection and fault-isolation capabilities of the test and diagnostic system.

Built-in Test.   In MIL-STD-2165A,[9] *built-in test* is defined as "the integral capability of the system or equipment which provides an automated test capability to detect, diagnose, or isolate failures."

Hardware which is identifiable as performing the built-in test function is called the *built-in test equipment* (BITE).

Diagnostics.   In MIL-STD-2165A,[9] *diagnostics* is defined as "The hardware, software, or other documented means to determine that a malfunction has occurred and to isolate the cause of the malfunction."

15.7.3 Testability Parameters

Testability requirements may be established at the system, subsystem, replaceable unit, or lower repair level. At the system and subsystem levels, they reflect the operational need to be able to detect failures as soon as practicable after they have occurred and, having detected a failure, to be able to isolate the failure to some reasonable subset of the system so that replacement and subsequent verification of the repair can be accomplished in a reasonable time. The three principal testability characteristics of modern equipments are

Fault-Detection Capability (FDC).   The fault-detection capability for a system is the percentage of failures that are automatically detected by the system.

FDC is estimated by dividing the failure rate for those portions of the system where failures can be detected by the test system by the system failure rate. This measure is also referred to as the *fraction of faults detected* (FFD). In a maintainability demonstration, FDC represents the number of real (or simulated) failures that are detected by the test system divided by the total number of real (or simulated) failures in the test.

Fault-Isolation Capability.   The *fault-isolation capability* (FIC) for a system measures the ambiguity associated with fault-isolation activities. Fault-isolation capability is frequently specified as follows: X_1 percent of the time the system shall be capable of isolating a fault to Y_1 or fewer line replacement units (LRUs); and X_2 percent of the time the system shall be capable of isolating a fault to Y_2 or fewer LRUs. In these statements X_1 and X_2 are typically 90 and 95 percent, Y_1 is typically 1 or 2 LRUs, and Y_2 is typically 2 or 3 LRUs.

In specifying FIC for a system, both built-in test capabilities as well as the use of external test equipment and manual procedures are permitted to perform isolation unless otherwise specified.

False-Alarm Rate.   The *false-alarm rate* (FAR) for a system is the rate at which the system declares the detection of a failure when, in fact, no failure has occurred. FAR is typically expressed as the ratio of the number of false alarms experienced divided by some temporal or operational interval for the system, for example, false alarms per month, per million transactions, or per thousand operating cycles.

15.7.4 Testability Design Guidelines

The maintainability design ground rules described in Section 15.6.1 should be reviewed to ensure that they adequately cover the test system as well as the host system.

MIL-STD-2165A contains a checklist with over 100 testability design criteria, a sample of which follows.

Mechanical design (for electronic functions):

- Is a standard grid layout used on boards to facilitate identification of components?
- Is enough spacing provided between components to allow for clips and test probes?

Test control:

- Can circuitry be quickly and easily driven to a known initial state?
- In microprocessor-based systems, does the tester have access to the data bus, address bus, and important control lines?

Test access:

- Are all high voltages scaled down within the item prior to providing test point access so as to be consistent with the tester capabilities?
- Is the measurement accuracy of the test equipment adequate compared to the tolerance requirement of the item being tested?

Parts selection:

- Have parts been selected which are well characterized in terms of failure modes?
- Is a single logic family being used? If not, is a common signal level used for interconnections?

Analog design:

- Is each test point adequately buffered or isolated from the main signal path?
- Does the design avoid or compensate for temperature-sensitive components?

RF design:

- Have suitable termination devices been employed in the off-line automatic test equipment (ATE) or BIT circuitry to accurately emulate the loading requirements for all RF (radio-frequency) signals to be tested?
- Do transmitter outputs have directional couplers or similar signal sensing or attenuation techniques employed for BIT or off-line test monitoring purposes, or both?

Electrooptical (EO) design:

- Have optical splitters/couplers been incorporated to provide signal accessibility without major disassembly?
- Do test routines and internal memories test pixels for shades of gray?

Digital design:

- Are all clocks of differing phases and frequencies derived from a single master clock?
- Do all buses have a default value when not selected?

Built-in test (BIT):

- Are on-board BIT indicators used for important functions? Are BIT indicators designed such that a BIT failure will give a "fail" indicator?
- Have means been established to identify whether hardware or software has caused a failure indication?

Performance monitoring:

- Have critical functions been identified which require monitoring for the system operation and users?
- Has the displayed output of the monitoring system received a human-engineering analysis to ensure that the user is supplied with the required information in the best usable form?

Sensors:

- Are pressure sensors placed very close to pressure sensing points to obtain wideband dynamic data?
- Have procedures for calibration of sensing devices been established?

Test requirements:

- For each maintenance level, has a decision been made on how BIT, ATE, and general-purpose test equipment will support fault detection and isolation?
- Is the planned degree of test automation consistent with the capabilities of the maintenance technician?

15.7.5 Design Analysis

At the LRU level, design for testability typically includes the development of failure modes and effect analyses for the LRUs and the identification of tests to be used to detect and isolate failed items. The test suite is then analyzed to identify optimal test sequences. This is especially important when individual tests may take a large amount of time (or be very expensive to implement, or require special maintenance skills). Simpson and Sheppard[10] provide detailed discussions, analyses, and examples of diagnostic modeling and its applications.

During the design and development phases, maintainability engineers will review the test system development effort as an integral part of the maintainability design review process to determine whether the testability approach is adequate for the system's maintainability needs, to estimate how the test system contributes to maintainability prediction times, and to determine how testability will be integrated into subsequent demonstration test efforts.

15.7.6 Testability Demonstration

Demonstration of the fault-detection and -isolation capabilities of a system can be performed concurrent with other maintainability demonstration activities. When actual or simulated faults are used to demonstrate maintainability, the fault detection and isolation activities and times that are recorded as part of the maintainability demonstration can be used to assess the system's conformance to testability requirements.

Maintainability demonstration is discussed in Sec. 15.9. In that section, Table 15.7 includes tests for fault detection and fault isolation which can be integrated with other formal maintainability demonstration tests.

However, since the simulation of false alarms is not possible during normal test or operational activities, the measurement of this parameter during a demonstration program (or in actual operation) depends on the collection of any false-alarm data that may occur during that time and the subsequent determination of the frequency of those events.

15.8 MAINTAINABILITY PREDICTION

Maintainability predictions are performed to assess whether a proposed design can meet a specified or allocated maintainability requirement and to evaluate the maintenance impact of design alternatives or other program decisions. Predictions also provide a method to measure a design's progress toward its goals and to identify opportunities for improving the maintainability characteristics of a design.

15.8.1 Predicting Corrective and Preventive Maintenance

MIL-HDBK-472[11] presents five procedures for estimating both corrective- and preventive-maintenance parameters, among others. These procedures, together with the parameters predicted by each procedure, are listed in Table 15.4. Procedures I through IV were initially published in the 1960s and predate the extensive use of built-in test. These procedures were developed for the maintenance of tube-type equipment, and the methodology and formulas associated with these procedures should be carefully reviewed before they are applied to modern equipment. On the other hand, procedure V, issued with Notice 1 in 1984, provides both early and detailed prediction methods that recognize the use of integrated diagnostic, isolation, and test capabilities. A simplified example of a maintainability prediction is presented in Sec. 15.8.3.

15.8.2 Estimating Maintenance Times

In developing maintenance-time predictions, it is necessary to develop estimates for the maintenance elements that constitute a maintenance task, such as removing bolts and panels or unplug-

TABLE 15.4 MIL-HDBK-472 Maintainability Prediction Methods

Prediction procedure	Application	Time of application	Predicted parameters
I	To predict flight-line maintenance of airborne electronic and electromechanical systems involving modular replacement at the flight line	After establishment of the design concept provided that appropriate data are available	Distribution of downtime for various elemental activities, maintenance categories, active repair times, and system repair time and system downtime
II	To predict the maintainability of shipboard and shore electronic equipment and systems; can also be used to predict the maintainability of mechanical equipments or systems if maintenance task times for applicable functional levels can be established	During the final design stage	Part A procedure: equipment repair time (ERT) (related to MTTR) Part B procedure: active maintenance in terms of (a) Mean corrective maintenance time in worker-hours (b) Mean preventive maintenance time in worker-hours (c) Mean active maintenance time in terms of mean worker-hours per maintenance action
III	To predict the mean and maximum active corrective-maintenance downtime for ground electronic systems and equipment	During the design and development stages	(a) Mean and maximum active corrective downtime (95th percentile) (b) Mean and maximum preventive maintenance (c) Mean downtime
IV	To predict the mean corrective, preventive, or total downtime of systems and equipment; applicable to all systems and equipment	Throughout the design and development stages with various degrees of detail	(a) Mean corrective downtime (b) Mean preventive downtime (c) Total mean downtime
V	To predict maintainability parameters of avionics, ground and shipboard electronics at the organizational, intermediate and depot levels of maintenance	Applied at any equipment or system level, at any level of maintenance, and for any maintenance concept pertinent to avionics, ground electronics, and shipboard electronics; may be applied to electromechanical or mechanical equipment or systems	(a) Mean time to repair (MTTR) (b) Maximum corrective maintenance time ($M_{max\,ct}$) (c) Mean maintenance man-hours per repair (MMH/repair) (d) Mean maintenance man-hours per operating hour (MMH/OH) (e) Mean maintenance man-hours per flight hour (MMH/FH) (f) Percent of faults isolatable to $\leq N$ replaceable items

ging cables. There are several sources for these data, including actual measurements on similar equipment, engineering judgment, simulations on equivalent equipments or mockups, or maintenance-time standards. Notice 1 of MIL-HDBK-472 provides maintenance-time standards for 57 common physical activities in the following categories: fasteners, latches, terminal connections, connectors, plug-in modules, and miscellaneous.

A maintainability engineer with an industrial engineering background which includes time measurements can also develop estimates for maintenance element times. Examples of element times prepared by this approach are shown in Tables 15.5 and 15.6. Table 15.5 provides maintenance element times for common maintenance tasks performed at the LRU level. Table 15.6 provides factors used to modify Table 15.5 times to compensate for different degrees of equipment access. The data in these tables have been validated by comparing predictions based on them with subsequent maintainability demonstration test data for the systems predicted.

These standards will not provide estimates for the times required for diagnostic or test programs to run. That information will be obtained from experience on similar equipments or from timing simulations of the diagnostic and test programs.

15.8.3 Maintainability Prediction: An Example

This section provides an example of a maintainability prediction for a multiprogrammer subsystem, which is part of a larger system. Assume that an earlier allocation has assigned a corrective-maintenance time (MTTR) of 0.50 h to this subsystem and that the maximum corrective-maintenance time ($M_{\max ct}$) at the 95th percentile should not exceed 1.0 h.

Equipment Description.   The multiprogrammer consists of 15 cards of five different types which plug into a mainframe that resides in a standard drawer in an equipment rack. The cards have the following functions:

Type/Quantity	Description
Type A/1	Digital-to-analog converter
Type B/2	Timer-pacer
Type C/10	Isolated digital input card
Type D/1	Interrupt card
Type E/1	Memory card

Maintenance Concept.   The system self-test automatically detects failures in the multiprogrammer and isolates to a failed card or to the mainframe in less than a minute. In the event of a mainframe failure, it will be removed and replaced with a spare mainframe, using the cards removed from the failed mainframe. In the event that a card fails, only the failed card will be removed and replaced. Verification of the maintenance action is accomplished by having the system run a "long test" for the type C cards (0.25 h in length) or a "short test" for the other cards and for the mainframe (0.17 h in length).

Sample Time Synthesis Analysis.   The first step in predicting the maintainability of the multiprogrammer is to analyze the design to identify the sequence of steps required to perform the necessary maintenance. Figure 15.4 identifies the failure rate for a type C card and lists the elemental activities (EAs) required to remove and replace a type C card. The third column of the figure records the data needed to estimate the elemental activity time. The fourth column lists the time, in hours, for the EAs taken from Table 15.5. An accessibility factor from Table 15.6 has been applied to these times to reflect physical workspace in which the equipment is located. Finally, the time for the "long" test is added to arrive at the MTTR for this card.

TABLE 15.5 Maintenance-Element-Time Standards

	Hours	
Maintenance element	First	Additional
Fasteners:		
1. Bolts or screws in threaded holes (hand tools)		
a. Install	0.0167	0.0083
b. Remove	0.0125	0.0042
2. Quick-acting—requires less than one rotation (hand tools)		
a. Install	0.0125	0.0042
b. Remove	0.0111	0.0028
3. Quick-acting—requires less than one rotation (no tool)		
a. Install	0.0042	0.0042
b. Remove	0.0028	0.0028
4. Door, hinged, with $\frac{1}{4}$-turn latch and T-handle		
a. Close	0.0042	0.0042
b. Open	0.0028	0.0028
Connectors:		
1. Three-prong power plug		
a. Plug in	0.0056	
b. Unplug	0.0028	
2. Bayonet-type connector		
a. Connect	0.0056	
b. Disconnect	0.0028	
3. Screw-shell-type connector		
a. Connect	0.0125	
b. Disconnect	0.0097	
4. Circuit card or card connector		
a. Connect	0.0069	
b. Disconnect	0.0042	
5. Rectangular connector with 2 screws		
a. Connect	0.0233	0.0139
b. Disconnect	0.0181	0.0098
6. Terminal connection (screw/loop terminal)		
a. Connect	0.0167	0.0083
b. Disconnect	0.0125	0.0056
7. Terminal connection (screw/spade terminal)		
a. Connect	0.0125	0.0028
b. Disconnect	0.0125	0.0028
8. Banana-jack connection		
a. Connect one only	0.0028	N/A
b. Connect 2 or more (pair jack/socket)	0.0056	0.0042
c. Disconnect	0.0028	0.0028
Operator or technician miscellaneous activities:		
1. Walk, per 10-ft increment	0.0028	
2. Handle part or assembly (one operator or technician)		
a. Replace on tracks in rack/position		
Light, small (<3 lb, <64 in³)	0.0056	
Medium (3–25 lb, <1 ft³)	0.0069	
Heavy, large (25–85 lb, size and shape manageable by one person)	0.0167	
b. Remove from tracks in rack		
Light, small	0.0028	
Medium	0.0035	
Heavy, large	0.0083	
c. Set aside or obtain part		
Light, small	0.0028	
Medium	0.0035	
Heavy, large	0.0083	

TABLE 15.6   Accessibility Factors

Access* type	Factor
1. *Good*—open, unrestricted movement; no obstacles to work motion	1.00
2. *Fair*—slightly restricted work motions such as working with special tools or with tools applied at awkward angles or where there are slight physical obstructions	1.05–1.20
3. *Poor*—restricted work conditions, cramped movements, and with physical obstructions	1.20–1.50
4. *Very poor*—"blind" work conditions, cramped and restricted, including physical obstructions, especially visual	2.00–5.00

*Access pertains to those physical conditions which must be overcome in gaining access to the element of interest.

Elemental Activity Analysis			
System or subsystem name: multiprogrammer LRU Name: card type C Quantity: 10 Maintenance task: remove and replace Failure rate: 150×10E-6			
EA no.	Elemental activity (EA)	Comments	EA time, h
		Disassembly activity	
1	Walk to rear of cabinet	Walk 20 ft at 0.0014 h per 10 ft	0.0028
2	Open door	Hinged door with ¼-turn latch and turn handle	0.0028
3	Disconnect power cable	3-prong connector	0.0028
4	Remove rear cover plate	1 screw at 0.0111 h, 3 screws at 0.0028 h	0.0195
5	Disconnect connector from failed card	Circuit card connector	0.0042
		Remove and replace activity	
6	Remove and replace failed card	Remove at 0.0028 h, plug in at 0.0028 h, aside at 0.0028 h	0.0084
		Reassembly activity	
7	Connect card connector	Circuit card connector	0.0069
8	Install rear cover plate	1 screw at 0.0125 h, 3 screws at 0.0042 h	0.0254
9	Connect power cable	3-prong connector	0.0056
10	Close rear door	As in EA2	0.0028
11	Walk to front of cabinet	As in EA1	0.0028
		Subtotal	0.0840
		Apply accessibility factor for "fair"	1.2
		Subtotal	0.1008
		Verification test activity	
12	Run verification test	"Long test"	0.25
		MTTR	0.3508

FIGURE 15.4   Maintainability prediction worksheet.

LRU	Qty (Q)	Total fail rate—$Q\lambda$ $(E-6)$	MTTR, h	$Q\lambda$ (MTTR), h	$(MTTR)^2$	$Q\lambda$ $(MTTR)^2$
Card A	1	33	0.2708	8.9364	0.073332	2.419977
Card B	2	26	0.2708	7.0408	0.073332	1.906648
Card C	10	150	0.3508	52.62	0.123060	18.45909
Card D	1	21	0.2708	5.6868	0.073332	1.539985
Card E	1	76	0.2708	20.5808	0.073332	5.573280
Mainframe	1	120	1.128	135.36	1.272384	152.6860
Totals		426		230.2248	1.688775	182.5850

Subsystem: Multiprogrammer
Allocated MTTR: 0.50 hour
Required $M_{\text{max ct}}$ (0.95): 1.00 hour

MTTR = 0.540 h $M_{\text{max ct}}(0.95) = 1.151$ h

FIGURE 15.5   Maintainability prediction summary worksheet.

Subsystem Prediction.   After the time synthesis analysis has been performed for each card type and for the mainframe, the resulting data are transferred to a maintainability prediction worksheet (Fig. 15.5). The worksheet contains a list of the LRUs for the multiprogrammer, and the quantity, failure rate, and MTTR of each LRU. Then, Eqs. (15.1) and (15.4) are used to calculate MTTR and $M_{\text{max ct}}$. In this example, the normal distribution has been used to calculate $M_{\text{max ct}}$.

Comparing Predicted Values to Allocations or Requirements.   The final step of the prediction process is to compare the predicted values to the allocations or requirements to determine whether the design meets them. In this example, neither maintainability requirement has been met. Furthermore, it is clear from Fig. 15.5 that the mainframe is the principal contributor to maintenance times. This provides the maintainability engineer with some information as to how the design could be modified to meet the quantitative requirements, such as, reduce the failure rate of the mainframe or simplify the physical efforts required to replace the mainframe when it does fail. The latter approach could involve improving the accessibility of the mainframe in the operating environment.

15.9   MAINTAINABILITY VERIFICATION, DEMONSTRATION, AND EVALUATION

Maintainability verification includes the maintainability engineering efforts that are conducted to determine the accuracy of maintainability analyses and predictions, to identify maintainability deficiencies before a design goes into production, and, in general, to gain assurance that the maintainability requirements of a system can be achieved and demonstrated later in a program. The usual method of verifying maintainability is by analysis. The most common methods of analysis are prediction, review of designs, and tradeoff analyses that are described in Sec. 15.6.

As prototype equipment and/or early production models become available, maintainability demonstrations can be planned and conducted to show that quantitative maintainability requirements have been met. Demonstrations will frequently use formal statistical test methods such as those defined in MIL-STD-471A[12] (see Table 15.7) and will be conducted in an environment which simulates, as closely as practicable, the equipment's operational and maintenance environment.

When maintainability tests are performed on development hardware, they are usually planned as part of the overall development testing. The major problem encountered in collecting maintenance data during this period is that, when a failure occurs, the maintenance team's first inclination is to perform a failure analysis for the situation and not to replace or repair the failed unit. Great care and planning are required to get valid maintainability data from these efforts. Also, only a small number of items are typically available for development testing, which makes it very difficult to establish confidence in the maintainability results. At best, it may be possible to identify major maintainability deficiencies that were not previously known or it may be possible to quantify and confirm the efforts required to accomplish portions of the maintenance tasks that occur during this time (e.g., assembly and disassembly times).

TABLE 15.7 MIL-STD-471A Maintainability Test Methods

Test method	Test index	Assumptions and comments
1	Mean	Test A: lognormal distribution; prior knowledge of variance Test B: distribution free; prior knowledge of variance
2	Critical percentile	Lognormal distribution; prior knowledge of variance
3	Critical maintenance time or worker-hours	None
4	Median	Lognormal distribution
5	Chargeable maintenance downtime per flight	Sample size large enough for the central limit theorem to apply (recommend >50 tasks)
6	Worker-hours per flight hour	Uses actual repair data during test operations
7	Worker-hours per flight hour	Similar to method 6, but uses simulated faults
8	Mean and percentile (90th or 95th) or dual percentiles	Lognormal distribution
9	Mean corrective-maintenance time, mean preventive-maintenance time, or mean maintenance time, and a maximum percentile of repair time	Sample size large enough for the central-limit theorem to apply (minimum = 30 tasks)
10	Median or maximum maintenance time	Acceptance criteria developed for sample size of 50 tasks
11	Mean preventive-maintenance time and maximum preventive-maintenance time (at any percentile)	All preventive-maintenance tasks must be performed
Notice 2	Proportion of faults detected Proportion of faults isolated to a given level of ambiguity	Depends on level being tested; at organization level, will be comparable with number required for maintainability demonstration
Notice 2	False-alarm rate	Use available data to compare observed false alarms with expected

Maintainability demonstrations are typically performed through dedicated tests and are therefore better suited to produce the data necessary to determine whether the design meets its quantitative maintainability requirements. These demonstrations are frequently performed on high-fidelity mockups or on preproduction or production hardware and software at the producer's or customer's facility by customer personnel and under conditions closely approximating those expected in the operational use. When carefully planned, staffed, and supported, maintainability demonstrations are an accurate method of verifying that the maintainability requirements have (or have not) been met.

The usual steps required to prepare for, perform, and evaluate the results of maintainability demonstrations are as follows:

1. Select the specific method(s) to be used. The choice depends on the system or equipment characteristics (e.g., whether it is made up of many duplicate subsystems or subassemblies or only a few), and the parameters to be demonstrated. Table 15.7 presents several test methods that range from formal statistical tests of hypotheses (method 1) to exhaustive accomplishment of preventive maintenance tasks (method 11). Table 15.7 also identifies methods for addressing testability parameters.

2. Establish accept/reject criteria and retest procedures should the accept criteria not be met.

3. Prepare a maintainability demonstration plan and detailed test procedures. The plan will address the personnel, facilities, documentation, training, and equipment required for the test and the schedule of test efforts.

4. Select the population of maintenance tasks from which the maintainability test sample will be selected.

5. Conduct pretest preparation. This includes assembling the test hardware, test support equipment, documentation, facilities needed for the test, and identifying and training, if necessary, the personnel who will perform the test and who will collect test data. If induced failures are to be used during the test, provisions should be made incorporating induced failures into the equipment and for restoring the equipment to an acceptable condition after the test.

6. Conduct of the maintainability demonstration test(s).

7. Perform posttest tasks, such as restoring test hardware to its original condition and verifying that it is acceptable for use on production items, if applicable, and returning test equipment and facilities to a pretest condition.

8. Analyze test data. This includes a determination of whether the acceptance criteria are met, analysis of the maintenance strengths and weaknesses of the equipment, supporting documentation, support equipment, training, and other aspects of the maintenance environment.

9. Recommend corrective actions and retest, if required, and, in any case, improvements which were identified during the test and analysis activities.

10. Prepare the demonstration test report.

Additional information on planning and conducting maintainability demonstrations is found in MIL-STD-471A[12] and the *Reliability Engineer's Toolkit.*[13]

REFERENCES

1. Retterer, B. L.; and Kowalski, R. A., "Maintainability: A Historical Perspective," *IEEE Transactions on Reliability,* **R-33** (1), April 1984, pp. 56–61.

2. Gryna, F. M., Jr.; McAfee, N.; Ryerson, C. M.; and Zwerling, S., eds., *Reliability Training Text,* 2d ed., Institute of Radio Engineers, New York, 1960.

3. Byron, J.; Deight, L.; Stratton, G., *RADC Testability Notebook,* Rome Air Development Center, TR-82-198, June 1982 (AD-A118881L).

4. MIL-STD-721C, *Definitions of Terms for Reliability and Maintainability,* June 12, 1981.

5. MIL-STD-470B, *Maintainability Program for Systems and Equipment,* May 30, 1989.

6. MIL-HDBK-338-1A, *Electronic Reliability Design Handbook,* Vol. I of II, Oct. 12, 1988.

7. MIL-HDBK-1472D, *Human Engineering Design Criteria for Military Systems, Equipment, and Facilities,* March 14, 1989.

8. MIL-HDBK-761A, *Human Engineering Guidelines for Management Information Systems,* Sept. 30, 1989.

9. MIL-STD-2165A, *Testability Program for Systems and Equipments,* Feb. 1, 1993.

10. Simpson, R.; Sheppard, J., *System Test and Diagnosis,* Kluwer Academic Publishers, Boston, 1994.

11. MIL-HDBK-472 (Notice 1), *Maintainability Prediction,* Jan. 12, 1984.

12. MIL-STD-471A, *Maintainability Verification/Demonstration/Evaluation,* March 27, 1973.

13. *Rome Laboratory Reliability Engineer's Toolkit,* Systems Reliability Division, Rome Laboratory, Air Force Material Command (AFMC), Griffiss Air Force Base (AFB), New York (NY 13441-4505), April 1993.

CHAPTER 16
COMPONENT RELIABILITY

Fred Watt
SAV-SOFT PRODUCTS
MILPITAS, CALIF.

16.1 INTRODUCTION

The task of a design engineer is to identify the necessary components and know how to connect them to achieve a higher-level function that we call an assembled product. When a product fails, the cause of failure can usually be traced (with the proper diligence and techniques) to a failed component or an interaction between components which failed to perform their intended function.

What general and specific component issues, then, does a design engineering team need to consider to design a product that has the potential to be reliable when properly manufactured? What are the specific tasks to be performed, and how are these best integrated into the design process?

16.1.1 General Principles

Before we start it is necessary to lay down some general principles which we can all understand and build from. These reliability axioms (see Fig. 16.1) are basic to the principles discussed and elaborated on in this chapter.

---

1. No amount of good manufacturing can fix a poor design.
2. On the other hand, poor manufacturing can ruin the best of engineering designs.

Therefore, there are three *requirements for achieving a reliable product*:

1. *The design must have margin* with respect to the stresses to which it will be subjected during manufacturing and actual field use.
2. The manufacturing *process must be stable* and completely documented. Variations from the "tried and true" must be considered experimental until proved.
3. There must be an *effective feedback and corrective-action system* which can identify and resolve problems, quickly, in engineering, manufacturing, and the field.

FIGURE 16.1   Reliability axioms.

Margin.   *Margin* is defined as the difference between the stress applied to a component and its strength with respect to that stress. This is also known as a *safety margin.* If the safety margin is slim, there is a greater probability that the part will fail. If there is a wide safety margin, many things can go wrong before the final straw breaks the camel's back.

The definition of stresses includes all the possible life-cycle stresses to which the component will be subjected during its entire life. The component's life begins back at the supplier's facility and extends to the customer's end-use environment. These stresses include handling, testing, soldering, shipping, and storage of the raw materials; of the partially finished component; of the finished component, the component in your factory; and of the component in your product in the customer's hands. The component has to make it through all these stresses with no damage.

Process Stability.   In order to produce reliable products every time, it is necessary to have a stable and predictable manufacturing process. It is very easy to be fooled into thinking that you "know how to make" your product and miss this very important step. If there are adequate margins, in general, the product will be reliable until something goes wrong in manufacturing. In many instances a product has been in manufacturing for a year or two with no problems. Then a part substitution is made or a tool begins to wear, and nobody knows what the consequences were until the customer calls.

There are two requirements for stability.

The First Requirement for Stability.   The first requirement for stability is that the manufacturing process be completely specified and documented. This means that there are adequate source-controlled documents defining exactly what is to be purchased and what requirements each component must meet. It means that there is a process which verifies that the components, subassemblies, and finished goods meet their respective requirements. Those requirements must be documented in adequate and correct assembly drawings, assembly instructions, and inspection and test instructions. The soldering times, temperatures, preheat, and cooldown cycles must be specified and monitored. The same goes for the cleanliness levels in the solder and cleaning solutions. And when the ionic contaminant levels in the solder wave or the cleaning solution exceed the specified levels or simply change too quickly, the process stops and the solder or cleaning solution is not just changed, but the question is asked why it changed. In other words, what was it about that last lot of boards that made the cleaning solution get dirty? That is, what dissolved off those boards? Was it the connectors, or what? "The drawing was wrong" or "the test program didn't test that condition" or "the *electrostatic discharge* (ESD) station ground was not connected" may be a reason there were failures, but not an excuse. A completely specified and documented process gives a baseline from which to measure the state of the process and make changes, if necessary. When changes are made to an undocumented process, the changes only last as long as the personalities involved remember to do it that certain way.

The Second Requirement for Stability.   The second requirement for stability comes from feedback control theory and is so important that it is really the third requirement for reliable product.

Effective Feedback and Corrective-Action System.   There must be an effective feedback and corrective-action system which can identify and resolve problems, quickly, in engineering, manufacturing, and the field.

The two key words in this statement are "effective" and "system." Often field service or board repair collects lots of failure data but it is either not in a form that can be used or it is not communicated clearly. The system must effectively collect data and feed it back to the appropriate department so that the correct corrective action can be implemented. How do you know that it is the correct corrective action? By identifying and correcting the root cause of the failure rather than working around the symptom. If the same problem is being solved over and over again, the root cause has not been identified. There must be a complete failure reporting, analysis, and corrective-action system (known as FRACAS), and it must be a system. Most problems smolder for months before the flames reach the rooftops, and, unfortunately, most companies wait until the flames reach the rooftops (or the boardroom) before reporting, analyzing and trying to correct the problems. A formal FRACAS system can see the signs of approaching doom and head off the apocalypse. That is, there should be a systematic process to look at everything that fails and ask "Is there some corrective action which can be put in place such that this problem will never appear again?"

16.1.2 The Role of Engineering

Few companies have the procedures to achieve what was just described. However, engineering has a specification to meet in the reliability area. If engineering does not have the tools to verify whether the specification can be met, then engineering can only guess at how to meet it. That might not be too bad if the new product only had to be a little better than the last product. The real problem comes in because the customers are demanding that the next generation of product be 2 to 10 times better than the last generation of product, and to accomplish that requires a total company effort. It is the role of engineering to drive the company toward that goal, because engineering's role is to drive the technology of the company. If engineering sees that something is not getting done, then it is engineering's responsibility to see that it is done either by releasing a specification that requires that it be done or by ensuring that another department's operating procedures require that it be done.

16.1.3 The Role of Manufacturing

Manufacturing's role is to "build to print." This means, of course, that the prints must be accurate and debugged and that processes exist which can build the finished product without damaging any of the components in the process. It is manufacturing's responsibility to develop processes and procedures which will produce a product that meets all its specifications including reliability specifications.

Manufacturing must have a process for the orderly development and qualification of tooling, assembly processes, handling procedures, and test and inspection procedures. So often, these processes are quickly thrown together to "get product out" as quickly as possible and are never adequately documented, refined, and qualified.

Manufacturing must have procedures to verify that the process is in control and that the product meets its reliability goals. These usually take the form of statistical process control and reliability monitor testing. Manufacturing also is the focal point for and should drive the failure analysis and corrective-action process.

16.2 DESIGNING FOR RELIABILITY

Figure 16.2 is a checklist of tasks required for an effective design-for-reliability program. These tasks can be applied to any design effort and are necessary to achieve reliability goals generally. They are put in the context of component and supplier selection here. The rest of this section describes the tasks and the reliability issues surrounding them.

16.2.1 Setting System Specifications for Reliability

It is very important to adequately specify the system specifications for life and mean time between failures (MTBF) and the environmental conditions under which the product must perform or withstand damage. Input for the specifications comes from customers, marketing, field failure reporting, failure analysis, and other sources. Far too many companies specify an MTBF only with no mention as to what temperature or other environmental conditions the number applies. There is usually no indication as to how it will be measured—whether from test or field data.

There are two specifications which must be considered: (1) all components must survive for the expected life of the system—that is, any wearout mechanisms must not dominate prior to the end of the useful life of the product; and (2) the rate at which the system fails must be acceptable

☐ Set system specifications for life and MTBF and the environmental conditions under which the product must perform and withstand damage.
☐ Allocate failure rates to subassemblies.
☐ Decide reliability design and component selection rules and ground rules.
☐ Design the system and select specific components.
☐ Evaluate supplier reliability data for the components and packages.
☐ Evaluate the stress margins on all components.
☐ Predict the failure rate of each component.
☐ Assign the failure rate required from each component so that the subassembly meets its allocation.
☐ Assign required wear, fatigue, shelf, or other life to applicable components to meet the system life specification.
☐ Compare required versus expected actual for failure rates and life and margins.
☐ Develop reliability critical-components list outlining areas of concern or uncertainty.
☐ Develop component-specific action plan to address any deficiencies or inadequacies.
☐ Qualify the components.
☐ Perform a reliability growth test on engineering-built systems.
☐ Perform a reliability demonstration test on manufacturing-built systems.
☐ Ensure that there is a working mechanism for measuring actual field reliability and field-use conditions.

FIGURE 16.2 Design-for-reliability checklist.

to the customer. This implies that the MTBF and life specifications must apply to the end-use environment in which the equipment will operate.

In order to gain a better understanding of these concepts and how they apply to components, it is important to understand the difference between MTBF and life.

The Bathtub Curve. Referring to Fig. 16.3*a*, if a population of components or machines is put on test or shipped to the field and cumulative failures are plotted against time, usually a number of units will fail rather quickly. Then, only a few will fail occasionally for (hopefully) a long

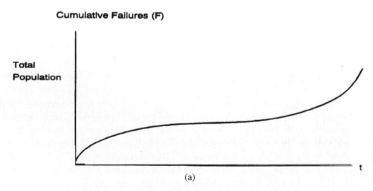

(a)

FIGURE 16.3 Reliability curves.

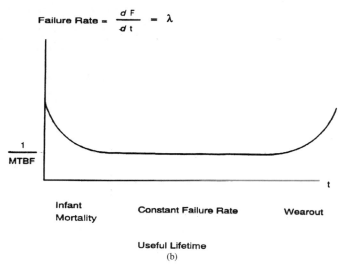

Failure Rate = $\dfrac{dF}{dt}$ = λ

Infant Mortality

Constant Failure Rate

Wearout

Useful Lifetime

(b)

FIGURE 16.3   (*Continued*) Reliability curves.

period of time. And then as various components wear out, more will fail until, finally, all components or machines fail.

The time derivative of this curve is the familiar "bathtub" curve (Fig. 16.3*b*), which graphs the failure rate λ versus time. This curve shows a period of relatively high failure rate (called the "infant mortality" period) decreasing to a relatively constant failure rate. When wearout mechanisms begin to dominate, the failure rate increases again. Useful lifetime may be defined from time zero if the product is a calculator or a watch, but, if the product is a pacemaker or satellite, useful life does not begin until infant mortality failures have been removed from the population and the product is at or near its inherent or random average failure rate as indicated by the flat portion of the bathtub curve.

Mean time between failures (MTBF) is defined as the reciprocal of the failure rate during the flat portion of the failure-rate curve. It says nothing about either the infant-mortality period or the wearout period. MTBF indicates neither how high the failure rate begins nor how long the infant-mortality period will last. MTBF has no correlation with when wearout will begin.

The unit of failure rate is (the number of) failures per hour. The unit of MTBF is hours per failure. Most statements of MTBF are in terms of hours, so it is easy to confuse MTBF and life. Be careful. The fact that a product has an MTBF of one million hours does not imply that it will last one million hours. Take the example of the incandescent light bulb. It has a failure mechanism where the filament incandesces or boils away after about 1000 h. During the period between 10 and 900 h very few burn out. In fact, the MTBF approaches one million hours. However, between 900 and 1200 h, all will burn out. So here is an example where the component has a one-million-hour MTBF (i.e., MTBF = 1×10^6 h), but a life of 1000 h.

So what does a *one-million-hour MTBF* mean? Referring to Fig. 16.4, the commonly used unit of failure rate for components is FITS. The definition of a "FIT" is one failure per billion hours. One thousand FITS is one failure per million hours or a one-million-hour MTBF. If the top and bottom (numerator and denominator) of the fraction are multiplied by 10^{-3}, the failure-rate expression becomes 0.1 percent per 1000 h. In other words, the rate at which failures occur will be 0.1 percent of the total population during every 1000-h period of operation. Or, for 42 days of continuous 24-h/day operation, one out of every 1000 units can be expected to fail on average.

Units of measure:

$$1 \text{ FIT} = 1 \text{ failure in } 10^9 \text{ h}$$

$$1000 \text{ FITS} = 1 \text{ failure in } 10^6 \text{ h} = 1{,}000{,}000\text{-h MTBF}$$

What does a *one-million-hour MTBF* mean?

$$\frac{1 \text{ failure}}{10^6 \text{ h}} = \frac{0.001 \text{ failure}}{10^3 \text{ h}} = \frac{0.1\% \text{ failure}}{1000 \text{ h}}$$

In other words, for every 1000-h period, 0.1 percent of the population can be expected to fail.

FIGURE 16.4   What does MTBF mean?

Specifying the Environment.   Table 16.1 lists the environmental conditions (and the general types of failures due to these stresses) which are usually specified for systems and equipment. A system specification should list the conditions under which the system will perform its intended function. Most systems will not meet either the life specification or the MTBF specification in the worst case of some or all of these environmental conditions. In fact, very few systems will meet their reliability specifications at only the hottest temperature specified for system operation. Therefore, it is important (although rarely done) to specify under what set of conditions the reliability specifications apply. Only when this is done can appropriate tests be designed to verify that the system is capable of meeting its specifications.

TABLE 16.1   Types of Reliability-Dependent Stresses to Be Defined at System and Component Level for Design and Manufacturing Processes

Stress	Some causes of failure
Temperature	Exceed component ratings; exceed material ratings
Voltage	Dielectric breakdown; overheating
Altitude	Temperature; arcing; air pressure
Humidity	Corrosion; electrical short-circuiting; drying of lubricants, inks, etc.
Thermal cycling	Storage environment; on/off cycling
Thermal shock	Soldering process; shipping environment; manufacturing stress screening
Mechanical vibration	Shipping environment; manufacturing stress screening; operating environment
Mechanical shock	Shipping environment; mishandling
Wear	Moving parts; vibration
Friction	Heat generation; particle generation
Handling	In manufacturing assembly and test; by the customer; electrostatic discharges

The set of environmental conditions in the specification should usually relate to the typical field conditions. However, when field returns are analyzed, most of the units which fail are those that are operated at the environmental extremes. These factors must be considered when establishing the specifications for MTBF and life.

16.2.2   Allocating Failure Rates to Subassemblies and Components

Since a system is composed of subassemblies and components, the next step is to allocate the failure rates to those subassemblies and components so that each major part has a reliability budget.

This process should be both a top–down and a bottom–up analysis. The results of this activity both establish and drive the basic design approach which must be taken for each subassembly.

For wearout mechanisms, each component must have a life equal to or greater than the system life. On the other hand, each component must have a failure rate much smaller than the system failure rate because failure rates are additive, whereas wear lives usually run in parallel.

The Probability of Success.   Referring to Fig. 16.5, during the constant failure-rate portion of the bathtub curve, the probability of success is given by the first term from the Poisson distribution which is the probability of zero failures or the probability of success P_s. The probability of success, then, is the exponential e, raised to the power $-\lambda t$ or e to the minus t/MTBF. Plotting P_s

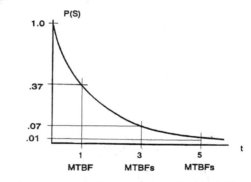

The probability of success

$$P_s = e^{-t/\mathrm{MTBF}} = e^{-\lambda t}$$

where

Failure rate $(\lambda) = 1/\mathrm{MTBF};$ $e = 2.718$

Therefore, after 1 MTBF, the probability for zero failures is 37 and 60 percent at 0.5 MTBF.

This assumes a constant failure rate over time. Individual components, infancy, and wearout may need to be modeled in other ways such as with Weibull or normal distributions.

FIGURE 16.5   Understanding failure rates.

versus time as in Fig. 16.5, it can be seen that, at time zero, the probability of success is 100 percent. When $t =$ one MTBF worth of time, the probability of success is one over e or 0.37. That is, after one MTBF worth of time 63 percent of the units will have experienced at least one failure and only 37 percent will not have experienced any failures. Unfortunately, most engineers believe that, if a system has a 10,000-h MTBF, at least most of the systems will last that long. And worse, customers believe that their systems will last that long! However, a customer may have received the system which failed after only 10 h or, maybe, the system which never fails.

The MTBF is not the mean of a normal distribution where 50 percent of the population lies on either side of the mean, but, rather, the mean of the exponential distribution. And actually, MTBF is more like a time constant. So, after three time constants, the probability of success is down to 7 percent and after five time constants, P_s is at 99 percent of its final value, or 99 percent of the units will have experienced at least one failure. The time-weighted average, then (63 percent failing between zero and one MTBF, 30 percent failing between one and three MTBFs, 6 percent failing between three and five MTBFs, and 1 percent lasting longer) is one MTBF.

The exponential distribution is valid only during the constant failure-rate portion of the bathtub curve. Infant mortality is usually modeled with a Weibull distribution, while wearout is usually modeled with a normal or lognormal distribution.

Series and Parallel Systems and Subassemblies

Series Systems. A *series system* is defined as a system in which, if any one component fails, the whole system fails. If the probabilities of success of each component or subassembly are independent of the other components or subassemblies, then the probability of success of the whole system is the product of the probabilities of success of each component or subassembly. Because the failure rate λ is in the exponent, the failure rate of the whole system is simply the sum of the exponents. This is illustrated in Fig. 16.6a.

Therefore, the system failure rate needs to be allocated to each subassembly such that the sum of the failure rates of all the subassemblies equals the system specification. This provides a reliability budget for each subassembly. Similarly, the failure rate of each subassembly needs to be allocated to each component in the subassembly.

Parallel Systems. With a *parallel redundant system* it is not so simple to allocate failure rates. Here it is important that the allocation be performed by someone who is well versed in working with probabilities. The simplest form of a parallel redundant system is the standby redundant system illustrated in Fig. 16.6b. Here, if one subsystem fails, the failure is detected and the standby system is switched in and performs the function.

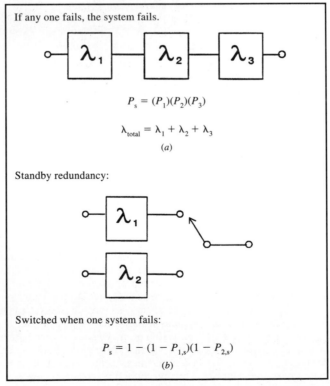

FIGURE 16.6 Series (a) and parallel (b) systems.

If the failure rate of the switch and the detection circuitry is neglected for the moment, an expression for the probability of success can be derived by noting that the probability of success plus the probability of failure is 1.0 or 100 percent. The system will either work or not. The probability that the switch will have to be thrown is the probability that system 1 will fail or $1 - P_{1,s}$. Similarly, the probability that system 2 will fail is $1 - P_{2,s}$. Therefore, the probability that both systems have failed is the product of the two probabilities of failure. And the probability of success of the parallel combination is 1 minus the probability that both systems will fail.

If the failure rates of the two subsystems are the same, then at one MTBF worth of time, the probability of success is up to 60 percent. Therefore, the effective failure rate of the system has been greatly improved. However, since the switch is still in series with the parallel combination, this configuration is heavily dependent on the failure rate of the switch and the detection circuitry.

For example, for a memory system with error detection and correction, an individual memory chip can be removed and the system will detect and correct it. In fact, the failure rate of the individual memory chips becomes relatively unimportant as the failure rate of the memory system is dominated by the *error-correction circuitry* (ECC) and any sense amplifiers and address drivers in series with the data path.

16.2.3 Defining Reliability and Component Selection Rules

It is necessary to adopt a set of reliability design rules at the beginning of each development project to assist designers in achieving a reliable design. These take several forms. First, there are decisions as to the technologies which will be used in the product. Then a set of derating guidelines must be adopted so that the parts are not overstressed. Next, there must be guidelines as to how new parts are to be evaluated before they can be released into manufacturing. And finally, the ground rules must be established for how the worst-case analysis is to be performed.

Technology Decisions.   Each new product faces a decision as to whether to stay with the "tried and true" or use a new promising technology. There are two types of new technologies.

There is the new technology that is new to the industry. It is built with processes which are new and unproven, but promises gains in performance and/or cost. It is important that both engineering and manufacturing understand its limitations and features and how these are different from the technologies that the company is familiar with. Often, new processes will need to be established and debugged. Old processes for designing and handling may not apply. Most users of this type of new technology recognize this and take appropriate steps to minimize the reliability risks.

The second type of new technology is the technology which has been available in the industry (perhaps for many years), but this is the first time your company has used the technology. The issues discussed in the previous paragraph apply here, too. For example, suppose all your previous products have been built using transistor-transistor logic or emitter-coupled logic (TTL or ECL) technologies. Someone decides to use a complementary metal oxide semiconductor (CMOS) in the new products, but forgets to establish a sound ESD control program. CMOS simply cannot be handled like bipolar circuits.

Technology decisions are far-reaching and cannot be made lightly. It is important that all functions in the company know what has to be done to use the technology reliably.

Derating Guidelines.   What is a *derating guide,* and how is it used? A *derating guide* is a standardized set of safety-margin definitions to be used by designers in order to avoid field and factory problems due to overstressed parts. It is best used only as a guideline, not as a law. Some companies try to lay it down as a law on designers only to find out that they have to make exceptions for various situations or products. As a guideline, designers are being told "Try to stay below these stress levels. If, in a given situation, these levels must be exceeded, then discussions will occur and decisions made as to whether the resulting level of reliability is acceptable." In other words, the derating guide should be used as a vehicle to generate discussion as to what makes a given design reliable and how reliable is reliable enough (for this product).

It is important to recognize that there is no one set of numbers that is right for all products and all marketplaces, and there will always be controversies as to what are the right numbers. There are no right or wrong numbers. The derating numbers are only right or wrong with respect to a particular product in a particular marketplace. For example, the numbers in the derating guide for junction temperatures for semiconductors are too low for automobile engine and transmission-mounted components or "down hole" in an oil well. They are also too high for parts implanted in the body such as pacemakers and other medical products. Therefore, it is important for each new development project to reevaluate the derating guide to establish the numbers that are "right" for this product and this marketplace. In this way, the designers also have a stake in the numbers and have agreed to abide by these "flags."

An Example of a Derating Guide.   The derating guide presented in Table 16.2 consists of two columns. The "Normal" column is typically used for office and computer room environments, while the "High-reliability" (or "Hi-Rel") columns is typical for medical, space, or financial environments. Most equipment has some parts that require higher reliability than do other parts, so these classifications are to be considered general, not specific.

In general, you should not require derating levels less than 50 percent for high reliability (sub)systems or greater than 70 percent for normal (sub)systems. There are exceptions to this. For example, if designers must derate propagation delays to 70 percent of the specification, they will laugh and simply will not do it, arguing that the system cost or performance objectives cannot be met with such a restriction. These guidelines must be balanced with the product and marketplace needs. For a relay or switch, the inductive load can damage the contacts unless it is heavily derated from the DC or resistive load current rating. An incandescent lamp simply will not glow bright enough at 70 percent of rated voltage.

Some numbers require further explanation. The power supply voltages for most digital and linear integrated circuits are already derated. That is, the normal operating voltage is already less than the absolute maximum. Typically the absolute maximum is 6.5 to 7 V, and the maximum operating voltage is 5.25 V.

Sometimes (such as for temperatures) both a percentage and a number are listed. In these cases, try to stay below the number, but below the percentage of absolute maximum is also acceptable, particularly for parts that have a higher-than-normal absolute maximum temperature.

Sometimes the ratios divide into the specification and sometimes they multiply by the specification. For example, propagation delays for semiconductors and dissipation factor for capacitors are divided by the derating ratio while a hold-time or voltage-level specification is multiplied by the derating factor.

For capacitors, the dissipation factor is given on the specification at 25°C and one frequency only. Therefore, the supplier's typical curves must be obtained for dissipation factor versus frequency and dissipation factor versus temperature in order to obtain the typical value for dissipation factor at your operating temperature and frequency. The derating factor is then applied to this number. In addition, when the temperature rise in the capacitor has been calculated, the maximum operating temperature must be below the temperature derating limit.

Component Selection Guidelines

Reliability Risks Associated with the "It Works, Doesn't It?" Approach.   It should be obvious (but rarely is) that, just because a part works in a prototype or test unit, it does not follow that it will work in manufacturing over the range of production tolerances. However, this kind of thinking prevails all too often in both engineering and manufacturing. In fact, it is promulgated by the supplier's salespeople whenever they ask you to "try it in your system and see if it works."

It is very important to be clear about what "it works" means. Does it work in the test machine over all specified voltage, temperature, vibration, etc. limits, or was this test simply at nominal conditions? Does "it works" mean that it will work over all production tolerances and specified environmental limits? Usually this is not the case. Therefore, a company definition of "it works" must be stated which includes all possible combinations of production tolerances and specified environmental conditions. Maintaining this definition as the "company culture" required constant

TABLE 16.2   Component Derating Guidelines

Device type	Parameter	Normal	High-reliability
Digital integrated circuits	Power supply and input voltages	Already derated	
	Output current (load, fanout)	0.8	0.8
	Junction temperature, TTL	120°C	100°C
	Junction temperature, CMOS	100°C	90°C
	Speed	0.9	0.8
Linear integrated circuits	Power supply voltage	Already derated	
	Input voltage	0.8	0.7
	Junction temperature (higher of the following):	0.75 or 100°C	0.65 or 70°C
Silicon (general-purpose) transistors	Collector current	0.7	0.5
	Voltage V_{ce}	0.75	0.6
	Junction temperature (higher of the following):	0.75 or 100°C	0.65 or 80°C
Silicon (power) transistors and thyristors	Collector current	0.75	0.6
	Voltage V_{ce}	0.75	0.6
	Voltage V_{be} (reverse bias)	0.8	0.7
	Junction temperature (higher of):	0.75 or 110°C	0.65 or 90°C
Silicon diodes (general-purpose and power), zener	Currents and voltages	0.8	0.5
	Junction temperature	120°C	100°C
	Power	0.8	0.6
Resistors (all types)	Operating temperature	Rated −20°C	Rated −40°C
Metal or carbon film	Power dissipation	0.7	0.5
Carbon composition	Power dissipation	0.7	0.6
Wirewound power	Power dissipation	0.7	0.5
Capacitors (all types): ceramic, mica, glass, porcelain, plastic film/foil, aluminum, and wet and dry tantalum	Voltage	0.7	0.6
	Operating temperature	Rated −20°C	Rated −40°C
	Dissipation factor	0.7	0.6
Polarized caps	Reverse voltage	−0.2 V	−0.1 V
Relays and switches	Resistive or capacitive current	0.8	0.5
	Inductive current	0.3	0.2
Motors	Voltage	0.8	0.6
	Temperature (windings and bearings)	0.8	0.6
Incandescent lamp	Voltage	0.9	0.8
Transformers and inductors	Voltage and current	0.8	0.7
Wire, cable, and power cords	Current and voltage	0.8	0.7
Connector signal pins	Contact current	0.7	0.5
	Voltage	0.6	0.5
Connector power pins	Current	0.5	0.4

reinforcement so that people stop saying "it works" and begin saying "it works over all production tolerances and specified environmental conditions."

The First Look—A Minimum Level of Part Evaluation.   Each company needs to have guidelines for design engineering which state what needs to be done before a new component and/or supplier is designed into and released in a new product. Most companies have no guidelines, either formal or informal. The implicit assumption usually made is that, if a company is selling a product, then it must be reliable enough.

However, when a wide variety of similar parts are being offered by other companies, it follows that there are also a wide variety of prices, reliabilities, qualities, etc. Therefore, the component life-cycle–asset-management viewpoint requires that the various offerings for parts be evaluated to gain an understanding of the benefits and risks associated with using the part and of the differences between the various offerings.

A minimum level of parts control is required for long-term process control and competitiveness. First, part numbers and suppliers must be identified on an *approved-source list*; and, second, steps must have been taken to ensure that the parts are compatible with the design, manufacturing, and field processes.

The minimum guidelines must start with a system specification against which all evaluations are made, and there must be a checklist for supplier and part evaluation.

The minimum component part evaluation should consist of a physical analysis, an assembly-level functional test, and an assembly-level characterization test. The minimum supplier evaluation should consist of gathering data such as part specification, annual report, reliability test data, and field data and analyzing it for anomalies.

This minimum evaluation (summarized in Fig. 16.7) will usually provide a basis for deciding whether or not to place a component on the critical-parts list for further qualification actions. Parts

---

☐ Obtain five samples from each supplier.
☐ Test each one in the assembly over the assembly's specified operating conditions. Operating conditions should include temperature, voltage, and any other parameter deemed important for the application.
☐ Determine at what voltage or other condition the parts fail when the conditions are varied outside normal operating conditions.
☐ Perform an internal visual examination. Look for weak points and strong points.
☐ Obtain, study, and evaluate specifications, application notes, and reliability data from the supplier.

FIGURE 16.7   Basic component evaluation guidelines.

---

on the critical-parts list will usually need some combination of additional qualification procedures and tests as outlined in Sec. 16.2.13.

Physical Analysis.   Form dictates function. In other words, the way that a component is put together and the materials used determines how and under what conditions it will be weak, how it will be strong, what parameters or performance characteristics it will easily achieve, and what parameters it will have difficulty achieving. Therefore, it is important to understand just how a given supplier's part is built, and the best way to do this is to dissect it. Pictures and drawings can never provide the insights into the component that are possible with a hands-on dissection. This is especially important with all electromechanical components such as relays, switches, and connectors, but is also important for semiconductors and other components, if for no other reason than to see when the supplier has made a change.

The construction analysis can help determine how to design stress tests to determine how weak the weak points are and how strong the strong points are.

Assembly-Level Functional Test.   This test asks the question "Does the part work in the assembly over all specified assembly operating conditions?" This involves obtaining five parts from each supplier and testing each one in the assembly over specified voltage, temperature, vibration, or other system operating conditions. Often, the design engineer gets five samples from the supplier and plugs only one into the system, and the other four go into the desk drawer as "spares," resulting in an unacceptable sample of one. This "blind faith" in the supplier's ability to supply a part that meets the specification in all respects can wreak havoc later on in manufacturing.

If you suspect that the supplier is "handpicking" the sample parts, then buy five more from a distributor and see whether or how they are different.

Assembly-Level Characterization.   This test asks the question "How well does the part work in the assembly?" Here the assembly is tested outside its specified operating conditions to ascertain at what point the system stops working. In this way one can determine how much margin the part has in the specified operating environment. In general, the more margin, the better.

For example, suppose samples have been obtained from several suppliers for a particular part and all parts seem to work in the assembly at 4.75, 5.0, and 5.25 V. Take each part, one at a time, and vary the power supply above and below its limits. Be careful, of course, not to exceed the absolute maximum voltage for the devices. For supplier A, all five samples work at 6.0 V and all the way down to 4.1 V. Supplier B's parts work at 6.0 V but fail at 4.6 V. Supplier C's parts all work at 6.0 V, and one works at 4.0 V, while two fail at 4.4 V and two fail at 4.6 V. Which supplier would you feel most comfortable using in manufacturing?

Worst-Case Analysis Guidelines.   It is not possible to test all possible combinations of stresses and parameters or to obtain worst-case devices to test. Because of this, a theoretical analysis of the system and the interactions between components is necessary to ensure that the system will function over all specified conditions and at the parameter limits of components. The guidelines for this activity need to be established so that all designers working on the project perform this analysis in the same manner. A checklist is presented in Fig. 16.8.

□ Determine minimum and maximum input conditions and environmental conditions.
□ Calculate peak voltage and peak and average currents for all circuit nodes.
□ Calculate average and peak power for all components.
□ Calculate capacitive loads for all circuit nodes. Determine worst-case rise time, fall time, and propagation delay for each circuit element.
□ Calculate propagation delays for each circuit path and verify that input timings for each component are met with the margin as defined in the derating guidelines.
□ Determine maximum cycling rates for relays, switches, connectors, motors, and all other electromechanical and mechanical elements.

FIGURE 16.8   Worst-case analysis guidelines.

The theoretical analysis should verify that the system will work with margin over all possible combinations of voltage, temperature, and any other appropriate general conditions when all the components are at the extremes of their specifications. This includes both a DC and an AC analysis.

16.2.4   Selecting Specific Components and Suppliers

Once the basic design rules have been established, the design of the system can proceed. Selection of components will be based on their suitability to perform their intended function and their ability to meet the selection guidelines. Final component and supplier selection may also depend on meeting the next eight items in Fig. 16.2, "Design for reliability checklist," which are detailed in Secs. 16.2.5 through 16.2.14.

16.2.5   Evaluating Supplier Reliability Data for the Components and Packages

Each supplier must be asked to provide reliability test data so that the component's reliability can be evaluated for your system. If the supplier does not have reliability data or a planned test to gather it, look for another supplier. There is no excuse for a supplier not having a reliability measurement program.

Reliability data may be presented in many forms. The data may be from test, or field data, or by prediction. It is all right to accept a prediction at the beginning of a design project, but there should be hard test data before the part is fully qualified.

Test Data.   There are many ways that test data can be taken. Usually, components will be tested at a higher than specified temperature in order to accelerate any temperature-dependent failure modes which may be present. This is generally called *burnin*. Other tests may need to be performed to evaluate reliability. These are discussed in Sec. 16.2.14. Here, we will discuss only temperature-accelerated life tests because this must always be performed and it is easy to use the χ^2 (Chi^2) distribution to calculate the failure rate.

The Arrhenius Equation and Acceleration Factors.   Figure 16.9 gives the Arrhenius equation for acceleration factors and describes the temperature dependency of the failure rate. This equation states that the failure rate at the normal operating temperature T_1 is less than the failure rate

Arrhenius equation for temperature-dependent reactions:

$$AF = \frac{\lambda_2}{\lambda_1} = \exp\left[\left(\frac{E_A}{K}\right)\left(\frac{1}{T_1} - \frac{1}{T_2}\right)\right]$$

where λ_1, λ_2 = failure rate at the component's internal temperature T_1, T_2, in kelvins
E_A = activation energy (0.3 to 1.4 eV)
K = Boltzmann constant (8.625×10^{-5} eV/K)

FIGURE 16.9   Acceleration factors.

at the (higher) test temperature T_2. How much less is determined by the activation energy E_A, which is an experimentally determined constant that relates to the speed at which a failure mode, chemical reaction, or physical process proceeds as a function of temperature.

Note that a large activation energy gives a large acceleration factor, which means that the failure mode is highly temperature-dependent. An example of this might be a microcrack in ceramic or silicon which has an activation energy of 1.3 eV (see Table 16.3). Thermal expansion will tend to propagate the crack very quickly. However, a crack will propagate much quicker with temperature cycling, vibration, or thermal shock. The Arrhenius equation says nothing about these other stresses.

A small activation energy gives a small acceleration factor, which means that the failure mode, chemical reaction, or physical process is not very temperature-dependent. An example of this is an oxide defect on an integrated circuit which has an activation energy of 0.3 eV. Oxide defects, of course, are highly accelerated by voltage. The Arrhenius equation says nothing about voltage.

TABLE 16.3   Failure Mechanisms and Activation
Energies for Integrated Circuits

Defect type	E_A (in eV)
Oxide defects	0.3
Silicon defects	0.3
Electromigration	0.6
Contact metallurgy	0.9
Surface charge	0.5–1.0
Slow trapping	1.0
Plastic chemistry	1.0
Polarization	1.0
Microcracks	1.3
Contamination	1.4

So, life tests which calculate a failure rate based on a constant-temperature life test will give a
"best case" failure rate based on only one of the many stresses which may cause failure.

Figure 16.10 plots the Arrhenius equation for an activation energy of 0.7 eV in terms of the
difference between the two temperatures, T_1 and T_2. Suppose that the components are burned in
at 130°C, and it is desired to know what the failure rate is at 50°C. The top curve represents the
temperature difference of 80°C. Therefore, the failure rate measured at 130°C will be about 146
times higher than the failure rate at 50°C.

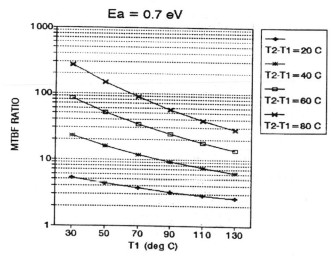

FIGURE 16.10   Acceleration factor versus temperature difference between
test temperature and specification temperature.

Another way to plot the Arrhenius equation is to assume that the normal system operating temperature is 30°C and plot the acceleration factor versus the burnin temperature for several activation energies. This is shown in Fig. 16.11. Note that it makes a big difference what value of
activation energy is used to calculate the normal operating temperature failure rate. Suppose the
burnin was done at 110°C and a failure rate of 10,000 FITS was measured. The failure rate is then
divided by the acceleration factor. If the activation energy used is 0.4 eV, the failure rate at 30°C
will be 400 FITS. If the activation energy used is 0.7 eV, the failure rate at 30°C will be about 50
FITS. And if the activation energy used is 1.0 eV, then the failure rate at 30°C will be about 3 FITS.

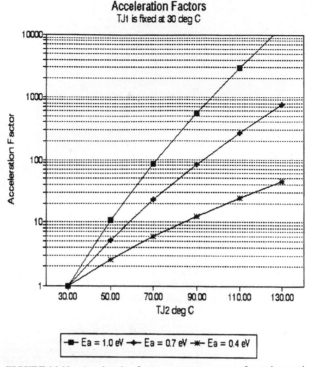

FIGURE 16.11   Acceleration factor versus temperature for various activation energies.

How Do You Know What the Right Activation Energy Is?   It cannot be assumed that the activation energy used by the supplier is the correct activation energy for the dominant failure modes of the component. Often, the supplier will use 1.0 eV simply because it makes the data look better. Obtain the actual test results which list how many devices were tested and for how long, and what failure modes were seen during the test. Be satisfied that the activation energy used is the weighted average of all the failure modes seen during the test. Be sure to determine whether the test was long enough to see the low-activation-energy failure modes. For example, if many devices were burned in, but for only 168 h, failure modes like oxide defects and metal migration will not have enough time to activate and the failure analysis of test failures will reflect only the high-activation-energy failure modes. Also, determine how heavily loaded the devices were during the test.

Table 16.4 lists the activation energies used by MIL-HDBK-217F to predict failure rates for various semiconductor types. These are fairly conservative, and other sources may use other activation energies, so this table is to be used as a guideline only. Also, suppliers vary as to the activation energy they use for similar processes, so be sure to compare various suppliers' data and assumptions.

Field Data.   Field data is usually the best data to have. However, the data collection and evaluation process may be flawed.

First, many component suppliers simply do not get back all (or any) failed components. Second, the only parts or modules received back may be those which failed during the warranty period. Since very few failures are reported after the warranty is up, the failure rate looks very good.

TABLE 16.4   Activation Energies for Semiconductors

Component type	E_A (in eV)
TTL, ASTTL, FTTL, ECL ICs	0.4
BiCMOS, LSTTL, ALSTTL ICs	0.5
NMOS, CMOS ICs (digital)	0.35
Linear MOS and CMOS ICs	0.65
Memories (all types)	0.6
Diodes (except zener)	0.26
Zener diodes	0.166
NPN and PNP transistors	0.18
FETs	0.166
Thyristors, SCRs, triacs	0.266
LEDs and other optoelectronic parts	0.24

Source: MIL-HDBK-217F, Notice 1.

With all data based on field returns, there will be some assumption as to the average number of hours per day that the components or modules were operating. You should examine this for applicability to your intended use conditions. Also, most companies experience about a 40 percent "no trouble found" rate. These parts were usually returned because replacing them allowed the system to work again. It must be determined whether the field diagnostic procedures are simply inadequate to properly isolate the component that failed or whether there are connector problems.

Predictions.   Reliability predictions are useful during the design phase to determine the relative merits of several different design approaches. Predictions for systems will be wildly different depending on which procedure was used and which revision of the procedure was used. Reliability predictions should not be relied on too heavily, but rather simply considered as another data point.

Often companies will modify the prediction method to meet the specification. Be sure to ask for backup data to support those modifications.

Specification, If Any.   Often, a supplier will give you a specification only with no backup data to support it. Be sure to ask for backup data or reliability test plans and results.

Problems with Reliability Data.   Figure 16.12 is a checklist for reading and studying supplier burnin and other qualification data. This list points to some of the inconsistencies and incompletenesses which appear in data. Often there is no clear answer to these questions and, so, the design engineer must evaluate all the data and make the best engineering judgment.

16.2.6   Evaluating the Stress Margins on All Components

Electronic Parts and Circuits Tolerance and Thermal Analysis.   A theoretical and experimental thermal analysis of each subassembly is necessary to determine the actual operating temperature ranges which the components will experience. Then, an examination of the effects of tolerance and parasitic effects over the specified operating temperatures can be performed. Tolerances must be examined with respect to both initial tolerances and end-of-life tolerances of components. These analyses verify that, given reasonable combinations of within-spec AC and DC characteristics and tolerance buildups, the system will perform within specification and the components will still meet the derating guidelines. These analyses should be performed according to the worst-case-analysis guidelines discussed earlier in this chapter.

☐ What is the source of the data presented? Prediction? Test? Or field data?
☐ If it is a prediction:
    What method and revision was used?
    What modifications to the procedure were made?
    What data supports those modifications?
    What process controls support the choice of the quality factor used?
☐ If the data is from test:
    What were the details of the test procedure and test results?
    What activation energy was used, and what is the justification for that value?
    Was the test long enough to see the low-activation-energy failure modes?
    What thermal resistance was used to calculate the junction temperature during test? Was a socket used, or were the parts soldered onto the burnin board? Was the airflow in the oven accounted for?
    What junction temperature was used for the failure rate in your system?
    What is the confidence level of the failure rate presented?
☐ If the data is from field data:
    What is the confidence that all failed parts were returned?
    How were "no trouble founds" handled?
    How complete was the failure analysis process?
    What was the estimated usage duty cycle in the field?
    How much time were the failed units in the field?
    Are there enough units in the field to see the variety of failure modes?
    Were the effects on present reliability factored in for design and/or process changes which fixed previous reliability problems?
☐ Were tests performed to analyze the effects of stresses other than temperature, such as temperature cycling, thermal shock, voltage, mechanical shock and vibration, handling, and chemical or humidity effects? How appropriate were the stress levels to the actual intended-use environment?

FIGURE 16.12   Checklist for reading and studying supplier burnin data.

Thermal Calculations.   All components have a specified temperature range of operation and an absolute maximum internal temperature beyond which damage may occur. It is important to have margin with respect to this absolute maximum temperature. Therefore, the internal operating temperature for each component needs to be calculated to ensure that, under worst-case conditions, no component damage will occur.

Figure 16.13 shows the equation for calculating the internal temperature of a semiconductor device. This equation is valid for calculating the internal temperature of any component where power is dissipated across a nonideal thermal conductor which is modeled as a thermal resistance Θ. In this case, there are two thermal resistors in series. The junction-to-case thermal resistance, Θ_{JC}, plus the case-to-ambient thermal resistance Θ_{CA}, yields the total junction-to-ambient thermal resistance Θ_{JA}. This is the simplest representation. Actually every heat path has a thermal resistance, and a more complete model would consider the thermal resistance from junction to leads and the thermal resistance of the printed-circuit board (PCB). There may be other heat paths as well.

The temperature rise of a component, then, is the power multiplied by the thermal resistance. Adding ambient temperature to this gives the actual internal temperature in a particular ambient-temperature environment.

The primary method of heat flow from junction to case is conduction. Θ_{JC} is a function of the thermal conductivity of the die, the die size, the die attach material, the package material, the package size, the number of leads, and the lead material. In order to obtain a lower Θ_{JC}, these packaging parameters must be lowered. Die size and die thermal conductivity generally cannot be changed. Gold eutectic die attach provides better thermal resistance that either thermally conductive epoxy or glass die attach. Ceramic or metal has a lower thermal resistance than plastic. However, plastic-packaged integrated circuits (ICs) with copper slugs and copper lead frames can have lower thermal resistance than ceramic or metal packages with alloy 42 or kovar lead frames.

Heat flows off the surface of a package mainly as a result of either convection or radiation. Θ_{CA} is a function of the package surface area, color, and surface finish. There are basically two ways to lower Θ_{CA}. One is to use airflow. The typical effect of airflow is shown in Fig. 16.14. The

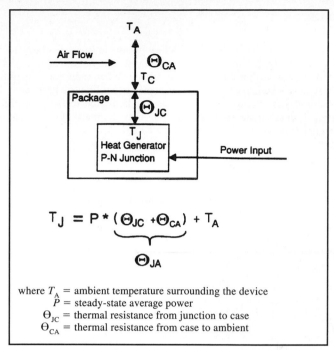

$$T_J = P * (\Theta_{JC} + \Theta_{CA}) + T_A$$

$$\Theta_{JA}$$

where T_A = ambient temperature surrounding the device
P = steady-state average power
Θ_{JC} = thermal resistance from junction to case
Θ_{CA} = thermal resistance from case to ambient

FIGURE 16.13   Thermal resistance basics.

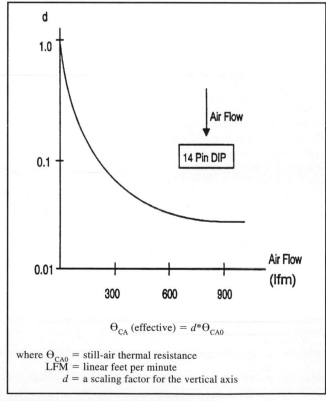

$$\Theta_{CA} \text{ (effective)} = d * \Theta_{CA0}$$

where Θ_{CA0} = still-air thermal resistance
LFM = linear feet per minute
d = a scaling factor for the vertical axis

FIGURE 16.14   Airflow effects on thermal resistance.

other is to use a heat dissipator. A heat dissipator increases the effective package surface area providing a surface from which to both convect and radiate heat. Radiation will be more with a black surface than with a shiny or white surface. When radiation constitutes a significant portion of the heat flow mechanism, thermal resistance is not linear with power. The thermal resistance will be lower at higher power levels because the hotter the surface, the higher the heat flow due to radiation. This is illustrated in Fig. 16.15. Also, the thermal resistance of a device in a socket will be higher than when it is soldered onto a PCB.

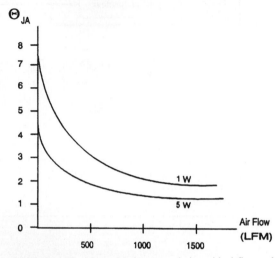

FIGURE 16.15   Thermal resistance variation with airflow and power dissipated for a transistor mounted on a heat sink where heat flow due to radiation is significant.

In a high-pin-count IC package, the thermal resistance from junction to leads may be more important than the thermal resistance because of heat flow off the package surface. A four-layer PCB will carry the heat away better than will a two-layer board. When thermal resistance data is given by the supplier, the test is usually performed using a *Joint Electronic Device Engineering Council* (JEDEC) standard procedure which specifies a two-layer PCB. These numbers may be higher than what is actually the case in your system.

The thermal resistances listed in Tables 16.5 and 16.6 were compiled from many different suppliers' data books to illustrate the variations which are possible from supplier to supplier and

TABLE 16.5   Typical Thermal Resistances for Discrete Transistors

Package type	Θ_{JC}, °C/W	Θ_{CA}, °C/W
	Transistors	
T0-3	0.7–4	25–35
T0-220	3–5	45–60
T0-5	15–25	110–135
T0-39 medium can (2N2219)	30–85	135–200
T0-18 small can (2N2222)	70–90	175–200
T0-92 plastic	100–125	150–200

TABLE 16.6 Typical Thermal Resistances for Integrated Circuits

Package type	Θ_{JC}, °C/W	Θ_{CA}, °C/W
Plastic DIP ICs (copper lead frame) $\pm30\%$		
8 pin	50	50
14 pin	48	38
16 pin	42	41
24 pin	23	30
28 pin	23	30
Cerdip $\pm30\%$		
8 pin	30	80
14 pin	30	80
16 pin	30	70
24 pin	26	34
28 pin	27	30
Surface-mount (Cu lead frame) $\pm30\%$		
SO-14 plastic gull wing	80	105
SO-16 plastic gull wing	65	80
SO-20 plastic gull wing	35	50
SO-24 plastic gull wing	30	50
100-pin PQFP with Cu slug	12	40
176-pin PQFP with Cu slug	10	35
208-pin PQFP with Cu slug	8	28

package type to package type. These are to be used as a guide only. Notice that for the 2N2219 (and all other types) there is a wide difference between different supplier's numbers. This translates to wide differences in operating temperature and, hence, wide differences in failure rates for a given level of power dissipation for parts from different suppliers. This is a primary reason to control suppliers with a qualified suppliers list.

Another item to keep in mind is that the thermal resistance increases as altitude increases or air pressure decreases. This is shown in Fig. 16.16. Note that if the system is specified to operate from 0 to 10,000 ft, the thermal resistance from case to ambient is about 20 percent higher at 10,000 ft than it is at sea level. Supplier data is usually given at sea level.

When calculating the power dissipation, both peak power and average power need to be calculated. Peaks of power cannot allow the junction temperature to exceed the absolute maximum. For this calculation, you may need to consider the thermal time constant of the package. Average power is used to calculate the junction temperature for use in the Arrhenius equation and other failure-rate calculations.

Timing Analysis.   When performing a worst-case timing analysis, it is important to recognize that the maximum or minimum number in the supplier's data sheet is not necessarily the maximum or minimum number in your circuit. All numbers in the data sheet are at a specific set of conditions. Be sure to determine what the correct minimum and maximum numbers are over voltage, temperature, and capacitive load. A commonly overlooked issue is the effect on propagation delays of the input rise and fall times.

One example of the pitfalls associated with simply saying that "it works" relates to the issue of minimum propagation delays. If the supplier's data sheet leaves the minimum column blank, then the designer must assume that the minimum can be zero+. Manufacturers do not like to spec-

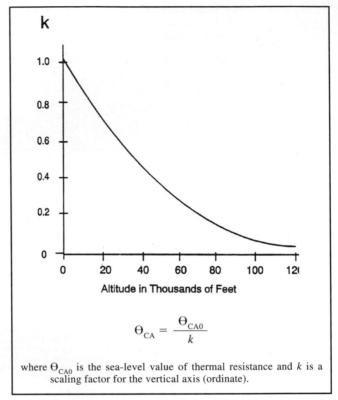

where Θ_{CA0} is the sea-level value of thermal resistance and k is a
scaling factor for the vertical axis (ordinate).

FIGURE 16.16   Thermal resistance variation with altitude.

ify these minimums because process improvements and die shrinks will lead to faster parts. Therefore, if the designer has not designed to zero+ minimum delays, the system will fail as parts get faster or someone decides that a faster part (or a better-than-spec part) can be substituted because it "should" work. This problem is particularly common with memories and *programmable array logic/generic array logic* (PALs/GALs) where the designer has relied on a minimum data hold time or simply assumed some number in place of the blank column.

Other Stresses Depending on Environment (Vibration, Humidity, etc.).   Worst-case analysis must be performed with respect to all stresses which a component may see. Some other obvious stresses are voltage and current levels, but there are others such as mechanical and chemical stresses which are usually not well analyzed from a theoretical standpoint. Often the design in these areas is more intuitive than analytical, and the designers rely on testing of the finished product to gain confidence that the system and components will survive under these other conditions. This approach does not guarantee any margin for the worst-case systems and environments.

16.2.7 Predicting the Failure Rate of Each Component

Failure-rate predictions describe the long-term failure rate after the infant-mortality period has passed. The prediction for each component provides another data point in order to evaluate whether the components can meet their reliability budget for the subassembly and whether the system is capable of meeting the reliability specification.

The Anatomy of a Prediction Procedure.   The purpose of a prediction procedure is to establish and maintain consistent and uniform methods for estimating the inherent or long-term reliability for particular types of equipment in particular marketplaces. For the military marketplace, the procedure is MIL-HDBK-217. For telecommunications equipment, the procedure is Bellcore TR-NWT-000332. Most companies use one of these two procedures or a modified version of them.

For each component type, a failure-rate model is developed to account for the various dominant stresses to which the part is likely to be subjected. The largest variable in these procedures is the quality-factor multipliers. These are based on your and your supplier's adherence to specified quality and reliability specifications and procedures.

All mathematical models have limitations. Some of these for reliability predictions are

1. The failure-rate models are empirically derived and based on available data which may be incomplete or inaccurate.

2. All failure modes relevant to the particular use environment may not be accounted for in the model. Wide variations in failure rate can occur because of differences in the system application and manufacturing environment.

3. Failure-rate models are not accurate at high stress levels. The methods assume that the parts are well derated.

4. A reliability prediction predicts the incidence of hard failures and does not include adjustment failures or soft or noise-induced failures. A hard failure requires the replacement of a failed module or part to get the unit working again.

5. New processes and products are seldom adequately modeled.

The prediction procedures do not cover mechanical parts, and, although there are failure rates published for some of these parts in various sources, the numbers are not to be considered accurate because they depend so much on the system application.

Most mechanical, electromechanical, and chemical components have wearout mechanisms which will determine their life.

The Benefits and Uses of Reliability Predictions.   There are many ways of using and misusing reliability predictions; some of these are

1. To evaluate alternative designs and design approaches.

2. To identify the most likely highest contributors to failure.

3. To identify overstressed parts.

4. To determine the dependency on the failure rate of various stress changes, such as temperature, voltage, frequency, and other parameters.

5. To measure the effectiveness of your manufacturing process by comparing actual field reliability with the predicted reliability.

6. To determine whether a MTBF specification is achievable.

7. To advertise a MTBF number. Be careful here.

MIL-HDBK-217F.   All prediction procedures were originally derived from MIL-HDBK-217 and subsequently modified to provide more "realistic" numbers for other end-use environments. MIL-HDBK-217 provides fairly accurate failure rate predictions for companies who follow the military quality specifications and use military-qualified parts.

The major issue with MIL-HDBK-217 is the failure-rate multiplier, called the *quality factor* π_Q, which is set at a high number for commercial parts. This effectively keeps commercial parts out of military equipment.

The issue here is that commercial components have unknown and/or varying levels of reliability because the processes are seldom standardized or controlled. In fact, there is a wide variability in the effective quality factor achieved by commercial companies based on how they qualify, purchase, inspect, and test components and how much care they take and the process controls they have as they build the components into their equipment.

An average commercial company is loosely defined as a manufacturing company with no special quality and reliability systems beyond what an "average" commercial company would have. The best commercial companies can experience field failure rates which are two to three times better than "average." A company with few component or process controls may experience field failure rates two to three times worse than "average." It is precisely this variability in commercial processes that must be carefully accounted for.

The major benefit of using MIL-HDBK-217 is that the failure rate models take into account actual device internal temperatures. No other prediction procedures do a good job of this. Quality factors and other parameters can then be adjusted on the basis of properly collected field data for similar equipment and manufacturing processes. To get an accurate number for a particular company's effective quality factor, a prediction on a similar machine in present production is compared and the quality factors are scaled to the actual field data. This will involve a detailed analysis of the field data for those machines and an analysis of manufacturing processes. This will yield a quality factor "tuned" to the particular manufacturing process and will allow the most accurate and believable prediction of all the various methods.

Bellcore.   The Bellcore procedure is required by suppliers to the "Baby Bells." It provides generally better failure rates than does MIL-HDBK-217F, which is supposedly based on field experience in communications equipment.

There are three quality levels described in Bellcore. A major limitation of this procedure is the way it treats the temperature dependence of the failure rate. "The device operating temperature is the unit operating temperature of the unit in which the device resides." With this definition, there will be no difference in the failure rate of an integrated circuit running at 125°C junction temperature versus one running at 90°C junction temperature. This procedure will understate the failure rate of motor drivers and any other hot-running devices.

How to Perform Predictions.   Two predictions need to be performed. One is per MIL-HDBK-217F and is done "by the book." When dealing with customers and other "outside" people, this is the number that should be used. The MIL-HDBK-217 by-the-book prediction should be considered worst-worst case. In some cases, the by-the-book prediction is so off-base as to make it unusable. This is the case for hybrids, field-effect transistors (FETs), relays, and motors and (to a lesser extent) for aluminum capacitors, PIN diodes, varactors, and fast-recovery diodes.

The other prediction (called a "realistic commercial prediction") is what can be realistically achieved by your company with the process now used for manufacturing and design. As the product matures and field data is collected which shows that the product is doing better than the "217F by-the-book" prediction, then the field data can be used to substantiate the better level of reliability.

Many suppliers of commercial parts continue to make great strides toward improving the reliability of their products and manufacturers of commercial equipment also continue to make great improvements in the process control needed for higher-reliability equipment. Some of the major design and process issues which allow these improved levels to be achieved are component- and equipment-level qualification testing, source control of components and suppliers, a solid ESD control program, and an effective failure reporting, analysis, and corrective-action system (FRACAS).

16.2.8   Assigning the Failure Rate Required from Each Component

At this point, data has been gathered suggesting what the failure rates of each component will probably be based on supplier data and predictions. The final step in the allocation process, then, is to assign the failure rate required from each component so that the subassembly meets its allocation.

16.2.9   Assigning Required Failure Mechanisms to Meet the System Life Specifications

The component requirements can also be established for end of life based on the systems needs for the parts to perform their intended functions.

16.2.10   Comparing Required versus Expected Actual for Failure Rates, Life, and Margins

By comparing the required failure rates and life characteristics with the expected actuals as gathered from the supplier's data, field data, and predictions, the areas of concern can readily be identified. These components can then be listed on the *reliability critical-components list*.

16.2.11   Developing Reliability Critical-Components List Outlining Areas of Concern or Uncertainty

The *reliability critical-items list* is a subset of the critical-components list (CCL). This list identifies components which have known reliability problems or which must have high reliability, assists in the determination of how much testing is enough, and helps establish priorities. Each part on the CCL must have a formal action plan. This list also drives alternative design efforts.

The critical-parts list is a list of parts made at the beginning of each product design and reviewed monthly which lists those parts which may have design or production problems. Any part that someone feels uncomfortable with should go on the list so that the proper engineering effort can be made to evaluate and alleviate the risk. That effort may consist of robust design techniques, supplier technical or logistical assistance, and/or additional qualification testing.

Another way of saying this is that every design action taken should have a plan. And, potentially troublesome or critical components should have a formalized plan. Success in any development program requires that problems be anticipated and steps taken, in advance, to avoid them.

All new components need to be qualified for use in manufacturing. However, not all components can (or should) be qualified by extensive environmental tests. The amount of qualification testing done on a component will depend on how critical the part is determined to be.

There are several types of parts that must go on the critical-component list. These are new technology, mechanical parts, and parts with known or suspected supplier problems. Figure 16.17 is a *quick-part-evaluation checklist* to determine whether a part belongs on the critical-components list.

☐ Are there any known problems or deficiencies? Now? In the past?
☐ Is the specification preliminary, incomplete, or inadequate?
☐ Does the failure rate appear to be unachievable?
☐ Is the supplier or process new or unqualified?
☐ Does the part meet the "minimum" evaluation guidelines?
☐ Will the part "work" under worst-case conditions?
☐ Is the supplier sound and established?
☐ Is the supplier committed to the part?
☐ Does the intended use meet the derating guideline?
☐ Is the part to be used in a "critical" application?

FIGURE 16.17   Quick-part-evaluation checklist.

16.2.12   Addressing Deficiencies, Inadequacies, and Unknowns

Component qualification may consist of a variety of actions which are designed to address any deficiencies, inadequacies, or unknowns in the component or supplier. These will generally include a combination of supplier actions, design actions, production process actions, and test or evaluation actions.

Components can be qualified by one of the following methods.

Qualification by Previous Usage.   Parts which have been used successfully in manufacturing prior to initiating a formal qualification program, can, in general, be used in future products. However, parts which are classified as "restricted" or "not for new design" may need to go on the critical-parts list for further testing prior to release in a new product.

Qualification by Similarity.   If a part is not on the critical-parts list and has passed the minimum level of part evaluation as described in Sec. 16.2.3, it may be qualified by similarity to some other part from the same supplier, as long as the part is the same technology and its function does not represent a radical departure from previously qualified parts.

Qualification by Test.   If a part cannot be qualified by one of the two methods described above, then some amount of testing must be performed to verify its suitability for production use. How much testing is performed will depend on how critical it has been ranked and how much risk is acceptable.

16.2.13   Testing Components

Device-Level Functional Test.   Functional testing asks the question "Does the part meet its data sheet?" This is a functional test to verify that all data-sheet parameters are being met. The sample size can be as low as five parts. This test should be performed at the temperature and voltage extremes specified by the supplier.

Device-Level Characterization.   Characterization testing asks the question "How well does the part work?" It involves testing the parts outside the recommended operating ranges for temperature, voltage, and/or other parameters to determine how much margin the supplier has for the parts and process. Comparison of different suppliers' margins is necessary to determine which suppliers out of four or five choices should be fully qualified and which suppliers will be held for emergencies. Also, a part with a wide margin will be more reliable in a given application than a part with a narrow margin. Again, five parts is all that is needed here.

Device Level Reliability and/or Life Testing.   *Reliability qualification* asks the question "Can the supplier ship me at least one lot of 50 or 100 devices which can last through an extended life test with no failures?" For integrated circuits, this is typically a 125°C burnin for 1000 h. One failure may be allowed during the infant-mortality period (up to 168 h), but there should be no failures between 168 and 1000 h.

The proper extended life test to run depends on the component type and the concerns which placed it on the critical-parts list. For mechanical parts, a life test which determines the wear or fatigue life of the component is important. For electrical components, the life test will determine whether the rate of failure is acceptable and that any wearout mechanisms do not occur during the useful lifetime of the product.

Confidence Level.   Often a user of a part will want to measure the failure rate of a part without understanding the cost in parts and time to make such a measurement. Actually, the time depends on the extent to which the test can be accelerated (see Sec. 16.2.5) and the level of confidence desired. For instance, for a 100-part 1000-h life test at 125°C, if the primary failure modes are expected have a temperature-acceleration factor based on an activation energy of 0.7 eV and there are no failures, you can demonstrate that the failure rate is, at most, 30 FITS at 40°C with a 60 percent confidence and 90 FITS at 40°C with a 90 percent confidence. If, however, the activation energy for the primary failure modes is governed by an activation energy of 0.4 eV, then the failure rate is, at most, 390 FITS (at 40°C) with a 60 percent confidence and, at most, 963 FITS (at 40°C) with a 90 percent confidence.

A confidence level is the probability that the failure rate measured in the test is the same as the failure rate of the total production population. Since an accelerated test can never get at all the failure modes which you will see in the field, these results must be considered "best case." Also, the implicit assumption is that the failure modes seen at the accelerated test stress conditions are the same as will be seen in actual use or field stress conditions.

The Test Temperature.   The test temperature should be as high as possible to achieve the largest acceleration factor. Since we are usually not testing for functionality at the burnin temperature, the limiting temperature will be the lowest absolute maximum temperature listed on the data sheet of all the components under test. For semiconductors, this is usually 125°C; for PCBs and systems, this is usually the temperature rating of the capacitors or plastic parts. Note that an absolute maximum temperature is usually an internal temperature for the component. Therefore, power dissipation and airflow in the test chamber will need to be accounted for. Also, the thermal resistance of the part in a test socket is not the same as the thermal resistance when the part is soldered onto a PCB.

When testing at temperatures above 100°C, consider using both a high-temperature PCB material and a high-temperature solder.

Other Stress Options.   Although temperature is the most common stress applied during testing, other types of stresses are appropriate depending on the component type. Again, in no case, should the absolute maximum be exceeded.

Higher-than-rated voltage testing is appropriate for semiconductors and capacitors. Do not exceed the rated breakdown voltage.

Low-temperature testing is appropriate for components which will be used in systems operating at subnormal temperatures.

Many other types of stresses can be applied to the components such as temperature cycling, thermal shock, vibration, and mechanical shock. What tests are performed must always be related back to what is required from the component in the final system. Test procedures and methods for these and other tests can be found in MIL-STD-883, MIL-STD-750, and MIL-STD-202.

16.2.14   Performing a Reliability Growth Test

Traditional Growth Testing within Component Specification Limits.   *Traditional reliability growth testing* consists of testing a finished product or prototype at higher-than-specified temperatures, but not so high that the component ratings are exceeded. Whenever a failure occurs, a concerted effort is made to determine if there is some corrective action which can be taken such that failure will never occur again.

The number of units tested will vary depending on the cost of a unit and the number of units to be built in production. When the cost of the unit is high, there may be only one unit available for this kind of test. Try to put as many units on test as possible so that there will be a higher confidence that most of the possible problems which could appear will appear. If the total production run is low, say, 10 or 100 units, then these tests will need to be performed on those units. In other words, enough time will have to be put on each system in the factory to ensure that there are no latent problems.

The test time should be long enough to verify that the specified MTBF can be met. For this, use the χ^2 distribution to calculate the failure rate based on the number of units on test, the test time, and the number of failures which are chargeable to the MTBF. A failure that is chargeable against the reliability of the system is a failure for which there is no corrective action which can be implemented such that failure will never appear again. In addition to the χ^2 distribution, there will probably also be a temperature-acceleration factor based on the Arrhenius equation.

Testing Outside Component Specification Limits (STRIFE, HAST, etc.).   This type of testing tries to discover the weakest link in a system by stressing the product way beyond the specification limits. This is similar to margin testing except that it uses temperature, vibration, thermal shock, and other environmental screens as the stress. There are many flavors of these kinds of tests. *STRIFE* means stress to end of life. *HAST* means highly accelerated stress test. They are all basically similar in that the purpose is to discover failures and put in a corrective action so that the failure will not occur again at that stress level. This assumes, of course, that the stress which causes failure is close to the specified environmental stress levels.

Since the object of these tests is to find the point at which a product fails, it is not possible to "pass" this kind of test. It is possible only to have successfully completed the process.

There are several pitfalls in this kind of testing (or any accelerated testing). Namely, is the failure that occurred a failure that can reasonably be expected to occur in the actual field environment? Or is the failure a phenomenon that is unique to these high stress levels? These questions must be explored before implementing a corrective action which may cause problems with system cost or add complexity to the manufacturing cycle.

16.2.15   Performing a Reliability Demonstration Test

A *reliability demonstration test* is similar to the reliability growth test except that it is performed on manufacturing-built equipment built to released engineering documentation. When a failure occurs which can become nonchargeable when a particular corrective action is implemented, the test must be restarted after the corrective action is implemented. Because of this, usually, no corrective actions are implemented on the machines until there are so many failures that it is clear that the design cannot meet its specified MTBF.

When changes are made to the design or manufacturing process after the reliability has been demonstrated, those changes must also be demonstrated.

16.2.16   Ensuring That There Is a Working Mechanism for Measuring Actual Field Reliability and Field-Use Conditions

A *failure reporting, analysis, and corrective-action system* (FRACAS) allows the measurement of reliability and generates corrective actions which improve yields and product reliability. It is a formal system which actively seeks out the causes of every failure and attempts to identify corrective actions which will prevent the failure from recurring. Engineering must have the data from previous products in order to improve the new products.

Failure Reporting.   *Failure reporting* must occur during the engineering, production, and field-usage phases of each product program. The failed item or assembly must be conspicuously tagged and controlled to allow further analysis. The failure tag should state when and where the failure occurred, who witnessed the failure, and a description of the symptoms and conditions of failure.

Failure Analysis.   *Failure analysis* is the determination of the causes, mechanisms, and potential effects of failures and reporting those results in such a manner that the results serve as a basis for decisions about which corrective action to implement. Repair is not failure analysis. It could be, but most repair operations make no attempt to identify exactly what failed. Nothing is more frustrating than trying to make sense out of repair data when five components were replaced at the same time because the diagnostic test did not bother to identify exactly which one failed. Not only that, but the reliability of the board itself has been degraded because of heating up five component locations instead of only one.

Corrective Actions.   When the root cause of the failure has been determined, a *corrective action* needs to be developed, documented, coordinated, and implemented to eliminate or reduce the recurrence rate of the failure. This corrective action may be applied to the design, the process, or the field-usage instructions.

One common area where actions are taken on failed components is the *material review board*. Here decisions are made daily on whether to use discrepant material. The role of engineering on this board is to ensure that the margins are not used up by using parts which do not meet the specifications.

16.3   MANUFACTURING FOR RELIABILITY

As discussed earlier, the manufacturing process is a very important contributor to system reliability. There are many issues which must be attended to in order to produce consistently reliable equipment. To assist in sorting out these issues and developing and maintaining a consistently reliable process, Fig. 16.18 presents a checklist for manufacturing. The discussions in this section follow this checklist.

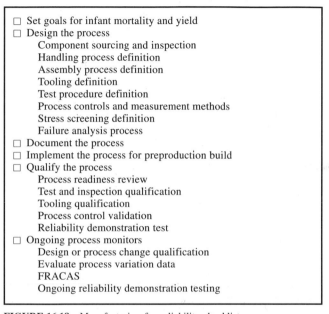

☐ Set goals for infant mortality and yield
☐ Design the process
    Component sourcing and inspection
    Handling process definition
    Assembly process definition
    Tooling definition
    Test procedure definition
    Process controls and measurement methods
    Stress screening definition
    Failure analysis process
☐ Document the process
☐ Implement the process for preproduction build
☐ Qualify the process
    Process readiness review
    Test and inspection qualification
    Tooling qualification
    Process control validation
    Reliability demonstration test
☐ Ongoing process monitors
    Design or process change qualification
    Evaluate process variation data
    FRACAS
    Ongoing reliability demonstration testing

FIGURE 16.18   Manufacturing-for-reliability checklist.

16.3.1   Setting Goals for Infant Mortality and Yield

The levels of infant mortality and yields are dependent on the level of the initial quality and reliability of the components received and how well they are put together in the assembly process. It is important to specify these goals so that the process can be designed to meet them. Processes must then be designed and qualified to meet these goals.

An infant-mortality specification needs to be related to the specified system failure rate, the end-use environment, and the desired warranty period failure rate. The failure rate during the infant-mortality period will be higher than the steady-state failure rate (see Fig. 16.3b). A typical process with no stress screening for components or assemblies could start out the bathtub curve between one and one-half and five times the steady-state failure rate. A rule of thumb is that after about one year in the field at normal operating temperatures, all the infant-mortality failures will have occurred. As with any rule of thumb, any given product will vary widely from this average.

For a system with a 1,000,000-h MTBF, 0.1 percent of the units in the field will fail during any given 1000-h period of operation after the infant-mortality period has passed. If an exponential approximation is used for the infant-mortality period failure rate, with one year (8760 h) of 24-h/day operation for the infant-mortality period, assume that the failure rate starts out at twice the steady-state failure rate; then about 1.03 percent of the units shipped will fail during their first year of operation.

16.3.2 Designing the Process

Once the process goals have been set, the next step is to design the process. Many companies simply use whatever process is readily available and miss this important step.

The complete process for manufacturing consists of receiving and handling the parts; assembling them into the product; and testing, inspecting, and screening actions to verify functionality and ensure that the product is capable of meeting its specifications.

Component Sourcing and Inspection.   A critical element of process control is the control of suppliers and the limiting of components used to only those which have been qualified for use in the product. It can be a fatal mistake to assume that just because a company is selling a product, it must be reliable.

This control is usually accomplished with an *approved-source list,* which lists the manufacturer and the manufacturer's part number including all prefixes and suffixes. It is also important to limit the number of suppliers, generally to three, so that the supplier's quality and performance can be adequately monitored. There should be at least two known good sources qualified for each part, and both sources should be purchased from on a regular basis. Many companies have had disruptions of production and reliability problems because only one of the qualified sources has been utilized; then when, for some reason, the preferred source has not been able to deliver, it is discovered that the second source cannot supply parts either. Then, there is a scramble to find any supplier who has a part that seems to work, and the quick evaluation demanded by production schedules creates an unnecessary risk to reliability.

Handling Process Definition.   Handling damage is probably the most significant cause of component failure. Electrostatic discharge damage to semiconductors and film resistors is one of the most pervasive causes of handling damage, but many other forms also exist.

Electrostatic discharges are insidious because the parts may not fail immediately, but simply be weakened so that they fail several months or several years after the product has been shipped to the customer. It is, therefore, necessary to design and properly implement a good, solid ESD control program in all parts of the factory including the receiving dock, receiving inspection, the storeroom, the kitting area, the assembly and rework areas, and all testing areas including final test.

It is important to recognize that components are delicate items which must be handled with care at all times, both as piece parts and as assemblies. Components should be received in protective packaging which minimizes the risk of electrical, mechanical, and chemical damage to the parts. The parts should be kept in their protective packaging as long as possible until they are assembled into the next level of assembly. And each level of higher assembly needs to be packaged to minimize the risk of damage. Components should be individually packaged on reels or trays so that they do not rub together. Use of bulk-packaged components (e.g., 1000 in a bag) should be avoided.

Many components and assemblies (particularly surface-mount PCBs) cannot withstand flexing and bending, and many parts are damaged while they are being placed in the protective packaging as well as just in picking them up carelessly.

Assembly Process Definition.   Material should flow smoothly through the manufacturing area such that it is difficult to mix tested and untested material. A process flowchart should be developed so that material flow is clear and logical for every step in the process. All assembly operations must be well defined and the test and inspection stations clearly identified and defined.

One of the most critical operations in electronics assembly is the soldering process. It is critical to reliability that this process be adequately defined and controlled whether the process is wave solder, infrared, convection, or vapor-phase reflow. The time and temperature profiles for preheat, reflow, and cooldown must be defined in terms of both the need to obtain a good solder connection and the ability of the components to handle the thermal shock of the soldering operation.

Tooling Definition.   The tooling and fixturing to be used to assemble the components and perform fabrication and assembly operations must be defined in such a way that these operations do not inadvertently damage the components. Many factory and field failures have been traced to poorly designed, worn, or broken tooling.

Tooling and fixturing needs to be a joint effort between design engineering and manufacturing engineering so that the tradeoffs between throughput, part function, and costs can be made properly and built into both the part and the tool.

Test Procedure Definition.   As with all other processes, the tests must be properly designed before they are implemented. All too often, the test is thrown together at the last minute with less than full test coverage and upgraded later when problems surface at later stages of the manufacturing process or in the field.

Receiving inspection is necessary to ensure that the components meet their specifications. It is not necessary that this be performed at your facility. It can be performed at the supplier's facility as an outgoing test if it can be assured that the parts being tested are the parts which will be received. If the parts are purchased through distribution, there are no guarantees that they were tested to your specifications. In this case some testing at receiving is in order.

For PCB testing, the preprogrammed test routines supplied by the manufacturer of the tester can damage components in some circuit situations. Therefore, the precise operation of each of these routines should be understood as they interact with the circuits under test before they are implemented.

Again the test engineers should work closely with the circuit designers to ensure that the proper test points are available and that any special circuitry is implemented on the board to test the board. One of the most basic items usually needed is a way to turn off the on-board clock and insert the tester clock.

Process Controls and Measurement Methods.   A key feature of manufacturing for reliability is to produce the same product, the same way, every time. This requires that the detailed process be documented so that a standard exists against which to monitor performance and make improvements when desired. The performance of the manufacturing process, then, needs to be monitored and controlled. Test data from functional test is simply not sufficient. Many things need to be measured and controlled.

At receiving inspection, critical parameters of parts should be measured to gain insight into the variability of the parts both over time from the same supplier and from different suppliers. These parameters need to be defined. Otherwise, receiving test will default to pass/fail testing, which gives no indication of variability. This should also be done at subassembly and final test.

In the soldering process, the actual time–temperature profile in the operation needs to be measured. Too often, the operators rely on the settings of the equipment and assume that, if the dial is set to some temperature, then the oven temperature must be what the dial says. This is rarely the case. There will be a variation across the oven as well as variations due to the size of the boards and components going through the ovens. There are many instances of reliability degradation due to the oven going over temperature.

Statistical process control is a technique where variations in parameters are measured and deviations from the norm are analyzed to determine why the changes have occurred and whether the change presents a reliability risk. It is important to measure the ionic contaminant levels in the cleaning solutions, and when they change, stop and ask why they changed. That is, what dissolved off the assembly and into the cleaning solution? Did it damage the assembly or components, and should the cleaning solution be changed for the next assembly lot?

Stress Screening Definition.   The purpose of environmental stress screening (ESS) is to cause the weakest parts in a lot to fail in the factory rather than at the customer's site. The failures that occur will generally be in the infant-mortality category. This type of testing will not have much effect on the long-term failure rate of the product unless the stresses are so high that otherwise good parts are weakened.

The ESS process can be applied to components, subassemblies, or systems. The stresses which can be applied without causing damage to good parts decrease as the level of assembly increases.

In other words, at the higher levels of assembly, the stresses which can be safely applied will not be able to weed out some types of component defects. Therefore, a balanced ESS program will identify which components need ESS and will perform that screening at the component level, and which subassemblies need ESS and will perform that screening at the subassembly level. Then, the only screening needed at the finished-product level will relate to those defects that manifest only at the finished-product level.

ESS is needed if failures are occurring at the next or higher levels of assembly or in the field which could have been forced to occur at lower levels of assembly. All too often, companies will perform ESS testing on assemblies and make no attempt to see if it is doing any good. The proper questions to ask here are "Are failures occurring as a result of the ESS test and not occurring at the next higher level? Or are failures occurring (or not occurring) as a result of ESS testing and still occurring at the next higher level of assembly or in early field life?" If an unacceptable level of early life failures is not occurring in the field, there is no reason to institute an ESS program. Yet, so often, an ESS program is instituted because someone (the boss or the customer) read in a magazine that other companies are performing ESS. It may be the right thing to do, but for the wrong reason.

The real purpose of ESS testing is to screen out process problems until the root cause can be identified and fixed. In most companies, no effort is made to analyze the failures to see if the weaknesses could have been prevented at an earlier stage in manufacturing. Or, even if the root problem does go away and the early life failures cease, ESS will continue to be performed because either the field data is not adequately collected and digested or the customer has been convinced that he should pay for this "premium" process.

Ideally, the type of ESS should be chosen to get at the primary failure modes of the device or assembly. However, high-temperature burnin is the most common screen used for semiconductors. The devices may not be functionally exercised at all during the burnin. This is called a *static burnin*. The devices may be exercised to one degree or another. This is called a *dynamic burnin*. Two types of tests were developed for discrete transistors: *high-temperature reverse bias* (HTRB) and *high-temperature operating life* (HTOL). HTRB tries to get at the failure modes associated with voltage stressing of thin oxides in FETs or to accelerate contamination failures by reverse-biasing the collector-base junction in bipolar transistors. HTOL tries to get at the failure modes associated with dissipating power in the transistors. For integrated circuits, then, the dynamic burnin is the best compromise between these tests, but this is also the most expensive to perform.

Other types of ESS such as thermal shock and vibration stress the mechanical connections inside components and between components. These are useful in detecting cold solder joints, poor crimp connections, bad wire bonds, and other mechanical connections. Again, there are many instances of companies that started thermal shock because "everyone else is doing it," only to discover new types of failures began to occur because the correct stresses and stress levels for the product were not properly thought out.

Failure Analysis Process.   The failure analysis for the specific product must be designed and integrated into the overall FRACAS program. There will be product-specific procedures and capabilities which should be in place before any failures occur. In this way, production will not have to wait for someone to figure out how or where to perform a specific analysis.

16.3.3 Documenting the Process

After engineering releases a product to manufacturing, manufacturing must develop process instructions so that the product can be put together in a step-by-step fashion. The two main types of process instructions are assembly instructions and inspection instructions. It is important that these be approved by design engineering as being consistent with the spirit and intention of the released assembly drawing. The manufacturing instructions need to be controlled such that no process changes can be implemented unless they are coordinated with the overall manufacturing process. When design changes are released, the manufacturing instructions need to be updated to reflect the changes. These are the product-specific process documents.

Another class of manufacturing documentation relates to general processes which are used across several products. These include the documentation describing the procedures and test/inspection limits for ESD control, kitting, soldering, cleaning, and shipping. Unless these processes are specified, there can be no process stability and reliability will, therefore, vary, also.

16.3.4 Implementing the Process for Preproduction Build

The next step is to implement the process which has been defined to verify that all the correct steps have been planned. The preproduction build may or may not be destined for customers. But, in any case, the customer should not be misled into believing that the process has been fully qualified. The customer should be informed that these are preproduction units which will be undergoing a full qualification test as described below.

16.3.5 Qualifying the Process

This important step verifies that the combination of design and process is capable of meeting the specified reliability goals. It consists of a combination of theoretical reviews and testing.

Process Readiness Review.   The *process readiness review* is a joint review with design engineering and manufacturing to verify that the process implemented is the correct process for building the product. Actually, several reviews should be conducted throughout the process definition stage and again after the process has been used for the preproduction build, and data should be available from the process controls for analysis.

Test and Inspection Qualification.   *Test qualification* involves measuring the signals applied to the components and assemblies during test to verify that the tests do not overstress the components.

Inspection qualification verifies that the inspection tools are adequate for locating defects and that the sampling plans provide a reasonable (for the product and marketplace) assurance that a defective lot will not be erroneously accepted. Keep in mind that in any sampling plan, there is a probability that a good lot will be rejected and that a bad lot will be accepted.

Tooling Qualification.   Measurements must be made on components while they are in the fixtures and tools to ensure that they are not being overstressed and that the parts are being assembled properly.

For example, the tooling for assembling connectors should be chosen or built to ensure that operator variability is minimized and the connections are made the same way every time. It is a big mistake to assume that just because purchasing found the lowest-cost cable assembly supplier that supplier has the correct tooling and is using it correctly. These things need to be verified. A "first article" inspection is rarely sufficient to accomplish this.

Another example that frequently occurs with surface-mounted PCBs is when a large board is assembled, soldered, and then broken up into smaller "finished" boards. Extreme care needs to be taken in separating the boards to ensure that the flexing specifications for ceramic capacitors, chip resistors, and leadless IC packages are not exceeded.

Process Control Validation.   The process controls, measurements made, and the selected control limits must be validated as being sufficient to raise flags when the product or process is varying unacceptably. This is a difficult step which involves a theoretical analysis as well as an analysis of ongoing measurements.

Reliability Demonstration Test.   This test is described in Sec. 16.2.15.

16.3.6 Ongoing Process Monitors

Ongoing process monitors verify that the manufacturing process is consistent and produces products which meet the specifications.

Design or Process Change Qualification.   Whenever a design change or a process change is implemented, it needs to be qualified. One of the most common causes of failure occurs when a design change is implemented without proper testing. Then the fix needs to be fixed. This is both frustrating and embarrassing.

Evaluate Process Variation Data.   The *statistical process control* (SPC) data needs to be constantly evaluated to ensure that the process is in control. Some companies gather lots of data, but do not analyze it until something goes wrong. The purpose of SPC is to prevent failures, not explain them after the fact.

FRACAS.   A failure reporting, analysis, and corrective action program is a companywide program. Usually, manufacturing is the driving force because most of the failure-analysis capability resides in manufacturing. This program is described in more detail in Sec. 16.2.16.

Ongoing Reliability Demonstration Testing.   At periodic intervals, finished product should be measured for failure rate and/or wear life. This process should continue throughout the manufacturing life of the product and the units to be tested accounted for in the product cost. If the units are not accounted for in the product cost and the testing planned for, this ongoing testing is unlikely to happen.

16.4   COMPONENT FUNCTION AND FAILURE SUMMARIES

In electrical or electronic systems, the semiconductors are one of the largest contributors to the failure rate of the system. Most field failures can be repaired by replacing a discrete semiconductor or an integrated circuit. The other large class of failures involves the connectors and interconnection points. Of course, all components are susceptible to handling damage.

16.4.1   Integrated Circuits

It is easy to get lost in the bewildering array of acronyms associated with the many different IC types and suppliers. The types of things which are important from a reliability viewpoint can be grouped into four categories. These are the process type, the process, the function or complexity, and the package.

Electrostatic damage is one of the greatest causes of semiconductor failure. Damage is not restricted to MOS and CMOS devices, but can occur in all semiconductors. Therefore, it is necessary to package and handle all semiconductors in a manner that does not allow ESD damage.

Process Types.   Bipolar process types include DTL, TTL, HTTL, AS, ALS, FAST, ECL, integrated injection logic (I^2L), and others. The MOS (metal oxide on semiconductor) process types include NMOS, PMOS, CMOS, MNOS, SOS, HC, HCT, FCT, ACT, BCT, and others. Gallium arsenide (GaAs) is also a MOS-type process. Bi-CMOS is a process which places both bipolar and MOS transistors on the same die. These are the general classes of process types which are available. If the type of process is not clearly spelled out on the data sheet, ask the supplier what it is and whether the process (as well as the particular device) is qualified.

Most of the basic failure modes encountered relate to failure modes that are general to bipolar and FET transistors (see Table 16.7). Other differences in process complexity, oxide thicknesses, and processing goals lead to differences in reliability between the various process types.

Bipolar IC Processes

DTL   Diode transistor logic
TTL   Transistor transistor logic

TABLE 16.7 General Failure Modes for Bipolar and FET Transistors and Integrated Circuits

Failure cause or mechanism	Description	Method of detection	Method of minimizing defect
Surface contamination	Ionic materials (Na$^+$, Ca^{2+}, Cl$^-$) which provide leakage paths and activate parasitic transistors	High-temperature reverse bias testing (HTRB)	Process control; proper cleaning
Metal-layer corrosion	Contaminants such as moisture or other corrosive gases and residues attack the metal	Not generally detectable	Cleanliness; process control; die passivation; dry out parts
Bulk defects	Die cracks; dislocations in the crystal which localize impurities	Thermal cycling; thermal shock; power aging	Visual inspection; difficult to eliminate all defects
Oxide defects	Cracks, holes and thin spots which can break down with voltage or provide leakage paths	HTRB	Process control; visual inspection
Die attachment	Voids between die and lead frame or package base increase thermal impedance	X-ray; thermal resistance	Process control
Metal-layer interconnect	Resistive contact due to oxide growth in the windows	High-temperature storage	Process control
Electrostatic discharge	High-voltage discharges rupture oxides and junctions	Failure analysis	Effective ESD control program
Excessive voltage and current	Causes localized hot spots which melt the die and cause short circuits; can melt bond wires	Failure analysis	Protection devices in the circuit design
Base-emitter reverse bias	Zenering the base emitter junction can cause the current gain beta to degrade	Electrical test	Protection in the circuit design
Assembly defects	Improper die orientation; misaligned bonds; excessive wire loops; surface scratches due to handling; excess die attach material; debris from wafer probing	X-ray; PIND (acoustic particle detection); constant acceleration; thermal resistance	Process control; process cleanliness; machinery maintenance
Wire bonds	Insufficient time, temperature, or pressure leading to incomplete bonds; gold-aluminum: intermetallic formations (purple plague), Kirkendall voids, nicked wire due to excess pressure, peeling of Al film under ball bond; aluminum-aluminum: microcracking at heel of bond; plastic packages: thermal intermittent due to mismatch of expansion coefficients	Bond-pull testing; high-temperature bake will aggravate plague and void failures; temperature cycling will aggravate Al-Al wedge cracks; thermal intermittent testing for plastic packages	Bond-pull testing; process control; observing and derating the temperature limits of plastic packaged devices
Degraded package	Bent leads; cracked package; moisture in surface-mount plastic packages; exposure to corrosives, finger acids, and greases; cracked or leaky seals	Failure analysis	Handling and humidity control

HTTL	High speed TTL (old style)
AS	Advanced shottky TTL
ALS	Advanced low power shottky TTL
FAST	Fast advanced shottky TTL
ECL	Emitter coupled logic
I^2L	Integrated injection logic

FET (Field Effect Transistor) Processes

MOS	Metal oxide FETs on a silicon substrate
NMOS	N-channel MOS
PMOS	P-channel MOS
CMOS	Complimentary MOS (included both P and N channel transistors)
SOS	Silicon on sapphire (a type of CMOS)
HC	High speed CMOS
HCT	High speed CMOS with TTL-compatible voltage levels
FCT	Same as HCT but faster
ACT	Similar to FCT
BCT	Similar to FCT
GaAs	A type of CMOS on a gallium arsenide substrate
BiCMOS	CMOS with some bipolar transistors

The Process.   The specific process which builds the circuit may have different feature sizes (between 0.3 and 2.0 μm), different mask design rules, different transistor parameters, or different oxide thicknesses or otherwise be optimized for specific functionality. It is this detailed process that is most important for process-related reliability.

Function or Complexity.   Again there is a bewildering array of acronyms for the many functions available in integrated circuits. What is usually important is the number of gates for digital devices, the number of transistors for linear circuits, and the number of bits for memories. Some examples of the acronyms relating to function are SSI, MSI, LSI, VLSI, ULSI (small-scale, medium-scale, large-scale, very-large-scale, ultra-large-scale integrated circuits), ROM, PROM, EPROM, EEPROM, EAROM (read-only memory; programmable, erasable programmable, erasable programmable ROM), flash memory, RAM, DRAM, SRAM, VRAM (random-access memory;), PAL, GAL, FPGA, FPLA, first-in first-out (FIFO), ASIC, and gate arrays. Output drive capability and the load is also very important.

Functional Acronyms

SSI	Small scale integrated circuits
MSI	Medium scale integrated circuits
LSI	Large scale integrated circuits
VLSI	Very large scale integrated circuits
ULSI	Ultra large scale integrated circuits
ROM	Read only memory
PROM	Programmable read only memory
EPROM	Erasable read only memory
EEPROM	Electrically erasable read only memory

EAROM	Electrically alterable read only memory
Flash	Similar to EEPROMs
RAM	Random access memory
DRAM	Dynamic random access memory
SRAM	Static random access memory
VRAM	Video random access memory (dual port)
PAL	Programmable array logic
GAL	Generic array logic
FPGA	Field programmable gate array
FPLA	Field programmable logic array
FIFO	First-in first-out memory buffer
ASIC	Application specific integrated circuit

The Package.   Again there are a wide variety of packages and acronyms. The important features for reliability are the number of pins, the thermal resistance, and the manner in which stresses between the board and the package are handled.

The major stresses which must be absorbed by the package leads are mechanical stresses due to flexing and the expansion and contraction stresses due to differences in the thermal coefficients of expansion between the package and the board substrate. Dual-in-line and gull-wing flat packages absorb these stresses very well. J-hook leads also absorb the stresses, but not as well as the DIPs (dual-in-line packages) and flatpacks. Leadless packages are the most susceptible to these stresses, and cracked solder joints are common with these types of packages. The pins of pin grid arrays do not comply to these stresses very well, but the holding force of the pins in the hole is generally strong enough to resist solder joint failure except for high stress levels.

The two major categories of packages are plastic and ceramic. There are also two major differences from a reliability standpoint.

First, the ceramic package is relatively much more hermetic than the plastic and can be used in humid environments. Second, the limit on the absolute maximum junction temperature for a die in a ceramic package is the limit allowed by the process, whereas the limiting factor for a molded plastic package is the glass transition temperature of the plastic. This is the temperature where the plastic starts to soften and leach impurities onto the die, causing contamination failures. Also, the thermal resistance changes at this temperature. When plastic devices are specified for an absolute maximum temperature greater than 150°C, the package is usually a cavity-type package where the plastic is not in direct contact with the surface of the die.

Some Reliability Rules for Bipolar Semiconductors.   When rise and fall times are greater than about 50 ns, the pullup and pulldown transistors are both turned on for a long enough time to cause extra heating in the device. This also slows down the switching time. In this case, it is best to use a Schmitt trigger device.

Voltages on the inputs greater than the maximum specification can zener the base-emitter junction of the multiple emitter input circuits. This causes a degradation of the current gain ß and slows down the device.

Some circuits, such as the 74LS74, 74LS244, and many others, use a PNP input structure which can saturate when the input goes negative. The device may then not switch until the excess charge has been removed from the base (the base storage time). Excess delays of 50 to 150 ns have been observed as a response to negative transients. If this is happening, an external diode is usually required to clamp the negative transient. The input clamp diodes are usually not fast enough.

During power up and power down the output of a logic circuit may glitch. Therefore, a filter circuit is necessary on the write-enable input to EEPROMs, battery-backed memories, and flash memories. Otherwise, bad data may be written into the memory.

Some Reliability Rules for CMOS Semiconductors.   As with bipolar circuits, rise and fall times should be kept as fast as possible to minimize overheating due to turning on both the upper and the lower transistors.

Unused inputs must be terminated. Otherwise, leakage currents will charge up the input to the threshold and turn on both the upper and lower devices which will overheat and fail.

Gate inputs which go off the board should be terminated with a pullup or pulldown resistor even if the final system will terminate the line through a connector or cable. ESD damage can occur as a result of handling of the loaded board. But that is not the only reason. During electrical test the line may not be terminated because of the spring-loaded test pin not making good contact. Then leakage currents can charge up the input and damage the device.

Some Reliability Rules for PROMs, EPROMs, EEPROMs, Flash Memories, and Other Programmable Devices.   Be sure that the programming equipment or method, the programming algorithm, and the timings are per the manufacturer's specifications. Minor variations from the specified procedures can cause the device to not program adequately. That is, a logic low may not be low enough and a logic high may not be high enough, or the memory may not meet its access time.

Derate the number of write/erase cycles for all reprogrammable devices. For electrically programmable devices which are written to during the course of normal operation, ensure that the software observes this specification.

16.4.2 Diodes and Transistors

The range of failure modes seen with diodes and transistors is similar to those seen with integrated circuits except that there are more opportunities to overstress the parts with either voltage or current.

A key feature of the assembled board test should be a low voltage, low current solder short circuit, or bridging test. By using this approach, the components will not be weakened by applying full power to the board and overstressing the parts because of a solder short.

LEDs require current limiting, or they will burn out. Also, the light output of LEDs degrades with time and degrades faster at high current levels. This can be a cause of failure for optically coupled devices. Another cause of failure for optically coupled devices is not supplying enough input current to cause the device to switch reliably for the luminosity life of the LED.

16.4.3 Resistors

A *resistor* is a device which resists the flow of current and in so doing dissipates power in the form of heat. The resistance of a uniform material with resistivity ρ is the resistivity times the length L divided by the cross-sectional area A (see Fig. 16.19). The cross-sectional area should be thick enough to handle the current; and the length, long enough to handle the voltage.

Table 16.8 lists the general failure modes seen in resistors. Other specific issues are discussed below for various resistor types.

A common failure caused by resistors which does not cause the resistor to fail occurs when the heat generated by the resistor overheats a nearby component such as an aluminum electrolytic capacitor. Figure 16.20 shows typical temperature specifications for various resistor types. For all resistors, two temperatures are specified: a temperature at which the resistor can dissipate 100 percent of its rated power and the hottest temperature it can see when no power is dissipated. This latter temperature is called the *maximum hot-spot* temperature. Both of these temperatures vary from supplier to supplier, so be careful.

Whenever the resistor is operated on the sloping portion of the graph, the resistor is at its maximum hot-spot temperature. Referring to Fig. 16.20a and using the maximum hot-spot temperature of 165°C, when the ambient temperature is 70°C, the temperature rise of the resistor is 95°C, due to dissipating 100 percent of its rated power. Figure 16.21a plots the temperature rise against percentage of rated power actually being dissipated. Notice that it does not matter whether the

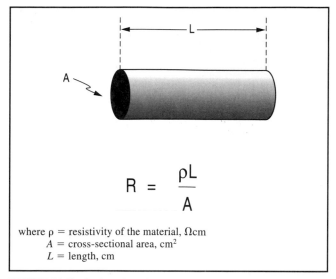

$$R = \frac{\rho L}{A}$$

where ρ = resistivity of the material, Ωcm
A = cross-sectional area, cm^2
L = length, cm

FIGURE 16.19   Basic resistor theory.

resistor is rated for 2 W or 0.1 W. At 50 percent of rated power, the surface temperature of even an 0805 0.1-W resistor will be 47.5°C above ambient. If the ambient temperature is 40°C and the part is too close to an 85°C-rated capacitor, the capacitor will fail prematurely. Figure 16.21b plots the temperature rise for power resistors.

Figure 16.22a gives the formula for calculating the failure rate of most resistors from MIL-HDBK-217F, Notice 1. The base failure rates are plotted in Fig. 16.22b for a RC-style carbon composition and a RN-style metal film resistor when operated at various percentages of rated power. From these graphs, it can be seen that the metal film resistors give better reliability performance at the higher temperatures, while the carbon composition resistor has a lower failure rate at lower temperatures.

Carbon-Composition Resistors.   A carbon-composition resistor is made from a compressed slug of finely ground carbon and silica (see Fig. 16.23). The leads are embedded in the slug, and the whole piece is then compressed and bonded under pressure. It is then molded in a case, cured in a bake cycle, and coated with a sealant to provide a moisture barrier. Unfortunately, the moisture barrier does not work very well, and humidity can cause the resistance to drift 10 to 15 percent from the specified value. The specified value and tolerance are valid only when the resistor is new and dry. The value of carbon-composition resistors will also change depending on the applied voltage. This is called the *voltage coefficient*.

All resistors will drift in value with time and stress. Typical values for various resistor types are given in Table 16.9. The sum of the percentages in each column can be used as a reasonable end-of-life tolerance.

Carbon-composition resistors are an old technology and have been replaced by films and chips in most applications. One area where no other resistor comes close to the carbon composition in performance is their ability to handle pulsed power in excess of the rated power for short pulse durations. Depending on pulse width, this can be 20 to 25 times the rated power. This ability is governed by the energy ratings for the resistor which must be obtained from the manufacturer.

The primary failure modes encountered with carbon-composition resistors are hot-spot breakdown, value drift, and separation of the lead attachment as a result of mishandling or not supporting the leads during the lead-forming operation.

TABLE 16.8 General Failure Modes for Resistors

Problem cause or mechanism	Description	Method of detection	Method of minimizing defect
Hot-spot breakdown	Exceeding the thermal limits of the resistor due to a combination of internal heating and external ambient	Visual inspection for carbon composition and/or failure analysis for thick- and thin-film resistors	Derate power and energy; improve heat flow paths
Cracked carbon slug	Intermittent resistance changes usually caused by rough handling or lead forming	Thermal intermittency test	Process design and control for parts handling
Cracked substrate	Creates open circuit for resistive element crossing crack; issue for all ceramic- or glass-based film resistors including chip resistors	Temp cycling; thermal shock	Process design and control for parts handling and soldering process temperature profile
Lead open circuit	Corrosion of the lead at the point of contact with the end cap on film resistors due to trapped contaminants; also can occur on wirewounds due to a weak weld of the wire to the lead; uncontrolled lead bending	PCB cleanliness inspection; failure analysis	Clean solder residues; process control; tooling control
End-cap open circuit	End cap peels from the resistor body for chips and films due to overheating, contaminants, or poor plating	Solderability test	Avoid excessive heat; sample-test incoming material
Resistance drift	Tests at rated temperature and rated power show that carbon resistors can drift 10% and metal resistor about 1.3%	Load life test	Derate power and tolerance
Resistance drift	Tests at 85°C and 85% relative humidity show metal resistors drifting about 1% and carbon resistors 10–15%	Humidity test	Derate tolerance
Corona breakdown	Voltage breakdown in high-voltage resistors which create a permanent low-resistance path	Visual inspection for tracks	Derate voltage; ensure cleanliness
Film buildup on pots	Oxide film buildup on wiper contacts can prevent current from flowing through the wiper	Turning the pot causes it to work	Minimize the use of pots; use "solvent-tolerant" pots; hand-solder and clean
Electrostatic damage	Can vaporize thin films (both leaded and chip) causing a change in resistance	Failure analysis	Handling process control
Thermal runaway in NTC thermistors	Caused by improper circuit design or noise	Failure analysis	Current limit

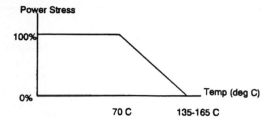

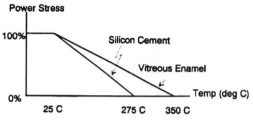

FIGURE 16.20   Resistor power versus temperature.

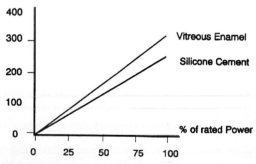

FIGURE 16.21   Temperature rise of resistors.

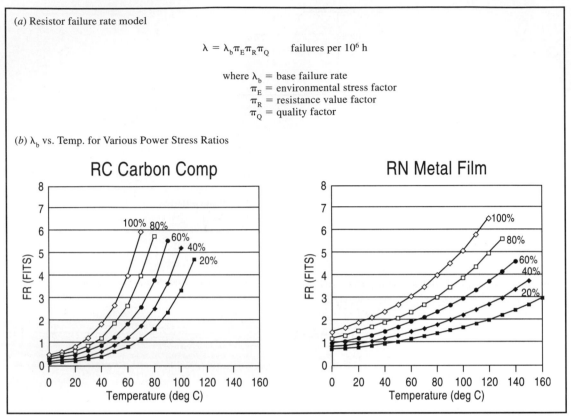

(a) Resistor failure rate model

$$\lambda = \lambda_b \pi_E \pi_R \pi_Q \qquad \text{failures per } 10^6 \text{ h}$$

where λ_b = base failure rate
π_E = environmental stress factor
π_R = resistance value factor
π_Q = quality factor

(b) λ_b vs. Temp. for Various Power Stress Ratios

FIGURE 16.22   Resistor failure rates from MIL-HDBK 217F, Notice 1.

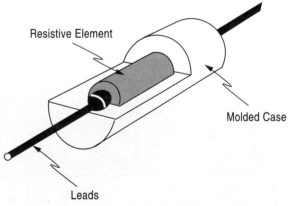

FIGURE 16.23   Carbon composition resistor construction.

TABLE 16.9 Resistance Changes (in Percent, %) in Resistors Due to Life Testing

Stress test	Wirewound	Carbon composition and film	Metal film, metal oxide, Cermet
Load life	2	10	1.3
High temp.	1	5	1
Low temp.	1	2	.4
Moisture resistance	1	10	1

Metal Film, Metal Oxide, and Carbon Film Resistors. Film resistors are constructed on a glass or ceramic substrate (see Fig. 16.24). A resistive film with a specific resistivity and thickness is applied to the substrate. The resistive film can be metal, metal oxide, or carbon particles suspended in a paintlike solution. The film can be applied by spraying, dipping, evaporation, or sputtering. Then a conductive metal coating is applied to the ends. At this point the resistor could be a MELF (*metal electrode face*) -type surface mount. If it is to be a leaded resistor, an end cap with a lead welded on is then soldered or welded to the conductive metal end coating. To obtain the final resistor value, a spiral cut is made in the resistive film coating. The effect of this is to increase the length and decrease the cross-sectional area, thereby increasing the resistance value. The whole resistor assembly is then coated with a phenolic or epoxy coating to provide insulation resistance.

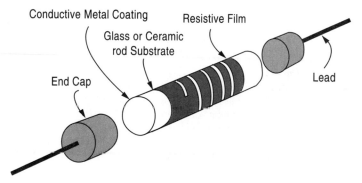

FIGURE 16.24 Film resistor construction.

The spiraling process creates an inductance that is in series with the resistance. This can be a problem for low values (generally $<100 \ \Omega$) at high frequencies.

Film resistors are capable of having a tight tolerance; a low temperature coefficient; and good stability over time, temperature, and humidity. In these respects the metals are better than the carbon films. Film resistors are susceptible to mechanical and thermal stresses which can crack the substrate. They generally are more reliable than carbon composition resistors when handled carefully.

Film debris in the spiral cut can create a resistance change, particularly over temperature or in humid environments. Some metal films are susceptible to corrosion. If the end caps and leads are not properly nickel-plated, they can corrode, also. Metal-based resistors generate less electrical noise than do the particle-based resistors such as carbon-composition and carbon film resistors. Film resistors are susceptible to electrostatic discharges and should be handled accordingly.

The primary failure modes seen with film resistors are hot-spot breakdown, cracked substrates, and peeling of the conductive coating from the resistive element. Leaded resistors can short-circuit to a trace if the overcoating is thin. The lead connection to the end cap can corrode if contaminants get trapped at the lead connection and the plating is not good.

Surface-Mount Film Resistors.   *Surface-mount resistors* are film resistors and have characteristics similar to those of the films. The construction of a typical surface-mount resistor is shown in Fig. 16.25. Instead of a spiral cut, a zigzag cut is made with a laser to obtain the various values. There is no inductance associated with the zigzag cut.

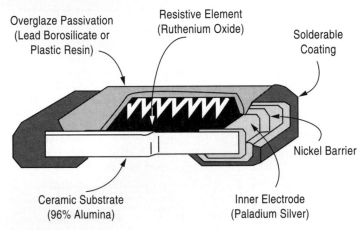

FIGURE 16.25   Chip resistor construction.

The resistors can be purchased as shown in Fig. 16.25 or without the solderable coating or without either the solderable coating and the nickel barrier. If the resistor is not purchased as shown, there will usually be solderability problems due to oxidation of the palladium-silver inner electrode or the nickel corrosion barrier.

The PCB layout is critical to prevent cracking or tombstoning during the cooldown cycle after reflow. PCB pads should allow even cooling of both ends during this cycle. Both pads should be the same size and plated through holes should be away from the pads. Even if these rules are observed, too rapid cooling can cause cracking of the resistor body or solder fillet or tombstoning because the PCB will contract more than the resistor. These are also issues for surface-mount ceramic capacitors.

Wirewound Resistors.   *Wirewound resistors* are usually designed for power, but precision wirewound resistors are also available.

The standard power wirewound resistor is shown in Fig. 16.26a. A resistance wire is wound around a high-thermal-conductivity cylindrical form, and the wires are welded to the leads. The form can be cardboard, aluminum oxide ceramic, beryllium oxide, or other material. The primary heat-flow path is out through the leads and into the mounting (for all resistors), although significant heat also flows off the surface of the resistor.

The common failure modes include hot-spot breakdown, separation of the welded wire ends from the leads, and open circuits caused by stretching the wire too tightly over the form, causing the wire cross-sectional area to be too small to handle the current flow.

The standard wirewound has a high inductance. This, coupled with the capacitance between adjacent wires, severely limits their high-frequency performance. There are two ways to limit the

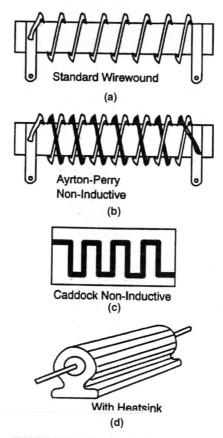

Standard Wirewound

(a)

**Ayrton-Perry
Non-Inductive**

(b)

Caddock Non-Inductive

(c)

With Heatsink

(d)

FIGURE 16.26   Wirewound resistor construction.

inductance. Figure 16.26*b* shows the Ayrton-Perry method of winding where two wires are wrapped in opposite directions so that the magnetic flux cancels out, producing a noninductive wirewound. Figure 16.26*c* shows the patented method used by Caddock. It consists of a metal clad plastic sheet etched with the pattern shown. The sheet is then wrapped around a cylindrical form. Current then flows first in one direction, then the other, allowing the magnetic fields to cancel.

Chassis-mounted resistors, such as the style shown in Fig. 16.26*d,* are usually rated for power only when mounted on a large heatsink with thermally conductive grease between the resistor and the heatsink. When operated in free air or mounted on a PCB, they may be good for only 40 percent of their rated power.

Use of tapped power resistors should be avoided. If the tap is a wire welded to the windings, the weld may weaken the wire. If the tap is a sliding type, it can cause wire wear, or particles may become trapped, causing an open circuit. Also the sliding tap is actually a connector, which is susceptible to film buildup or galvanic corrosion.

Variable Resistors.   The number 1 reliability rule for potentiometers is *do not use them.* If they must be used, be sure that the wipers can be turned occasionally to remove the films which will inevitably build up or which are deposited during the cleaning cycle.

Variable-resistor elements can be carbon-based, wirewound, or cermet. The carbon elements will wear and shed particles. Wirewound elements are noisy when the wiper jumps from one wire to the next. Cermet elements are the most reliable and consist of a metalized ceramic glaze.

16.4.4 Capacitors

A *capacitor* is a device which stores charge when a voltage is applied between two parallel plates on opposite sides of an insulating dielectric material. This is illustrated in Fig. 16.27. An electrolyte is necessary to carry current from the plate to the surface of the dielectric material. When the dielectric is in direct contact with the plate, there is no need for an electrolyte. The dielectric

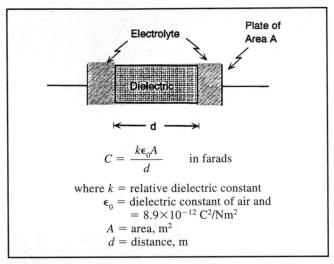

$$C = \frac{k\epsilon_0 A}{d} \qquad \text{in farads}$$

where k = relative dielectric constant
ϵ_0 = dielectric constant of air and
$\qquad = 8.9 \times 10^{-12}$ C^2/Nm^2
A = area, m^2
d = distance, m

FIGURE 16.27   Capacitance between two parallel plates.

material determines most of the properties of a capacitor. Table 16.10 lists the relative dielectric constants of the common materials used for capacitor dielectrics. These numbers will vary somewhat based on impurities.

The electrolyte, when present, affects the dissipation factor, the temperature range of operation, and the life expectancy. For all capacitors, with or without electrolyte, the life expectancy is also a function of the applied voltage. The wearout mechanisms related to capacitor life are due to drying or loss of the wet electrolytes and time-dependent breakdown of the dielectric material.

TABLE 16.10   Dielectric Constants of Common Materials

Material	k
Dry air	1.0
Window glass	4.2–8.0
Mica	2.5–6.0
Nylon	3.4–22.4
Paper	2.0
Polystyrene	2.5
Impregnated paper	2.0–2.6
Teflon	2.0
Mylar	3.0
Polypropylene	2.5
Aluminum oxide	7.0
Tantalum pentoxide	26
Ceramic	20–12,000

Figure 16.28*a* shows the typical effects of voltage on the life of a metalized polyester capacitor, and Fig. 16.28*b* shows the typical effects of temperature on the life of an aluminum electrolytic. In both cases, *life* is defined as the capacitors drifting out of specification for value, dissipation factor, or leakage current.

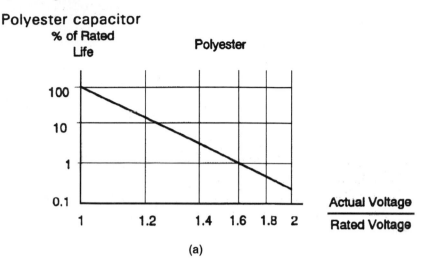

Polyester capacitor

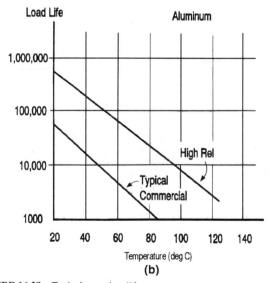

Aluminum capacitor

FIGURE 16.28   Typical capacitor life curves.

Dissipation factor is a measurement of the DC and AC losses in a capacitor. A small dissipation factor means that the capacitor is very close to an ideal capacitor and has very low impedance. The equivalent-series resistance is derived from the dissipation factor, and both are a function of temperature and frequency.

Table 16.11 shows the general failure modes associated with capacitors. Figure 16.29 shows capacitor failure rates from MIL-HDBK-217F, Notice 1 for a CK-style ceramic capacitor and a solid tantalum capacitor as a function of temperature and voltage stress ratio.

TABLE 16.11 General Failure Modes for Capacitors

Problem cause or mechanism	Description	Method of detection	Method of minimizing defect
Metal migration	Aluminum migrating into plastic film's pinhole defects for metalized plastic; usually self-healing, cracks in ceramic allowing plate metal migration creating high-resistance short circuits; sufficient energy or voltage in the charging circuit will clear the short; dendrite growth in silver mica	Random voltage spikes for self-healing plastic metal film; high leakage current for other short circuits	Use metal film plastics where spike does not affect circuit; for ceramics, process control, soldering control, shock, vibration, and flexing control; for silver mica, control temperature, humidity, and voltage
Dielectric breakdown	Fast charge or discharge of tantalums causing breakdown of oxide, momentary overheating and short-circuiting (will self-heal with high series impedance); reverse voltage for polarized aluminum and tantalums; deterioration of aluminum oxide dielectric with storage life, then applying power; damaged seal allowing water to get in and dissolve the dielectric layer for aluminum	High leakage current	Limit tantalum charge current; use nonpolarized aluminum and tantalums if reverse voltages are possible; recondition aluminum that have been stored for long periods; process and handling control for aqueous cleaning
Electrolyte loss	Loss of electrolyte caused by dissolving rubber seals with solvents, high temperature, or low pressure; loss of wet electrolyte due to damaged seals	High ESR	Verify compatibility of solvents; minimize handling damage to seals
Overheating	Too much power plus ambient temperature; improper time–temperature profile for surface mount (especially aluminum and plastic)	High ESR	Derate temperature and AC power; process control for soldering process
Delamination of electrode pickups	Too much heat for surface mount; contamination when electrode was laminated	Impedance test or visual inspection	Process control for soldering process, especially rework; process control at supplier
Damaged package	Lead corrosion due to trapping contaminants from soldering process or atmosphere; breaking glass to metal seals for ceramics and tantalums; wet slug tantalums can leak acidic electrolyte	Open circuit or high-resistance contact; visual inspection of seals	Use clean cleaning solutions; control lead-bending operations; avoid wet tantalums
Parameter changes	Aluminum and Class II ceramics have large temperature coefficients and large end-of-life drifts on most parameters	Capacitance vs. temperature test	Derate
Voltage	Class II ceramics have a large voltage coefficient	Capacitance vs. voltage test	Derate
Humidity	Causes capacitance drift in plastics	Humidity test	Derate
Shock or vibration	Loose welds of pickup tabs for low-ESR aluminum and foil tantalums increase ESR; causes cracking of ceramic- and glass-based parts; breakage of other internal and external connections; the larger the part (for its type) the worse the damage	Open circuit usually; short circuit if ceramic crack allowed metal migration	Process control; shock, vibration, and flexing controls

(*a*) Failure rate model

$$\lambda_P = \lambda_b \pi_E \pi_{CV} \pi_{SR} \pi_Q \pi_C \text{ per } 10^6 \text{ h}$$

where $\lambda_b = f$ (temperature and voltage stress)
π_E = environmental factor
π_{CV} = capacitance value factor
π_{SR} = a factor for solid tantalums only
π_C = construction factor for nonsolid tantalums
π_Q = quality screening factor

(*b*) λ_b vs. temperature for various voltage stress ratios.

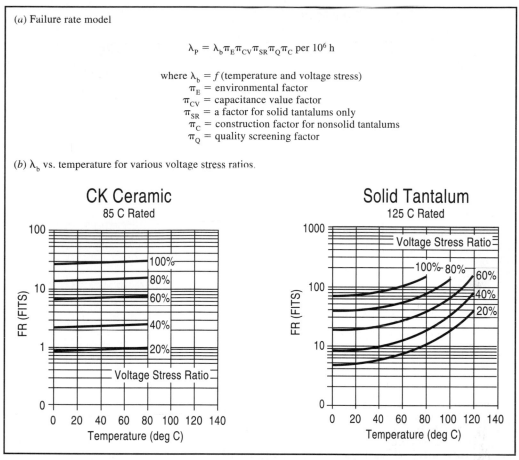

FIGURE 16.29 Capacitor failure-rate curves from MIL-HDBK 217F, Notice 1.

Ceramic Capacitors. A *ceramic capacitor* (see Fig. 16.30) consists of thin layers of a ceramic dielectric sandwiched between electrode plates. The plates are staggered to allow attachment of the electrode pickups after firing. The electrode pickup is then nickel plated to provide a corrosion barrier. For a surface-mount capacitor, a solder plating is applied to the nickel. For a leaded capacitor, the lead is welded or soldered to the pickup. Then the capacitor is dipped or molded with a phenolic or epoxy coating.

Ceramic capacitors are manufactured in three basic types of dielectric material, generally referred to as *low-, medium-,* and *high-dielectric-constant* types. The low-dielectric constant capacitors are the most stable with time, temperature, voltage, and frequency. The high-dielectric-constant capacitors can vary greatly with these stresses.

A common error made in selecting ceramic capacitors is to confuse the capacitance tolerance with the temperature coefficient. A capacitor may have a +80, −20 percent tolerance and a +20, −80 percent temperature coefficient. It is important to be clear whether the tolerance is specified over the whole temperature range or just at one temperature. The voltage coefficient is seldom specified, and can range from zero for the low-dielectric-constant capacitors to as much as −80 percent for the high-dielectric-constant capacitors.

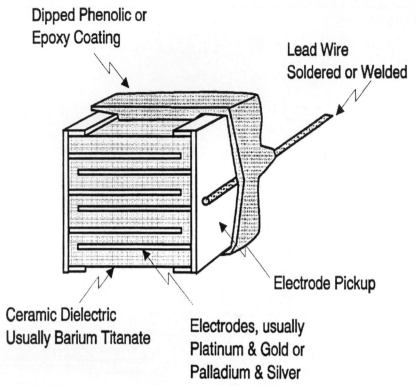

Dipped Phenolic or
Epoxy Coating

Lead Wire
Soldered or Welded

Electrode Pickup

Ceramic Dielectric
Usually Barium Titanate

Electrodes, usually
Platinum & Gold or
Palladium & Silver

FIGURE 16.30   Construction of the multilayer ceramic capacitor

The major failure mode associated with multilayer ceramic capacitors is related to voids and cracks in the dielectric which may be caused by an initial manufacturing defect or handling shock or, for surface-mounted capacitors, by flexing the PCB or cooling the board too quickly after soldering. When a path is opened up between the two plates and a voltage is applied, plate metal will migrate through the crack and cause a high-resistance short circuit. If there is enough current available from the applied voltage source, the metal will vaporize and clear the short circuit. If insufficient energy is available to clear the short circuit, the leakage current will continue to increase until the circuit fails.

Plastic Capacitors.   Plastic capacitors have replaced paper dielectric capacitors in most applications where paper used to be used.

The metalized plastic capacitor (see Fig. 16.31) consists of a sheet of very pure plastic film. Aluminum is sputtered or deposited on both surfaces to form the plates. Then leads are attached to contact the outside circuitry.

Even though the plastic is very pure, it still has pinholes which fill with aluminum during the metal deposition process. The capacitor may be short-circuited when it is new, in which case, when a voltage is applied, and if a sufficiently high current flows, the metal will vaporize and clear the short circuit. This process is called *clearing* or *self-healing*. Consider the case where the pinhole is not short-circuited when the capacitor is new. When a voltage is applied, a high electric field causes a dielectric breakdown of the remaining plastic. Metal again migrates through the hole and creates a short circuit. Current then flows and clears the short circuit. This causes a voltage spike which can be between 0.1 and 1 V in amplitude. This spike can create unwanted effects in circuits connected to the capacitor. Therefore, metalized plastic capacitors should not be used

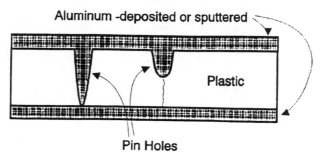

FIGURE 16.31   Construction of the metalized film plastic capacitor.

in circuits that are sensitive to transient voltage spikes. This is not a problem with the foil versions of plastic capacitors because the metal does not fill the pinholes and the electric fields are not high enough at normal plate separation to allow metal migration. This problem can never be adequately screened out even with burnin at elevated voltage because there will always be some pinholes left.

Because there is no need for an electrolyte (the plates are in direct contact with the dielectric), these capacitors can carry high-AC currents. Here, the temperature rise caused by dissipating power across the equivalent-series resistance is limited by the temperature limits of the plastic.

Every type of plastic known has been tried as a dielectric for these capacitors. Different plastics have different characteristics. The most common is the polyester capacitor because it is the least expensive and fairly reliable. Be sure to select the right capacitor for the job.

Most catastrophic failures of plastic capacitors are caused by overheating. This is particularly an issue for surface-mount devices, where the whole device sees the soldering temperature, versus through-hole devices, where only the leads see the solder temperature.

Aluminum Electrolytic Capacitors.   Aluminum capacitors provide the largest capacitance for their size of any of the capacitor types. They are generally not as reliable as the solid tantalum capacitors, but when used properly provide adequate reliability for most applications.

The aluminum capacitor starts with a large-area sheet of very pure (≥99.99% or better) aluminum foil. One size is then oxidized to form a very thin dielectric layer of aluminum oxide (see Fig. 16.32c). Because the foil can never be perfectly pure, no oxide forms over impurities. These areas allow leakage current to flow. Aluminum capacitors have relatively high levels of leakage current which is highly temperature-dependent. A porous-paper separator is then impregnated with a wet electrolyte and sandwiched between the two sheets of aluminum foil. The separator is needed so that the two foils do not rub off the very thin layer of aluminum oxide. The electrolyte is needed to carry current from the unoxidized plate to the surface of the aluminum oxide dielectric. The assembly is then rolled up, an anode tab and a cathode tab are then welded or stitched to the plates (Fig. 16.32a), and the whole thing is sealed in a metal case (Fig. 16.32b).

The case needs to be sealed to keep moisture out and electrolyte in. If moisture gets in, it will dissolve the aluminum oxide into the electrolyte, allowing large currents to flow. This will boil the electrolyte, causing the capacitor to explode. A weak spot in the header is provided to force the electrolyte out the header rather than explode the metal case.

Some capacitors have a venting rubber seal which will give way when the internal pressure builds up too high because of overheating. Most of the carbon-based rubbers and plastics will degrade to one degree or another by various cleaning solvents used in PCB cleaning. The halogenated hydrocarbon cleaning solutions are particularly bad in this respect, but all cleaning operation may have problems. Each supplier of aluminum electrolytics publishes a list of which solvents are safe to use with the rubber vents. It is very important that the list of acceptable solvents for all plastic and rubber parts be developed for those suppliers on the approved-source list. If any electrolyte leaks out, the capacitor life is severely shortened. If moisture is forced in during high-pressure aqueous cleaning, the life will also be shortened.

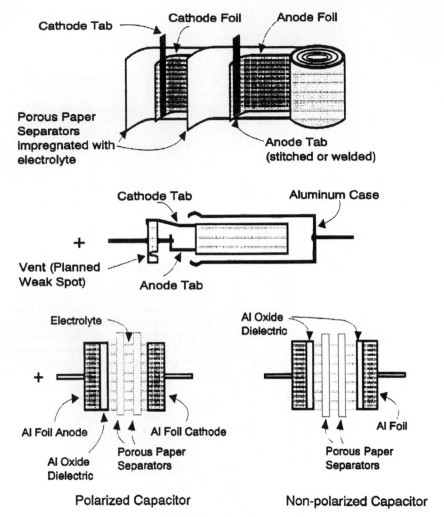

FIGURE 16.32 Construction of the aluminum electrolytic capacitor.

This is a polarized capacitor, and a reverse voltage will also dissolve the oxide back into the electrolyte and result in overheating. If no voltage is applied, the oxide will again dissolve into the electrolyte, particularly when stored at high temperatures. A capacitor which has been stored for a long period will exhibit an increased leakage current. These parts can be reconditioned by applying the proper polarity of voltage with current limiting to regrow the oxide until the leakage current is back within specification.

A nonpolarized aluminum capacitor is built by oxidizing both sheets of aluminum foil. Here there is always an insulating layer of dielectric to limit DC current flow. However, if the voltage polarity is only rarely changed and mainly in one direction only, the capacitor may revert to being a polarized capacitor. For this application a nonpolarized solid tantalum capacitor may be the better choice.

A low-equivalent-series resistance (low-ESR) capacitor is built by attaching many anode tabs and many cathode tabs so that, when it is rolled up, there is a bundle of anode tabs and a bundle of cathode tabs. This reduces the resistance of the large plates. A common failure mode associated with this part type is when the spot welds come loose because of poor process control at the manufacturer, resulting in paying for a premium capacitor but getting an ordinary capacitor.

Tantalum Capacitors.   Tantalum capacitors are available in wet and dry electrolyte and foil and slug versions. Tantalum capacitors have generally more stable characteristics over time, temperature, and voltage and are more reliable than are aluminum capacitors. The dry-slug tantalum is the most stable and the most reliable of all the tantalums. The wet-foil tantalum can be obtained in higher voltage levels than the other tantalums, but it is also the least reliable. These capacitors do not have a storage-life problem. They are available in both polarized and nonpolarized versions.

   The dry-slug tantalum capacitor is pictured in Fig. 16.33. The tantalum pellet is porous and has a relatively large surface area which is oxidized with a thin layer of tantalum pentoxide for the dielectric. The rest of the layers are shown. The coated pellet is then encased in a tin brass case. A tantalum wire is embedded in the pellet and is welded to a tinned nickel wire which can withstand soldering temperatures.

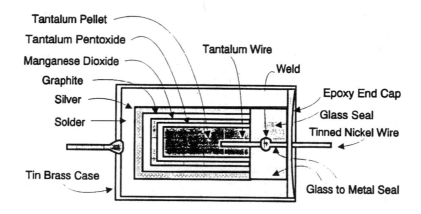

Layers (from inside to outside): Ta, Ta_2O_5, MnO_2 (cathode), graphite, silver, solder, case.

FIGURE 16.33   Construction of the tantalum capacitor.

   Reverse voltages will break down the tantalum pentoxide, creating a short-circuited capacitor, which will usually explode with a loud bang. A nonpolarized tantalum is made by connecting two polarized capacitors back-to-back and packaging them in the same case.

   Tantalum capacitors cannot withstand high surge currents so must be current limited with an external resistor for charging and discharging if the source impedance is less than one ohm. What happens is that the tantalum pentoxide breaks down and a local area overheats. If the capacitor is not permanently damaged or short-circuited, the heat converts the manganese dioxide to manganese oxide, which is not as electrically conductive. So the capacitor will self-heal but with a higher leakage current than before. However, the life will be degraded. There is some evidence that the lower the leakage current, the longer the life of solid tantalums.

   A wet-slug tantalum is similar to the dry-slug tantalum except the pellet is held with a plastic boot and a wet electrolyte replaces the solder.

   Foil tantalums are built similar to the aluminum electrolytic except with tantalum foil electrodes.

16.4.5   Connectors

Connectors usually cause more failures than any other type of components. Many of these failures are not reported because they can be "fixed" by reseating the connector. Table 16.12 lists the general failure modes for connectors.

The single greatest cause of connector failure from a design standpoint is the defeating of or loss of connector float. *Connector float* is the ability of the connector pin to self-center in the socket. If the pin binds against the socket, loss of contact area results and the connection can overheat and open up.

The greatest cause of connector failure for tin-plated connectors is fretting corrosion caused by micromovements of the contact interface, which allows tin oxide deposits to build up and eventually prevent current flow. This can be minimized by the use of contact lubricants, but these can also attract dust and dirt. It is important that the contact pair have a high enough normal force, a large enough contact area, and a wiping action to break through the films.

Most other connector failures are caused by mishandling either during assembly or in the field. In order to obtain reliable connections, the proper tooling must be used in the assembly process.

16.4.6   Inductors, Transformers, and Other Wound Components

Inductors, transformers, and other wound components have many failure modes in common relating to the windings. Winding failures usually occur because of abrasion in the winding process which causes short-circuiting to the coil form, if metal, or to other windings. The other major class of failures occurs because of overheating in the windings, which breaks down the insulation, causing short circuits. All wound components have a thermal resistance from windings to ambient which allow a calculation to be made using the equation given earlier (in Fig. 16.13). It is important that the winding temperature be less than the temperature rating of the winding insulation.

Motors also have a thermal resistance from bearings to ambient, and it is important that the temperature of the bearings does not cause loss of lubricant. If this happens, the motor will pull more current to try to overcome the increased frictional load. This increases the power and the temperature rise until the windings burn up.

One of the major problems with these components is lack of supplier process control. Many suppliers are small and do not have the resources to properly characterize and qualify their parts and processes. Table 16.13 lists the failure modes common to all wound components.

16.4.7   Batteries

The most common failure mode for batteries is electrolyte leakage due to either handling damage to the case, overheating from too rapid charging, or excessively high ambient temperature. However, a large number of batteries will simply corrode the casing with time and normal temperatures. For this reason, batteries should not be soldered into a PCB, so that they can be replaced before they leak and damage the board. Table 16.14 lists the general failure modes seen with batteries.

16.4.8   Switches and Relays

Switches and relays have many failure modes in common with connector contacts. Relay coils are wound components and have many of the failure modes shown in Table 16.13. Table 16.15 summarizes the failure modes seen for these types of components.

16.4.9   Crystals and Oscillators

The most common failure modes seen with crystals and oscillators occur when the seals are broken and contaminants corrode the crystal connections. Unfortunately, there are many dented and damaged crystals available from distributors.

TABLE 16.12 General Failure Modes for Connectors

Problem cause or mechanism	Description	Method of detection	Method of minimizing defect
Fretting corrosion	Oxide and film buildup at connection point due to small movements of the contacts; causes high resistance or open circuit, particularly for currents <100 μA and voltages <10 V. Condition accelerated by temperature and humidity	Intermittent failure; fixed by unplugging and reconnecting; usually, debris visible with microscope	Minimize environmental vibrations; investigate use of lubricants or inhibitors; ensure plating metals do not readily oxidize
Galvanic corrosion	A galvanic potential at the interface between two dissimilar metals hastens the growth of intermetallic formations which lead to high resistance or open circuit	Visual inspection; "purple plague"-type growths	Use similar metals on both sides of the contact
Los of contact float	Loss of the ability of the pin to self-center in the spring contact; causes loss of contact area, overheating and disconnecting of the connection	Intermittent failure will occur more often at higher temperatures; sometimes charring or discoloration of the housing	Ensure proper connector mounting and cable harnessing; ensure PCB warpage and mounting method; maintain and/or create float
Corrosion	Longer-term failure due to corroded base metals hastened by contamination from improper cleaning or the environment	Visual inspection	Ensure that ionic or corrosives are cleaned from the connector; design for the end-use environment
Plating peeling on contacts	Peeling or flaking of nickel plate over base material; usually caused by contamination during plating or overheating	Visual inspection	Supplier process control
Poor crimp or solder connection	Loss of contact caused by poor crimp or stress on the crimp; common failure with non-strain-relieved insulation-displacement connectors	Intermittent failure may go away when cable is gently pulled	Use proper crimp and other tooling; use strain reliefs on all connections
Contact wear	Wearing of either lubricant layer or corrosion protection layer due to too many insertions or withdrawals; contact rides on high spots and loses contact area and allows films to build up	Visual inspection of contact area for wear	Derate insertions or withdrawals
No contact wipe	Connection forces are not strong enough to break through the films; open circuit for low currents and voltages	Visual inspection for wear marks at connection interface	Proper connector design must include wipe
Loss of normal force	The force pressing the spring against the pin maintains the gastight connection and provides a lower resistance based on surface area; normal force can be lost if a connector overheats, is not plugged in all the way, after too many insertion withdrawals, if the housing is cracked, warped, or cold flows, or if the spring or pin is damaged	Visual inspection; very difficult to measure normal force for most connectors; loss of insertion or withdrawal force is indicator	Use latching connectors; derate temperature and insertion/withdrawal cycles; maintain contact float; observe "delicate handling" rules for kitting, inspection, assembly, and test
Overheated contact	Caused by too much current plus ambient temperature or loss of connection area caused by loss of normal force due to movement of the plastic housing	Charred or discolored housing; temperature measurements on the housing	Derate temperature and contact current; maintain contact float relationship

TABLE 16.13 General Failure Modes for Inductors, Coils, Transformers, and Motor and Solenoid Windings

Problem cause or mechanism	Description	Method of detection	Method of minimizing defect
Thermal breakdown of insulation	Winding insulation breakdown caused by exceeding the temperature rating of the insulation due to a combination of internal heating and external ambient; internal heating can be generated by ohmic losses in the windings and/or core losses, particularly if the core is saturated	Burned or discolored case; open circuit or low winding resistance	Derate power and temperature; controlled motor and transformer loads
Broken winding	Open circuit caused by broken winding wire or wire that was kinked or nicked during the winding process and then overheated; poor bonding of the winding to the lead	Open-circuit winding	Process control
Breakdown of insulation	Dielectric breakdown of insulation caused by too much voltage or thin spots in the insulation; insulation can be abraded off during winding of toroidal windings	Low breakdown voltage	Derate voltage; process control
Mechanical damage	Cracked cores, housings, and terminations due to mishandling, shock, or vibration; core delamination due to poor construction, temperature cycling, or overheating	Visual inspection	Process and handling control
Chemical damage	Corrosion of metal parts and deterioration of potting compounds and insulation due to trapped contamination from the manufacturing process or the environment	Visual inspection	Process and cleaning control; environmental control

TABLE 16.14 General Failure Modes for Batteries

Problem cause or mechanism	Description	Method of detection	Method of minimizing defect
Electrolyte leakage	Loss of electrolyte due to broken or ruptured seals	Visual inspection	Replace batteries regularly; charge properly
Film buildup on terminations	Loss of battery connection due to oxide formation or corrosives from the environment	Open circuit; usually fixed by replacing battery in socket	Cleanliness
Mechanical damage	Broken terminations; damaged case; cracks in solid electrolytes; rupture of case due to gas buildup caused by discharging or charging too quickly	External or internal visual inspection	Handling control
Electrode and electrolyte depletion	Low or no voltage due to depletion of the active material in the battery caused by normal usage or self-discharge due to temperature-dependent leakage currents; can be reversed a limited number of times in rechargeable batteries by recharging	Low voltage	Limit temperature, number of charge cycles, and usage
Overheating	Discharging or charging too rapidly causes overheating and loss of electrolyte and rupture; high storage temperatures discharge the battery prematurely	Visual inspection	Proper charging and storage

TABLE 16.15 General Failure Modes for Switches and Relays

Problem cause or mechanism	Description	Method of detection	Method of minimizing defect
Solvent damage to plastic housings	Softening or dissolving of plastic pieces due to cleaning solvents of contamination	Visual inspection; measuring contaminant and polymer levels in cleaning solutions	Handling control; verify solvent compatibility with components
Contact contamination	Film buildup on contacts of unsealed components caused usually by cleaning solutions or particles getting into the contact area	Works after several switching cycles	Use sealed components; hand-insert and clean
Fretting corrosion	Oxide or film buildup on contacts due to vibration and oxidation inhibits the contact's ability to carry low currents	Works after several switching cycles	Minimize vibration and corrosive gasses; select proper contact material
Contacts pitted, burned, or welded closed	Contact overheating usually caused by arcing; can occur with hot-switching or inductive loads; lamps and power supplies have high inrush currents; the switch should actuate fast enough to minimize arcing; high-altitude operation allows arcing at smaller contact separations	Visual inspection of contacts	Derate contact currents; choose proper material for current levels and load; derate cycle life; use arc suppressor
Contact wear	Wear of the contact plating due to exceeding the specified cycle life	Visual inspection of contacts	Derate cycle life
Actuator wear	Failure of the actuating mechanism due to loss of lubricant at high temperatures or exceeding the cycle life	Failure analysis	Derate temperature; derate cycle life

TABLE 16.16 General Failure Modes for Surface-Mount Assemblies

Problem cause or mechanism	Description	Method of detection	Method of minimizing defect
Moisture in plastic packages	Moisture in plastic-encapsulated semiconductors and other parts boils during assembly and delaminates internal structures or explodes the package	Visual inspection and X ray	Dry out parts before soldering; handle parts in humidity-controlled environments
Compression fracture	Chipping of ceramic capacitors and chip resistors due to cooling the board too fast after reflow; can also fracture along an existing microcrack.	Visual inspection; electrical test (may not catch it); high leakage current in capacitors	Control soldering temperature profile
Cracked solder fillet	Cracking of solder connection due to cooling the PCB too rapidly	Visual inspection; electrical test (may not catch it)	Control soldering temperature profile
Peeling terminations	Separation of end-cap terminations due to too much heat (usually during rework), or poor plating during component manufacture	Electrical test (may not catch it); visual inspection	Solderability test of lot sample, control of soldering, and rework temperatures
Metal migration	Dendrite growth between fine line traces and component leads due to temperature, humidity, and voltage; high-resistance short circuits in ceramic capacitors along microcracks created during initial manufacture or by handling shock or flexing the finished board	Electrical test; leakage current tests on capacitors	Clean boards after soldering; control of handling and packaging of finished board
Corrosion	Corrosion of leads and traces due to ionic contaminants, finger grease, or poor plating of component leads and traces	Visual inspection	Control of handling; clean boards
Overheating during assembly	Damage to components caused by excessive temperatures during preheat, reflow, and/or cooldown; may cause field failures	Electrical test (may not catch it)	Temperature control during soldering
High temperature rise	Small surface-mount packages have higher thermal resistance than DIPs, causing the parts to run hotter and have a higher failure rate than equivalent DIP parts	Thermal scan during development	Derate temperature; worst-case analysis

16.59

A *crystal oscillator* is actually a hybrid with an integrated circuit in the same package as the crystal. Many hybrid suppliers are small and do not have the process controls to guarantee good reliability. Be sure to properly evaluate these suppliers.

16.4.10   PCB Layout, Fabrication, and Assembly

Printed-circuit boards have become more important as surface mounting has become more widespread and fine-pitch packages have become more widely available. Close traces invite solder bridging, which can damage components if not removed before power is applied. Controlled impedances are necessary for today's fast-rise-time integrated circuits, and careful attention needs to be paid to the operating temperature of components which are packed as close together as possible, raising the ambient temperature around the components due to secondary heating effects. Proper aspect ratios must be observed for plated through-holes to ensure their reliability because they are not filled with solder as they are in a wave-soldered board. Traces and pads should be as wide as possible to conduct heat away from components.

When the board is fabricated, contaminants must be removed prior to applying the solder mask to prevent them from corroding components and traces. Solderable surfaces should not be gold-plated, to avoid embrittlement of the connection.

Assembly needs to be done carefully (as always) and in a controlled manner. Table 16.16 lists some of the common failure modes seen with surface-mount PCBs. Boards should be cleaned to remove contaminants which can cause corrosion and leakage paths.

Boards should be mounted so that they are not flexed, particularly during service and repair.

16.5   SUMMARY

To obtain reliable equipment, it is necessary to pay attention to detail in the design, manufacturing, and field-use phases of a product life cycle. A formal reliability program helps focus the tasks and ensures that the right things get done at the right time.

In design, the components must be selected carefully, and their reliability must be determined to be consistent with the product's reliability, functionality, and cost goals. In manufacturing, a process needs to be used which does not damage the components and allows unreliable parts to be identified and removed from the product before they get into the customer's site. A field reliability measurement program is essential to measure how well both the design and manufacturing efforts are doing and to measure the actual field environments in which the product must function.

Only when these things are attended to properly can a company's products meet the demands of several times reliability improvement required by today's customers.

16.6   BIBLIOGRAPHY

1. Anderson, R. T., *Reliability Design Handbook,* IIT Research Institute, 1975.

2. Adler, R. B.; Smith, A. C.; and Longini, R. L., *Introduction to Semiconductor Physics,* Wiley, New York, 1964.

3. Arcidy, M., "Maximizing Chip Resistor Performance and Reliability for Surface Mount Applications," *Evaluation Engineering,* Sept. 1986.

4. Baily, R. A.; and Gilbert, R. A., "STRIFE Testing," *Quality,* Nov. 1982.

5. Balaban, H. S., "Reliability Improvement," *Quality,* Nov. 1978.

6. *Bell Communications Research Reliability Manual,* Bellcore, SR-TSY-000385.

7. Biancomano, V., "Film Capacitors," *Electronic Design,* Nov. 13, 1986.

8. Bourns, Inc, Application Notes, *Best of the Trimmer Primers.*

9. Campbell, M., "Monitored Burn-in Improves VLSI IC Reliability," *Computer Design,* April 1985.

10. Clemente, S.; Lidow, D.; and Lidow, A., "Armed with a Battery of Data, Designers Can Reinforce MOS-FET Circuit Reliability," *Electronic Design,* June 6, 1985.

11. *Component Reliability Assurance Requirements for Telecommunications Equipment,* Bellcore, TR-TSY-000357.

12. *Component Reliability Analysis: General Findings and Trends,* Bellcore, SR-TSY-001630.

13. Costlow, T., "Electrolytic Capacitors," *Electronic Design,* Jan. 24, 1985.

14. Denton, D. L., "Finding the Failure Rate," *Quality,* April 1980.

15. Denton, D. L., "Quality and Reliability Screens Impact Cost of Owning IC's," *Electronic Design,* Nov. 20, 1986.

16. Doty, L., *Reliability for the Technologies,* ASQC Press, 1989.

17. Dummer, G.; and Winton, R., *Elementary Guide to Reliability,* Pergamon Press, New York, 1986.

18. Epland, Donald; and Liddane, Ken, "Selecting Capacitors Properly," *Electronic Design,* June 21, 1977.

19. *Failure Mode/Mechanism Distributions,* Reliability Analysis Center, Rome, New York, Air Development Command.

20. Farrell, John; and Fuqua, Norman, *Parts Selection, Application and Control,* Reliability Analysis Center, Rome, New York, Air Development Command.

21. Furlow, W., "High Reliability Systems (Special Report)," *Computer Design,* April 1985.

22. Giesse, A., "Influence of Aluminum Layer Thickness on the Self-Healing Properties of Metalized Polycarbonate Capacitors," *Insulation Circuits,* Nov. 1978.

23. Jones, T., *Electronic Component Handbook,* Reston Publishing, Reston, Va., 1978.

24. Kay, H., "Junction-Temperature Planning Boosts Product Reliability," *EDN,* April 18, 1985.

25. Klinger, David J.; Nakada, Yoshinao; and Menendez, Maria A., *AT&T Reliability Manual,* Van Nostrand Reinhold, New York, 1990.

26. Koppel, R., "RAM Reliability in Large Memory Systems—Significance of Predicting MTBF," *Computer Design,* Feb. 1979.

27. Krishnamoorthi, K. S., *Reliability Methods for Engineers,* ASQC Press, Milwaukee, Wis., 1992.

28. Laffin, J. P., *Guidelines to Semiconductor Thermal Management,* International Electronic Research Corp., Burbank, Calif.

29. Lowrie, R. W., "Relating Reliability Improvement Costs to Field Repair Costs," *Electronics Test,* Nov. 1983.

30. Mazda, F., *Discrete Electronic Components,* Cambridge University Press, Cambridge, U.K., 1981.

31. Michaels, I., "Resistor Networks," *Electronic Design*, Nov. 13, 1986.

32. MIL-HDBK-108, *Quality Control and Reliability.*

33. MIL-HDBK-217, *Reliability Prediction Procedure for Electrical and Electronic Systems.*

34. MIL-HDBK-978, *NASA Parts Application Handbook.*

35. MIL-STD-202, *Test Methods for Electronic and Electrical Components.*

36. MIL-STD-750, *Test Methods for Semiconductors.*

37. MIL-STD-883, *Test Methods for Microelectronics.*

38. Mullin, W., *ABC's of Capacitors,* Howard Sams, Indianapolis, 1978.

39. O'Connor, P., *Practical Reliability Engineering,* 2nd ed, Wiley, New York, 1985.

40. Oppenheimer, C. P., "Reliable Designs Begin with the Basics," *Computer Design,* Aug. 1983.

41. Prasad, Ray P., *Surface Mount Technology,* Van Nostrand Reinhold, New York, 1989.

42. Rauhe, M. L., *Thermal Control for High Density Packaging,* IBM Corp., Oswego, N.Y.

43. *Reliability Design Handbook,* Reliability Analysis Center, Rome Air Development Command, Griffiss AFB, New York (NY 13441-5700).

44. *The Reliability Handbook,* National Semiconductor Corp., Vol. 1, April 1982.

45. *Reliability Prediction Procedure for Electronic Equipment,* Bellcore, TR-TSY-000332.

46. Sokal, N. O., "Use of Check List Prevents Problems with TTL Systems," *EDN,* Nov. 13, 1986.

47. Sokal, N. O., "Check Lists Help You Avoid Trouble with MOS and Memory ICs," *EDN,* Nov. 27, 1986.

48. Taylor, M. A., "Consider How TTL Outputs Work during Power-downs," EDN, November 27, 1987.

49. Tersteeg, Douglas, "Reliability Demonstration Testing and Strife Testing—Is There a Correlation?," presented at the Power Electronics Conference, 1990.

50. Turner, T., "IC Failure Rate Calculations Evaluate Components Realistically," *EDN,* Apr. 15, 1981.

51. *Voltage Regulator Handbook,* National Semiconductor, Santa Clara, Calif.

52. "VRCI Checks—Impact of Wave Soldering on Reliability of Trimmers," *Evaluation Engineering,* May/June 1983.

53. Warring, R., *Electronic Components Handbook,* TAB Books, Blueridge Summit, Pa., 1983.

54. Watt, Fred, *Component Technology and Reliability Seminar Manual,* Sav-Soft Products, P.O. Box 360974, Milpitas, CA 95036, 1992.

55. Watt, Fred, *Reliability Prediction Methodologies Seminar Manual,* Sav-Soft Products, 1992.

56. Westerman, W., "More Reliable Plastic Film Capacitors," *Evaluation Engineering,* April 1983.

57. Winward, H., "Focus on Surface Mounted Electrolytic Capacitors," *Electronic Design,* Aug. 7, 1986.

CHAPTER 17
THERMAL MANAGEMENT AND RELIABILITY OF ELECTRONICS

Vivek Mansingh
FUJITSU COMPUTER PACKAGING TECHNOLOGY
SAN JOSE, CALIF.

17.1 STRUCTURE OF ELECTRONIC SYSTEMS

Semiconductor devices are used in almost all types of electronic systems these days. In common semiconductor devices, also called *electronic chips,* the desired pattern of microscopic electronic circuits are formed on a piece of semiconductor. These electronic chips are generally housed in packages which protect chips from the atmosphere, provide signal and power connections to the chips, as well as facilitate the dissipation of heat from the chips.

Several chip packages are mounted on a printed-circuit board (PCB), where layers of conductor networks are fabricated to connect these packages electrically. An electronic system is generally composed of a number of electronic chips mounted on PCBs, several PCBs connected by a mother PCB, a power supply, and accessories. Figure 17.1 shows the four main packaging levels of a typical electronic system.

17.1.1 Operating-Temperature Effects on Semiconductor Device Reliability

In semiconductor devices, the input power that is not converted to perform a useful electromagnetic function is dissipated in the form of heat, and the heat generated is inversely proportional to the efficiency of the circuit.

As the performance demands from the semiconductor devices and electronic chips have been increasing over the years, the number of circuits (transistors or diode emitter-base junctions) per chip has also been increasing steadily. Because of the increase in the number of circuits per chip and the frequency of operation of these chips, the heat dissipation from chips has also been increasing significantly (Table 17.1).

The heat dissipation from the semiconductor devices elevates the operating temperature of these devices. Unfortunately, most electronic devices are prone to failure at elevated temperatures, and the reliability of each device is affected by the device's operating temperature. Virtually all the failure mechanisms are enhanced by the increased temperature. Most common failures due to increased operating temperature result from:

1. Thermal-coefficient-of-expansion (TCE) mismatch in different materials of chip and package
2. Creep in the bonding materials
3. Corrosion

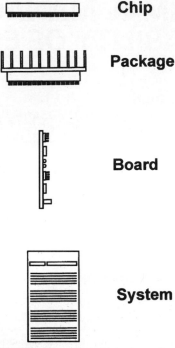

Chip

Package

Board

System

FIGURE 17.1 Typical structure of an electronic system.

TABLE 17.1 Chip Packaging Trends

Year	1992	1995	1998	2001	2004	2007
Feature size, μm	0.5	0.35	0.25	0.18	0.12	0.1
Number of transistors, millions	3	7	20	50	100	200
Chip frequency, MHz	120	200	350	500	750	1000
Chip power, W	15	20	30	40	120	200

4. Electromigration

5. Diffusion in the devices

Figure 17.2 shows the effect of operating temperature on the failure rates of some typical electronic devices. These curves are adopted from MIL-HDBK-217F,[1] and the figures are qualitative only. However, these figures clearly show a very strong dependence of reliability on the operating temperature. It has been shown in some studies that up to 75 percent of the failure in electronic devices can be temperature-related.[2]

17.1.2 Principles of Thermal Management

It is, therefore, extremely important for proper and reliable operation of electronic systems that the electronic devices that make up a system are maintained below their maximum allowable temperatures at the worst-case operating conditions. The aim of thermal management, therefore, is to properly control the operating temperatures of the electronic devices that make up the system by

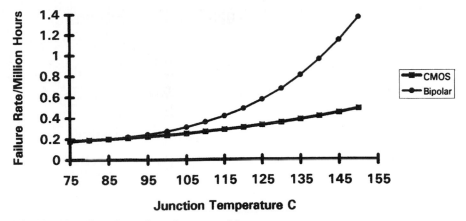

FIGURE 17.2   Effect of operating temperature on failure rate.

safe, efficient, and economical removal of heat. This goal can be achieved only by applying proper thermal management techniques at all of the four levels of packaging described earlier.

A common industry practice is to maintain average device temperatures (also called *junction temperatures*) below 100°C at worst operating conditions. However, in systems requiring very high reliability, junction temperatures, as low as 85°C may be desirable. This maximum allowable junction temperature is generally suggested by the manufacturer of the component on the basis of its power dissipation and reliability requirement. In an electronic system, the cooling technique and the system cooling configuration is based on the maximum allowable component temperatures, heat dissipation rates, and environmental specifications.

In the section that follows, a typical thermal design process and the fundamentals of heat transfer are reviewed. Analysis techniques for cooling of microelectronics systems, boards, and packages are then illustrated by examples. Experimental techniques used for thermal design are then presented in the next section. Then, some CAD tools used for the thermal design of electronic systems are briefly discussed.

17.2   THERMAL DESIGN PROCESS

A key to good thermal design practice is to start thermal analysis early in the system design process, preferably right from the conceptual stage of a product. A recommended thermal design process flow diagram is shown in Fig. 17.3. After the conceptual product design is agreed on, the best estimates of power dissipation at system, board, and package levels, along with the geometric parameters and environmental specifications, should be used to do a quick (back-of-the-envelope) analysis. The purpose of this analysis is to select a proper cooling technique and make sure that the conceptual design is feasible. Then, a more detailed analysis preferably aided by spreadsheet or simple network tools, should be performed. Once again, the purpose of this analysis is to further confirm that the conceptual design is feasible.

A simplified numerical analysis using CAD tools should then be carried out. The numerical analysis results will help with not only the sanity check of the design but also the design of experiments. While the experiments are being designed and performed, a more detailed numerical analysis should be carried out. After the experimental results are available, they should be compared with the numerical results, and the numerical model should be improved based on the input from the experimental results. Once a good numerical model is available, all the optimization and design improvement work should be done using the numerical model. After an optimized design is avail-

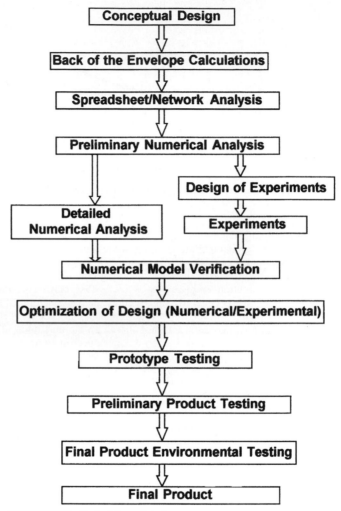

FIGURE 17.3 Typical thermal design process.

able, prototype experimental testing should be carried out, followed by in situ product testing. In the end, a thorough environmental testing should always be done before releasing the product.

17.3 HEAT-TRANSFER FUNDAMENTALS

When a temperature gradient exists in a medium, whether in a solid or in a fluid, heat transfer takes place from the hotter body to a cooler body. The heat-transfer processes can be categorized into three basic types of modes: conduction, convection, and radiation. However, a problem of practical interest generally involves all of the three modes of heat transfer taking place simultaneously.

17.3.1 Conduction Heat Transfer

The *conduction heat transfer* in solids involves the transfer of heat from one part of a body at a higher temperature to another part of a body at a lower temperature, or from one body at a higher temperature to another body at a lower temperature but in physical contact to each other.

The conduction-heat-transfer-rate equation is known as *Fourier's law.* In case of a solid plane wall, as shown in Fig. 17.4, it is given by

$$Q = \frac{-kA(T_2 - T_1)}{L} \tag{17.1}$$

where Q = heat flow, W
$T_2 - T_1$ = temperature difference (°C) across the conducting surface of thickness L, m
k = thermal conductivity of material, W/m°C
A = heat transfer area, m^2

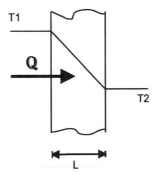

FIGURE 17.4 Heat conduction through solid plane wall.

Thermal conductivity is a property of the material. The thermal conductivity of a number of materials, some common liquids and gases, is given in Table 17.2. Generally, metals are better heat conductors than nonmetals, solids are better conductors than liquids, and liquids are better conductors than gases.

Following the analogy between heat flow and electrical current flow governed by Ohm's law, when the heat flow is steady and one-dimensional, the conductive thermal resistance can be defined as

$$R = \frac{T_1 - T_2}{Q} = \frac{L}{kA} \tag{17.2}$$

where R is thermal resistance, °C/W; and Q is heat flow, W.

A typical example of thermal conduction in electronic components is the heat flow from chip to the package, package to board, and package to heatsink.

17.3.2 Thermal Radiation

In conduction and convection heat transfer, a material medium is involved to facilitate the heat transfer. However, heat can also be transferred through a perfect vacuum. Energy transferred by electro-

TABLE 17.2 Thermal Conductivity of Electronics Packaging Materials at 26°C

Material	Conductivity, W/m°C	Material	Conductivity, W/m°C
Solids		Solids	
Aluminum, pure	216	Phosphor, bronze	52
Alumina (96% Al_2O_3)	29	Platinum	69
Alumina (90% Al_2O_3)	13	Polyimide	0.33–0.4
Aluminum 2024 T_4	121	Quartz	3.0–5.0
Aluminum 6061 T_6	155	RTV	0.31
Aluminum nitride	150–220	Sapphire	30
Beryllium	164	Silicon	84
Beryllia, 95% pure	155	Silicon carbide	220
Beryllia, 99% pure	242	Silicon dioxide	10
Brass, red	110	Silicone rubber	0.19
Brass, yellow	95	Silver	418
Copper	398	Solder (Pb-In)	36
Diamond	2000–2300	Solder 80-20 Au-Sn	52
Epoxy (conductive)	0.35–0.87	Stainless steel	15
Epoxy (dielectric)	0.23	Steel kovar	16.6
Epoxy glass (PCB)	0.24–0.3	Steel SAE 1020	55
Eutectic bond	68	Tantalum	50
Gallium arsenide	50	Teflon	0.25
Germanium	60	Titanium	22
Gold	297	Tungsten	170
Iron, cast	55		
Iron, pure	74	Liquids and gases	
Iron, wrought	58		
Lead	34	Water	0.6
Mica	0.5	FC77	0.065
Molybdenum	140	Air	0.028
Nickel	92	Helium	0.15
Nylon	0.24		

magnetic waves from a surface, with or without the presence of any medium, is by a mode called *thermal radiation.* Heat transfer by radiation is governed by the Stefan-Boltzmann law and is given by

$$Q = \sigma \varepsilon AF(T_1^4 - T_2^4) \tag{17.3}$$

where σ = Stefan-Boltzmann constant with the value of 5.669×10^{-8} W/m²K⁴
ε = emissivity of the radiating surface
A = surface area, m²
T_1 = radiating surface temperature, kelvins (K)
T_2 = surrounding surface temperature, K
F = view factor

Emissivity is the property of the radiating surface, and its value lies between 0 and 1. For electronic components and PCBs, except for shiny metal surfaces, ε is in the range of 0.8 to 0.9. The emissivity of a number of different surfaces is given in Refs. 3, 5, and 6.

As electromagnetic radiation travels in straight lines, not all radiation leaving a radiating surface reaches the receiving surface. The *view factor* takes this into account; it is defined as the fraction of energy leaving a radiating surface that reaches a receiving surface. Its value also lies between 0 and 1. View factors for different configurations are given in Refs. 3, 5, and 6.

In electronic systems, thermal radiation typically takes place between components and boards, between different boards, and from boards to the cabinet walls. However, heat transfer by radiation can generally be ignored if the temperature difference between radiating surface and receiving surface is small. For example, if two PCBs next to each other are at about the same temperature, then radiation effects are small and can be ignored. But, when the components and boards are facing a cabinet wall which is at a much lower temperature, radiation effects can be important.

For natural-convection-cooled systems, up to 50 percent of the heat can be dissipated by radiation, whereas for forced-convection-cooled systems, radiation heat transfer is generally small and can be ignored, at least in the preliminary analysis.

A quick way to estimate the heat dissipation by radiation is to use the Stefan-Boltzmann equation while assuming a view factor of 1 and emissivity of about 0.9. If it is then found that the heat transfer by radiation is significant, more accurate calculations should be performed. Additional background on thermal radiation can be found in Refs. 3–6.

17.3.3 Convection

Convection heat transfer deals with heat exchange between a solid surface and a moving viscous fluid. It is expressed by the Newton law of cooling and is given by

$$Q = h A (T_w - T_a) \tag{17.4}$$

where Q = heat flow, W
   h = convection heat-transfer coefficient, W/m²°C
   A = heat-transfer area, m²
   T_w = surface temperature, °C
   T_a = fluid free-stream temperature, °C

If the fluid motion is caused by a fan, the process is called *forced convection*. If there is no fan, and the fluid motion is caused by density differences due to heating of fluid, the process is called *natural* or *free convection*.

The convective thermal resistance is defined as

$$R = \frac{1}{hA} \tag{17.5}$$

where R is thermal resistance, °C/W.

The convective heat-transfer coefficient for a surface depends on the fluid velocity, surface geometry, temperature difference between the fluid and the surface, gravitational forces, and fluid properties. Although the heat-transfer coefficient h is a simple term to use in the preceding equations, it is extremely difficult to calculate h for objects of different geometries in practical situations. Typical values of heat-transfer coefficients for some common objects are given in Table 17.3. The heat-transfer coefficient is generally calculated using experimental correlations available in the literature.[3–5,15,16] These correlations are given in terms of the following nondimensional numbers.

Nondimensional Numbers

Grashoff Number.   The *Grashoff number* is used for natural convection. It is defined as

$$Gr = \frac{\rho^2 g L^3 \beta (T_s - T_a)}{\mu^2} \tag{17.6}$$

TABLE 17.3 Approximate Values of Heat-Transfer Coefficients For Some Common Configurations

Mode of heat transfer	h, W/m^2
Natural convection	
$T_s = 100$, $T_a = 55$, air	
Vertical plate 0.2 m high	5
Vertical plate 0.02 m high	9
Circular cylinder, vertical, 0.02 m high, 0.01 m diameter	6.5
Horizontal plate 0.2 m long	5
Horizontal plate 0.02 m long	8
Circular cylinder, horizontal, 0.02 m long, 0.01 m diameter	8.5
Vertical plate 0.3 m high in water	3,500
Forced convection	
$T_s = 100$, $T_a = 55$, air	
Flat plate 0.3 m high, 1 m/s	7
Flat plate 0.3 m high, 2 m/s	10
Flat plate 0.3 m high, 3 m/s	12
Flat plate 0.3 m high, 4 m/s	14
Flat plate 0.02 m high, 1 m/s	25
Flat plate 0.02 m high, 2 m/s	38
Flat plate 0.02 m high, 3 m/s	47
Circular cylinder 0.01 m diameter, 2 m/s	45
Flat plate 0.03 m high, 3 m/s in water	9,000
Boiling water	25,000–100,000

where $T_s - T_a$ = temperature difference between the solid surface and the free-stream fluid, °C or K

ß = coefficient of thermal expansion, K^{-1}

L = characteristic length, m, varies for different geometries given in Table 17.4

ρ = fluid density, kg/m^3

g = acceleration due to gravity, m/s^2

μ = fluid dynamic viscosity, kg · m/s

TABLE 17.4 Characteristic Lengths of Different Surfaces for Natural Convection

Surface	Position	Characteristic length
Flat plate	Vertical	Height
Flat plate, nonrectangular	Vertical	Area/horizontal width
Cylinder	Vertical	Height
Flat plate	Horizontal	2 (length × width)/ (length + width)
Cylinder	Horizontal	Diameter

A higher Grashoff number generally means that the flow is turbulent. For natural convection in air over a vertical flat plate, the flow is considered turbulent if the Grashoff number is higher than approximately 4×10^8.

Reynolds Number.   The Reynolds number used for forced convection is defined as

$$\text{Re} = \frac{\rho U L}{\mu} \quad \text{or} \quad \frac{UL}{\nu} \tag{17.7}$$

where L = characteristic length, m, varies for different geometries given in Table 17.5
  U = fluid velocity, m/s
  ν − fluid kinematic viscosity, m²/s

TABLE 17.5   Empirical Constants for Forced-Convection Correlations for Fully Developed Flow

Surface	Characteristic length	Re	C	n	m
Flat plate	Length	$<5\times10^5$	0.664	0.5	0.3
		$>5\times10^5$	0.023	0.8	0.33
Tube flow, circular	Diameter	>2000	0.023	0.8	0.3
Tube flow, noncircular	$Dh = (4\times A)/P$	>2000	0.023	0.8	0.3
Cylinder	Diameter	40–4000	0.683	0.46	0.33
		4×10^3–4×10^4	0.19	0.61	0.33

When the Reynolds number is small, generally the flow is laminar, whereas for high Reynolds numbers the flow is considered turbulent. For internal flow such as pipe flow or flow through a duct, the Reynolds number at which the flow becomes turbulent from laminar, called the *critical Reynolds number,* is about 2300. For external flow such as flow over a flat plate, the critical Reynolds number is approximately 5×10^5.

Prandtl Number.   The Prandtl number is based on the fluid properties. It is given by

$$\text{Pr} = \frac{\mu \, C_p}{k} \tag{17.8}$$

where C_p is fluid specific heat, J/kg°C; and k is thermal conductivity of the fluid, W/m°C.
  The Prandtl number has a value of 0.7 for air and 7.0 for water at 25°C.

Nusselt Number.   The nondimensional form of convective heat-transfer coefficient is defined as the Nusselt number. It is given by

$$\text{Nu} = \frac{hL}{k} \tag{17.9}$$

where L is the characteristic length (same as used for calculating Grashoff or Reynolds numbers); and k is thermal conductivity of fluid, W/m°C.
  To calculate the heat-transfer coefficient for an object of interest, first the nondimensional numbers such as Reynolds number or Grashoff number and Prandtl number are calculated. Then the Nusselt number is calculated using experimental correlation for that particular object. Knowing the thermal conductivity of the fluid and the characteristic length, one can then calculate the convective heat transfer coefficient from the Nusselt number.

17.3.4 Extended Surfaces

As can be seen from the convective heat-transfer equation [Eq. (17.4)], convection heat transfer from a solid body can be increased by increasing the convective heat-transfer coefficient h or by increasing the surface area A. One way to increase h for air-cooled equipment is to increase the air velocity. However, increasing the velocity increases the pressure drop through the system, requiring larger fans. Increasing the velocity also generally increases the acoustic noise problem. Therefore, sometimes a more effective method of increasing the convection heat transfer is to increase the effective surface area of heat dissipation by using enhanced surfaces called *fins*. Fins are available in a variety of shapes. Two commonly used fins for electronics cooling applications are straight rectangular fins and pin fins, as shown in Fig. 17.5. A heatsink generally consists of a number of fins.

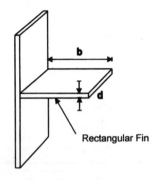

Rectangular Fin

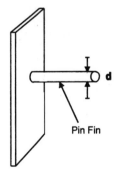

Pin Fin

FIGURE 17.5 Commonly used fin geometries.

Because the temperature of a fin decreases as the distance from fin base increases, a fin does not dissipate as much heat as it could have, had it been at the base temperature. To account for this effect, a term called *fin efficiency* η is defined as η = (actual heat dissipation from the fin surface)/(heat dissipation from the fin if the entire fin were at the base temperature).

Fin efficiency can be calculated using the following equation:

$$\eta = \frac{\tanh(mb)}{mb} \tag{17.10}$$

where $m = (4h/kd)^{1/2}$ for pin fins and $(2h/kd)^{1/2}$ for rectangular straight fins
  b = fin length, m
  d = fin thickness or fin diameter, m
  k = thermal conductivity of fin material, W/m°C
  h = fin convective heat-transfer coefficient, W/m²°C

Once the fin efficiency is calculated, the thermal resistance of a fin can be calculated by

$$R_{\text{fin}} = \frac{1}{\eta h A} \qquad (17.11)$$

where A is surface area of the fins, m².
  A more comprehensive discussion of enhanced surface and fins is given in Refs. 3–6.

17.4 NATURAL CONVECTION IN ELECTRONIC SYSTEMS

In the air-cooled electronic systems which do not use fans, natural convection and radiation are the main modes of heat transfer at component, board, and system levels. The main advantages of natural-convection cooling are low cost, no need for maintenance (for cooling parts such as fans), and no acoustic noise. However, relatively low-power-dissipation components and systems can be cooled using natural convection.
  To calculate natural-convection heat transfer from a component surface, PCB, or cabinet wall, the heat-transfer coefficient is first calculated. Then Eq. (17.4) can be used to calculate the total heat dissipation. The heat-transfer coefficient in natural convection is governed by the following relationship:

$$\text{Nu} = C\,(\text{GrPr})^n \qquad (17.12)$$

where C and n are empirical constants that depend on the temperature difference between the surface and the fluid, the geometry of the heat-dissipating surfaces, and whether the fluid flow is laminar or turbulent.
  Table 17.6 presents the values of C and n to be used in Eq. (17.12) for some common surfaces in electronic systems. Equations for the heat-transfer coefficient for various other surfaces and flow conditions are given elsewhere.[3,4,6,15]

TABLE 17.6 Empirical Constants for Natural-Convection Correlations

Surface	GrPr	C	n
Vertical plate	10^4–10^9	0.59	0.25
	10^9–10^{13}	0.10	0.33
Horizontal plate	10^4–10^7	0.54	0.25
Hot side up	10^7–10^{11}	0.15	0.33
Hot side down	10^5–10^{11}	0.27	0.25
Vertical cylinder	10^4–10^7	0.54	0.25
	10^7–10^{11}	0.15	0.33
Horizontal cylinder	10^3–10^9	0.53	0.25

The following example illustrates calculation procedure for natural-convection heat transfer from a PCB. Heat transfer from component surfaces and cabinet walls can also be calculated in a similar manner.

EXAMPLE 17.1   A PCB, 0.2 m high and 0.1 m wide, is located in a cabinet with an ambient temperature of 55°C. It dissipates power from one side of the board. If the board maximum temperature is 100°C, what is the maximum power that can be dissipated from the board? Also find its convective thermal resistance. Neglect the heat transfer by radiation.

Solution   The mean film temperature is

$$T_f = \frac{T_s + T_a}{2} = \frac{100 + 55}{2} = 77.5°C$$

The properties of air at 77.5°C[6] are $\rho = 0.998$ kg/m^3, $g = 9.8$ m/s^2, $k = 0.03$ W/m°C, and $\mu = 2.08 \times 10^{-5}$ kg/ms, Pr = 0.7. So

$$\beta = 1/T_f = \frac{1}{273 + 77.5} = 0.00285 \text{ K}^{-1}$$

$$L = 0.2 \text{ m}$$

Therefore, the Grashoff number is

$$Gr = \frac{\rho^2 g L^3 \beta (T_s - T_a)}{\mu^2}$$

$$= 2.32 \ 10^7$$

and

$$GrPr = 1.62 \ 10^7$$

Using Eq. (17.12) with constants selected from Table 17.6, we obtain

$$Nu = 0.59(1.62 \ 10^7)^{0.25} = 37.4$$

$$But Nu = \frac{hL}{k}$$

so $h = 37.4 \times 0.03/0.2 = 5.6$ W/m^2°C. The area of the board is $A = 0.2 \times 0.1 = 0.02$ m^2. So

$$Q = hA(T_w - T_a)$$

$$= 5.6 \times 0.02 \times (100 - 55)$$

$$= 5 \text{ W}$$

and

$$R_{conv} = 1/hA = 1/(5.6*0.02) = 8.9°C/W$$

17.5   *FORCED CONVECTION IN ELECTRONIC SYSTEMS*

In air-cooled electronic systems that use fans or blowers, forced convection is generally the main mode of heat transfer at component, board, and system levels. Compared to natural convection, forced convection allows a higher level of heat dissipation. However, it adds cost, weight, and acoustic noise to the system. Therefore, forced-convection-cooled design should only be used if natural-convection-cooled design cannot dissipate the required amount of heat.

To calculate the heat that can be dissipated from component surfaces, PCBs, or cabinet walls by forced convection, we must first calculate the convective heat-transfer coefficient. Then we can use Eq. (17.4) to calculate the heat dissipation.

The general equation used for calculating the heat-transfer coefficient for forced convection is

$$\mathrm{Nu} = C \mathrm{Re}^n \mathrm{Pr}^m \qquad (17.13)$$

where C, m, and n are the empirical constants. These constants mainly depend on the geometry of the heat-dissipating surfaces, and fluid flow characteristics determining laminar or turbulent flow. Table 17.5 presents values of C, m, and n to be used in Eq. (17.13) for some common surfaces in electronic systems. Equations for the heat-transfer coefficient for various other surfaces and flow conditions are given in Refs. 3, 4, 6, and 15.

Example 17.2 shows the calculation procedure of forced-convection heat transfer from a board. Heat transfer from component surfaces and cabinet walls can also be calculated in the same way.

EXAMPLE 17.2   A PCB, 0.4 m high and 0.1 m wide, is located in a cabinet with the ambient temperature of 27°C. It is cooled by air flowing at 2 m/s over the board. It dissipates power from one side of the board. If the board maximum temperature is 60°C, find the maximum power that can be dissipated from the board. Neglect the heat transfer by radiation.

Solution   The mean film temperature is $T_f = (T_s + T_a)/2 = (60 + 27)/2 = 43.5°C$. The properties of air at 43.5°C are $v = 17.36 \ 10^{-6} \ \mathrm{m^2/s}$, $k = 0.0274 \ \mathrm{W/m°C}$; $\mathrm{Pr} = 0.7$.

Assuming the board as a flat plate with the length of the board $L = 0.4$ m, the Reynolds number is $\mathrm{Re}_L = UL/v = (2*0.4)/(17.36 \ 10^{-6}) = 46,082$. Using Eq. (17.13) with constants from Table 17.6,

$$\mathrm{Nu}_L = 0.664 \times (46082)^{0.5} \times 0.7^{0.33}$$

$$\mathrm{Nu}_L = 126.5$$

$$\mathrm{But \ Nu} = hL/k$$

Therefore $h = 126.5 \times 0.0274/0.4 = 8.7 \ \mathrm{W/m^2°C}$.

The area of the board is $A = 0.4 \times 0.1 = 0.04 \ \mathrm{m^2}$. So, heat transfer from the board is

$$Q = hA \ (T_s - T_a) = 11.48 \ \mathrm{W}$$

17.6   *FAN SELECTION*

For forced-convection cooled electronic systems, one critical thermal design task is the selection of a fan or fans. Before a fan can be selected, however, the airflow required to dissipate the heat generated from the system and the pressure drop characteristics of the system have to be found.

To calculate the airflow, both the amount of heat to be dissipated and the density of the air must be known. The energy balance equation used is

$$Q = C_p m(T_o - T_i) \qquad (17.14)$$

where Q = amount of heat to be dissipated, W
C_p = specific heat of air, J/kg°C
T_i = inlet temperature of air, °C
T_o = outlet temperature of air, °C
m = mass flow of air, kg/s

It is given by

$$m = V\rho \qquad (17.15)$$

where V is airflow, m³/s; and ρ is density of air, kg/m³.

If Q, C_p, ρ, and inlet−outlet temperature difference, are known, airflow can be calculated. Experience has shown that the air inlet−outlet temperature difference should be maintained at about 15°C. It should also be noted from Eq. (17.14) that the mass of air, not its volume, governs the amount of cooling. This becomes important if the system has to be operated at high altitudes, because density of air decreases as the altitude increases.

17.6.1   Determining System Pressure Drop

After the airflow needed has been determined, the amount of system pressure drop at this airflow must be found. This resistance to flow, called *system pressure drop,* is a function of airflow rate. A typical system pressure drop curve, in most electronic systems, follows what is called the *square law,* which means that the static pressure changes as a square function of changes in the airflow rate. Figure 17.6 shows a typical pressure-drop curve for a system. Because of the complexity of geometry of an electronic system, static pressure of a system cannot be easily calculated. It is generally measured in a prototype or in a real system. The measurements are performed in wind tunnels. The measurement procedure is described in the experimental measurement section later in the chapter. Some fan manufacturers also provide the measurement and testing services.

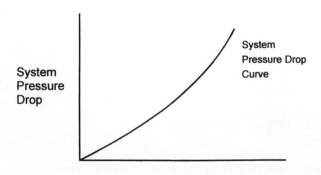

FIGURE 17.6   Typical system pressure-drop curve.

17.6.2   System Flow

The governing principle in fan selection is that any given fan can deliver only one flow rate at one pressure drop in a given system. Once the volume of air and the static pressure drop of the system are known, a fan can be selected. The fan performance curve, supplied by the fan manufacturers, and the system pressure-drop curves are superimposed on each other as shown in Fig. 17.7. The point of intersection shows the flow rate that the fan under consideration can provide for the system. As long as the fan flow-rate capacity is more than the needed airflow, the fan will work well in the system. If the fan flow-rate or pressure-drop capacity is less than the needed flow rate or pressure drop, either a larger fan or multiple fans have to be used.

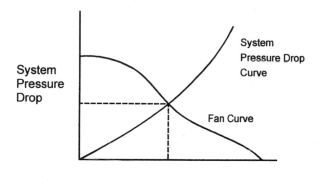

FIGURE 17.7   Fan selection graph.

It is a good practice, however, to analyze the system for possible reduction in the overall pressure drop, before increasing the number of fans or using a larger fan. This can be achieved by increasing the board spacing, using different filters or grills, and increasing the size of the cabinet, to name only a few alternatives.

17.6.3   Series and Parallel Operation

Combining fans in series or parallel can achieve the desired airflow in a given system without greatly increasing the fan size. *Parallel operation,* which is defined as two or more fans blowing together side by side, is used to increase the flow rate through a given system. The performance of two fans in parallel will result in doubling the volume flow, but only when the system has no pressure drop. The higher the system pressure drop, the less would be the increase in total flow rate with parallel fan operation. *Series operation* can be defined as using multiple fans in a push or pull arrangement. By staging two or more fans in series, the static pressure capability at a given airflow can be increased.

17.7   *PACKAGE THERMAL RESISTANCE*

Once the environment of a package is determined by the system and board-level thermal analysis or measurements, the junction temperature of the components can be calculated using the package junction-to-air or junction-to-package thermal resistance. These thermal resistances include a number of conduction, spreading, contact, convective, and radiative resistances which are present in a component.

The *junction-to-air thermal resistance* of a package is defined as

$$R_{ja} = \frac{T_j - T_a}{P}$$ (17.16)

and *junction-to-case thermal resistance* is defined as

$$R_{jc} = \frac{T_j - T_c}{P}$$ (17.17)

where P = power dissipation from the chip, W
T_j = is average chip junction temperature, °C
T_c = package case temperature, °C
T_a = ambient or air temperature adjacent to the package, °C

Also

$$R_{ja} = R_{jc} + R_{ca}$$ (17.18)

where R_{ca} is the case-to-ambient thermal resistance.

Although package thermal resistance can be calculated by analytical or numerical analysis, because of the complex geometry of the component, unknown contact, spreading and other resistances, it is generally measured experimentally. A package manufacturer generally supplies thermal resistance data for the package at various air velocities, or in the form of a curve as shown in Fig. 17.8.

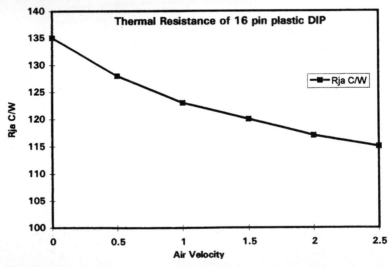

FIGURE 17.8   Thermal resistance of a DIP package.

Several types of packages are used in electronic systems. Some commonly used packages are dual inline package (DIP), small-outline package (SOP), quad flat package (QFP), plastic leaded chip carrier (PLCC), and pin-grid array (PGA). More details on different types of packages are available in Ref. 7. Typical thermal resistance values for some common packages are given in Table 17.7. These values should be used for reference purposes only. Thermal resistance data should be obtained from the package or component manufacturer.

TABLE 17.7 Typical Thermal-Resistance Values for Some Common Packages

Package	R_{ja} natural convection, °C/W	R_{ja} forced convection at 2 m/s°CW
CDIP 8L	130	78
CDIP 24L	65	45
CDIP 40L	48	32
SOP 8L	165	138
SOP 28L	68	50
PQFP 44L	58	45
PQFP 128L	38	30
PQFP 160L	25	15
CQFP 120L	35	26
CQFP 208L	27	20
PLCC 20L	100	72
PLCC 44L	60	43
PLCC 84L	36	23
PPGA 175L	27	16
CPGA 68L	45	37
CPGA 129L	35	20
CPGA 407L	13	6

Even the manufacturers' data for thermal resistance should be used with caution. The reason for this lies in the fact that thermal resistance of a package also depends on its environment, method of attachment, PCB material, and neighboring components. Therefore, along with the thermal resistance values, the manufacturer should also supply all the details of the test. If the package environment in the real system is quite different from the test situation, it is advisable to measure thermal resistance of the packages in the environment and on the board where the component will be used.

It is important to note that T_a is the package ambient temperature and not the board or system ambient temperature. If the component under consideration is located downstream of some power-dissipating component, the average T_a for the package can be either measured experimentally or calculated using CAD tools, discussed later. A simple but approximate analytical way of determining T_a is by using the following energy-balance equation:

$$T_a = T_i + T_p = T_j + \frac{P_b}{mC_p} \tag{17.19}$$

where T_i = board air inlet temperature, °C
T_p = board preheat temperature, °C
P_b = board power dissipation upstream of the component under consideration, W
m = air mass flow, kg/s

As mentioned earlier, if T_a, R_{ja}, and power dissipation P are known, T_j can be calculated as shown in Example 17.3.

EXAMPLE 17.3 A 16-pin plastic DIP package is located on a board cooled by air at a velocity of 1.5 m/s. The board inlet air temperature is 40°C, while the air preheat temperature due to heat dissipated by the components upstream of the dip is 15°C. Assuming that the power dissipation by the dip is 1 W, calculate the junction temperature of the chip. For package junction-to-air thermal resistance, use the chart shown in Fig. 17.8.

Solution   From the chart in Fig. 17.8, at the air velocity of 1.5 m/s,

$R_{ja} = 120°C/W$
$T_i = 40°C$, T_p (preheat temperature) = 15°C
$T_a = T_i + T_p = 40 + 15 = 55°C$
$T_j = T_a + R_{ja}P$
$T_j = 55 + 120 \times 1$
$T_j = 175°C$

The calculated junction temperature of 175°C is much higher than the generally acceptable 100°C. In such situations, the designer can do one or a combination of the following: reduce power dissipation from the package, use a better package with lower thermal resistance, improve the package environment by lowering air temperature and/or increasing air velocity, or use a heatsink.

17.8 HEATSINK FOR ELECTRONIC SYSTEMS

If the junction temperature of a chip exceeds the maximum allowable chip temperature, a heatsink may be used to improve heat transfer from the package. Heatsinks enhance heat transfer from the package by providing added heat-dissipation area through fins. Figure 17.9 shows some heatsinks

FIGURE 17.9   Heatsinks used in electronic systems. (*Courtesy of Wakefield Engineering Inc.*)

used for electronic components. Heatsinks are attached to a package using conductive epoxy, or using thermal grease with some mechanical clamping device.

EXAMPLE 17.4   In the problem of Example 17.3 above, find the thermal resistance of a heatsink that can keep the junction temperature below 100°C. The package to case thermal resistance R_{jc} is 40°C/W.

Solution   As per Eq. (17.17), $R_{ja} = R_{jc} + R_{hs}$ or $R_{hs} = R_{ja} - R_{jc}$. But

$R_{ja} = (T_j - T_a)/P = (100 - 55)/1 = 45°C/W$
$R_{jc} = 40°C/W$
$R_{hs} = (T_j - T_a)/P - R_{jc}$
$R_{hs} = 45 - 40 = 5°C/W$

Therefore, the package needs a heatsink with a thermal resistance (including the epoxy thermal resistance) of at least 5°C/W.

A number of heatsink suppliers, such as Thermalloy, Wakefield, and Aavid, sell hundreds of off-the-shelf heatsinks through their catalogs. These can typically be purchased at a cost ranging from $0.50 for simple stamped heatsinks to $25.00 for high-performance heatsinks. Most of the heatsink vendors can also provide support for selecting an appropriate heatsink for specific applications.

17.9 EXPERIMENTAL MEASUREMENTS AND TESTING

Experimental measurements and testing are a very important part of the thermal design process. Depending on the purpose of testing, measurements are carried out at component, board, and system levels in a wind tunnel, system prototype (mockup), or a real system. A rigorous final environmental testing must always be carried out before the system release. These tests, generally carried out in specially designed environmental chambers, should reflect the use of the electronic system in its worst-case environment.

The measurements generally include: heat dissipation, airflow rate, air velocities, static pressure drop in the system, as well as ambient, package, heatsink, and chip temperatures. Some of the measurement techniques used for thermal measurements and testing of electronic systems are briefly described below. More information on experimental measurements is available in Ref. 8.

17.9.1 Air-Velocity Measurements

Air-velocity measurements are carried out at package, board, and system levels. Velocities at various points in a prototype or real system are measured to determine the airflow distribution in the system. Board- and package-level velocity measurements are performed in prototypes or real systems to make sure that all the packages are getting the airflow needed to cool them to their proper junction temperatures.

Velocity measurements are generally carried out using a hot-wire anemometer. The hot-wire anemometer is inserted into the system, and the tip of the hot wire is placed at a point where measurement is required. The air velocity can then be read directly on an analog or digital meter attached to the hot wire.

The basic technology behind the hot-wire anemometers is constant-temperature anemometry. The sensor element generally contains a heated wire or a thermistor. An electronic control circuit maintains the sensor at a constant temperature. As the air passes by the sensor, it cools the sensor and more power is required to maintain the sensor temperature. Therefore, the power dissipated by the sensor is directly related to the air velocity. Once the sensor is calibrated against a known velocity, it can be used independently to measure velocity.

For system-level air-velocity measurements using a hot-wire anemometer, one generally has to drill holes through the sheet metal and PCBs to measure velocity at different locations in the system. This process is very time-consuming and tedious. It is also not very accurate because of probe intrusion, especially if the board spacing is small. A new air-velocity measurement system has recently been introduced which uses small-size velocity sensors connected to a computer with flexible cables (Fig. 17.10). A number of velocity sensors can be installed at various locations in the system, and simultaneous velocity measurements can be performed at all the points. More details of this instrument can be found in Ref. 9.

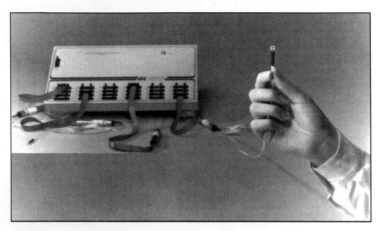

FIGURE 17.10 Multichannel airflow and temperature monitor. (*Courtesy of Cambridge Aeroflo, Inc.*)

17.9.2 Airflow Rate and Pressure-Drop Measurements

As explained earlier, airflow rate and pressure-drop measurements are important for fan selection. Before selecting a fan, one has to know the system pressure drop at various flow rates. These measurements are carried out in wind tunnels such as shown in Fig. 17.11. A wind tunnel typically has a variable-speed blower, flow straighteners, an orifice plate, and a settling chamber. The system to be tested is mounted on the settling chamber. The pressure drop across the orifice plate is measured using a manometer. Knowing the pressure drop and the type of orifice plate, one can find the flow rate through the wind tunnel. Another manometer attached at the settling chamber outlet

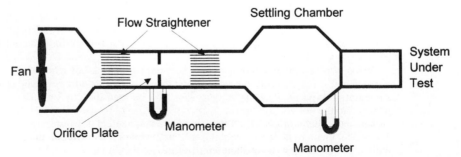

FIGURE 17.11 Wind tunnel layout for system pressure-drop measurements.

shows the pressure drop through the system at a particular flow rate. Once pressure drop measurements at three or four flow rates are measured, a system pressure drop curve such as shown in Fig. 17.6 can be created.

17.9.3 Temperature Measurement

Temperature of PCBs, packages, and chips are measured during a system thermal design and testing. A number of instruments and sensors are used for measuring these temperatures. Some of the commonly used temperature measurement techniques are described below.

Thermocouples.   Surface and air temperatures are commonly measured using thermocouples. A *thermocouple* is a temperature-measurement sensor that consists of two dissimilar metals joined together at one end, called a *junction,* and attached to a measurement device at the other end. This produces a thermoelectric voltage that is proportional to the temperature of the junction.

To measure component surface temperatures, the thermocouples are generally attached to the surface using a high-thermal-conductivity epoxy. Thermocouples at the other end are connected to a dedicated meter, a multimeter, or a data-acquisition system which converts the analog voltage into temperature and shows the surface temperature. The desirable features of thermocouples are accuracy, wide operating range, fast response time, sturdiness, and low cost.

Infrared Thermography.   *Infrared thermography* is another common technique used for measuring PCB, package, and chip surface temperatures without contact. An infrared system uses energy radiated from hot surfaces at infrared wavelength to determine the surface temperature. Infrared thermography uses a photocell which produces a DC output proportional to the surface temperature of the object. It is important to note that the infrared system requires visibility of the surface for measurements. Infrared systems can be a single-point measurement probe, or a full-screen camera system that can be used to measure full board temperature in one shot. The temperatures can be read on a dedicated meter or a computer screen.

17.9.4 Chip-Junction-Temperature Measurement

A *chip junction temperature* is generally measured to evaluate the thermal resistance of a package. Knowing junction temperature, ambient temperature, package case temperature, and package power dissipation, one can find the package thermal resistances.

Two measurement techniques are commonly used to determine the chip junction temperature. The first, called *direct technique,* uses IR thermography to measure chip temperatures. In order to use the IR thermography approach to measure chip temperature, the semiconductor package has to be modified to allow a direct line of sight to the top surface of the chip. The advantages of this approach are that it gives a complete map of temperature distribution, which is difficult to measure using an indirect technique (Fig. 17.12). However, the disadvantages are that it requires altering the package, thus possibly altering the chip temperature; it is also difficult to use for in situ measurements in a real system, and allows chip back temperature measurement only for back-bonded chips.

The second chip-junction-measurement technique, called the *indirect technique,* makes use of diodes or resistors on a test chip to determine the chip temperature. The test chips are generally supplied by the semiconductor vendors. These test chips contain heaters for heat dissipation from the chip and diodes or resistors for temperature sensing. Typical chip layouts for test chips are shown in Fig. 17.13.[10] These test chips are put in a package for which thermal resistance is to be measured. Diodes are preferred for measuring the local temperatures at different points on a chip, whereas resistors are used for measuring an average chip temperature.

Before the thermal tests, the temperature-dependent parameter on the chip is calibrated against a known temperature in a bath or oven without supplying power to the chip heaters. For resistors as sensors, resistance is measured as a function of temperature, while for diodes as sensors, a very

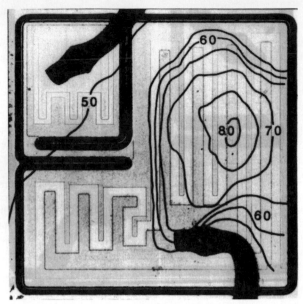

FIGURE 17.12   Temperature distribution on a chip.

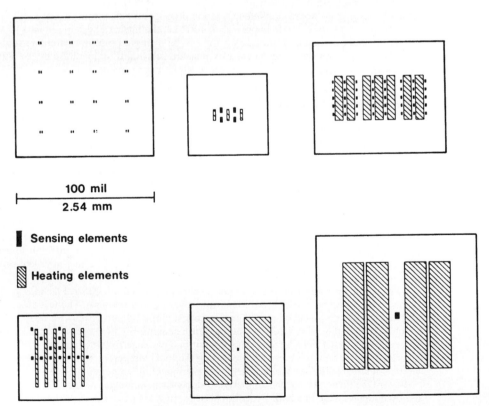

100 mil

2.54 mm

▮ Sensing elements

▨ Heating elements

FIGURE 17.13   Heating and sensing areas of a test chip.

small constant current is passed through the emitter-base junction, and the forward voltage drop is measured as a function of temperature. After measurements at a few temperatures, a diode voltage or resistor resistance and temperature relationship is developed. Then, during actual measurements, power is supplied to the package heaters and the chip temperature is measured using the calibrated temperature-dependent parameter.

17.10 CAD TOOLS FOR THERMAL MODELING

For most electronic systems, the product design cycle times, as well as the product development team sizes have shrunk to half of what they were 5 years ago. Therefore, with shorter product design cycle times and limited workforce, product design can be extremely difficult to complete without the use of appropriate computer-aided-design (CAD) tools.

In the thermal design, these tools are used for numerically modeling thermal-fluid characteristics in systems, PCBs, packages, and chips. These models can be used for checking the design feasibility, exploring different design scenarios, and most importantly for the optimizing of the design.

17.10.1 Benefits of Thermal Modeling

Some of the main benefits of thermal modeling are as follows:

- Thermal modeling generally takes less time and can be performed fairly quickly compared to measurements.
- In most cases, the cost of modeling is significantly lower than the cost of a corresponding experimental investigation.
- The most important benefit of modeling is that it allows optimization of the design. A designer can study the implications of several parameters in a short time and choose the optimum design. On the other hand, a corresponding experimental investigation is time-consuming and is generally much more expensive.
- A thermal model gives much more information in the solution domain than does a corresponding experimental study. It can provide the values of all the relevant variables (such as velocity, pressure, temperature) at any point in the model. Such complete information is very helpful in understanding the design bottlenecks and improving the design.
- In a thermal model, realistic conditions, such as zero gravity, high-temperature environment, and high-altitude effects, can be easily calculated. Experimental measurements in these conditions can be extremely expensive.

17.10.2 Drawbacks of Thermal Modeling

Some of the main drawbacks of thermal modeling are

- Realistic models, typically system-level, are difficult and time-consuming to create. In most cases, thermal models have to be created using mechanical drawings, which is a very tedious process. Some of the modeling tools are developing links to mechanical drawing tools so that drawings can be read directly into a modeling tool. However, such links are still at early stages of development.
- Models have some errors due to numerical techniques and assumptions made to simplify the models.

- For very complex problems, such as calculating the velocity distribution and overall pressure drop for a large electronic system, modeling can also be time-consuming and expensive.
- Computing resources are still limiting factors for complex models. Although most of the tools can now be run on a personal computer or a workstation, complex problems can take several hours or days to solve.
- In most of the cases of practical significance, some level of experimental verification is necessary.

17.10.3 Thermal Modeling Techniques

For thermal analysis in electronic systems, modeling tools are based mainly on one of the following three techniques.

Thermal Network Technique. The *thermal network technique* uses the analogy between the heat transfer and electric charge to create an equivalent electrical network. The thermal problem is modeled with lumped thermal resistances, as described in Sec. 17.3. Once a networked model is ready, it can be solved using any circuit simulation program, such as SPICE, a common program from the University of California, Berkeley, Calif., or thermal network solver programs such as Sauna.[11] The accuracy of the solution depends on the accuracy of lumped resistances compared to the actual thermal resistances.

Finite Element Method. The *finite element method* involves the solution domain to be divided into a finite number of discrete elements. Then, the governing equations of heat transfer and fluid flow (equations of conservation of mass, momentum, and energy, called *Navier-Stokes equations*) are integrated over each element, generally using a weighted residual method. The advantage of using the finite element method over the other methods is its ability to model complex boundaries. Therefore, if possible, applications involving complicated geometric boundaries should be solved using modeling tools based on finite element methods.

Finite Difference Method. The *finite difference method* involves the solution domain to be discretized into a finite number of points called *nodes*. Then, the governing heat transfer and fluid-flow equations (governing equations of conservation of mass, momentum, and energy, called *Navier-Stokes equations*) are replaced with finite difference algebraic equations. These algebraic equations are then solved using matrix inversion or iterative techniques.

17.10.4 Commercial Thermal Modeling Tools

A number of finite-difference- and finite-element-technique-based commercial tools are available which can be used to solve a variety of heat-transfer and fluids problems in electronic systems. In a broad sense, these can be divided into two groups.

Integrated Board-Level Thermal Modeling Tools. These tools are generally based on finite element technique, and are integrated with structural analysis, reliability analysis, and board routing tools using a common database. Most popular commercial tools available for the board-level thermal modeling are AUTOTHERM from MENTOR Graphics, Thermax from Cadence, and Pacific Numerics. In these tools, the models are easy to create using interactive geometry creation and editing commands. An extensive parts library is also available for model generation. Models can include PCBs, components, and heatsinks.

The finite element mesh is created automatically. Typical boundary conditions include fluid velocities, heat loads, material properties, component thermal resistances, and heatsink thermal resistances. These programs run on workstations, and a typical board with 100 to 200 components can be analyzed in a couple of minutes. Models can solve conduction, convection, and radiation heat-transfer problems for steady-state and transient conditions. All these programs use experimental correlations for calculating convective heat-transfer coefficients for component and board surfaces. The results, such as junction temperature and board temperature maps, are available in graphical form. These tools are very helpful in thermally optimizing a board layout. They can also be used for a simplified package-level thermal analysis which gives package thermal resistance values.

CFD Thermal Modeling Tools.   These thermal modeling tools use computational fluid dynamics (CFD) techniques, generally based on finite element or finite difference methods, to solve a wide variety of fluid-flow and thermal problems. Although these tools can be used to model package-, system-, and board-level problems, the board-level tools described earlier are easier and faster for modeling board-level thermal problems.

A number of commercial CFD modeling programs are available which can be used to model fluid-flow and heat-transfer problems in electronic systems. Some of the main commercially available CFD programs are Fluent, Phoenics, Fidap, Flotran, Flow 3D, and Fire. In the last few years, some CFD programs designed especially for thermal modeling of electronic systems have also become available. The main ones are Flotherm, Autoflow, and Hot Box. All these programs are available for personal computers, workstations, and mainframe computers.

In these programs, the model is generated with appropriate boundary conditions using an interactive preprocessor. The model is then discretized with mesh that is generated manually. Then, the governing equations are solved using finite element or finite difference technique. After the equations are solved, the results of appropriate variables such as velocity, pressure, and temperature are available in graphical form. A system-level model showing flow distribution through a computer system is shown in Fig. 17.14.[12] More details about CFD modeling techniques are given in Ref. 13.

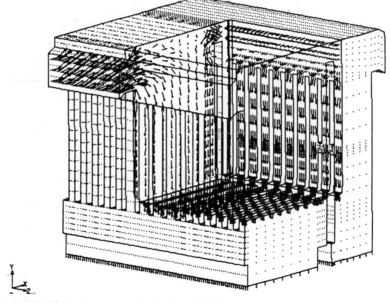

FIGURE 17.14   Temperature distribution of a PCB.

17.11 *CONCLUDING REMARKS*

Thermal management can play a very key role in improving the reliability of the electronic systems. In this chapter, a brief overview of the thermal design process is presented, followed by thermal analysis, modeling, and measurement techniques. For advanced analysis techniques, readers should consult Refs. 14–16. These techniques, when applied from the early stages of design of an electronic system, can improve the reliability significantly.

REFERENCES

1. MIL-HDBK-217F, *Reliability Prediction of Electronic Equipment,* Jan. 1990.

2. Udo, Alt, "Thermal Analysis and Environment of Flotherm," *Proceedings of Flotherm International Users Conference,* Guildford, U.K., Sept. 1991.

3. Holman, J. P., *Heat Transfer,* McGraw-Hill, New York, 1991.

4. Bar Cohen, A.; and Kraus, A. D., *Thermal Analysis and Control of Electronic Equipment,* Hemisphere Publishing Co., 1983.

5. Incropera, F. P.; and DeWitt, D. P., *Fundamentals of Heat Transfer,* Wiley, New York, 1985.

6. Rohsenow, W. M.; Hartnett, J. P.; and Ganic, E. N., *Handbook of Heat Transfer Fundamentals,* McGraw-Hill, New York, 1985.

7. Rymaszewski, E. J.; and Tummala, R. R., *Microelectronics Packaging Handbook,* Van Nostrand Reinhold, New York, 1988.

8. Beckwith, T. G.; Buck, N. L.; and Marangoni, R. D., *Mechanical Measurements,* Addison-Wesley, Reading, Mass., 1982.

9. *Is This How You Are Still Performing Thermal Measurements of Your Electronic Systems?,* Product Information Brochure, Cambridge Aeroflo, Shirley, Mass.

10. Oettinger, F. F.; and Blackburn, D. L., *Thermal Resistance Measurements,* NIST Publication 400 86.

11. *Sauna Thermal Analysis,* Tatum Labs, Inc., 3917 Research Park Drive, B-1, Ann Arbor, Mich.

12. Mansingh, V.; and Misegades, K., "Thermal-Fluids Modeling of Computer System," paper presented at 3rd ASME-JSME Thermal Engineering Conference at Reno (Nev.), March 17–22, 1990.

13. Patankar, S. V., *Numerical Heat Transfer,* Hemisphere Publishing Co., New York, 1980.

14. Bar Cohen, A.; and Kraus, A. D., *Advances in Thermal Modeling of Electronic Components and Systems,* Vol. 1, Hemisphere Publishing Co., New York, 1988.

15. Bar Cohen, A.; and Kraus, A. D., *Advances in Thermal Modeling of Electronic Components and Systems,* Vol. 2, ASME Press, New York, 1990.

16. Bar Cohen, A.; and Kraus, A. D., *Advances in Thermal Modeling of Electronic Components and Systems,* Vol. 3, ASME Press, New York, 1993.

CHAPTER 18
MECHANICAL STRESS AND ANALYSIS

Richard L. Doyle
DOYLE AND ASSOCIATES
LA JOLLA, CALIF.

18.1 INTRODUCTION

Reliability, just like other important parameters, is established by the design and manufacturing process. Actual reliability (time to failure) rarely exceeds desired reliability no matter how good the production, quality, or maintenance program may be. The design disciplines and analyses that are discussed in the next three chapters (Chaps. 18 to 20) do nothing more than identify and focus attention on design weaknesses so that they may be corrected, protected against, or accepted after consideration. They also provide a means of assuring that a design will meet specified time-to-failure (reliability-to-life) requirements prior to a production commitment.

It is generally true that you cannot practically demonstrate reliability by test, with any degree of statistical confidence, prior to a production commitment. There normally isn't the time nor the number of test samples available to demonstrate the high mean time between failures (MTBF) or time between overhauls required for a new design. It is also generally true that people don't pay any attention to requirements they don't have to demonstrate. The normal inclination is to spend time and effort on design parameters that can be readily demonstrated, such as performance, with a resulting deemphasis on everything else. Thus, as reliability requirements become more and more stringent and, therefore, harder or impossible to demonstrate cost-effectively, emphasis tends to *decrease* rather than increase on the reliability design parameter. In addition, by the time reliability problems become apparent by test, it is often too late to do anything about them, so even then they are ignored and production proceeds. It becomes necessary, therefore, to translate reliability requirements into design criteria, against which compliance can be analytically measured prior to a production commitment.

Although high reliability itself cannot be effectively measured, the design parameters that determine reliability can be, and the control and analytical verification of these parameters will lead to the attainment of a reliable design.

There are two fundamental bases for failure. One type of failure results from higher stresses than the part can stand; the second type of failure is that the part literally wears out. Both types of failures are covered in this handbook. However, the second type of failure is the most intriguing since wearout comes in many different forms. Wearout can be due to fatigue, such as a spring failing after being subjected to the same load for many years. It can be due to abrasion, such as a railroad car carrying coal for many years. The coal car side plates wear thin and finally rupture. Wear and wearout also are caused by friction, thermal cycling, and other factors. Generally, all mechanical parts have a finite life. It can be very long, like the pyramids of Egypt, or very short, like the tires on your car. We will attempt to quantify the wearout time for all parts based on stresses and wear in these next three chapters. This way one can predict the MTBF and when to perform maintenance, whether it is maintenance on an aircraft carrier or on the Statue of Liberty. Even buildings and automobile bridges have a design life based on the fatigue life of the structural steel or the design life of the reinforcing steel in the concrete.

The first step in the development of a reliable system is the translation of threat-based operational requirement, such as *probability of being available and reliable* (PAR) or *mission reliability* (MTBF or time between overhauls) into a parameter that means something to designers. The MTBF values can be translated into safety factor guidelines since stress causes failures (see Fig. 18.1). Compliance with safety-factor guidelines can be verified by a mechanical stress analysis. Stress levels determined by the stress analysis can be used to determine mechanical failure rates. Failure rates, in turn, can be used to determine the need for overhaul at certain intervals which, in turn, can be used in conjunction with the math model of the system and the mission profile to predict operating parameters such as availability. Figure 18.2 shows how the mechanical failure rates can be used to establish overhaul requirements. Stage I (infant mortality) is a very short time period. As a result, it is not desirable to test all mechanical components for extended lengths of time.

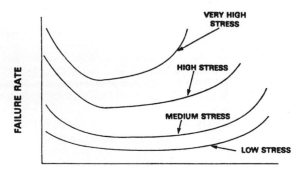

FIGURE 18.1 Effect of stress levels on mechanical failure rates.

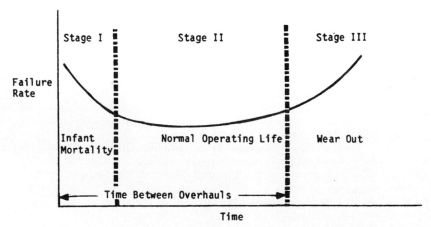

FIGURE 18.2 Mechanical-component life cycle.

The majority of a mechanical component's life is spent in stages II and III, which show a gradual increase in failure rate during stage II and an accelerated failure rate during stage III. This is brought about by the fact that mechanical components have continuous wear and fatigue throughout their operating life. An excellent method of selecting the time between overhauls is to determine the time (number of operating hours) when the failure-rate curve begins to accelerate the sharpest, sometimes called the knee of the *hazard-rate*.

Up to this point, the failure rate is somewhat constant with time, and can be modeled as a constant failure rate. Field data indicates that mechanical components demonstrate a roughly constant failure rate when the time between overhauls is sufficiently short to keep them from having to operate past the knee of the curve, shown in Fig. 18.2.

A simplified information flow is shown graphically in Fig. 18.3. This figure shows how operational requirements are translated into design requirements, how compliance with design requirements can be determined by analysis (in order to provide an assessment of the design), and how the results can be compared to the operational requirement. (In Fig. 18.3 FMEA represents *failure modes and effects analysis*.)

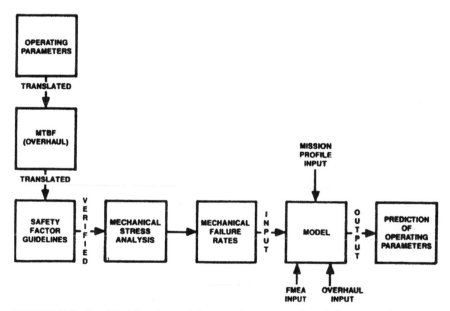

FIGURE 18.3 Simplified flow for translating operating parameters into design processes.

18.2 MECHANICAL STRESS ANALYSIS

As can be seen from Fig. 18.3, mechanical stress analysis is an important design discipline and should be an integral part of all mechanical designs. The analysis is a procedure by which the maximum actual stresses induced on a material in its application are identified. The primary purpose is to ensure that a *safety margin* is designed in.

The optimum time to perform a mechanical stress analysis is concurrent with the detailed design. Selection of a material or component for a specific application is primarily based on that material or component meeting the performance requirements and being able to withstand the external and internal applied stresses. Hence, stress analysis is normally conducted as an integral part of the material and component design selection process.

- *Life-cycle phase:* Stress analysis is normally conducted during full-scale development as the detailed design definition progresses.
- *Purpose:* The results of conducting stress analysis provide
  1. Verification that the stresses are identified and controlled, thereby maximizing life.
  2. A confirmation of compliance with safety margins.
- *Objective:* The primary objective of stress analysis conducted on mechanical systems and equipment is to identify any life-limiting stresses. This is performed by comparing stress-strength characteristics of parts and materials with internal and/or external loading so that when the design is exposed to stress levels experienced in service, its life meets the requirements for that equipment. Stress analysis requirements should be invoked as a design technique to reveal system/equipment design deficiencies.

18.2.1 Material Stress-Strength Comparison

A material's strength or capability of handling a given stress varies from lot to lot and manufacturer to manufacturer. This variation for all materials of the same type can be represented by a statistical distribution of material strength. Similarly, the stress applied to a material changes from one point in time to another, with instantaneous changes occurring in temperature and mechanical stresses and transients, vibration, shock, and other deleterious environments. At a random point in time the environmental effects can combine, reaching stress levels beyond the material's strength, resulting in failure of the material.

This strength-stress relationship can be described graphically by two overlapping probability density distributions, as shown in Fig. 18.4. For materials rated at the mean distribution used in an environment with an average stress equal to the mean of the stress distribution, the probability of failure is equal to the product of the areas of the two distributions where overlapping occurs (i.e., the probability of the stress being greater than the minimum strength times the probability of the strength being lower than the maximum stress). This probability of failure is represented by the solid dark area of Fig. 18.4.

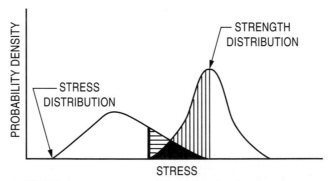

FIGURE 18.4 Two overlapping statistical probability-density distributions.

To reduce the probability of failure, one of two approaches must be taken: (1) reduce the potential stress levels to a point where there is a very small probability of the stress exceeding the material's strength (shift stress distribution to the left) or (2) increase the material's strength so that the probability of the combined stresses reaching or exceeding this strength is very small (shift strength distribution to the right). In most instances, the stresses cannot be reduced and the only approach is to increase the material's strength. This is accomplished by using a different or stronger material stressed to only a small percentage of its capability. This technique can also reduce material internal temperatures, decreasing the rate of chemical reaction.

18.2.2 Procedure for Mechanical Stress Analysis

This chapter explains the fundamental mechanical stress analysis principles and shows the methods and formulas necessary for calculating mechanical stresses, safety factors, and margins of safety associated with mechanical designs. These calculations are used in the mechanical stress analysis in two ways:

1. They ensure that the structure has sufficient strength to operate in its service environment and under the loads specified.
2. They ensure that there is no appreciable excess of material or overdesign, which could result in higher costs and burden the program with excessive weight, mass, volume, etc.

Mechanical stress analysis is a detailed and complex engineering discipline. However, for this chapter, simplifying assumptions will be made which provide equations for evaluating the approximate values of the stress, loads, margins of safety, and safety factors.

Several important factors must be thoroughly understood before performing a mechanical analysis. These include the following:

1. Important parameters of a mechanical analysis are the loads, the material properties, and the actual calculated stresses.
2. Factors to be considered in the mechanical analysis include the size, weight, material joining techniques, welding, riveting, loads, material properties, thermal environment, and thermal limitations of the material.
3. It is important to use standard terminology and the proper equations for mechanical analysis so that the calculations can be understood and verified.
4. Environmental stresses such as corrosion and temperature must be determined and applied to the mechanical properties of the material.
5. Safety factors either derived or specified must be applied. The safety factor is used in calculating the margins of safety.
6. Actual stress values must be calculated under steady-state loading conditions.
7. Stress amplification factors must be calculated under dynamic conditions.
8. Actual strain energy must be calculated under shock loading.

18.2.3 Symbols and Abbreviations

Symbols and abbreviations used in this chapter are defined per U.S. MIL-HDBK-5E. These symbols and abbreviations will be used throughout this section, with the exception of special statistical symbols that are presented in Chap. 19. To go further than this information, the reader should be familiar with the symbols and abbreviations used in mechanical structural analysis.

18.3 STANDARD DEFINITIONS, NOMENCLATURE, AND FUNDAMENTAL EQUATIONS

All mechanical systems, while in operation, are subjected to external loading. These applied loads must be transmitted through components and between components to the supports, which provide the necessary reactions to maintain equilibrium. In the process, forces and moments are developed in each component of the system that resist the externally applied loads, providing equilibrium for the components. The internal forces and moments at any point within the system are called *stress resultants*. These internal forces and moments can be categorized in terms of the deformations that are produced.

18.3.1   Loading and Stress Resultants

Normal Force.   The uniform bar shown in Fig. 18.5 is subject to a force P which must be transmitted to the support. Under this loading the bar will increase in length and produce internal stress resultants, called *normal forces,* denoted by P. Normal forces are also associated with mechanical components that shorten in length under applied compression loads.

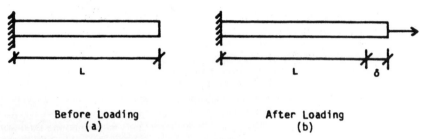

Before Loading
(a)

After Loading
(b)

FIGURE 18.5   Uniform bar before (*a*) and after (*b*) loading.

Shear Force.   The uniform square plate shown in Fig. 18.6 is subjected to a uniformly distributed force f on three sides of the plate. Again these forces must be transmitted to the support, but the deformations are quite different from those related to normal force. This form of deformation is called *shear deformation* and is associated with an internal stress resultant called *shear force,* denoted by S.

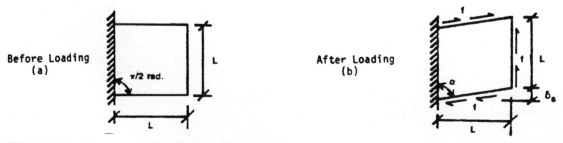

FIGURE 18.6   Uniform square plate before (*a*) and after (*b*) loading.

Bending Movement.   The uniform beam shown in Fig. 18.7 is subjected to a couple C and produces the deformations illustrated. Notice that the beam is no longer straight, but has become bent. Deformations that bend mechanical components are associated with internal stress resultants called *bending moments,* denoted by M.

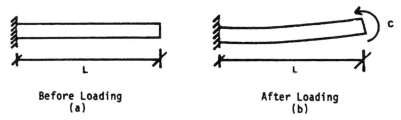

Before Loading
(a)

After Loading
(b)

FIGURE 18.7   Uniform beam before (*a*) and after (*b*) loading.

Torque.   The uniform circular shaft shown in Fig. 18.8 is subjected to a couple C. Although the deformations are difficult to show in a drawing, it should be apparent that the shaft must twist as the couple is transmitted to the support. Deformations that twist mechanical components are associated with internal stress resultants called *torque,* denoted by the vector C.

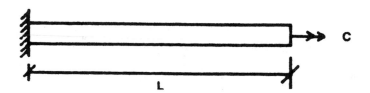

FIGURE 18.8   Uniform circular shaft.

18.3.2 Fundamental Stress Equations

Internal stress resultants represent the total force or moment acting at any point within a mechanical system. In reality these stress resultants are distributed over surfaces within the components, and their magnitudes per unit area are called *stress*. Stress is defined as force per unit area. Therefore, bending moments and torques must be described by equivalent static force systems made up of distributed normal or shear forces, in order to evaluate their effect in terms of stress.

Normal and Shear Stress.   Stresses acting perpendicular or normal to a surface are called *normal stresses,* while stresses acting in the plane of the surface are called *shear stresses.* Internal stress resultants, once reduced to force resultants, do not necessarily act normal or in the plane of the surface in question. For convenience, in mechanical stress analysis, internal force resultants are always represented in component form so that stresses are either normal stresses or shear stresses.

Consider an incremental internal force resultant ΔR acting on an incremental surface area ΔA. The vector sum of ΔP (normal force) and ΔS (shear force) may be used to replace ΔR. The normal force per unit area and the shear force per unit area are then called *normal stress f_n*, and *shear stress f_s*, respectively. Therefore, average values are defined by

$$f_{n(avg)} = \frac{\Delta P}{\Delta A}; \qquad f_{s(avg)} = \frac{\Delta S}{\Delta A} \tag{18.1a,b}$$

Notice that the correct stress involves more than magnitude and area. Normally, stress involves direction, sense, and location in addition to magnitude and area. Normal stresses acting away from a surface are considered positive and are called *tensile stress,* while normal stresses acting into the surface are considered negative and are called *compressive stress.* Unfortunately, there is no universal sign convention for shear stress.

18.3.3 Fundamental Strain Equations

All mechanical components or mechanical systems deform under load. In the first part of Sec. 18.3, the internal stress resultants were categorized in terms of the deformations produced. This section will investigate deformations in terms of change in size, change in shape, or both with respect to the original size and shape.

Normal Strain. Normal strain e_n is defined as a change in length per unit of original length. Recall that the uniform bar (Fig. 18.5) increased in length by δ when subjected to the normal force P. Therefore, average values for normal strain are defined by

$$e_{s(avg)} = \frac{\delta}{L} = \frac{L_{final} - L_{initial}}{L_{initial}} \tag{18.2a}$$

Shear Strain. Shear strain e_s is defined as change in angle between any two originally perpendicular lengths, measured in radians. Recall that the uniform square plate (Fig. 18.6) changed in shape when subjected to the shear force S. The lower-left corner forms perpendicular lengths, which changed under load to form the angle α. Therefore, average values for shear strain are defined in radians by

$$e_{s(avg)} = \frac{\pi}{2} - \alpha \tag{18.2b}$$

When the deformations are small and the angle change is also small, average shear strain can be defined in terms of the shear deformation δ_s:

$$e_{s(avg)} \approx \tan[e_{s(avg)}] = \frac{\delta_s}{L} \tag{18.3}$$

18.4 COMMONLY USED FORMULAS

18.4.1 Formulas and Equations

The formulas in the following sections are listed for reference purposes and are extracted directly from MIL-HDBK-5E. The sign conventions generally accepted in their use are as follows. Quantities associated with tensile action (load, stress, strain, etc.) are considered as positive and quantities associated with compressive action are considered negative. When compressive action is of primary interest, however, it is sometimes convenient to drop the minus sign and consider the associated quantities to be positive.

18.4.2 Simple Unit Stresses

$$f_t = \frac{P}{A} \qquad \text{(tension)} \tag{18.4}$$

$$f_c = \frac{P}{A} \qquad \text{(compression)} \tag{18.5}$$

$$f_b = \frac{My}{I} = \frac{M}{Z} \tag{18.6}$$

$$f_s = \frac{S}{A} \qquad \text{(average direct shear stress)} \tag{18.7}$$

$$f_s = \frac{SQ}{I_b} \qquad \text{(longitudinal or transverse shear stress)} \tag{18.8}$$

$$f_s = \frac{Ty}{I_p} \qquad \text{(shear stress in round tubes due to torsion)} \tag{18.9}$$

$$f_s = \frac{T}{2} \, At \qquad \text{(shear stress due to torsion in thin-walled structures of closed section)} \tag{18.10}$$

[Note that A in Eq. (18.10) is the area enclosed by the median line of the section.]

$$f_H = \frac{Pr}{t} \qquad \text{(hoop stress in thin-walled structures)} \tag{18.11}$$

$$f_A = Bf_H; \qquad f_T = Bf_L \tag{18.12}$$

18.4.3 Combined Stresses

$$f_n = f_c + f_b \qquad \text{(compression and bending)} \tag{18.13}$$

$$f_{s,max} = \sqrt{f_s^2 + \left(\frac{f_n}{2}\right)^2} \qquad \text{(compression, bending torsion)} \tag{18.14}$$

$$f_{n,max} = \frac{f_n}{2} + f_{s,max} \tag{18.15}$$

18.4.4 Deflections (Axial)

$$e = \frac{\delta}{L} \qquad \text{(unit deformation or strain)} \tag{18.16}$$

$$E = \frac{f}{e} \tag{18.17}$$

(This equation applies when E is to be found from tests in which f and e are measured.)

$$\delta = eL = \left(\frac{f}{E} \right) L \qquad (18.18)$$

$$\delta = \frac{PL}{AE} \qquad (18.19)$$

(This equation applies when the deflection is to be calculated using a known value of E.)

18.4.5 Deflection (Bending)

$$\frac{dt}{dx} = \frac{M}{EI} \qquad (18.20)$$

(This expresses change of slope per unit length of beam, radians per unit length.)

$$f_2 = i_1 + \int_{x_1}^{x_2} \left(\frac{M}{EI} \right) dx = \text{slope at point 2} \qquad (18.21)$$

(The integral denotes the area under the curve of M/EI plotted against x, between the limits of x_1 and x_2.)

$$y_2 = y_1 + i_1(x_2 - x_1) + \int_{x_1}^{x_2} \left(\frac{M}{EI} \right)(x_2 - x_1) dx = \text{deflection at point 2} \qquad (18.22)$$

(The integral denotes the area under the curve having ordinates equal to M/EI multiplied by the corresponding distances to point 2, plotted against x, between the limits of x_1 and x_2.)

$$y_2 = y_1 + \int_{x_1}^{x_2} i \, dx = \text{deflection at point 2} \qquad (18.23)$$

(The integral denotes the area under the curve of i plotted against x, between the limits x_1 and x_2.)

18.4.6 Deflections (Torsion)

$$\frac{d\Phi}{dx} = \frac{T}{GJ} \qquad (18.24)$$

(This expresses change of angular deflection or twist per unit length of member, radians per unit length.)

$$\Phi = \int_{x_1}^{x_2} \left(\frac{T}{GJ} \right) dx = \text{total twist over a length from } x_1 \text{ to } x_2 \qquad (18.25)$$

(The integral denotes the area under the curve of T/GJ plotted against x, between the limits of x_1 and x_2.)

$$\Phi = \frac{TL}{GJ} \qquad (18.26)$$

(This equation is used when torque T/GJ is constant over length L.)

18.4.7 Biaxial Elastic Deformation

$$\mu = \frac{\text{unit lateral deformation}}{\text{unit axial deformation}} \tag{18.27}$$

(This is Poisson's ratio in uniaxial loading.)

$$Ee_x = f_x - \mu f_y \tag{18.28}$$

$$Ee_y = f_y - \mu f_x \tag{18.29}$$

$$E_{\text{biaxial}} = \frac{E}{(1-\mu B)} \qquad \text{(biaxial elastic modulus)} \tag{18.30}$$

18.4.8 Basic Column Formula

$$F_c = \pi^2 E_t \left(\frac{L'}{\rho} \right)^2 \qquad \text{where} \qquad L' = L\sqrt{c} \qquad (c \text{ is end-fixity coefficient}) \tag{18.31}$$

18.5 *MECHANICAL FAILURE MODES*

This section will briefly present types of mechanical failure modes which can occur and which should be evaluated to ensure that there is sufficient material strength and adequate margins of safety.

- *Tensile-yield-strength failure:* This type of failure occurs under pure tension. It occurs when the applied stress exceeds the yield strength of the material. The result is permanent set or permanent deformation in the structure. This is seldom a catastrophic condition.
- *Ultimate tensile-strength failure:* This type of failure occurs when the applied stress exceeds the ultimate tensile strength and causes total failure of the structure at this cross-sectional point. This is a catastrophic condition.
- *Compressive failures:* Compressive failures are similar to the preceding tensile failures only under compressive loads. They result in permanent deformation or total compressive failure, causing cracking or rupturing of the material.
- *Failures due to shear loading:* Yield and ultimate failures occur when the shear stress exceeds the strength of the material when applying high torsion or shear loads. These failures generally occur on a 45° axis with respect to the principal axis.
- *Bearing failures:* Bearing failures are similar to compressive failures. However, they are generally caused by a round, cylindrical surface bearing on either a flat or a concave surface, such as roller bearings in a race. Both yield and ultimate bearing stresses should be determined.
- *Creep and stress rupture failures:* Long-term loads generally measured in years cause elastic materials to stretch even though they are below the normal yield strength of the material. If the load is maintained continuously and the material stretches (creeps), it will generally terminate in a rupture. Creep accelerates at elevated temperatures. This results in plastic strain due to

creep even though the load is in the elastic region. Creep should always be checked under conditions where high loading for long periods of time are anticipated.

- *Fatigue failures:* Fatigue failures result from repeated loading and unloading of a component. This is a major reason for reducing the design-allowable stresses below those listed in the various design tables for static properties. The amount of reduction in the design load is based on complex curves and test data. The amount of load cycling with respect to the maximum load and the number of cycles are the predominant variables. These variables are generally described in Moody diagrams (these are also called *S-N* curves, which relate stress to number of life cycles).

- *Metallurgical failures:* Metallurgical failures are the failures of materials due to extreme oxidation or operation in corrosive environments. Certain environmental conditions accelerate these metallurgical failures. The environmental conditions are (1) heat, (2) erosion, (3) corrosive media, and (4) nuclear radiation. Therefore, depending on the environment, these conditions should be considered.

- *Brittle fracture:* Certain materials which have little capacity for plastic flow and are generally brittle, such as glass or some of the harder steels, are extremely susceptible to surface flaws and imperfections. The material is elastic until the fracture stress is reached. Then the crack propagates rapidly and completely through the component. These fractures can be analyzed using fracture mechanics.

- *Bending failures:* A *bending failure* is a combined failure where an outer surface is in tension and the other outer surface is in compression. The failure can be represented by tensile rupture of the outer material.

- *Failures due to stress concentration:* This occurs when there is an uneven stress "flow" through a mechanical design. This stress concentration generally takes place at abrupt transitions from thick gauges to thin gauges, at abrupt changes in loading along a structure, at right-angle joints, or at various attachment conditions.

- *Failures due to flaws in materials:* This is generally due to poor quality assurance or improper inspection of materials, weld defects, fatigue cracks, or small cracks and flaws. These reduce the allowable strength of the material and result in premature failure at the flawed location.

- *Instability failures:* Instability failures occur in structural members such as beams and columns particularly those made from thin material and where the loading is generally in compression. The failure may also be caused by torsion or by combined loading including bending and compression. The failure condition is generally a crippling or complete failure of the structure.

All the preceding failure modes should be analyzed as necessary to confirm the adequacy of the mechanical design. It should be noted that some of the analysis is very involved, particularly in the case of combined loads or complex structures. These may require the use of computer programs which are described in Chap. 20.

18.6   SAFETY FACTORS AND MARGINS OF SAFETY

To design a mechanical component or system to perform a given function in a reliable manner, the designer must understand all the possible ways the component or system could lose its ability to perform that given function. This requires a complete understanding of how the system or component responds to externally applied loads. With the aforementioned modes of failure established, reasonable limits on load can be defined.

The *safety factor* (SF) is a strength design factor defined by the ratio of a critical design strength parameter (tensile, yield, etc.) to the anticipated operating stress under normal operating conditions. For example, let F be the strength of the material and f_{mw} be the maximum allowable working stress. Then the factor of safety becomes

$$SF = \frac{F}{f_{mw}} \tag{18.32}$$

The *safety margin* or *margin of safety* (MS) is usually expressed as the maximum allowable working stress (f_{mw}) divided by the applied stress f minus 1, as shown in the following equation:

$$MS = \frac{f_{mw}}{f} - 1.0 = \frac{F/SF}{f} - 1.0 \tag{18.33}$$

Any negative MS value indicates that the structure will fail because the applied stress is in excess of the allowable material strength. This is only for unidirectional stresses; biaxial and triaxial stresses require further analysis.

CHAPTER 19
MECHANICAL RELIABILITY

Richard L. Doyle
DOYLE AND ASSOCIATES
LA JOLLA, CALIF.

19.1 RELIABILITY PREDICTION (PRESENT TECHNIQUES)

Chapter 18 was devoted to determining the various stress levels or loads on mechanical systems. The lower the stress level with respect to capability or strength, the better the reliability. This chapter provides currently available techniques for determining the reliability of mechanical components.

With electronic hardware, manufacturers produce millions of similar parts that are then incorporated into thousands of different types of electronic equipment. Billions of hours of operation are accumulated, which are used to guess at electronic part failure rates. Because of the many different uses, it is possible to accumulate extensive data, with respect to failure rate, under various environmental, temperature, and electrical stress conditions.

With the exception of nuts-and-bolts-type hardware, mechanical equipment is designed for a specific configuration and use. There is usually insufficient quantity of any specific device to accumulate the hours necessary to establish a basic failure rate. This does not consider the many failure-rate variations caused by different stress levels, environment, and temperature extremes.

Sections 19.9 through 19.11 present a *probabilistic stress and strength analysis,* which provides techniques for determining structural reliability through the computation of the probability of failure. Although this technique provides a good reliability estimate based on stress and strength, it will be found that to use this method for every mechanical part of a mechanical system would be a long and costly effort.

One method for limiting the effort involved in probabilistic stress and strength analysis is through the use of a *failure modes, effects, and criticality analysis.* This procedure eliminates those failure mechanisms which are not critical to equipment performance, or will not otherwise affect the performance of a prescribed mission. Another method is to establish a reasonable safety factor to minimize the risk of part failure.

There are several methods, both in use and in the development stages, that can be used to determine the reliability of a mechanical system. These methods may vary from precise to simple estimates. The precise methods require extensive computations for each element that is used in the mechanical system. The simple approach uses generic data for each of the parts to obtain an estimated reliability, and as such, does not consider stress levels, temperature extremes, and material strengths.

19.2 BASIC RELIABILITY DEFINITION

To use standard exponential reliability formulas, it is necessary to assume that the failure rates of all parts within a given unit are constant, that is, that the failure rates are constant and do not increase or decrease with time. Those parts which exhibit a wearout mechanism or otherwise have a limited life, are replaced prior to any appreciable increase in failure rate.

As an example, a shutoff valve has a constant failure rate of 6.5 failures per one million cycles. The mean life is 90,000 cycles. The valve is used in equipment that is to have a 10-year life. On an average, the valve will be activated 10 times per hour throughout the equipment's life. Therefore, wearout will occur at:

$$\frac{90,000 \text{ cycles}}{10 \text{ cycles/h}} = 9000 \text{ h}$$

For the 6.5 failures per one million cycles for this device to be valid over the 10-year life of the equipment, this device must be scheduled for replacement every year (8760 h), or sooner.

19.3 PROPOSED METHODS OF DETERMINING MECHANICAL SYSTEM RELIABILITY

To date, no single method or technique has been accepted as the standard for computing the reliability of mechanical systems. The following paragraphs briefly describe two procedures that have been used in the past for computing mechanical systems reliability. Sections 19.5 through 19.8 provide the final implementation of these early practices.

19.3.1 U.S. MIL-STD-XXX, Procedures for Performing a Reliability Stress Analysis of Mechanical Equipment

In 1980, a proposed military standard was assembled entitled *Procedures for Performing a Reliability Stress Analysis of Mechanical Equipment*. This proposed document, which has never been approved for use, recognized the lack of standard failure rates and the problems associated with their establishment. The details in this proposed standard make note of the fact that there are considerable variations in failure rates for similar mechanical parts. This document provides a good description of the steps that should be taken during the process of computing reliability. Because of the lack of standard data, the difficulties in using these procedures were apparent and should become more apparent to the reader in the later paragraphs regarding data sources.

The concept of relating mechanical stress analysis to failure rate in this proposed standard is based on first determining the critical failure modes, and then limiting the stress analysis to those failure modes which are determined to be critical to equipment operation. This is a generally accepted procedure to limit the depth of the mechanical stress analysis.

For those failure modes considered to be critical, the standard requires

> An evaluation of the stress/strength processes for each part analyzed. This evaluation may be entirely deterministic, entirely probabilistic, or a combination thereof depending on the criticality of the evaluation results. A method for translating stress/strength relationships such as safety factors, into a reliability value shall be provided or probabilistic techniques shall be used.

This proposed standard also defines the procedures for the performance of a failure modes, effects, and criticality analysis (FMECA). These procedures are "form and format" or U.S. MIL-STD-1629A, *Procedures for Performing a Failure Modes, Effects and Criticality Analysis.* From the results of this type of analysis, direction is obtained with respect to limiting detailed stress analyses to those critical parameters of parts found to be critical to the performance of the entire system.

19.3.2 TTCP Committee for Reliability Prediction and Design Techniques

The Technical Cooperation Program (TTCP), an international committee representing the United States, United Kingdom, Canada, Australia, and New Zealand, has issued several documents relating to reliability and predictions of mechanical equipment. This provides a ballpark estimate at best.

In these documents, recognition is given to the problem of predicting the reliability of mechanical equipment:

> Many problems are encountered in trying to predict the reliability of mechanical components and systems, because of the unique characteristics of mechanical designs. Mechanical designs are often one of a kind so that historical data from "similar" equipment is not helpful in predicting performance reliability of the new equipment.

Table 19.1, taken from TTCP *Reliability Prediction of Mechanical Equipment,* presents typical variations in the failure rates for similar mechanical hardware from various failure-rate sources.

At this time, failure-rate models are being determined and test data derived for 18 standard

TABLE 19.1   Reliability Prediction Problems

Component	Range of published failure rates in failures per million hours	Maximum/minimum value ratio
Pump, medium oil	4.2–324	77
Valve, general	0.004–225	56,000
Valve, regulating	1.7–400	235

Source: TTCP *Reliability Prediction of Mechanical Equipment.*

building blocks for mechanical equipment. The technique and results are discussed in Secs. 19.5 through 19.8.

19.4   USE OF EXISTING GENERIC FAILURE-RATE DATA

Earlier paragraphs have briefly described efforts being made to establish methods for the determination of the reliability of mechanical equipment. This provides a ballpark estimate at best.

The easiest and most direct approach to computing the reliability of mechanical hardware is to use the failure rates of generic parts. As an example, it is possible to find the generic failure rate of a hydraulic pressure valve from several sources. Many different valve configurations may fit this general or generic description. However, this valve is probably only one part in a much larger mechanical system. Therefore, by using generic failure rates for all the parts that make up the system, it is possible to determine which parts will probably be the highest contributors to the system's unreliability. This directs emphasis and design attention to the "weak link" in the system.

There are several sources of failure rate information with respect to mechanical hardware. Data sources include:

GIDEP   Government Industry Data Exchange Program

NPRD-3   *Nonelectronic Parts Reliability Data,* Fall 1985, Reliability Analysis Center, Rome Air Development Center

AVCO   D. R. Earles, *Failure Rates,* AVCO Corp.

The following paragraphs briefly describe and provide examples of each of these data sources.

19.4.1 GIDEP Failure-Rate Sources

GIDEP publishes many documents which are continually being updated to provide government and industry with an in-depth perspective of what is currently taking place in practically every technology. Members contribute test data, reports, and other nonrestricted research data and reports. One benefit of GIDEP is that if one member has a question concerning a particular subject matter, GIDEP circulates the question, either as a simple or urgent request, to other GIDEP members. If any member has information in that area, it is normally provided within a relatively short period of time. GIDEP does not have the resources to carefully screen all of the data that is entered into their data base.

Figure 19.1 presents an example of one of the GIDEP data bases. In this data base, the reliability of mechanical components is provided. This GIDEP summary is issued on an annual basis. The primary drawback to using GIDEP as a source of failure-rate data for a part in a specific environment is the lengthy search through computer databases to obtain needed backup data. This is required when environmental conditions differ from the listed environment, or there is an appreciable difference in part application.

19.4.2 NPRD-3

NPRD-3, *Nonelectronic Parts Reliability Data Notebook 3,* published in Fall 1985 by the Reliability Analysis Center at Rome Air Development Center, is a report that is organized into five major sections. It presents reliability information based on field operation, dormant state, and test data for more than 250 major nonelectronic part types. The five sections are "Background," "Generic Data," "Detailed Data," "Failure Modes," and "Mechanisms." Each device type contains reliability information in relation to specific operational environments. The data presented in this reliability publication is intended for use as failure-rate data, not part replacement data (scheduled maintenance). Only verified failures were used in the calculations of the failure rates.

NPRD-3 provides failure-rate and failure-mode information for mechanical, electromechanical, electrical pneumatic, hydraulic, and rotating parts. The data utilized in the development of NPRD-3 represents equipment-level experience under field conditions in military, industrial, and commercial applications. In order to use this notebook, it is necessary to accept the assumption that the parts display a constant failure rate (exponential distribution). This assumption is necessary due to virtual absence of data containing individual times or cycles to failure.

Figure 19.2 presents an example of the type of information provided by NPRD-3 with respect to failure-rate information. It is noteworthy that no attempt has been made to either (1) develop environment factors that allow the use of these failure rates for environmental conditions other than those indicated, (2) develop temperature history, or (3) modify these failure rates for various safety factors (SFs) or materials.

19.4.3 AVCO Published Failure Rates

Although the failure-rate handbook prepared as part of a Reliability Engineer Data Series by D. R. Earles for the AVCO Corporation is outdated (1962), it provides an excellent source of information with respect to mechanical part failure rates. These generic failure rates are also easily computed for different environmental conditions.

Figure 19.3 presents an example of the types of generic failure-rate data provided in this handbook. It will be noted that in addition to providing failure rates in failures per million hours, it also provides failure rates in failures per million cycles, where applicable, as well as providing average life of parts either in hours or cycles.

G-DEP RELIABILITY-MAINTAINABILITY DATA SUMMARIES
PERMANENT ISSUE OCTOBER, 1984
REELS F027-F032 ANALYZED FAILURE RATE DATA

MAJOR SUBJECT CATEGORIES (NOMENCLATURE)								90% GROUP CONFIDENCE INTERVAL		
MICROFILM ACCESS NUMBER	ORIG CODE	MFR AND PART NUMBERS	OBSERVED ENVIRONMENT AND MODIFIER	APPLICATION	STRESS LEVEL SCR E-T-M CLS	OBSERVATN DATES BEGIN/END F-A-I-L-U-R-E	MEAN REPAIR HOURS M-O-D-E-S	TOTAL ITEMS TESTED FAILED	# OF ITEMS FAILED	MILLION ITEM-UNITS OF MEASURE	UNIT OF MEAS	MILLION FAILURE RATE PER MILLION LOWER / MEAN / UPPER

TRANSISTORS, POWER, NPN (CONTINUED FROM PREVIOUS PAGE)

| F027-1724 | N1 | MOTOROLA GM7515H | DOR STORAG | MISSILES | - - | 7512/7601 | | 110 | | 0.2088 HRS | | 4.790 |
| F027-1724 | N1 | FAIRCHILD G657071 | DOR STORAG | MISSILES | - - | 7310/7601 | | 10 | | 0.1994 HRS | | 5.010 |

TRANSISTORS, SWITCHING, SILICON

| F027-1724 | N1 | TELEDYNE 2N2907, TM33117 | DOR STORAG | MISSILES | - - | 7207/7601 | | 80 | GROUP INTERVAL NOT COMPUTED | 2.4367 HRS | | 0.410 |

TRANSISTORS, SWITCHING, SILICON

| F031-1361 | J7 | ---- 2N2907A JAN | GND CNTRLD | RECEIVER/TRANSMITTER | -M- | 7007/7807 | | 9 | 0 | 8.5811 HRS | | <0.107 |
| F031-1361 | J7 | ---- 2N897 JAN | GND CNTRLD | RECEIVER/TRANSMITTER | -M- | 7007/7807 | | 1 | 0 | 0.9457 HRS | | <0.969 |

VALVES, FLOW CONTROL, WATER

| F031-0598 | FY | CLA-VAL 882240204 | SHP SEBSD | FIRE MAIN | - - | 7504/7701 | 1.8 | 58 | GROUP INTERVAL NOT COMPUTED | 0.8960 HRS | | <1.023 |

VALVES, PILOT, RELEASE

| F031-0598 | FY | FIGGIE INT SHP SEBSD 1H 4210-00-575-9297, 88224D028 | FIRE MAINS/SPRINKLER | | - - | 7504/7701 | 5.8 | 90 | GROUP INTERVAL NOT COMPUTED | 1.4400 HRS | | <0.836 |

FIGURE 19.1 GIDEP reliability summary. (*Source: 1991 AR&MS Tutorial Notes.*)

VALVE, HYD.

COMPONENT PART TYPE	APPL ENV	USER CODE	POINT ESTIMATE	60% UPPER SINGLE-SIDED	20% LOWER INTERVAL	80% UPPER INTERVAL	# OF RECS	# OF FAIL	OPERATING HRS (E6)
FUEL	AUF	M	24.450	-----	19.787	30.271	1	20	0.818
FUEL	AUT	C	3.056	-----	2.487	3.762	7	21	6.872
FUEL	DOR	M	-----	0.127	-----	-----	1	0	7.220
FUEL	GF	M	-----	8.327	-----	-----	1	0	0.110
FUEL	HEL	M	40.000	-----	16.271	86.462	1	2	0.050

*The table above is headed "FAILURE RATE PER MILLION HRS. *"*

VALVE, HYD.

COMPONENT PART TYPE	APPL ENV	USER CODE	POINT ESTIMATE	60% UPPER SINGLE-SIDED	20% LOWER INTERVAL	80% UPPER INTERVAL	# OF RECS	# OF FAIL	OPERATING HRS (E6)
GATE	A	M	32.448	-----	24.092	43.695	1	11	0.339
GATE	GF	?	8.471	-----	8.075	8.889	13	328	38.722
GATE	GF	M	1.336	-----	0.975	1.829	4	10	7.484
GATE	HEL	M	71.429	-----	44.179	113.510	1	5	0.070

The table above is headed "FAILURE RATE PER MILLION HRS."

VALVE, HYD.

COMPONENT PART TYPE	APPL ENV	USER CODE	POINT ESTIMATE	60% UPPER SINGLE-SIDED	20% LOWER INTERVAL	80% UPPER INTERVAL	# OF RECS	# OF FAIL	OPERATING HRS (E6)
GLOBE	GF	?	3.616	-----	3.397	3.853	17	194	53.646
GLOBE	GF	M	0.173	-----	0.034	0.526	2	1	5.784
GLOBE	GM	M	-----	1.104	-----	-----	1	0	0.829

The table above is headed "FAILURE RATE PER MILLION HRS."

*Obviously 5 significant places far exceed the factor of 2 prediction capability.

FIGURE 19.2 NPRD-3 example: failure-rate information. (*Source: 1991 Annual Reliability and Maintainability Symposium.*)

Also included in Fig. 19.3 are the environmental factors that are applicable to the failure rates in this handbook. There is also a table of factors for use in computing the average life of parts with respect to the part application and environment. Because of the age of this handbook, it may be difficult to obtain a copy. Also, there are a large number of mechanical components with no failure rates or average life provided. However, it is still widely used today in spite of its inadequacies.

Generic Failure Rate / Generic Life Expectancy Table

Component or Part	Generic Failure Rate (1/10⁶ hrs, cyc)			Generic Life Expectancy (10⁶ ops)		
	Lower Limit	Mean	Upper Limit	Lower Limit	Mean	Upper Limit
Turbine	3.33	10.0	16.67	0.008	0.011	0.015
Valves, Air	0.112	5.1	14.8	0.0001cy	0.15cy	1.2?cy
Valves, Ball	0.112	0.212	2.29	50.0cy	60.0cy	100.0cy
Valves, Blade	1.11	4.6	7.7	0.07cy	0.1cy	0.2cy
Valves, Bleed	1.08	4.6	7.4	0.07cy	0.1cy	0.2cy
Valves, Butterfly	3.112	5.7	18.94	0.05cy	0.09cy	0.15cy
Valves, By-Pass	1.33	3.4	8.13	0.08cy	0.15cy	0.25cy
Valves, Check	0.16	2.24	4.7			
Valves, Control	0.24	1.9	2.2	0.04cy	0.08cy	0.125cy
Valves, Drain					10.0cy	
Valves, Filler	0.1	0.22?	1.12			
Valves, Float	5.50	8.0	11.2			
Valves, Fuel	1.24	6.4	37.2			
Valves, G		0.112				
Valves, Gate				0.05cy	0.08cy	1.2?cy
Valves, Jettison	0.112	5.5	32.5	0.0001cy	0.067cy	0.2cy
Valves, Pressure	0.165	6.3	14.8			
Valves, Priority		37.2			0.087cy?	
Valves, Propellant		0.56			0.035	
Valves, Regulator	0.224	5.7	14.1	0.05cy	0.1cy	0.25cy
Valves, Relief	0.224	3.92	12.5		0.093	
Valves, Relief, Pressure	5.6	6.4	12.3			
Valves, Relief, Thermal	2.70	6.88	16.8			
Valves, Reservoir .7Tl	3.70	16.0	19.7			
Valves, Selector	0.57	1.10	2.14	0.02cy	0.05cy	0.08cy
Valves, Selector, Manual	2.22	3.0	4.16			
Valves, Wing Flap	2.10	4.6	91.0			
Valves, Sequence	16.8	30.0	56.0	0.045cy	0.09cy	0.12cy
Valves, Servo	0.112	6.5	10.2			
Valves, Shut-off		0.112				
Valves, Shut-off, Manual	0.56	3.36	2.86		0.4	
Valves, Shut-off, Motor Drive		0.56				
Valves, Shut-off, Press. Drive		1.12				
Valves, Shuttle		1.12		0.06cy	0.15cy	0.3cy
Valves, Slider	2.27	11.0	19.7		0.052cy	
Valves, Solenoid	1.87	4.6	7.41		50.0cy	
Valves, Solenoid, 3 Way	1.81	4.6	7.22			
Valves, Solenoid, 4 Way	2.99	6.9	9.76			
Valves, Spool	0.26	0.5	2.24	0.05cy	0.1cy	0.02?cy
Valves, Transfer	3.41	5.7	1.531		0.05	
Valves, Vent and Relief				0.012	0.015	0.02
Variac						
Vibrators	0.4	0.5	1.2			
Wave Guides, Fixed	0.59	1.5	454			
Wave Guides, Flexible	0.59	1.1	1.92			
Wave Guides, Irid	1.133	2.64	4.54			

Operating Mode Table

OPERATING MODE	Average Kf
Satellite (In-Orbit)	1
Laboratory Computer	1
Ground Equipment	8
Shipboard Equipment	15
Rail-Mounted Equipment	22
Aircraft Equipment (Bench Test)	30
Missile Equipment (Bench Test)	40
Aircraft Equipment (In-Flight)	90
Missile Equipment (In-Flight)	400
Nose Cone Compartment (In-Flight)	800
Between Tanks, Stage II (In-Flight)	900
Sustainer Engine Compartment (In-Flight)	1000
Between Tanks, Stage I (In-Flight)	1250
Booster Engine Compartment	1500

Kel Table

Installation Environment	Kel			
	All Eqpt.	Electronic and Electrical Eqpt.	Electro-Mechanical Eqpt.	Dynamic Mechanical Eqpt.
Satellites	2.50	2.60	2.40	2.10
Laboratory Computer	1.00	1.00	1.00	1.00
Bench Test	0.54	0.55	0.51	0.50
Ground	0.30	0.31	0.26	0.23
Shipboard	0.19	0.21	0.17	0.15
Aircraft	0.16	0.18	0.14	0.12
Missiles	0.15	0.17	0.13	0.11

FIGURE 19.3 AVCO example: failure-rate information. (*Source: 1991 AR&MS Tutorial Notes.*)

19.4.4 Comparison of Generic Failure-Rate Data

Because the GIDEP failure-rate data source is more complex to use, this comparison has been limited to NPRD-3 and the AVCO handbook. Table 19.2 presents a comparison of five different generic part types. The O-ring gasket is the only part whose failure rate is in agreement. There is about a 9:1 difference in failure rates for ball bearings; AVCO is more conservative.

TABLE 19.2 Failure-Rate Comparison: NPRD-3 to AVCO Handbook

Part description	Environment	Failure rate/million (M), hours or cycles	
		NPRD-3	AVCO
Ball bearings	Naval-sheltered	1.148/M h	9.75/M h
O-ring gasket	Naval-sheltered	0.530/M h	0.30/M h
Fuel pump	Ground-fixed	176.4/M h	18.0/M h
Pressure regulator	Ground-fixed	2.435/M h	34.0/M cycles
Ball valve	Ground-fixed	0.647/M h	36.8/M cycles

Note: To convert failure rate per million cycles to failure rate per million hours, multiply the number of cycles per hour that the device operates times the cycle failure rate.

EXAMPLE 19.1   A pressure regulator failure rate is 34.0 failures per million cycles. It will operate at 200 cycles per hour. Therefore, 34.0 failures/1,000,000 cycles $\times$ 200 cycles/h and

$$\text{Failure rate} = 6800 \text{ failures per million hours}$$

The large discrepancy in fuel pumps is probably due to the type of fuel pump(s) from which the data was derived. The pressure regulator and the ball valve have a different basis for failure rates, namely, hours and cycles. Notice that NPRD-3 is in failures per million hours, while AVCO is in failures per million cycles. Failures per million cycles is probably more realistic, since cyclic devices are generally designed for a life expectancy specified in number of cycles. Mechanical life is more dependent on actual cycles than operating hours.

19.4.5 Computing Mechanical Systems Reliability

The method of computing the reliability of a mechanical system, either MTBF or probability of success, will depend on personal preference. The reliability can normally be determined, in descending order of preference by

1. Failure-rate models developed in Secs. 19.5 through 19.8
2. Probabilistic stress and strength analysis for each part (Secs. 19.9 through 19.11)
3. Probabilistic stress and strength analysis for each critical part, with generic data for all others
4. Parts-count technique for all parts using generic data
5. GIDEP data for all parts

Because of costs and time constraints, technique 2 is rarely imposed. For critical mechanical equipment, technique 1 should be required. For the majority of mechanical systems, technique 1 or 3 is usually acceptable.

The use of the parts-count technique greatly simplifies the reliability computations of mechanical systems. This technique, basically, is to list all mechanical parts used, determine the generic failure rate for each part in a specific environment, add up the failure rates, and then compute the reliability, either MTBF or probability of success. One should realize that predicting a value does not guarantee that the value will be within a factor of 10 of the actual test results, but it will provide trend data and show which are the critical parts.

19.4.6 Example Reliability Calculation

Table 19.3 presents an example listing of various mechanical parts, quantity used, cyclic use (if applicable), failure rate per million hours for a fixed-ground environment, and the source of the failure-rate data. These failure rates are then summed together for a total failure rate of this example mechanical system. Inverting the total failure rate provides the Mean Time Between Failures (MTBF).

TABLE 19.3 Example of Mechanical Parts and System Failure Rates

Part description	Quantity	Cyclic use	Generic failure rate per million hours	Total failure rate per million hours	Data source
Heavy-duty ball bearing	6	N/A	14.4	86.4	AVCO
Brake assembly	4	N/A	16.8	67.2	AVCO
Cam	2	$1\ h^{-1}$	0.016	0.032	AVCO
Pneumatic hose	1	N/A	29.28	29.28	AVCO
Fixed displacement pump	1	N/A	1.464	1.464	NPRD-3
Manifold	1	N/A	8.80	8.80	AVCO
Guide pin	5	N/A	13.0	65.0	AVCO
Control valve	1	$40\ h^{-1}$	15.20	15.20	AVCO
Total assembly failure rate				273.376	

MTBF = 1/0.000273376 = 3657.9 h.

For a mission time of 500 h, the probability of success P_s is

$$P_s = \exp\left(-500 \times .000273376\right) = .872$$

Remember, generic failure-rate sources assume that parts display a constant failure rate.

The utilization of generic data sources (NPRD-3 and AVCO) provides a relatively fast but somewhat inaccurate computation of a mechanical equipment's reliability. However, it generally isolates those elements that contribute the greatest degree to the equipment's unreliability. In other words, this technique provides a fast, easy method of determining reliability as well as providing direction as to where concentration should be placed in the design to achieve a more reliable mechanical system.

19.5 RELIABILITY PREDICTION (USING PRODUCT MULTIPLIERS)

19.5.1 Introduction

At the present time there is no commonly accepted method for predicting the reliability of mechanical equipment that is equivalent to that for electronic equipment. Documents such as the *Nonelectronic Part Reliability Data Notebook* (Sec. 19.4) were developed from existing failure-rate databases and contain a listing of failure rates for various mechanical components. The accuracy of a reliability prediction using the database approach cannot be determined because of the wide dispersion of failure rates which occur for apparently similar components. Variations in failure rates for mechanical equipment are the result of the following:

- *Multiple functions:* Individual mechanical components such as valves and gearboxes often perform more than one function, and failure data for specific applications of nonstandard components are seldom available.

- *Nonconstant failure rate:* Failure rates of mechanical components are seldom described by a constant failure rate because of wear, fatigue, and other stress-related failure mechanisms resulting in equipment degradation. Data gathering is complicated when a constant failure rate cannot be assumed, and individual times to failure must be recorded in addition to total operating hours and total failures.

- *Stress history:* Mechanical equipment reliability is more sensitive to loading, operating mode, and utilization rate than is electronic equipment reliability. Failure-rate data based on operating time alone is usually inadequate for a reliability prediction of mechanical equipment.

- *Criticality of failure:* Definition of failure for mechanical equipment depends on the application. For example, failure due to excessive noise or leakage cannot be universally established. Lack of such information in a failure-rate database limits its usefulness.

These variables associated with acquiring failure-rate data for mechanical components demonstrate the need for reliability prediction models that do not rely solely on existing failure-rate databases. Useful models must provide the capability of predicting the reliability of all types of mechanical components by specific failure modes considering the operating environment, the effects of wear, and other potential causes of degradation. The models developed are based on the identified failure modes and their causes. The first step in developing the models was to derive equations for each failure mode from design information and experimental data contained in published technical reports and journals. These equations were simplified to retain those variables affecting reliability as indicated from field-experience data. The failure-rate models utilize the resulting parameters in the equations, and modification factors were compiled for each variable to reflect its effect on the failure rate of individual component parts. The total failure of the component is the sum of the failure rates for the component parts. Failure-rate equations for each component part, the methods used to generate the models in terms of failures per hour or failures per cycle, and the limitations of the models are presented. The models are being validated with laboratory testing results.

The following mechanical components and parts have been identified for which failure-rate models are to be developed:

- Group A
  Valve, regulator
  Actuator, cylinder
  Seal, O-ring, gasket
  Spring
  Bearing

- Group B
  Slider-crank mechanism
  Gear, spline
  Impacting device
  Solenoid
  Connector, fitting

- Group C
  Drive unit, gearbox
  Clutch, brake assembly

Coupling, universal joint

Structural joint

- Group D

Motor, pump, compressor

Transducer, sensor

Accumulator, reservoir

Filter assembly

19.5.2 Failure-Rate Models for Valve Assemblies

This chapter contains a failure-rate model of a seal for a fluid valve assembly. The technique can be used to support the development of a proposed design modification. The model is intended to focus attention on design analyses which should be accomplished to provide a reliability estimate of the valve in its intended operational service environment.

Failure-rate models should be based on identified failure modes of the individual parts. Failure modes and their failure causes and effects have been developed for a valve assembly as presented in Table 19.4.

TABLE 19.4   Failure Modes of a Valve Assembly

Failure mode	Failure cause	Local effect
Seal leakage	Embrittlement, installation damage, wear, surface damage, distortion, dynamic instability	Internal or external valve leakage
Worn or damaged poppet seat	Wear of poppet or seat assembly, contaminants	Poppet not seating properly, causing internal leakage and low or erratic pressure drop
Sticking valve piston in main valve body	Contaminants, loss of lubrication, air entrapment; excessively high temperature; structural interference	Low erratic pressure drop; slow operating response, valve immobile
Broken spring or damaged spring ends	Fatigue	Unable to adjust or maintain pressure
Inoperative solenoid assembly	Open coil winding, misalignment of solenoid with respect to spool or poppet stem	Valve fails to open or close
Cracked connector or housing	Fatigue, external shock, vibration	External leakage

After the failure rates are determined for each component part, the results are summed to determine the failure rate of the total valve assembly (for statistically independent failures):

$$\lambda_p = \lambda_1 + \lambda_2 + \lambda_3 + \lambda_4 + \lambda_5 + \lambda_6 \tag{19.1}$$

where λ_p is the failure rate of the total valve assembly in failures per million cycles and λ_1 through λ_6 are the failure rates for the individual component parts in failures per million cycles. Failure mechanisms resulting in part degradation and the failure rate as a function of time are considered statistically independently in each failure-rate model.

19.6 *FAILURE-RATE MODEL FOR SEALS*

A *seal* is a device placed between two surfaces to restrict the flow of fluid from one region to another. Static seals, such as gaskets and O-rings, are used to prevent leakage through a mechanical joint when there is no relative motion of mating surfaces other than that induced by changes in the environment. A *dynamic seal* is a mechanical device used to control leakage of fluid from one region to another when there is relative motion between the sealing interface. For valve applications, relatively low rotary or linear speeds are involved. The effectiveness of a seal design is determined by the capability of the seal interfaces to maintain good mating over the seal operating environment.

The primary failure mode of a seal is leakage. Table 19.5 is a list of failure mechanisms and causes of seal leakage. Other failure mechanisms and causes should be identified for a specific application to ensure that all considerations of reliability are included in the prediction.

TABLE 19.5   Failure-Rate Considerations for Seals

Failure mode	Failure mechanism	Failure causes
Leakage	Wear	Contaminants, inadequate lubrication
	Elastic deformation	Extreme temperature
	Seal distortion	Misalignment; shaft out-of-roundness; seal eccentricity
	Surface damage	Inadequate lubrication
	Dynamic instability	Contaminants; fluid-seal incompatibility; idle periods of component use

The first step in determining the failure rate of a seal is to isolate each leakage path. In many cases backup seals are used in a valve design, and the failure rate is not necessarily proportional to the total number of seals in the valve. Each seal design must be evaluated as a potential internal or external leakage path.

A review of failure-rate data suggests that the following characteristics be included in the failure rate model for seals:

- Leakage requirements
- Material characteristics
- Amount of seal compression
- Surface irregularities
- Extent of pressure pulses
- Fluid viscosity
- Fluid-material compatibility
- Static versus dynamic conditions
- Fluid pressure
- Seal size
- Quality control and manufacturing processes
- Contamination level

The failure rate of a seal or gasket material will be proportional to the ratio of actual leakage to that allowable under conditions of usage. This rate can be expressed as follows:

$$\lambda_1 = \lambda_{b1} \left(\frac{Q_a}{Q_f} \right)$$ (19.2)

where λ_1 = failure rate of seal or gasket, failures per million cycles
λ_{b1} = base failure rate of seal or gasket, failures per million cycles
Q_a = actual leakage rate tendency, in³/min
Q_f = leakage rate considered to be a device failure, in³/min

The allowable leakage Q_f is determined from design drawings, specifications, or knowledge of component applications. The actual leakage rate Q_a for an O-ring is determined from a standard equation for laminar flow around two curved surfaces:

$$Q_a = \frac{\pi(P_s^2 - P_0^2)}{24vP_0} \left[\frac{r_2 + r_1}{r_2 - r_1} \right] H^3$$ (19.3)

where P_s = system pressure, lb/in²
P_0 = standard atmospheric pressure or downstream pressure, lb/in²
v = absolute fluid viscosity, lb · min/in²
r_1 = inside radius of circular interface, in
r_2 = outside radius of circular interface, in
H = conduction parameter, in

For flat seals or gaskets, the leakage can be determined from the following equation:

$$Q_a = \frac{2\pi r_1(P_s^2 - P_0^2)}{24vLP_0} H^3$$ (19.4)

where r_1 is inside radius and L is contact length. The conduction parameter H is dependent on contact stress, material properties, and surface finish. To find H, the fluid pressure P across the seal is determined and the apparent contact stress (load/area) calculated. Next, the ratio of contact stress to Meyer hardness of the softer interface material is computed and the surface finish of the harder material determined. The conduction parameter is computed from the empirically devised formula

$$H^3 = 10^{-1.1(C/M)} f^{1.5}$$ (19.5)

where M = Meyer hardness (or Young's modulus) for rubber-resilient materials
C = apparent contact stress, psi (lb/in²)
f = surface finish, in

The surface finish will deteriorate at a rate dependent on the rate of fluid flow and the number of contaminants according to the following relationship:

$$Z \approx f(\alpha \eta Q dT)$$ (19.6)

where Z = seal degradation
α = contaminant wear coefficient, (in³/particle)²
η = number of contaminant particles, in⁻³
Q = flow rate, in³/min
d = ratio of time the seal is subjected to contaminants under pressure versus no pressure
T = temperature of operation, °F

The *contaminant wear coefficient* is a sensitivity factor for the valve assembly based on performance requirements. The number of contaminants includes those produced by wear in upstream components after the filter and those ingested by the system. Combining and simplifying terms provides the following equation for the failure rate of a seal:

$$\lambda_1 = \lambda_{b1} xf\left[\frac{(P_s^2 - P_0^2)}{Q_f \nu P_0}\left(\frac{r_2 + r_1}{r_2 - r_1}\right)(H^3)(\alpha \eta dT)\right] \qquad \text{[for circular seals (O-rings)]} \quad (19.7a)$$

or

$$\lambda_1 = \lambda_{b1} xf\left[\frac{(P_s^2 - P_0^2)r_1}{Q_f \nu L P_0}(H^3)(\alpha \eta dT)\right] \qquad \text{(for seals and gaskets)} \quad (19.7b)$$

where λ_b is the base failure rate of a seal due to random cuts, installation errors, etc., based on field experience data.

$$\lambda_1 = \lambda_{b1} C_P C_Q C_D C_H C_f C_V C_T C_N C_W \qquad (19.8)$$

where λ_1 = failure rate of a seal in failures/million cycles
λ_{b1} = base failure rate of seal
C_P = multiplying factor which considers the effect of fluid pressure on the base failure rate
C_Q = multiplying factor which considers the effect of allowable leakage on the base failure rate
C_D = multiplying factor which considers seal size
C_H = multiplying factor which considers the effect of surface hardness and other conductance parameters on the base failure rate
C_f = multiplying factor which considers seal smoothness
C_V = multiplying factor which considers the effect of fluid viscosity on the base failure rate
C_T = multiplying factor which considers the effect of temperature on the base failure rate
C_N = multiplying factor which considers the effect of contaminants on the base failure rate
C_W = multiplying factor which considers the effect of flow rate on base failure rate

Many of the parameters in the failure-rate equation can be located on an engineering drawing, by knowledge of design standards or by actual measurement. Other parameters which have a minor effect on reliability are included in the base failure rate, as determined from field-performance data.

19.7 RELIABILITY PREDICTION OF BEARINGS

19.7.1 General

From the standpoint of reliability, bearings are by far the most important gearbox components, since they are among the few components that are designed for a finite life. Bearing life is usually calculated using the Lundberg-Palmgren method. This method is a statistical technique based on the subsurface initiation of fatigue cracks through hardened air-melt-bearing material. Bearing life is generally expressed as the L_{10} life, which is the number of hours at a given load that 90 percent of a set of apparently identical bearings will complete or exceed. There are a number of other factors that can be applied to the L_{10} life so that it more accurately correlates with the observed life. These factors include material, processing, lubrication film thickness, and misalignment.

The mean time to failure (MTTF) of a system of bearings can be expressed as

$$\text{MTTF}_{\text{B}} = \frac{5.45L_{10}}{N^{0.9}} \qquad\qquad (19.9a)$$

where MTTF_{B} = design MTTF for the system of bearings
$\quad\quad L_{10}$ = design life for each individual bearing
$\quad\quad N$ = number of bearings in system

Most handbooks express the L_{10} life of a bearing to be a Weibull distribution as follows (refer to *Marks' Handbook,* 8th ed., pp. 8–138):

$$L_{10} = 16{,}700/\text{rpm} \left(\frac{C}{P} \right)^k \qquad\qquad (19.9b)$$

where L_{10} = rated life in revolutions
$\quad\quad C$ = basic load rating, lb (lookup value from handbooks)
$\quad\quad P$ = equivalent radial, load, lb
$\quad\quad K$ = constant: 3 for ball bearings; 10/3 for roller bearings
$\quad$ rpm = shaft rotating velocity, rev/min

The preceding expression assumes a Weibull failure distribution with the shape parameter equal to 10/9, the value generally used for the number N of rolling-element applicable bearings. The life L_{10} given in Eq. (19.9b) includes the effects of lubrication, material, and misalignment under design conditions in addition to the common parameters. This value will usually suffice during early design. Typical material processing factors used for vacuum-melt aircraft-quality bearings are 6.0 for ball bearings and 4.0 for roller and tapered bearings. These material processing factors should be incorporated into the basic MTTF_{B} to determine a representative life of these bearings. These factors should be multiplied times the MTTF_{B} to get the expected MTTF.

If the bearings being considered are low-speed (<100 rpm) the MTTF_{B} should be multiplied by 2 to account for the fact that the lubrication factor is very low.

19.7.2 Ball Bearings

Ball bearings are generally used where there is likely to be excessive misalignment or shaft deflection. They are also used, especially in duplex arrangements, where accurate axial positioning is required in the presence of thrust loads, such as with bevel gear shafts. Ball bearings are not as common in the main drive train of more recent designs because of advancements made with tapered roller bearings. Ball bearings are, however, often used on lightly loaded accessory shafts. Higher bearing life is easily achieved in these applications because of the very small loads, thus the installation is simplified. Also since a ball bearing is nonseparable, it requires no special setup procedures.

19.7.3 Cylindrical Roller Bearings

Cylindrical roller bearings are used to support pure radial loads. They are often used at one end of highly loaded gear shafts with either tapered roller bearings or multiple-row matched ball bearings at the other end. Roller bearing life is drastically reduced by excessive misalignment or deflection; hence, when using roller bearings, the stackup of tolerances contributing to misalignment and the shaft or housing deflections should be carefully considered. To compensate for misalignment or deflection and to carry heavy radial loads, roller bearings are crowned to prevent the phenomenon known as *end-loading* (see Fig. 19.4).

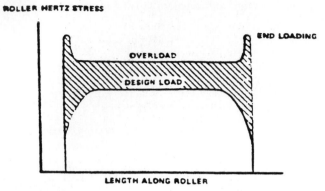

FIGURE 19.4   End-loading of cylindrical roller bearings.

End-loading invariably leads to a drastic reduction in bearing life. The amount of crown to be used should be based on maximum continuous power. At lower powers, the crown will not appreciably change bearing life, and at higher powers transients load the bearings for short durations only.

19.7.4   Internal Clearance

Internal clearance is an important consideration in the design of ball and roller bearings, since improper internal clearance can drastically shorten the life of a bearing. Too little internal clearance limits the amount of misalignment that can be tolerated and can lead to heavily preloaded bearings, particularly at low temperatures. Excessive internal clearance will cause the load to be carried by too few rolling elements. The best practice is to ensure that under all conditions there will be a small internal clearance. Usually, the most appreciable factors to consider when determining the mounted internal clearance of the bearing are reduction of internal clearance due to shaft or housing fits, and the effect of temperature on the magnesium housing and outer race interface diameters.

19.7.5   Tapered Roller Bearings

Tapered roller bearings are being used increasingly in modern drive systems, since they can react to both thrust and radial loads and can offer the greatest load-carrying capacity in the smallest possible envelope. Although early tapered roller bearings were speed-limited, these restrictions have been removed by utilizing bearings with special lubrication features. However, on very high-speed shafts, the use of tapered roller bearings may be precluded because of their inability to operate for required time intervals under survivability (on/off) conditions. Tapered roller bearings, unlike single-row ball and cylindrical roller bearings, require spacers or shims to give these bearings the proper amount of preload or end play for proper operation. Usually it is desirable to have a light preload, although a small amount of end play is often acceptable. As with internal clearance, extremes in end play or preload should be scrupulously avoided.

19.7.6   Bearing Failure Modes

The most prevalent mode of bearing failure is *spalling,* which is defined as chipping or breaking of the bearing surface. The failure is usually caused by poor lubrication (premature failure) or by fatigue or wearout due to the repeated stress and strain of a lifetime of usage. Another mode of bearing failure is *creeping* or *spin,* caused by an improper fit between the bearing and the shaft or outer surface of the bearing. This allows movement of the race in relation to the housing or shaft. The surfaces then wear or score, thereby damaging the surfaces and preventing a firm, fixed contact.

Roller and tapered bearings have an additional failure mode defined as *hard lines* or *scuffing* of the bearing surfaces. This failure mode is usually caused by bearing exposure to an excessive load for an extensive period of time. The surfaces of the moving parts are scored or scratched, increasing the roughness of the surfaces, setting up stress concentrations, and increasing friction. The scoring also interferes with the normal lubricant film and increases the metal-to-metal contact during use.

19.7.7 Bearing Failure-Rate Prediction

The basic failure rate for bearings sets can be determined by using Eq. (19.9) to obtain

$$L_{10} = \frac{N^{0.9}(\mathrm{MTTF_B})}{5.45} \tag{19.10}$$

MTTF would be the average life of the bearings excluding other components. In spite of incorrect mathematics, people often invert the MTTF to get a pseudo–failure rate. This pseudo–failure rate of bearings in failures per million hours of operation can be expressed as

$$\lambda_{b_1} = \frac{10^6 N^{0.9}}{5.45 \, L_{10}} \tag{19.11}$$

and this can be expressed as λ_{b_1}:

$$\lambda_{b_1} = \frac{10^6}{(\mathrm{MTTF_B})} \qquad \text{(by definition)} \tag{19.12}$$

where λ_{b_1} is pseudo–failure rate per million hours of operation.

Few mechanical systems are utilized precisely as the bearing manufacturer envisioned; therefore, some adjustment factors must be used to approximate the failure rate of the bearings under specific conditions. The following is an expression for predicting the failure rate of bearings:

$$\lambda_b = \lambda_{b_1} \left(\frac{L_A}{L_S} \right)^y \left(\frac{A}{0.006} \right)^{2.36} \left(\frac{v_0}{v_L} \right)^{0.54} C^{2/3} \tag{19.13}$$

where λ_b = predicted failure rate of bearing, using actual conditions, per million hours of operation
  y = 3.33 for roller bearing; 4.0 for ball bearings
  L_A = actual load, psi
  L_S = specification load, psi
  A = alignment error, radians
  v_0 = specification lubricant viscosity, lb · min/in²
  v_L = operating lubricant viscosity, lb · min/in²
  C = (actual contamination level, µg/m³)/(60 µg/m³)

All the base parameters can be obtained from the manufacturer's drawings or specifications. The environmental parameters can either be measured at the operating site or predicted from a knowledge of the operating conditions.

To estimate the failure rate of a specific bearing, the basic conditions and the operating conditions can be substituted in Eqs. (19.11) to (19.13).

19.8   RELIABILITY PREDICTION OF OTHER WEAROUT PARTS

19.8.1   Introduction

Mathematical models provide the capability of predicting the reliability of all types of mechanical components. These are categorized by specific part types and failure modes. They also consider the operating environment, the effects of wear, and other potential causes of degradation. The models developed are based on the derivation of equations for each failure mode from design information and experimental data contained in published technical reports and journals. These equations are simplified to only retain those variables affecting reliability as indicated from field experience data. The failure-rate models use the parameters in the equations and factors as compiled for each variable to reflect its effect on the failure rate of individual parts. The total failure of the component is the sum of the failure rates of the parts. This represents a series reliability model where no redundancy is considered. Failure-rate equations for each part, the methods used to generate the models in terms of failures per hour or failures per cycle, and the limitations of the models are presented in the U.S. Navy *Handbook of Reliability Prediction Procedures for Mechanical Equipment* (NSWC-92/L01, May 1992). The models are being validated with laboratory testing results.

The following are other mechanical parts (and assemblies) and the basis for the mathematical model. Most of the following mathematical and failure-rate models have already been developed, but the reader can tailor these models to fit any design problem for which a solution is needed.

19.8.2   Failure-Rate Models for Parts

Probabilistic design can be determined for each type of part. This is determined by taking the basic equation that describes the stresses in the part and performing a Taylor's series expansion of the basic equation as described in Sec. 19.11. For example, bending stress in a beam would be calculated using the equation MY/I [from Chap. 18, Eq. (18.6)]. However, since I and Y are related (not statistically independent) one must use the section modulus Z. It is very important to use statistically independent variables or the technique will not be correct.

One can also incorporate time varying parameters such as wear and wearout. To do this, it is necessary to quantify the wear rate. An example given earlier (Chap. 18) was a railroad car carrying coal for many years. The coal car side plates wear thin and finally rupture. Plate thickness as a function of time can be determined and placed into the statistical plate equation. This will provide a time to wear out and a probability failure rate about that wearout failure. This same technique works for fatigue life. The only difference is that strength of the material is changing as a function of stress cycles (time and magnitude of load fluctuation). This can be factored into the equation for determining the fatigue life. For a more complex problem one can determine the fatigue life of a coal car (fatigue- and wear-related failures).

The following constitute the basis for the fundamental part failures. Each of these is relatively simple. The complex equations should be simplified by holding constant the less important parameters. This will provide ease of calculation.

Beam	Based on standard beam stress equations
Shaft	Based on standard shaft stress equations
Plate	Based on standard plate stress equations
Spring	Based on standard spring stress equations
Connector, fitting	Based on standard fitting stress equations
Structural joint	Based on standard fitting stress equations
Accumulator, reservoir	Based on pressure vessel stress equations
Bearing	As described in Sec. 19.7

Gear, spline	Based on standard gear stress equations
Seal, O-ring, gasket	As described in Sec. 19.6
Solenoid	Based on standard failure-rate data

19.8.3 Failure-Rate Models for Assemblies

Probabilistic design and life can be determined for subassemblies and major assemblies, also. This is determined by taking the failure rates of each part in the assembly and combining these failure rates. Extreme care must be taken when combining failure rates to ensure that only series reliability failures occur. Otherwise one must consider the redundancy equations, which are an order of magnitude more complex.

The following assemblies have one or more of the fundamental parts listed above which make up the system. Also, there may be multiple loading conditions and multiple load paths. This may require a more complex stress analysis such as using sophisticated tools (finite element models) as described in Chap. 20. However, these parts generally are nothing more than the sum of their pieces. Therefore, the reliability value is simply determined by summing the failure rates of each of the failure modes. One can quickly see that the reliability basis is exactly "the sum of all failure rate data."

Drive unit, gearbox	Based on sum of all failure-rate data
Clutch, brake assembly	Based on sum of all failure-rate data
Coupling, universal joint	Based on sum of all failure-rate data
Filter assembly	Based on sum of all failure-rate data
Transducer, sensor	Based on sum of all failure-rate data
Motor, pump, compressor	Based on sum of all failure-rate data
Valve, regulator	Based on sum of all failure-rate data
Actuator, cylinder	Based on sum of all failure-rate data
Impacting device	Based on sum of all failure-rate data
Slider-crank mechanism	Based on sum of all failure-rate data

The ultimate design goal would be to manufacture a product that fails slightly past its useful life. The tolerance of failure, say, 3 sigma (3σ), would be added onto its useful life and that would be the design MTBF. More on this topic is covered in Chap. 20.

19.9 PROBABILISTIC STRESS AND STRENGTH ANALYSIS

19.9.1 Introduction

Probabilistic design can be a very important evaluation tool for predicting the probability of failure of mechanical systems operating in a random loading environment. Ships, airplanes, missiles, and similar vehicles experience random forces which must be considered by the designer during the conceptual phases of the design. The ultimate design goal would be to manufacture a product that never fails. This, however, is rarely achievable because of other considerations, such as cost and weight, which must be integral parts of any design. On the other hand, it may be possible to design a product (or system) that has a low probability of failure (i.e., high reliability) which does not appreciably impact other design constraints.

Conventional design and stress analysis studies do not recognize the fact that material properties, design specifications, and applied loads can be characterized as having a nominal value and some variation. Knowledge of the variables, as well as their statistical properties, will enable the designer to estimate the probability that the part will not fail.

In Sections 19.10 and 19.11 it is assumed that all random variables are statistically independent and are normally distributed. Assumptions of dependent variables and underlying distributions other than normality are complex to analyze. These nonnormal distributions seldom enhance the validity of the analysis since we don't know what the distribution is.

19.10   *STATISTICAL CONCEPTS AND PRINCIPLES*

The following paragraphs discuss concepts and principles which must be understood in order to effectively address the topic of probabilistic design with respect to the stress-strength relationship.

19.10.1   Normal Distribution

The normal (or gaussian) distribution is one of the most important continuous probability distributions available to the designer. It has practical applications in numerous disciplines and can be used as a design tool for predicting stress-strength relationships.

The *normal distribution* is a bell-shaped curve characterized by a mean μ and standard deviation σ. Once the mean and standard deviation have been specified, the curve is completely determined. Figure 19.5 illustrates a typical normal curve for a random variable X having mean μ and standard deviation σ.

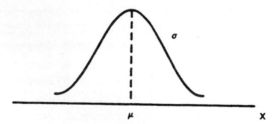

FIGURE 19.5   The normal curve.

In Fig. 19.5, the area under the curve for all values of X where $-\infty < X < \infty$ is unity. Consequently, the probability of X lying between two values, for example, X_1 and X_2, can be found by determining the area under the curve from X_1 to X_2. This probability would be represented by and the shaded area of Fig. 19.6.

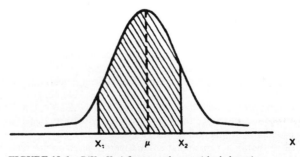

FIGURE 19.6 $P(X_1 < X < X_2)$ for normal curve (shaded area).

19.10.2 Standard Normal Distribution

In practical applications, a linear transformation is always made to a random variable Z where

$$Z = \frac{X - \mu}{\sigma} \tag{19.14}$$

This transformation converts any normally distributed random variable X having mean μ and standard deviation σ to a random variable Z having a mean of zero and a standard deviation of 1. Therefore, if $X_1 < X < X_2$, then $Z_1 < Z < Z_2$, where

$$Z_1 = \frac{X_1 - \mu}{\sigma} \qquad \text{and} \qquad Z_2 = \frac{X_2 - \mu}{\sigma} \tag{19.15a,b}$$

The probability that X lies between X_1 and X_2 can be found by determining the probability that Z lies between Z_1 and Z_2.

Table A.4 in App. A gives the area under the standard normal curve from $-\infty$ to Z. To find $P(Z_1 < Z < Z_2)$ it is necessary to determine the area under the curve up to Z_2 and then subtract the area under the curve up to Z_1.

EXAMPLE 19.2   *GIVEN:* X distributed normally with $\mu = 25$ and $\sigma = 5$. *Determine:* the probability that X assumes values between 19 and 28.

Solution   Making a transformation to Z, Z_1 and Z_2 can be determined as

$$Z_1 = \frac{19 - 25}{5}$$

$$= -1.20$$

and

$$Z_2 = \frac{28 - 25}{5}$$

$$= +0.60$$

thus

$$P(19 < X < 28) = P(-1.20 < Z < 0.60)$$

$$= P(Z < 0.60) - P(Z < -1.20)$$

$$= .7257 - .1151$$

$$= .6106$$

This answer is obtained using the appropriate values for Z_1 and Z_2 found in Table A.4.

19.10.3 Mean and Variance of *n* Statistically Independent Variables

In solving practical problems involving one or more random variables, it is necessary to define a mathematical relationship expressing how the variables are related.

For simplicity, let the variable W (not to be confused with the Z transformation for the standard distribution) be a function of these random variables. Inasmuch as each random variable is assumed to be normally distributed with a mean μ_i and a standard deviation σ_i, it follows that W is also distributed normally and has a mean and variance that must be determined. Since it is assumed that the random variables are statistically independent, it follows that there is no linear correlation between them, thus eliminating the need to estimate a statistical parameter called the correlation coefficient.

Consider the sum of two statistically independent normally distributed random variables, for example, X and Y. To determine the mean and variance (σ^2) of their sum, that is, $X + Y$, let $W = X + Y$ be the random variable representing the sum. Let the mean and standard deviations of X and Y be represented by μ_X, σ_X, μ_Y, and σ_Y, respectively. It follows that W is also normally distributed, having mean μ_W and standard deviation σ_W. It can be shown that $\mu_W = \mu_X + \mu_Y$ and that $(\sigma_W)^2 = (\sigma_X)^2 + (\sigma_Y)^2$. It is possible to determine the probability that W is greater than some value, lies between two specified values, is less than some value, etc., depending on the problem to be solved.

There are, of course, other ways in which random variables can be combined. Table 19.6 contains various combinations of the random variables X and Y. This table will provide the basic relationships required for one method of determining appropriate means and variances in the remainder of the chapter.

TABLE 19.6 Means and Standard Deviations of Statistically Independent Random Variables X and Y

Algebraic function, W	Mean	Standard deviation
1. $W = ax$	$a\mu_x$	$a\sigma x$
2. $W = ax \pm by$	$a\mu_x \pm b\mu_y$	$\sqrt{a^2\sigma_x^2 + b^2\sigma_Y^2}$
3. $W = axy$	$\cong a\mu_x\mu_y$	$\cong a\sqrt{\mu_x^2\sigma_y^2 + \mu_y^2\sigma_x^2}$
4. $W = ax/by$	$\cong a\mu_x/b\mu_y$	$\cong \dfrac{a}{b\mu_y}\sqrt{\dfrac{\mu_y^2\sigma_x^2 + \mu_x^2\sigma_y^2}{\mu_y^2}}$
5. $W = x^2$	$\cong \mu_x^2 + \sigma_x^2$	$\cong \sqrt{4\mu_x^2\sigma_x^2 + 2\sigma_x^4}$
6. $W = 1/x$	$\cong 1/\mu_x$	$\cong \sigma_x/\mu_x^2$

Notes: (1) a and b are constants representing real numbers—they can be negative quantities, but the absolute values of a and b must be used for the standard deviation; (2) it is almost impossible to estimate the standard deviation accurately even for large samples.

19.10.4 Stress-Strength Relationship

Both stress and strength exhibit statistical variability, each with a mean and standard deviation. Failure occurs whenever the stress applied exceeds the strength. Knowledge of the statistical parameters defining the strength and stress enables the calculation of the probability of failure. In addition, the reliability can be defined as one minus the probability of failure. Possibly the easiest way to illustrate how the probability of failure can be estimated is through an example.

EXAMPLE 19.3 *Given:* a structural member is under a load. The strength of the member is a random variable having a mean and standard deviation of 27,000 and 3200 psi, respectively. The stress is also a random variable having a mean and standard deviation of 18,000 and 1500 psi, respectively.

Determine: the probability of failure.

Solution   The design does not fail as long as the strength of the beam exceeds the stress imposed on it. Let F represent a random variable associated with the strength and f a random variable associated with the stress. It is assumed that F and f are statistically independent and normally distributed random variables. Thus, $\mu_F = 27{,}000$, $\sigma_F = 3200$, $\mu_f = 18{,}000$ and $\sigma_f = 1500$. Setting up the random variable D where $D = F - f$, the failure occurs when f exceeds F, or conversely $D < 0$. In order to determine the $P\,(D < 0)$, it is necessary to know that it is a normal distribution, and its mean and variance. Referring to Table 19.6, if $W = X - Y$, then its mean and variance are given as $\mu_x - \mu_y$ and $\sigma_x^2 + \sigma_y^2$, respectively. However, there are no variables X and Y. By letting F and f replace X and Y, respectively, one obtains $\mu_0 = \mu_F - \mu_f$ and $\sigma_D^2 = \sigma_f^2 + \sigma_f^2$. The graph defining the random variable D would be depicted as shown in Fig. 19.7.

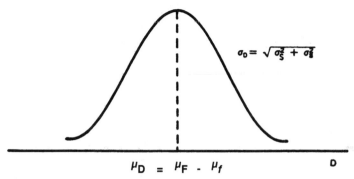

FIGURE 19.7   Random variable.

Since it is known that D is normally distributed, transformation can be made to the standard normal curve for computing probabilities involving D. Letting

$$Z = \frac{D - \mu_D}{\sigma_D} \tag{19.16}$$

Then

$$p(D < 0) = p\left(Z < -\frac{\mu_D}{\sigma_D}\right)$$

$$= p\left(Z < -\frac{\mu_F - \mu_f}{\sqrt{\sigma_F^2 + \sigma_f^2}}\right) \tag{19.17}$$

The shaded area of the standard normal curve shown in Fig. 19.8 represents the probability of failure. Substituting appropriate values:

$$p(Z < 0) = p\left(Z < -\frac{27{,}000 - 18{,}000}{\sqrt{(3200)^2 + (1500)^2}}\right) \tag{19.18}$$

$$= p\left(Z < -\frac{9000}{3534}\right)$$

$$= p(Z < -2.547)$$

$$= .0054$$

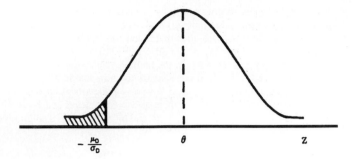

FIGURE 19.8   Probability of failure.

Therefore, the probability of failure is estimated to be .0054 and the reliability, to be .9946.

Suppose that the specification has a reliability requirement of .998, keeping σ_p, μ_p, and σ_F the same. What would have to be the value of μ_F to attain this reliability? The solution would be to determine the value of Z such that the area to the left of that value is .0002. Referring to Table A.4, $Z = -3.49$, and the area under the curve to the left is .0002. Thus

$$-3.49 = -\frac{\mu_S - 18,000}{\sqrt{(3200^2) + (1500^2)}}$$

(19.19)

Solving for μ_S gives a value of 30,300 psi.

19.11   TAYLOR'S SERIES FOR DETERMINING MEANS AND VARIANCES OF RANDOM VARIABLES

Determination of the mean and variance of functions which are nonlinear in terms of their components can be accomplished by using a Taylor's series. Basically, the function is approximated by a linear function in the region of interest. For this study, the region of interest is considered to be the mean values of each component. Functions of n variables, for example, $f(x_1, x_2, x_3, \ldots, x_n)$, can be expanded into a Taylor's series about a particular point, giving

$$f(x_1, x_2, \ldots, x_n) = f(\mu_1, \mu_2, \ldots, \mu_n) + (x_1 - \mu_1)\left.\frac{\partial f}{\partial x_1}\right|_{\mu_1, \mu_2, \ldots, \mu_n}, + \cdots + (x_n - \mu_n)$$

$$\left.\frac{\partial f}{\partial x_n}\right|_{\mu_1, \mu_2, \ldots, \mu_n} \qquad \text{(excluding terms of higher order)}$$

(19.20)

If the higher terms are neglected and the variables are assumed to be statistically independent, the mean of the function, denoted by μ_f is

$$\mu_f = f(\mu_1, \mu_2, \ldots, \mu_n)$$

(19.21)

The variance is approximately given by

$$\sigma_f^2 = \sigma_{x1}^2 \left(\left.\frac{\partial f}{\partial x_1}\right|_{\mu_1, \mu_2, \ldots, \mu_n}\right)^2 + \cdots + \sigma_{xn}^2 \left(\left.\frac{\partial f}{\partial x_n}\right|_{\mu_1, \mu_2, \ldots, \mu_n}\right)^2$$

(19.22)

To demonstrate how the preceding discussion would be used in solving a practical problem, assume that a random variable Z is expressed in terms of the product of two variables, such as x_1 and x_2. Furthermore, let $x_1 \sim (\mu_1, \sigma_1)$ and $x_2 \sim (\mu_2, \sigma_2)$. Then

$$W = f(x_1, x_2) = x_1 x_2 \tag{19.23}$$

The mean and variance of Z is desired. Thus, using Taylor's expansion

$$f(\mu_1, \mu_2) = \mu_1 \mu_2 \tag{19.24}$$

$$\frac{\partial f}{\partial x_1} \bigg|_{\mu_1, \mu_2} = x_2 \big|_{\mu_1, \mu_2} = \mu_2 \tag{19.25}$$

and

$$\frac{\partial f}{\partial x_2} \bigg|_{\mu_1, \mu_2} = x_1 \big|_{\mu_1, \mu_2} = \mu_1 \tag{19.26}$$

Thus, W can be expressed as

$$W = \mu_1 \mu_2 + (x_1 - \mu_1)\mu_2 + (x_2 - \mu_2)\mu_1 \tag{19.27}$$

The mean and variance of the function are given as

$$\mu_z = \mu_1 \mu_2 \tag{19.28}$$

and

$$\sigma_z^2 = \mu_2{}^2 \sigma_1{}^2 + \mu_1{}^2 \sigma_2{}^2 \tag{19.29}$$

These equations are identical to the equations presented in $W = xy$ equations found in Table 19.6.

The Taylor's series expansion about a point greatly facilitates computation of means and variances encountered in mechanical stress analysis. It is sometimes difficult to determine which are the appropriate equations. However, all that is required is for the designer to define the stress equation. The mean and variance of this function can then be estimated using the Taylor's expansion.

EXAMPLE 19.4   *GIVEN:* a lug, depicted in Fig. 19.9, subjected to a certain tensile load.

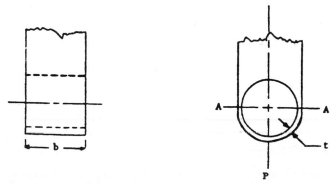

FIGURE 19.9   Lug subjected to a tensile load.

Determine: the reliability of the lug when considered in light of the following statistics and their respective values.

Parameter	Symbol	Value
Ultimate tensile strength mean S	μ_F	65,000 psi
Standard deviation of ultimate tensile strength	σ_F	3,600 psi
Load P	μ_P	21,200 lb
Standard deviation of load	σ_P	2,120 lb
Radial thickness mean t	μ_t	0.250 in
Radial thickness standard deviation	σ_t	0.005 in
Width b	μ_b	0.8625 in
Standard deviation of width	σ_b	0.005 in

As shown in Fig. 19.9, the potential failure mode of this lug is prevalent in section A–A. The potential failure governing the stress is action at A–A and can be expressed mathematically as

$$f = \frac{\text{load}}{\text{area}} = \frac{P}{A} \tag{19.30}$$

where A represents the area of section A–A. Furthermore, A can be expressed mathematically as

$$A = 2bt \tag{19.31}$$

Substitution of the relationship defining A in Eq. (19.31) into Eq. (19.30) gives f as

$$f = \frac{P}{2bt} \tag{19.32}$$

Determination of the mean and standard deviation of the stress will enable the designer to estimate the reliability.

Taking the partial derivatives of f with respect to each of the random variables P, b, and t gives

$$\frac{\partial f}{\partial p} = \frac{1}{2bt} \tag{19.33}$$

$$\frac{\partial f}{\partial b} = -\frac{P}{2b^2 t} \tag{19.34}$$

and

$$\frac{\partial f}{\partial t} = -\frac{P}{2bt^2} \tag{19.35}$$

Expanding the function describing f in a Taylor's series gives

$$f = \frac{\mu_P}{2\mu_b\mu_t} + \frac{(P - \mu_P)}{2\mu_b\mu_t} - \frac{(b - \mu_b)\mu_P}{2\mu_t\mu_b^2} - \frac{(t - \mu_t)\mu_P}{2\mu_b\mu_t^2} \tag{19.36}$$

It follows that the mean value of f, say, μ_f is always given by the first term in the expansion and is

$$\mu_f = \frac{\mu_P}{2\mu_b\mu_t} \approx \frac{P}{2bt} \tag{19.37}$$

Substituting the appropriate values of P, b, and t into Eq. (19.37) gives

$$\mu_f = \frac{21{,}200}{2(.8625)(.25)} = 49{,}159 \tag{19.37a}$$

The variance of f is also directly obtained from squaring the Δ values in Taylor's expansion for f and is given as (these are statistically independent)

$$\sigma_f^2 = \left(\frac{\partial f}{\partial P}\right)^2 \sigma_P^2 + \left(\frac{\partial f}{\partial b}\right)^2 \sigma_b^2 + \left(\frac{\partial f}{\partial t}\right)^2 \sigma_t^2 \tag{19.38}$$

Substituting the appropriate values for the partial derivatives in this equation gives

$$\sigma_f^2 = \frac{(2120)^2}{4(.8625)^2(.25)^2} + \frac{(21{,}200)^2(.005)^2}{4(.25)^2(.8625)^4} + \frac{(21{,}200)^2(.005)^2}{4(.8625)^2(.25)^4} \tag{19.39}$$

Taking the square root of $(\sigma_f)^2$ gives the standard deviation of the stress as 5020 psi.

Determining the reliability of the lug is relatively straightforward once μ_F, σ_F, μ_f, and σ_f are known. Setting up the D statistic where

$$D = F - f \tag{19.40}$$

and proceeding in the same manner gives

$$p(D < 0) = P\left(Z < -\frac{\mu_F - \mu_f}{\sqrt{\sigma_F^2 + \sigma_f^2}}\right)$$

$$= p\left(Z < -\frac{65{,}000 - 49{,}159}{\sqrt{(3600)^2 + (5020)^2}}\right)$$

$$= p(Z < -2.57)$$

$$= .0051 \ (0.51\%)$$

Therefore, the reliability is $(1 - .0051) = .9949$.

See also Tables 20.1 and 20.2 for variations in metal strength and in manufacturing tolerances.

CHAPTER 20
DESIGN FOR MECHANICAL RELIABILITY

Richard L. Doyle
DOYLE AND ASSOCIATES
LA JOLLA, CALIF.

20.1 INTRODUCTION

Reliability, just like other important parameters, is established by the design and manufacturing process. Design for reliability (time to failure) and design for manufacturability will provide the highest reliability, but will be dependent on how good the production, quality, or maintenance program may be. The design disciplines and analyses that are discussed in this chapter identify and focus on design weaknesses so that they may be corrected, protected against, or accepted after consideration. They also provide a means of assuring that a design will meet specified time-to-failure requirements prior to a production commitment.

The purpose of this chapter is to establish and define *design for mechanical reliability* (DFMR) rules and guidelines for the design of mechanical parts, assemblies, and mechanical systems no matter how large or small they are. The integrated use of DFMR rules can help reduce new product introduction time and eliminate most delays caused by repeated design iterations and rework.

The DFMR guidelines presented in this chapter are intended to assist design engineers with a rules-driven system and focus on an increased level of awareness on the constraints of downstream fabrication, assembly, and the operating environmental stresses. Increasing both design simplicity and component reliability are the thrust behind higher levels of component life and the necessity of getting the design right the first time. Without DFMR rules in place, it is almost certain that poor reliability will result and defeat the high-yield, low-cost assembly goals.

It is generally difficult to demonstrate reliability with a high degree of statistical confidence, prior to actual usage. There normally is neither the time nor the number of test samples available to demonstrate a high mean time between failures (MTBF) for a new design. Thus, as reliability requirements become more stringent and harder to demonstrate cost-effectively, emphasis tends to increase on the mechanical design parameters. By the time production has started, it is often too late to identify reliability problems. It becomes necessary, therefore, to translate reliability requirements into design criteria, against which compliance can be analytically measured prior to a production commitment. Although high reliability itself cannot be efficiently measured, the design parameters that determine reliability can be, and the control and analytical verification of these parameters will lead to the attainment of a reliable design.

There are two fundamental bases for failure: (1) failures resulting from higher stresses than the part can withstand and (2) failures due to the part actually wearing out. Both types of failures are covered in this chapter. However, the second type of failure is the most intriguing, since wearout comes in many different forms. Wearout can be due to fatigue, like a spring failing after being subjected to the same load for many years. It can be due to abrasion, like a railroad car carrying coal for many years. The coal car side plates wear thin and finally rupture. Wear and wearout also are caused by friction, thermal cycling, etc. Generally all mechanical parts have a finite life. It can be very long like the pyramids of Egypt or very short like the tires on an automobile. We will attempt to quantify the wearout time for all parts on the basis of stresses and wear in this chapter. This way one can predict the MTBF and when to perform maintenance, whether it is maintenance on an aircraft carrier or on the Statue of Liberty. Even buildings and automobile bridges have design lives, based on the fatigue life of the structural steel or the design life of the reinforcing steel in the concrete.

20.1.1   Design Objectives

This section covers seven important design objectives to achieve high mechanical reliability and system performance:

1. *Mechanical failure modes (structural):* The different types of failures that can occur and how they can be prevented.
2. *Safety factors and margins of safety:* The reliability that will be built into a part. You set the reliability value that you want.
3. *Alternate load paths (structural redundancy):* When the structure fails, is it a disaster, or does the design reliability of the structure (backup load paths) provide a safe failure (noncatastrophic)?
4. *Mechanical system redundancy:* When an entire subsystem fails, is it a disaster, or does the reliability of the system provide an alternate system and results in a safe failure (noncatastrophic)?
5. *Mechanical system failures:* When an entire system fails, is it a disaster, or does the reliability of the system provide a safe failure (noncatastrophic)?
6. *Warranty period and availability:* When any part of a system fails, it must be repaired or replaced. This section considers who is responsible for the cost of repair and what the effects are on system availability.
7. *Mechanical failure-rate distributions:* Different types of failure distributions and how they impact the mechanical design are discussed.

The first step in the development of a reliable system is to analyze the types of failure that may occur and determine the safety factor or the probability of success. Success may be measured in many ways, being available and reliable (PAR) or mission reliability (MTBF or time between overhauls). Reliability must be built in during the design and manufacturing phase by the designers and producers of the product.

20.1.2   Mechanical Failure Modes

This section will briefly present the types of mechanical failure modes which can be avoided by good design practices. The designer should consider each of these modes to ensure that there is sufficient material strength and adequate margins of safety to meet the reliability goals.

Tensile-Yield-Strength Failure.   This type of failure occurs under pure tension. It occurs when the applied stress exceeds the yield strength of the material. The result is permanent set or permanent deformation in the structure. This is seldom a catastrophic condition. Recommended safety factors on this type of failure can be as low as 1 (design to yield) or high (4 or 5) in the case of bridges and pressure vessels.

Ultimate Tensile-Strength Failure.   This type of failure occurs when the applied stress exceeds the ultimate tensile strength and causes total failure of the structure at a cross-sectional point. This is a catastrophic condition. Recommended safety factors on this type of failure can be as low as 1 (design to fail) or high (5 or 6) in the case of bridges and pressure vessels. The entire cross section is highly stressed and consequently is sensitive to flaws. This generally requires a good surface finish, and good manufacturing controls to minimize internal flaws.

Compressive Failures.   Compressive failures are similar to the preceding tensile failures, except under compressive loads. They result in permanent deformation or total compressive failure causing cracking or rupturing of the material. This is a catastrophic condition. Recommended safety factors on this type of failure can be as low as 1 (designed to fail in compression for some controlled crushing, not buckling) or high (5 or 6) in the case of bridges.

Failures Due to Shear Loading.   Yield and ultimate failures occur when the shear stress exceeds the strengths of the material when applying high-torsion or high-shear loads. These failures generally occur on a 45° axis with respect to the principal axis. The outer surface is highly stressed, and consequently is sensitive to surface flaws. This generally requires a good surface finish.

Bearing Failures.   Bearing failures are similar to compressive failures. However, they are generally caused by a round, cylindrical surface bearing on either a flat or a curved surface like roller bearings in a race. Both yield and ultimate bearing stresses should be determined. It is critical to consider fatigue failures if repeated loading occurs.

Creep and Stress Rupture Failures.   Long-term loads, generally measured in years, cause elastic materials to stretch even though they are below the normal yield strength of the material. If the load is maintained continuously and the material stretches (creeps), it will generally terminate in a rupture. Creep accelerates at elevated temperatures. This results in plastic strain due to creep even though the load is in the elastic region. Creep should always be checked under conditions where high loading for long periods of time are anticipated. The material tends to anneal and weaken over a long period of time or at an elevated temperature.

Fatigue Failures.   Fatigue failures result from repeated loading and unloading (or partial unloading) of a component. This is a major reason for reducing the design-allowable stresses below those listed in the various design tables for static properties. The amount of reduction in the design load is based on complex curves and test data. The amount of load cycling with respect to the maximum load and the number of cycles are the predominant variables. These variables are described in S-N (stress versus number of life cycles) diagrams. The actual material used must be evaluated or the part may fail before the expected wearout time. The choice of materials is very important to prevent this type of failure. Steel will outlast aluminum under cyclic loading. The S-N curves for steel extend much further than aluminum and tend to flatten out (S-N curve dog leg). An example of the best materials for fatigue endurance is to look at good spring material. You do not find many aluminum springs because the fatigue characteristics are poor.

Metallurgical Failures.   *Metallurgical failures* are the failures of materials due to extreme oxidation or operation in corrosive environments. Certain environmental conditions accelerate these metallurgical failures. The environmental conditions are (1) heat, (2) erosion, (3) corrosive media, and (4) nuclear radiation. Therefore, depending on the environment, these conditions should be considered. They are all a function of time and magnitude of the corrosion rate. Both variables can be determined and are very predictable.

Brittle Fracture.   Certain materials which have little capacity for plastic flow and are generally brittle, such as glass or some of the harder steels, are extremely susceptible to surface flaws and imperfections. The material is elastic until the fracture stress is reached. Then the crack propagates rapidly and completely through the component. These brittle fractures can be predicted on the basis of the surface finish and porosity, using fracture mechanics. Reliability generally requires a good surface finish and very few internal flaws.

Bending Failures.   A *bending failure* is a combined failure where an outer surface is in tension and the other outer surface is in compression. The failure can be represented by tensile rupture of the outer material. The outer surfaces are highly stressed, and consequently the tensile side is sensitive to surface flaws. Reliability generally requires a good surface finish.

Failures Due to Stress Concentration.   These failures occur when there is an uneven stress "flow" through a mechanical design. Stress concentration generally occurs at abrupt transitions from thick gauges to thin gauges, at abrupt changes in loading along a structure, at right-angle joints, or at various attachments and connections. Stress concentration factors are discussed in Sec. 20.4.

Failures Due to Flaws in Materials.   This is generally due to poor quality assurance or improper inspection of materials, weld defects, fatigue cracks, or small cracks and flaws. These reduce the allowable strength of the material and result in premature failure at the flawed location.

Instability Failures.   Instability failures occur in structural members such as beams and columns, particularly those made from thin material where the loading is generally in compression. The failure may also be caused by torsion, or by combined loading including bending and compression. The failure condition is generally a crippling or complete failure of the structure. Since these failures are very predictable but have a large dispersion from sample to sample, it is advisable to use a generous safety factor. Usually a safety factor of 2 is sufficient, but never use less than 1.5 because failures of instability will sometimes occur at less than the predicted value.

Conclusion.   All the preceding mechanical failure modes should be analyzed as necessary to confirm the adequacy of the design. It should be noted that some of the analysis is very involved, particularly in the case of combined loads or complex structures. These may require the use of computer programs which are described in Sec. 20.13.

20.1.3 Safety Factors and Margins of Safety

In order to design a mechanical component or system to perform a given function in a reliable manner, the designer must understand all the possible ways the component or system could fail. This requires a complete understanding of how the system or component responds to externally applied loads. With the aforementioned modes of failure established, reasonable limits on load can be defined.

As mentioned in Chap. 18, the *safety factor* (SF) is a strength design factor defined by the ratio of a critical-design strength parameter (tensile, yield, etc.) to the anticipated operating stress under normal operating conditions. The factor of safety is

$$SF = \frac{F}{f_{mw}} \tag{20.1}$$

where SF = safety factor
      F = strength of the material
      f_{mw} = maximum allowable working stress

The safety factor must always be greater than 1.0. Any value less than one indicates that the structure will fail, because the applied stress is in excess of the material strength.

Taking the derivatives of the terms and evaluating the mean and standard deviation provides an uncertainty value for the safety factor. This value can be used to determine whether the safety factor is sufficient or if it should be adjusted upward (or downward).

$$\Delta SF = \frac{\Delta F}{f_{mw}} - \frac{F}{f_{mw}^2} \Delta f_{mw} \tag{20.2}$$

Combining the squares of the standard deviation yields the standard deviation for the safety factor as follows:

$$\sigma_{SF} = \sqrt{\left(\frac{\sigma_F}{f_{mw}}\right)^2 + \left(\frac{F}{f_{mw}^2}\right)^2 (\sigma_{f_{mw}})^2} \tag{20.3}$$

where the σ_F is determined per Sec. 20.3.2 and $\sigma_{f_{mw}}$ is determined per Sec. 19.11. The mean value for the safety factor is determined by using the mean values in Eq. (20.2).

The *margin of safety* (MS) is usually expressed as the maximum allowable working stress divided by the applied stress minus 1, as shown in the following equation:

$$MS = \frac{f_{mw}}{f} - 1.0 = \frac{F/SF}{f} - 1.0 \tag{20.4}$$

where MS = margin of safety
f_{mw} = maximum allowable working stress
f = applied stress
F = strength of the material

The margin of safety must always be greater than zero. Any value less than zero (negative) indicates that the structure will fail because the applied stress is in excess of the allowable material strength.

Taking the derivatives of the terms and evaluating the mean and standard deviation provides an uncertainty value for the margin of safety. This value can be used to determine whether the structure is sufficient or should be redesigned.

Combining the squares of the standard deviation yields the standard deviation for the margin of safety as follows:

$$\sigma_{MS} = \frac{1}{SF} \sqrt{\left(\frac{\sigma_F}{f}\right)^2 + \left(\frac{F}{f^2}\right)^2 (\sigma_f)^2} \tag{20.5}$$

where the σ_F is determined per Sec. 20.3.2 and σ_f is determined per variations in stresses as described in Sec. 19.11. The mean value for the margin of safety is determined by using the mean values in Eq. (20.4).

20.1.4 Alternate Load Paths (Redundancy)

When possible, the structure should contain alternate load paths. These paths may be lightly loaded until a yield or ultimate failure occurs elsewhere in the structure, after which the load is redistributed throughout the structure. If properly designed, the structure will withstand the load for at least a short time, so that the operator may safely shut down the failed equipment or operate at a lower loading condition. A redundant structure will always have multiple paths, so that no single link will cause a complete or catastrophic failure.

Structural redundancy is provided using many different designs. Included are

1. *Safety tie strap:* This is used for some highway bridges. If the span is pulled off its bearing support, the tie strap stops it from plunging to total destruction.

2. *Alternate load-carrying members:* If the load is higher than anticipated, the primary structure is stretched beyond its allowable limit. If this limit is the yield point, permanent distortion occurs in the primary structure. At this point, more of the load will be redistributed through nonyielded structures if present. These nonyielded structures will carry a larger percentage of the load and will eventually carry 100 percent of the load if the primary structure goes to ultimate failure. The secondary structure may also fail if it cannot carry the total load (zipper failure). To prevent this, a rip stop is sometimes designed into the structure. You find this type of design in hull plating on aircraft fuselages.

3. *The part fails but does not fly loose:* This design is used on structures that might break but the pieces should not hit personnel, such as flying projectiles during the crash of an aircraft or automobile.

4. *Redundant load paths such as tandem or dual wheels on a truck:* Multiple fan belts or multiple gears in a drive train will also provide power-train redundancy.

5. *Laminated structures will provide a more homogeneous material:* Knot holes or discontinuities in a portion of the wooden structure will not allow a total failure since the next layer will not have this same discontinuity.

20.1.5 Mechanical System Redundancy

When possible, a mechanical system should contain alternate paths for power and control. These paths may equally share the load until a failure occurs in one of the paths; then the control or power function is redistributed through the remainder of the operating system. If properly designed, such a mechanical system can withstand the loss of any single component for at least a short time, so that an operator may safely shut down the failed equipment. This type of design would always have multiple paths so that no single link will cause a complete or catastrophic failure.

Equipment redundancy is provided in many different ways. Included are the following:

1. Redundant power trains such as twin props on a boat, multiple engines on an aircraft, or tandem or dual wheels on a truck will provide multiple power or load paths. Also multiple gears in a drive train will provide power-train redundancy.

2. Multiple check valves or regulating valves will provide redundancy so that no single failure will cause a catastrophe.

3. Multiple control systems, such as two and three hydraulic systems on an aircraft for controlling the flight-control system, brakes, engine controls, and other critical single-link structural controls, will provide redundancy.

20.1.6 Mechanical System Failures

We can all benefit from past mistakes. Therefore, when possible, the mechanical system should be reviewed against previous failures or similar equipment that has had major failures. This information may be illuminating, by suggesting design criteria that will eliminate or minimize the recurrence of a failure in the system you are responsible for. This could be as simple as placing a speed control limit on a hydraulic ram, such that when a valve sticks (and it will), there will be no catastrophic results. Other possibilities include looking at failures in seals, blow-by, or high leakage. This would indicate a marginal design, and will not provide the high reliability desired in the system.

If properly designed, the new mechanical system will withstand the high loads that caused any previous failures. All single-component failures should be backed up so that the structural redundancy or system redundancy will prevent any recurrence of a common failure or previous failure. A good design will always have multiple paths so that no single link will cause a complete or catastrophic failure.

20.1.7 Warranty Period and Availability

When any part of a system fails, it must be repaired or replaced. A third option, not discussed here, is that the failure is allowed to exist and equipment continues to operate, possibly in a degraded operating condition (a fail-safe mode). The failed part is not replaced until it is absolutely necessary to repair the system. This section considers the cost of repair, what the effects are on system availability, and who is responsible for the repair.

Warranty Calculations.   New and completely redesigned mechanical systems require a detailed analysis to ensure that the design performance desired is the performance that will be achieved. For the first few years of providing a new product, the company may observe high field failure rates (approximately 6 percent per year per major subsystem) and low production yields (5 to 30 percent reject rate, depending on past production experience with similar products). The following are the highlights of an analysis to provide cost guidelines for pricing warranty costs. As the design matures, the warranty costs and failures will go down. However, for new products (first 6 months of production), the warranty costs could be three to four times the initial predictions.
   The following topics are discussed in detail:

1. Warranty costs
2. Potential problems
3. Service contract
4. Field failures
5. Failure rate
6. System availability

Each is discussed in its own corresponding paragraph.

1. *Warranty costs:* Expected warranty costs should be determined early in the program and used as a "field support" cost for the product. This cost can be predicted for each system sold and should be factored into the sales cost of the product. Warranty costs are divided into three areas as follows: (*a*) cost of each repair, (*b*) frequency of each repair (failure rate), and (*c*) period of guaranteed free repair by the manufacturer (warranty period). Areas *a* and *b* are rarely known during preliminary design and can only be estimated. Area *c* is to be determined based on best information available. Remember that the warranty period should not equal the expected life of the product (lifetime guarantee) without considering wearout time.
    a. *Cost of each repair:* Since some subsystems may be repaired easily, and others may have to be totally replaced, we will use an average cost of repair or (R\$), where R\$ — *x*% (a small percent) of the cost of the total system. This does not include repair costs for on-site repair. See paragraph 4, *field failures,* for further comments on travel costs.
    b. *Frequency of each repair (failure rate):* Assume a constant failure rate of 0.5 percent per month (6 percent per year) for each subsystem in each system, and a 3σ variance of 0.17 percent failure per month. Constant failure rate, MTBF, and reliability distributions are discussed in Sec. 20.1.8. System failure rate, spares, and redundancy are discussed in other parts of this chapter.
    c. *Period of guaranteed free repair by the manufacturer (warranty period):* Since we will only consider that all subsystems have a constant failure rate (FR) and constant cost of repair (R\$), the warranty costs are the same year after year. The warranty cost per year is failure rate per year times the average cost of repair ($W\$ = \text{FR} \times R\$$). This key fact allows one to price out any length of warranty. A 10-year warranty would be just 10 times the price of a 1-year warranty. The only difference is the present value of money received at time of sale for repair work promised 10 years from now. One factor not considered in the equation is mechanical wearout. If the part wears out before the warranty period is up, then it must be replaced free of charge. Therefore, the wearout life must be known or accurately predictable.
2. *Potential problems:* Potential problems during engineering development and during production development make the warranty period a high-risk period for approximately the first 2 years. After this the design and fabrication techniques are streamlined, yields are up, failure

rates are down, and the product is highly successful. Hopefully this transition period is shorter than 2 years, but we are talking about very complex systems.

3. *Service contract:* One option to writing a long warranty period would be to sell a yearly service contract for repair of returned subsystems or for on-site repairs. This contract could be renewable and at the same price year after year. A constant failure rate means that the system will have the same number of failures in the tenth year as in the first year. Therefore, the tenth year of the service contract would be the same price as the first year. The only increase in costs would be due to inflation, and this might be mitigated by ease of repair after 10 years of experience. If the part wears out before the service contract is up, it must be replaced free of charge. Therefore, the wearout life must be known, predictable, or a good estimate.

4. *Field failures:* Field failures need to be discussed from the following perspectives: (*a*) system downtime, (*b*) site maintenance, (*c*) subsystem replacement by on-site personnel, (*d*) spares available at operating location, (*e*) failed parts returned to manufacturer for repair or replacement, and (*f*) repairs not covered by warranty.

   a. *System downtime:* Downtime will be minimized by (1) system redundancy, (2) several levels of spares, and (3) good training so that on-site personnel can troubleshoot and replace inoperative subsystems efficiently (so that no redundant system goes down) and quickly.

   b. *Site maintenance:* The on-site personnel will also have other tools available to them. These tools might include (1) training of service personnel, (2) good service manuals, (3) use of built-in testing, (4) proper inspection periods, and (5) proper servicing periods. A tool in the near future is a high-speed modem which will provide manufacturer's technical support. This may also monitor the system for potential failures and recommend part changes before the part fails. Another option is to provide field services at the operating site. This is not recommended because of high transportation costs and limited tools available once the manufacturer's repairman arrives at the site. This is particularly expensive for overseas installations.

   c. *Subsystem replacement by on-site personnel:* The on-site personnel will exchange failed subsystems with local spares and send the failed subsystems to the manufacturer or service center for repair or replacement.

   d. *Spares available at operating location:* It may be necessary to stock two spares of each subsystem (based on its rate of failure). This would provide one spare after a failure, since the failed subsystem will be en route to the manufacturer or being repaired at their facility. A bad second option would be for the site to have only one spare. At first failure, the spare is used. Then, the operator requests a second spare from the manufacturer, which could take up to 3 days to arrive. It is highly unlikely that a second failure of the same subsystem would occur, but not impossible (and this would result in a very unhappy customer). Therefore, we recommend two spares, particularly in high-failure-rate areas or where the system downtime is extremely critical.

   e. *Subsystem returned to the manufacturer for repair or replacement:* The manufacturer will repair or replace all returned subsystems as soon as possible. However, this could typically take 1 to 3 months in a normal repair cycle. Therefore, the customer should have sufficient spares on site to survive for that period of time. The repaired subsystem would continue to live out its warranty period, depending on the time left on the subsystem. Spares will also use up their warranty period sitting on the shelf at a customer's site. (This is a small argument against stocking spares.)

   f. *Repair or replacement not covered by warranty:* If the subsystem is out of warranty or damaged by customers in a way that is not covered by warranty, customers will have to pay the manufacturer for any repair service. The other option is for customers to repair the subsystem themselves or send it to a third-party repair facility.

5. *Failure rate:* Initially, we may observe high field failure rates (approximately 6 percent per year per subsystem) and low production yields (5 to 30 percent reject rate). As the design matures, the warranty costs and failure rates will usually decrease. For new products, the failure rates could be three to four times what is predicted.

6. *System availability:* The predicted system reliability should be greater than 1 year if a failure might require major repairs. The system must be repaired if there is a *single-point failure,* a

condition where no redundancy exists, or a failure in the equipment that switches the redundant system into action. Other failures may occur, but these will be invisible because of redundancy, spares, and backup systems. The failed subsystems will be diagnosed as defective and switched on the fly, while the system is fully operational (or during normal maintenance). This will also provide a high system availability.

The intent of this discussion is to improve the warranty cost estimates. Generally a long warranty period will have the effect of improving the reliability of the product (lower failure rates) while ensuring customer satisfaction and minimizing failures during the period of the warranty. This occurs because the manufacturer generally tries to lower all costs, including warranty. The customer perceives a long warranty period as indicating a high-quality product, so a long warranty period is sometimes demanded. The one concern "How reliable is each subsystem?" should be resolved initially, particularly where some of these subsystems may be extremely complex or expensive. We will also have to consider good wearin (breakin) and screening to minimize field failures. Accelerated life tests will provide some information about weak components and high-failure-rate subsystems.

20.1.8 Mechanical Failure-Rate Distributions

This section presents mechanical failure-rate predictions and distributions based on accelerated and extended life tests. This example shows a 4-month test of 20 systems with no failures.

The reliability estimate shows that the overall system failure rate is less than 15.6 failures per million hours (MTTF > 7.3 years, where MTTF is mean time to failure). This is based on no failures in 80 unit-months of testing (20 units × 4 months). It is based on the upper 60 percent confidence limit for the failure-rate estimate, using χ^2 (chi-square) statistics (see "Confidence in a Failure-Rate Prediction" below).

A summary of this example shows that there should be no more than 17 percent failures in the first 5 years of life of the mechanical system. This is based on the failures fitting a normal (gaussian) distribution, which is a further subject of this section. Other distributions show higher failure rates, but these are not easily justified.

Reliability Testing.   This failure rate is high and still requires more test time. The intent of additional test time is to improve the MTTF while improving the confidence in the predicted value. This should also lower the failure rate. Caution must be taken so that one does not predict reliability out past the wear-out region.

Reliability Estimate.   The reliability estimate is based on the mechanical system having a failure rate of less than 15.6 failures per million hours. Based on this failure rate λ (lambda), the MTTF is $1/\lambda$ = 88 months (7.3 years). The MTTF is defined as the mean time to first failure. This would be represented as 50 percent of the parts failing in the first 7.3 years, or 1 failure out of two units tested for 7.3 years, but it depends on which distribution function is used and what confidence factor is chosen. Actual test results might provide zero failures in 80 unit-months of test. This is the condition used in the following example.

χ^2 ***Distribution.***   This section explains how to estimate the failure rate using the life-test summary data. The failure rate resulting from the life test is an estimate of the expected failure rate for the mechanical product and process technology. This calculation is made for the upper 60 percent confidence limit for the failure-rate estimate using χ^2 statistics. The following results are based on the maximum failure rate or worst-case condition:

	Confidence, %	χ^2 value	MTTF, years
1.	60	1.83	7.3
2.	50	1.4	9.5
3.	40	1.0	13.4

This table shows the correlation of percent confidence in the MTTF with the predicted minimum MTTF. We will use the standard 60 percent confidence limit in the following example.

Failure-Rate Models.   The actual failure rate depends on many different parameters: the quality of parts, quality of manufacturing processes, use environment, vibration levels, and other internal and external stresses. Four different failure distributions for modeling failure rate have been analyzed. All are based on an assumed MTTF (mean value) of 7.3 years.

Distribution	Failures in first 5 years, %	Standard deviation σ, years	Comments
1. Poisson	26	2.7	σ^2 = mean
2. Normal	17	2.4	3σ = mean
3. Weibull	31	3.8	$a = 8.25, b = 2$
4. Exponential	50	7.3	σ = mean

This table shows the direct correlation of percent failures in the first 5 years and the standard deviation. For the Poisson and exponential distributions the standard deviation is a function of the mean value. For the Weibull distribution, it is a function of both the mean value and the beta (β) value chosen.

Failure-Rate Distributions.   The failure-rate distributions are derived from functions that always have an integrated area of one. This means that, after a long time, the probability of failure will be one (100 percent).

For failure-rate modeling, the most commonly used distribution is the exponential. This is because it has a constant failure rate (hazard rate = 1), and is mathematically simple. The failure rate is constant based on the number of surviving parts. As fewer parts are left, fewer parts fail (exponential decay). Many people dislike this distribution because of the high failure rate at the start, and this does not even consider infant mortality. This characteristic makes the system look unreliable initially.

The normal distribution may be the next best choice for failure predictions (or the lognormal, for special applications), but the biggest unknown then becomes the choice of the standard deviation value. Three sigma is generally used for manufacturing tolerances. This has been selected as the time-zero starting time for this example. This is the preferred distribution for this example, because of the possibly optimistic results that it provides. It could be considered a wearout distribution, where the example chosen approximates the average life of an automobile.

The Weibull distribution is probably the distribution of choice for those who do not like the exponential. However, the shape parameter β is a very critical term. When $\beta = 1$, the Weibull distribution is identical to the exponential. It is best to determine this value from life-test data, but who can wait 10 to 20 years to get life-test data? The value used for this example was $\beta = 2$. Research of specific failure data would probably show a better value to use, but this will have to be performed by the user of the data.

The Poisson distribution is the least attractive of the contenders, but is sometimes suggested since it follows the binomial distribution so closely. The binomial distribution is probably the most popular discrete distribution, and is used frequently in determining pass/fail criteria for sample inspection of large lots.

Confidence in a Failure-Rate Prediction.   This section explains how to calculate the failure rate (estimate) by using the life-test summary data. The failure rate resulting from the life test is an average, or estimate, of the typical expected failure rate for a typical mechanical product and process technology. This calculation is made for the upper 60 percent confidence limit for the failure-rate estimate using χ^2 statistics. The following formula predicts the maximum failure rate or worst-case condition:

$$\lambda_{(max)} = \frac{\chi^2(1-a)}{2t} \qquad \text{with } 2(r+1) \text{ degrees of freedom (dof)} \qquad (20.6)$$

where $\chi^2 = \chi^2$ distribution value
r = number of failures
dof = degrees of freedom
t = device hours
a = statistical error expected in estimate (for 60% confidence, $a = 0.4$ or $1 - a = 0.6$)
$\lambda_{(max)}$ = maximum failure rate

The value $(1-a)$ can be interpreted to mean that we can state with statistical confidence of $1 - a$ (i.e., 60 percent) that the actual failure rate is equal to or less than the calculated maximum $\lambda_{(max)}$ failure rate.

To calculate the total number of device hours, multiply the total samples tested (survivors) by the total hours of test. Please note that the equations can also be used for total cycles or total miles of test.

Suppose that we continue to run a life test on 20 samples. So far no devices have failed. The test has been running continuously for 4 months. The test results represent the actual use conditions; therefore, no temperature corrections are required for the data. This data does not require any correction using a generalized acceleration equation. The acceleration factor is considered to be 1 for this data.

To find the maximum failure rate under normal operating conditions with 60 percent confidence, using Eq. (20.6), we have

$\chi^2 = 1.83$ (for a total of 0 failures, $\chi^2 = 1.83$ with 60 percent confidence; χ^2 values are listed in App. A.6 or can be calculated using a computer program)
$r = 0$
dof = 2
$t = \frac{1}{3}$ year $\times$ 8766 h/year $\times$ 20 units = 58,440 h
$a = 0.4$ [for 60 percent confidence, $a = 0.4$ for Q failures, since $1 - a = 0.6 = 60$ percent for P success (probability of success)]

$$\lambda_{(max)} = \frac{1.83}{2 \times (58,440 \text{ h})}$$

$$= 15.6 \times 10^{-6} \text{ failures per hour}$$

or

$$\lambda_{(max)} = 15.6 \text{ failures per million hours}$$

$$\text{MTBF} = \frac{1}{\lambda_{(max)}} = 64,100 \text{ h} = 7.3 \text{ years}$$

20.2   *RELIABILITY BY DESIGN: REVIEW OF ANALYSIS*

This section is designed to provide the reliability engineer with the tools for reviewing a mechanical stress analysis. To review and adequately comment on a mechanical stress analysis report, an engineer must fully comprehend how to conduct such a stress analysis, and understand the purpose of the equipment, the overall mission, and the operating environment. This is where the design failures will occur. By looking for these potential errors early in the design, one can obtain a higher reliability of the system. This section addresses the following techniques: (1) tricks of the trade, (2) pitfalls of a mechanical stress analysis, and (3) step-by-step review process. There are certain methods that should be followed in an orderly fashion to effectively and efficiently evaluate the stress analysis. These methods are discussed in the following paragraphs.

20.2.1   Tricks of the Trade

The reliability engineer reviewing a mechanical stress analysis should be an experienced mechanical and materials engineer. In the absence of experience, the reliability engineer should rely on the fundamental equations as presented in Chap. 18, whenever a process is unclear or unfounded. In some instances the approach may be based on previous assumptions which are not explicitly stated. Therefore, one must be cautious and complete in the review. One area of primary concern should be any highly stressed areas. These will be easily identified as having margins of safety close to zero.

20.2.2   Environments

The environmental conditions are critical to the stress analysis and to the structural design. These environments should be reviewed and carefully checked. The loads and operating conditions used for the analysis should appear realistic considering the function the equipment is to perform. Environments to account for include (1) nominal operating temperature (possibly an elevated operating temperature), (2) shock loading, (3) vibration inputs and resonant frequencies, (4) thermal cycling, (5) corrosive atmosphere, and (6) load cycling (fatigue). Environments 1 to 3 are most common. Thermal cycling must be considered for such items as high-temperature piping, equipment and controllers near furnaces and boilers, rocket nozzles, gun barrels, and other high-temperature applications. One characteristic often overlooked is the effect of materials at high temperatures. They change phase or the grain structure, and become annealed or have altered strength characteristics. This is extremely risky if the structure continues to see high loads, since the material no longer has full strength. One should always use the lowest-strength properties for the material, particularly if thermal cycling or high-temperature operation is required.

20.2.3   Structural Failure Modes

All structural failure modes should be analyzed, and one should determine which failure modes are the most critical in the design. These failure modes include (1) tensile yield, (2) tensile ultimate failure (after yield), (3) compressive yield and ultimate failure, (4) structure operating near resonant frequencies, (5) fatigue failure, (6) column or beam buckling, (7) plate or shell buckling, and (8) local beam buckling (flange or web).

Generally, only one or two of these failure modes are analyzed. Sometimes, all may be critical, depending on the loading conditions. While none should be ignored, a simple analysis will quickly verify whether any of these failure modes are critical, because the analysis will provide positive or negative margins of safety.

20.2.4   Stress Review Checklist

A stress review checklist should be prepared and included as part of the review procedures. This checklist should contain the following detailed sections: (1) identifying all possible loading conditions, (2) evaluating the allowable strength including safety factors for all materials used, (3) determining all the combined stresses present in the various critical or weak points of the structure, (4) ensuring that all computer analyses are complete and accurate, and (5) ensuring that all detailed hand calculations are complete.

This checklist will aid in evaluating the stress analysis and at the same time will provide continuity for the analysis. Be sure that all highly stressed parts have been analyzed.

20.2.5 Pitfalls

Sometimes, fundamental pitfalls that are relatively simple and straightforward escape scrutiny. The reliability engineer can avoid these common pitfalls by considering fundamental review questions such as

1. Has the operating load been accurately assessed by a comprehensive load study?
2. What was assumed to be the maximum loading condition?
3. Have part tolerances been included in the stress analysis? In some instances tolerances may build to such a large value that holes may be on extreme edges of plates, etc.
4. Have the margins of safety been properly determined? Primarily, what allowable stresses are used in welded regions, and has the weld weakened the apparent material?
5. Has the stress analysis included all of the critical parts and has it included all of the critical loads?
6. Does the stress analysis include all the system, or is it incomplete?
7. Is it possible that the design analyzed is not the current design (latest drawings) that will be built?
8. Do the applied stresses fall within the allowable safety factors of the material strengths, or are some parts overstressed and shown to fail?
9. Has the stress analysis addressed all the mechanical failure modes, or does it address only the critical one for each part, mentioning nothing about the other types of failures, or why they weren't considered?

Failure to recognize and avoid these common pitfalls can result in a poor design and an inaccurate stress analysis. This may result in the use of overstressed parts in the final design of the equipment. A comprehensive explanation of these pitfalls and how to avoid them is discussed in the following section.

20.2.6 Step-by-Step Review Process

The review of a mechanical stress analysis can be broken down into six distinct verification processes as follows:

- Mechanical design
- Applied loads
- Material strength properties
- Margins of safety due to static loading
- Margins of safety due to shock loading
- Resonant frequencies, mode shapes, and dynamic loading

20.2.7 Mechanical Design Review

This is an important step in the review process. The reliability engineer should verify that the mechanical design is accurately defined and that it will perform the mechanical functions for which it was designed, and should also ensure that the parts are correct and the points where the loads are applied are correct—which may be accomplished through a detailed check of current drawings and verification of all critical dimensions. The steps include (1) obtain a complete set of drawings for the mechanical layout (this should be part of the stress analysis package), (2) randomly select a portion or several items on the drawings and review these parts accurately, and (3) perform a dimensional analysis on these sample parts to ensure that tolerances, minimum gauges, and other critical dimensions have been properly accounted for. If the results closely agree with those of the stress analysis, then proceed to the next step of the review process. If there are major differences in this initial analysis (particularly if there are significant errors), then a more in-depth review of the mechanical values may be necessary. You may have to reduce the predicted reliability of the system based on the findings in the stress analysis. However, at this stage, generally just comments on the inaccuracies in the design should be sufficient.

20.2.8 Verification of Applied Loads

Another step of the review process is to verify that the loading analysis is correct. To perform this verification the reliability engineer must determine the loads. This may be a classic case where the statistical probability of loads is not properly considered. The reliability engineer should know more about the probability of loads occurring than the stress analysts. The loading conditions may vary as a function of weight and inertia of the equipment. These might be verified through actual measurements. Also, the combined loading conditions should have been considered. Loads in two or more axes generally provide the greatest stresses, and should be resolved into principal stresses.

If shock loads are to be determined, the proper analysis must be performed. This can be done using a detailed computer program to evaluate the transient loading conditions or by equivalent steady-state loading conditions using a shock design factor (SDF).

If dynamic loads have been considered, the reliability engineer should ensure that the proper frequency spectrum and power spectrum have been applied. Included in this evaluation is a proper review of the structure from a dynamics viewpoint, including load amplification factors and the damping or SDF factors for the structure.

If, after reviewing the mechanical loads, it can be determined that the specified maximum loading conditions were not utilized in the analysis, or that the combined loading condition is greater than what was utilized, then there is evidence that the stress analysis is underestimating the stresses and the reliability of the structure. You may have to reduce the predicted reliability of the system based on these findings.

20.2.9 Verification of Material-Strength Properties

Another verification process in the review consists of examining the design to determine what materials are used in the structure. If several materials are specified for one complex part, one should remember that the lowest strength material should be evaluated first, since it is the most likely to fail first. An example is high-strength steel bolts holding two aluminum plates together, which will quite likely fail at the bolt holes rather than at the bolts. Also, one should use the lowest strengths available for the material present in that structure. This includes the lower strength of a welded zone (heat-affected zone) in a material. Also, one should use a minimum-strength value for the material or the mean and standard deviation, not a nominal-strength value for the material. Minimum strength is based on the A basis or B basis, or some lower statistical limit in the strength of the material. The A basis is recommended, which is the minimum strength of the material where only 1 percent of the material will be weaker, and 99 percent of the material will exceed this strength value. The B basis is not recommended since it has a minimum strength where 10 percent of the material will be weaker, and 90 percent of the material will exceed this strength value.

Another area associated with material-strength values is the use of dissimilar materials, such as cadmium-plated steel bolts with aluminum structures. This causes a corrosive effect within the aluminum and will eventually weaken the aluminum in the region of the attachments. One should provide sufficient strength for the material to ensure that corrosion does not greatly affect the strength of the overall assembly.

If the calculated part strengths seem to be comparable with the analysis, the reliability engineer can move on to the next step. If the results are not similar, especially if the reliability analysis shows significantly higher part stresses, then the predicted reliability of the system may have to be reduced. (*Remember:* You cannot properly calculate the structural margin of safety unless the material strength has been accurately determined. You should further consider the strength at elevated temperatures, or use complex temperature profiles if this is an operating condition.)

20.2.10 "Margin of Safety" for Static Loads

To perform this verification, the reliability engineer would perform all the structural analysis calculations associated with the applied stresses in the part. This is very time-consuming. It is therefore recommended that you concentrate only on those parts with margins of safety of less than 0.5 or 0.2. These are the critically stressed parts, which will fatigue first or fail at high temperature. These are the parts that will determine the reliability prediction of the structure. Obviously, parts with a margin of safety of 5 or 10 are unlikely to fail when connected directly to a part with a safety margin of 0.1. As discussed previously, the pitfall here is that the previous analysis may have considered only a few of the possible failure conditions and may have allowed some of the critical failure conditions to go unnoticed. Therefore, the margin of safety may really be much lower or even negative (impeding failure).

To determine bending stresses, one must first determine the cross-sectional area properties, including the area moment of inertia, and for complex cross sections this is where most errors occur. Therefore, the reliability engineer should use a computer program and spot-check some of the complex cross-sectional designs, to confirm that the area moments of inertia are calculated properly. Once the stress has been calculated for the part, one should verify that this is the stress for the maximum load and should use the appropriate safety factor to determine the ratio of the allowable working stress to the applied stress. One should further determine the margin of safety. The combined loads are necessary for determining the complete stress tensor on a structural element. One must determine the principal stresses and compare these to the allowable working stress. If the principal stresses are greater than the maximum working load, then the margin of safety is negative and the structure will fail.

20.2.11 Margins of Safety for Shock Loads

Verification of the structure-borne stresses when subjected to shock loading is accomplished by comparing the actual stresses incurred during the shock loading to the maximum allowable shock stresses for the specified material. This verification can also be accomplished on a sample basis. Be certain, however, that the sample includes the review of each shock-loading condition and that critical stressed areas are analyzed for the shock-loading conditions.

For example, if a ship can encounter a shock load due to a flat wet deck slamming into a high-sea state, the load is transmitted throughout the bow structure of the ship. Any attachments or brackets attached to the decks or bulkheads will see this high shock load and should be appropriately evaluated for this loading condition.

The major pitfall is that some stress analyses include only the steady-state loading conditions, and fail to analyze the transient loading conditions. These transients are sometimes the extreme loading condition and are much higher than the static loading condition. If the stress analysis only includes the static loading conditions, then you may have to reduce the predicted reliability of the system. It must be reduced if there is a dynamic environment in which the part must survive, especially if the reliability analysis shows significantly higher part stresses. (*Remember:* You cannot properly calculate the structural margin of safety unless the maximum loads and stresses have been accurately determined.)

20.2.12 Determining the Resonant Frequencies, Mode Shapes, and Dynamic Loads

The final step in the verification of the baseline design is to determine the frequencies, mode shapes, and dynamic loads. One should verify that the dynamic loads do not excite resonant frequencies, since undamped oscillations (structural members have very little damping, usually 2 to 6 percent) lead to eventual catastrophic failure. The best method to evaluate the resonant frequencies and mode shapes is to evaluate the model using a finite element computer program. If the stress analysis does not include this, the reliability engineer should at least perform a modal analysis or spectrum analysis using mechanical test equipment and dynamic (vibration) analyzers. Short of this type of analysis, there is no way to verify what the response of the structure will be to various input frequencies. The results could be an extreme resonant condition resulting from the rotation of a propeller on a ship or aircraft, or the buffeting of a missile, due to the vortex impinging on a control surface.

If the stress analysis does not provide a detailed dynamic analysis of the structure and compare the dynamic loading input to the structure, you may have to reduce the predicted reliability of the system. It must be reduced if there is a dynamic environment in which the part must survive, especially if the reliability analysis shows that the structure is operating near a primary resonant frequency.

Remember that it is your equipment—you have to ensure that it is reliable and safe. You have to maintain it, repair it, and replace worn-out parts. You do not have to agree with the stress engineer's justification and an inadequate dynamic analysis.

20.2.13 Review Process Summary

The overall objectives of reviewing a mechanical stress analysis are to (1) verify that the critical areas are adequately addressed and (2) ensure that the methods of analysis were correct and complete. Verification of the stresses in the critical design areas are, in fact, the overall objective of the design process. Comparison of these stresses with the allowable stresses for the material is necessary to confirm that the structure will withstand the input loads. (*Remember:* High stresses cause failures.)

If there is any uncertainty that the structure has insufficient strength, the reliability prediction must be reduced drastically. In fact, you should either do an initial analysis to verify that the structure is properly designed, or recommend that the design be modified to withstand the loading conditions specified. As a reliability engineer, you should use your understanding of stress analysis methods, tricks of the trade, and pitfalls to check the correctness of the analysis. Do not hesitate to question the stress analysis on unclear results. The stress analysts will expect you to ask some questions based on the fact that they have the best understanding of the hardware, and have the best understanding of the analysis performed.

20.3 *MATERIAL AND DIMENSIONING TOLERANCES*

20.3.1 Tolerances

A design specification for a dimension is usually given as a nominal value plus or minus a tolerance, e.g., 1.530 ± 0.003 in. If asked exactly what these plus and minus limits mean, as the designer, you will usually answer that all parts should fall within these limits. If pressed further, you will admit that you don't really expect all, but almost all, of the items to fall within these dimensions. If this statement is questioned, the answer will finally result in a statement that no more than a given fraction of the items produced will fall outside these limits. In other words, this dimension is a random variable having a probability distribution with the property that no more than the given fraction will fall outside the lower and upper specification limit.

In practice, the nominal value plus or minus the tolerance value is often interpreted to mean that 99.72 percent of all manufactured parts should meet the design specifications. The percentage is based on the dimension fitting a normal distribution. To achieve this high probability, the variability in the manufacturing process must be such that three times the standard deviation (σ) must equal the tolerance value. Thus, if the tolerance is 0.003 in, then $3\sigma = 0.003$ in, giving $\sigma = 0.001$ in. Consequently, whenever a designer specifies a nominal value plus or minus a tolerance, there is a probability of .9972 of a part selected at random having dimensions within the upper and lower specifications.

20.3.2 Variations in Materials and Manufacturing Processes

Representative mean and standard deviation of static strengths for several widely used metal alloys are contained in Table 20.1. Variability of tolerances for different types of manufacturing processes are contained in Table 20.2. The data contained in these tables can be of value to the designer in estimating probabilistic strength and stress analysis problems.

TABLE 20.1 Typical Variations in the Strength of Metals

Material	Condition	Ultimate tensile strength S_u, ksi		Tensile yield strength S_y, ksi	
		μ	σ	μ	σ
2021-T3 aluminum	(Bare sheet and plate) <0.250 in	73	3.7	54	2.7
6061-T6 aluminum	$\frac{1}{2}$-in sheet	46	1.9	42	2.9
7075-T6 aluminum	$\frac{1}{4}$-in plate; bare	85	3.4	76	3.8
A231B-0 magnesium	$\frac{1}{4}$-in plate; longitudinal	41	3.8	24	3.8
Ti-SA1-4V titanium alloy	Sheet and bar; annealed	135	6.8	131	7.5
304 stainless steel	Round bars; annealed (0.50–4.6 in)	85	4.1	38	3.8
ASTM-A7 Steel	$\frac{1}{2}$-in plate	66	3.3	40	4.0
AISI 1018	Cold-drawn round bar (0.75–1.25 in)	88	5.7	78	5.9

TABLE 20.2 Variability in Manufacturing Tolerances

Process	Tolerance, in ($\pm$)	Standard deviation σ, mils
Flame cutting	0.060	20.0
Sawing	0.020	6.7
Shaping	0.010	3.3
Broaching	0.005	1.7
Milling	0.005	1.7
Turning	0.005	1.7
Drilling	0.010	3.3
Reaming	0.002	0.7
Hobbing	0.005	1.7
Grinding	0.001	0.3
Lapping	0.0002	0.1
Stamping	0.010	3.3
Drawing	0.010	3.3

If no statistical data is available for the material, then one might obtain the mean and standard deviation from other statistical data readily available for the material (MIL-HDBK-5 or other material-strength tables). The statistical data available is generally the *A* basis, where the strength of the material exceeds this value for 99 percent of the population and 1 percent of the population will fail. The second set of data which is also necessary is the *B* basis, where the strength of the material exceeds this value for 90 percent of the population and 10 percent of the population will fail. Since the *A* basis has a 1 percent failure or a *Z* value equal to -2.33σ and a *B* basis has a 10 percent failure or a *Z* value equal to -1.28σ, one can determine what a 1σ value is. The difference between the *A* basis and the *B* basis values represents a 1.05σ ($+2.33\sigma - 1.28\sigma$). Therefore, by multiplying the difference in material strengths by 0.95 ($= 1/1.05\sigma$), one obtains the 1σ value. The mean value is determined by shifting the *B*-basis value 1.28σ to the right (adding 1.28σ to the *B*-basis value). One can also use the *A* basis to find the mean value, but it must be shifted further to the right ($+2.33\sigma$). Both results should be the same. One point of caution—you are using a small difference in two large numbers, so the error in the statistical properties could be greater than desired.

20.4   STRESS CONCENTRATION FACTORS

These are the stress concentration factors due to cutouts and other interruptions in the stress flow through the base material. Also, when parts are not aligned, the stress cannot flow smoothly in one direction. This causes higher stresses at the point where the stresses and forces make a transition around a corner.

This section presents a method for evaluating the stress concentration factors in a mechanical design. It is generally at these points that the structure is most susceptible to failure, and, therefore, it is important to evaluate in detail any stress concentration. There are several amplification factors associated with the high stress values at these locations.

The maximum stresses in a structure or machined part will generally occur at the location where a notch, abrupt change of cross section, or structural transition point occurs. The stresses at these locations will be greater than the stresses calculated by the elementary formulas presented in Chap. 18. The ratio of the maximum expected stress to the nominal calculated stress is expressed as the stress concentration factor *K*.

The stress concentration factor is based on the shape of the structural material and the shape of the transition zone and is independent of the material, provided that it is isotropic. This does not infer that anisotropic materials, such as fiberglass, do not have stress concentrations at structural transitions. The stress concentration factors for anisotropic materials are much more complex to evaluate, and will be discussed at the end of this section. The stress concentration factors are generally determined empirically, on the basis of many tests. However, they are also explained and analyzed through the theory of elasticity.

Stress concentration will cause failure in a brittle material if the concentrated stress is larger than the ultimate strength of the material. In ductile materials, the concentrated stress will generally produce a yield failure causing local plastic deformation and a redistribution of the stresses over the more uniform load path. Stress concentrations in ductile materials are subject to additional failures if cyclical loading occurs.

20.4.1   Stress Concentration Factors in Flat Plates

The stress concentration factors in flat plates are analyzed for (1) circular holes in plates, (2) semicircular cutouts, and (3) fillets or width/thickness transitions. Two loading conditions are analyzed. The first to be presented includes various configurations in tension and compression. The second set will be analyzed for bending stresses (Sec. 20.4.2). The transition in a plate thickness or a plate width causes a stress concentration at that transition point. The actual working stress or applied stress is the calculated stress at the thinner cross section, and amplified by the stress concentration factor, which is always greater than 1. The stress concentration factors for this condition are presented in Figs. 20.1 and 20.2. If these concentration factors are applied to brittle materials, unstable or catastrophic failures will originate at the preexisting stress transition area.

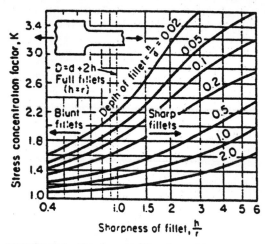

FIGURE 20.1   Flat plate with fillets, in tension.

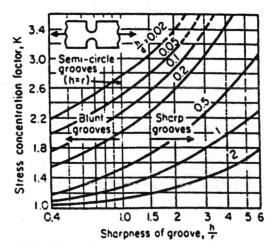

FIGURE 20.2   Flat plate with grooves, in tension.

The second stress concentration factor is due to a semicircular groove or cutout in a flat plate, again loaded in tension. This is shown in Fig. 20.2. A description of the variables is presented as follows:

r = the radius of the fillet or cutout
D = the height of the full material or maximum dimension in the plane of the cutout
d = the minimum height (after all cutouts have been removed from the maximum height) in the plane of the cutouts
h = the height of the cutout

20.4.2   Stress Concentration Factors in Bending

The following stress concentration factors should be used when a plate is loaded in bending and where the cutouts are in the plane of bending. The first stress-concentration-factor curve is for the transition in plate thickness or plate width, as shown in Fig. 20.3. The other condition, a stress concentration factor due to a semicircular groove, is presented in Fig. 20.4. It should be noted that stress concentration factors in bending loads are about 30 percent less than stress concentration

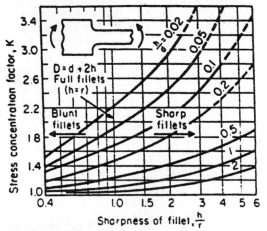

FIGURE 20.3   Flat plate with fillets, in bending.

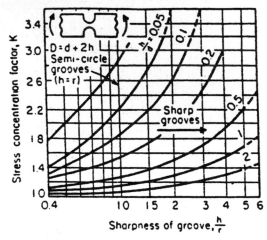

FIGURE 20.4   Flat plate with grooves, in bending.

factors in tension. It is believed that the compression side of the bending helps relieve the problem of stress concentration, since it is the tensile load at a cutout where crack propagation is initiated.

Stress concentration factors due to tension, compression, and bending have been summarized for special conditions in Fig. 20.5. The special conditions are that all cases ignore the depth of the groove as being a variable. The depth of the groove is equal to the radius of the cutout. In other

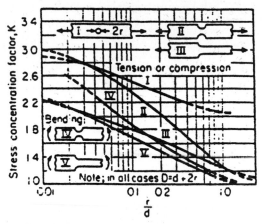

FIGURE 20.5 Flat plate with semicircular fillets and grooves, or with holes, in tension or compression.

words, all cutouts are exactly semicircular and do not extend deeper than half the diameter of the circle. The curve is composed of all $h = r$ values from the previous four curves plus case I (in Fig. 20.5), which is a circular cutout (hole drilled in the plate). This differs from case II, which has the same notch-equivalent cutout, but in which the internal hole is less critical at the higher r/d ratios. As r approaches $2d$, there is hardly any material left to maintain the integrity of the structural member. When $r = d$, exactly half of the cross section has been removed due to the hole.

20.4.3 Stress Concentration Factors Due to Torsion Loads

This section evaluates the notch sensitivity due to a transition in a torsion structure. We consider only round shafting for this brief analysis, and the definition of the variables is the same as that for flat plates. Dimensions refer to continuous diameters rather than cutouts in a single plane. Figure 20.6 shows a complete transition from a thick shaft to a thin shaft with a fillet of radius r.

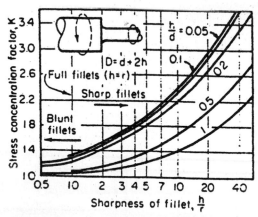

FIGURE 20.6 Stress concentration factors for filleted shaft, in torsion.

A notch in a shaft has a stress concentration factor as shown in Fig. 20.7. In summary, torsion sensitivity to stress concentration is less than the bending condition, which is less than tensile loading. Also, the notch cutout has a stress concentration factor higher than that for a transition from a thick to a thin member.

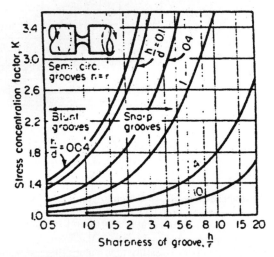

FIGURE 20.7 Stress concentration factors for grooved shaft, in torsion.

Since the notch has the higher stress concentration factor, we will present one more curve on this subject (Fig. 20.8) for notch sensitivity versus the angle of the notch. This opening allows the notch to be more tapered and reduces the stress concentration factor. If the sides of the notch are approximately 90°, the stress concentration factor is reduced to 82 percent of the original stress concentration factor. This applies to angular notches in tension and bending of flat plates, but does not apply to the torsion condition. Notch sensitivity of a specific type of material is determined by the Charpy V-notch (CVN) impact test, or other tests.

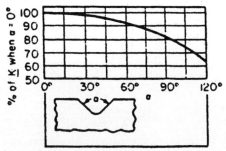

FIGURE 20.8 Flat plate with angular notch, in tension or bending.

20.5 FITTINGS, FASTENING, AND JOINTS

This section concentrates on the various attachment methods in mechanical design. It is generally at these points where most structural failures occur, and, therefore, it is important to evaluate in detail any critical joint design. Several amplification factors are associated with the stress values in joint designs: (1) the stress concentration factors due to cutouts as described in the preceding section; (2) any interruption in the stress flow through the base material; and (3) when parts are not aligned, the stress cannot flow in a continuous direction. This causes higher stresses at the point where the stresses and forces make a transition around a corner.

Two basic areas of fittings, fastening, and joining are presented in this section: (1) bolted and riveted fastening and the evaluation of the joints associated with this type of fastening and (2) welding and the evaluation of welded joints. These two subjects will be developed in the following paragraphs.

20.6 BOLTED AND RIVETED JOINTS

20.6.1 Fasteners

Many types of fasteners are used to distribute the loads to improve mechanical strength, including conventional steel bolts, inserts, pin fasteners, rivets, eyelets, staples, retaining rings, and quick-operating fasteners, and washers. As stated earlier, all fasteners must be analyzed with respect to their strength capabilities. The loads on these parts are generally quite high, and the stress concentration factors are also quite high. A simple bolt may have cyclic loading or a complex load path. The advantage of fasteners is that they are generally made of extremely high-strength materials. In this section, only bolts will be analyzed, including bolt torques and the joint strength generated by the bolted connection. However, much of this analysis is applicable to the other types of fasteners as well.

20.6.2 Bolt Strength

Bolt strength varies drastically depending on the material and the grade of the bolt. Typically, steel bolts are the most common in the industry. However, super-high-strength bolts are available, including titanium bolts used in military airframes. Titanium bolts may be used for the following: (1) weight savings (50 percent the weight of steel), (2) high strength, (3) ability to provide high strength at high temperature, and (4) excellent corrosion resistance. Other materials include popular stainless steels, which are required and must be used for fastening aluminum because of the electrolytic action. If one uses steel or cadmium-plated steel in contact with aluminum, the effects of battery action tend to dissolve the aluminum. Stainless steel (passivated) is inert and presents no problem to the aluminum. Other materials for fastening devices include the nonmetallic materials such as nylon, Teflon, and fiberglass.

Steel bolts may be coated to prevent corrosion. These coatings include electrodeposited zinc, tin, nickel, and chromium plating; nickel coatings such as the zinc or phosphate alloys; and chromates such as zinc chromate. Hot-dip coatings may be used as an inexpensive coating on high-strength ferrous fasteners. The most popular is the hot-dip zinc (galvanized). One warning on these coatings is that the parent material must be extremely clean prior to the coating, or a major adhesion problem will occur on the underside of the coating. Also, a coating that should always be included is the standard painting or finishing. This further protects the material and is the primary protection on structures exposed to harsh environments. Figure 20.9 shows the typical markings on the head of a bolt to indicate its strength value. These strength values are translated into a proof load and minimum tensile strength in Table 20.3.

SAE Grade 1   ASTM A 307
   Low-carbon steel suited for cold and hot heading. Used for large upsets such as square-head bolts, carriage bolts, etc.

SAE Grade 2
   Low-carbon steel, bright finish. Primarily intended for cold-headed products. Cold working increases strength. Widely used for cold-headed hexagon bolts. Sometimes necessary to stress relieve fasteners with large upset heads.

SAE Grade 5.2
   A quenched-and-tempered, low-carbon, martensitic steel which has been treated with boron.

SAE Grade 5   ASTM A 449
   A quenched and tempered medium-carbon steel bolt. Recommended where high preloading of the bolt is practical. This grade is considered the most economical on a highest-clamping-force per dollar-of-bolt-cost basis.

SAE Grade 7
   A quenched-and-tempered, medium-carbon, alloy steel. Threads are cold rolled after heat treatment for improved fatigue strength.

SAE Grade 5.1
   A quenched and tempered, medium-carbon steel specifically intended for use in preassembled washer-and-screw assemblies.

SAE Grade 8   ASTM A 354, Grade BD
   Quenched-and-tempered medium-carbon alloy steel which has a higher strength than Grade 7 material.

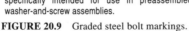

FIGURE 20.9   Graded steel bolt markings.

   Bolts of grade 5 or better are generally recommended for military use, but there is a tendency to use lower-strength bolts in the less critical areas, to minimize cost.
   Screw thread inserts made of high-strength materials are useful for providing increased thread strength and life. Soft or ductile materials with inserts exhibit improved load-carrying capacity under both static and dynamic loading conditions. Holes in which threads have stripped out or have been otherwise damaged can generally be restored through the use of thread inserts.
   Holes for thread inserts are drilled oversize and specially tapped to receive the insert selected to mate with the threaded fastener used. The standard material for inserts is 18-8 stainless steel, but other materials are available, such as phosphor bronze and Inconel. Recommended insert lengths are given in Table 20.4.

TABLE 20.3 Physical Requirements for Threaded Fasteners (Bolts and Cap Screws)

SAE grade	Bolt size diameter, inches	Proof load, psi	Ultimate strength	Hardness Brinell	Rockwell
0	All sizes				
1	All sizes		55,000	207 max	95 B max
2	$\leq\frac{1}{2}$	55,000	69,000	241 max	100 B max
	$>\frac{1}{2}-\frac{3}{4}$	52,000	64,000		
	$>\frac{3}{4}-1\frac{1}{2}$	28,000	55,000	207 max	
3	$\leq\frac{1}{2}$	85,000	110,000	207–269	95–104 B
	$>\frac{1}{2}-\frac{5}{8}$	80,000	100,000		
5	$\leq\frac{3}{4}$	85,000	120,000	241–302	23–32 C
	$>\frac{3}{4}-1$	78,000	115,000	235–302	22–32 C
	$>1-1\frac{1}{2}$	74,000	105,000	223–285	19–30 C
5.1	$\leq\frac{3}{8}$	85,000	120,000	241–375	23–40 C
7	$\leq1\frac{1}{2}$	105,000	133,000	269–321	28–34 C
	$\leq1\frac{1}{2}$	120,000	150,000	302–352	32–38 C

TABLE 20.4 Screw Thread Insert Lengths

Shear strength of parent material, ksi	Bolt material ultimate tensile strength, psi (length in terms of nominal insert diameter)				
	60	90	125	170	220
15,000	$1\frac{1}{2}$	2	$2\frac{1}{2}$	3	
20,000	1	$1\frac{1}{2}$	2	$2\frac{1}{2}$	3
25,000	1	$1\frac{1}{2}$	2	2	$2\frac{1}{2}$
30,000	1	1	$1\frac{1}{2}$	2	
40,000	1	1	$1\frac{1}{2}$	$1\frac{1}{2}$	2
50,000	1	1	1	$1\frac{1}{2}$	$2\frac{1}{2}$

20.6.3 Bolt Torques

The joint strength is affected more by the clamping force than by the related tensile strength of the bolt. The *clamping force* is the force exerted by the bolt when it is tightened to a specified torque value. The equations for determining the torque value are fairly complex and will not be developed here, but the suggested tightening torque values for various-grade bolts are listed in Table 20.5. Also, the tensile strength and the stress area are provided for evaluating the allowable stress in the bolt. This is the actual stress on the bolt due to a specific torque value. This also allows for calculating the safety margin and reliability values for each bolt.

For a bolt being loaded cyclically, the amount of initial stress (preload) will change the fluctuating stresses to the extent that the bolt has an acceptable value for a given safety factor. Generally, if the preload is set to the maximum cyclic load, the bolt will not be subjected to cyclic fatigue and should last longer (static load condition).

Effect of Spring Washers and Gaskets. If a spring washer or a gasket is used in a bolt joint, an analysis must be made to determine the exact effect, since this may make the assembly either weaker or stronger depending on the loading conditions. The important factor in a joint is to achieve sufficient clamping to prevent separation in the joint or unloading of the bolt during the load cycle. The clamping load (which is a function of bolt torque) should never be exceeded during any loading condition.

TABLE 20.5 Corresponding Bolt Clamping Loads

Size	Bolt diameter D, in	Tensile stress area A_s, in²	SAE grade 2 bolts					SAE grade 5 bolts					SAE grade 7 bolts		
			Tensile strength, min psi	Proof load, psi	Clamp* load P, lb	Tightening dry, $K=0.2$	Torque† lub., $K=0.15$	Tensile strength, min psi	Proof load, psi	Clamp* load P, lb	Tightening dry, $K=0.2$	Torque† lub., $K=0.15$	Clamp* load P, lb	Tightening dry, $K=0.2$	Torque† lub., $K=0.15$
						lb · in	lb · in				lb · in	lb · in		lb · in	lb · in
4–40	0.1120	0.00604	74,000	55,000	240	5	4	120,000	85,000	380	8	6	480	11	8
4–48	0.1120	0.00661			280	6	5			420	9	8	520	12	15
6–32	0.1380	0.00909			380	10	8			580	16	12	720	20	15
6–40	0.1380	0.01015			420	12	9			640	18	13	800	22	17
8–32	0.1640	0.01400			580	19	14			900	30	22	1,100	36	27
8–36	0.1640	0.01474			600	20	15			940	31	23	1,160	38	29
10–24	0.1900	0.01750			720	27	21			1,120	43	32	1,380	52	39
10–32	0.1900	0.02000			820	31	23			1,285	49	36	1,580	60	45
¼–20	0.2500	0.0318			1,320	66	49			2,020	96	75	2,500	120	96
¼–28	0.2500	0.0364			1,500	76	56			2,320	120	86	2,860	144	105
						lb · ft	lb · ft				lb · ft	lb · ft		lb · ft	lb · ft
5⁄16–18	0.3125	0.0524			2,160	11	8			3,340	17	13	4,120	21	16
5⁄16–24	0.3125	0.0580			2,400	12	9			3,700	19	14	4,560	24	18
3⁄8–16	0.3750	0.0775			3,200	20	15			4,940	30	23	6,100	40	30
3⁄8–24	0.3750	0.0878			3,620	23	17			5,600	35	25	6,900	45	30
7⁄16–14	0.4375	0.1063			4,380	30	24			6,800	50	35	8,400	60	45

Size	D	Tensile stress area, in²		Tensile strength				Tensile strength			Clamp load		
7/16–20	0.4375	0.1187		4,900	35	25		7,550	55	40	9,350	70	50
1/2–13	0.5000	0.1419		5,840	50	35		9,050	75	55	11,200	95	70
1/2–20	0.5000	0.1599		6,600	55	40		10,700	90	65	12,600	100	80
9/16–12	0.5625	0.1820		7,500	70	55		11,600	110	80	14,350	135	100
9/16–18	0.5625	0.2030		8,400	80	60		12,950	120	90	16,000	150	110
5/8–11	0.6250	0.2260		9,300	100	75		14,400	150	110	17,800	190	140
5/8–18	0.6250	0.2560		10,600	110	85		16,300	170	130	20,150	210	160
3/4–10	0.7500	0.3340		13,800	175	130		21,300	260	200	26,300	320	240
3/4–16	0.7500	0.3730		15,400	195	145		23,800	300	220	29,400	360	280
7/8–9	0.8750	0.4620	60,000	11,400	165	125		29,400	430	320	36,400	520	400
7/8–14	0.8750	0.5090	33,000	12,600	185	140		32,400	470	350	40,100	580	440
1–8	1.0000	0.6060		15,000	250	190		38,600	640	480	47,700	800	600
1–12	1.0000	0.6630		16,400	270	200		42,200	700	530	52,200	860	680
1 1/8–7	1.1250	0.7630	105,000	18,900	350	270		42,300	800	600	60,100	1120	840
1 1/8–12	1.1250	0.8560	74,000	21,200	400	300		47,500	880	660	67,400	1260	940
1 1/4–7	1.2500	0.9690		24,000	500	380		53,800	1120	840	76,300	1580	1100
1 1/4–12	1.2500	1.0730		26,600	550	420		59,600	1240	920	84,500	1760	1320
1 3/8–6	1.3750	1.1550		28,600	660	490		64,100	1460	1100	91,000	2080	1560
1 3/8–12	1.3750	1.3150		32,500	740	560		73,000	1680	1260	104,000	2380	1780
1 1/2–6	1.5000	1.4050		34,800	870	650		78,000	1940	1460	111,000	2780	2080
1 1/2–12	1.5000	1.5800		39,100	980	730		87,700	2200	1640	124,000	3100	2320

*Clamp load is also known as *preload* or *initial load* in tension on bolt. Clamp load (lb) is calculated by arbitrarily assuming that usable bolt strength is 75 percent of bolt proof load (psi) times tensile stress area (in²) of threaded section of each bolt size. Higher or lower value of clamp load can be used depending on the application requirement, and the judgment of the designer.

†Tightening torque values are calculated from the formula $T = K D P$, where $T =$ tightening torque, lb · in; $K =$ torque friction coefficient; $D =$ nominal bolt diameter, in; and $P =$ bolt clamping load developed by tightening, lb.

‡Tensile strength of all grade 7 bolts is 133,000 psi. Proof load is 105,000 psi.

A spring washer is used to provide additional flexibility in the joint under repeated loading. However, the spring should provide the additional flexibility required without ever increasing the load in the bolt; that is, it should never reach the limit of the spring travel. One note is that a compressed lock washer is not a spring washer. The lock washer provides no additional flexibility once it has bottomed out. Gaskets provide this springiness in an interface such that any surface discontinuities are removed and there is equal force between the two surfaces along the entire area of contact.

20.6.4 Joint Strength

Joint strengths are analyzed as follows. One must take a cross section to include all the holes where the bolts occur. The effective cross-sectional width is generally the width minus the number of bolts times the diameter of each bolt. This effective cross-sectional width is multiplied by the thickness t of the material to determine the stressed area in the plate. The following is the equation for the stresses in the plate:

$$f_{tu} = \frac{P_u}{(W - nD)t} \tag{20.7}$$

where f_{tu} = plate tensile stress (ultimate)
P_u = plate tensile load (ultimate)
W = plate width
n = number of bolts
D = diameter of bolt hole
t = plate thickness

Here the plate tensile stress f_{tu} is compared to the ultimate strength of the material F_{tu} to ensure adequate strength.

Bearing Strength.   The bearing strength of the plate must be evaluated where the bearing stress f_b is

$$f_b = \frac{P_b}{A_b} \tag{20.8}$$

For one bolt:

$$f_b = \frac{P_b}{tD} \tag{20.9}$$

where f_b = plate bearing stress
P_b = plate bearing load
D = diameter of bolt
t = plate thickness

Note that D is the diameter of the bolt, not the bolt hole.

Joint Efficiency.   A bolted or riveted joint cannot be as strong as a solid plate or sheet. The reason is that the holes in the plate reduce the effective width. The *joint efficiency* is the ratio of the strength of the joint to the strength of the thinnest sheet or plate being joined. To determine the strength of the joint, calculations must be performed for bearing, shear, and tension stresses. Approximate efficiencies of well-designed riveted joints are listed in Table 20.6.

TABLE 20.6 Approximate Efficiencies
of Riveted Joints

Type of joint	Efficiency, %
Riveted butt joint	
Single	60–70
Double	75–83
Triple	80–89
Riveted lap joint	
Single	50–60
Double	60–70
Triple	70–80

Joint strength is analyzed for the following three forms of joint attachment: bolting, riveting, or spot welding. However, spot-welding failures are not quite the same as the first two, since a spot weld has a stress limit based on its peel strength in addition to the tensile and shear stresses. The typical failures that occur in a riveted joint are shown in Fig. 20.10.

The simple shear load for a bolted or riveted joint is shown in Fig. 20.11.

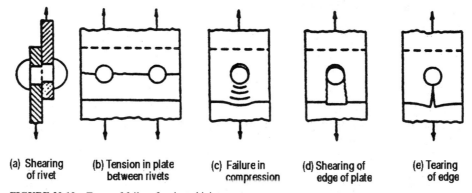

(a) Shearing of rivet   (b) Tension in plate between rivets   (c) Failure in compression   (d) Shearing of edge of plate   (e) Tearing of edge

FIGURE 20.10   Types of failure for riveted joints.

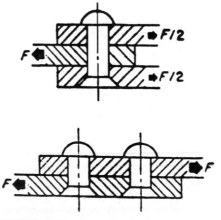

FIGURE 20.11   Simple shear load.

$$f_s = \frac{P}{An}$$ (20.10)

where f_s = fastener shear stress, psi
P = fastener shear load, lb
A = fastener shear area, in^2
n = number of shear planes

20.7   WELDING AND WELDED JOINTS

20.7.1   Welding Processes

There are many welding processes available for different types of metallic materials. These methods include impact welding, friction/inertia welding, ultrasonic welding, gas welding, resistance welding, and electric-arc welding. These various processes are briefly described as follows.

Impact Welding.   *Impact welding* is a process that employs an explosion or high pressure on two parts to force these parts to join in a molecular bond at the interface. This is a very specialized welding procedure.

Friction/Inertia Welding.   *Friction/inertia welding* is a spin-forming process that brings the part in contact with a fixture. The spinning motion of one part against the other, stationary, part generates extreme amounts of heat due to friction, which creates a molten state, resulting in an inertia weld as the rotation stops. The part is held in place as the molten metal cools down and forms its permanent weld.

Ultrasonic Welding.   *Ultrasonic welding* is a rubbing process that brings the two parts in contact. The rubbing motion of one part against the other, stationary, part generates heat due to friction, which creates a small molten area resulting from the ultrasonic energy as the motion stops. The part is held in place as the molten metal cools down and forms its permanent weld. This method is used to weld extremely small parts such as welding bond wires to the die pads on an integrated circuit chip.

Gas Welding.   Burning a mixture of oxygen and combustible gas generates heat to melt the material locally where a weld is to be formed. This method of welding is used when portability of welding equipment is necessary (for remote areas), when welding extremely thin materials, or where the welding of materials at relatively low temperatures is required. The limitations are that the process tends to be labor-intensive, requires a nonoxidizing or a slow-oxidizing material, and is slower than other welding methods.

Resistance Welding.   *Resistance welding* employs electricity to generate a current through the parts to be welded. The current flows through the metal and creates a melting of the metal locally, causing it to weld together. This is used for spot welding, but may also be used for seam welding and other forms of welding. The spot becomes a weld point and is considered to be similar to a bolted or riveted joint (Sec. 20.6).

Electric-Arc Welding.   *Electric-arc welding* is a broad group of welding processes where heating the base material and the filler material is provided by an electric spark gap between the electrode and the base material. The most common processes are (1) shielded-metal-arc welding (SMAW), (2) submerged-arc welding (SAW), (3) gas-tungsten-arc welding (GTAW or TIG), (4) gas-metal-arc welding (GMAW or MIG), and (5) arc spot welding (ASW). Processes using inert gas are necessary when welding aluminum and other rapidly corroding materials. The welding process without the inert gas causes undesirable oxidation of the basic material, creates large imperfections, and causes serious degradation in the weld. Electric-arc welding lends itself to automatic seam tracking, and is used with robotic welding systems. However, if production is small, hand welding with various types of electric arc systems is most practical.

20.7.2 Materials for Welding

The various high-quality structural materials to be welded include steel, nickel, copper, aluminum, magnesium, and alloys of these materials.

The filler metals chosen for the welding rod are generally the same material; however, the alloy is slightly different to provide both a higher strength and better flow characteristics. Filler metals are specified in various forms, including material type, electrode diameter or wire size, the recommended amperage or current, and the rate at which the wire should be fed for that amperage. The AWS structural welding code (D 1.1-72) permits the same values of stress for the base metal and the filler metal except that special formulas are provided for fatigue loading. There is an area near the weld which is referred to as the *heat-affected zone*. Generally this is the area immediately adjoining the filler material and is the most critical to fracture. Since the filler material is sized to be as strong as the base material, the weld zone has full strength. It is the heat-affected zone that is weak, as a result of an annealing process that reduces the yield and ultimate strength of the material. Any analysis performed on the welded seam should be based on the strength of the welded properties of the base material and the properties in the heat-affected zone. Typical electrode mechanical properties are shown in Table 20.7.

TABLE 20.7   Typical Mechanical Properties of Mild Steel Electrodes

Weld rod material	E601U, E6011	E6012, E6013	E6UZO, E6021	E7014, E7024	E7U15 & E7016, E7018 & E7028
Tensile strength, min psi	62,000	67,000	62,000	72,000	72,000
Yield point, min psi	50,000	55,000	50,000	60,000	60,000
Elongation in 2 in, min %	22	17	25	17	22

20.7.3 Weld Design

The various designs for the process of joining two members are presented in this section. The fundamental methods of joining include the butt joint, corner joint, edge joint, tee joint, and lap joint (Fig. 20.12). Generally, the welded area should be at least as thick as the base material and may exceed that thickness by a factor of 1.4 to 2, to ensure that the strength is adequate to overcome any stress concentration factors in the joint. All welded plates should be properly prepared with appropriate chamfers prior to the welding, to ensure maximum strength in the joint. This preparation is covered by the AWS structural welding code and various other specifications. These include the appropriate single and double V and U grooves, single and double bevel, and J grooves.

20.7.4 Strength of Fillet Welds

Fillet welds loaded in both shear and tension are evaluated to determine whether the strength of the weld is adequate on the basis of the diagrams and equations in Fig. 20.13.

Stresses in tension and shear of a butt joint in the fillet weld are identical to the stresses of the base material. However, in a tee weld with the weld on only one side, the stresses increase by the square root of 2. This is because the weld throat, which is a cross-sectional area at 45° to the baseplate, is less than the principal weld dimension h.

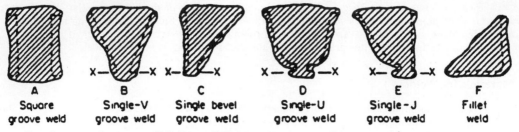

A	B	C	D	E	F
Square groove weld	Single-V groove weld	Single bevel groove weld	Single-U groove weld	Single-J groove weld	Fillet weld

Double welds for types B, C, D and E formed by constructing symmetrically about x - x

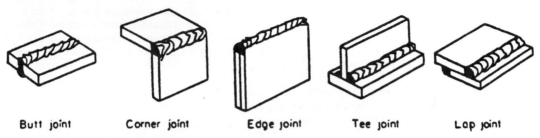

Butt joint   Corner joint   Edge joint   Tee joint   Lap joint

FIGURE 20.12 Types of welds (upper) and joints (lower).

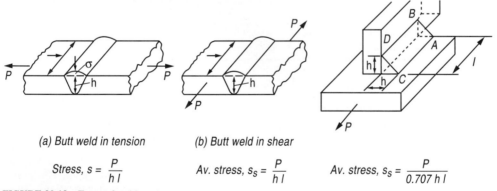

(a) Butt weld in tension

$$\text{Stress, } s = \frac{P}{h\,l}$$

(b) Butt weld in shear

$$\text{Av. stress, } s_s = \frac{P}{h\,l}$$

$$\text{Av. stress, } s_s = \frac{P}{0.707\,h\,l}$$

FIGURE 20.13 Types of welds and average values for stresses.

In a lap joint one must also use the same square-root-of-2 factor, which is illustrated in Fig. 20.14.

The S_i term is the stress along the action plane (direction of P). However, it is noted that $S_s = P/hl = S_t$ which are the shear and tensile stresses, respectively. Since both shear and tensile stresses are present in this weld joint, we have to analyze the combined loading condition. The resultant normal stress in the lap weld S_i is given by the following equation:

$$S_i = 0.5S_t + \sqrt{(0.5S_t)^2 + S_s^2} = 1.618S_t = 1.618\,\frac{P}{hl} \tag{20.11}$$

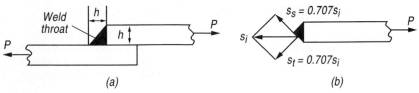

FIGURE 20.14   Fillet weld, in tension.

The following is a typical example of a fillet weld loaded in shear.

EXAMPLE 20.1   *GIVEN. A pulley is 15 in in diameter with a $3\frac{1}{2}$-in-OD (outer-diameter) hub. The weld plate has a $\frac{1}{4}$-in fillet weld on each side. The belt transmits 30 hp to the shaft, which turns at 1200 rpm. Determine:* the value of the torsion shearing stress in the welds.

Solution

$$\text{Torque: } T = 63{,}000 \text{ hp/N} = 63{,}000 \,(30/1200) = 1575 \text{ in} \cdot \text{lb}$$

$$\text{Radius to center of throat: } r_1 = \frac{3.5}{2} + \frac{0.25}{4} = 1.81 \text{ in}$$

$$\text{Force carried by welds: } P = \frac{1575}{1.81} = 870 \text{ lb}$$

$$\text{Length of both welds: } L = 2 \times 2\,\pi\, r_1 = 4\pi \times 1.81 = 22.8 \text{ in}$$

$$f_s = \frac{P}{0.707hL} = \frac{870}{0.707 \times 0.25 \times 22.8} = 216 \text{ psi}$$

20.7.5 Stress Concentration in Welds

Generally there is an abrupt change in the stress flow at the point of weld. Therefore, the stress concentration factor present at that transition must be considered. For welds, the stress concentration effects can be minimized for static or steady loads; however, for fluctuating loads, it is very important to include a stress concentration factor. Also, the weld metal and the plates should be thoroughly fused together with the minimum amount of imperfections and discontinuities, as these flaws in the weld cause additional stress concentrations to occur.

Fillet welds have stress concentrations at the toe and heel points of a weld, and particularly for a lap joint the forces are not directly opposed. The stress concentration factors that should be applied to various welds are listed in Table 20.8.

TABLE 20.8   Stress Concentration Factors K for Welds

Location	K
Reinforced butt weld	1.2
Toe of transverse fillet weld	1.5
End of parallel fillet weld	2.7
T-butt joint with sharp corners	2.0

20.7.6 Eccentrically Loaded Welds

Eccentrically loaded welds are offset from the line of force and create shear and tension loads. When a joint consists of a number of welds, it is customary to assume that the stress at any point is proportional to the distance from the center of gravity to the group of welds. Therefore, a stress moment s acts perpendicular to the radius r from the center of the force out to the weld. The torque equation for this is as follows:

$$T = \int sr \, da = \int \left(\frac{s}{r} \right) r^2 \, da = \frac{s}{r} \int r^2 \, da = \frac{sJ}{r}; \quad \text{or} \quad s = \frac{Tr}{J} \tag{20.12}$$

It is important to calculate the proper area moment of inertia J for this weld stress calculation. The area moment of inertia includes both the moment of inertia of the cross-sectional area I_{xx} plus the offset distance from the center of the area to the force point r_1, and this inertia term is $I' = Ar_1^2$, where A is the area of the welded section.

20.8 FRACTURE MECHANICS

Fracture mechanics is an analytical method for analyzing the notch sensitivity of various materials. This is typically applied to ultra-high-strength materials, brittle materials, and anisotropic materials. The analysis is very complex, so computer programs are required along with detailed analysis. Fracture-mechanics problems also extend into the elastic/plastic region. The exact type of deformation prior to a catastrophic failure must be determined.

20.9 THERMAL STRESSES

This section presents thermal stresses that are encountered during the operational life of various types of structures. These thermal stresses can be extremely high and, if the structures are not properly designed, can destroy the structure without any other mechanical loading. Thermal stresses *must* be evaluated and determined acceptable for the design.

Thermal stresses are the result of temperature differences; constant temperature will not create thermal stresses. Transient conditions cause expansion and contraction of the material. If multiple materials are present, the rate of expansion of each is different, and stresses are created at the interfaces. Also, for a single material, the heat flow may cause a temperature differential across a cross section of the structure. Under a transient heat pulse, the surface of the structure may become extremely hot while the heat has not yet penetrated the structure to the inner portion. If the inner portion is relatively cool, the expansion of the outer elements may be excessive and cause fracturing. If there is no heat flowing into or out of the structure, then the structure has reached a steady-state condition and there are no further changes in stresses due to thermal effects.

20.9.1 Thermal Stresses in a Long Bar

The thermal stresses in a long bar with free expansion is zero unless the bar is constrained. The expansion dL would be the coefficient of thermal expansion α times the delta temperature dT times the length of the beam L or

$$dL = \alpha dTL \tag{20.13}$$

This same beam may be compressed, where the spring constant K is

$$K = \frac{AE}{L} \tag{20.14}$$

where A = cross-sectional area
E = modulus of elasticity
L = length of the beam

The force in the beam to constrain the dL distance is $F = KdL$. Combining the equations results in

$$dL = \frac{F}{K} = \frac{FL}{AE} = \alpha dTL \tag{20.15}$$

$$s_t = \frac{F}{A} = \alpha dTE \tag{20.16}$$

where $dT = (T_{new\ ambient} - T_{old\ ambient})$.

20.9.2 Thermal Stresses in Circular Rings and Pipes

To determine the tensile force P in a ring and the hoop stress in that ring, we will evaluate the ring as follows:

$$s_r = \frac{P}{A} = \frac{P}{tL} \tag{20.17}$$

where t is the thickness of the circular ring and L is the length. If this ring is forced to expand with an increase in temperature, but is constrained because of an outer shell, stresses are developed in the internal shell and possibly the outer shell.

20.9.3 Thermal Stresses of Bimetallic Structure

An example of a bimetallic structure might be a strip of Monel metal and steel layered together. Both have equal cross sections and will bend if a temperature change occurs because of the differences in thermal expansion of the two materials and the differences between the flexural rigidity E of the steel and Monel. When the temperature rises, the Monel metal, having a greater coefficient of expansion, will become a longer beam and will subject the steel beam to tension as the Monel beam experiences compression. Each beam element has a tensile or compressive force and a bending moment due to the clamping effect of the two beams. The moments are calculated as

$$M_1 = \frac{E_s I_s}{R} \quad \text{and} \quad M_2 = \frac{E_m I_m}{R} \tag{20.18}$$

The tensile force P in one beam is equal to the compressive force in the other beam. If H is the sum of the thickness of the two equal materials, then the sum of the moments, $M_1 + M_2 = PH/2$.

By combining equations and simplifying, we resolve the many moments and stresses in the beams as follows:

$$s_{max} = \frac{HE}{3R} \qquad (20.19)$$

and for the temperature change

$$s_{max} = \tfrac{1}{2} EdT(\alpha_m - \alpha_s) \qquad (20.20)$$

20.10 DESIGN STANDARDS

A mechanical design is acceptable only if it is based on the actual loading and the minimum strength required to safely carry that load. Design standards to establish the design criteria are established by U.S. agencies and international organizations. Design codes are established by ASTM and ANSI for piping and pressure vessels. They also provide standard material strengths for steel, aluminum, etc. Proof loads may be required for high loading conditions such as for high-strength bolts, piping, pressure vessels, and other highly stressed parts. Safety factors were discussed in detail in Secs. 20.1.2 and 20.1.3, so they will not be reviewed here. However, safety factors will be presented as they are applied to various design criteria. This section concentrates primarily on pressure-vessel design.

20.10.1 Pressure-Vessel Design Standards

Pressure-vessel designs are based on many criteria, including the following:

1. Vessels under internal pressure must include stresses in the cylindrical shell, pressure of fluid, pressure and temperature ratings of the steel for the vessel, and pipe flanges.
2. Vessels under external pressure must include the minimum thickness of cylindrical shells to prevent buckling, for both cylindrical and spherical vessels.
3. Design of tall towers utilizes special design criteria and includes variables such as wind load, weight of the vessel, seismic load, vibration load, eccentric load, elastic stability, deflection, anchor construction, and base ring design.
4. Large horizontal vessels supported by two saddles must be analyzed for the stiffener ring requirements for the horizontal vessel, for the design of the saddles, and for the thermal expansion and contraction of the horizontal vessel.
5. The design for openings and cutouts must be analyzed considering
   a. Reinforcing pads and simple cutouts without reinforcing pads
   b. Any extension of the opening
   c. Strength of the attachment joining the opening to the vessel
   d. Length of the coupling in the pipes attached to the opening
6. Welding the pressure vessel includes analyzing all welded joints—butt welds, lap welds, etc.
7. Regulations and specifications related to pressure vessels depend on
   a. The type of service
   b. The type of plate material and its thickness
   c. The classification of the contents of the vessel, including flammable and combustible liquids
   d. Material-strength properties
   e. Specific design specifications for pressure vessels

Classification of Shell Stresses.   The primary stresses of interest in a pressure vessel for externally applied loads and internal pressure (per the Welding Research Council) are primarily the membrane stresses. These stresses are the result of forces applied at a cutout in the pressure vessel and include both axial loads and bending moments applied in a direction perpendicular to the cutout. We will refer to the cutout as the *nozzle portion,* since this is where the high fluid flow exists. This is generally where a boss is welded and piping extends from the pressure vessel at this point.

A convenient way to identify stresses is by using a stress concentration factor K. A subscript "m" represents membrane stresses, and a subscript "b" indicates bending stresses. There also are secondary stresses from the externally applied load. These secondary stresses are the result of axial and bending loads applied on the nozzle. Bending stresses will generate tension on one side and compression on the other side of the pressure vessel wall. A third group of stresses are the stress concentrations due to notches or sharp corners at this interface.

These nozzle stresses are produced by external loads and internal pressures. Generally, the critical stress in the nozzle will be found in that region near the nozzle where it is joined to the shell. It is this area that is recommended for local reinforcement by ASME codes. The stress in the nozzle due to external loads is calculated assuming a cantilever beam to be fixed at the point where the nozzle joins the shell. The combined loading condition must be considered and should include the cantilever bending and the combined compression or tension load, depending on the direction of the axial load on the nozzle. The force from the internal pressure inside the pipe nozzle must be included, because it adds to the axial force. Finally the pressure stresses in the nozzle wall must be evaluated. These are calculated using the hoop stress and a stress concentration multiplier where the nozzle joins the pressure vessel. It is recommended that the designer use the ASME codes, Section VIII and Section III, Classes 2 and 3. These are the standard pressure vessel codes and include various protrusions and nozzles attached thereto.

The pressure stress coefficient applied to hoop stress for cylindrical vessels I_v is obtained from the *Pressure Vessel Handbook.* From the pressure in a pressure vessel the hoop stress f_p is

$$f_p = \frac{PRI_v}{t} \tag{20.21}$$

where f_p = hoop stress
P = pressure
R = radius of cylindrical
I_v = pressure stress coefficient
t = wall thickness of cylindrical

20.11   COMPONENT DESIGN: SHAFTS AND SPRINGS

This section provides an introduction to the various component designs and design considerations for shafts and springs, including the fundamental equations required to perform mechanical stress analysis and reliability factors. One should analyze the loads and stresses to ensure that the components are designed so that they will not yield during their normal lifetime. This section provides the reader with a basic knowledge of designing some of the more critical components that are attached to a primary structure. The tools (numerical analysis) presented in this section support the overall structural analysis tools presented in the preceding sections. However, where the preceding sections have been general, the tools here are more specific to the application of the attachment structure or energy transmission.

Various components are reviewed using standard design practices to ensure sufficient strength and endurance. These components include shafts and springs, as described in detail in the following paragraphs.

20.11.1 Shaft Design

Shafts are common in most machinery and mechanical equipment. They generally transmit rotating energy. The elementary theory of a circular shaft with a static torsion load is the most general case. However, shafts are also subjected to fluctuating loads, including bending and torsion with various degrees of stress concentration. The primary type of failure for shafts is due to fatigue, caused by fluctuating loads. Design considerations should also include the keyway and the coupling. In general, the operating speed should not be close to the critical vibration frequency of the shaft; otherwise, an additional safety margin is required.

A standard shaft in pure torsion causes all the structural elements to be loaded in shear, where the highest shear stresses are at the outer surface of the shaft. The torsional stiffness K of a shaft (in foot-pounds per radian) is calculated as follows:

$$K = \frac{T}{\theta} = \frac{JG}{L} \tag{20.22}$$

where the variables are torque T, polar moment of inertia J, modulus of rigidity or shear modulus G, length of the shaft L, and angle of twist θ. The stress for the shaft is f_s:

$$f_s = \frac{Tr}{J} \tag{20.23}$$

where r is the radius.

The shear stress values for various odd shaft forms are presented in books on shaft design. From the equations polar moment of inertia, J can be determined. Also included is the angle of twist θ. The torsional stiffness K may then be determined from the ratio

$$K = \frac{T}{\theta} \tag{20.24}$$

The following examples are for circular shafts.

EXAMPLE 20.2   *GIVEN:* A hollow shaft must carry a torque of 30,000 in · lb at a shearing stress of 8000 psi. The inside diameter is to be 0.65 of the outside diameter. *Determine:* the value of the outside diameter.

Solution

$$d_i = 0.65\, d_o$$

$$J = \left(\frac{\pi}{32}\right)(d_o^4 - 0.65^4\, d_o^4) = 0.08065\, d_o^4$$

$$= \frac{(30{,}000 \times 0.5\, d_o)}{8000} = 0.08065\, d_o^4 \qquad \left(J = \frac{Tr}{s}\right)$$

$$d_o^3 = 23.2486$$

$$d_o = 2.854 \text{ in}$$

Since most shafts have cyclical fluctuating loads, the ASME code for design of transmission shafting is used to determine the maximum shear stress. The equation is as follows (subscript "yp" denotes yield point):

$$S_{s\,max} = \frac{0.5 S_{yp}}{FS} = \sqrt{\frac{C_m S^2}{2} + (C_t S_s)^2} = \frac{16}{\pi d^3}\sqrt{(C_m M)^2 + (C_t T)^2} \tag{20.25}$$

where C_m is the numerical combined shock and fatigue factor to be applied in every case to the computed bending moment and C_t is the corresponding factor to be applied to the computed torque. The recommended values for shock and fatigue factors are given in Table 20.9.

TABLE 20.9   Values for Shock and Fatigue

Nature of loading	Values for	
	Cm	Ct
Stationary shafts		
Gradually applied load	1.0	1.0
Suddenly applied load	1.5–2.0	1.5–2.0
Rotating shafts		
Gradually applied or steady load	1.5	1.0
Suddenly applied loads, minor shocks only	1.5–2.0	1.0–1.5
Suddenly applied loads, heavy shocks	2.0–3.0	1.5–3.0

There are more elaborate equations for evaluating fluctuating loads, such as using the maximum shear theory and the Mises-Hencky theory for shafting. These should be used for complex loading problems. Keyways are necessary for fastening hubs to transmit torque from shafts to gears, pulleys, levers, etc. Various methods of transmitting the torque from the shaft to the hub are illustrated in Fig. 20.15.

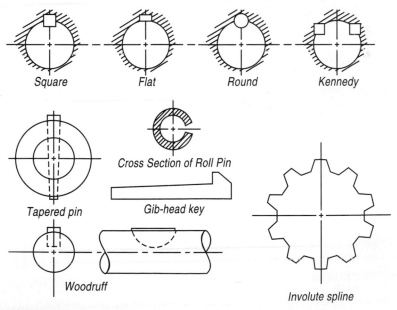

FIGURE 20.15   Types of keys.

These are areas of high stress concentration that should be carefully evaluated to ensure that the keyway and the shaft are adequate. Some recommended dimensions for square keys are presented in Table 20.10.

TABLE 20.10   Dimensions (in Inches) of Square Keys

Shaft diameter	Key size	Shaft diameter	Key size	Shaft diameter	Key size
$\frac{1}{2}-\frac{9}{16}$	$\frac{1}{8}$	$1\frac{7}{16}-1\frac{3}{4}$	$\frac{3}{8}$	$3\frac{3}{8}-3\frac{3}{4}$	$\frac{7}{8}$
$\frac{5}{8}-\frac{7}{8}$	$\frac{3}{16}$	$1\frac{13}{16}-2\frac{1}{4}$	$\frac{1}{2}$	$3\frac{7}{8}-4\frac{1}{2}$	1
$\frac{15}{16}-1\frac{1}{4}$	$\frac{1}{4}$	$2\frac{5}{16}-2\frac{3}{4}$	$\frac{5}{8}$	$4\frac{3}{4}-5\frac{1}{2}$	$1\frac{1}{4}$
$1\frac{5}{16}-1\frac{3}{8}$	$\frac{5}{16}$	$2\frac{7}{8}-3\frac{1}{4}$	$\frac{3}{4}$	$5\frac{3}{4}-6$	$1\frac{1}{2}$

The force distribution on the key is dependent on the fit of the key in the groove, and other variables. One method of evaluating the keyway is to assume that the stresses are calculated as a shear force at the radius of the outer surface of the shaft:

$$F = \frac{T}{r} \tag{20.26}$$

where the shear stress (S_s) is

$$S_s = \frac{F}{A} \tag{20.27}$$

However, a sufficiently large safety factor should be employed, since the stresses are not linear on the key—they are generally much higher at the ends of the key. A stress concentration factor of approximately 2 should be used for the keyway area.

EXAMPLE 20.3   *GIVEN:* A $3\frac{7}{16}$-in-diameter shaft is made from material with a yield-point value of 58,000 psi. A $\frac{7}{8} \times \frac{7}{8}$-in key is to be used of material with a yield-point value of 48,000 psi. Assume a safety factor of 2 is required and $S_{syp} = 0.5\ S_{yp}$. *Determine:* the required length of key based on the torque value of the gross shaft.

Solution

Shaft:

$$S_{yp} = 58,000 \text{ psi, working stress } S = \frac{58,000}{2} = 29,000 \text{ psi}$$

$$S_{syp} = 29,000 \text{ psi, working stress } S_s = \frac{29,000}{2} = 14,500 \text{ psi}$$

Key:

$$S_{yp} = 48,000 \text{ psi, working stress } S_s = \frac{48,000}{2} = 24,000 \text{ psi}$$

$$S_{syp} = 24,000 \text{ psi, working stress } S_s = \frac{24,000}{2} = 12,000 \text{ psi}$$

$$J = \frac{\pi d^4}{32}$$

Torque in shaft:

$$T = S_s \frac{J}{r} = \frac{14,500 \times 13,708}{1.719} = 115,650 \text{ in} \cdot \text{lb}$$

Force at shaft surface:

$$F = \frac{T}{r} = \frac{115,650}{1.719} = 67,280 \text{ lb}$$

Based on bearing on shaft:

$$L = 67,280/29,000 \times 0.438 = 5.30 \text{ in}$$

Based on bearing key:

$$L = 67,280/24,000 \times 0.438 = 6.41 \text{ in}$$

Based on shear in key:

$$L = 67,280/12,000 \times 0.875 = 6.41 \text{ in}$$

Other critical designs for shafting include (1) coupling analysis, (2) the bending loads in two planes on a shaft, (3) a shaft with three or more rigid supports, (4) crank shafts and the analysis associated with them, (5) critical speed of rotating shafts, and (6) stresses and deflections in shafts with nonuniform diameter and irregular shapes, including noncircular shafts.

20.11.2 Spring Design

When flexibility or deflection is required in a mechanical system, a spring may be used. The steel spring has many advantages, such as its capability to endure high cycle life prior to fatigue failure, and its relative insensitivity to the environment. Steel has a much greater application than elastomeric materials, which tend to operate in the plastic range. They wear out much more rapidly and have a significant aging factor.

Helical Springs and Coil Spring Design.   These springs can be represented as a shaft in torsion. The equations are similar to those presented in the preceding section, where the torsional shearing stresses are

$$S_s = \frac{Tr}{J} = \frac{16\, Pr}{\pi D^3} \tag{20.28}$$

This equation is applicable for a solid round (straight) bar loaded in pure torsion.

After a coil has been formed from a straight round shaft, the cross section has an additional stress resulting from the transverse shear. The transverse shear stress is

$$S_{ts} = \frac{1.23\, P}{A} \tag{20.29}$$

20.12   *VIBRATION ANALYSIS*

Thus far we have considered mechanical components and mechanical systems subjected to static loads. In this section the basic concepts of dynamic equilibrium and behavior of mechanical systems undergoing vibration are introduced. All bodies possessing mass and stiffness are capable of vibration. Most mechanical systems experience some form of vibration, and their design may require the investigation of additional stresses and strains associated with these vibrations, since this is the major cause of fatigue failure. For the analysis of such systems, it is often advantageous to represent the system by a model formed from lumped masses, springs, and dashpots (dampers).

A structural dynamic analysis of various subsystems is performed in this section. However, it is first necessary to establish some fundamental principles and to derive some basic equations which are used throughout the dynamic stress analysis.

20.12.1   Single Degree of Freedom System

A *single degree of freedom* within a dynamic system is generally described as having one mass and one spring. This may be more complex if a damping term or a forcing function is added. Also, the motion can be either translation or rotation. If both motions are involved, the system is generally described as having two or more degrees of freedom. Consider the system in Fig. 20.16, with a single mass and a single spring of negligible weight and spring constant K (lb/in). This spring constant is defined as the force produced in a unit displacement of the spring.

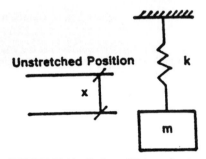

FIGURE 20.16   Static equilibrium of spring and mass.

In the absence of externally applied loads and vibration, the spring will stretch a distance and come to a position of static equilibrium. If the mass is disturbed from the static equilibrium position, vibration will begin. This type of vibration is called *free vibration* if damping is negligible (it is generally very small in mechanical systems). It approaches 2 to 5 percent in welded structures and 5 to 10 percent in bolted and riveted structures. This motion can be defined by Newton's Second Law. All dynamic problems are basically solved by this natural law, which is $F = ma$. This same equation can be used for multiple degrees of freedom and unit forcing functions, as long as all components are properly accounted for. In the case of a spring-mass system, the mass provides the m for the equation, a for acceleration, and the spring exerts a force F which is equal to the spring rate k times the displacement of the spring x. This could also be expressed in terms of a rotating spring mass system where $T = I\alpha$, where the torque T equals the torsion spring rate k times the angular displacement θ, where θ is measured in radians. The moment of inertia I is the rotating moment of inertia or polar moment of inertia. The angular acceleration α is equal to the second derivative of angular displacement with respect to time $(d^2\theta/dt^2)$.

20.12.2 Free Vibration without Damping

A single degree of freedom system with free vibration and no damping is simply written as

$$mx'' + kx = 0 \tag{20.30}$$

The general solution to this equation can be written as

$$x = C_1 \sin \sqrt{\frac{K}{M}}\, t + C_2 \cos \sqrt{\frac{K}{M}}\, t \tag{20.31}$$

where $\sqrt{K/M}$ = natural circular frequency in radians per second and C_1 and C_2 are constants of integration and are the conditions of initial displacement and initial velocity. The solution can be verified by differentiating it twice and then substituting into the fundamental equation.

If we want to solve this equation for the spring-mass system, initially the mass is extended such that the spring has a force of $k\,x$, the extension is x_0, and $x' = 0$, since there is no velocity. It is just being stretched to an initial condition. By substituting the solution into the general equation, one can find that $C_2 = x_0$. Since we have two constants, the second condition is that $x' = 0$. Taking the derivative of the general equation, we find x where $C_2 = x_0$, C_1 must equal 0. A specific solution for the boundary condition is

$$x = x_0 \cos \sqrt{\frac{K}{M}}\, t \tag{20.32}$$

This represents the undamped vibration. One cycle occurs when the cosine term goes through 360°, or 2π radians. The time of one cycle is the period t. From the preceding equation we have

$$\sqrt{\frac{K}{M}} \cdot t = 2\pi \tag{20.33}$$

The natural frequency f_n is found by the following substitution:

$$f_n = \frac{1}{t} = \frac{1}{2\pi} \sqrt{\frac{K}{M}} \tag{20.34}$$

This same equation can be derived using the conservation-of-energy principle, where work is being done over the distance 0 to x as follows:

$$W = Fx = \int_0^x kx\, dx = \tfrac{1}{2} kx^2 \tag{20.35}$$

The kinetic energy, also a work term in inch-pounds, is equal to $\tfrac{1}{2}\, mv^2$. If one assumes that the displacement follows a simple sinusoidal path, then $x = x_0 \sin \omega t$.

Taking the derivative of this provides us with $v = x_0\, \omega \cos \omega t$. Equating the work term at maximum displacement to the kinetic energy at zero displacement provides us with $\tfrac{1}{2} kx^2 = \tfrac{1}{2} mv^2$. Since the cosine of $0° = 1.0$, and $v = x_0\, \omega \cos \omega t$, at maximum displacement $v = x_0\, \omega \cos p/2 = x_0\, \omega$. Substituting these relationships into the energy equation, one obtains

$$\tfrac{1}{2} kx_0^2 = \tfrac{1}{2} m\, \omega^2 x_0^2 \tag{20.36}$$

Canceling like terms provides us with the simple equation:

$$\omega^2 = \frac{K}{M}$$ (20.37)

This energy solution is known as the *Rayleigh method,* and is used when the differential equations become complex.

The single-degree-of-freedom idealization can be used to approximate the natural frequencies of many real mechanical systems. Excellent accuracy is achieved when the mass of the elastic portion of the system is small compared with the lumped mass of the system or when the distributed mass is properly equated to an equivalent lumped mass.

EXAMPLE 20.4   *GIVEN:* the beam shown in Fig. 20.17 with mass *m.*

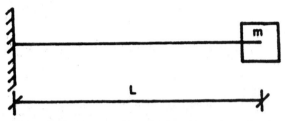

FIGURE 20.17   Example of free vibration without viscous damping.

Determine: the natural circular frequency and the natural frequency if

$M = 10\ \text{lb}_f \cdot \text{s}^2/\text{in}$
$L = 100\ \text{in}$
$I = 250\ \text{in}^4$
$E = 29 \times 10^6\ \text{psi}$

Solution   From standard deflection equations $K = 3\ EI/L^3$:

$$\omega_n = \sqrt{\frac{K}{M}} = \sqrt{\frac{3EI}{ML^3}}$$

$$= \sqrt{\frac{3(29)(10^6)(250)}{10(100)^3}} = 46.64\ \text{rad/s}$$

$$f = \frac{\omega_n}{2\pi} = \frac{46.64}{2\pi} = 7.42\ \text{cycles/s}$$

20.12.3   Free Vibration with Viscous Damping

In all free-vibration systems, energy is dissipated and the maximum amplitude cannot be maintained. The analytical representation of damping is extremely difficult, because it may be a function of displacement, velocity, acceleration, stress, or other factors. Many mathematical approximations have been proposed, but the simplest and most widely used is viscous damping. Free vibration with standard viscous damping is dependent on the constant damping term C, and

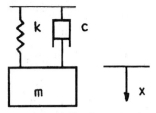

FIGURE 20.18 Free vibration with viscous damping.

is multiplied by the velocity term x' in the equations of motion (see Fig. 20.18). It is called *viscous damping* because it represents a fairly constant value (linear) as a function of velocity and can be represented as the damping in a dashpot when using oil.

Applying Newton's Second Law to the damped single-degree-of-freedom model in Fig. 20.18 generates the following equation of motion:

$$mx'' + cx' + kx = 0 \tag{20.38}$$

This second-order homogeneous differential equation is solved by assuming a solution of the form

$$x = e^{st} \tag{20.39}$$

which, by substitution, yields

$$(ms^2 + cs + d)e^{st} = 0 \tag{20.40}$$

The two solutions to this equation are

$$S = \frac{-c}{2m} \pm \sqrt{\frac{c^2}{4m^2} - \frac{k}{m}} \tag{20.41}$$

The general solution to this equation is written

$$x = c_1 e^{s_1 t} + c_2 e^{s_2 t} \tag{20.42}$$

where c_1 and c_2 are again arbitrary constants of integration. There are three types of solutions to this equation. These different solutions are dependent on the results inside the square-root sign. If $-k/m$ is greater than the damping term, the square root of a minus sign indicates that the answer is imaginary, and the exponential function to an imaginary exponent results in an oscillatory motion. This is termed *underdamped*.

If the value under the square-root sign is real, the solution is a decaying exponential and is called *overdamped*. The transition point (where the square root term is equal to zero) is the critical damped condition, and is equal to

$$C_c = 2m \sqrt{\frac{K}{M}} = 2m\omega_n \tag{20.43}$$

An oscillation characteristic of an underdamped system is shown in Fig. 20.19.

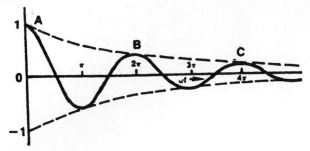

FIGURE 20.19   Free vibration with damping less than critical.

A damped oscillation of this nature decays each succeeding peak by the factor

$$e^{-(\pi C/mq)} \tag{20.44}$$

where q equals

$$q = \sqrt{\frac{k}{m} - \frac{c^2}{4m^2}} \tag{20.45}$$

The ratio between two consecutive peaks is constant. The amplitude decreases in a geometric series. Let the value

$$\Delta = \frac{\pi C}{mq} \tag{20.46}$$

where Δ is known as the *logarithmic decrement.* For small damping values, we have Δ approximately equal to

$$\Delta \cong \frac{2\pi c}{C_c} \tag{20.47}$$

The frequency also shifts as a result of damping. The ratio of the damped to undamped natural frequency is

$$\frac{q}{\omega_n} = \sqrt{1 - \frac{c^2}{C_c^2}} \tag{20.48}$$

20.12.4   Forced Vibration without Damping

The next important dynamic condition is a forced natural vibrating condition where no damping exists, but where we have an oscillating forcing function. The equation for the dynamic forces is

$$m\ddot{x} + k\dot{x} = P_0 \sin \omega t \tag{20.49}$$

One may find that a solution for this equation is

$$x = x_0 \sin \omega t \tag{20.50}$$

Taking derivatives of this equation and substituting it into the fundamental equation results in the following:

$$x_0 = \frac{P_0}{k - m^2} = \frac{P_0/k}{1 - m\omega^2/k} = \frac{P_0/k}{1 - (\omega/\omega_n)^2} \tag{20.51}$$

$$x = \frac{P_0/k}{1 - (\omega/\omega_n)^2} \sin \omega t \tag{20.52}$$

One can substitute this into the previous equation to find the value of x at any instant in time. Also, the expression P_0/k in the numerator represents a static deflection of the spring under the constant load P_0. Therefore $x_{st} = P_0/k$. Thus, the solution becomes

$$\frac{x}{x_{st}} = \frac{1}{1 - (\omega/\omega_n)^2} \sin \omega t \tag{20.53}$$

This is a specific solution, and represents the undamped forced vibration. Other terms in the complete solution represent the undamped free vibration similar to that discussed in Sec. 20.12.1.

20.12.5   Forced Vibration with Viscous Damping

Finally, the complete system can be expressed as

$$m\ddot{x} + c\dot{x} + kx = P_a \sin \omega t \tag{20.54}$$

This general equation provides the basis for all two-degree-of-freedom and multiple-degree systems. However, the analysis generally assumes that the damping can be neglected unless we are providing isolation or a shock damping analysis. The specific solution for the forcing function is

$$x_0 = \frac{P_0}{\sqrt{(c\omega)^2 + (k - m\omega^2)^2}} \tag{20.55}$$

The following four variables apply to this analysis:

1. $\omega_n = \sqrt{\dfrac{K}{M}} =$ undamped natural frequency

2. $q = \sqrt{\dfrac{K}{m} - \dfrac{c^2}{4m^2}} =$ damped natural frequency

3. The frequency of the maximum forced amplitude is sometimes referred to as the *resonant frequency*

4. The damping coefficient, zeta (ζ) = c/C_c

EXAMPLE 20.5   *GIVEN:* an automobile has a body weighing 3000 lb mounted on four equal springs which sag 9 in under the weight of the body. Each of the four shock absorbers has a damping coefficient of 7 lb for a velocity of 1 in per second. The car is placed with all four wheels on a test platform which is moved up and down at resonant frequency with an amplitude of 1 in. *Determine:* the amplitude of the car body on its springs, assuming the center of gravity to be in the center of the wheel base.

solution   The natural frequency is

$K = 3000 \text{ lb}/9 \text{ in} = 333$
$M = 3000 \text{ lb}/386 = 7.8$
$\omega_n = 2\pi f_n = (333/7.8) = (386/9) = 6.6 \text{ rad/s}$

The damping of the system (four shock absorbers) is

$$c = 4 \times 7 = 28 \text{ lb/in} \cdot \text{s}$$

The differential equation governing the motion has (at resonance) a disturbing force of

$$\sqrt{(ka_0)^2 + (ca_0)^2}$$

Here $k = 3000 \text{ lb}/9 \text{ in} = 333 \text{ lb/in}$; $a_0 = 1 \text{ in}$; $c = 28 \text{ lb/in} \cdot \text{s}$ and $\omega = \omega_n = 6.6 \text{ rad/s}$. Therefore

$$\sqrt{(ka_0)^2 + (ca_0)^2} = \sqrt{(333)^2 + (185)^2} = 380 \text{ lb}$$

The amplitude of the car body is found:

$$x_0 = \frac{P_0}{c\omega} = \frac{380}{28(6.6)} = 2.06 \text{ in}$$

20.12.6   Two-Degree-of-Freedom Systems

Free Vibration without Damping.   The most general undamped two-degree-of-freedom system can be reduced to that shown in Fig. 20.20. This consists of two masses (M_1 and M_2) and either two or three springs (depending on the model), where K_1 is sometimes neglected. A typical example of this type of spring system would be heavy equipment M mounted on springs (motor mounts) attached to the primary structure attached to the ground or a deck of a ship (M_2) with a spring stiffness of K_2.

Any motion present in the primary structure or forces on the primary structure would tend to excite both the structural platform and any equipment spring-mounted on the platform. These compound frequencies must be carefully analyzed. Disregarding the K_1 spring, the following two frequencies occur:

$$\omega_a^2 = \frac{K_3}{M_1} = \text{the natural frequency of the equipment} \tag{20.56}$$

$$\omega^2 = \frac{K_2}{M_2} = \text{the natural frequency of the main system} \tag{20.57}$$

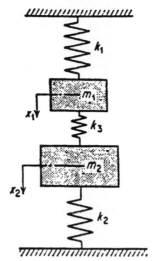

FIGURE 20.20 Undamped two-degree-of-freedom system with spring coupling.

$$\mu = \frac{M_1}{M_2} = \text{the mass ratio} \tag{20.58}$$

The solution is

$$\frac{x_1}{x_{st}} = 1 - \frac{\dfrac{\omega^2}{\omega_a^2}}{\left(1 - \dfrac{\omega^2}{\omega_a^2}\right)\left(1 + \mu - \dfrac{\omega^2}{\omega_a^2}\right)} - \mu \sin \omega t \tag{20.59}$$

Self-Excited Vibrations.   So far we have basically discussed free vibration and forced vibration, which accounts for a majority of the cases to be analyzed. However, some disturbances have been observed which belong to a fundamentally different class known as *self-excited vibrations*. An example is an engine running at its motor-mount frequency. In a self-excited vibration, the alternating force that sustains the motion is created and controlled by the motion itself. When the motion stops, the alternating force disappears. If the motion is caused by an external force, then stopping the motion does not stop the vibratory force, and vice versa. Another method of describing a self-excited vibration is to consider free vibration with negative damping. The reader may have already discovered that this type of condition is unstable, and each successive oscillation is larger and larger. Typically, self-excited vibration lends itself to instability and eventually to destruction. The frequency of the self-excited vibration is the natural frequency of the system. The frequency is relatively easy to calculate depending on the complexity of the system. In a truly linear self-excited system, the amplitude will become infinitely large, given sufficient time. During each cycle more energy is put into the system.

20.12.7 Instability Caused by Friction

Generally, sliding friction is an unstable friction force since the tendency of two moving surfaces to stick will result in a higher force than the actual sliding friction. A sliding body has a tendency to stick and then let go, and this jerking motion tends to create problems. Other examples of unstable friction force include shafts whirling in dry bearings, and variations in pressure versus flow rate on fan curves, particularly if they have a point of negative slope. Also, hysteresis may cause instability. Galloping of electrical transmission lines and the characteristics of Von Karman vortices and vortex shedding are also cases of self-excited vibration and major problems of instability. When a fluid flows around a cylindrical obstacle, the wake behind the obstacle is no longer regular, but is found to have distinct vortices. This phenomenon has been studied experimentally, and it has been found that there is a relationship between the frequency f, the diameter D of a cylinder in a flow field, and the velocity V, expressed by the following equation:

$$\frac{fd}{V} = 0.22 \tag{20.60}$$

This equation would indicate that a cylinder will move forward about $4\frac{1}{2}$ diameters (1/0.22) during one period of vibration. The 0.22 is dimensionless, and is known as the *Strouhal number.* Vortex shedding on alternate sides of the cylinder causes a harmonically varying force on the cylinder, in a direction perpendicular to the flow stream. Maximum intensity expressed in standard aerodynamic terms is as follows:

$$F_k = C_k \cdot \frac{1}{2} V^2 A \sin \omega t \tag{20.61}$$

where "k" stands for Karman. Generally $C_k = 1$ is a good approximation for large Reynolds numbers, from 10^2 to 10^7. Steel smokestacks regularly show this resonant vibration at wind speeds of 30 knots or more, and vibrations become worse if riveted stacks are replaced by welded stacks having less internal damping.

20.13 COMPUTER-AIDED ENGINEERING

This section provides a brief overview of the various elements of computer-aided engineering and how they can be applied by the engineer, at the design and analysis levels, to verify design integrity. This section provides you with a knowledge of the various software programs available for your use.

The tools presented here and referenced elsewhere in this handbook are used for preprocessing the data prior to analysis or to reduce the data and optimize the results for final correlation and presentations.

The hardware associated with the computer programs presented in this section include microcomputers compatible with MS-DOS. The programs mentioned herein generally are readily available on various media, including floppy disk and CD-ROM. The programs can be obtained from major software houses, and some are offered as shareware through various computer clubs or bulletin boards.

20.13.1 Mechanical Reliability Software

Various types of mechanical reliability analyses are performed on a microcomputer. These include

1. General mechanical engineering analysis
2. Finite element analysis
3. Special analysis of bar linkage, etc.
4. Thermal analysis

5. Mechanical reliability prediction
6. Units conversion program

Only the last two subjects are discussed in this section. The others may be obtained from references already mentioned.

20.13.2 Mechanical Reliability Program

The mechanical reliability program conforms with the theory presented in Chaps. 18 and 19. The program and an operator's guide can be obtained from U.S. Army Belvoir Research, Development, and Engineering Center, For Belvoir, Virginia 22060 (Clarence Meese). Also, many of the reliability software houses have their own version of a mechanical reliability program. Many of these are listed regularly in the *IEEE Reliability Society Newsletter.*

20.13.3 Units Conversion Program

The units conversion program can convert any quantity of the 200 different units. This would be approximately 2001 types of conversions that can be performed using this program. The program may be obtained from the author of this chapter. It is also up on some of the Reliability BBS. Try the IEEE Reliability Chapter BBS in Los Angeles, if you cannot find it locally.

CHAPTER 21
SYSTEM RELIABILITY

Henry A. Malec
BIOSCIENCE LABORATORIES, INC.
BOZEMAN, MONTANA

21.1 INTRODUCTION

Reliability is defined as the ability of an item to perform a required function under stated conditions for a stated period of time. When one groups together a number of units and forms a system, reliability is defined differently. A definition in this context for systems is that *system reliability* is the control of failures. If one can accurately predict how often the system will fail and in the exact manner in which the system will fail, controls can be put in place to minimize any failure effects that the customer or user experiences. This is the thrust of system reliability which allows one to add units to the system in the many different forms of redundancy or either add fault-tolerance or error-correction techniques. System reliability is identified with a state of knowledge as averages or probabilities are determined, and not the exact time prediction for a specific event.

21.1.1 Customer System-Reliability Process

It is very important from all reliability aspects to focus on the customer who uses the system. It is necessary to understand what the system is to be used for, how it is to be used, and what the customer expects for the reliability performance. System reliability focuses on the expected performance that the customer desires. A four-step process for determining which parameters are crucial to the customer is presented in Fig. 21.1.

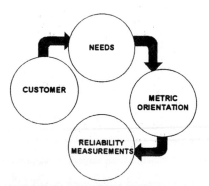

FIGURE 21.1 Four-step process for customer system reliability requirements.

With an international market base for product and often challenged with multinational development teams, the system reliability engineer has to have a broad base of mathematical, design, and manufacturing knowledge. The customers for product have often attained expertise in system-reliability estimation and many have extensive databases that summarize operational reliability performance.

21.1.2 Customer System Reliability Process Overview

The steps of this process are as follows:

1. *List all the system-reliability customers:* There are two types of customers: internal customers and external customers. The *external customers* purchase or use the product or service that will be provided. Cellular telephone service and mainframe computers are examples of products or services in which external customers have a vested interest in terms of systems reliability. An automobile or an individual phone has reliability requirements; however, generally no forms of redundancy are incorporated in these products and they are defined with *item reliability.* Internal customers for system reliability activities are personnel from the company or organization that is producing the product or service. They have an interest in how the system will perform in terms of reliability. They may be design engineers, program management, marketing personnel, or similar.

2. *Define their specific measurement needs:* The needs have a wide spectrum as a design engineer is interested in the characteristics of the design, and the customer support personnel are interested in the modes of failure. System users may need to know how the system will perform in their manufacturing plant environments or how it will perform under overload conditions.

3. *Obtain their specific metric orientation: Metric orientation* is defined as *the individual or organization interest in a particular type of measurement.* These can be of the form of go/no-go, software-phase details, test results, benchmark data, or statistical process control characteristics. Designers and customers are both interested in system performance measurements. Program management responsible for developing the system is interested in go/no-go measurements that will indicate when the system can move to the next design phase or when it is ready to ship to the customer.

4. *List the specific system reliability measurements:* There are many different reliability measurements. The more frequently used are system availability, mean time between system failures, and fault density.

21.1.3 Customer System-Reliability Process Examples

Inherent in the definition of customer, both internal and external customers are considered in the four-step customer system reliability process. Table 21.1 presents a selection of external customers with their measurement needs, and Table 21.2 presents a selection of internal customers with their measurement needs.

TABLE 21.1 External Customer System Reliability Measurements

Customer:	Needs	Orientation	Measurement
External A	Continuous operation	System performance	System availability
External B	Maintainability	Failure perspective	Failure rate or MTBF
External C	Product maturity	Relative measure	Time to inherent reliability

TABLE 21.2   Internal Customer System Design Reliability Measurements

Customer	Needs	Orientation	Measurement
Hardware engineering	Design characterization	Performance	Failure rate
Software engineering	Design heuristics	Design characteristics	Complexity
Reliability engineering	Model validation	Parametric data	Availability
Program management	Gate review	Go/no go	Trend index
Customer support	Service cost estimation	Failures	Fault density

21.2   SYSTEM-RELIABILITY MEASURE OF AVAILABILITY

Customer A, an external customer from Table 21.1, wants to evaluate several systems that must operate for 24 h each day for every day of the year. Two examples are telephone switching systems and computer systems operating automatic teller machines (ATMs), where customers are oriented toward the overall system performance. These two example systems have some type of redundancy built in, resulting in a single failure, not disabling operations. The measure of interest is system availability, which is measured as a simple number less than or equal to one. In this first example, steady-state availability is used as the interest of customer A. The presentation of availability is either in a numerical form or is associated with time. For example, an availability of .99995 in terms of hours per year would be 8,759.562 h, given 8760 h in a year.

21.2.1   Steady-State Availability

The most commonly defined type of availability is *steady-state availability* or *limiting availability*. Steady-state availability A_s is expressed as

$$A_s = \lim_{t \to \infty} \frac{1}{t} \int_0^t A(t) dt \tag{21.1}$$

where $A(t)$ is the instantaneous availability and t is the time of interest.

21.2.2   Instantaneous Availability

Instantaneous availability is the probability that a system is operating at a selected time t. Instantaneous availability $A(t)$ is expressed as

$$A(t) = \frac{\mu}{\mu + \lambda} + \frac{\lambda}{\mu + \lambda} e^{-(\mu + \lambda)t} \tag{21.2}$$

where μ = unit repair rate
λ = unit failure rate
t = any point in time

21.2.3   Interval Availability

The third type of availability is interval availability, which is also known as *mission availability*. *Interval availability* is the proportion of time that the system is available in an interval t_1 to t_2 and is expressed as

$$A_m(t) = \frac{1}{t_2 - t_1} \int_{t_1}^{t_2} A(t)\, dt \tag{21.3}$$

where t_1 is the start of mission; usually $t_i = 0$, and t_2 = end of mission.

21.2.4 Unavailability

From a practical viewpoint, unavailability is used more often than availability as it allows one to plan for the average total time in a given year a system would not be available. *Unavailability* is defined as

$$\text{Unavailability} = 1 - \text{availability} \tag{21.4}$$

In this example given the availability of .99995, the unavailability = .00005, and the time that the system is not available is 8760 times .00005, which is equal to 0.438 h or about 26 min per year. Thus external customer A has only to plan for the system not being available for 26 min each year.

21.3 SYSTEM-RELIABILITY MEASURE OF FREQUENCY OF FAILURE: MTBF

Customer B, an external customer from Table 21.1, is interested in evaluating two different system maintenance scenarios. The two possibilities are (1) the permanent hiring of a maintenance person for the system being purchased and (2) the contracting for maintenance service by a third-party vendor. Two examples of products where such considerations are made are the high-speed laser printers for a computer processing center and a surface-mount pick-and-place machine for the manufacturing shop floor printed-wire-board process. Customer B's interests are in equipment maintenance; this customer specifically wants to know the amount and frequency of repairs due to failure and the amount of preventive maintenance associated with the system. Both of these maintenance items are based on the failure rates of the component parts or equipment modules of these systems. The preventive maintenance is based on periodic replacement, cleaning, adjustment, etc., helping to prevent a catastrophic failure of the system. The unplanned failures are generally specified in terms of an MTBF (mean time between failures) for the system, where MTBF is a measure for the reliability during the useful life and is the mean time that elapses between two consecutive failures.

System MTBF calculations for a system often include redundant equipment failure sequences and the chance that the system may fail while maintenance is being performed for a failure that doesn't affect operation. In this example, Customer B is concerned with the total amount of failures that will need to be fixed. With a redundant system, the total number of items that fail is much greater than the number of system failures. MTBF can be stated as the number of operations or in the amount of elapsed time. There are four types of MTBFs for repaired equipment: observed MTBF, assessed MTBF, extrapolated MTBF, and predicted MTBF. In our discussions, observed MTBF will be used as the interest of customer B.

21.3.1 Observed MTBF

The *observed mean time between failures* is defined as the mean value of the length between consecutive failures computed as the ratio of the cumulative observed time to the number of failures:

$$\text{Observed MTBF} = \frac{\text{cumulative time}}{\text{number of failures}} \tag{21.5}$$

21.3.2 Observed Failure Rate

The observed MTBF is often specified in terms of a failure rate. Thus, the *observed failure rate* is defined as the ratio of the total number of failures to the cumulative observed time:

$$\text{Observed failure rate} = \frac{\text{number of failures}}{\text{cumulative time}} = \lambda \qquad (21.6)$$

since

$$\text{Observed MTBF} = \frac{1}{\text{observed failure rate}} \qquad (21.7)$$

or

$$\text{MTBF} = \frac{1}{\lambda} \qquad (21.8)$$

21.3.3 Observed MTBF Example

External customer B, as shown in Table 21.1, has contacted another user of the high-speed laser printers referenced by the manufacturer that has three high-speed laser printers installed. The data provided indicated that the units are used 60 h per week and need preventive maintenance on each unit once every 3 months. Over the past 2 years, there have been a total of 36 failures for the three units. Thus, the observed MTBF is calculated from Eq. (21.5). The cumulative time is determined by using 52.14 weeks per year times 60 h of operation per week times 3 systems, which equals 9385.2 system operational hours. The three systems had a total of 36 failures. Therefore:

$$\text{Observed MTBF} = \frac{9385.2}{36} = 260.7 \text{ h} \qquad (21.9)$$

The MTBF is 260.7 h of operation for a single unit. In other words, if customer B intends to use a single unit for 40 h per week, a failure every 6.5 weeks can be expected.

21.3.4 MTBF Acceleration Factor

If customer B intends to use a single unit described in Sec. 21.3.2 for 60 h per week, a failure every 4.4 weeks can be expected. Originally, the analysis indicated that if customer B intended to use a single unit for 40 h per week, a failure every 6.5 weeks can be expected. The different times between failures is the *acceleration factor,* which transforms the base failure rate to one indicative of the customer application stress. Given that the site in this example with three printers operates them at 200 printed pages per minute, if customer B operates a single unit at a significantly lower printing rate, the MTBF would be increased; whereas if customer B operated at a higher printing rate, the MTBF would be reduced. A high-speed laser printer or a surface-mount pick-and-place machine have dominant failure modes that are mechanical adjustments which are a direct function of machine use. If customer B operates the one machine at 100 printed pages per minute, the calculated 6.5 weeks of MTBF would be increased to 13 weeks. In addition, the time between preventive-maintenance actions would be extended, but most likely at a factor of less than 2 (e.g., printed-page-count reduction of 200 to 100/min).

21.4 *SYSTEM-RELIABILITY MEASURE OF INHERENT RELIABILITY*

Customer C, an external customer from Table 21.1, is interested in selecting a large mass-storage tape-library unit and is interested in selecting a mature product operating at the inherent reliability level. In the context of system reliability, there are several categories of MTBF, of which the inherent MTBF is one. Table 21.3 presents a tier-level concept for MTBF. In general, the higher the tier level, the higher the MTBF. Most often MTBF measurements are not equal when measured in the field or in a test in the manufacturing plant. There are several reasons for this, including different operating conditions and different definitions of what constitutes a failure.

TABLE 21.3   MTBF Tier Levels

Tier	MTBF	Description
1	Minimum	Minimum ship MTBF (statistical test)
2	Specified	Specified MTBF which meets the market needs and support service pricing
3	Design	Design goal MTBF (calculation)
4	Inherent	Inherent MTBF (plant measurement)
5	Field	Field MTBF measurement

With the tier concept, products can receive first-customer-ship (FCS) status at the minimum MTBF. This level is the minimum allowable level for product shipment or initialization of service. Manufacturing has a commitment to drive a product's reliability to the tier 2 level to achieve a product that is profitable. The raising of the MTBF up to the specified MTBF is supported by design engineering. Emphasis is placed on continuous improvement by meeting or exceeding the MTBF design goal. At product maturity, the inherent MTBF is reached as measured in the plant or service facility, and it is desirable that the field measurements and plant or service measurements are equivalent.

Ideally the MTBFs of tiers 2 through 5 would be equal; however, the accuracy of field MTBF measurements are a continuing challenge. Thus a three- to five-tier approach is practical. Customer C, as an example, is interested in the number of failures per exchanges of the storage tape cartridges. There are many forms of inherent reliability, and for this specific application it is in terms of failures per million moves. Figure 21.2 presents the requested data by the customer, which indicates that the product has reached maturity in the fourth quarter. This satisfies the customer concerns that the product is mature.

FIGURE 21.2   Product maturity.

21.5 *MODELS FOR SYSTEM RELIABILITY*

In supporting design and determining the measurements of failure rate for hardware engineering and availability for reliability engineering, there are many models and modeling techniques that will provide such measurements.

TABLE 21.4   System Reliability Analysis Methods

Method	Symbolic representation	Quantitative analysis
Parts count	List	Components and system failure rates
Block diagram	Block diagram	System reliability and availability
Markov analysis	System state diagram	System reliability and availability
Fault tree	Failure tree	System reliability and availability
Petri-Net	Petri-Net graph	System reliability, availability, and performance

Table 21.4 presents the most commonly used models with their characteristics. Certain models are appropriate for specific products or services, and experience ensures success. This table is a guide to match the requirements of the desired system-reliability analysis with a model or prediction technique.

21.6   PARTS-COUNT METHOD

The *parts-count method* is a bottom-up method used early in a given design program to estimate the frequency of part failures. A list of component parts that are to be used in the system along with their stresses and environment are needed. The method is based on the assumption that any component failure will cause the system to fail. The advantages are that the cost is low and the completion time is minimal. There is no mathematics involved in this method. The disadvantages are many. Repair and maintenance considerations are impossible, and time-sequential failure and events cannot be included.

In performing a reliability prediction for a set of components in series, the following assumptions are made:

1. Failure of any one of the components gives rise to failure of the complete unit.
2. Failures are completely independent of each other from a statistical viewpoint.
3. The components as well as the overall unit (set of components) follow the traditional *reliability bathtub curve.*
4. The unit reliability bathtub curve is the combination of the individual component reliability curves.

In order that the unit function at time *t,* all components of the unit must be functional at that time *t.* Thus, the probability of survival of the unit at an instant of time is equal to the product of the probabilities of survival of its components at the exact same instant. This is illustrated in Eq. (21.10).

$$R_{\text{unit}} = R_{\text{component 1}} \times R_{\text{component 2}} \times \cdots \times R_{\text{component } n} \tag{21.10}$$

This can be written as Eq. (21.11):

$$R_{\text{unit}} = e^{-\lambda_1 t} \times e^{-\lambda_2 t} \times \cdots \times e^{-\lambda_n t} \tag{21.11}$$

where λ is failure rate per hour and *t* is time.

The following can be written for the system MTTF (mean time to failure), which for nonrepairable units is preferred over MTBF (mean time between failures), which is the usual nomenclature for repairable systems of units:

$$\text{MTTF} = \int_0^\infty R(t)\, dt = \int_0^\infty e^{-(\lambda_1 + \lambda_2 + \cdots + \lambda_n)t}\, dt \tag{21.12}$$

where $R(t)$ = probability of survival at time t
λ = failure rate per hour
t = time

Thus it is possible to sum the component failure rate to obtain the reliability of the unit as shown in Eq. (21.13) as derived from Eq. (21.12):

$$\text{MTTF} = \frac{1}{\lambda_1 + \lambda_2 + \cdots + \lambda_n} \tag{21.13}$$

Failure rates are normally specified in FITS (failures per 10^9 operating hours). Sometimes they will appear in percent per 1000 h. Given a component with a failure rate of 0.002 percent per 1000 h, this is equivalent to 20 FITS. In the early days of electronic components, vacuum tubes often had failure rates of 40 percent per 1000 h. In recent years, integrated circuits (ICs) have demonstrated a failure rate of 0.00002 percent per 1000 h.

In terms of calculation, fewer errors are made with less zeros; thus, the reliability practitioners have adopted the FIT nomenclature as the IC would have a failure rate of 2 FITS. Some resistors presently have failure rates less than 1 FIT; however, it is not expected that the reliability practitioners will adopt a failure rate based on 10^{12} operating hours.

Given a unit is composed of four components as presented in Fig. 21.3, the summation rate of the failure rates yields 400 FITS. Thus, from Eq. (21.13), the MTTF is 2,500,000 h as shown in the following equations:

$$\text{MTTF} = \frac{1}{200 \times 10^{-9} + 150 \times 10^{-9} + 48 \times 10^{-9} + 2 \times 10^{-9}} \tag{21.14}$$

$$\text{MTTF} = \frac{1}{400 \times 10^{-9}} = \frac{10^9}{400} = 2,500,000 \text{ h} \tag{21.15}$$

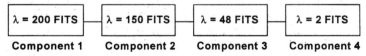

λ = 200 FITS	λ = 150 FITS	λ = 48 FITS	λ = 2 FITS
Component 1	Component 2	Component 3	Component 4

FIGURE 21.3 Unit composed of four components.

21.7 BLOCK-DIAGRAM METHOD

The *block-diagram method* is a top-down method which is opposite to the parts-count method. The resultant diagram pictures the organization of the system allowing for functional organization into subsystems. The mathematical analysis can utilize truth tables, boolean techniques, or path and cut-set analysis. The main advantage of this method is that it allows for parallel, redundant, and standby paths and can be constructed in either a failure or success basis. The method results in a clear and concise diagram for the entire system. The method will provide models for the prediction of system reliability in probabilistic terminology. The main disadvantage is that the method is not able to integrate complex repair strategies.

21.7.1 Active Redundancy

One of the most common forms of active redundancy is two or more units in parallel. In the electronics world, power-supply units are able to be paralleled in a manner such that if one or more

Unit A

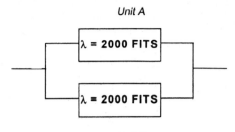

Unit B

FIGURE 21.4   Active redundancy: two units in parallel.

units fail, the remaining units can provide the required power for the load. Figure 21.4 illustrates the basic form of two units in parallel.

The MTTF of the individual 2000-FIT units is 500,000 h using Eq. (21.13). To calculate the system MTTF, probability theory will be used. The sum of the probability of survival and the probability of failure of the total system equals 1 as illustrated in Eq. (21.16):

$$R_S + Q_S = 1 \tag{21.16}$$

where R is reliability and Q is unreliability. Four possible scenarios in the operation of these units can occur:

1. Unit A and B are operational: $R_A R_B$.
2. Unit A is operational and unit B is failed: $R_A Q_B$.
3. Unit A is failed and unit B is operational: $Q_A R_B$.
4. Unit A and B are failed: $Q_A Q_B$.

The probability that at least one power unit will be operational is provided in Eq. (21.17) as

$$R_S = R_A R_B + R_A Q_B + Q_A R_B \tag{21.17}$$

Rearranging yields Eq. (21.18) with a final desired form presented in Eq. (21.19):

$$R_S = R_A R_B + R_A(1 - R_B) + (1 - R_A)R_B \tag{21.18}$$

$$R_S = R_A + R_B - R_A R_B \tag{21.19}$$

Substituting for the original R values yields Eq. (21.20):

$$R_S = e^{-\lambda_A t} + e^{-\lambda_B t} - e^{-\lambda_A t} \times e^{-\lambda_B t} \tag{21.20}$$

where λ is failure rate per hour and t is time.

The MTTF of the power unit system can be calculated from Eq. (21.21):

$$\text{MTTF}_S = \int_0^\infty R_S \, dt \tag{21.21}$$

Substituting for R_S yields Eq. (21.22), and the standard form is shown in Eq. (21.23):

$$\text{MTTF}_S = \int_0^\infty (e^{-\lambda_A t} + e^{-\lambda_B t} - e^{-\lambda_A t} \times e^{-\lambda_B t})\, dt \tag{21.22}$$

where λ is failure rate per hour and t is time.

$$\text{MTTF}_S = \frac{1}{\lambda_A} + \frac{1}{\lambda_B} - \frac{1}{\lambda_A + \lambda_B} \tag{21.23}$$

Equation (21.23) is the standard form for two units in parallel in which the units can have different failure rates. In the case shown in Fig. 21.4, units A and B have identical failure rates; thus, $\lambda_A = \lambda_B = \lambda$. With this equivalence, Eq. (21.23) transforms into Eq. (21.24):

$$\text{MTTF}_S = \frac{1}{\lambda} + \frac{1}{\lambda} - \frac{1}{2\lambda} = \frac{3}{2\lambda} \tag{21.24}$$

Thus, the MTTF of the system of two units operating in active redundancy as identical units result in a system MTTF that has a 50 percent gain over the individual unit MTTFs. Specifically, each unit had a 500,000-h MTTF as shown in Fig. 21.4, and with their parallel operation, the resultant is a 750,000-h MTTF for the system.

21.8   MARKOV ANALYSIS METHOD

The *Markov analysis method* is a bottom–up method that allows the analysis of complex systems and repair strategies including self-repair analysis. The method is based on the *theory of Markov chains*. The mathematical basis for the analysis is that the particular system elements be in an operating, failure, or repair state at specific points in time with the analysis incorporating all elements in a time sphere. Thus, the advantage is that the system behavior can be studied on a state of wellness or partial failure with complex, dependent repair strategies. The main disadvantage is the large number of system states encountered in the analysis is not easily absorbed by the design or reliability personnel who are trying to understand the results.

21.8.1   State-Space Diagram

A *state-space diagram* is defined as a graphical representation of the reliability states of a system. An analysis of such a system gives the dynamic reliability performance of the system. This integral part of the Markov analysis method uses a standard set of symbols to represent the reliability parameters for failure occurrences, error detection and correction, and maintenance actions. Figure 21.5 presents two identical units in parallel that can be repaired after failure with a failure rate λ and a repair rate μ.

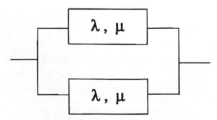

FIGURE 21.5   Two identical units in parallel.

The system in Fig. 21.5 can exist in the following three states:

1. State 0 (S0): both units are operational.
2. State 1 (S1): one of the units is operational and the other unit is failed.
3. State 2 (S2): both units are failed.

Since the units can be repaired, Fig. 21.6 presents the state diagram for two identical units in parallel.

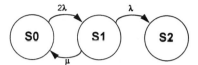

FIGURE 21.6   State diagram for two identical units in parallel.

These three states are mutually exclusive, which means that the system can be in only any one of the states at any one time. Since only one event can occur during a small instant of time Δt, transitions between states correspond to only one failure event or one repair event. In addition, these transitions must be capable of taking place while the systems are operational. Thus the three transitions shown in Fig. 21.6 are presented in Fig. 21.7a–c.

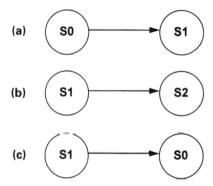

(a)

(b)

(c)

FIGURE 21.7   Transition (a) caused by a failure of one of the units; (b) caused by the failure of the remaining unit; (c) effected by the repair of the failed unit.

Each of these three transitions is characterized by transition rates. Since either of the two units in Fig. 21.5 can fail, the transition rate is $\lambda + \lambda = 2\lambda$, which is the transition rate from S0 to S1 as shown in Figs. 21.6 and 21.7. If the failed unit is repaired, the transition rate from S1 to S0 is μ. Once there is one failed unit, the system is in S1. If the remaining unit fails, a transition occurs from S1 to S2 with a transition rate of λ. The state diagram as shown in Fig. 21.6 contains these transitions and their corresponding rates. To complete a state diagram, loops are drawn at the operative states: S0 and S1 of Fig. 21.6. These loops measure the stability of the state and are characterized by a number equal to minus the sum of the values on the transition flow lines leaving the specific state. Figure 21.8 presents this complete transition state diagram.

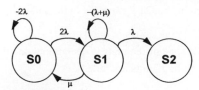

FIGURE 21.8   Transition state diagram for two identical units in parallel.

21.8.2   Transition Matrix

The transition matrix A can be constructed from a transition state diagram such as the one shown in Fig. 21.8. In this case of two identical units in parallel, the resultant transition matrix will be a 3×3 matrix. The elements of the matrix are determined in a form a_{ij} with a_{ij} *defined as the value on the* transition flow line pointing from state j to state i. If no transition line exists, the corresponding matrix element is zero. In the example of Fig. 21.8, $i = 1, 2, 3$ and $j = 1, 2, 3$, and the resultant transition matrix is presented in Fig. 21.9, where

1. The matrix is a square matrix.
2. The sums of the columns are equal to zero.
3. The diagonal elements of the matrix correspond to the state loops in the transition state diagram of Fig. 21.8.

	State 0	State 1	State 2
State 0	-2λ	μ	0
State 1	2λ	$-(\lambda+\mu)$	0
State 2	0	λ	0
Sums	0	0	0

FIGURE 21.9   Transition matrix for two identical units in parallel with repair.

21.8.3   MTBF of Two Units in Parallel with Repair

Given the two identical units in parallel as described by the block diagram in Fig. 21.5 and the transition state diagram of Fig. 21.8, the MTBF can be determined by the solution of equations. Let $P_0(t)$, $P_1(t)$, and $P_2(t)$ be the probability of being in states 0, 1, and 2 at a time t. The probability of being in a given state is $= (1 - \text{probability of leaving})$. Then at a time $t + \Delta t$:

$$P_0(t + \Delta t) = P_0(t) - P_0(t)\, 2\lambda\, \Delta t + P_1(t)\, \mu\, \Delta t \qquad (21.25)$$

where $P_0(t)\, 2\lambda\, \Delta t$ = probability of going to state 1 in Δt

$P_1(t)\, \mu\, \Delta t$ = probability of going by repair to state 0 in Δt

and

$$P_1(t + \Delta t) = P_0(t)\, 2\lambda\, \Delta t + P_1(t) - P_1(t)\, (\mu + \lambda)\, \Delta t \qquad (21.26)$$

$$P_2(t + \Delta t) = P_1(t) \lambda \Delta t + P_2(t) \tag{21.27}$$

The reliability $R(t)$ of the two units in parallel with repair since the system is operational in states 0 and 1 is

$$R(t) = P_0(t) + P_1(t) \tag{21.28}$$

where $P_0(t)$ = probability of being in state 0 at time t

$P_1(t)$ = probability of being in state 1 at time t

The MTBF of the two units in parallel with repair is

$$\text{MTBF} = \int_0^\infty R(t) \, dt = \int_0^\infty P_0(t) \, \Delta t + \int_0^\infty P_1(t) \, \Delta t \tag{21.29}$$

where $R(t)$ = probability of survival at time t
λ = failure rate per hour
t = time
$P_0(t)$ = probability of being in state 0 at time t
$P_1(t)$ = probability of being in state 1 at time t

As $\Delta t \to 0$, Eqs. (21.25) to (21.27) for $P_0(t + \Delta t)$, $P_1(t + \Delta t)$, and $P_2(t + \Delta t)$ can be restated as Eqs. (21.30) to (21.32):

$$P_0'(t) = -2\lambda \, P_0(t) + \mu \, P_1(t) \tag{21.30}$$

$$P_1'(t) = 2\lambda \, P_0(t) - (\lambda + \mu)P_1(t) \tag{21.31}$$

$$P_2'(t) = \lambda \, P_1(t) \tag{21.32}$$

where the expressions $P_0'(t)$, $P_1'(t)$, and $P_2'(t)$ are the first-order derivatives of $P_0(t)$, $P_1(t)$, and $P_2(t)$, respectively. Equations (21.30) to (21.32) form a set of three first-order differential equations which can be solved with general-purpose mathematical computer programs or by obtaining a solution using Laplace transforms. Since the exact formula is of interest, these equations will be solved.

Equations (21.30) to (21.32) can be written in matrix form:

$$
\begin{vmatrix} P_0'(t) \\[4pt] P_1'(t) \\[4pt] P_2'(t) \end{vmatrix}
=
\begin{vmatrix} -2\lambda & \mu & 0 \\[4pt] 2\lambda & -(\lambda + \mu) & 0 \\[4pt] 0 & \lambda & 0 \end{vmatrix}
\cdot
\begin{vmatrix} P_0(t) \\[4pt] P_1(t) \\[4pt] P_2(t) \end{vmatrix}
$$

The elements of the middle matrix are exactly the same elements as those of Fig. 21.9 (which were derived from the transition state diagram. To solve for the MTBF of the two units in parallel with repair, Eq. (21.29) is written as Eq. (21.33):

$$\text{MTBF} = \int_0^\infty P_0(t)\,\Delta t + \int_0^\infty P_1(t)\,\Delta t = T_0 + T_1 \tag{21.33}$$

The values of T_0 and T_1 can be obtained by solving the following:

$$\int_0^\infty \begin{vmatrix} P'_0(t) \\ P'_1(t) \\ P'_2(t) \end{vmatrix} dt = \int_0^\infty \begin{vmatrix} -2\lambda & \mu & 0 \\ 2\lambda & -(\lambda + \mu) & 0 \\ 0 & \lambda & 0 \end{vmatrix} \cdot \begin{vmatrix} P_0(t) \\ P_1(t) \\ P_2(t) \end{vmatrix} dt \tag{21.34}$$

The 3×3 matrix is constant, and Eq. (21.34) can be written as

$$\int_0^\infty \begin{vmatrix} P'_0(t) \\ P'_1(t) \\ P'_2(t) \end{vmatrix} dt = \begin{vmatrix} -2\lambda & \mu & 0 \\ 2\lambda & -(\lambda + \mu) & 0 \\ 0 & \lambda & 0 \end{vmatrix} \cdot \int_0^\infty \begin{vmatrix} P_0(t) \\ P_1(t) \\ P_2(t) \end{vmatrix} dt \tag{21.35}$$

Equation (21.35) can also be written as

$$\int_0^\infty \begin{vmatrix} P'_0(t) \\ P'_1(t) \\ P'_2(t) \end{vmatrix} dt = \begin{vmatrix} -2\lambda & \mu & 0 \\ 2\lambda & -(\lambda + \mu) & 0 \\ 0 & \lambda & 0 \end{vmatrix} \cdot \begin{vmatrix} \int_0^\infty P_0(t) \\ \int_0^\infty P_1(t) \\ \int_0^\infty P_2(t) \end{vmatrix} dt \tag{21.36}$$

or as

$$\begin{vmatrix} \int_0^\infty P'_0(t)\,dt \\ \int_0^\infty P'_1(t)\,dt \\ \int_0^\infty P'_2(t)\,dt \end{vmatrix} = \begin{vmatrix} -2\lambda & \mu & 0 \\ 2\lambda & -(\lambda + \mu) & 0 \\ 0 & \lambda & 0 \end{vmatrix} \cdot \begin{vmatrix} \int_0^\infty P_0(t)\,dt \\ \int_0^\infty P_1(t)\,dt \\ \int_0^\infty P_2(t)\,dt \end{vmatrix} \tag{21.37}$$

or as

$$
\begin{vmatrix} P_0(\infty) - P_0(0) \\ P_1(\infty) - P_1(0) \\ P_2(\infty) - P_2(0) \end{vmatrix} = \begin{vmatrix} -2\lambda & \mu & 0 \\ 2\lambda & -(\lambda + \mu) & 0 \\ 0 & \lambda & 0 \end{vmatrix} \cdot \begin{vmatrix} T_0 \\ T_1 \\ T_2 \end{vmatrix} \qquad (21.38)
$$

If the assumption is made that the system is fully operational at time $= 0$, then

$$P_0(0) = 1$$

$$P_1(0) = 0$$

$$P_2(0) = 0$$

At time $t = \infty$, the system will have failed; then

$$P_0(\infty) = 0$$

$$P_1(\infty) = 0$$

$$P_2(\infty) = 1$$

On the basis of these criteria, Eq. (21.38) can be reduced to

$$
\begin{vmatrix} -1 \\ 0 \\ 1 \end{vmatrix} = \begin{vmatrix} -2\lambda & \mu & 0 \\ 2\lambda & -(\lambda + \mu) & 0 \\ 0 & \lambda & 0 \end{vmatrix} \cdot \begin{vmatrix} T_0 \\ T_1 \\ T_2 \end{vmatrix} \qquad (21.39)
$$

Changing from matrix notation to equation form by matrix multiplication, a set of three equations is derived:

$$-1 = -2\lambda\, T_0 + \mu\, T_1 \qquad (21.40)$$

$$0 = 2\lambda\, T_0 - (\lambda + \mu)\, T_1 \qquad (21.41)$$

$$1 = \lambda\, T_1 \qquad (21.42)$$

Solving the set of Eqs. (21.40) to (21.42) yields values for T_0 and T_1:

$$T_0 = \frac{\lambda + \mu}{2\lambda^2} \qquad (21.43)$$

$$T_1 = \frac{1}{\lambda} \tag{21.44}$$

where μ is unit repair rate and λ is unit failure rate.

Equation (21.33) formulated the MTBF for the two identical units in parallel with repair and can be solved using Eqs. (21.43) and (21.44):

$$\text{MTBF} = T_0 + T_1 = \frac{\lambda + \mu}{2\lambda^2} + \frac{1}{\lambda} = \frac{3\lambda + \mu}{2\lambda^2} \tag{21.45}$$

21.8.4 Availability of Two Units in Parallel with Repair

Given the two identical units in parallel as described by the block diagram in Fig. 21.5, a transition state diagram can be drawn to include the repair from state 2 as shown in Fig. 21.10:

1. State 0 (S0): both units are operational.
2. State 1 (S1): one of the units is operational and the other unit is failed.
3. State 2 (S2): both units are failed.

There are two different repair rates in Fig. 21.10: μ_1 and μ_2. The repair rate for transitions of the system from state S2 to S1 might take additional repair time than the transition from state S1 to S0. Thus, two different repair rates are considered for this analysis. In addition, if one compares Fig. 21.6, which was symbolic of the determination of the system MTBF, to Fig. 21.10 for system

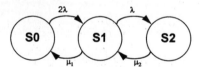

FIGURE 21.10 State diagram for two identical units in parallel.

availability, one can note that the availability system model allows for continuous operation of the system down and the system operational. The MTBF model essentially stops with a system failure. To complete a state diagram, loops are drawn at the operative states, namely, S0, S1, and S2 of Fig. 21.10. These loops measure the stability of the state and are characterized by a number equal to minus the sum of the values on the transition flow lines leaving the specific state. Figure 21.11 presents this complete transition state diagram.

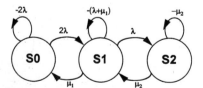

FIGURE 21.11 Transition state diagram for two identical units in parallel.

The solution for the availability follows a method very close to the solution method for the MTBF as derived in Sec. 21.8.3. Let $P_0(t)$, $P_1(t)$, and $P_2(t)$ be the probability of being in states 0,

1, and 2 at a time t. The probability of being in a given state is $= (1 - \text{probability of leaving})$. Then at a time $t + \Delta t$

$$P_0(t + \Delta t) = P_0(t)(1 - \lambda \Delta t)^2 + P_1(t)(1 - \lambda \Delta t)\mu_1 \Delta t \tag{21.46}$$

$$P_1(t + \Delta t) = P_0(t) 2\lambda \Delta t(1 - \mu_1 \Delta t) + P_1(t)(1 - \lambda \Delta t)(1 - \mu_1 \Delta t) + P_2(t) \mu_2 \Delta t \tag{21.47}$$

$$P_2(t + \Delta t) = P_1(t) \lambda \Delta t(1 - \mu_1 \Delta t) + P_2(t)(1 - \mu_2 \Delta t) \tag{21.48}$$

As $\Delta t \to 0$, Eqs. (21.46) to (21.48) for $P_0(t + \Delta t)$, $P_1(t + \Delta t)$, and $P_2(t + \Delta t)$ can be restated as Eqs. (21.49) to (21.51):

$$P'_0(t) = -2\lambda P_0(t) + \mu_1 P_1(t) \tag{21.49}$$

$$P'_1(t) = 2\lambda P_0(t) - (\lambda + \mu_1)P_1(t) - \mu_2 P_2(t) \tag{21.50}$$

$$P'_2(t) = \lambda P_1(t) - \mu_2 P_2(t) \tag{21.51}$$

where the expressions $P'_0(t)$, $P'_1(t)$, and $P'_2(t)$ are the first-order derivatives of $P_0(t)$, $P_1(t)$, and $P_2(t)$, respectively. Equations (21.49) to (21.51) form a set of three first-order differential equations. Equations (21.49) to (21.51) can be written in matrix form:

$$
\begin{vmatrix} P'_0(t) \\ P'_1(t) \\ P'_2(t) \end{vmatrix}
=
\begin{vmatrix} -2\lambda & \mu_1 & 0 \\ 2\lambda & -(\lambda + \mu_1) & \mu_2 \\ 0 & \lambda & -\mu_2 \end{vmatrix}
\cdot
\begin{vmatrix} P_0(t) \\ P_1(t) \\ P_2(t) \end{vmatrix}
$$

As the time t increases, the value of the probability function $P_i(t)$ tends to the constant value P_i. Thus, the first-order derivatives $P'_0(t)$, $P'_1(t)$, and $P'_2(t)$ become equal to zero for large values of time t. For large times, which is the normal situation for availability predictions for systems, the matrix can be reduced to

$$
\begin{vmatrix} 0 \\ 0 \\ 0 \end{vmatrix}
=
\begin{vmatrix} -2\lambda & \mu_1 & 0 \\ 2\lambda & -(\lambda + \mu_1) & \mu_2 \\ 0 & \lambda & -\mu_2 \end{vmatrix}
\cdot
\begin{vmatrix} P_0 \\ P_1 \\ P_2 \end{vmatrix}
$$

This set of equations is indeterminate; thus, one of the equations needs to be replaced with Eq. (21.52), which makes use of the fact that the system must be in one of the states S0, S1, or S2:

$$P_0 + P_1 + P_2 = 1 \tag{21.52}$$

Replacing the third equation results in the modification of the matrix to the following one:

$$
\begin{vmatrix} 0 \\ 0 \\ 1 \end{vmatrix}
=
\begin{vmatrix} -2\lambda & \mu_1 & 0 \\ 2\lambda & -(\lambda+\mu_1) & \mu_2 \\ 1 & 1 & 1 \end{vmatrix}
\cdot
\begin{vmatrix} P_0 \\ P_1 \\ P_2 \end{vmatrix}
$$

Changing from matrix notation to equation form by matrix multiplication, we can derive a set of three equations:

$$0 = -2\lambda\,P_0 + \mu_1\,P_1 \tag{21.53}$$

$$0 = 2\lambda\,P_0 - (\lambda+\mu_1)\,P_1 + \mu_2\,P_2 \tag{21.54}$$

$$1 = P_0 + P_1 + P_2 \tag{21.55}$$

Solving the set of Eqs. (21.53) to (21.55) yields values for P_0 and P_1:

$$P_0 = \frac{\mu_1\mu_2}{\mu_1\mu_2 + 2\lambda\mu_2 + 2\lambda^2} \tag{21.56}$$

$$P_1 = \frac{2\lambda\,\mu_2}{\mu_1\,\mu_2 + 2\lambda\mu_2 + 2\lambda^2} \tag{21.57}$$

where μ is unit repair rate and λ is unit failure rate.

The availability of the system A_s incorporates the probabilities of reliability and maintainability. Operation in states S0 and S1 produce satisfactory operation and the operation in state S2 is undesirable. *Availability* is the probability that a system is operating satisfactorily and will not have any failures which cannot, through maintainability, be restored to operational service within a permissible time constraint. Thus, Eq. (21.58) can be written for the availability of two units in parallel with repair:

$$\text{Availability} = A_s = P_0 + P_1 = \frac{(2\lambda+\mu_1)\,\mu_2}{\mu_1\,\mu_2 + 2\lambda\mu_2 + 2\lambda^2} \tag{21.58}$$

where μ is unit repair rate and λ is unit failure rate.

From a practical viewpoint, *unavailability* is used more often than *availability* as it allows one to plan for the average total time in a given year that a system would not be available. Unavailability, U_s is defined as

$$\text{Unavailability} = U_s = 1 - \text{availability} = P_2$$

$$= \frac{2\lambda^2}{\mu_2(2\lambda+\mu_1) + 2\lambda^2} \tag{21.59}$$

where μ is unit repair rate and λ is unit failure rate. If $\mu_1 = \mu_2$, then Eqs. (21.58) and (21.59) can be written as

$$\text{Availability} = A_s = \frac{(2\lambda + \mu)\,\mu}{(2\lambda + \mu)\,\mu + 2\lambda^2} \tag{21.60}$$

$$\text{Unavailability} = U_S = 1 - \text{availability} = P_2$$

$$= \frac{2\lambda^2}{\mu\,(2\lambda + \mu) + 2\lambda^2} \tag{21.61}$$

where μ is unit repair rate and λ is unit failure rate. If μ is very large with respect to λ, an approximation can be made for the unavailability as follows:

$$\text{Unavailability} = U_s \cong \frac{2\lambda^2}{\mu^2} \tag{21.62}$$

21.9 FAULT-TREE METHOD

The *fault-tree method* is a top–down method for identifying the factors and conditions which would cause an undesirable main failure or catastrophe. Starting with the critical event, the failure causes or modes are determined on the next-lower functional level, and this is repeated progressively down until the desired level is reached. Methods such as cut-set analysis and boolean analysis can be applied to estimate reliability and availability based on the established fault tree with appropriate failure rates, modes, repair rate, etc., for the tree elements. The advantages are the ability to analyze parallel, redundant, and standby paths, which can accommodate several cross-linked systems. This method is very useful when only one or two major issues need analysis. The disadvantages are that often large trees are necessary to describe accurately the situation, and the method requires a separate tree for each concern, and complex repair strategies are not easily dealt with.

The top event needs to be the focus of the entire analysis. Generally, fault trees are applied in the safety critical applications such as the nuclear power generation stations, airplanes, and communications networks. The event to be defined is the onset or existence of a dangerous condition or subsystem failure of system failure. It is possible to develop fault trees that focus on more than one event. Fault trees are pictured by graphics. There are five basic steps in fault-tree analysis:

1. *Define the top event with a measure:* Usually the top event is the undesirable event or primary reason why the fault-tree analysis is being conducted. The top event should be defined in physically measurable units.

2. *Establish limits for the measure:* It is important to understand whether the measure is a 100 percent loss or if even a certain level of partial loss is considered a catastrophe.

3. *Understand the system operation:* The fault tree is a diagram in which events are linked by logic gates, with each gate having one output but one or more inputs. It is necessary to understand the logic of the diagram and the timing of the events.

4. *Construct and analyze the fault tree:* Fault trees are drawn either vertically or horizontally. Symbols used are (*a*) event description box, (*b*) fault-tree logic symbol (AND, OR, EXCLUSIVE OR, INHIBIT, M out of N, etc.), (*c*) gate input connection line, (*d*) transfer out, and (*e*) basic event (terminating symbol).

5. *Report on any corrective action necessary.* Often several solutions can be implemented and the effectiveness of these solutions should be quantified.

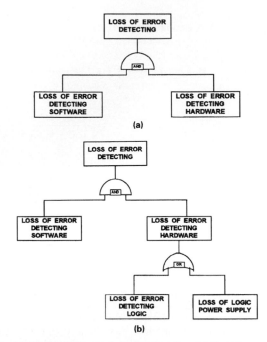

FIGURE 21.12   Loss of error detection: (*a*) first hierarchical level; (*b*) second hierarchical level.

As an example of the method, the first hierarchical level for a fault tree for the loss of error detection is presented in Fig. 21.12*a*.

In Fig. 21.12*a*, one would need to lose both the error-detecting software and the error-detecting hardware to cause a loss of error detection. Since the error-detecting software consists of two major items and loss of error-detecting hardware will result if either of these items fails, Fig. 21.12*b* presents the second hierarchical level with an OR gate for that combination.

After developing a fault tree, one will evaluate the probability of an occurrence of the top event. This can be accomplished for the example in Fig. 21.12*b* with the use of the basic rules of probability given the logic presented. The values for the units of Fig. 21.12*b* are presented in Table 21.5.

TABLE 21.5   Fault-Tree Reliability Analysis of Error-Detecting Units

Name of unit	Reliability = R	Unreliability = $Q = (1-R)$
Error-detecting software (EDS)	.9969	.0031
Error-detecting logic (EDL)	.9920	.0080
Logic power supply (LPS)	.9990	.0010

In Fig. 21.12*b*, the reliability of the error-detecting hardware (EDH) is

$$R_{EDH} = R_{EDL} \times R_{LPS} = .9920 \times .9990 = .991 \tag{21.63}$$

and

$$Q_{EDH} = 1 - R_{EDH} = .009 \tag{21.64}$$

The *unreliability of error detection* (ED) is a function of the unreliability of the error-detecting hardware and the unreliability of the error-detecting software:

$$Q_{ED} = Q_{EDH} \times Q_{EDS} = .009 \times .0031 = .000279 \qquad (21.65)$$

and

$$R_{ED} = 1 - Q_{ED} = .999721 \qquad (21.66)$$

Thus, the reliability of error detection is .999721.

21.10   PETRI-NET METHOD

The *Petri-Net method* is a bottom–up method that utilizes a symbolic language. A Petri-Net structure consists of places and transitions and the Petri-Net graph allows the user to represent the actual system functions and use markings to assign tokens to the net. One can study the blockages or failures that affect the system while monitoring the performance and reliability levels.

With the pictorial representation, the elements can be easily modified. With the use of a computer program, results can be obtained and element changes studied. It is often important to know what performance will occur with one or more units or subsystems operating with some failures present. In summary, *Petri-Nets are a tool for the study of combined system performance and reliability.*

A Petri-Net has four basic parts that allow construction of a Petri-Net graph:

1. A set of *places* labeled "P" and represented by a *circle,* ○
2. A set of *transitions* labeled "T" and represented by a *vertical bar,* |
3. An *input function* labeled "I"
4. An *output function* labeled "O"

A Petri-Net graph, which can represent a system, is composed of

1. A set of *places* represented by *circles,* ○
2. A set of *transitions* represented by a *bar,* |
3. A set of *directed arcs* represented by *arrows,* →, which connect places to transitions and transitions to places

Places can contain *tokens* represented by *dots,* ·. Tokens can be considered resources. The Petri-Net *marking* is defined by the number of tokens resident at each place, which is also designated as the *state* of the Petri-Net. In a *timed* Petri-Net, transitions have time values associated with them. Tokens reside in places and control the execution of the transitions of the Petri-Net. Many applications of the timed model include finding the expected delay in a complex set of actions, average throughput capacities of parallel computers, or the average failure rates for fault-tolerant designs.

There are several variances of Petri-Nets used in system reliability analysis. A *stochastic Petri-Net* (SPN) is obtained with the association of a *firing time* with each transition. The firing of a transition causes a change of state. A *generalized stochastic Petri-Net* (GSPN) allows timed transitions with zero firing times and exponentially distributed firing times. The *extended stochastic Petri-Net* (ESPN) allows the firing times to belong to an arbitrary statistical distribution.

The evaluation of reliability and performance of complex systems lends itself to modeling using *stochastic Petri-Nets.* The state space Markov process is defined by a set of system states, transition probabilities, and an initial state of the system. It is known that Petri-Nets are isomorphic to discrete state-space Markov processes. Thus, the Petri-Net model will have its making graph equivalent to the desired state diagram. The symbolic language of the Petri-Net method allows one to describe the system in the language of Petri-Nets and use a computer program to translate the model into a state diagram equation format and have the resultant set of equations automatically solved and the results easily understood.

CHAPTER 22
SOFTWARE RELIABILITY AND THE DEVELOPMENT PROCESS

Samuel J. Keene
STORAGE TECHNOLOGY CORPORATION
LOUISVILLE, COLORADO

22.1 INTRODUCTION

This chapter addresses the two main focuses of software reliability. First the emphasis is on the proactive steps that one can take to assure the development of more reliable code. Then a discussion is given as to know when the software is good enough to ship, specifically, when the code meets its specified reliability requirements. The code is typically tested to its specified operating conditions. Failures are detected and their source is tracked down and eliminated. The code is successively refined by the removal of bugs that cause failures. The code then exhibits reliability growth. This process normally continues until the code behaves sufficiently well to ship to the customer.

A survey is made of the field of software reliability. Much emphasis is placed on the early phases of product development in setting the framework for a high-quality product. This can best be accomplished by an awareness of the early developmental vulnerabilities.

Many lessons learned on a software development program point back to the early development phase. Typical citations include

- Attempt to deliver too much function at once
- Inadequate up-front engineering
- Need to identify and resolve interoperability issues
- Inadequate definition of the operations concept
- Need to recognize and handle off-nominal program inputs
- Product requirements not complete or validated
- Inconsistencies between the developmental hardware and the target hardware
- Need consistent set of agreed-to tools and operational conventions (naming modules)
- Need to define configuration management process up front
- Requirements team not composed of the "right people"
- Lack of adequate architectural concept, problem definitions, processes, methods, and guidelines

A set of proactive guides are given to guard-band these deficiencies.

22.2 *TERMINOLOGY*

Software reliability is the probability that the software operates successfully, without failure, for a given operational time period in a specified operational environment or under a specified operational profile. Software reliability and software maintenance are typically the major reliability concerns in today's newly developed systems. Further the extent of software used in our products is growing exponentially over time as shown in Fig. 22.1.

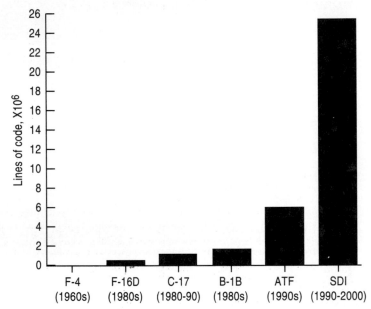

FIGURE 22.1   Growth in software size.

Operational profile is a representation of the actual use environment such that when the software is evaluated for its reliability characteristics or life, the test results obtained are indeed representative of the actual end-use application. A host application might have a word-processor function as well as a function that calculates Bessel functions, for instance. The former feature will likely be used much more than the Bessel function. So the testing environment should place more emphasis on the word-processing application than the Bessel function one.

Design errors are created by the system engineers and the designers and programmers. Their genesis is

- *Incorrectly defined user requirements:* These errors can originate with the user not knowing or recognizing the needed system requirements. The user requirements can evolve or change over the period of development leaving the user wanting when the product is delivered. There can also be misunderstanding of the requirements by the developer, or, finally, these requirements may be subject to misinterpretation or mistranslation throughout the development process.

- *Semantic errors:* These are logical errors. An example of a semantic error could be the erroneous assumption for an algorithm that the sum of the exponential of a set of variables is the same as the exponential of the sum. This is mathematically incorrect.

- *Syntax error:* These are typically grammatical errors, such as a wrong delimiter in an expression or a delimiter in the wrong place in an expression or computer statement. A second example would be the placement of a O (letter O) where a 0 (zero) was desired. This happens quite frequently and can be very difficult to isolate. Sometimes the programmer will place a set of temporary markers or flags in the code to see how much of it will compile before the failure is experienced. Narrowing the flag separation down and recompiling the program allows the programmer to isolate and identify the fault.

- *Nonrobust design:* Robust designs will do the correct response in the face of errant or off-nominal input conditions. An example of this is demonstrated in the following simple arithmetic expression:

$$A + B = C \tag{22.1}$$

This operation is deterministic when the inputs A and B are real numbers. What should happen when one of the inputs is negative, or imaginary, or alphabetic? These would constitute erroneous inputs. The equation would function best if the program recognized the aberrant condition, flagged it, and isolated the error to the user. This code is then said to be robust in its design.

Faults are the embedded tendencies to fail in the code. These are the shipped defects in the delivered code. These are also called the *latent defects*. These faults are weak or susceptible points that are subject to spawning a failure. The first time the newly developed F-16 fighter plane flew south of the equator, the navigational system of the plane caused it to flip over. This part of the navigational software had never been exercised prior to this trial. As long as the plane stayed in the Northern Hemisphere, there was no problem. A fault is merely a susceptibility to failure; it may never be triggered and thus a failure may never occur. One can readily imagine parallel examples of faults in terms of legal exposures or medical threats to our well-being. Some of us are more susceptible to disease than others.

Faults are a design measure of reliability. The typical measure is in fault per thousand lines of source code (KSLOC). One limitation of this approach to measuring reliability is the difficulty of agreeing on what constitutes a line of code. Usually the definition is limited to the executable lines of code and the commentary lines of source code are left out. Typical levels of faults found in the code are shown in Table 22.1.

TABLE 22.1   Code Defect Rates (Faults/KSLOC)

	Total development	At delivery

Traditional development
  Bottom–up design,
  unstructured code,
  removal through testing
	50–60	15–18

Modern practice
  Top–down design,
  structured code,
  design and code
  inspections,
  incremental releases
	20–40	2–4

Future improvements
  Advanced software
  engineering,
  verification practice,
  reliability measurement
	0–20	0–1

Space Shuttle code	6–12	0.1

The difference between the amount of faults injected into the code and those that survive to be shipped to the field reflects the goodness of the development process.

Defects are removed by a number of development quality processes such as inspections, reviews, and tests. Eighty-five percent of all faults are removed prior to system test phase on the highly successful Space Shuttle program. The better the development process, the fewer defects escape (detection and removal) and go to the field as latent defects.

The so-called traditional code is the "spaghetti code." This was a custom design code that was optimized for speed and function; however, it lacked a sense of structure and was very difficult to understand. It has been replaced largely by structured coding techniques that follow a set of design rules, have a limited set of commands, and are more controllable in terms of entrances and exits to the module.

Failures are a deviation of performance from the customer's expectations, requirements, and needs. When a word-processing system locks up and doesn't respond, that is a failure in the eyes of the customer. The customer makes the call as to what a failure is. The occurrence of a failure signals the presence of an underlying fault that permitted the failure to occur. A point should be made here, that is, that faults are a logical concept and that they may often be corrected or defended in more than one way, with each being effective but different. This is well illustrated when software changes are used to patch or fix a hardware problem. The Hubble telescope had early vibration problems that were damped by software modifications.

Fault triggers can originate from erroneous inputs made to the system by the user, operator, or maintainer. The user might direct the system to do two opposing things at once: go up and go down simultaneously. These aberrant or off-nominal input conditions are a test of the code robustness. These conditions are the ones likely to trigger the faults and produce a failure. The *exception-handling* aspects of the code are more complex than the operational code. Its inputs are more varied and the combination of inputs are potentially explosive in number, making exception code critical from a reliability standpoint.

Fault triggers can also arise from program navigation down untested operational paths or sequences that are found stressful for the code. These paths could be sensitive to program timing. In some instances the timing would be such as to fail the operation of the code. The failure could also be sensitive to the path taken to get to a particular branch of the executable code. There are so many sequences of paths through a program that it is typically said to be *combinatory-explosive*. Ten paths taken in random sequence lead to 2^{10} possible path combinations or 1024 combinations. Practical testing considerations limit the testing to the major path combinations. There will always be some risk on the other path combinations. The relationship between errors, faults, and failures is shown in Fig. 22.2.

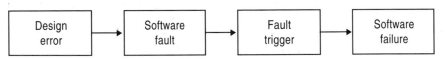

FIGURE 22.2   Path to failure.

Fault tolerance is a misnomer. It is closely related to robustness of the code. The code is not tolerant of faults. It is really tolerant of operational errors. Fault-tolerant code continues to work in the face of off-nominal input conditions and the various operating environments under which it is expected to operate. It requires a different design approach to achieve a fault-tolerant design.

Designers must consider the possibility of off-nominal conditions that their codes must successfully handle. This is a proactive or defensive design approach. For example, in numeric intensive algorithms, there is always the possibility of divide-by-zero situations where the formula would go unbounded. When many calculations are involved or when one is dealing with inverting matrices, a divide-by-zero happens too frequently. The code can defend itself by checking the denominator to ensure that it is nonzero before dividing. There is a performance cost to

TABLE 22.2   Some Typical Fault-Tolerance Questions for the Programmer

1. How will you detect a failure in your module? Programmed exception handling or abort with default exception handling?
2. Will you detect out-of-range data?
3. How will you detect invalid commands?
4. How will you detect invalid addresses?
5. What happens if bad data is delivered to your module?
6. What will prevent a divide-by-zero from causing an abort?
7. Will there be sufficient data in a log to trace a problem with the module?
8. Will the module catch all its own faults? (What percentage detection and unambiguous isolation?)
9. Do you have dependencies for fault detection and recovery on other modules or the operating system?
10. What is the probability of isolating faults to the module? Line of code?
11. Does your module handle "exceptions" by providing meaningful messages to allow operator-assisted recovery without total restart?

do this operation. It is best when the designer is aware of the exposures and the tradeoffs. A sample list of fault-tolerance considerations that can be applied to the design process is shown in Table 22.2.

22.3   SOFTWARE-RELIABILITY EXAMPLES

Software is the embodiment of the programmer's thought processes. It does exactly what it is told to do even though the circumstances under which it acts might not have been anticipated by the author of the code. A planetary probe satellite was programmed to switch to its backup power system if it did not receive any communications from the ground over a 7-day period. It was believed that such lack of communication would signal a disruption in the power system and switching to the backup would restore power operations. This was a built-in algorithm in the software, a piece of artificial intelligence that would mimic the actions of an on-board service person. In the second decade following launch this enabling condition of 7 days' silence was experienced, and the system switched to backup power. The mainline system was still working, and unfortunately the standby system was not. The logical error led to driving the satellite into a failed state. The satellite was just doing what it was told to do more than 10 years earlier. Given the operational environment, an error in the satellite operating requirements had led to a failed state.

There are many everyday analogies to software faults and failures. Any transaction that we embark on that has residual risks and uncertainties has a counterpart in software reliability. Look at the potential risks in a real estate contract. This could include a flaw in the title, the house might encroach on an easement, be in a floodplain, be flawed in its construction or condition, etc. A prudent person wants to mitigate these faults or their likelihood and builds defenses against their occurrence. Likewise, a prudent programmer will think proactively and deal with contingent conditions and mitigate any adverse effects they might present.

Some more failure examples or examples of the lack of reliability are discussed in the following paragraphs. These failures typically arise from some abnormal or limiting condition that the designer did not consider and account for in the design. Several examples are enumerated here.

22.3.1   Get Main, Free Main

This is the sequence of obtaining an allocation of main memory and then freeing up the same memory when the application is finished with it. This is part of the operating-system code. If the memory is not freed up, the resource dwindles and programs cannot find sufficient memory to properly execute. This condition also occurs as memory is used and released but left in a chopped-up array without being contiguously accessible. A practical example of how memory can affect the user reliability occurs with word-processing programs. As the program size grows as it is used, and more data is linked, the author has found the keyboard to lock up. The program would then no longer respond. The best defense is to periodically back up the active work and also to restart the program so that memory can be reinitialized and organized.

22.3.2 Minor Code Changes

These create a special problem because the programmers and testers discount their impact and they feel that it is too expensive to study these changes in the detail required or to test them as much as the original code was tested. This process of retesting to the original test conditions is called *regression testing* of the code.

When this critical step is omitted, disastrous consequences can occur. In 1990 a sequence of major phone networks began to fail in the large metropolitan areas of the United States. The problem was traced to a change of four lines of code in a system containing more than one million lines of code. The regression testing would have taken 13 weeks. The decision was made to skip it. Disastrous consequences followed in phone network disruptions.

22.3.3 Resource Misallocation

This can disrupt system operations. An operating system for a computer network has preprogrammed intelligence in it. It is instructed beforehand as to how to handle resources, priorities, and application programs. The operating system has to manage and balance the needs of the system components. It is not always successful in this regard. In several instances a lowly printer can tie up the system. The printer can get in a loop requesting CPU (central processing unit) resources. The CPU has been known to lose all control to the print requests and have its other resources and applications time out, bringing the system to failure.

22.3.4 High-Severity System Failure Modes

An infamous failure of a high-energy X-ray machine was labeled "Malfunction 54.[1]" This machine had two modes of operation. One was a high-energy electron accelerator, and the second was a gamma-ray (γ-ray) machine. When the machine was operated in the gamma mode, high-energy electron rays impinged on a target. This produced a secondary emission of gamma rays. Under a malfunction mode the operator screen displayed "Malfunction 54." As the operator keyed in commands to clear the failure, high-energy X rays directly impinged on the patient. The aberrant dosage was 100 times over the allowable limit. Of course, the transpiring events are not immediately comprehended. Several patients died before the problem was sufficiently diagnosed.

The point here is that the *embedded software* in the machine was taking a highly undesirable course of action for this medical equipment in the presence of a failure. This was also the result of the operator attempting to navigate past the problem with the machine. The failure mode was lethal and should have been avoided. This failure susceptibility could have been avoided by systematically analyzing the system's behavior in the face of potential failures and erroneous inputs such as the operator had done.

The safest way to preclude this type of failure is to incorporate hardware interlocks to ensure that the high-energy electron rays could not be emitted without the target being in place. This type of analysis can be systematically done by one of two methods. One is *failure modes and effects analysis* (FMEA), which logs the effects of failures starting at the lowest level of analysis in the system. For hardware this is usually a field replaceable unit such as a circuit board. For software it typically is performed at the module level. Likely failures in the module are postulated on the basis of experience. They could be

* No output from the module
* Wrong timing on the output: too slow or too fast
* Wrong output provided

FMEA provides a systematic survey and documentation of the effect of faults throughout the design. *Fault trees* are the second systematic fault-tracking system.

Fault trees start at the top level of the system and find the root causes for high-level failures that lead to disastrous consequences. The programmer and/or designer is concentrating on making the system work. That covers one face of the design problem. Ferreting out the probable failure modes and their consequences can lead to a more robust and safe design. These failure-mapping tools are a great asset to apply to the system design.

22.4   COMPARING HARDWARE AND SOFTWARE RELIABILITY

The United States Air Force system's experience reported that mean time between failures (MTBF) on IBM central processors (3080 class machines) had grown to 10 years in their application by 1987.[2] By contrast, they found newly developed software had an MTBF of 160 to 200 h, although an estimated 90 percent of these were not considered fatal. Thus they did not interrupt operations by bringing the system down. Discounting the nonfatal flaws, this data still indicates software to be 40 to 50 times more prone to failure than hardware. Recently the author studied a system that had been fielded for over two decades and found that the software is 10 times more likely to fail than hardware. The point is that software is an increasingly critical component of system reliability. As a component of system cost, software maintenance is an increasingly high component as shown in Fig. 22.3.

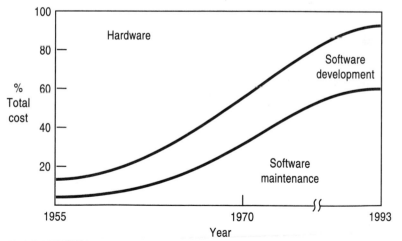

FIGURE 22.3   Hardware and software costs.

Classifying the origin of intermittent failure in hardware or software is not easy. In his book, John Musa talks about the unknown origins of failure.[3] His recommendation is "that you classify a failure as not-software if it doesn't reoccur when you run the program with all the software and non-software the same."[3] Jim Gray, on the other hand, has found intermittent failures a predominant aspect of software.[4] His experiments have shown that broken code recovers 99 percent of the time by warm starting and retrying the faulty operation. The *warm start* restructures the state environment and facilitates recovery.

There is one compensating factor about software failures vis-à-vis hardware failures. Software failures are typically remedied by restarting the code. This takes only a few minutes. Hardware failures, on the other hand, take more than an hour on average. So software fails an order of magnitude more often, but its system interruption is an order of magnitude less. The total downtime may be comparable from software and hardware. This was observed on 1970s and 1980s hardware. Hardware is constantly improving, and this aggravates the relative contribution of the software.

Why do software faults dominate over hardware faults? Software development is very labor-intensive. The problems found in system and operational test are often related to interface anomalies; software technology today does not have the level of integration that hardware has. This means more unverified software interfaces than hardware interfaces. Also hardware has more restricted inputs between its functional component areas. Further, the human operator is more interactive with the software than with the hardware.

In comparison to software, hardware design is more modularized and standardized. It is more testable, and its quality is more verifiable. Better tools exist for validating the reliability of hardware than for validating the reliability of software. In personal communication with Harlan Mills, who has been one of the foremost leaders in developing reliable software, he expressed his belief that software design today still remains an art. A comparison of software and hardware reliability characteristics is shown in Table 22.3.*

22.5   SOFTWARE RELIABILITY AND THE DEVELOPMENT PROCESS

There is more recognition today as the determinant power of the process on the final product reliability. This is the foundation of TQM, ISO 9000, and Malcolm Baldrige activities. Product reliability usually has to be built in. In software this is especially true. So the development process for software is critical to the qualities of software reliability, schedule, and cost. The earliest software model used (de facto) was the "test and fix" model, sometimes called TAF. After so many fixes the code structure degraded and subsequent fixes became increasingly difficult to implement and expensive to incorporate. Frequently there was a poor match of code to user's needs and the so-called spaghetti code resulted.

22.5.1   Waterfall Model

The TAF model has been largely replaced by the more disciplined model called the "waterfall model," shown in Fig. 22.4.

This model recognizes successive requirements decomposition phases followed by a series of integration phases to put together the final system configuration. Initially the requirements are broken down to put them into handleable size pieces. First the functionality is allocated between the hardware and the software. Then the software is further broken down into functional elements and then to modules. This is the programmer design level. Modules usually contain several hundred lines of code. Modules should have a high cohesiveness of function and relative independence of other modules. Greater independence of modules within their boundaries simplifies the design and therefore increases software reliability and maintainability. The workable programmable level is the module. Once the modules are designed, built, and tested, the integration phases start.

---

*Derived from material developed by Patrick D. T. O'Connor and published in "Practical Reliability Engineering," J. Wiley and Son, New York, 1981, p. 234, used with permission.

TABLE 22.3   Comparison of Hardware and Software Characteristics

Hardware	Software
1. Failures can be caused be deficiencies in design, production, use, and maintenance.	1. Failures are due primarily to design faults. Repairs are made by modifying the design to make it robust against the condition that triggered the failure.
2. Failures can be due to wear of other energy-related phenomena. Sometimes warning is available before failure occurs (systems can become noisy indicating degradation and impending failure).	2. There is no wearout phenomenon. Software failures occur without warning, although very old code can exhibit an increasing failure rate as a function of errors induced into the code while making functional code upgrades.
3. Repairs can be made that might make the equipment more reliable. This would be the case with preventive maintenance where a component is restored to a like-new condition.	3. There is no repair. The only solution is redesign (reprogramming), which, if it removes the error and introduces no others, will result in higher reliability.
4. Reliability can depend on burnin or wearout phenomena; that is, failure rates can be decreasing, constant, or increasing with respect to operating time.	4. Reliability is not so time-dependent. Reliability improvement over time may be affected, but this is not an operational time relationship. Rather, it is a function of reliability growth of the code through detecting and correcting errors.
5. Reliability is time-related, with failures occurring as a function of operating (or storage) time.	5. Reliability is not time related. Failures occur when a program step or path that contains the fault is executed and triggers a failure.
6. Reliability is related to environmental factors (temperature, vibration, humidity, etc.).	6. The external environment does not affect reliability, except insofar as it might affect program inputs.
7. Reliability can be predicted in theory from knowledge of design, usage, and environmental stress factors.	7. Reliability cannot be predicted from any physical bases, since it entirely depends on human factors in design. Some a priori approaches exist based on the development process used and the extent of the code.
8. Reliability can usually be improved by redundancy. The successful use of redundancy presumes ready detection, isolation, and switching of assets.	8. Reliability cannot be improved by redundancy if the parallel program paths are identical, since if one path fails, the other will have the same error. It is possible to provide redundancy by having parallel paths, each with different programs written and checked by different teams.
9. Failures can occur in components of a system in a pattern that is, to some extent, predictable from the stresses on the components and other factors. Reliability critical lists are useful to identify high-risk items.	9. Failures are rarely predictable from analyses of separate statements. Errors are likely to exist randomly throughout the program, and any statement may be in error. Reliability critical lists of failures are not appropriate.
10. Hardware interfaces are visual; one can see a 10-pin connector.	10. Software interfaces are conceptual rather than visual.
11. Hardware products use standard components as basic building block.	11. There are no standard parts in software, although there are standardized logic structures. Also, software reuse is being deployed, but on a limited basis.

Module integration starts after unit testing. Independent modules and compilable units are tested. This is usually performed by the programmer. These working units are then combined through a process called *software integration and test* (SWIT), in which successively larger blocks of code are compiled and executed. When the total software system is completed, system testing begins. This is the first time the entire system of hardware and software can be evaluated. The high-level

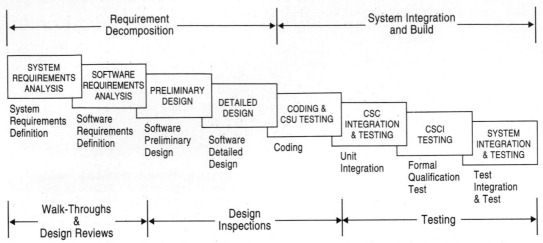

FIGURE 22.4 The waterfall model, where CSU = Computer Software Units, CSC = Computer Software Component, CSCI = Computer Software Configuration Item.

requirements were decomposed to get down to a containable level for the programmer. Then the system was rebuilt by verifying that the modules and larger integrated structures worked together satisfactorily.

Each phase of the waterfall model has its own assurance check points. The decomposition and build steps (coding) are all checked by walk-throughs, inspections, and reviews. These are paper analyses. Once SWIT starts, the emphasis is on testing.

The code is tested and refined by the removal of bugs found. In the superreliable Space Shuttle program, the steps prior to test were sufficiently rigorous to remove 85 percent of all the defects injected into the code. Most of these defects are removed at code reviews following high-level design, low-level design, and code development.

22.5.2 Code Reviews

- *Design:* The *design* category encompasses all errors which are a carryover from the design phase. These errors include poor interface design, functionality which clearly needs to be implemented to support the code but which was not specified in the design, bad or missing error and recovery messages, serious performance problems due to problems with the system design, and ambiguities resulting from the design of the user interface with the system.

- *Logic:* The *logic* category is distinguished by defects that center on problems with program structure and execution. Errors of this kind frequently show up in pseudocode. They are typified by serialization and concurrence problems, control logic, algorithmic problems, recovery logic, loop control, signal handling, recovery logic, error-handling logic, test and branch, if/else, switch and case incorrectly implemented (logic problem), and functionality specified in the design but incorrectly implemented. Typically, this code works but has an incorrect result.

- *Coding mechanics:* Defects in the *coding mechanics* category encompasses violations of project or language standards. Violations of this type may not prevent the code from functioning, but make it difficult to maintain. Examples of this kind of defect are violations of naming conventions for source files, functions, subroutines, and modules; unmaintainable language constructs such as indiscriminate use of complex compound statements, indiscriminate use of GOTOs, excessively deep nesting levels, constants not symbolically defined, depth or size of if/else dictates use of case/switch, nesting levels, hidden declarations, unused variables, linkage, nonstandard register usage, and unused variables. The only coding standards that can be brought up during the code review are those that are part of the operating agreement.

- *Documentation:* The *documentation* errors include incorrect program or module prologues, incomplete or missing inline comments or block comments, commentary that doesn't match the code, and missing commentary needed to understand a complex coding construct. Other types of documentation errors are pseudocode or block comment that doesn't match the code or a lack of white space. These errors are distinguished from those under the standard coding practice category in that they are not executable statements.

The faults found in the stages of the development process have been found to follow a Rayleigh curve.[5] The Rayleigh curve is shown in Fig. 22.5.

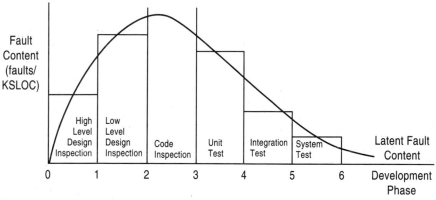

FIGURE 22.5   Typical fault-discovery profile.

A fundamental difficulty with the waterfall model has been its emphasis on a fully elaborated documentation as completion criteria for the early requirements and design stages. This elaboration is performed even when the design is still "fuzzy" Document-driven standards have pushed projects to document poorly understood interfaces and then to follow that with poorly written code. The main problem is that if requirements are not correctly fathomed in the planning and requirements stages, they propagate problems throughout the design. Most requirements problems are not found until they are in the field. Then the customer finally checks out the designed system. The cost of discovered errors is shown in Fig. 22.6. The final cost of faults that hit the field could easily be upward of $50,000, as shown in Fig. 22.6.

Steps can be taken to prevent or minimize a large number of the model limitations mentioned above. These include

- Keeping the customer involved throughout the development process to get customer reviews, inputs, and feedback.
- Level-setting the customer's expectation of the product. A computer manufacturer was beta testing its product to a limited number of customers. These customers were eager to buy this product. The manufacturer recognized that the current reliability level was still below its goal. The manufacturer's disclosure didn't dampen the customers' desire to have the product before final release. They knowledgeably bought into the current status and proved to be satisfied customers.
- Building prototypes, simulations and graphic representations of the design. This is to avoid the vagaries of natural language (that which we speak). A picture is worth a thousand words, and a working model must be worth a million words. The customer will rapidly refine requirements after working with the initial prototypes. Everyone benefits from this learning curve.

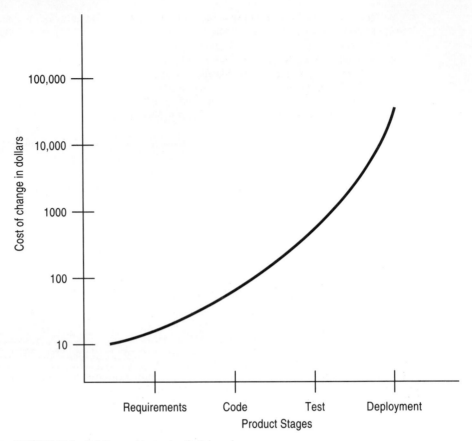

FIGURE 22.6   Relative cost to repair a fault in code.

22.6   *SOFTWARE-RELIABILITY MODELS*

22.6.1   Predevelopment Phase

Typically, system reliability-maintenance-availability models must be created before the beginning of software design (e.g., at the proposal stage). At that point, the functions of each module and the approximate amount of code are defined. Using this information, one can create a failure-mode effect criticality analysis (FMECA) with a reasonable degree of accuracy. The FMECA should identify the impacts to system performance (e.g., failure or degraded mode) that all components, hardware units, and software modules will have. A reliability block diagram can also be created to show single points of failure redundancy.

The next step is to gather data on mean time between failures (MTBF) and mean time to repair (MTTR) for each component. A model is needed to calculate the MTBF for a piece of software from the amount of code.

Because little knowledge of the code is available at this stage, fault-content projections are based on past experience. Typically, fault densities are measured in terms of faults per thousand source lines of code (KSLOC). As discussed by Koss,[2] the number of faults per KSLOC is dependent on the development process used, including the inspection process, the amount and types of testing, and the length of the development cycle.

The clean-room software development process has driven the latent fault content of the code to less than one fault per KSLOC. The clean-room development is strongly structured and is aided by random, well-structured testing techniques and formal statistical controls. However, the emphasis on fault removal should not wait until testing, but should occur as early in the development cycle as possible. This not only provides for more reliable code but also drastically reduces the cost of software development.

22.6.2 Development Phase

During the development of the code, more information regarding fault content can be gained. This information can be used to refine the initial estimates (predevelopment phase) of latent fault density. Tractenberg[6] has shown that the number of faults per KSLOC during development and testing follows a Rayleigh curve; Gaffney[7,8] has taken this concept and data gathered on a number of software development projects to create a model that predicts the expected, maximum, or minimum number of faults for each remaining phase of development, including the latent faults delivered to the customer. Inspection (high-level, low-level, and code) and test (unit, integration, and system) fault-discovery data are collected for these projects. Fault-discovery profiles can be fitted against experience curves or Rayleigh models to form the basis for the fault-content prediction model.

The model predicts future fault densities for newly developed code by matching current fault-content data collected to the Rayleigh curves. The latent defect rate to be shipped to the field is then projected. These Rayleigh curves provide a signature of the development process. Better processes detect and remove faults earlier in the development cycle. As processes change, the Rayleigh curves project the overall effect on the reliability of the shipped product.

Once the projected number of latent faults is estimated, other models are used to estimate the detection (failure) rate of the code once it becomes operational. One method, derived by Gaffney,[6,7] is based on statistics gathered from numerous projects. He proposes that the cumulative percentage of faults experienced F can be projected by the following formula:

$$F = 100(1 - 0.56^t) \qquad (22.2)$$

where t is years after release of the code. This equation indicates that 90 percent of the faults in the code will be discovered after 4 years. Simple in nature, the equation does not consider factors such as the relative usage or complexity of the model, the execution environment, and multiple failures from the same fault. Because of its simplicity, this code is ideal for use early in development. Future studies that could account for the other factors without requiring much data would enhance the field of software reliability considerably.

22.6.3 Testing Phase

During early testing, the test group validates that the product meets its functional requirements. Next, test subjects the product to a set of extreme input variations to measure the robustness of the product. Finally, the product is operationally tested in a simulated customer environment. In this setting, the reliability of the product can be determined.

Figure 22.7 shows how reliability grows over the operational test period as bugs are formed and removed. The code continues to improve to its targeted shipment level. The additional test time required to refine the code to a specified level can be estimated (see Fig. 22.7).

During operational testing, faults are discovered through the failures they manifest. The underlying faults are then fixed. Although most faults will be from the software, hardware faults will also be removed. The reliability growth profile follows a saturating profile with time that closely fits an exponential curve. If it is assumed that all faults are equally likely to be detected, then the discovery and removal of faults ΔF over time Δt will be

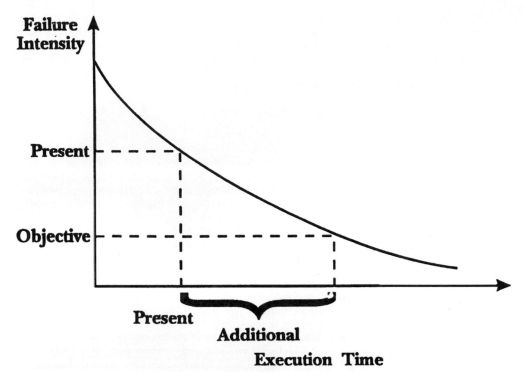

FIGURE 22.7 Execution-time component.

$$\frac{\Delta F}{\Delta t} \propto F \tag{22.3}$$

or

$$\frac{\Delta F}{\Delta t} = -kF \tag{22.4}$$

Because faults F decrease in proportion to the number of faults in the system, the greater the number of faults, the more likely they will be detected and removed.

Integrating Eq. (22.4) with respect to time t

$$\log F = -kt \tag{22.5}$$

and changing to the exponential form and normalizing, we obtain

$$F = F_0 e^{-kt} \tag{22.6}$$

which is the anticipated reliability growth profile.

The question of how time is measured should be raised. Once testing begins, failures-per-unit-time data can be collected. During this phase, Musa advocates using program execution time rather than calendar time as a metric, because the code will run in test for only short and irregular periods of time.[3]

Kan used the basic exponential to model defect data during test on the IBM AS/400 system.[5] The biggest limitation he acknowledged was the restriction of measuring test time as calendar time. This limitation was mitigated by the testing efforts remaining consistently high and homogeneous. Actual field defects have been tracked following shipment, and field data validated Kan's test projections.

Second, Musa's model is a logarithmic Poisson model that has certain advantages[3]:

- Measures product reliability in user terms of MTBF rather than design terms of faults per thousand lines of source code (KSLOC).
- Estimates current MTBF at any point during test.
- Projects the additional test resources (test time, personnel, CPU resources) to meet a specified target reliability.
- Bounds estimates with confidence intervals.
- Uses CPU execution time as the independent variable rather than calendar time. This removes the noise from nonhomogeneous testing in estimating reliability.

Third, Jelinski-Moranda developed a model called a *geometric deeutrophication model.* In their model, the failure rate decreases geometrically, which is a more realistic assumption than the notion that all faults have the same size. They reasoned that faults removed early in the process were more obvious, and had higher impact. The remaining faults were more subtle and occurred less often during test or operations.

22.6.4 SMERFS Model

There is a set of software reliability models that have found wide acceptance. It was developed by software experts at the Naval Surface Warfare Center, Dahlgren, Virginia (USA). It is called *statistical modeling and estimation of reliability functions for software* (SMERFS).

This package has a good representation of popular software reliability models. Some of the models in the SMERFS package are

- *Generalized Poisson model:* This is more specifically called the *nonhomogeneous Poisson process model* (NPPM). Poisson processes are based on an underlying exponential failure probability. The nonhomogeneous modifier means that the failure probabilities are changing over time; e.g., as faults are discovered and removed, the code failure rate diminishes over time.
- *Yamada's S-shaped reliability model:* The exponential model assumes the peak of the defect arrival is at the beginning of the test or field deployment phase. The S-model has a delayed peak defect arrival time.
- *Jelinski/Moranda deeutrophication model:* Most models assume all the errors to be the same size. This model assumes that the failure rate is decreasing in a geometric order.
- *Littlewood's model:* This model assumes that

  1. The failure rates of the errors still in the program are independently identically distributed.
  2. The program failure rate is the sum of the individual failure rates.
  3. No new errors are introduced during fixes.

SMERFS has tools to compare the accuracy of the models to measure the particular data set. Through these analyses, users can easily determine which models are good candidates for their data sets. The SMERFS package can be obtained at a nominal cost. For information, contact:

Dr. William H. Farr (B10)
NSWCDD
17320 Dahlgren
Dahlgren, VA 22448-5000

22.7 *CONCLUSIONS*

Proactive design for reliability requires a strong focus on communications, first to get the requirements correct, then to assure design coordination among the team members. The design can be made more robust by systematically examining the reaction of the program to off-nominal input conditions.

Finally, the operational reliability of the code can be measured during system test. This presumes that the code is following an operational profile or sequence that simulates the real operating condition. Code reliability grows through test and fix of discovered defects. So present code reliability can be estimated along with the rate of future improvements.

REFERENCES

1. Joyce, E. J., "Software Bugs: A Matter of Life and Liability," *Datamation,* May 15, 1987, pp. 88–92.

2. Koss, Edward W., "Software Reliability Metrics for Military Systems," *Proceedings of 1988 Reliability and Maintainability Symposium,* Los Angeles, Calif., pp. 190–194.

3. Musa, John D., *Software Reliability, Measurement, Prediction and Application,* McGraw-Hill, New York, 1987, 615 pp.

4. Gray, Jim, "Why Do Computers Stop and What Can Be Done About It?," *Proceedings of the Fifth Symposium on Reliability in Distributed Software and Data Base Systems,* 1986, pp. 3–12.

5. Kan, S. H., "Modeling and Software Development Quality," *IBM Systems Journal,* **30** (3), 1991, pp. 351–362.

6. Tractenberg, M., "Discovering How to Ensure Software Reliability," *RCA Engineering,* Jan./Feb., 1982, pp. 53–57.

7. Gafney, John, "Software Error Estimation Program," (Sweep), user manual SPC-92017-CMC Software Productivity Consortium, Herndon, Va. 22070.

8. Gafney, B. J., "Estimation of Software Code Size Based on Quantitative Aspects of Function," *Journal of Parametrics,* Sept. 1984, pp. 23–33.

CHAPTER 23
SUPPLIER RELIABILITY AND QUALITY ASSURANCE

David L. Nelson
THE BOEING COMPANY
SEATTLE, WASHINGTON

23.1 INTRODUCTION

For many engineering-intensive companies, a substantial proportion of the finished product consists of components produced by suppliers. Any effective program for product quality must therefore include a means of assuring supplier quality and reliability.

As the means of assuring quality and reliability has evolved in many companies' own systems (e.g., more reliance on statistical process control and less on end item inspection), it is also evolving with respect to customer-supplier relationships. A customer's typical concern used to be "Have you, my supplier, sent me components that are within tolerance, except for a predetermined level of allowable defectives?" This concern is rapidly being replaced by "Have you, my supplier, employed reliable, stable, and efficient processes to create components near the nominal specification, with low variability?"

The latter question sounds a bit more complicated, and carries a sense of the customer getting much closer to a supplier's operations. The implications of this change are far-reaching in terms of the kind of quality a customer can expect, the supplier's and the customer's responsibilities to ensure high quality, and the price the customer will have to pay for it. Surprisingly, higher quality and reliability achieved through modern quality-assurance methods usually results in lower, not higher, costs.

Both quality and reliability are full-spectrum issues; that is, they require a comprehensive set of disciplines and approaches that encompass the entire product chain from subcontractors through the ultimate customer. In many arenas, such full-spectrum issues are being addressed under the heading of *total quality*. Since total quality includes reliability considerations, and since the same disciplines tend to be applied to reliability improvement and to quality improvement, this chapter will emphasize assurance of quality.

The following sections describe modern quality-assurance methods, how they work, and how they have been employed at major manufacturers, principally The Boeing Company.

23.2   OBJECTIVES FOR SUPPLIER QUALITY AND RELIABILITY

23.2.1   Quality Operating System

A quality system (or quality *operating* system) is a combination of organizational structure, responsibilities and authority, methods, documents, and records whose purpose is to assure quality in processes and products. Every company has some quality system operating, or else product delivery would be virtually impossible. Surprisingly few companies, however, have an effective supplier quality system.

23.2.2   Objectives for Suppliers

The primary objectives in instituting quality systems within suppliers are increased quality, reduced costs, and reduced cycle time for the customer. These objectives are best achieved through a concurrent reduction of defects in the supplier's manufacturing processes and reduced reliance on mass inspection to achieve delivered quality. Reducing defects and mass inspection can happen only if variation in manufacturing processes is decreased dramatically, so that the risk of escape of a defect to the customer is minimal.

To see why this is so, consider a customer whose supplier's manufacturing processes allow a great deal of product variability. The supplier who is aware that product variability spills beyond customer specifications will spend time and energy inspecting products, to sort the deliverable ones from the defectives. The cost of reworking or scrapping the defectives will be a factor in determining the price the customer pays. Since excess variability usually results in increased manufacturing time to create acceptable product, schedules are vulnerable.

Since inspection is always imperfect, some defective product will reach the customer. The customer will have to order early and in perhaps excessive quantity to ensure the delivery of sufficient quantities of acceptable product on time. The customer will probably back up the supplier's inspection systems with receiving inspection. The customer will tend to keep an excess inventory for contingencies.

Each activity mentioned above is expensive, yet all these activities add no value to the product itself. Taken in total, these activities can overwhelm the actual cost of manufacture. A customer who takes no steps to reduce supplier product variability is simply paying too much for the supplier's product. The only choices become to recover those costs from downstream customers or go out of business.

On the other hand, if variation in suppliers' products can be substantially reduced, then products will tend to meet specifications, even without the expense of mass inspection. Delivery schedules will be reliable, scrap and rework minimal, the need for receiving inspection small, and the keeping of excess inventory a thing of the past. *Just-in-time* (JIT) delivery, itself the subject of many books and articles, can succeed only where variation is reduced to minuscule levels.

23.2.3   Key Characteristics

In the midst of considering the objectives described above, it is crucial to realize that variation reduction is more important for some features of a deliverable than for others. For this reason, there must be a mechanism between the supplier and the customer to identify those features of the delivered product for which variation has the greatest impact on quality. Boeing and many other companies call these features *key characteristics*.

Once key characteristics are determined, the manufacturing processes that contribute most to the variation of these key characteristics are identified, usually through a scientific analysis. Sources of variation in the manufacturing processes must be systematically removed. An employee training program that is both effective and measurable must usually be implemented to impart the necessary skills. These new activities must be budgeted and led by top management.

A number of related objectives for supplier quality are inherent in these statements. The following sections provide the necessary details on how they can be realized.

23.3   SUPPLIER QUALITY SYSTEMS

The need for effective supplier quality-assurance programs has been borne out recently through several important initiatives: the skyrocketing impact of the ISO 9000 series of quality-system documents, the emphasis on customer satisfaction represented in the Malcolm Baldrige National Quality Award criteria in the United States, the supplier quality-enhancement programs of the automotive industry, and standards established by other companies and government agencies to drive the use of modern quality-assurance practices into their supplier base. Many companies have used some combination of these criteria in an attempt to increase supplier quality and reliability without increasing costs.

The most successful of these supplier quality programs seem to be based on a two-tiered approach: (1) establish a basic quality system that is well founded, sound and documented, and provides needed information about the status of product and process quality; and (2) employ statistical techniques to improve the quality of products and processes and to reduce their cost. The first tier is referred to as a *basic quality system* in the remainder of this chapter, while the second is called a *statistically based improvement system.*

23.4   BASIC QUALITY SYSTEMS

Important objectives are addressed by basic quality systems, including management commitment and employee training. Other positive effects include improved planning and configuration control of processes and information, all of which tend to reduce cost and cycle time.

23.4.1   MIL-Q-9858A

A forerunner of many of today's basic quality-system requirements is MIL-Q-9858A, *Quality Program Requirements,* first published by the U.S. Department of Defense in 1963.[2] It describes requirements for

- Quality management and organization
- Planning
- Work instructions
- Records
- Corrective action
- Costs
- Drawings and documentation
- Measuring and test equipment
- Tooling
- Inspection and testing
- Purchasing
- Materials
- Production processes
- Handling, storage, and delivery

- Disposition of nonconforming material
- Inspection stamps

23.4.2 ISO 9000 Series

Borrowing from MIL-Q-9858A, as well as basic quality standards in the nuclear industry and elsewhere, the International Organization for Standardization (ISO) in 1987 published five documents that together make up the ISO 9000 series.[4–9] The ISO 9000 document itself provides guidance on the selection of the appropriate requirements document to serve as a standard between a given customer and supplier. ISO 9001, 9002, and 9003 are the requirements documents, only one of which should be selected. ISO 9001 is the most comprehensive among them, describing a basic quality system with all the elements noted above, along with additional standards for design. ISO 9002 is closest to the topic list of MIL-Q-9858A, covering quality assurance in production, installation, and servicing. ISO 9003 covers quality assurance for final inspection and test only (see Fig. 23.1). ISO 9004 provides an elaboration on the principles of managing for quality embodied in the other ISO 9000 series documents. Bossert[10] reproduces the ANSI/American Society for Quality Control equivalents of the ISO quality standards. (Much of the ISO 9000 series was revised extensively and republished in 1994.)

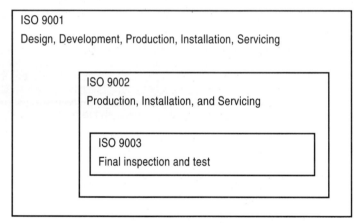

ISO 9001

Design, Development, Production, Installation, Servicing

ISO 9002

Production, Installation, and Servicing

ISO 9003

Final inspection and test

FIGURE 23.1 Relationship of ISO quality requirements documents.

23.4.3 Malcolm Baldrige National Quality Award

The Malcolm Baldrige National Quality Award, made a reality by the U.S. Congress in 1987, also establishes basic quality system standard for those who choose to pursue the award. Except for rare instances where a customer might require suppliers to pursue the award, it is entirely voluntary. Only a limited number of companies can win the award each year. Even though the award criteria have evolved, the basic quality categories of leadership, information and analysis, strategic quality planning, human resource utilization, quality assurance, quality results, and customer satisfaction have remained. The award criteria call for detailed descriptions of how a company achieves certain goals, rather than requirements for those goals. Nevertheless, many companies have used the criteria as implicit requirements as they pursue the award. Countless others have performed a self-assessment, using the award criteria as guidelines for quality improvement. Supplier quality per se is covered as a subcategory of quality results. As mentioned earlier, a company could choose to impose Baldrige criteria on its suppliers itself or require suppliers to pursue the award.

23.4.4 Industry-Specific Quality Systems

Ford, General Motors, and Chrysler have recently drafted a joint supplier requirements document based on ISO 9001, with additional requirements that are common to the three manufacturers and several that are unique to a single manufacturer. A number of other automotive assemblers in the United States have subscribed to this supplier standard. In a similar fashion, the Boeing *Advanced Quality System*[1] contains basic requirements for supplier quality systems, adapted from ISO 9002.

Although basic quality systems are an important ingredient in the worldwide movement toward quality improvement, one limitation is that they alone cannot assure that quality happens; when it does not occur, however, the incidence and nature of defects should be well recorded.

23.5 STATISTICALLY BASED IMPROVEMENT SYSTEMS

To meet the remainder of the quality system objectives outlined in Sec. 23.2.2, it is necessary to implement additional requirements. As mentioned earlier, these requirements will be described under the heading of *statistically based improvement systems.*

23.5.1 Development in Industry

Beginning in the 1920s with Walter Shewhart's work at Western Electric,[3] methods for assuring quality through statistics-based methods came into common use. Statistical process control and designed experimentation often provide an advantageous substitute for acceptance sampling, which is described, for example, in military standards 105, 414, and 1235.

From the 1950s through the present day, the major influence of these methods has been felt in Japan. It was largely through the efforts of such eminent American quality proponents as W. Edwards Deming and Joseph Juran that Japan embraced statistical process control as the foundation of its manufacturing quality. Many companies in the rest of the world are now seeing the powerful influence it can have on quality and reliability, when applied both in-house and within a supplier base.

There are benefits of statistical assurance of quality to both customers and suppliers. Customer benefits include higher quality at a lower price, because costly process problems in the supplier base tend to be corrected. (Using nonstatistical approaches, higher quality of the delivered product tends to be achieved by incorporating tighter specifications or more stringent inspections, both of which add to cost.) Benefits to the supplier include better information about what the customer views as important, resulting in less guesswork; better control of processes, resulting in more predictable outgoing product quality; less inspection, particularly of end items; fewer rejections, both in-process and postdelivery; and lower warranty costs.

23.5.2 Statistically Based Improvement System Example

Overview. In 1991, Boeing revised its quality system for suppliers, which before that time had been based solely on a MIL-Q-9858A-like standard. The new standard includes requirements for application of statistical methods to achieve manufacturing process control and capability. The new document, called D1-9000[1] (no relationship to ISO 9000), still describes requirements for a basic quality system. However, the bulk of the document consists of both statistical tools and a process flow within which to employ them.

The simplest description of the process is shown in Fig. 23.2. In each purchase order under the new standard, key characteristics must be identified. Key characteristics are the most important features of the product delivered by the supplier in terms of variation. If a feature's variation from the nominal specification has a significantly adverse effect on the fit, performance, or service life

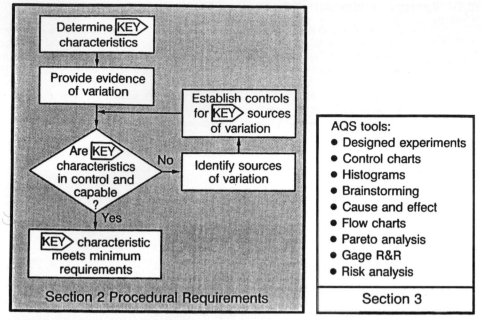

FIGURE 23.2   D1-9000 procedural flow.

of the product, then that feature should be identified as a key characteristic. A well-designed product should have few key characteristics—in other words, it should arise from a design that is robust against variation in manufacturing processes, environmental conditions, and usage.

Boeing identifies the key characteristics within many of its engineering specifications. If Boeing chooses not to do so, the supplier is obligated to identify the key characteristics, using the best information available concerning variation, downstream assembly problems, and rejections. For any key characteristics identified, the supplier must establish measurement and monitoring using appropriate statistical process control techniques.

Once statistical control has been achieved, the supplier must compute the capability of the processes creating the key characteristics, by comparing normal process output to the engineering specifications through a process capability ratio called C_{pk}.* If a process fails to be either in control or capable, D1-9000 requires the supplier to take corrective action, which continues until both control and capability are established.

Boeing prescribes a comprehensive process for corrective action, including the conduct of gauge-variation studies, investigating process sources of variation, and correcting them through the use of such tools as statistically designed experiments. The D1-9000 process flow is described in detail in Sec. 23.6.

Approximately 1000 Boeing suppliers have incorporated D1-9000 in their quality systems and are actively using it in their manufacturing processes. For some, D1-9000 constitutes their only quality system, and is used for all products for all customers. In addition, many Boeing suppliers impose virtually the same requirements to D1-9000 on *their* suppliers.

---

*For a detailed discussion on the definition and usage of C_p and C_{pk}, see Sec. 23.6.3, and V. E. Kane, "Process Capability Indices," *Journal of Quality Technology,* ASQC, 1986, Milwaukee, Wis., vol. 18, pp. 41–52.

23.6 BOEING ADVANCED QUALITY SYSTEM: DETAILS OF PROCESS FLOW

D1-9000, the *Advanced Quality System* used by Boeing suppliers (and many Boeing internal manufacturing organizations as well), is just one way of employing statistical techniques to improve quality. Nevertheless, it has been highly effective in organizing improvement efforts chronologically, with an emphasis on meeting customer and design requirements. The process flow is generally applicable, in a variety of product and service environments.

This section steps through the process flow diagram of Fig. 23.2 in detail, showing how a supplier can improve products and the processes that create them, in order to meet customer needs.

23.6.1 Determine Key Characteristics

The most important party to satisfy with regard to quality and reliability is the customer. If the customer provides any information about what is important, particularly with regard to variation control, that information should be recorded and incorporated into improvement strategies.

If the customer does not identify key characteristics, or if the supplier believes that there is an advantage in identifying more than those provided by the customer, the supplier should use a formal process to identify key characteristics. That process should begin with the compilation and analysis of data concerning the product, including its design, its use, and the processes and methods used to manufacture it. Any of these areas may provide clues as to how the product is affected by variation.

Any data used in key characteristic selection should be analyzed in a team environment, with all knowledgeable and affected parties represented (e.g., design, manufacturing, customer services, facilities, purchasing, tooling, quality assurance). Information on which features of the product are most affected by variation is extremely helpful. As mentioned earlier, variation in fit, performance, or service life is of particular interest, since controlling those kinds of variation is most closely allied with quality and reliability of product. The information may be charted, as in a Pareto diagram (see Fig. 23.3) showing which features of the same or similar products have caused problems in the past.

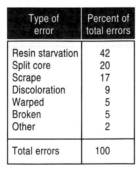

Reasons for
Rejected Crushed
Core Panels

Type of error	Percent of total errors
Resin starvation	42
Split core	20
Scrape	17
Discoloration	9
Warped	5
Broken	5
Other	2
Total errors	100

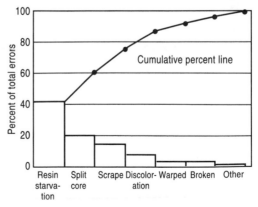

FIGURE 23.3 Example Pareto diagram.

Once data is available, it is important to analyze it dispassionately. One means of selecting key characteristics (called *special characteristics*) has been in use at Ford Motor Company for some time. A *risk-priority number,* based on scales of *severity, occurrence,* and *detection,* is derived using a form of failure modes and effects analysis (FMEA). Hewlett-Packard and Boeing, among others, use similar approaches for prioritizing characteristic selection.

23.6.2   Provide Evidence of Variation

Designation of a feature as a key characteristic means that the feature is to be measured, charted, and controlled. It also means that variation reduction of the key characteristic receives emphasis in the manufacturing environment. The most effective time to measure a key characteristic, and to address any excess variation it might display, is at the earliest possible point in the manufacturing process. Early detection of process problems facilitates early correction. Boeing encourages its suppliers to train operators on how to collect data and how to take corrective action on out-of-control or out-of-specification conditions.

Key characteristics are to be plotted on statistical process control charts. The most effective of such control charts plot *variable data,* measured on a numerical scale, rather than go/no go or pass/fail data, called *attribute data.* Many references provide guidance for the appropriate selection of control charts, but D1-9000 also contains a fairly thorough tutorial on their proper use.

It isn't enough simply to designate a measurement that must be taken on a key characteristic. Control charts are typically derived from measurements of samples, not of the entire production output. How often to sample and how many items to measure in each sample must be decided, usually on the basis of production throughput and historical levels of variation.

The ideal person to collect measurements and maintain control charts is the one who first completes the key characteristic. Charts should be available, however, to quality assurance, manufacturing management, and engineering, since they should follow through on improvement actions suggested by production conditions displayed on the charts.

23.6.3   Are Key Characteristics in Control and Capable?

Statistical control means that the process being measured is operating consistently within a band that is considered to be typical for that process, given historical data. A control chart consists of plotted data values, a *centerline* that represents the average of historical process output, and *control limits* that are usually placed three standard deviations above and below the center line (see Fig. 23.4). (Standard-deviation estimates are computed differently for various types of control charts; see a standard statistical reference for details.) As long as the process continues to operate within the control limits, with no discernible nonrandom patterns, it is considered to be in a state of statistical control, or *in control.* Another way of putting it is that a process in control is stable,

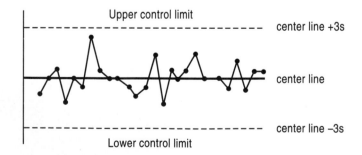

s = standard deviation

FIGURE 23.4   Typical control chart, where each point is taken from a sample of product output.

reliable, and predictable. If there are plotted points outside the control limits, the process is called *out of control,* and is considered unstable. Unstable processes are at risk of producing unsatisfactory product, and typically cost too much to operate.

Even a process that is in control must also meet engineering specifications a high percentage of the time in order to satisfy the customer. By "minimum capability," Boeing means that the measured process output compares favorably to specifications, as determined by the ratios C_p and C_{pk}. C_p is defined as follows:

$$C_p = \frac{USL - LSL}{6s}$$

where USL and LSL are the upper and lower specifications, respectively, and s is the estimate of the standard deviation of the entire process output (see Fig. 23.5). Since C_p simply compares specification width relative to process width, it is insensitive to process shifts away from the nominal specification.

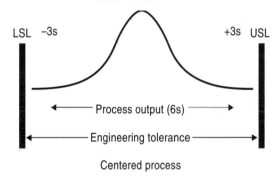

FIGURE 23.5   Process output compared to specifications (centered process).

A refinement of C_p, C_{pk} (discussed earlier in this section), does take process shifts into account. C_{pk} is defined as

$$C_{pk} = \text{minimum of} \left[\frac{USL - Avg}{3s} , \frac{Avg - LSL}{3s} \right]$$

where Avg is the average of process output (see Fig. 23.6). Boeing requires corrective action by the supplier whenever the capability ratio C_{pk} of a key characteristic is below 1.0. If the C_{pk} is 1.0 or greater with 90 percent confidence (confidence being attained by having a sufficiently large collection of measurements), the process is said to be *capable.* (Some other manufacturers using C_{pk} as a standard measure of process performance, notably U.S. and Japanese automobile manufacturers, require a higher value of C_{pk}.) At a C_{pk} of 1.0, the process fits neatly within the specification limits, except perhaps for a small amount in the tail regions of the product output distribution falling outside the specifications. If the process that produces the key characteristic is both in control and capable, supplier actions to improve the process are optional.

23.6.4   Identify Sources of Variation

When first starting to measure a process, it is rare to find it in a state of statistical control. At least some plotted points from the process typically fall outside the control limits. This process insta-

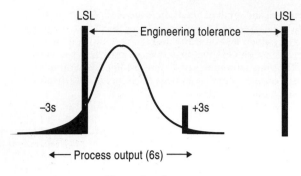

FIGURE 23.6   Process output compared to specifications (noncentered process).

bility usually has identifiable, or assignable, causes. The Boeing *Advanced Quality System* (AQS) process flow leads the supplier to investigate assignable causes for each out-of-control event.

23.6.5   Establish Controls for Key Sources of Variation

If special causes of variation can be assigned, the supplier must take corrective action to remove them. An example might be to change a cutter blade, add chemicals to a bath, replace a faulty autoclave thermostat, or use more reliable raw materials. The supplier's corrective action should ensure that the special cause of variation, once removed, will not recur.

Any time a process has been altered, it can be expected to result in somewhat different process output. The altered process should be measured for awhile to determine its distribution and state of statistical control. New control limits may have to be calculated.

It is important to establish that the measurement system itself has only a small amount of variation relative to the specifications. Measurement system variability can easily make an otherwise acceptable process appear to be delivering product characteristics that are out of control or not capable. It may be necessary to perform a *gauge-variation study,* which is typically conducted with a number of operators using measuring devices several times each on several items of product. The study is planned in such a way that measurement system variability due to the measuring devices, the ways in which operators use them, and the surrounding environment can be determined through statistical means. It is not unusual for a very high proportion of the engineering tolerance to be consumed by variability in the measuring process. Many world-class suppliers perform gauge studies prior to taking any measurements for statistical process control.

It is usually possible to reduce measurement variability substantially through the simple steps of standardizing the measurement process for all operators; keeping the part, gauges or tooling stationary; or calibrating the measurement devices more frequently. Occasionally it may be necessary to obtain more capable gauges.

If gauge variation is acceptable and the process still fails to be in control or capable, it is likely that something in the process producing the key characteristic is causing excessive variation. It is important to ascertain what the sources might be. From a list of potential sources, a team of knowledgeable individuals should use their expertise, available data, or a statistically designed experiment (DOE) to determine which process conditions are most likely to be sources of product variation, and to what degree. With a DOE, it should be possible also to establish process parameter settings that provide near-nominal levels of the key characteristic, with robustness against variation.

A *key process parameter* is one that has a significant effect on the level or the variability of a product key characteristic. Once it has been determined which process parameters are key, and what the nominal settings of those parameters need to be for the quality of the key characteristics, controls must be placed on the manufacturing process to hold those parameter settings. Often these controls are established in the form of operator instructions or automated control settings.

It is important that process improvement information be kept up to date. Whether the information is stored in a computer-accessible file or on paper, it should be readily available to those in charge of operating or improving the process.

23.6.6 Key Characteristic Meets Minimum Requirements

Once this stage is reached, there is no further obligation for the supplier to improve the processes that create the key characteristics. Boeing, however, as do other major manufacturers, sets the expectation that continuous improvement of product quality and reliability will be a way of life for its suppliers. Most suppliers who reach this stage realize the benefit to themselves, in productivity and customer satisfaction, of continuing to follow the improvement cycle embedded in the AQS process flow.

23.7 CUSTOMER INVOLVEMENT IN SUPPLIER QUALITY AND RELIABILITY SYSTEM

23.7.1 Specifications

Most of a customer's expectations and requirements for quality and reliability should ideally be expressed through contracts, with an emphasis on clarity. Whether the customer imposes one of the ISO 9000 series of standards, MIL-Q-9858A, the Boeing *Advanced Quality System,* the Chrysler-Ford-General Motors quality system standard, or any of a number of other quality operating systems on its suppliers, there is an obligation for the customer to describe requirements clearly and to participate appropriately in the realization of those requirements. Obviously, the requirements should cover both basic quality-system elements and those of a statistically based improvement system described in earlier sections.

General terms and conditions, purchase-order notes, and product and process specifications are often declared contractually. If the customer has a formal quality standard, it is usually referenced in these documents as part of a supplier's requirements. In addition, quality feedback mechanisms, such as the supplier's quality and reliability history and supplier quality rating, are important for the customer to establish. There should be a close relationship between the customer's procurement and quality-assurance departments, so that demonstrated supplier quality and life-cycle cost help form the basis for purchasing decisions. Achieving this goal may require fairly extensive education of the customer's procurement organization on the importance of the quality standard. It is important that the concepts of high quality and reliability, value, and life-cycle cost replace the criterion of low initial price in procurement.

23.7.2 Improvement Process

In addition to contractual terms, the customer should in every way communicate the importance of improving manufacturing systems so that they deliver a product that is close to the nominal specification, that the processes be maintained in a state of statistical control and capability, that action will be expected to reduce excessive variation, and that there be less outgoing and less receiving inspection, *all with no price increase.*

In choosing among suppliers, procurement should look for top-management commitment to quality; a comprehensive implementation plan, including identified production areas and a schedule for implementing steps to achieve improved quality and reliability; a designated manager responsible for implementation; effective training, and a means of assessing its effectiveness; documentation of the supplier's quality system; the existence of an accessible quality database;

flowdown of requirements to both internal organizations and to subcontractors; corrective-action processes; and documentation of quality status for review by both supplier management and customers. The expectations for these elements of a quality operating system should be clearly communicated to suppliers, and assessed for completeness and for the supplier's perceived ability to follow their own system.

23.7.3   Design for Quality and Reliability

As much as possible, the customer should drive quality considerations upstream into their own engineering design processes. It is too much to ask a supplier to make up in manufacturing excellence for faulty design. Quality must be driven into the design of both the product and the manufacturing processes that will produce it. Attention must be paid to robustness, so that variation in materials, environment, manufacturing conditions, or usage have little effect on quality. Engineering should also ensure that design specifications are consistent with real requirements for fit, performance, and service life of the product. Since suppliers generally cannot measure service life as part of their own in-process control, it may be important for the customer to help flow service life (reliability) requirements down to components for which the supplier can exercise control of quality.

23.7.4   Supplier Audit

A periodic audit of the supplier's quality system implementation is often essential to sustain improvement efforts. Reviewing C_{pk} information during these audits can help the customer understand the supplier's manufacturing capabilities. The customer should insist on corrective action to address any faulty or tardy implementation efforts.

23.7.5   Feedback

Timely feedback on quality problems associated with use of the supplier's product is also beneficial to the supplier. It is surprising how little information currently gets back to suppliers from either intermediate or end users. Particularly in areas of performance and service life (or reliability), the customer is usually in a much better position to collect information about the end user's problems with the product than the supplier is. This information should be regularly communicated to suppliers. World-class suppliers view such information as a gold mine for improvement opportunities.

23.7.6   Support by Customer

Some customers go quite a bit farther to ensure supplier quality. Free or low-cost training and training materials, seminars, workshops, videotapes, statistical software, quality handbooks, and the like are provided by many manufacturers to their suppliers. Supplier audits often include advice and counsel on how to improve manufacturing systems. Customer statisticians may travel to the supplier's site and provide consultation on such applications as data collection and analysis, control charts, sampling, designed experiments, and use of statistical software.

Suppliers are more likely to succeed in their quality improvement efforts if they are following a methodology the customer has proved in practice. Examples of applications that incorporate the use of the supplier's product are extremely valuable in showing what is important to the customer and how the supplier might improve it.

23.8   ASSURING RELIABILITY AT THE SUPPLIER

One key-characteristic category is that of service life. Service life is most highly correlated with a customer's typical needs regarding reliability. Indeed, the classic definition of reliability is as follows: "Reliability is the probability of a product performing without failure a specified function under given conditions for a specified period of time" (Juran,[11] p. 2–7).

Some common products for which reliability is often included in the specifications are in electronic, avionic, computing, hydraulic, and mechanical systems. Reliability requirements are worthy of separate treatment in such systems because they can seldom be exhaustively tested either in the laboratory or in the service environment by the supplier. Instead, suppliers and their customers usually depend on other means for assuring reliability.

23.8.1 Customer Requirements for Reliability

A common way of describing reliability requirements is through a reliability figure of merit, such as mean time between failures (MTBF). (Other common figures of merit include failure rate, mean time to failure, mean time to first failure, mean time between maintenance, and availability.) The customer will typically specify the desired figure of merit contractually in engineering documents, and will base it on system criticality or safety; reliability of available components; target cost; and repair, replacement, servicing, and maintenance costs. A figure of merit such as MTBF may be specified with maintenance assumed, without maintenance, or both.

Other customer specifications may include mandatory materials, use of specific components, required similarity to existing designs, or adherence to MIL-STD specs or equivalent commercial specifications. The customer may also require that testing of the end item or components shall consist of a certain number of cycles under certain test conditions (e.g., temperature, pressure, time, humidity, corrosiveness, and fatigue).

23.8.2 Supplier Design Responsibilities

Once agreed on, the customer's reliability requirements become the supplier's job to meet, through manufacturing, through verification tests, and possibly through additional design. An important responsibility of the supplier can be described as managing the relationship between component reliabilities and the desired reliability of the end product. The design document itself could be the result of collaboration between the customer and the supplier, in which case the supplier has already begun that task.

Important tools for the supplier at the early stages of reliability assurance are the design tree, fault trees, and failure modes and effects analysis (FMEA). The design tree shows how the functionality of the end item is assured and how detail components build up to the end product. The supplier may depend on existing components and subassemblies with known reliability characteristics for some or all of the design tree. Fault trees and FMEA are tools that describe how failures in components result in system failure, with associated probabilities. The goal is to determine those failure modes that are most likely to occur, so that they can be precluded by the use of higher-quality components or redundant subsystems. The customer is interested, at this stage, in evidence in the design that product reliability will be achieved. If the supplier is responsible for any part of design, then the parties should be meeting frequently to conduct design reviews, at which the meeting of reliability requirements is a prominent subject.

23.8.3 Reliability through Manufacturing Design

Employing a thorough knowledge of the product design, the supplier is responsible for design of the manufacturing process. The customer should expect evidence in the manufacturing process design that product reliability will be addressed, through use of appropriate and well-sequenced processing steps. The customer may review and/or approve the supplier's manufacturing plans from a reliability standpoint.

23.8.4 Test Plans and Failure Analysis

A prime ingredient of the supplier's planning must include test plans. As already mentioned, the supplier is in a poor position to prove the product via laboratory tests or in-service usage (besides which, the latter would be too late). Test plans usually include accelerated or stress testing of the product to gain reliability information prior to full-scale production. The supplier is responsible for developing the test plan in such a way that it can demonstrate that the design objectives will be achieved. As mentioned, the test plan should include desired numbers of tests under various environmental and stress conditions that at least cover the design envelope. The test plan may be submitted to the customer's engineering for approval.

Along with the test plan, the supplier should respond to any requirements for failure analysis, imposed either by the customer or by internal procedures. During testing, premature product, subsystem, or component failures are likely to occur. The steps for effective analysis for root causes and for corrective action should be spelled out completely in operating procedures approved by the customer.

23.8.5 Prototypes and First Articles

Prior to production, a series of prototypes may be developed. These prototypes should help verify design, manufacturing processes, failure analysis procedures, and test plans, and should possibly provide some indications of component and subsystem reliability and/or failure modes. The customer may require submission of prototypes for installation or other tests. At this stage, the customer is looking for physical evidence that design objectives will be achieved. For a relatively new development, the customer may submit a series of change requests following prototype delivery, resulting in redesign (e.g., components with higher reliability, or more system redundancy) or changes in the manufacturing process or test procedures. The supplier responds to these change requests, which converge to the final product design.

The customer may require that certain tests be conducted on the first production article(s), in order to verify that the final product design, manufacture, and testing are consistent with reliability specifications. Beyond the first articles, there are usually warranties for product reliability, agreed to with either the immediate customer or the end user. In-service failures often are analyzed exhaustively by the supplier to determine root causes, with the resulting information used to feed product improvement efforts.

23.8.6 Statistically Based Reliability Improvement

The tools described in Secs. 23.5 and 23.6 can be applied in a straightforward fashion to reliability requirements. If MTBF or a similar requirement is specified by the customer, it should be flowed down as a key characteristic to influential subassemblies and components that are important for system service life. The design tree, fault trees, and FMEA can supply useful information for this flowdown. Typically, component materials; dimensional characteristics; or chemical, electronic, or physical properties can be identified and designated as lower-level key characteristics. In addition, process parameters, such as those that might apply in heat treating, machining, or soldering, can be identified that have a significant influence on product reliability. These parameters can be designated as key process parameters. Sometimes key characteristics and key process parameters will have to be flowed down to those subcontractors who control them.

Once key characteristics and key process parameters have been identified, improvement steps such as those described in Sec. 23.6 can be applied, using statistical process control, capability analysis, measurement system analysis, designed experiments and corrective action to achieve stable and capable processes from a reliability standpoint.

For example, suppose the customer has specified a reliability requirement for an electronic lighting ballast in terms of an MTBF of at least 300,000 h. MTBF was chosen because of an inordinate number of in-service replacements of similar items. This service life key characteristic is not directly measurable on the end item prior to delivery. However, the requirement can be flowed down to inspectable and measurable components. In looking at the design tree and analyzing failure modes and probabilities, the supplier determines that the MTBF of the ballast is dominated by the MTBF of the printed circuit board (PCB). Service life of the PCB can be flowed down to com-

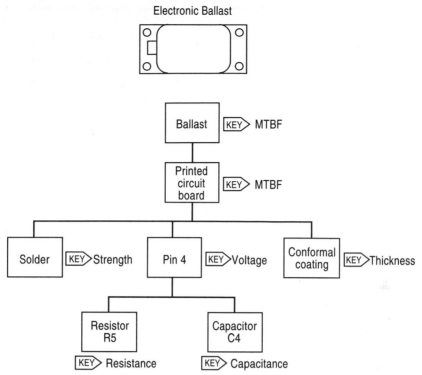

FIGURE 23.7 Flowdown of reliability (service life) key characteristics in an electronic lighting ballast.

ponent key characteristics, as shown in Fig. 23.7. The key characteristics of solder strength, pin 4 voltage, conformal coating thickness, resistance at resistor R5, and capacitance at capacitor C4 are measured and charted using SPC. Once the processes are determined to be in control, C_{pk} is computed to assess process capability. Other steps, such as those in the AQS process flow, are followed and improvements implemented until design reliability is reasonably assured. The effectiveness of the control of variation in the components is verified through failure analysis of test articles and, eventually, in-service systems.

REFERENCES

1. D1-9000, *Advanced Quality System,* The Boeing Company, 1991.

2. MIL-Q-9858A, *Quality Program Requirements,* U.S. Department of Defense, 1963.

3. Shewhart, Walter, *Economic Control of Quality of Manufactured Product,* 1931 (reprinted by American Society for Quality Control, 1980).

4. ISO 9000-1:1994, *Quality Management and Quality Assurance Standards,* Part 1, *Guidelines for Selection and Use,* International Organization for Standardization, 1994.

5. ISO 9000-2:1994, *Quality Management and Quality Assurance Standards,* Part 2, *Generic Guidelines for the Application of ISO 9001, ISO 9002, and ISO 9003,* International Organization for Standardization, 1994.

6. ISO 9001:1994, *Quality Systems—Model for Quality Assurance in Design, Development, Production, Installation and Servicing,* International Organization for Standardization, 1994.

7. ISO 9002:1994, *Quality Systems—Model for Quality Assurance in Production, Installation and Servicing,* International Organization for Standardization, 1994.

8. ISO 9003:1994, *Quality Systems—Model for Quality Assurance in Final Inspection and Test,* International Organization for Standardization, 1994.

9. ISO 9004:1987, *Quality Management and Quality System Elements—Guidelines,* International Organization for Standardization, 1987.

10. Bossert, James L., ed., *Procurement Quality Control,* 4th ed., ASQC Quality Press, 1988, Milwaukee, Wis.

11. Juran, J. M., ed.-in-chief, *Quality Control Handbook,* 3d ed., McGraw-Hill, New York, 1979.

CHAPTER 24
TECHNIQUES OF ESTIMATING RELIABILITY AT DESIGN STAGE

Kailash C. Kapur*

24.1   INTRODUCTION

Reliability is basically a design parameter and must be incorporated into the system at the design stage. Reliability is an inherent attribute of a system resulting from design, just as is the system's capacity, performance, or power rating. The reliability level must be established at the design phase, and subsequent testing and production will not raise the reliability without a basic design change. With increasing system complexity, reliability becomes an elusive and difficult design parameter to define and achieve. It also becomes more difficult to control, demonstrate, and ensure as an operational characteristic under the projected conditions of use by the customer. However, past history has demonstrated that where reliability was recognized as a necessary program development component, with the practice of various reliability engineering methods throughout the evolutionary life cycle of the system, reliability can be quantified during the specification of design requirements, can be predicted by testing, can be controlled during production, and can be sustained in the field.

The term *system effectiveness* is often used to describe the overall capability of a system to accomplish its mission or perform its intended function. System effectiveness is defined as the probability that the system can successfully meet an operational demand within a given time when operating under specified conditions.[1] For consumer products, system effectiveness is related to customer satisfaction. Our objective is to design and manufacture products that will meet the needs and expectations of the customer. Effectiveness is influenced by the way the system is designed, manufactured, used, and maintained, and thus is a function of all the life-cycle activities as well as system attributes such as design adequacy, performance measures, safety, reliability, quality, producibility, maintainability, and availability (see Fig. 24.1). Reliability is one of the major attributes determining system effectiveness.

The purpose of this chapter is to present some of the reliability methodologies and philosophies that are applicable during the design stage in the life cycle of a system. Figure 24.2 represents the complexity of balancing different design requirements[2] and tradeoffs between reliability, maintainability, cost, etc.

---

*Adapted from Ireson and Coombs, *Handbook of Reliability Engineering and Management*, McGraw-Hill, New York, 1988, Chap. 18.

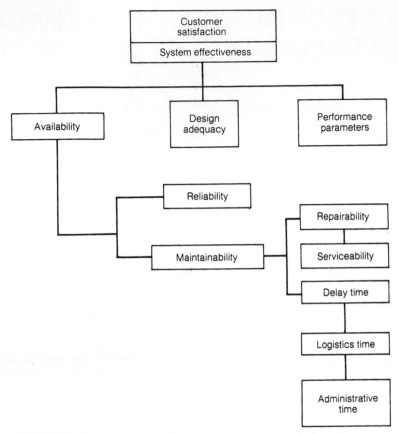

FIGURE 24.1 Concept of system effectiveness and customer satisfaction.

24.2 *SOURCES OF FAILURE DATA*

This section briefly explains some of the sources of failure data which designers can use for reliability analysis. Relatively few sources for failure-rate data are available. There are some data on electronic components, particularly in military applications. However, on mechanical components practically no good data are commercially available. The use of any existing database to obtain failure-rate data on any particular design should be done with great care. The applicability of any past history on failure rate to a current design depends on the degree of similarities in the design, the environment, and the definition of failure. With these words of caution, three sources of failure-rate information are covered in this section.

MIL-HDBK-217. This handbook is concerned with reliability prediction for electronic systems.[3] It contains two methods of reliability prediction, the *part-stress analysis* and the *parts-count methods*. These methods vary in complexity and in the degree of information needed to apply them. The part-stress analysis method requires the greatest amount of information and is applicable during the later design stages when actual hardware and circuits are being designed. To apply this method, a detailed parts list including part stresses must be available.

The parts-count method requires less information and is relatively easy to apply. The information needed to apply this method is (1) generic part type (resistor, capacitor, etc.) and quanti-

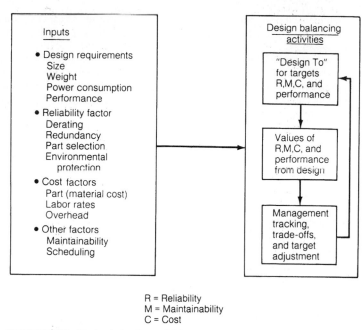

R = Reliability
M = Maintainability
C = Cost

FIGURE 24.2   Design balancing activities.

ty, (2) part quality levels, and (3) equipment environment. Tables are provided in the handbook to determine the factors in a failure-rate model that can be used to predict the overall failure rate. The parts-count method is obviously easier to apply and, one would assume, less accurate than the stress analysis method.

Nonelectronic Parts Reliability Data.   The *Nonelectronic Parts Reliability Data Handbook*[4] (1981) was prepared under the supervision of the Reliability Analysis Center at Griffiss AFB. This handbook is intended to complement MIL-HDBK-217 in that it contains information on some mechanical components. Specifically, the handbook provides failure-rate and failure-mode information for mechanical, electrical, pneumatic, hydraulic, and rotating parts. Again, little is known about the environment, design specifics, etc., that produced the failure-rate data in this handbook.

GIDEP.   The Government-Industry Data Exchange Program (GIDEP)[5] (1976) is a cooperative venture between government and industry that provides a means to exchange certain types of technical data. Participants in GIDEP are provided with access to various databases. The failure experience database contains failure information generated when significant problems are identified on systems in the field. These data are reported to a central database by the participants. The reliability-maintainability database contains failure-rate and failure-mode data on parts, components, and systems based on field operations. The data are reported by the participants. Here again, the problem is one of identifying the exact operating conditions that caused failure.

24.3   *DESIGN REVIEW*

The design review, a formal and documented review of a system design, is conducted by a committee of senior company personnel who are experienced in various pertinent aspects of product design, reliability, manufacturing, materials, stress analysis, human factors, safety, logistics, maintenance, etc. The design review extends over all phases of product development from conception to production. In each phase previous work is updated, and the review is based on current information.

A mature design requires tradeoffs between many conflicting factors such as performance, manufacturability, reliability, and maintainability. These tradeoffs depend heavily on experienced judgment and require continuous communication between experienced reviewers. The design review committee approach has been found to be extremely beneficial to this process. The committee adopts the system's point of view and considers all conceivable phases of design and system use to ensure that the best tradeoffs have been made for the particular situation (see Fig. 24.2). A complete design review procedure must be multiphased in order to follow the design cycle until the system is released for production. The product development cycle is presented in Table 24.1. A typical review committee is shown in Table 24.2, where the review phases are keyed to the product development cycle given in Table 24.1. Here the review process has been subdivided into six phases, and each phase is an update or more detailed analysis based on the latest knowledge.

TABLE 24.1 Product Design Cycle

   I. Marketing research
      Needs analysis
      Forecast sales
      Set broad performance objectives
      Establish program cost objectives
      Establish technical feasibility
      Establish manufacturing capacity

  II. Concepts
      Formulate project teams
      Refine project—broadly outline product
      Develop rough ideas of product
      Develop and consider alternatives

 III. Design
      A. Preliminary design
         Design calculations
         Rough drawings and sketches
         Pursue different alternatives
         Obtain manufacturing engineering cost estimates of design approaches
      B. Detailed design
         Design calculations
         Stress analysis
         Vibration considerations
         Complete and detailed design package (drawings)

 IV. Manufacturing engineering
      A. Process planning
      B. Quality system planning

  V. Prototype program
      A. Build components and prototypes
      B. Write test plans
         1. Component and subsystem tests
         2. System tests

 VI. Finalized design
      A. Design changes due to manufacturing engineering input and test input
      B. Freeze design
      C. Release to manufacturing

TABLE 24.2 Design Review Committee

Member	I	II	III	IV	V	VI	Responsibility
Chairperson (manager of product design function)	X	X	X	X	X	X	Ensure that review is conducted in an efficient fashion; issue major reports and monitor follow-up.
Design engineer (of this product)		X	X	X	X	X	Prepare and present design approaches with calculations and supporting data
Design engineer (not of this product)		X	X	X	X	X	Review and verify adequacy of design (may require more than one specialist)
Customer and/or marketing representative	X	X	X	X	X	X	Ensure that customer's viewpoint is adequately presented (especially at the design-tradeoff stage)
Reliability engineer	X	X	X	X	X	X	Evaluate design for reliability consistent with system goals
Manufacturing engineer			X	X	X	X	Ensure manufacturability at reasonable cost; check for tooling adequacy and assembly problems
Materials engineer			X				Ensure optimum material usage considering application and environment
Stress analysis			X				Review and verify stress calculations
Quality-control engineer			X	X	X	X	Review tolerancing problems, manufacturing capability, inspection, and testing problems
Human-factors engineer			X	X			Ensure adequate consideration to human operator, identification of potential human-induced problems, and person-machine interface
Safety engineer			X				Ensure safety to operating and auxiliary personnel
Maintainability engineer		X	X				Analyze for ease of maintenance, repair, and handling of field service problems
Test engineer			X		X	X	Present test procedures and results
Logistics			X				Evaluate and specify logistic support; identify logistics problems

Ultimately, the design engineer has the responsibility of investigating and incorporating the ideas and suggestions posed by the design review committee. The design review committee chairperson is responsible for adequately reporting all suggestions by way of a formal and documented summary. The design engineer then can accept or reject various points in the summary, but must formally report back to the committee stating reasons for any actions.

The basic philosophy of a design review is being presented here, and it should be recognized that considerably more thought and detail must go into developing the management structure and procedures for conduct in order to have a successful review procedure. It should be noted that this review procedure considers not only reliability but also all important factors in order to ensure that a mature design will result from the design effort.

A design review technique has proved effective in identifying failure situations early in the design cycle and before product testing. With proper organization and appropriate management support, the design review procedure is an effective means for promoting early product maturity. Further discussion on design review can be found in Refs. 6, 7, and 8. A brief discussion of some of the components for design review are given below.

24.3.1 Policy

An endeavor should be made to conduct design reviews on all new products and on major revisions of existing products which seriously affect cost, interchange ability, function, performance, or appearance.

24.3.2 Appointment of Chairperson

The task of the chairperson requires a high level of tact, broad understanding of the design requirements, and technical knowledge of the various disciplines involved. Each review committee will have its designated chair, and the person selected should neither be the designer of the product nor be associated with the design.

24.3.3 Participants

The various functional activities represented in a design review vary with the type of review. Generally, the participants will include personnel from engineering, manufacturing, tooling, marketing, purchasing, reliability, quality control, and cost reduction. In addition, it may be desirable to invite specialists from other activities to participate. A design review committee should consist of about a dozen people.

24.3.4 Agenda

The chairperson will have the agenda prepared for the meeting and will distribute it with the advanced information.

24.3.5 Checklists

Checklists will be developed to remind participants of all items which should be considered in each review. Such a list also aids designers by reminding them of the things that should not be overlooked. Some of the points to consider during design review are given below.

- Quality of products is defined by customers.
- Review adequately the needs and expectation of the customer.
- Review customer performance and environmental requirements.
- Confirmation on use of approved parts in an approved manner.
- For high reliability, make conservative choices of materials and components.
- Simplify design wherever possible.
- Wherever possible avoid new state-of-the-art use of materials and manufacturing processes.
- Wherever possible, use previously proven designs and components.
- Look for misapplication of components and materials.
- Check for the influence of environmental extremes on the system.

- Made provision for vibration and shock.
- Make provision for heat transfer.
- Search for potential sources of fluid leaks.
- Check all subsystem interfaces for failure problems.
- Minimize the requirement for assembly adjustments and selective fit requirements in production.
- Make provisions for testability and inspectability where required.
- Determine serviceability index, particularly for components with highest failure rates. Good maintainability may compensate for poor reliability.
- Analyze potential failure modes and their effect.
- Perform a worst-case analysis.
- Make sure that review checklists are based on product experience in design, production, and field use.

24.3.6 Conduct of Design Review Meetings

Introductory comments by the chairperson should set a constructive tone and climate for the meeting. The specific objectives of the design review should be stated and should relate to the overall objectives, namely, achieving optimum product design from the standpoint of reliability, function, cost, appearance, and other requirements of the customer.

An attendee will be appointed secretary and will take notes on useful ideas submitted and other pertinent comments. The person will also record when additional action is required and by whom. It may be a good idea to tape the entire design review committee meetings to make sure that the secretary has correct information for inclusion in the record. It may be useful to save the tapes and use them to resolve arguments in the future.

The design engineer or product manager should describe adequately the product being reviewed and include a comparison of customer requirements versus the expected performance of the product.

The chairperson should make sure that the discussion follows a systematic plan so that no major subject areas are omitted. The discussion should follow the prepared agenda. Checklists should be used to prevent omission of important design considerations.

24.3.7 Timing and Duration of Reviews

It is usually desirable to have design reviews at more than one point in the design and development cycle of a new product; e.g., a design review may be held when

1. Marketing or customer requirements are completed
2. Specifications and drawings are completed
3. After testing the prototypes and prior to manufacturing release
4. After production assessment tests

In addition, tool drawings should be similarly reviewed prior to their manufacture.

24.3.8 Scheduling of Reviews

The reliability manager and engineering manager will schedule all design reviews with the appropriate attendees and advise them of time, place, and subject. Specific time for the design reviews will be incorporated into the product design and development schedule and approved so that all participating functions can plan their efforts efficiently.

24.3.9 Advance Information

At least 10 days before the actual date of the review, information should be distributed or made available to the designated participants. This may be in the form of specification, a competitive cost analysis, preliminary layouts, etc. Distributing this information beforehand will help assure that the participants are well prepared to contribute constructively to the objectives of the review.

24.3.10 Follow-up

Documentation: The secretary is responsible for preparing and distributing minutes of the meeting, indicating the ideas generated and the action to be taken. The minutes should also note by whom the action is to be taken and when.

Utilization of ideas: The designer is responsible for investigating and incorporating into the design those ideas which will aid in achieving optimum product design. The design review chairperson will have the responsibility of following up on the utilization of the ideas proposed and of the assigned action items.

Final report: A final report should be issued covering the investigation of suggestions made and the reasons for their adoption or rejection.

24.3.11 Reliability Design Guidelines

There are some basic principles of reliability in design that are useful for the designer.[9] Each concept is briefly discussed in terms of its role in the design of reliable systems.

Simplicity.   Simplification of system configuration contributes to reliability improvement by reducing the number of failure modes. A common approach is called *component integration,* which is the use of a single part to perform multiple functions.

Use of Proven Components and Preferred Designs

1. If working within time and cost constraints, use proven components because this minimizes analysis and testing to verify reliability.
2. Mechanical and fluid system design concepts can be categorized and proven configurations given first preference.

Stress and Strength Design.   Use various sources of data on strength of materials and strength degradation with time as it relates to fatigue. The traditional and common use of safety factors does not address reliability, and new techniques like the probabilistic design approach should be used. The probabilistic design approach is explained in a following section.

Redundancy.   Redundancy sometimes may be the only cost-effective way to design a reliable system from less reliable components.

Local Environmental Control.   Severe local environment sometimes prevents achievement of required component reliability. The environment should be modified to achieve high reliability. Some typical environmental problems are as follows:

1. Shock and vibration
2. Heat
3. Corrosion

Identification and Elimination of Critical Failure Modes.   This is accomplished through failure mode, effect, and criticality analysis (FMECA) and also by fault-tree analysis. These are discussed later in this chapter.

Self-Healing.   A design approach which has possibilities for future development is the use of self-healing devices. Automatic sensing and switching devices represent a form of self-healing.

Detection of Impending Failures.   Achieved reliability in the field can be improved by the introduction of methods and/or devices for detecting impending failures. Some of the examples are

1. Screening of parts and components
2. Periodic maintenance schedules
3. Monitoring of operations

Preventive Maintenance.   Preventive-maintenance procedures can enhance the achieved reliability, but the procedures are sometimes difficult to implement. Hence, effective preventive maintenance procedures must be considered at the design stage.

Tolerance Evaluation.   In a complex system, it is necessary to consider the expected range of manufacturing process tolerances, operational environment, and all stresses, as well as the effect of time. Two tolerance evaluation methods, worst-case tolerance analysis and statistical tolerance analysis, are discussed later in this chapter.

Human Engineering.   Human activities and limitations can be very important to system reliability. The design engineer must consider factors which directly refer to human aspects, such as

1. Human factors
2. Person-machine interface
3. Evaluation of the person in the system
4. Human reliability

24.4 PROBABILISTIC APPROACH TO DESIGN

Reliability is basically a design parameter and must be incorporated into the system at the design stage. One way to quantify reliability during design and to design for reliability is the probabilistic approach to design.[10–13] The design variables and parameters are random variables and, hence, the design methodology must consider them as random variables. The reliability of any system is a function of the reliabilities of its components. In order to analyze the reliability of the system, we have to first understand how to compute the reliabilities of the components. The basic idea in reliability analysis from the probabilistic design methodology viewpoint is that a given component has certain strength which, if exceeded, will result in the failure of the component. The factors which determine the strength of the component are random variables as are the factors which determine the stresses or loads acting on the component. *Stress* is used to indicate any agency that tends to induce failure, while *strength* indicates any agency resisting failure. *Failure* itself is taken to mean failure to function as intended. It is defined to have occurred when the actual stress exceeds the actual strength for the first time.

Let $f(x)$ and $g(y)$ be the probability-density functions for the stress random variable X and the strength random variable Y, respectively, for a certain mode of failure. Also, let $F(x)$ and $G(y)$ be the cumulative distribution functions for the random variables X and Y, respectively. Then the reliability R of the component for the failure mode under consideration with the assumption that the stress and the strength are independent random variables is given by (see Fig. 24.3)[12–14]

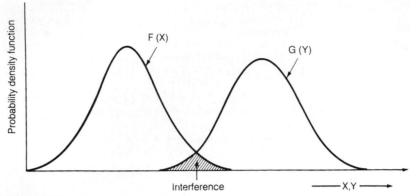

FIGURE 24.3   Stress-strength interference.

$$R = P\{Y > X\}$$

$$= \int_{-\infty}^{\infty} g(y) \left\{ \int_{-\infty}^{y} f(x)\,dx \right\} dy$$

$$= \int_{-\infty}^{\infty} g(y) F(y)\,dy$$

$$= \int_{-\infty}^{\infty} f(x) \left\{ \int_{x}^{\infty} g(y)\,dy \right\} dx$$

$$= \int_{-\infty}^{\infty} f(x)\{1 - G(x)\}\,dx \qquad (24.1)$$

For example, suppose the stress random variable X is normally distributed with a mean value of μ_X and standard deviation of σ_X and the strength random variable is also normally distributed with parameters μ_Y and σ_Y, then the reliability R is given by

$$R = \Phi\left[\frac{\mu_Y - \mu_X}{\sqrt{\sigma_Y^2 + \sigma_X^2}} \right] \qquad (24.2)$$

where $\Phi[.]$ is the cumulative distribution function for the standard normal variable.

EXAMPLE 24.1

$$\mu_Y = 40{,}000 \qquad \sigma_Y = 4{,}000$$
$$\mu_X = 30{,}000 \qquad \sigma_X = 3{,}000$$

Then, factor of safety = 40,000/30,000 = 1.33, and

$$R = \Phi\left[\frac{40{,}000 - 30{,}000}{\sqrt{(4{,}000)^2 + (3{,}000)^2}} \right] = \Phi[2]$$
$$= 0.97725$$

If we change μ_X to 20,000, increasing the factor of safety to 2, we have

$$R = \Phi\left[\frac{40,000 - 20,000}{\sqrt{(4,000)^2 + (3,000)^2}}\right] = \Phi[4]$$

$$= 0.99997$$

We now develop the relationship between reliability, factor of safety and variability of the stress, and the strength random variables. Let

$$V_Y = \text{coefficient of variation for the strength random variable } Y$$

$$= \frac{\sigma_Y}{\mu_Y}$$

$$V_X = \text{coefficient of variation for the stress random variable } X$$

$$= \frac{\sigma_X}{\mu_X}$$

$$n = \text{factor of safety}$$

$$= \frac{\mu_Y}{\mu_X}$$

Substituting these values in Eq. (24.2) for reliability R, when X and Y are normally distributed, we have

$$R = \Phi\left[\frac{n - 1}{\sqrt{V_Y^2 n^2 + V_X^2}}\right] \tag{24.3}$$

Thus, the above relation can be used to relate reliability, factor of safety, coefficient of variation for stress, and strength random variable. For example, let $n = 2.0$, $V_Y = 0.25$, $V_X = 0.15$. Then

$$R = \Phi\left[\frac{2 - 1}{\sqrt{(0.25)^2(2)^2 + (0.15)^2}}\right]$$

$$= \Phi[1.91] = 0.972$$

There are four basic ways in which the designer can increase reliability:

1. *Increase mean strength:*   Can be achieved by increasing size, weight, using stronger material, etc.
2. *Decrease average stress:*   Controlling loads, using higher dimensions.
3. *Decrease stress variations:*   This variation is harder to control, but can be effectively truncated by putting limitations on use conditions.
4. *Decrease strength variation:*   The inherent part-to-part variation can be reduced by improving the basic process, controlling the process, and utilizing tests to eliminate the less desirable parts.

The probabilistic design methodology is illustrated in Fig. 24.4.

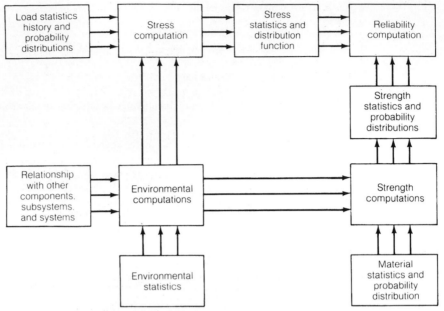

FIGURE 24.4 Probabilistic design methodology.

If we consider the total *design reliability program,* the steps can be summarized as follows:

1. Define the design problem.
2. Identify the design variables and parameters involved.
3. Conduct a failure modes, effects, and criticality analysis.
4. Verify the significant design parameter selection.
5. Formulate the relationship between the critical parameters and the failure-governing criteria involved.
6. Determine the failure-governing stress function.
7. Determine the failure-governing stress distribution.
8. Determine the failure-governing strength function.
9. Determine the failure-governing strength distribution.
10. Calculate the reliability associated with these failure-governing distributions for each critical failure mode.
11. Iterate the design to obtain the design reliability goal.
12. Optimize the design in terms of performance, cost, weight, etc.
13. Repeat optimization for each critical component.
14. Calculate system reliability.
15. Iterate to optimize system reliability.

24.5   *STATISTICAL VARIATION AND TOLERANCE ANALYSIS*

The reliability of an engineering design is a function of several design parameters and random variables. The design performance can be expressed as a function of these design variables and parameters. In this section we discuss how to combine several random variables. For example, a voltage distribution can be derived from the resistance and the current probability distributions. In a mechanical system, the stress distribution can be derived from the force and cross-sectional-area distributions. We have to develop mathematical variability models for the behavior of the system. The model must be accurate enough to simulate the behavior of the system over the range of operation. We discuss some of the techniques for variation analysis.

Let us assume that the performance of the system Y is a function of n random variables and/or parameters $X_1, X_2,..., X_n$, i.e.,

$$Y = f(X_1, X_2,..., X_n) \tag{24.4}$$

We wish to determine the properties of the random variable Y. If we know the probability-density functions of the random variables $X_1,..., X_n$, we may be able to find the probability-density function of random variable Y, but this may be difficult. In many design situations, only the first few moments of the random variables $X_1, X_2,..., X_n$ are known, and it is necessary to find the corresponding moments of the random variable Y. Computation of the moments of Y using the Taylor series approximation is discussed below.

24.5.1 Taylor Series Approximation to Find Moments

Let $Y = f(X_1, X_2,..., X_n)$ represent a general equation where a design variable Y is a function of other design variables $X_1, X_2,..., X_n$. Given

$$E[X_i] = \mu_i \qquad i = 1, 2,..., n$$

$$V[X_i] = \sigma_i \qquad i = 1, 2,..., n$$

We have the approximate values for μ_Y and σ_Y using Taylor series approximations as follows:

$$\mu_y \cong f(\mu_1, \mu_2,..., \mu_n) + \frac{1}{2} \sum_{i=1}^{n} \frac{\partial^2 f(X)}{\partial X_i^2}\bigg|_{x=\mu} V(X_i) \tag{24.5}$$

$$V[Y] \cong \sum_{i=1}^{n} \left\{ \frac{\partial f(X)}{\partial X_i}\bigg|_{x=\mu} \right\}^2 V(X_i) \tag{24.6}$$

where $X = (X_1,..., X_n)$
$\mu = (\mu_1,..., \mu_n)$

EXAMPLE 24.2 Let us consider two resistances in parallel as shown in Fig. 24.5.[13] The mean and standard deviation for each resistance R_1 and R_2 are given, and we wish to approximately find the mean and standard deviation of R_T, the terminal resistance. We have

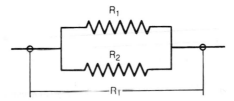

FIGURE 24.5 Two resistances in parallel.

$$R_T = f(R_1, R_2) = \frac{R_1 R_2}{R_1 + R_2}$$

$$\mu_{R_1} = 100 \ \Omega \qquad \sigma_{R_1} = 10 \ \Omega$$

$$\mu_{R_2} = 200 \ \Omega \qquad \sigma_{R_2} = 15 \ \Omega$$

Then

$$E[R_T] \cong f(100, 200) = \frac{100 \times 200}{100 + 200} = 66.7 \ \Omega$$

$$\frac{\partial f}{\partial R_1} = \frac{R_2^2}{(R_1 + R_2)^2} \qquad \frac{\partial f}{\partial R_2} = \frac{R_1^2}{(R_1 + R_2)^2}$$

$$\frac{\partial f}{\partial R_1}\bigg|_{R=\mu} = 0.444 \qquad \frac{\partial f}{\partial R_2}\bigg|_{R=\mu} = 0.111$$

Hence, using the Taylor series approximation [Eq. (24.6)], we have

$$\sigma^2_{RT} \cong (0.444)^2(10)^2 + (0.111)^2(15)^2$$

$$= 22.4858$$

or

$$\sigma_{RT} = 4.74 \ \Omega$$

If we use 3σ tolerance limits, we have $R_1 = 100 \pm 30$, $R_2 = 200 \pm 45$ and then $R_T = 66.7 \pm 14.2$.

If the random variables $X_1, \ldots, X_n$ are *not independent*, Eq. (24.6) for $V[Y]$ using the Taylor series approximation is given by

$$V(y) = \sum_{i=1}^{n} \left\{ \frac{\partial f(X)}{\partial X_i}\bigg|_{x=\mu} \right\}^2 V(X_i)$$

$$+ 2\sum_{i=1}^{n-1} \sum_{j=i+1}^{n} \left(\frac{\partial f}{\partial X_i}\bigg|_{x=\mu} \right)\left(\frac{\partial f}{\partial X_j}\bigg|_{x=\mu} \right) \text{cov}\, (X_i, X_j) \qquad (24.7)$$

24.5.2 Worst-Case Methods[10]

The worst-case method of variability analysis is a nonstatistical approach that can be used to determine whether it is possible, with given parameter tolerance limits, for the system performance characteristics to fall outside specifications. The answer is obtained by using system models in which parameters are set at either their upper or lower tolerance limits. Parameter values are chosen to cause each performance characteristic to assume first its maximum and then its minimum expected value. If the performance characteristic values fall within specifications, the designer can be confident that the system has high drift reliability. If specifications are exceeded, drift-type failures are possible, but the probability of their occurrence remains unknown.

Worst-case analysis is based on expressing the model for the performance variable Y as a function of design parameters $X_1, \ldots, X_n$ and expanding these functions in Taylor series about the nominal values. The design parameters include all pertinent part characteristics, inputs, loads, and environmental factors. Let the model for the performance variable Y_i be

$$Y_i = f_i X_1, \ldots, X_n \tag{24.8}$$

The linear relation which relates changes in Y_i to changes in design parameters $X_1, X_2, \ldots, X_n$ is

$$\Delta Y_i = \sum_{j=1}^{n} \left(\frac{\partial f_i}{\partial X_j} \bigg|_{x=\mu} \Delta X_j \right) \tag{24.9}$$

where vector μ represents the nominal value of the design parameter vector X and ΔX_j is the variation of design parameter X_j, $\Delta X_j = \mu_j - X_{j\min}$ or $\Delta X_{j\max} - \mu_j$, where μ_j is the nominal value for parameter j, $j = 1, \ldots, n$.

A set of these equations must be derived to relate all performance factors to all design variables. The partial derivatives of f with respect to each dependent variable X_j must be computed. One of the most important steps in a worst-case analysis is to decide whether to use a high or low parameter tolerance limit for each component part when analyzing a specific performance characteristic. If the slope of the function that relates a parameter to a performance characteristic is known, the selection of parameter limit is easy: when the slope of the parameter function is positive, the upper tolerance limit is chosen if the maximum value of the performance characteristic is desired. For parameter functions with negative slopes, the lower tolerance limit corresponds to the maximum performance characteristic value.

An important part of worst-case analysis is to determine the sensitivity of system performance to variations in input parameters. The sensitivity of a system essentially is measured as the effect of parameter variations on the system performance. In equation form, sensitivity can be expressed by

$$S_{ij} = \frac{\partial Y_i}{\partial X_j} \bigg|_{x=\mu} \quad \text{or} \quad \frac{\partial Y_i}{\partial X_j} \bigg|_o \tag{24.10}$$

where $S_{ij} =$ the sensitivity of the performance measure Y_i to the variation in the system design parameter X_j
$o =$ evaluated at nominal conditions, usually the mean values

Another equation that is used is

$$s_{ij} = \frac{\partial \ln Y_i}{\partial \ln X_j} \bigg|_{x=\mu} \approx \frac{\Delta Y_i / Y_{io}}{\Delta X_j / X_{jo}} \tag{24.11}$$

The forms of the variation equation which correspond to the two sensitivities are

$$\Delta Y_i = \sum_{j=1}^{n} S_{ij} \Delta X_j \tag{24.12}$$

$$\frac{\Delta Y_i}{Y_{io}} = \sum_{j=1}^{n} s_{ij} \frac{\Delta X_j}{X_{jo}} \tag{24.13}$$

If a design fails the worst-case analysis, look at the absolute values of the individual terms in Eq. (24.10) or (24.11). The ones which contribute the most ought to be reduced—they are the bottlenecks. It does little good to reduce the small terms because they have so little effect on the total variation. It is not unusual to have well over half the variation due to one or two parameters. If several performance parameters have too much variation, the major contributors ought to be listed for each. If a few parameters are causing most of the difficulty, attention can be devoted to

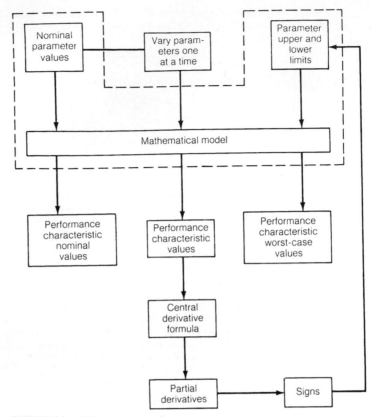

FIGURE 24.6   Worst-case method.

them. If not, an extensive redesign might be necessary. Figure 24.6 shows a block diagram of the worst-case method.

24.5.3 Monte Carlo Method[10,15,16]

In the Monte Carlo method a large number of replicas of the system are simulated by mathematical models. The values of the variables and parameters are randomly selected based on their probability distributions. The performance of the overall system can be compared to a given set of specifications. The large number of values of the system performance are used to develop a frequency distribution for the performance of the system. Figure 24.7 is a block diagram of the Monte Carlo method. Monte Carlo methods are very good for simulating the performance of the system but give very little help in identifying and correcting problems.

24.6 *ALLOCATION OF RELIABILITY REQUIREMENTS*

Reliability and design engineers must translate overall system performance, including reliability, into component performance, including reliability. The process of assigning reliability requirements to individual components to attain the specified system reliability is called *reliability allo-*

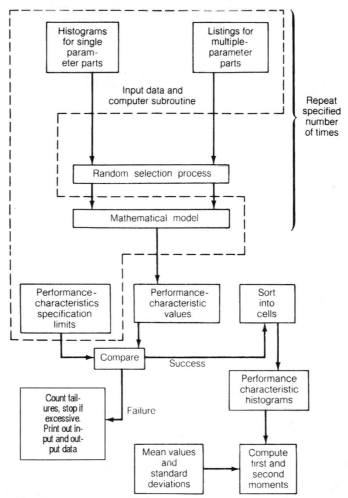

FIGURE 24.7   The Monte Carlo method.

cation. There exist many different ways that reliability can be allocated to components in order to achieve a specific system reliability.[1] The allocation problem is complex for several reasons, among which are (1) the role a component plays for the functioning of the system, (2) the methods available for accomplishing this function, (3) the complexity of the component, and (4) the changeable reliability of the component with the type of function to be performed. The problem is further complicated by the lack of detailed information on many of these factors early in the system design phase. However, a tentative reliability allocation must be accomplished in order to guide the design engineer. The typical decision process from a reliability allocation standpoint is illustrated in Fig. 24.8. A process such as this attempts to force all concerned to make decisions in an orderly and knowledgeable fashion rather than on an ad hoc basis.

Some of the advantages of the reliability allocation program are

1. The reliability allocation program forces system design and development personnel to understand and develop the relationships between component, subsystem, and system reliabilities.

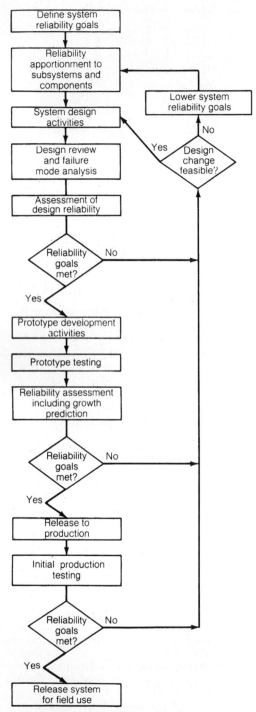

FIGURE 24.8   Reliability allocation process.

This leads to an understanding of the basic reliability problems inherent in the design.

2. The design engineer is obliged to consider reliability equally with other system parameters such as weight, cost, and performance characteristics.

3. Reliability allocation program ensures adequate design, manufacturing methods, and testing procedures.

The allocation process is approximate and the system effectiveness parameters, such as reliability and maintainability apportioned to the subsystems, are used as guidelines to determine design feasibility. If the allocated parameters for a system cannot be achieved using the current technology, then the system must be modified and the allocations reassigned. This procedure is repeated until an allocation is achieved that satisfies the system requirements.

The allocation of specified system reliability R^* to the component reliability requires solving the following inequality

$$f(R_1, R_2, \ldots, R_n) \geq R^* \tag{24.14}$$

where R_i = the reliability allocated to the ith unit
$\quad\; f$ = the functional relationship between the components and the system

For series and parallel systems, the functional relationship f is well known. This relationship is complex for other system configurations. If we are interested in reliability as a function of time, Eq. (24.14) may be generalized by considering R^* and R_i, $i = 1, \ldots, n$ as functions of time t.

Most of the basic reliability allocation models are based on the assumption that component failures are independent, the failure of any component results in system failure (i.e., the system is composed of components in series), and the failure rates of the components are constant. These assumptions lead to the following equation as a special case of Eq. (24.14):

$$R_1(t)R_2(t) \cdots R_n(t) \geq R^*(t) \tag{24.15}$$

Let λ_i = failure rate of the ith component and λ^* = failure rate of the system; then Eq. (24.15) becomes, using the exponential distribution,

$$e^{-\lambda_1 t} e^{-\lambda_2 t} \cdots e^{-\lambda_n t} \geq e^{-\lambda^* t}$$

or

$$\lambda_1 + \lambda_2 + \cdots + \lambda_n \leq \lambda^*$$

The purpose of the allocation model is to allocate $R^*(t)$ or λ^* to the components of the system. We will discuss some of the reliability allocation methods.

In the ARINC method,[1] we assume that the subsystems are in series with constant failure rates, that any subsystem failure causes a system failure, and that the subsystem mission times equal the system mission time. This apportionment technique requires the expression of the required reliability in terms of the failure rates. The objective is to choose λ_i^*'s such that

$$\sum_{i=1}^{n} \lambda_i^* \leq \lambda^* \tag{24.16}$$

where λ_i^* = the failure rate allocated to subsystem i
$\quad\quad i = 1, \ldots, n$
$\quad \lambda^*$ = the required system failure rate

The following steps summarize this technique:

1. Determine the subsystem failure rates λ_i from the past data, observed or estimated.

2. Assign a weighting factor ω_i to each subsystem according to the failure rates determined in step 1, where ω_i is given by

$$\omega_i = \frac{\lambda_i}{\sum\limits_{i=1}^{n} \lambda_i} \qquad i = 1,\ldots, n \tag{24.17}$$

Thus ω_i represents the relative failure vulnerability of the ith component and

$$\sum_{i=1}^{n} \omega_i = 1$$

3. Compute the subsystem failure rate requirements using

$$\lambda_i = \omega_i \lambda^* \qquad i = 1,\ldots, n$$

assuming quality holds in Eq. (24.16).

It is clear that this method allocates the new failure rates based on relative weighting factors that are functions of the past failure rates of the subsystems.

EXAMPLE 24.3[13]   Consider a system composed of three subsystems with the estimated failure rates of $\lambda_1 = 0.005$, $\lambda_2 = 0.003$, and $\lambda_3 = 0.001$ failure per hour, respectively. The system has a mission time of 20 h. A system reliability of 0.95 is required. Find the reliability requirements for the subsystems.
 Using Eq. (24.17), we compute the weighting factors:

$$\omega^1 = \frac{0.005}{0.005 + 0.003 + 0.001} = 0.555$$

$$\omega^2 = \frac{0.003}{0.005 + 0.003 + 0.001} = 0.333$$

$$\omega^3 = \frac{0.001}{0.005 + 0.003 + 0.001} = 0.111$$

We know that

$$R^*(20) = \exp[-\lambda^*(20)] = 0.95$$

or

$$\lambda^* = 0.00256 \text{ failure per hour}$$

Hence the failure rates for the subsystems are

$$\lambda_1 = \omega_1 \lambda^* = 0.555 \times 0.00256 = 0.00142$$

and similarly,

$$\lambda_2^* = 0.333 \times 0.00256 = 0.000852$$

$$\lambda_3^* = 0.111 \times 0.00256 = 0.000284$$

The corresponding apportioned reliabilities for the subsystem are

$$R_1^*(20) = \exp[-0.00142(20)] = 0.97$$

$$R_2^*(20) = \exp[-0.000852(20)] = 0.98$$

$$R_3^*(20) = \exp[-0.000284(20)] = 0.99$$

The AGREE allocation method[17] is more sophisticated than the previous method. This method is based on component or subsystem complexity and explicitly considers the relationship between component and system failure. The AGREE formula is used to determine the minimum mean time between failures (MTBF) for each component required to meet the system reliability. The components are supposed to have constant failure rates that are independent of each other and they operate in series with respect to their effect on system success.

Component complexity is defined in terms of modules and their associated circuitry. Examples of a module are an electron tube, a transistor, or a magnetic tape; a diode is considered a half-module. It is recommended that for digital computers (where the module count is high), the count should be reduced because failure rates for digital parts are generally far lower than for radio-radar types. The importance factor of a unit or subsystem is defined in terms of the probability of system failure if the particular subsystem fails. The importance factor of 1 means that the subsystem must operate for the system to operate successfully, and the importance factor of 0 means that the failure of the subsystem has no effect on system operation.

The allocation assumes that each module makes an equal contribution to system success. An equivalent requirement is that each module have the same failure rate. Making the observation that $e^{-x} \approx 1 - x$ when x is very small, the allocated failure rate to the ith unit is given by

$$\lambda_i = \frac{N_i[-\ln R^*(t)]}{N\omega_i t_i} \qquad i = 1, 2, \ldots, n \tag{24.18}$$

where t = mission time, or the required system operation time
t_i = time units for which the ith subsystem will be required to operate during t units of system operation $(0 < t_i \le t)$
N_i = number of modules in ith subsystem
N = total number of modules in the system = ΣN_i
ω_i = importance factor for the ith subsystem
  = P [system failure|subsystem i fails]
$R^*(t)$ = required system reliability for operation time t

The allocated reliability for the ith subsystem for t_i operating time units is given by

$$R_i(t_i) = 1 - \frac{1 - [R^*(t)]^{N_i/N}}{\omega_i} \tag{24.19}$$

The AGREE formula will lead to distorted allocation if the importance factor for a certain unit is very low. It is a good approximation if ω_i is close to one for each subsystem.

EXAMPLE 24.4[13]   A system consisting of four subsystems is required to demonstrate a reliability level of 0.95 for 10 h of continuous operation. Subsystems 1 and 3 are essential for the successful operation of the system. Subsystem 2 has to function for only 9 h for the operation of the system, and its importance factor is 0.95. Subsystem 4 has an importance factor of 0.90 and must function for 8 h for the system to function. Solve the reliability allocation problem by AGREE method using the data in Table 24.3. We have

$$N = \sum_{i=1}^{4} N_i = 20$$

TABLE 24.3   Data for Reliability Allocation

Subsystem number	Number of modules N_i	Importance factor ω_i	Operating time t_i
1	15	1.00	10
2	25	0.95	9
3	100	1.00	10
4	70	0.90	8

The minimum acceptable failure rates for the subsystems are given by Eq. (24.18), and these are

$$\lambda^1 = \frac{15(-\ln 0.95)}{(210)(1.0)(10)} = 0.000366$$

$$\lambda^2 = \frac{25(-\ln 0.95)}{(210)(0.95)(9)} = 0.000714$$

$$\lambda^3 = \frac{100(-\ln 0.95)}{(210)(1.0)(10)} = 0.002442$$

$$\lambda^4 = \frac{70(-\ln 0.95)}{(210)(0.90)(8)} = 0.002377$$

Thus, the allocated subsystem reliabilities are, using Eq. (24.19)

$$R_1(10) = 1 - \frac{1 - (0.95)^{15/210}}{1} = 0.99635$$

$$R_2(9) = 1 - \frac{1 - (0.95)^{25/210}}{0.95} = 0.99274$$

$$R_3(10) = 1 - \frac{1 - (0.95)^{100/210}}{1} = 0.97587$$

$$R_4(8) = 1 - \frac{1 - (0.95)^{70/210}}{0.90} = 0.98116$$

As a check, we have the system reliability as

$$R^* = (0.99635)(0.99274)(0.97587)(0.98116)$$

$$= 0.94723$$

which is slightly less than the specified reliability. This is due to the approximate nature of the formula and because the importance factors for subsystems 2 and 4 are less than 1.

REFERENCES

1. Aeronautical Research Incorporated, Engineering and Statistical Staff, in W. H. Von Alven, ed., *Reliability Engineering,* Prentice-Hall, Englewood Cliffs, N.J., 1964.

2. *Reliability Design Handbook,* Reliability Analysis Center, RDG-376, Griffiss Air Force Base, N.Y., March 1976.

3. MIL-HDBK-217, *Military Standardization Handbook, Reliability Prediction of Electronic Equipment,* U.S. Department of Defense, April 1979.

4. NPRD-Z, *Nonelectronic Parts Reliability Data Handbook,* Reliability Analysis Center, Rome Air Development Center, Griffiss AFB, N.Y., 1981.

5. MIL-STD-1556A, *Government-Industry Data Exchange Program (GIDEP),* U.S. Department of Defense, USAF, Feb. 1976.

6. Juran, J. M.; and Gryna, F. M., Jr., *Quality Planning and Analysis,* 2d ed., McGraw-Hill, New York, 1980.

7. MIL-STD-785B, *Reliability Program for Systems and Equipment Development and Production,* U.S. Department of Defense, Sept. 15, 1980.

8. MIL-STD-1543A, *Reliability Program Requirements for Space and Missile Systems,* U.S. Department of Defense, June 25, 1982.

9. *Quality Assurance-Reliability Handbook,* AMC Pamphlet No. 702-3, Headquarters, U.S. Army Materiel Command, Alexandria, Va., Oct. 1968.

10. *Engineering Design Handbook,* Part 2, *Design for Reliability,* AMC Pamphlet No. 706-196, U.S. Army Materiel Command, Alexandria, Va., Jan. 1976.

11. Kececioglu, D.; and Cormier, D., "Designing a Specified Reliability Directly into a Component," *Proceedings of the Third Annual Aerospace Reliability and Maintainability Conference,* 1968, pp. 520–530.

12. Haugen, E. G., *Probabilistic Approach to Design,* Wiley, New York, 1968.

13. Kapur, K. C.; and Lamberson, L. R., *Reliability in Engineering Design,* Wiley, New York, 1977.

14. Kapur, K. C., "Reliability and Maintainability," *Industrial Engineering Handbook,* chap. 5, Wiley, New York, 1982.

15. Mark, D. G.; and Stember, L. H., Jr., "Variability Analysis," *Electro-Technology,* July 1965.

16. NASA CR-1126, *Practical Reliability,* Vol. I, *Parameter Variation Analysis,* Research Triangle Institute, Research Triangle Park, N.C., July 1968.

17. Reliability of Military Electronic Equipment, Advisory Group on Reliability of Electronic Equipment (AGREE), Office of the Assistant Secretary of Defense, U.S. Government Printing Office, Washington, D.C., June 1957.

CHAPTER 25
MATHEMATICAL AND STATISTICAL METHODS AND MODELS IN RELIABILITY AND LIFE STUDIES

Kailash C. Kapur

25.1 ELEMENTS OF PROBABILITY THEORY

Reliability is the probability of a product performing its intended function satisfactorily for its intended life under specified operating conditions. Hence study of probability theory is essential for understanding the reliability of a product. The purpose of this section is to provide a basic knowledge of probability theory.

Probability is a method to model and describe the random variation in systems. Probability theory provides the mathematical foundation and language of statistics and helps us deduce from a mathematical model the properties of a physical process, whereas statistical inference determines the properties of the model from observed data. Statistics is the art and science of gathering, analyzing, and making decisions from the data and will be discussed in Sec. 25.3.

Table 25.1 gives a glossary of the symbols and abbreviations to be used in this chapter.

25.1.1 Fundamentals of Set Theory

A *set* is a collection of objects viewed as a single entity. The individual objects of the set are called the *elements* of the set. Sets usually are denoted by capital letters: $A, B, C, \ldots, Y, Z$; elements are designated by lowercase letters: $a, b, c, \ldots, y, z$. If a is an element of the set A, we write $a \in A$, and we write $a \notin A$ for a is not an element of A. A set is called a *finite* set when it contains a finite number of elements and an *infinite* set otherwise. The *null set* $\emptyset$ is the set that contains no elements. The *total* or *universal set* Ω is the set which contains all the elements under consideration.

We say a set A is a subset of set B if each element of A is also an element of B and write as $A \subseteq B$. The relation $\subseteq$ is referred to as *set inclusion*.

TABLE 25.1   Glossary of Symbols and Abbreviations

$P(\cdot)$	Probability of event $(\cdot)$
$\binom{n}{k}$	Combination of size k from n objects
R	Reliability
p.d.f.	Probability-density function
c.d.f.	Cumulative distribution function
r.v.	Random variable
$f(t)$	Probability-density function for random variable T
$F(t)$	Cumulative distribution function for random variable
$R(t)$	Reliability function for random variable T
$h(t)$	Failure rate
$E[x]$	Expected value for random variable x
μ	Mean of a random variable
σ	Standard deviation
ρ	Correlation coefficient
z	Standard normal variable
$\Phi(z)$	Cumulative distribution function for standard normal variable
$\Gamma(\cdot)$	Gamma function
MTBF	Mean time between failures
MTTR	Mean time to repair
$M(t)$	Maintainability function
χ^2	Chi-square distribution
F	F distribution
$\hat{\theta}$	Point estimator of parameter θ
β	Slope parameter for Weibull distribution

The Algebra Sets

Definition: The *union* of the two sets A and B, denoted by $A \cup B$, is the set of all elements of either set, that is, $c \in (A \cup B)$ means $c \in A$, or $c \in B$, or both.

Definition: The *intersection* of the two sets A and B, denoted by $A \cap B$, is the set of all elements common to both A and B, that is, $c \in (A \cap B)$ means $c \in A$ and $c \in B$.

Definition: The *complement* of a set A, denoted by $\overline{A}$, is the set of elements of the universal set that do not belong to A.

Definition: Two sets are said to be *disjoint* or *mutually exclusive* if they have no elements in common, i.e., $A \cap B = \emptyset$.

When considering sets and operations on sets, *Venn diagrams* can be used to represent sets diagrammatically. Figure 25.1*a* shows a Venn diagram for $A \cap B$ and Fig. 25.1*b* shows a Venn diagram for $A \cup B$. Figure 25.1*c* shows a Venn diagram with three sets A, B, and C.

25.1.2 Probability Definitions

There is a natural relation between probability theory and set theory based on the concept of a random experiment for which it is impossible to state a particular outcome, but we can define the set of all possible outcomes.

Definition: The *sample space* of an experiment, denoted by S, is the set of all possible outcomes of the experiment.

Definition: An *event* is any collection of outcomes of the experiment or subset of the sample space S. An event is said to be *simple* if it consists of exactly one outcome, and *compound* if it consists of more than one outcome.

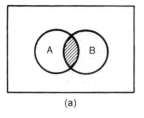
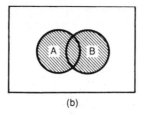

(a)   (b)

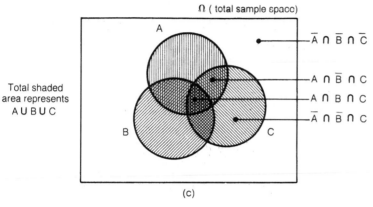

Ω (total sample space)

A

Total shaded
area represents
$A \cup B \cup C$

$\overline{A} \cap \overline{B} \cap \overline{C}$

$A \cap \overline{B} \cap C$

$A \cap B \cap C$

$\overline{A} \cap \overline{B} \cap C$

B

C

(c)

FIGURE 25.1   Venn diagrams: (*a*) $A \cap B$ (shaded); (*b*) $A \cup B$ (shaded); (*c*) three sets *A*,
B, and *C*.

The objective of probability is to assign to each event *A* of the sample space *S* associated with
an experiment a number *P(A)*, called the *probability of event A,* which will give a precise measure
of the chance that *A* will occur. The function $P(\cdot)$ has the following properties:

1. $0 \le P(A) \le 1$ for each event *A* of *S*
2. $P(S) = 1$
3. For any finite number *k* of mutually exclusive events defined on *S*,

$$P\left(\bigcup_{i=1}^{k} A_i \right) = \sum_{i=1}^{k} P(A_i) \tag{25.1}$$

If $A_1, A_2, A_3,...$ is a denumerable or countably infinite sequence of mutually exclusive events
defined on *S*, then

$$P(A_1 \cup A_2 \cup A_3 \cup \cdots) = P(A_1) + P(A_2) + P(A_3) + \cdots \tag{25.2}$$

We can also use the concept *relative frequency* to develop the function $P(\cdot)$. If we repeat an experiment *n* times and event *A* occurs n_A times, $0 \le n_A \le n$, then the value of the relative frequency f_A
$= n_A/n$ approaches *P(A)* as *n* increases to infinity.

Properties of Probability

1. If $\emptyset$ is the empty or null set, then $P(\emptyset) = 0$
2. $P(\overline{A}) = 1 - P(A)$ $\tag{25.3}$

3. $P(A \cup B) = P(A) + P(B) - P(A \cap B)$ (25.4)

4. $P(A_1 \cup A_2 \cup \cdots \cup A_n) = \sum\limits_{i=1}^{k} P(A_i) - \sum\limits_{i=1}^{n-1} \sum\limits_{j=i+1}^{n} P(A_i \cap A_j)$

$$+ \sum\limits_{i=1}^{n-2} \sum\limits_{j=i+1}^{n-1} \sum\limits_{k=j+1}^{n} P(A_i \cap A_j \cap A_k)$$

$$+ \cdots + (-1)^{n+1} P(A_1 \cap A_2 \cap \cdots \cap A_n)$$ (25.5)

Conditional Probability.   We will frequently be interested in evaluating the probability of events where the event is *conditioned* on some subset of the sample space.

Definition: The *conditional probability* of event A given event B is defined as

$$P(A/B) = \frac{P(A \cap B)}{P(B)} \qquad if\ P(B) > 0$$ (25.6)

This statement can be restated to what is often called the *multiplication rule*; that is,

$$P(A \cap B) = P(A/B)P(B) \qquad P(B) > 0$$ (25.7)

$$P(A \cap B) = P(B/A)P(A) \qquad P(A) > 0$$ (25.8)

Definition: A and *B* are called *independent* events if and only if

$$P(A \cap B) = P(A)P(B)$$ (25.9)

This definition leads to the following statement. If A and B are independent events, then

$$P(A/B) = P(A) \qquad \text{and} \qquad P(B/A) = P(B)$$

Partitions, Total Probability, and Bayes' Theorem.   If $A_1, \ldots, A_n$ are disjoint subsets of S (mutually exclusive events) and if $A_1 \cup A_2 \cup \cdots \cup A_n = S$, then these subsets are said to form a *partition* of S. The *total probability* of any other event B is given by

$$P(B) = \sum\limits_{i=1}^{n} P(B/A_i)P(A_i)$$ (25.10)

Another important result of total probability is *Bayes' theorem*. If $A_1, A_2, \ldots, A_k$ constitute a partition of the sample space S and B is an arbitrary event, then Bayes' theorem states that

$$P(A_i/B) = \frac{P(A_i \cap B)}{P(B)} = \frac{P(B/A_i)P(A_i)}{\sum\limits_{i=1}^{n} P(B/A_i)P(A_i)} \qquad i = 1, 2, \ldots, n$$ (25.11)

where $P(A_i) > 0$ for $i = 1, \ldots, n$ and $P(B) > 0$.

Permutations and Combinations.   In some situations we will have to resort to the relative frequency concept and successive trials to *estimate* probabilities and hence have to enumerate the number of outcomes favorable to event A.

Definition: Suppose that a set consists of ordered collections of k elements (k-tuples) and that there are n possible choices for the first element; for each choice of first element there are n_2 possible choices of the second element, and so for each possible choice of the first $(k - 1)$ elements, there are n_k choices of the kth element. Then the product rule states that there are n_1, $n_2, ..., n_k$ possible k element (k-tuple) sets.

Definition: Any ordered sequence of k objects from n distinct objects is called a *permutation* of size k of the objects. The number of permutations of n objects taken k at a time is denoted by P_k^n and is given by

$$P_k^n = n(n - 1)\cdots(n - k + 1) = \frac{n!}{(n - k)!} \tag{25.12}$$

where

$$n! = n(n - 1)(n - 2)\cdots(2)(1) = \prod_{i=1}^{n} i$$

$n!$ is read n factorial; also $0! = 1$.

Definition: Given a set of n distinct objects, any unordered subset of size k of the objects is called a *combination*. The number of combinations of size k which can be formed from n distinct objects is denoted by $\binom{n}{k}$ and is given by

$$\binom{n}{k} = \frac{n!}{(n-k)!k!} \tag{25.13}$$

Further details on probability concepts can be found in Refs. 1 and 2.

EXAMPLE 25.1: System Reliability Computation   In the following sections, applications of probability concepts discussed before are presented to compute reliability of a system given reliabilities of the components. A reliability block diagram has to be developed based on a careful analysis of the manner in which the system operates. Let R_i be the reliability of the ith component, $i = 1, ..., n$. We present reliability equations for different system configurations.

series structure   The structure is called a *series* structure when the system functions if and only if all the n components of the system function (see Fig. 25.2*a*). The components are assumed to fail or function independently of one another. Then the reliability R_S of the system is given by

$$R_S = \prod_{i=1}^{n} R_i \tag{25.14}$$

parallel structure   The structure is said to be *parallel* when the system functions if at least one of the n components of the system functions (see Fig. 25.2*b*). The reliability of the system is given by

$$R_S = 1 - \prod_{i=1}^{n} (1 - R_i) \tag{25.15}$$

combination structure   We can compute reliability of a system which consists of both series and parallel subassemblies. Figure 25.2*c* shows such a structure. The numbers in each block are the reliabilities of the components. The system reliability is given by

$$R_S = [1 - (1 - 0.95)(1 - 0.99)](0.98)[1 - (1 - 0.99 \times 0.97)(1 - 0.90)]$$

$$= 0.9995 \times 0.98 \times 0.99603$$

$$= 0.97562$$

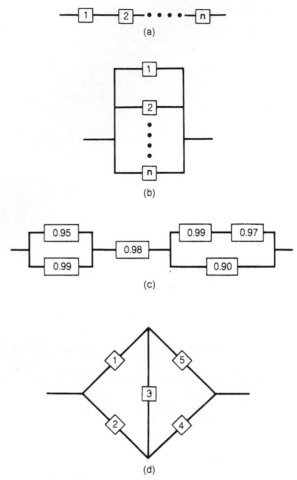

FIGURE 25.2   Structures: (*a*) series; (*b*) parallel; (*c*) combination; (*d*) bridge.

the *k*-out-of-*n* structure   For this structure, the system works if and only if at least *k* components out of the *n* components work; $1 \leq k \leq n$. For the case when $R_i = R$ for all *i,* we have

$$R_S = \sum_{i=k}^{n} \binom{n}{i} R^i (1 - R)^{n-i} \tag{25.16}$$

Let us now consider an example of the above system.

A military planner has eight helicopters available to perform a mission. At least six helicopters are required to perform the mission successfully. Each helicopter has a reliability of 0.80 for the duration of the mission. The helicopters work or fail independently of each other. What is the reliability of this mission, i.e., what is the probability that the mission will be a success? We have, using Eq. (25.16),

$$R_{\text{mission}} = \sum_{k=6}^{8} \binom{8}{k} (0.8)^k (1 - 0.8)^{8-k}$$

$$= 0.79692$$

coherent systems   The reliability block diagrams for many systems cannot be represented by the above three configurations. In general, the concept of coherent systems can be used to determine the reliability of any system.[3] The performance of each of the n components in the system is represented by a binary indicator variable x_i which takes the value 1 if the ith component functions and 0 if the ith component fails. Similarly, the binary variable $\varnothing$ indicates the state of the system, and $\varnothing$ is a function of $x = (x_1, \ldots, x_n)$ and $\varnothing(x)$ is called the *structure function* of the system. The structure function is represented by using the concept of minimal path and minimal cut. A *minimal path* is a minimal set of components whose functioning ensures the functioning of the system. A *minimal cut* is a minimal set of components whose failures cause the system to fail. Let $\alpha_j(x)$ be the jth minimal path series structure for path $A_j, j = 1, \ldots, p$ and $\beta_k(x)$ be the kth minimal parallel cut structure for cut $B_k, k = 1, \ldots, s$. Then, we have

$$\alpha_j(x) = \prod_{i \in Aj} x_i \tag{25.17}$$

$$\beta_k(x) = 1 - \prod_{i \in B_k} (1 - x_i) \tag{25.18}$$

and

$$\varnothing(x) = 1 - \prod_{j=1}^{p} [1 - \alpha_j(x)] \tag{25.19}$$

$$= \prod_{k=1}^{s} \beta_k(x) \tag{25.20}$$

For the bridge structure (Fig. 25.2d), we have

$$\alpha_1 = x_1 x_5 \qquad \beta_1 = 1 - (1 - x_1)(1 - x_2)$$

$$\alpha_2 = x_2 x_4 \qquad \beta_2 = 1 - (1 - x_4)(1 - x_5)$$

$$\alpha_3 = x_1 x_3 x_4 \qquad \beta_3 = 1 - (1 - x_1)(1 - x_3)(1 - x_4)$$

$$\alpha_4 = x_2 x_3 x_5 \qquad \beta_4 = 1 - (1 - x_2)(1 - x_3)(1 - x_5)$$

Then, the reliability of the system is given by $R_S = P[\varnothing(x) = 1]$.
The reliability R_S for the bridge structure is given by the following expression

$$R_S = R_1 R_5 + R_1 R_3 R_4 + R_2 R_3 R_5 + R_2 R_4$$

$$- R_1 R_3 R_4 R_5 - R_1 R_2 R_3 R_5 - R_1 R_2 R_4 R_5$$

$$- R_1 R_2 R_3 R_4 - R_2 R_3 R_4 R_5 + 2 R_1 R_2 R_3 R_4 R_5$$

If all $R_i = R = 0.9$, we have

$$R_S = 2R^2 + 2R^3 - 5R^4 + 2R^5$$

$$= 0.9785$$

The exact calculations for R_S are generally very tedious because the paths and the cuts are dependent, since they may contain a same component. Bounds on system reliability are given by

$$\prod_{k=1}^{s} P[\beta_k(x) = 1] \le P[\emptyset(x) = 1] \le 1 - \prod_{j=1}^{p} \{1 - P[\alpha_j(x) = 1]\}$$

Using these bounds for the bridge structure (Fig. 25.2d), we have when $R_i = R = 0.9$

$$\text{Upper bound on } R_S = 1 - (1 - R^2)^2(1 - R^3)^2$$

$$= 0.9973$$

$$\text{Lower bound on } R_S = [1 - (1 - R)^2]^2[1 - (1 - R)^3]^3$$

$$= 0.9781$$

The bounds on system reliability can be improved, and details are given in Barlow and Proschan.[3]

25.1.3 Discrete and Continuous Random Variables

Definition: Let S be the sample space associated with experiment ε. Let X be a function that assigns a real number $X(e)$ to every outcome $e \in S$, then $X(e)$ is called a *random variable* (r.v.).

Discrete Random Variables.   If the range space R_X of the random variable X is either finite or countably infinite, then X is called a *discrete random variable*.

Continuous Random Variables.   If the range space R_X of the random variable X is an interval or a collection of intervals, then X is called a *continuous random variable*.

Probability-Density Function.   For a continuous random variable X, we define

$$P[a \le X \le b] = \int_a^b f_x(x)\, dx \tag{25.21}$$

where the function f_X called the *probability-density function* (p.d.f.), satisfies the following properties:

1. $f_X(x) \ge 0$ for all $x \in R_X$

2. $\int_{R_x} f_X(x)\, dx = 1$

Probability-Mass Function.   For a discrete random variable, we associate a number $p_X(x_i) = P(X = x_i)$ for each outcome x_i in R_X, where the numbers $p_X(x_i)$ satisfy the following properties:

1. $p_X(x_i) \ge 0$ for all i

2. $\sum_{\text{all } i} p_X(x_i) = 1$

Cumulative Distribution Function.   The cumulative distribution function (c.d.f.) of a random variable X is denoted by $F_X(\cdot)$ and is defined by

$$F_X(x) = P[X \le x] \qquad \text{for all } x$$

$$F_X(x) = \sum_{\text{all } i \text{ such that } x_i \le x} p(x_i) \qquad X \text{ discrete random variable} \tag{25.22}$$

$$= \int_{-\infty}^x f(u)\, du \qquad X \text{ continuous random variable} \tag{25.23}$$

Properties of the cumulative distribution function are

$$F(-\infty) = \lim_{x \to -\infty} F(x) = 0$$

$$F(+\infty) = \lim_{x \to +\infty} F(x) = 1$$

$$F(x_2) \le F(x_1) \qquad \text{for all } x_2 \le x_1$$

EXAMPLE 25.2: Reliability Function   Let T be the time-to-failure random variable. Then reliability at time t, $R(t)$, is the probability that the system will not fail by time t, or

$$
\begin{aligned}
R(t) &= P[T > t] \\
&= 1 - P[T \le t] \\
&= 1 - F(t) \\
&= 1 - \int_0^t f(\tau)\, d\tau
\end{aligned}
\tag{25.24}
$$

where $f(t)$ and $F(t)$ are the probability-density function and cumulative distribution function for the T, respectively. For example, if the time-to-failure T is exponentially distributed, then

$$f(t) = \lambda e^{-\lambda t} \qquad\qquad t \ge 0, \lambda > 0$$

$$F(t) = \int_0^t \lambda e^{-\lambda \tau}\, d\tau = 1 - e^{-\lambda t} \qquad t \ge 0$$

and

$$R(t) = e^{-\lambda t} \qquad\qquad t \ge 0$$

Figure 25.3 shows the shapes of these functions.

EXAMPLE 25.3: Failure Rate   If we have a large population of the items whose reliability we are interested in studying, then for replacement and maintenance purposes we are interested in the rate at which the items in the population, which have survived at any point in time, will fail. This is called the *failure rate* or *hazard rate* and is given by the following relationship:

$$h(t) = \frac{f(t)}{R(t)} \tag{25.25}$$

The failure rate for most components follows the curve shown in Fig. 25.4a, which is called the *life characteristic curve*.[4] Figure 25.4b gives three types of failures, namely quality failures, stress-related failures, and wearout failures. The sum total of these failures gives the overall failure rate of Fig. 25.4a. Figure 25.4b is also referred to as the "bathtub curve." The failure rate curve in Fig. 25.4a has three distinct periods. The initial decreasing failure rate is termed "infant mortality" and is due to the early failure of substandard products. Latent material defects, poor assembly methods, and poor quality control can contribute to a high initial failure rate. A short period of in-plant product testing, termed *burnin,* is used by manufacturers to eliminate these early failures from the consumer market. The flat, middle portion of the failure-rate curve represents the design failure rate for the specific product as used by the consumer market. During the useful-life portion, the failure rate is relatively constant. It might be decreased by redesign or restricting usage. Finally, as products age they reach a wearout phase characterized by an increasing failure rate.

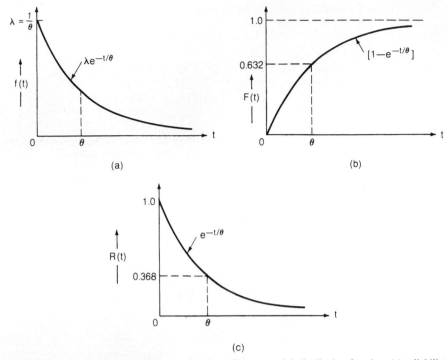

FIGURE 25.3 (*a*) Exponential density function; (*b*) exponential distribution function; (*c*) reliability function for exponential distribution.

The hazard rate is defined as the limit of the instantaneous failure rate given no failure up to time t, and is given by[5]

$$h(t) = \lim_{\Delta t \to 0} \frac{P[t < T \le t + \Delta t \mid T > t]}{\Delta t}$$

$$= \lim_{\Delta t \to 0} \frac{R(t) - R(t + \Delta t)}{\Delta t \cdot R(t)}$$

$$= \frac{1}{R(t)} \left[-\frac{d}{dt} R(t) \right]$$

$$= \frac{f(t)}{R(t)}$$

If the time-to-failure distribution is exponential, then using Example 25.2, we have

$$h(t) = \frac{\lambda e^{-\lambda t}}{e^{-\lambda t}} = \lambda$$

Thus, exponential distribution has a constant failure rate. Also,

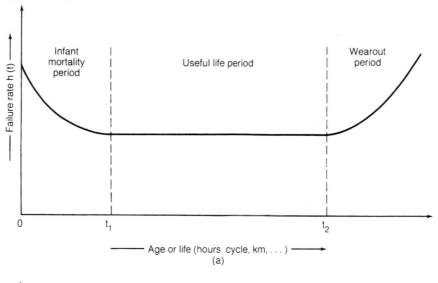

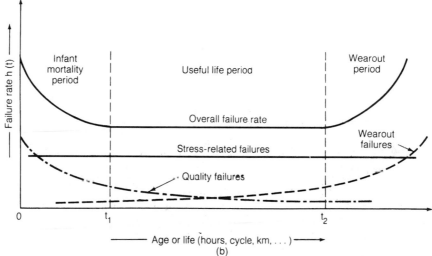

FIGURE 25.4   (*a*) Failure rate-life characteristic curve; (*b*) failure rate based on components of failure.

$$f(t) = h(t)\, exp\left[-\int_0^t h(\tau)\, d\tau\right] \tag{25.26}$$

and thus

$$R(t) = exp\left[-\int_0^t h(\tau)\, d\tau\right] \tag{25.27}$$

The term *mission reliability* is the reliability of a product for a specified time period t^*, i.e.,

$$\text{Mission reliability} = R(t^*) = P[T > t^*] \tag{25.28}$$

25.1.4 Expected Value

Definition: The expected value of a function $g(X)$ of a discrete or continuous random variable is

$$E[g(X)] = \sum_{\text{all } x} g(x)p(x) \qquad X \text{ is discrete r.v.}$$

$$= \int_{-\infty}^{\infty} g(x)f(x)\, dx \qquad X \text{ is continuous r.v.} \qquad (25.29)$$

Some of the properties of expected value are

$$E[c] = c \qquad c \text{ is a constant}$$

$$E[cg(X)] = cE[g(X)]$$

$$E[g_1(X) + g_2(X)] = E[g_1(X)] + E[g_2(X)]$$

Definition: The kth moment of the random variable X about origin is defined as

$$v_k = E[X^k] = \sum_{\text{all } x} x^k p(x) \qquad X \text{ is discrete random variable}$$

$$= \int_{-\infty}^{\infty} x^k f(x)\, dx \qquad X \text{ is continuous random variable} \qquad (25.30)$$

The mean or expected value of the random variable X is

$$v_1 = E[X] \qquad (25.31)$$

$E[X]$ is also denoted by μ or μ_1.

A measure that is frequently used as an indirect indicator of system reliability is called the *mean time to failure* (MTTF), which is the expected or mean value of the time-to-failure random variable. Thus, the MTTF is theoretically defined as

$$\text{MTTF} = E[T] = \int_0^{\infty} tf(t)\, dt = \int_0^{\infty} R(t)\, dt \qquad (25.32)$$

Sometimes the term *mean time between failures* (MTBF) is also used to denote $E[T]$. For the exponential distribution

$$E[T] = \int_0^{\infty} t\lambda e^{-\lambda t}\, dt = \frac{1}{\lambda} = \theta \qquad (25.33)$$

Definition: The kth moment of the random variable X about mean v_1 is defined as

$$\mu_k = E[(x - v_1)^k] \qquad k \geq 2$$

$$= E[(x - \mu)^k] \qquad k \geq 2 \qquad (25.34)$$

We have

$$\mu_k = \sum_{i=0}^{k} \binom{k}{i} (-1)^i v_{k-i} v_1^i \qquad (25.35)$$

$\mu_2 = E[(x - \mu)^2] = \sigma^2$ is called the *variance* of the random variable X and *standard deviation* σ is the positive square root of the variance. Other measures[6] for a random variable are

$$\text{Coefficient of variation} = \eta = \frac{\sigma}{\mu}$$

$$\text{Coefficient of skewness} = \alpha_3 = \frac{\mu_3}{\mu_2^{3/2}}$$

$$\text{Coefficient of kurtosis} = \alpha_4 - \frac{\mu_4}{\mu_2^2}$$

Definition: The moment-generating function (MGF) of a random variable X is the expected value of e^{tX} and is given by

$$M_X(t) = E[e^{tX}] \tag{25.36}$$

Whenever MGF exists, it is unique and completely determines the distribution. MGF can be used to find all the moments about origin v_k of the random variable X. The moments are given by

$$v_k = \left. \frac{\partial^k M_X(t)}{\partial t^k} \right|_{t=0} \tag{25.37}$$

Thus MGF is used to find moments and study the form of a probability distribution.

25.1.5 Bivariate Probability Distributions

If X_1 and X_2 are discrete random variables, joint probability distribution is given by

$$p(x_1, x_2) = P[X_1 = x_1, X_2 = x_2] \tag{25.38}$$

where the function P has the properties:

$$p(x_1, x_2) \geq 0$$

and

$$\sum_{\text{all } x_1} \sum_{\text{all } x_2} p(x_1, x_2) = 1$$

For continuous random variables X_1 and X_2, the joint density function $f(x_1, x_2)$ has the properties

$$f(x_1, x_2) \geq 0 \qquad \text{for all } x_1, x_2$$

$$\int_{x_2} \int_{x_1} f(x_1, x_2) \, dx_1 \, dx_2 = 1$$

The marginal distributions of X_1, and X_2 are given by

$$f_1(x_1) = \int_{-\infty}^{\infty} f(x_1, x_2) \, dx_2 \tag{25.39}$$

$$f_2(x_2) = \int_{-\infty}^{\infty} f(x_1, x_2)\, dx_1 \tag{25.40}$$

Definition: If $[X_1, X_2]$ is a two-dimensional random variable, the *covariance,* denoted by σ_{12} is

$$\text{cov}\,(X_1, X_2) = \sigma_{12} = E[(X_1 - E(X_1))(X_2 - E(X_2))]$$

and the correlation coefficient, denoted by ρ, is

$$\rho = \frac{\text{cov}\,(X_1, X_2)}{\sqrt{V(X_1)}\sqrt{V(X_2)}} = \frac{\sigma_{12}}{\sigma_1 \sigma_2} \tag{25.41}$$

The value of ρ will always be in the interval $[-1, +1]$. X_1 and X_2 are independent random variables if and only if

$$f(x_1, x_2) = f_1(x_1) f_2(x_2) \tag{25.42}$$

If X_1 and X_2 are independent, then $\rho = 0$.

25.2 DISTRIBUTIONS USED IN RELIABILITY

In this section we will summarize some of the important distributions used in reliability. Table 25.2 summarizes some of the discrete distributions. The table gives equations for mean, variance, and moment-generating functions for different distributions.

We will briefly discuss the applications of the discrete distributions given in Table 25.2. The *Bernoulli* distribution is extensively used for situations where an individual experiment, trial, or test has only two possible outcomes, such as success or failure, go or no-go, defective or nondefective, or pass or fail. The n Bernoulli trials are called a *Bernoulli process* if the trials are independent and each trial has only two possible outcomes, and probability of success remains constant from trial to trial. The number of successes in n Bernoulli trials has a *binomial* distribution. Applications of the binomial distribution will be discussed in this chapter. Some of the properties of the binomial distribution are given below:

1. $p(x)$ has maximum value for values of x for which $p(n + 1) - 1 \le x \le p(n + 1)$.
2. As $n \to \infty$ and $p \to 0$ in such a way that np remains constant, the binomial distribution approaches the Poisson distribution with parameter $\alpha = np$.
3. As $n \to \infty$, the binomial distribution approaches the normal distribution with $\mu = np$ and $\sigma^2 = np(1 - p)$. This approximation is good when $p = \frac{1}{2}$ and poor for $p < 1/(n + 1)$, $p > n/(n + 1)$, and outside 3σ limits for the random variable.

The *geometric* distribution is related to a sequence of Bernoulli trials except that the number of trials is not fixed and the random variable of interest is defined to be the number of trials required to achieve the first success. An interesting and useful property of the geometric distribution is that it has no memory, that is,

$$P(X > x + k \mid X > k) = P(X > x)$$

This property is analogous to the memoryless property of the exponential distribution. The geometric distribution is the only discrete distribution having this property. The *Pascal* distribution is an extension of the geometric distribution where the random variable X is the Bernoulli trial on

TABLE 25.2 Summary of Discrete Distributions

Distribution	Parameters	Probability function $p(x)$	Mean	Variance	Moment-generating function
Bernoulli	$0 \leq p \leq 1$	$p(x) = p^x q^{1-x} \quad x = 0, 1$ $= 0 \quad$ otherwise	p	pq	$pe^t + q$
Binomial	$n = 1, 2, \dots$ $0 \leq p \leq 1$	$p(x) = \binom{n}{x} p^x q^{n-x} \quad x = 0, 1, 2, \dots, n$ $= 0 \quad$ otherwise	np	npq	$(pe^t + q)^n$
Geometric	$0 < p < 1$	$p(x) = pq^{x-1} \quad x = 1, 2, \dots$ $= 0 \quad$ otherwise	$1/p$	q/p^2	$pe^t/(1 - qe^t)$
Pascal (negative binomial)	$0 < p < 1$ $r = 1, 2, \dots$ $(r > 0)$	$p(x) = \binom{x-1}{r-1} p^r q^{x-r} \quad x = r, r+1, r+2, \dots$ $= 0 \quad$ otherwise	r/p	rq/p^2	$\left[\dfrac{pe^t}{1 - qe^t}\right]^r$
Hypergeometric	$N = 1, 2, \dots$ $n = 1, 2, \dots, N$ $D = 1, 2, \dots, N$	$p(x) = \dfrac{\binom{D}{x}\binom{N-D}{n-x}}{\binom{N}{n}} \quad x = 0, 1, 2, \dots, \min(n, D)$ $= 0 \quad$ otherwise	$n\left[\dfrac{D}{N}\right]$	$n\left[\dfrac{D}{N}\right]\left[1 - \dfrac{D}{N}\right]\left[\dfrac{N-n}{N-1}\right]$	—
Poisson	$\alpha > 0$	$p(x) = e^{-\alpha}(\alpha)^x/x! \quad x = 0, 1, 2, \dots$ $= 0 \quad$ otherwise	α	α	$\exp[\alpha(e^t - 1)]$

Source: Ref. 1.

25.15

which the rth success occurs, where r is an integer. If r is not an integer, then the distribution given in Table 25.2 is called a *negative binomial distribution*. The *hypergeometric distribution* is used to model sampling without replacement from a population. Let the universe consist of N elements, D of which possess a given property. Then the probability that a random sample of size n, without replacement, will contain exactly x elements which possess this property is given by a hypergeometric distribution. As $n \to \infty$, the hypergeometric distribution approaches the binomial distribution with parameters n and $p = D/N$. The approximation is satisfactory for $10n < N$. The *Poisson distribution* is a useful approximation to the binomial and hypergeometric distributions. It also arises when the number of possible events is large but the probability of occurrence of the event over a given area or interval is small, e.g., number of defects, number of failures in a given time interval, and problems in maintainability and availability.

Now we summarize the probability-density function, cumulative distribution function, reliability function, hazard function, and MTBF for some of the well-known continuous distributions that are used in reliability.

Exponential Distribution

$$f(t) = \lambda e^{-\lambda t} \qquad t \geq 0 \tag{25.43}$$

$$F(t) = 1 - e^{-\lambda t} \qquad t \geq 0 \tag{25.44}$$

$$R(t) = e^{-\lambda t} \qquad t \geq 0 \tag{25.45}$$

$$h(t) = \lambda \tag{25.46}$$

$$\text{MTBF} = \theta = \frac{1}{\lambda} \tag{25.47}$$

Thus, the failure rate for the exponential distribution is always constant.

Normal Distribution

$$f(t) = \frac{1}{\sigma \sqrt{2\pi}} \exp\left[-\frac{1}{2} \left(\frac{t - \mu}{\sigma} \right)^2 \right] \qquad -\infty < t < \infty \tag{25.48}$$

$$F(t) = \Phi\left(\frac{t - \mu}{\sigma} \right) \tag{25.49}$$

$$R(t) = 1 - \Phi\left(\frac{t - \mu}{\sigma} \right) \tag{25.50}$$

$$h(t) = \frac{\phi[(t - \mu)/\sigma]/\sigma}{R(t)} \tag{25.51}$$

$$\text{MTBF} = \mu$$

$\Phi(Z)$ is the cumulative distribution function and $\phi(Z)$ is the probability-density function for the standard normal variate Z. The failure rate for the normal distribution is a monotonically increasing function.

Lognormal Distribution

$$f(t) = \frac{1}{\sigma_t \sqrt{2\pi}} \exp\left[-\frac{1}{2}\left(\frac{\ln t - \mu}{\sigma} \right)^2 \right] \qquad t \geq 0 \tag{25.52}$$

$$F(t) = \Phi\left(\frac{\ln t - \mu}{\sigma} \right) \tag{25.53}$$

$$R(t) = 1 - \Phi\left(\frac{\ln t - \mu}{\sigma} \right) \tag{25.54}$$

$$h(t) = \frac{\phi[(\ln t - \mu)/\sigma]}{t\sigma R(t)} \tag{25.55}$$

$$\text{MTBF} = \exp\left(\mu + \frac{\sigma^2}{2} \right)$$

The failure rate for the lognormal distribution is neither always increasing nor always decreasing. It takes different shapes depending on the parameters μ and σ.

Weibull Distribution

$$f(t) = \frac{\beta(t - \delta)^{\beta - 1}}{(\theta - \delta)^\beta} \exp\left[-\left(\frac{t - \delta}{\theta - \delta} \right)^\beta \right] \qquad t \geq \delta \geq 0 \tag{25.56}$$

$$F(t) = 1 - \exp\left[-\left(\frac{t - \delta}{\theta - \delta} \right)^\beta \right] \tag{25.57}$$

$$R(t) = \exp\left[-\left(\frac{t - \delta}{\theta - \delta} \right)^\beta \right] \tag{25.58}$$

$$h(t) = \frac{\beta(t - \delta)^{\beta - 1}}{(\theta - \delta)^\beta} \tag{25.59}$$

$$\text{MTBF} = \theta\Gamma\left(1 + \frac{1}{\beta} \right) \tag{25.60}$$

The failure rate for the Weibull distribution is decreasing when $\beta < 1$, is constant when $\beta = 1$ (same as the exponential distribution), and is increasing when $\beta > 1$.

Gamma Distribution

$$f(t) = \frac{\lambda^{\eta}}{\Gamma(\eta)} \, t^{\eta-1} e^{-\lambda t} \qquad t \geq 0 \tag{25.61}$$

$$F(t) = \sum_{k=\eta}^{\infty} \frac{(\lambda t)^k e^{-\lambda t}}{k!} \qquad \text{when } \eta \text{ is an integer} \tag{25.62}$$

$$R(t) = \sum_{k=0}^{\eta-1} \frac{(\lambda t)^k e^{-\lambda t}}{k!} \qquad \text{when } \eta \text{ is an integer} \tag{25.63}$$

$$h(t) = \frac{f(t)}{R(t)} \qquad [\text{using Eqs. (25.61) and (25.63)}]$$

$$\text{MTBF} = \frac{\eta}{\lambda} \tag{25.64}$$

We will now briefly discuss the application of some of the above continuous distributions in reliability analysis of products. *Exponential distribution* is a good model for the life of a complex system which has a large number of components. Because the exponential distribution has a constant failure rate, it is a good model for the useful life of many products after the end of the infant mortality period. Some applications for the exponential distribution are electrical and electronic systems, computer systems, and automobile transmissions. The *normal distribution* is used to model various physical, mechanical, electrical, or chemical properties of systems. Some examples are gas molecule velocity, wear, noise, chamber pressure from firing ammunition, tensile strength of aluminum alloy steel, capacity variation of electrical condensers, electrical power consumption in a given area, generator output voltage, and electrical resistance. The *lognormal distribution* is a skewed distribution and can be used to model situations where large occurrences are concentrated at the tail (left) end of the range. Some examples are amount of electricity used by different customers, downtime of systems, time to repair, light intensities of bulbs, concentration of chemical process residues, and automotive mileage accumulation by different customers. The *two*-parameter *Weibull distribution* can also be used to model skewed data. When $\beta < 1$, the failure rate for the Weibull distribution is decreasing and hence can be used to model infant mortality or debugging period or for situations when the reliability in terms of failure rate is improving or for reliability growth. When $\beta = 1$, the Weibull distribution is the same as the exponential distribution, and all of the previous comments for the exponential distribution are applicable. When $\beta > 1$, the failure rate is increasing, and hence it is a good model for determining wearout and end-of-useful life period. Some of the examples are corrosion life, fatigue life, antifriction bearings, transmission gears, and life of electronic tubes. The *three*-parameter Weibull distribution is a good model when we have a minimum life and the odds of the component failing before the minimum life are close to zero. Many strength characteristics of systems do have a minimum value significantly greater than zero. Some examples are electrical resistance, capacitance, and fatigue strength.

The failure rate for the gamma distribution is decreasing when $\eta < 1$, is constant when $\eta = 1$, and is increasing when $n > 1$.

EXAMPLE 25.4: Normal Distribution The time to failure for a component is normally distributed with an expected value of 20 hours and a standard deviation of 3 h. We wish to answer the following questions:

1. What is the reliability of the component at 25 h of operating time?
2. What is the probability that the component will fail between 25 and 28 h?
3. What is the failure rate of the component at 25 h of operating time?

For question 1, we have, using Eq. (25.50),

$$R(25) = 1 - F(25) = 1 - \Phi\left(\frac{25 - 20}{3}\right) = 1 - \Phi(1.667)$$

$$= 1 - 0.95221 = 0.04779$$

For question 2, we have, using Eq. (25.49),

$$F(28) - F(25) = \Phi(2.667) - \Phi(1.667)$$

$$= 0.99617 - 0.95221 = 0.04396$$

For question 3, we have, using Eq. (25.51),

$$h(25) = \frac{\Phi[(25 - 20)/3]}{\sigma R(25)} = \frac{0.09949}{3 \times 0.04779} = 0.6939 \text{ failures/h}$$

EXAMPLE 25.5: Weibull Distribution   A component has a Weibull failure distribution for the time to failure with the following parameters:

$$\delta = 1000 \qquad \beta = 4.5 \qquad \theta = 3000$$

One hundred components are put on test at time zero, and we wish to answer the following questions.

1. What is the expected number of components functioning at 2000 time units?
2. What is the expected number of failures in the time interval (2000, 2100)?
3. What is the failure rate at 2000 time units?

We have, using Eq. (25.58),

$$R(2000) = \exp[-(2000 - 1000/3000 - 1000)^{4.5}] = 0.9567$$

$$R(2100) = \exp[-(2100 - 1000/3000 - 1000)^{4.5}] = 0.9344$$

Hence the expected number of components functioning at 2000 is

$$E[N_S(2000)] = 95.67$$

Similarly,

$$E[N_S(2100)] = 93.44$$

and hence the expected number of failures in the interval [2000, 2100] = 2.23. The failure rate is [using Eq. (25.59)]

$$h(2000) = \frac{4.5(1000)^{3.5}}{(2000)^{4.5}} = 0.00019887 \text{ failures/h}$$

EXAMPLE 25.6: Maintainability Analysis   Maintainability is one of the system design parameters which has a great impact on the effectiveness of the system. Failures will occur no matter how reliable a system is made. The ability of a system to be maintained, i.e., retained in, or restored to, effective usable condition is often as important to system effectiveness as is its reliability. Maintainability is a characteristic of systems and it is designed just like reliability. It is concerned with such system attributes as accessibility to failed parts, diagnosis of failures, repairs, test points, test equipment and tools, maintenance manuals, displays, and safety. *Maintainability* can be defined as a characteristic of design and installation which imparts to a system a great inherent ability to be maintained, so as to lower the required maintenance labor-hours, skill levels, tools, test equipment, facilities, and logistics costs and thus achieve greater availability.

maintainability measures   Maintainability is the probability that a system in need of maintenance will be retained in, or restored to, a specified operational condition within a given period of time. Thus, the underlying random variable is the maintenance time. Let T be the repair time random variable. Then the maintainability function $M(t)$ is given by

$$M(t) = P[T \le t]$$

If the repair time T follows the exponential distribution with mean time to repair (MTTR) of $1/\mu$, where μ is the repair rate, then

$$M(t) = 1 - \exp\left(-\frac{t}{\text{MTTR}}\right)$$

Various other distributions such as lognormal, Weibull, and normal are used to model the repair time. In addition, other time-related indices such as median (50th percentile) and M_{max} (90th or 95th percentile) are used as maintainability measures. The lognormal probability-density functions with a median time to repair of 15 minutes but with different values for standard deviation are given in Fig. 25.5, and Fig. 25.6 shows the associated maintainability functions. Distribution 1 has the least variability, and distribution 3 has the highest variability. From the maintainability function plot, different percentiles, such as 90th, can be easily read. Distribution 3 has the highest value for the 90th percentile. In other instances, the maintenance labor-hours per system operating hour or maintenance ratio (MR) can be specified and maintainability design goals then derived from such specifications.

MTTR, which is the mean of the distribution of system repair time, can be evaluated by

$$\text{MTTR} = \frac{\sum\limits_{i=1}^{n} \lambda_i t_i}{\sum\limits_{i=1}^{n} \lambda_i}$$

where n = number of components in the system
λ_i = failure rate of the ith repairable component
t_i = time required to repair the system when the ith component fails

In addition, other quantities, such as mean active corrective-maintenance time and mean active preventive-maintenance time, are used to measure maintainability. Some components of the corrective maintenance tasks are

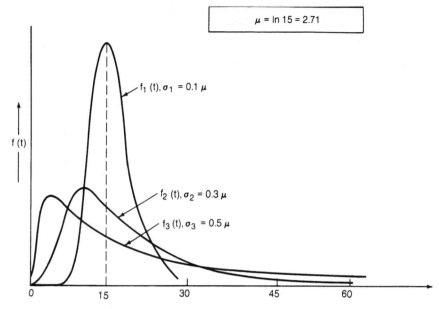

FIGURE 25.5   Lognormal probability density function.

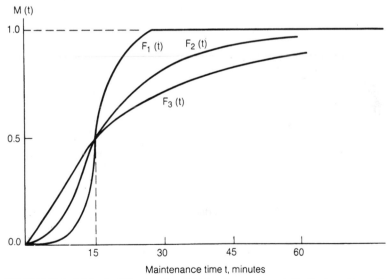

FIGURE 25.6   Maintainability functions $M(t)$ based on lognormal distribution with median = 15 min.

Localization: determining the location of a failure to the extent possible, without using accessory test equipment

Isolation: determining the location of a failure by the use of accessory test equipment

Disassembly: disassembling the equipment to gain access to the item being replaced

Interchange: removing the failed item and installing the replacement

Alignment: performing any alignment, testing, and adjustment made necessary by the repair action

Checkout: performing checks or tests to verify that the equipment has been restored to a satisfactory operating condition

As an example of MTTR computation, assume a communication system consists of five assemblies with the data given in Table 25.3. Column 2 gives the number of units n_i for assembly *i*. Column 3 indicates the failure rate per thousand hours for each unit. Thus, column 4 gives us

TABLE 25.3 Worksheet for MTTR Prediction

1	2	3	4	5	6
Assemblies	n_i	$\lambda_i \times 10^3$	$n_i\lambda_i \times 10^3$	t_p h	Repair time per 10^3 h, $n_i\lambda_i t_i$
1	4	10	40	0.10	4.0
2	6	5	30	0.20	6.0
3	2	8	16	1.00	16.0
4	1	15	15	0.50	7.5
5	5	12	60	0.50	30.0
			$\Sigma = 161$		$\Sigma = 63.5$

the total failure rate for an assembly *i*. Column 5 gives the average time to perform all the maintenance actions discussed before. Then, MTTR is given by

$$\text{MTTR} = \frac{\Sigma\, n_i\lambda_i t_i}{\Sigma\, n_i\lambda_i}$$

$$= \frac{63.5}{161} = 0.394 \text{ h}$$

EXAMPLE 25.7: Gamma Distribution The failure distribution of a component has the form of the gamma distribution with parameters $\eta = 2$ and $\lambda = 0.001$. We wish to determine the reliability of the component and the hazard function after an operation of 100 time units and also the mean life of the component. Here $\eta = 2$, $\lambda = 0.001$, $t = 100$, and hence $\lambda t = 0.1$. Using Eq. (25.61), we have

$$f(t) = \frac{(0.001)^2}{\Gamma(2)}\, te^{-0.001t} \qquad t \geq 0$$

and

$$f(100) = \frac{(0.001)^2}{1} \times 100e^{-0.1} = 9 \times 10^{-5}$$

Also from Eq. (25.63)

$$R(t) = \sum_{k=0}^{\eta-1} \frac{(\lambda t)^k e^{-\lambda t}}{k!}$$

Hence

$$R(100) = \frac{(0.1)^0 e^{-0.1}}{1} + \frac{(0.1)^1 e^{-0.1}}{1} = 0.995$$

Also, we have

$$h(100) = \frac{f(100)}{R(100)} = \frac{9 \times 10^{-5}}{0.995} = 9.09 \times 10^{-5} \text{ failures/time unit}$$

Mean life = $2/0.001$ = 2000 time units [using Eq. (25.64)].

25.3 ELEMENTS OF STATISTICAL THEORY

In order to evaluate the reliability of a system, we have to observe and collect data. We only collect data for a part of the population. Statistics is the science of drawing conclusions about a population based on an analysis of the sample data from the population. A subset of observations selected from a population is called a *sample*. There are many different ways to take samples from a population. Furthermore, the conclusions that we can draw about the population will depend on how the sample is selected. We want the samples to be *representative* of the population and we want the sample to be random. A *random sample* is one such that every member of the population has an equal chance of being in the sample.

The process of drawing conclusions about populations based on sample data makes considerable use of statistics. A *statistic* is any function of the observations in a random sample that does not depend on any unknown parameters. For example, if $X_1, X_2, \ldots, X_n$ is a random sample of size n, then the sample mean $\overline{X}$ and the sample variance S^2 are statistics, where

$$\overline{X} = \frac{\sum_{i=1}^{n} X_i}{n} \tag{25.65}$$

and

$$S^2 = \frac{\sum_{i=1}^{n} (X_i - \overline{X})^2}{n-1} \tag{25.66}$$

The statistical procedures require that we understand the probabilistic behavior of the statistic. In general, we call the probability distribution of a statistic a *sampling distribution*. There are several important sampling distributions used in reliability estimation and we briefly discuss some of them.

25.3.1 Sampling Distributions[6,7]

For the sampling distributions of $\overline{X}$ [Eq. (25.65)], we have

$$\mu_{\overline{X}} = \mu_X \quad \text{and} \quad \sigma_{\overline{X}}^2 = \frac{\sigma_X^2}{n} \tag{25.67}$$

Central-Limit Theorem.   If a population has a mean μ and a finite variance σ^2, then the sampling distribution of means approaches the normal distribution with mean μ and variance σ^2/n as n increases, i.e., the sampling distribution of mean is asymptotically normal.

Chi-Square Distribution.   Let $Z_1, Z_2,..., Z_\nu$ be normally and independently distributed random variables, with mean $\mu = 0$ and variance $\sigma^2 = 1$. Then the random variable

$$\chi^2 = Z_1^2 + Z_2^2 + \cdots + Z_\nu^2 \tag{25.68}$$

has χ^2 distribution with ν degrees of freedom with the probability-density function

$$f_{\chi^2}(u) = \frac{1}{2^{\nu/2}\Gamma(\nu/2)}\, u^{(\nu/2)-1}e^{-u/2} \qquad u > 0 \tag{25.69}$$

The percentage points of χ_ν^2 distribution are given in the Appendix.

$$P[\chi_\nu^2 \geq \chi_{\alpha,\,\nu}^2] = \alpha \tag{25.70}$$

For example, $P[\chi_{18}^2 \geq \chi_{0.05,\,18}^2] = P[\chi_{18}^2 \geq 18.31] = 0.05$. When ν is large, χ^2 can be approximated by normal distribution and the approximation is given by

$$\chi_{\alpha,\nu} \approx \frac{[Z_\alpha + \sqrt{2\nu - 1}]^2}{2} \tag{25.71}$$

For example, we wish to find $\chi_{0.05,\,170}^2$. We have

$$\chi_{0.05,\,170}^2 \approx \frac{[1.645 + \sqrt{2(170) - 1}]^2}{2} = 201.10$$

Theorem.   If $X_1,..., X_n$ are random samples from a normal population with mean μ and variance σ^2, then $(n-1)S^2/\sigma^2$ is distributed as chi-square with $(n-1)$ degrees of freedom.

Goodness-of-Fit Test.   The test statistic is given by

$$\chi^2 = \frac{\sum\limits_{i=1}^{k}(f_{oi} - f_{ei})^2}{f_{ei}} \tag{25.72}$$

where k = number of classes
$\quad f_{oi}$ = observed frequency of ith class
$\quad f_{ei}$ = expected frequency of the ith class (should be greater than 5)

The larger value of χ^2 [Eq. (25.72)], the greater the discrepancy. The degrees of freedom are given by $\nu = k - 1 - m$, where m is the number of population parameters used in the computation of expected frequencies.

Student's t Distribution.   Let Z be normally distributed with mean 0 and variance 1, and V be a χ^2 with ν degrees of freedom. If Z and V are independent, then the random variable

$$T = \frac{Z}{\sqrt{V/k}} \tag{25.73}$$

has the probability-density function

$$f(t) = \frac{\Gamma[(\nu + 1)/2]}{\sqrt{\pi\nu}\Gamma(\nu/2)}\left(1 + \frac{t^2}{\nu}\right)^{-(\nu + 1)/2} \qquad -\infty < t < \infty \tag{25.74}$$

and is said to follow the t distribution with ν degrees of freedom and abbreviated t_ν.

The cumulative distribution functions for the t distribution are given in the Appendix.

If X_i, $i = 1,\ldots, n$ are independently normally distributed random variables with mean μ and variance σ^2, then $Z = (\overline{X} - \mu)/(\sigma/\sqrt{n})$ is $N(0, 1)$. Furthermore, $t = (\overline{X} - \mu)/(S/\sqrt{n})$ has a t distribution with $(n - 1)$ degrees of freedom.

F Distribution. Given two independently distributed chi-square variables χ_1^2 with ν_1 degrees of freedom and χ_2^2 with ν_2 degrees of freedom, the random variable

$$F = \frac{\chi_1^2/\nu_1}{\chi_2^2/\nu_2} \tag{25.75}$$

has an F distribution with ν_1 and ν_2 degrees of freedom. The probability-density function of F is given by

$$f(F) = \frac{\Gamma[(\nu_1 + \nu_2)/2]}{\Gamma(\nu_1/2)\Gamma(\nu_2/2)}\left(\frac{\nu_1}{\nu_2}\right)^{\nu_1/2}\frac{F^{(\nu_1/2) - 1}}{[1 + (\nu_1/\nu_2)F]^{(\nu_1 + \nu_2)/2}} \qquad 0 < F < \infty \tag{25.76}$$

The percent points of the F distribution are given in App. A.7. For example, $P[F \geq F_{0.05,5,10}] = P[F \geq 3.33] = 0.05$. We also have

$$F_{(1 - \alpha), \nu_1, \nu_2} = \frac{1}{F_{\alpha, \nu_1, \nu_2}}$$

Consider two random samples, one from each of two normal populations; then the statistic $(s_1^2/\sigma_1^2)/(s_2^2/\sigma_2^2)$ is distributed as an F distribution with $(n_1 - 1)$ and $(n_2 - 1)$ degrees of freedom.

25.3.2 Statistical Inference

Statistical inference is the process by which information from sample data is used to draw conclusions about the population from which the sample was selected. The techniques of statistical inference can be divided into two major areas: parameter estimation and hypothesis testing.

Parameter Estimation. One of the important problems in reliability is to estimate the parameters of the life distributions. For example, we wish to know the average life of a light bulb. An *estimator* is some function of the sample values which provides an estimate of the population parameter. A single-valued estimate of the population parameter is called a *point estimate*. Let $\hat{\theta} = h(X_1,\ldots, X_n)$ be a point estimator of the parameter θ. Some of the properties of the estimators are given below.

1. $\hat{\theta}$ is an *unbiased* estimator of θ if $E(\hat{\theta}) = \theta$.
2. $\hat{\theta}$ is a *consistent* estimator of θ if $\hat{\theta}$ converges to θ in probability as sample size increases.

3. If $\hat{\theta}_1$ and $\hat{\theta}_2$ are two different unbiased estimators of θ and if $E[\hat{\theta}_1 - \theta)^2] < E[\hat{\theta}_2 - \theta)^2]$, then $\hat{\theta}_1$ is a more *efficient* estimator of θ than $\hat{\theta}_2$.

4. $\hat{\theta}$ is a *sufficient* estimator if no other independent estimate based on the sample is able to yield any further information about the parameter which is being estimated.

To construct an interval estimator of the unknown parameter θ, we find two numbers θ_L and θ_U such that

$$P[\theta_L \le \theta \le \theta_U] = 1 - \alpha$$

and the resulting interval, $\theta_L \le \theta \le \theta_U$, is called a $100(1 - \alpha)$ percent confidence interval for θ. The interpretation of the confidence interval is that if many random samples are collected, and a $100(1 - \alpha)$ percent confidence interval on θ is computed from each sample, then $100(1 - \alpha)$ percent of these intervals will contain the true value of θ.

Tests of Hypotheses.   Many reliability problems require that we decide whether or not a statement about some parameter of the life distribution is true or false. This statement is usually called a *hypothesis.* For example, we want to say that the mean life of a component which follows exponential distribution is at least 1000 h. The *null hypothesis* (H_0) is the hypothesis that we are interested in. Any hypothesis which differs from this is called the *alternate hypothesis,* designated by H_1. Type I error, designated α (also called *producer's risk*), is the error made in rejecting a hypothesis that is true. 100α is the *significance* level of the test. Type II error (see Table 25.4), designated β (also called *consumer's risk*), is the error made in accepting a hypothesis which is false. $(1 - \beta)$ is called the *power* of the test.

TABLE 25.4   Decisions in Hypothesis Testing

	H_0 is true	H_0 is false
Accept H_0	No error $(1 - \alpha)$	Type II error or β error
Reject H_0	Type I error or α error	No error $(1 - \beta)$

A graph of $(1 - \beta)$ versus values of H_1 is called the *power curve.* A graph of β versus values of H_1 is called the operating characteristic (OC) curve. Thus the OC curve gives us a Type II error as a function of different values of the underlying parameter for which we have formulated a hypothesis. Since reliability is a function of time as well as of the parameters of the underlying distribution, it is not practical to provide OC curves in terms of actual reliability. OC curves are mostly provided in terms of the unknown parameter of the assumed distribution. In practice, the problem of selecting a test procedure on the basis of operating characteristics can be formulated as that of selecting two points on the OC curve and then finding a test plan which meets this specification. (See Sec. 25.9 for illustration.)

25.4   *THE EXPONENTIAL DISTRIBUTION*

The exponential distribution is a very popular and easy-to-use model to represent time to failure. Selection of the exponential distribution as an appropriate model implies that the failure rate is constant over the range of predictions. For certain failure situations and over certain portions of product life, the assumption of a constant failure rate may be appropriate. The probability density function for an exponentially distributed time-to-failure random variable T is given by [see Eq. (25.43)]

$$f(t, \lambda) = \lambda e^{-\lambda t} \qquad t \geq 0 \tag{25.77}$$

where the parameter λ is the failure rate. The reciprocal of the failure rate ($\theta = 1/\lambda$) is the mean or expected life. For products that are repairable, the parameter θ is referred to as the mean time between failures (MTBF) and for nonrepairable products θ is called the *mean time to failure* (MTTF). The parameter λ must be known (or estimated) for any specific application situation.

The reliability function is given by

$$R(t) = e^{-\lambda t} \qquad t \geq 0 \tag{25.78}$$

The relationship between $f(t)$ and $R(t)$ is illustrated in Fig. 25.7. For any value of t the quantity $R(t)$ provides the chance of survival beyond time t. If the reliability function is evaluated at the MTBF, it should be noted that $R(\theta) = 0.368$, or there is only a 36.8 percent chance of surviving the mean life.

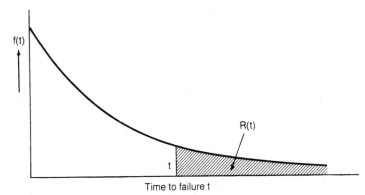

FIGURE 25.7   The exponential distribution.

25.4.1 Point Estimation[5,7]

The estimator for the mean life parameter θ is given by

$$\hat{\theta} = \frac{T}{r} \tag{25.79}$$

where T = the total accumulated test time considering both failed and unfailed items, and r = the total number of failures. It is reasonable to take the estimator for λ as $\hat{\lambda} = 1/\hat{\theta}$. The estimator $\hat{\theta}$ is termed a *maximum likelihood* estimator. It has the properties of *unbiasedness, minimum variance, efficiency,* and *sufficiency.*

The reliability is estimated by

$$\hat{R}(t) = e^{-\hat{\lambda} t} \qquad t \geq 0 \tag{25.80}$$

and if one wants to estimate a time for a given reliability level R, this is obtained by

$$\hat{t} = \hat{\theta} \ln\left(\frac{1}{R}\right) \tag{25.81}$$

Confidence-Interval Estimates.[8–11]   The confidence intervals for the mean life or for reliability depend on the testing situation. A life-testing situation where n devices are placed on test and it is agreed to terminate the test at the time of the rth failure ($r \leq n$) is called *failure-censored* and is termed *Type II life testing*. A time-censored life test is one where the total accumulated time T is specified; this is termed *Type I life testing*. So, in Type I testing T is specified and r occurs as a result of testing, whereas in a Type II situation, R is specified and T results from testing. The confidence limits are slightly different for each situation.

Failure-Censored Life Tests.   In this situation the number of failures r at which the test will be terminated is specified with n items placed on test. The $100\,(1 - \alpha)$ percent two-sided confidence interval for θ, the mean life, is

$$\frac{2T}{\chi^2_{\alpha/2,\,2r}} \leq \theta \leq \frac{2T}{\chi^2_{1 - \alpha/2,\,2r}} \tag{25.82}$$

The quantities $\chi^2_{\delta,\,v}$ are the $(1 - \delta)$ percentiles of the chi-square distribution with v degrees of freedom.

The one-sided lower $100(1 - \alpha)$ percent confidence limit is given by

$$\frac{2T}{\chi^2_{\alpha,\,2r}} \leq \theta \tag{25.83}$$

The confidence limits on reliability for any specified time can be found by substituting the above limits on θ into the reliability function.

Time-Censored Life Tests.   In this situation the accumulated test time T is specified for the test and the test produces r failures. The $100(1 - \alpha)$ percent two-sided confidence interval on the mean life is given by

$$\frac{2T}{\chi^2_{\alpha/2,\,2(r + 1)}} \leq \theta \leq \frac{2T}{\chi^2_{1 - \alpha/2,\,2r}} \tag{25.84}$$

If only the one-sided $100(1 - \alpha)$ percent lower confidence interval is desired, this is given by

$$\frac{2T}{\chi^2_{\alpha,\,2(r + 1)}} \leq \theta \tag{25.85}$$

For a number of failure-censored life tests, the termination time T_r is a random variable with

$$E[T_r] = \theta \sum_{i=1}^{r} \frac{1}{n - i + 1} \tag{25.86}$$

and

$$V[T_r] = \theta^2 \sum_{i=1}^{r} \frac{1}{(n - i + 1)^2} \tag{25.87}$$

EXAMPLE 25.8   A designer wants to use a certain electronic counter in the design of a piece of equipment. Results from a previous test on 50 counters shows that none failed during a 3000-h test. Based on this test and a 90 percent one-sided confidence limit, is it realistic to expect that the probability of survival of this counter will be at least 85 percent for a one-year period of operation? Assume exponential distribution as the underlying failure distribution. For this example, we have

$$\text{Total test time} = 50 \times 3000 = 150{,}000 \text{ h}$$

Hence, using Eq. (25.85), we have

$$\frac{2T}{\chi^2_{0.1,\,2}} \le \theta \quad \text{or} \quad \frac{2 \times 150{,}000}{4{,}605} \le \theta \quad \text{or} \quad 65{,}146 \le \theta$$

Hence,

$$R(365 \times 24) = R(8760) = e^{-8760/65{,}146} = 0.8742$$

Hence it is realistic to expect that the probability of survival of this counter will be at least 85 percent for a one-year period of operation.

25.4.2 Hypothesis Testing

Hypothesis testing is another approach to statistical decision making. Whereas confidence limits offer some degree of protection against meager statistical knowledge, the hypothesis testing approach is easily misused. To use this approach one should be very familiar with the inherent errors involved in applying the hypothesis testing procedure.

Consider a Type II testing situation where n items are placed on test and the test is terminated at the time of the rth failure. An MTBF goal θ_g is to be verified. The hypotheses for this situation are

$$
\begin{aligned}
H_0: & \quad \theta \le \theta_g \\[2mm]
H_1: & \quad \theta > \theta_g
\end{aligned}
\tag{25.88}
$$

If H_0 is rejected, it will be concluded that the goal has been met.

The statistic calculated from test data is

$$\chi^2_c = \frac{2T}{\theta_g} \tag{25.89}$$

and the decision criteria is to reject H_0 and assume that the goal has been met if $\chi^2_c > \chi^2_{\alpha,\,2r}$. Here the level of significance is taken as α.

The probability of accepting a design, that is, concluding that the goal has been met where the design has a true MTBF of θ_1, is

$$P_1 = P\left[\chi^2_{2r} \ge \frac{\theta_g}{\theta_1}\, \text{X}^2_{\alpha,\,2r}\right] \tag{25.90}$$

and can be looked up in χ^2 tables.

EXAMPLE 25.9 The reliability of an electronic device is under investigation. The MTBF goal for this device for a particular application is 2000 h.

Ten units are placed on test and the test is truncated at the time of the second failure. The failure times are

$$t_i: \qquad 187 \text{ h}; \qquad 462 \text{ h}$$

The estimated mean life would be

$$\hat{\theta} = \frac{187 + 462 + 8(462)}{2} = 2172.5 \text{ h}$$

The calculated χ^2 value is

$$\chi_c^2 = \frac{2(4345)}{2000} = 4.345$$

Using $\alpha = 0.10$, the critical value as obtained from χ^2 tables is

$$\chi_{0.10, 4}^2 = 7.779$$

So, in this case we would not conclude that the goal has been met.

Now let us investigate the Type II error associated with this procedure. Suppose the true mean was $\theta_1 = 3000$ h; then the probability of concluding that the goal was met by this testing procedure is

$$P_1 = P\left[\chi_4^2 \geq \frac{2000}{3000} \ (7.779) \right] = 0.24$$

Or, even though the goal is 2000 h, the decision procedure has only a 24 percent chance of accepting a design with a true MTBF of 3000 h. Obviously the decision procedure may not be acceptable in a practical situation. The procedure can be improved by increasing r, the number of failures. For example, if we used $r = 8$, then

$$P_1 = P\left[\chi_{16}^2 \geq \frac{2000}{3000} \ (7.779) \right] = 0.995$$

Or, we now only have a 0.5 percent chance of concluding that the design has not met the MTBF goal.

25.5   THE WEIBULL DISTRIBUTION

The Weibull distribution[11-13] is considerably more versatile than the exponential distribution and can be expected to fit many different failure patterns. However, when applying a distribution, the failure pattern should be carefully studied and the mixture of failure modes noted. The selection and application of a distribution should then be based on this study and on any knowledge of the underlying physical failure phenomena.

Graphical procedures for the Weibull distribution are attractive in that they provide practitioners with a visual representation of the situation. Although the graphical approach will be covered in the following material, it should be recognized that there are better statistical estimation procedures; however, these procedures require the use of a computer. Different computer programs are available in the open literature and should be used wherever possible.

The reliability function for the three-parameter Weibull distribution is given by

$$R(t) = \exp\left[-\left(\frac{t - \delta}{\theta - \delta}\right)^{\beta}\right] \qquad t \geq \delta \tag{25.91}$$

where δ = the minimum life ($\delta \geq 0$)
  θ = the characteristic life ($\theta > \delta$)
  β = the Weibull slope or shape parameter ($\beta > 0$)

The two-parameter Weibull has a minimum life of zero, and the reliability function is

$$R(t) = \exp\left[-\left(\frac{t}{\theta}\right)^{\beta}\right] \qquad t \geq 0 \tag{25.92}$$

where θ and β are as previously defined ($\theta > 0$). The term characteristic life resulted from the fact that $R(\theta) = 0.368$; or, there is a 36.8 percent chance of surviving the characteristic life for any Weibull distribution.

The hazard function for the Weibull distribution is given by

$$h(t) = \frac{\beta}{\theta^{\beta}} t^{\beta - 1} \qquad t \geq 0 \tag{25.93}$$

It can be seen that the hazard function will decrease for $\beta < 1$, increase for $\beta > 1$, and remain constant for $\beta = 1$.

The expected or mean life for the two-parameter Weibull is given by

$$\mu = \theta\Gamma\left(1 + \frac{1}{\beta}\right) \tag{25.94}$$

where $\Gamma(\cdot)$ is a gamma function and its value can be found in gamma tables. The standard deviation for the Weibull distribution is

$$\sigma = \theta\sqrt{\Gamma\left(1 + \frac{2}{\beta}\right) - \Gamma^2\left(1 + \frac{1}{\beta}\right)} \tag{25.95}$$

25.5.1 Graphical Estimation[9,11]

The Weibull distribution is very amenable to graphical estimation. This procedure will now be illustrated.

The cumulative distribution for the two-parameter Weibull is given by

$$F(t) = 1 - \exp\left[-(t/\theta)^{\beta}\right] \tag{25.96}$$

Hence by rearranging and taking logarithms one can obtain

$$\ln\left(\ln\frac{1}{1 - F(t)}\right) = \beta \ln t - \beta \ln \theta \tag{25.97}$$

Weibull paper is scaled such that t_j and $p_j = F(t_j)$ can be plotted directly and a straight line fitted to the data. A convenient way to assign values to p_i for plotting is to calculate

$$p_j = \frac{j - 0.3}{n + 0.4} \tag{25.98}$$

where j is the order of magnitude of the observation and n is the sample size. This is essentially median rank plotting and tables for median ranks are available[11] and can also be used.

EXAMPLE 25.10   The design for an aluminum flexible-drive hub on computer disk packs is under study. The failure mode of interest is fatigue. Data from 14 hubs placed on an accelerated life test follow (see Table 25.5). The plotted data with a visually fitted line are shown on the Weibull paper in Fig. 25.8. The slope of this line provides an estimate of β, which in this case is about 2.3. Most commercially available Weibull papers have a special scale for estimating β.

TABLE 25.5   Data for Aluminum Flexible-Drive Hub

j	Cycles to failure	p_j*
1	93,000	5.6
2	147,000	13.5
3	192,000	21.7
4	214,000	29.8
5	260,000	37.9
6	278,000	46.0
7	297,000	54.0
8	319,000	62.1
9	349,000	70.1
10	388,000	78.2
11	460,000	86.3
12	510,000	94.4

*The p_j values were calculated using Eq. (25.98).

The characteristic life can be estimated by recalling that $R(\theta) = 0.368$; or that $F(\theta) = 0.632$. So, one can locate 63.2 percent on the cumulative probability scale, project across to the plotted line, and then project down to the time-to-failure axis. In Fig. 25.8, the characteristic life is about 330,000 cycles.

Weibull paper offers a quick and convenient method for analyzing a failure situation. The population line plotted on the paper can be used to estimate either percent failure at a given time or the time at which a given percentage will fail. Also, a concave plot is indicative of a nonzero minimum life.[9–11]

Confidence Limits for Graphical Analysis.   Lower and upper confidence limits for graphical analysis can be computed using the following equations.[11]
Lower limit:

$$w_\alpha = \frac{j/(n - j + 1)}{F_{1 - \alpha, 2(n - j + 1), 2j} + j/(n - j + 1)} \qquad \alpha \geq 0.50 \tag{25.99}$$

Upper limit:

$$w_\alpha = \frac{[j/(n - j + 1)]F_{\alpha, 2j, 2(n - j + 1)}}{1 + [j/(n - j + 1)]F_{\alpha, 2j, 2(n - j + 1)}} \qquad \alpha < 0.50 \tag{25.100}$$

Use $\alpha/2$ for two-sided limits.

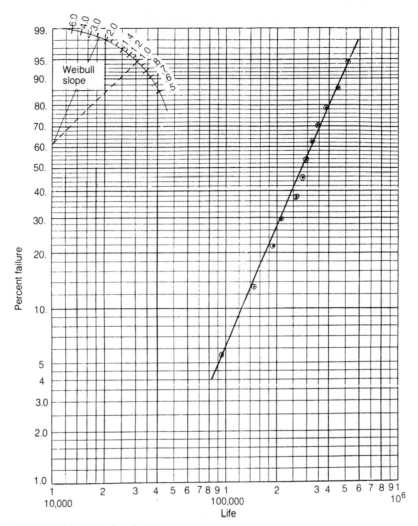

FIGURE 25.8   Weibull probability paper.

25.6   SUCCESS/FAILURE TESTING

Success/failure testing[9,10] describes a situation where a product (component, subsystem, etc.) is subjected to a test for a specified length of time T (or cycles, stress reversals, miles, etc.). The product either survives to time T (i.e., survives the test) or fails prior to time T.

Testing of this type can frequently be found in engineering laboratories where a test "bogey" or target has been established and new designs are tested against this bogey. The bogey will specify a set number of cycles in a certain test environment and at a predetermined stress level.

The probability model for this testing situation is the binomial distribution given by

$$P(y) = \binom{n}{y} R^y (1 - R)^{n-y} \qquad y = 0, 1, 2, \ldots, n \qquad (25.101)$$

where R = the probability of surviving the test
n = the number of items placed on test
y = the number of survivors

The value R is the reliability, which is the probability of surviving the test. Procedures for estimating product liability R based on this testing situation will now be covered.

25.6.1   Point Estimate

The point estimate of reliability is simply calculated as

$$\hat{R} = \frac{y}{n} \tag{25.102}$$

25.6.2   Confidence-Limit Estimate

The $100(1 - \alpha)$ percent lower confidence limit on the reliability R is calculated by

$$R_L = \frac{y}{y + (n - y + 1)F_{\alpha,\, 2(n - y + 1),\, 2y}} \tag{25.103}$$

where $F_{\alpha,\, 2(n - y + 1),\, 2y}$ is obtained from F tables. Here again n is the number of items placed on test and y is the number of survivors. The $100(1 - \alpha)$ percent upper confidence limit on R is given by

$$R_U = \frac{(y + 1)F_{\alpha,\, 2(y + 1),\, 2(n - y)}}{(n - y) + (y + 1)F_{\alpha,\, 2(y + 1),\, 2(n - y)}} \tag{25.104}$$

The F tables that are usually available are somewhat limited. Therefore, it is convenient to have an approximation for the lower confidence limit that uses the standard normal distribution. The lower confidence on reliability can be approximated by

$$R_L = \frac{y - 1}{n + z_\alpha \sqrt{n(n - y + 1)/(y - 2)}} \tag{25.105}$$

where z_α = the standard normal variate as given in Table 25.6
y = the number of successes
n = the sample size

It should be noted that $P[Z \geq z_\alpha] = \alpha$ where Z is the standard normal variable. Values given in Table 25.6 can be read from cumulative distribution tables for standard normal variables given in App. A.4.

TABLE 25.6   Standard Normal Variates

Confidence level, $1 - \alpha$, %	Confidence factor, z_α
95	1.645
90	1.281
80	0.841
70	0.525
50	0.0

EXAMPLE 25.11 A weapon system has completed a test schedule. The test is equivalent to 60 missions. Dividing the test schedule up into 60 missions results in seven failed missions. Let us estimate the mission reliability.

In this case the number of successes (y) is

$$y = 60 - 7 = 53 \text{ successful missions}$$

out of $n = 60$ missions. Then the point estimate for mission reliability is

$$\hat{R}_m = 53/60 = 0.883$$

Let us now find a 75 percent lower confidence limit. The exact lower 75 percent limit is found by using an F value of

$$F_{0.25, 8, 106} = 1.31$$

and substituting into the confidence limit, Eq. (25.103) gives

$$R_L = \frac{53}{53 + (8 \times 1.31)} = 0.835$$

Or, the 75 percent lower confidence limit on mission reliability is $0.835 \leq R_m$. If the normal approximation was used, the lower limit's value would be

$$R_L = \frac{52}{60 + 0.675\sqrt{[60(60 - 53 + 1)/51]}} = 0.838$$

As can be seen, this approximation provides limits that are reasonably close to the exact values.

25.6.3 Success Testing

In receiving inspection and sometimes in engineering test labs one encounters a situation where a no-failure ($r = 0$) test is to be specified. The concern is usually to ensure that a reliability level has been achieved at a specified confidence level. A special adaptation of the confidence-limit formula can be derived for this situation.

For the special case where $r = 0$ (i.e., no failures), the lower $100(1 - \alpha)$ percent confidence limit on the reliability is

$$R_L = \alpha^{1/n} \tag{25.106}$$

where α is the level of significance and n is the sample size (i.e., number placed on test). Then with $100(1 - \alpha)$ percent confidence, we can say that

$$R_L \leq R$$

where R is the true reliability.

If we let $C = 1 - \alpha$ be the desired confidence level (0.80, 0.90, etc.), then the necessary sample size to demonstrate a desired reliability level R is

$$n = \frac{\ln (1 - C)}{\ln R} \qquad (25.107)$$

For example, if $R = 0.80$ is to be demonstrated with 90 percent confidence, we have

$$n = \frac{\ln 0.10}{\ln 0.80} = 11$$

Thus, we must place 11 items on test and allow no failures. This is frequently referred to as *success testing*.

25.6.4 Conversion to the Exponential Distribution

In some cases, where it is reasonable to assume an exponential time-to-failure distribution, it is desirable to obtain the reliability to predict failure over time. If

n = number placed on test
r = number failing
y = number of successes

then a point estimate of the fraction failing the fixed length T test is

$$\hat{p} = \frac{r}{n}$$

Or, the reliability is estimated as

$$\hat{R} = \frac{y}{n}$$

From the exponential distribution, the probability of failing prior to time T is given by

$$F(T) = 1 - e^{-T/\theta}$$

So, an estimate of the MTBF can be obtained from

$$\frac{r}{n} = 1 - e^{-T/\theta}$$

or

$$\hat{\theta} = \frac{T}{\ln [1/(1 - r/n)]} \qquad (25.108)$$

Confidence limits can also be placed on the MTBF. If we denote the lower confidence limits on the reliability by

$$R_L \leq R$$

then, one can also say

$$R_L \leq e^{-T/\theta}$$

or

$$\frac{T}{\ln\left(1/R_L\right)} \leq \theta \qquad (25.109)$$

In the case of success testing where no failures are allowed, the lower confidence limit on the MTBF is [using Eq. (25.106)]

$$-\frac{nT}{\ln \alpha} \leq \theta$$

Or, for any specified time t, the $100(1 - \alpha)$ percent lower confidence limit on reliability is

$$\exp\left[-\frac{t(-\ln \alpha)}{nT}\right] \leq R(t) \qquad (25.110)$$

25.7   SYSTEM-RELIABILITY MODELS

Dynamic system reliability models are an extension of the static models (Example 25.1) where time-dependent reliability functions are used for each subsystem. Consider the following notation:

$$R_i(t) = \text{the reliability function for the } i\text{th system}$$

$$R_s(t) = \text{the system reliability function}$$

These will be used to reformulate series and parallel systems considered in Example 25.1.

25.7.1   Series Configuration

The system reliability is given by

$$R_s(t) = \prod_{i=1}^{n} R_i(t) \qquad (25.111)$$

where n = the number of subsystems.
    The failure rate $h_s(t)$ for the series system is given by

$$h_s(t) = \sum_{i=1}^{n} h_i(t) \qquad (25.112)$$

where $h_i(t)$ is the failure rate for the ith subsystem.
    If all subsystems have an exponentially distributed time to failure, then

$$h_s(t) = \sum_{i=1}^{n} \lambda_i \qquad (25.113)$$

where λ_i is the failure rate for the ith subsystem.

25.7.2   Parallel Configuration

The system reliability for a parallel configuration where all parallel subsystems are activated when the system is turned on is given by

$$R_s(t) = 1 - \prod_{i=1}^{n} [1 - R_i(t)] \tag{25.114}$$

If the time to failure for each subsystem is exponential, then

$$R_s(t) = 1 - \prod_{i=1}^{n} [1 - \exp(-\lambda_i t)] \tag{25.115}$$

If $\lambda_i = \lambda = 1/\theta$ for all i, then the expected system life is given by

$$E_s(T) = \sum_{i=1}^{n} \frac{\theta}{i} \tag{25.116}$$

A standby parallel redundant system is depicted in Fig. 25.9. In this system the standby unit is activated by the switching mechanism S. For a system with two units (i.e., one standby unit), the system reliability is

$$R_s^2(t) = R_1(t) + \int_0^t f_1(t_1)R_2(t - t_1)dt_1 \tag{25.117}$$

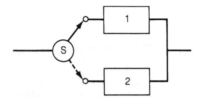

FIGURE 25.9   Standby redundant configuration.

where it is assumed that the switch cannot fail. Here $f_1(t)$ is the probability-density function for device 1 and $R_2(t)$ is the reliability function for device 2. For the special case where each subsystem is identical with an exponential time to failure, the system reliability is

$$R_s^2(t) = e^{-\lambda t}(1 + \lambda t) \qquad t \geq 0 \tag{25.118}$$

Similarly, if we have three subsystems, the reliability of the system is given by

$$R_s^3(t) = e^{-\lambda t}\left[1 + \lambda t + \frac{(\lambda t)^2}{2}\right] \qquad t \geq 0 \tag{25.119}$$

For the general case, when we have n subsystems,

$$R_s^n(t) = e^{-\lambda t}\sum_{i=0}^{n-1} \frac{(\lambda t)^i}{i!} \tag{25.120}$$

If the switch has a reliability of P_s, then the reliability for the two-unit system is

$$R_s^2(t) = R_1(t) + P_s \int_0^t f_1(t_1)R_2(t - t_1)dt_1 \tag{25.121}$$

For identical subsystems we have an exponential time to failure of

$$R_s^2(t) = e^{-\lambda t}(1 + P_s\lambda t) \qquad t \geq 0 \tag{25.122}$$

25.8   *BAYESIAN RELIABILITY IN DESIGN AND TESTING*

Probability is a mathematical concept used in conjunction with random events. The concept of relative frequency is widely used to define probability. Suppose we are interested in the probability of an event A associated with a random experiment ε. We perform the experiment N times and event A occurs N_A times, $0 \le N_A \le N$. Then, the probability of event A is defined by

$$P(A) = \lim_{N \to \infty} \frac{N_A}{N}$$

Thus probability is associated with "long-run" percentages. Another popular use of probability is as "degree of belief." The term *Bayes' probability* has become associated with degree-of-belief probability. The degree-of-belief probability is also called *subjective probability*. If a person is prudent, the degree-of-belief probability is the same as the long-run percentage, when that percentage becomes known.

Bayes' formula provides a means of converting the degree-of-belief probability we had before any test data or objective data were obtained to the degree-of-belief probability after the test data or objective data are obtained. Thus, Bayes' formula provides the mathematics by which a rational person has the degree-of-belief probability changed by evidence or some outcomes of a random experiment. Thus, the degree-of-belief or subjective probabilities are dynamic in nature. The Bayesian approach is based on the work by Reverend Thomas Bayes in the eighteenth century. His work was republished in *Biometrika*.[14] The Bayesian approach is illustrated in Fig. 25.10.

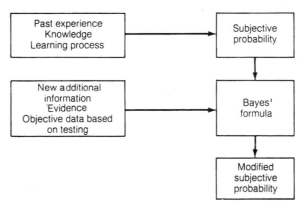

FIGURE 25.10   Bayesian approach.

The mathematics underlying the Bayesian approach is not controversial. The approach is controversial in terms of interpretation of equations concerning knowledge as "prior" and what constitutes reasonable prior knowledge. Also, it is difficult to develop experimental or analytical methods for the quantification of belief in the performance of new systems based on past experience.

EXAMPLE 25.12   We are concerned with the reliability level of a new but as yet untested system. We feel that the system might have one of two possible reliability levels denoted by R_1 and R_2. (It can be easily generalized for a continuous range of reliability.) Based on past experience we believe that the system may have a reliability level $R_1 = 0.95$, but if the design engineer has miscalculated on a certain factor in question, then the reliability level will be at a lower level $R_2 = 0.75$.

We will express our confidence in the system designer by assigning an 80 percent chance that level R_1 has been attained, thus leaving a 20 percent chance that level R_2 has been attained.

In this example, the confidence probabilities are subjective in nature and are based on past experience with similar systems. Objective data are obtained by testing. Now suppose that we test one system and find that it operates successfully. We now wish to find the probability that level R_1 has been attained. New information should modify our belief in the reliability of the system.

Bayes' theorem is based on the concept of conditional probabilities. We have

$$P(A|B) = P(A) \frac{P(B|A)}{P(B)} \tag{25.123}$$

$P(A)$ is the *prior* probability of the event A before information about B becomes available. In the Bayesian approach, $P(A)$ is generally a degree-of-belief or subjective probability. $P(A|B)$ is called the *posterior* probability, which is the probability of event A modified by the objective data about event B.

The conditional probability can be extended to the following version of Bayes' theorem. Let A_1, $A_2, \ldots, A_k$ be a partition of the sample space associated with the random experiment which is used to collect objective data. Then, Bayes' theorem is [see also Eq. (25.11)]

$$P(A_i|B) = \frac{P(A_i)P(B|A_i)}{\sum\limits_{i=1}^{k} P(A_i)P(B|A_i)} \tag{25.124}$$

Applications to Example 25.12   Let us apply the previous result to Example 25.12. Let us define

R_i = the event that reliability level R_i has been attained
S_i = the event that the ith system results in a success

We want to find

$$P(R_1|S_1) = \frac{P(R_1)P(S_1|R_1)}{P(R_1)P(S_1|R_1) + P(R_2)P(S_1|R_2)}$$

$$= \frac{0.80 \times 0.95}{0.80 \times 0.95 + 0.20 \times 0.75}$$

$$= 0.835$$

Thus, our confidence that level R_1 has been achieved goes up from 80 to 83.5 percent.

Let us assume that the second system is tested, and it is also successful. Then we wish to compute

$$P(R_1|S_1 \cap S_2) = \frac{P(R_1)P(S_1 \cap S_2|R_1)}{P(R_1)P(S_1 \cap S_2|R_1) + P(R_2)P(S_1 \cap S_2|R_2)}$$

$$= \frac{0.80 \times (0.95)^2}{(0.80)(0.95)^2 + (0.20)(0.75)^2}$$

$$= 0.865$$

We can see how the probability of events R_1 is updated by application of Bayes' theorem as new information becomes available. If we had tested one system and it resulted in a failure, then, as expected, our confidence would go down.

$$P(R_1|F_1) = \frac{P(R_1)P(F_1|R_1)}{P(R_1)P(F_1|R_1) + P(R_2)P(F_1|R_2)}$$

$$= \frac{0.80 \times 0.05}{0.80 \times 0.05 + 0.20 \times 0.25}$$

$$= 0.4444$$

25.8.1 Bayes' Theorem for Continuous Random Variables

This theorem enables prior knowledge about an unknown parameter θ (for example, true reliability or failure rate or MTBF) to be combined with subsequent test data to produce an updated or posterior knowledge about the unknown parameter θ about which we have some prior knowledge in the form of a probability-density function.

Bayes' theorem is

$$k(\theta|y) = \frac{h(\theta)g(y|\theta)}{\int \theta\, h(\theta)g(y|\theta)\, d\theta} \tag{25.125}$$

where $h(\theta) =$ prior probability-density function that expresses our degree of belief in θ
$g(y|\theta) =$ conditional probability-density function for y (hard or actual test data or statistic) given θ
$k(\theta|y) =$ posterior probability-density function for θ given y

EXAMPLE 25.13   Let us consider the continuous version of Example 25.12, and say that our belief in the reliability R of the system is expressed by a prior probability-density function

$$h(R) = 4R^3 \qquad 0 \leq R \leq 1$$

Thus, the expected value of reliability for the system is

$$E[R] = \int_0^1 R \cdot 4R^3\, dR = 0.80$$

Let us assume that we test one system and it is a success. Then

$$k(R|S_1) = \frac{h(R)g(S_1|R)}{\int_0^1 h(R)g(S_1|R)\, dR}$$

$$= \frac{4R^3 \cdot R}{\int_0^1 4R^3 \cdot R\, dR}$$

$$= 5R^4 \qquad 0 \leq R \leq 1$$

Thus, the posterior expected value for reliability is

$$E[R] = \int_0^1 R \cdot 5R^4\, dR = \frac{5}{6}$$

Similarly, we have the posterior reliability as given below when we test one system and it fails.

$$k(R|F_1) = \frac{h(R)g(F_1|R)}{\displaystyle\int_0^1 h(R)g(F_1|R)\,dR}$$

$$= \frac{4R^3(1-R)}{\displaystyle\int_0^1 4R^3(1-R)\,dR}$$

$$= 5(4R^3 - 4R^4) \qquad 0 \le R \le 1$$

The posterior expected value for reliability now is less than before, as expected, and is given by

$$E[R] = \int_0^1 R \cdot 5(4R^3 - 4R^4)\,dR = \frac{2}{3}$$

25.9   *SEQUENTIAL LIFE TESTING*

Sequential testing[15] is a situation where one continually reassesses test results to arrive at a decision with the minimum amount of testing. So, at decision points in the test, the alternatives that are available are

1. Accept the product
2. Reject the product
3. Continue testing—not enough information to make a decision

The OC curve (see Sec. 25.3.2) for a sequential test must be determined beforehand by specifying four values. These values are illustrated in the OC curve in Fig. 25.11. The values R_0, R_1, α, β must be specified in order to design the sequential test. Once the OC curve has been determined, then a sequential graph can be constructed that will have three regions as shown in Fig. 25.12.

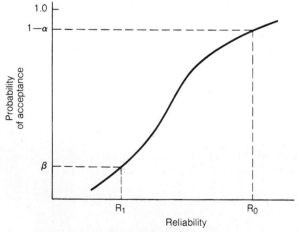

FIGURE 25.11   Operating-characteristic curve.

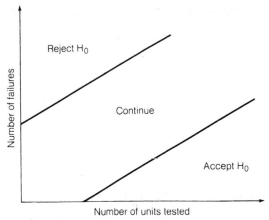

FIGURE 25.12   Sequential life testing.

25.9.1 Success/Failure Testing: Sequential Life Testing for Binomial Distribution[11]

In this situation each product tested either survives or fails the test. Let p be the probability of failure for a product. Let the null hypothesis be

$$H_0: \quad p \le p_0$$

and the alternate hypothesis be

$$H_1: \quad p > p_0$$

Here p_0 is the value of p such that if $p = p_0$, then the probability of accepting H_0 is $(1 - \alpha)$. Also let p_1 be a value of p such that $p_1 > p_0$, and for $p = p_1$, the probability of accepting H_0 is β. These quantities $(\alpha, p_0, p_1, \text{and } \beta)$ determine the sequential test.

Let $y =$ the total number of failures in n trials. Then the sequential test plan is given by (the continue region)

$$\frac{n}{D} \ln\left(\frac{1 - p_0}{1 - p_1}\right) - \frac{1}{D} \ln\left(\frac{1 - \alpha}{\beta}\right) < y < \frac{n}{D} \ln\left(\frac{1 - p_0}{1 - p_1}\right) + \frac{1}{D} \ln\left(\frac{1 - \beta}{\alpha}\right) \quad (25.126)$$

where

$$D = \ln\left[\left(\frac{p_1}{p_0}\right)\left(\frac{1 - p_0}{1 - p_1}\right)\right] \quad (25.127)$$

Let the inequality in Eq. (25.126) be represented as

$$A_n < y < B_n \quad (25.128)$$

Now by our procedural rules we accept H_0 if $y \le A_n$, reject H_0 if $y \ge B_n$, and take an additional observation if $A_n < y < B_n$.

The boundaries of Eq. (25.128) graph as parallel straight lines, and a sequential testing graph can be constructed similar to Fig. 25.12. The test results can then be plotted on this graph to provide a visual representation of the test progress.

The entire OC curve for this test can also be determined. The probability of accepting H_0 when p is the true fraction failing is

$$Pa(p) = \frac{B^h - 1}{B^h - A^h} \qquad (25.129)$$

where

$$p = \frac{1 - [(1 - p_1)/(1 - p_0)]^h}{(p_1/p_0)^h - [(1 - p_1)/(1 - p_0)]^h} \qquad A = \frac{\beta}{1 - \alpha}, \, B = \frac{1 - \beta}{\alpha} \qquad (25.130)$$

To obtain the OC curve for this test, Eq. (25.130) is first used with an arbitrarily selected value of h to compute p. Then Eq. (25.129) is used to calculate the probability of acceptance for the value of p. Of course, for $p = p_0$, $Pa(p_0) = \alpha$ and for $p = p_1$, $Pa(p_1) = \beta$.

The expected number of observations needed to reach a decision is given by

$$E(p, n) = \frac{Pa(p) \ln A + [1 - Pa(p)] \ln B}{p \ln (p_1/p_0) + (1 - p) \ln [(1 - p_1)/(1 - p_0)]} \qquad (25.131)$$

where $Pa(p)$ is given by Eq. (25.129).

25.9.2 Exponential Distribution

Let us consider the sequential test procedure for the following test of a hypothesis:

$$H_0: \quad \theta = \theta_0$$

$$H_1: \quad \theta = \theta_1 \qquad \theta_1 < \theta_0$$

In order to design the test, the following requirements must be specified:

1. Determine an acceptable mean life θ_0.
2. Determine the magnitude of an α risk which may be tolerated, where α is the probability of rejecting H_0 if $\theta = \theta_0$ or $P(H_1/H_0) = \alpha$.
3. Determine an unacceptable or limiting mean life θ_1 where $\theta_1 < \theta_0$.
4. Determine the magnitude of a β risk which may be tolerated, where $P(H_0/H_1) = \beta$ or β is the probability of accepting H_0 if $\theta = \theta_1 < \theta_0$.

Let us consider the situation where items are tested to failure in a sequential (one-at-a-time) fashion. Let t_i be the time to failure of the ith item. The sequential decision criteria are

If $h_1 + ns < V(t) < h_0 + ns$, continue the test
If $V(t) \geq h_0 + ns$, stop the test and accept H_0
If $V(t) \leq h_1 + ns$, stop the test and reject H_0

where $V(t)$ is the total time on test $\sum_{i=1}^{n} t_i$, and n is the number of failures by total time $V(t)$ and where

$$h_0 = \frac{\ln [(1 - \alpha)/\beta]}{1/\theta_1 - 1/\theta_0} \tag{25.132}$$

$$h_1 = \frac{\ln [\alpha/(1 - \beta)]}{1/\theta_1 - 1/\theta_0} \tag{25.133}$$

$$s = \frac{\ln (\theta_0/\theta_1)}{1/\theta_1 - 1/\theta_0} \tag{25.134}$$

The OC curve for the test is given by

$$P(\theta) = \text{probability of accepting } H_0 \text{ when } \theta \text{ is the true parameter}$$

$$= \frac{B^h - 1}{B^h - A^h} \tag{25.135}$$

where

$$\theta = \frac{(\theta_0/\theta_1)^h - 1}{h(1/\theta_1 - 1/\theta_0)}$$

$$A = \frac{\beta}{1 - \alpha} \qquad B = \frac{1 - \beta}{\alpha}$$

Parameter h can be any real number, and meaningful selections are made by trial and error.

The five points which can be easily found and enable us to sketch the OC curve are given in Table 25.7. The expected number of failures required to reach a decision, $E_\theta(R)$, is dependent on the lot mean life θ and may be found as follows:

$$E_\theta(R) = \frac{- h_1 - (h_0 - h_1)P(\theta)}{s - \theta} \qquad s \neq \theta \tag{25.136}$$

$$E_\theta(R) = \frac{- h_0 h_1}{s^2} \qquad s = \theta \tag{25.137}$$

where the random variable R is the number of failures required to reach a decision. If $\theta = \theta_1$ we have

$$E_\theta(R) \approx \frac{\beta \ln [\beta/(1 - \alpha)] + (1 - \beta) \ln [(1 - \beta)/\alpha]}{\ln (\theta_0/\theta_1) - [1 - (\theta_1/\theta_0)]} \tag{25.138}$$

and when $\theta = \theta_0$, we have

$$E_\theta(R) \approx \frac{(1 - \alpha) \ln [\beta/(1 - \alpha)] + \alpha \ln [(1 - \beta)/\alpha]}{\ln (\theta_0/\theta_1) - [(\theta_0/\theta_1) - 1]} \tag{25.139}$$

The expected waiting time to reach a decision, $E_\theta(T)$, is a function of θ and is found by (n is the number of items put on test at a time)

$$E_\theta(T) = \frac{\theta}{n} E_\theta(R) \qquad \text{for replacement test} \tag{25.140}$$

or

$$E_\theta(T) \approx \theta \ln\left[\frac{n}{n - E_\theta(R)}\right] \qquad \text{for a nonreplacement test} \tag{25.141}$$

EXAMPLE 25.14   We want to test a new product to see if it meets a standard of $\theta_0 = 1000$ h for its MTBF with $\alpha = 0.05$. We decide $\theta_1 = 500$ h with $\beta = 0.10$. Then, we have

$$h_0 = \frac{\ln[(1 - 0.05)/0.10]}{1/500 - 1/1000} = 2251.29$$

$$h_1 = \frac{\ln[0.05(1 - 0.10)]}{1/500 - 1/1000} = -2890.37$$

$$s = \frac{\ln(1000/500)}{1/500 - 1/1000} = 693.15$$

Hence, continue region is given by

$$693n - 2890 < V(t) < 693n + 2251$$

Accept H_0 when $V(t) \geq 693n + 2251$

Reject H_0 when $V(t) \leq 693n - 2890$

The OC curve can be plotted using Table 25.7 and we have

$$P(500) = 0.10$$

$$P(1000) = 0.95$$

TABLE 25.7   OC Curve for Sequential Plan

θ	$P(\theta)$
0	0
θ_1	β
s	$\dfrac{\ln[(1 - \beta)/\alpha]}{\ln[(1 - \beta)/\alpha] - \ln[\beta/(1 - \alpha)]}$
θ_0	$1 - \alpha$
α	1

and

$$P[\theta = 693] = \frac{\ln\,[(1 - 0.1)/0.05]}{\ln\,[(1 - 0.1)/0.05] - \ln[0.10/(1 - [0.05)]}$$

$$= \frac{2.89037}{2.89037 - (-2.25129)}$$

$$= 0.56214$$

The expected number of failures to reach a decision is given by [using Eq. (25.137) to (25.139)]

$$E_{500}(R) = 12.31 \qquad E_{693}(R) = 13.54 \qquad E_{1000}(R) = 6.50$$

Similarly, the expected time to reach a decision is given by [using Eqs. (25.140) and (25.141)]

$$E_{500}(T) = 500 \times 12.31 = 6155 \text{ h}$$

$$E_{693}(T) = 693 \times 12.54 = 9385 \text{ h}$$

$$E_{1000}(T) = 1000 \times 6.50 = 6500 \text{ h}$$

25.10   CONCLUSION

In this chapter we have presented probability theory and statistical methods that can help us model and estimate reliability of the product. Reliability is defined in terms of probability, and hence probability theory plays an important role in reliability. Probability theory is concerned with the methods of analysis that can be used to study random phenomena associated with the life of a product. A failure distribution describes and models mathematically the life of a product. There are several physical or other causes that may be responsible for the failure of a device, and generally it is not possible to isolate these causes and mathematically account for all of them. Thus the selection of a failure distribution is based on statistical studies about the life of a product and the data accumulated on the product. The failure data and an understanding of the physical causes for failure should complement each other in the selection of a failure distribution. Some of the statistical procedures to estimate the parameters of the failure distribution are given in this chapter. There is tremendous interplay between probability, statistics, and reliability.

REFERENCES

1. Hines, W. W.; and Montgomery, D. C., *Probability and Statistics in Engineering and Management Science,* 2d ed., Wiley, New York, 1980.

2. Feller, W., *An Introduction to Probability Theory and its Applications,* 3d ed., Wiley, New York, 1968.

3. Barlow, R. E.; and Proschan, F., *Statistical Theory of Reliability and Life Testing,* Holt, Rinehart and Winston, New York, 1975.

4. *Reliability Design Handbook,* Reliability Analysis Center, RDG-376, Griffiss Air Force Base, N.Y., March 1976.

5. Mann, N. R.; Schafer, R. E.; and Singpurwalla, N. D., *Methods for Statistical Analysis of Reliability and Life Data,* Wiley, New York, 1974.

6. Bowker, A. H.; and Lieberman, G. J., *Handbook of Industrial Statistics,* Prentice-Hall, Englewood Cliffs, N.J., 1955.

7. Duncan, A. J., *Quality Control and Industrial Statistics,* 4th ed., Irwin, Homewood, Ill., 1974.

8. Kapur, K. C., "Reliability and Maintainability," Chap. 8.5 in G. Salvendy, ed., *Industrial Engineering Handbook,* Wiley, New York, 1982.

9. Kapur, K. C.; and Lamberson, L. R., "Reliability," in H. A. Rothbart, ed., *Mechanical Design and Systems Handbook,* McGraw-Hill, New York, 1985.

10. Kapur, K. C.; and Lamberson, L. R., "Reliability in Product Design and Testing," Lecture Notes, Detroit, Michigan, copyrighted 1982.

11. Kapur, K. C.; and Lamberson, L. R., *Reliability in Engineering Design,* Wiley, New York, 1977.

12. Weibull, W., "A Statistical Distribution Function of Wide Applicability," *Journal of Applied Mechanics,* 1951, pp. 293–296.

13. Weibull, W., *Fatigue Testing and Analysis of Results,* Macmillan, New York, 1961.

14. Bayes, T., "An Essay Towards a Problem in the Doctrine of Chances," *Biometrika,* **45,** 1958.

15. Wald, A., *Sequential Analysis,* Wiley, New York, 1947.

CHAPTER 26
LIFE DISTRIBUTIONS AND CONCEPTS

Wayne Nelson
PRIVATE RELIABILITY CONSULTANT
SCHENECTADY, NEW YORK

26.1 INTRODUCTION

Almost every major company spends millions of dollars annually on product reliability. Much management and engineering effort goes into evaluating risks and liabilities, predicting warranty costs, evaluating replacement policies, assessing design changes, identifying causes of failure, and comparing alternate designs, vendors, materials, manufacturing methods, and the like. Major decisions are based on product life data, often from a few units. This chapter presents life distributions for modeling life-test and field data. Such models are successfully used for a great variety of products by many who have just a working knowledge of basic statistics from a first course.

This chapter presents basic concepts and theory for product life distributions, used as models for the life of products, materials, people, television programs, and many other things. It presents the commonly used exponential, normal, lognormal, and Weibull distributions. Chapter 27 shows how to use them to analyze data graphically. This chapter presents the series-system model for products with a number of failure modes. This chapter (which is based on Wayne Nelson,[8] *How to Analyze Reliability Data,* copyright American Society for Quality Control, Inc. 1983, Milwaukee, Wis.; reprinted by permission) also presents the Poisson and binomial distributions, which are models for the observed numbers of failures or defectives. For more detail and other distributions, consult Nelson,[7,9] O'Connor,[10] and Tobias and Trindade.[12]

26.2 BASIC CONCEPTS AND THE EXPONENTIAL DISTRIBUTION

The *cumulative distribution function $F(t)$* for a continuous distribution represents the population fraction failing by age t. Any such $F(t)$ has the mathematical properties:

1. It is a continuous function for all t.
2. $\lim_{t \to -\infty} F(t) = 0$ and $\lim_{t \to +\infty} F(t) = 1$.
3. $F(t) \leq F(t')$ for all $t < t'$.

The *exponential cumulative distribution function* for the population fraction failing by age t is

$$F(t) = 1 - e^{-t/\theta}, \qquad t \geq 0$$

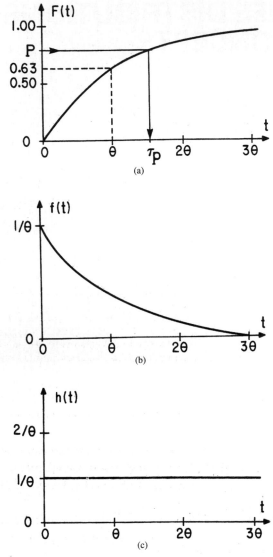

FIGURE 26.1   (a) Exponential cumulative distribution; (b) exponential probability density; (c) exponential hazard function.

$\theta > 0$ is the mean time to failure. θ is in the same measurement units as t, for example, hours, months, cycles, etc. Figure 26.1a shows this cumulative distribution function. In terms of the "failure rate" $\lambda \equiv 1/\theta$,

$$F(t) = 1 - e^{-\lambda t}, \qquad t \geq 0$$

The parameters θ and λ are estimated from data as shown in Chap. 27.

Engine Fan Example.   The exponential distribution with a mean of $\theta = 28{,}700$ h was used to describe the hours to failure of a population of fans on diesel engines. The population failure rate is $\lambda = 1/28,700 = 34.8$ failures per million hours. For the engine fans, the population fraction failing on an 8000-h warranty is $F(8000) = 1 - \exp(-8000/28,700) = 0.24$. Management decided to replace the fan design.

The *reliability function $R(t)$* for a life distribution is the probability of survival beyond age t:

$$R(t) \equiv 1 - F(t)$$

The *exponential reliability function* is

$$R(t) = e^{-t/\theta}, \qquad t \geq 0$$

For the engine fans, reliability for 8000 h is $R(8000) = \exp(-8000/28,700) = 0.76$; thus, 76 percent of such fans survive warranty.

The *100Pth percentile* of a distribution $F(\ )$ is the age t_P by which a proportion P of the population fails. It is the solution of

$$P = F(t_P)$$

In life-data work, one often wants to know low percentiles such as the 1 and 10 percent points, which correspond to early failure. The 50 percent point is called the *median* and is commonly used as a "typical" life.

The *100Pth exponential percentile* is

$$t_P = -\theta \ln(1 - P)$$

For example, the mean θ is roughly the 63d percentile of the exponential distribution. For the diesel engine fans, median life is $t_{.50} = -28,700 \ln(1 - 0.50) = 19,900$ h. The 1st percentile is $t_{0.01} = -28,700 \ln(1 - 0.01) = 288$ h. The *probability density* of a cumulative distribution function is

$$f(t) \equiv \frac{dF(t)}{dt}$$

It corresponds to a histogram of the population life times. The *exponential probability density* is

$$f(t) = \left(\frac{1}{\theta}\right) e^{-t/\theta}, \qquad t > 0$$

Figure 26.1*b* depicts this probability density. Also,

$$f(t) = \lambda e^{-\lambda t}, \qquad t \geq 0$$

The *mean μ* of a distribution with probability density $f(t)$ is

$$\mu \equiv \int_{-\infty}^{\infty} t\, f(t)\, dt$$

The integral runs over all possible outcomes t. The mean is also called the *average* or *expected life*. It corresponds to the arithmetic average of the lives of all units in a population. It is used as still another "typical" life.

The *mean of the exponential distribution* is

$$\mu = \int_{0}^{\infty} t\left(\frac{1}{\theta}\right) e^{-t/\theta}\, dt = \theta$$

This shows why θ is called the "mean time to failure" (MTTF). Also, $\mu = 1/\lambda$ for this distribution and no others. For the diesel engine fans, the mean life is $\mu = 28,700$ h. Some repairable products are assumed to have exponentially distributed time *between* failures, particularly after most components have been replaced a few times. Then θ is called the *mean time between failures* (MTBF). The Poisson model (Sec. 26.6) describes such products. Chapter 27 and Nelson[6,8] present graphical analyses of repair data.

The *hazard function* $h(t)$ of a distribution is defined as

$$h(t) \equiv \frac{f(t)}{1 - F(t)} = \frac{f(t)}{R(t)}$$

It is the *(instantaneous) failure rate* at age t. That is, in the short time Δ from t to $t + \Delta$, a proportion $h(t) \cdot \Delta$ of the population that reached age t fails. $h(t)$ is a measure of proneness to failure as a function of age. It is also called the *hazard rate* and the *force of mortality*. In many applications, one wants to know whether the population failure rate of a product increases or decreases with product age.

The *exponential hazard function* is

$$h(t) = \frac{(1/\theta)e^{-t/\theta}}{e^{-t/\theta}} = \frac{1}{\theta}, \qquad t \geq 0$$

Figure 26.1c shows this constant hazard function. Also, $h(t) = \lambda$, $t \geq 0$. Only the exponential distribution has a constant failure rate, a key characteristic. That is, for this distribution only, an old unit and a new unit have the same chance of failing over a future time interval Δ. For example, engine fans of any age will fail at a constant rate of $h(t) = 34.8$ failures per million hours.

A decreasing hazard function during the early life of a product is said to correspond to "infant mortality." Figure 26.2 shows this near time zero. Such a failure rate often indicates that the product suffers from manufacturing or design defects. Some products, such as some semiconductor devices, have a decreasing failure rate over their observed life.

An increasing hazard function during later life of a product is said to correspond to statistical *wearout* failure. This often indicates that failures are due to the product physically wearing out. Figure 26.2 shows this feature in the latter part of the curve. Many products have an increasing failure rate over the entire range of life.

The Bathtub Curve.   A few products show a decreasing failure rate in the early life and an increasing failure rate in later life. Figure 26.2 shows such a hazard function, called a "bathtub

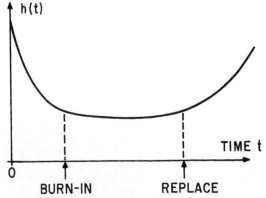

FIGURE 26.2   Bathtub hazard (failure rate) curve.

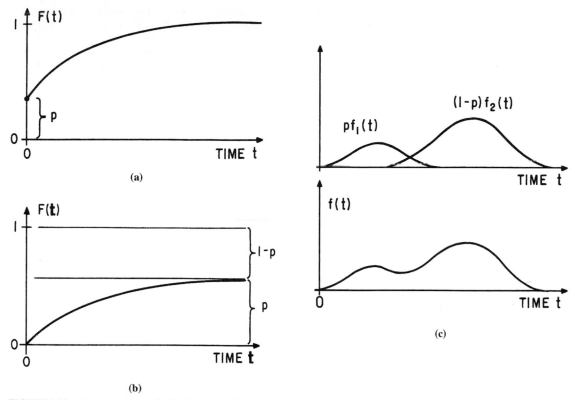

FIGURE 26.3   (*a*) A cumulative distribution with a fraction *p* failed at time zero; (*b*) a cumulative distribution with eternal survivors; (*c*) a mixture of distribution densities.

curve." Some products with an initially decreasing failure rate, such as high-reliability capacitors and semiconductor devices, are subjected to a burn-in. Then surviving units that go into service have a lower failure rate. Jensen and Petersen[4] comprehensively treat planning and analysis of burn-in, including the economics. Also, for products with an increasing failure rate as the population ages, units are removed from service before wearout starts. Then units are in service only in the low-failure-rate portion of their life. This increases their reliability in service.

Distributions with Failure at Time Zero.   A fraction of a population may already be failed at time zero. All of us have encountered products that do not work when first used. The model for this consists of the proportion *p* failed at time zero and a continuous life distribution for the rest. Such a cumulative distribution appears in Fig. 26.3*a*. The sample proportion failed at time zero is used to estimate *p,* and the failure times in the remainder of the sample are used to estimate the continuous distribution.

Distributions with External Survivors.   Some units may never fail. This applies to (1) the time to death from a disease when some individuals are immune, (2) the time to redemption of trading stamps (some stamps are lost and never redeemed), (3) the time to product failure from a particular defect when some units lack that defect, and (4) time to warranty claim on a product whose warranty applies only to original owners, some of which sell the product before failure. Figure 26.3*b* depicts this situation.

Mixtures of Distributions. A population may consist of two or more subpopulations. Figure 26.3c depicts distribution densities for this situation. Units from different production periods may have different life distributions due to differences in design, raw materials, environment, etc. It is often important to identify such a situation and the production period, customer, environment, etc., that has poor units. Then suitable action may be taken on that portion of the population. A mixture should be distinguished from competing failure modes, described in Sec. 26.8.

26.3   NORMAL DISTRIBUTION

This section presents the well-known normal distribution. Its hazard function increases. Thus it may describe products with wearout failure.

The *normal probability density* is

$$f(t) = (2 \pi \sigma^2)^{-1/2} \exp\left[-\frac{(t - \mu)^2}{2\sigma^2} \right], \qquad -\infty < t < \infty$$

Here, μ is the population mean and may have any value; σ is the population standard deviation and must be positive; μ and σ are in the same measurement units as t, for example, hours, months, cycles, etc. These parameters are estimated from data as shown in Chap. 27. Figure 26.4a depicts this probability density, which is symmetrical about the mean μ. The figure shows that μ is the distribution median and σ determines the distribution spread.

The range of t is from $-\infty$ to $+\infty$. Life must, of course, be positive. Thus the distribution fraction below zero must be small for this distribution to be a satisfactory approximation in practice.

The *normal cumulative distribution function* for the population fraction failing by age t is

$$F(t) = \int_{-\infty}^{t} (2 \pi \sigma^2)^{-1/2} \exp\left[-\frac{(x - \mu)^2}{2\sigma^2} \right] dx, \qquad -\infty < t < \infty$$

Figure 26.4b depicts this function. This can be expressed in terms of the standard normal cumulative distribution function $\Phi(\ )$ given in Table A.4 (App. A) as

$$F(t) = \Phi\left[\frac{(t - \mu)}{\sigma} \right], \qquad -\infty < t < \infty$$

Many tables of $\Phi(z)$ give values only for $z \geq 0$. One then uses $\Phi(-z) = 1 - \Phi(z)$.

Transformer Example. A normal life distribution with $\mu = 6250$ h and $\sigma = 2600$ h was used to represent life of a transformer population. The fraction of the distribution with negative life times is $F(0) = \Phi[(0-6250)/2600] = \Phi(-2.40) = 0.0082$. This small fraction is ignored hereafter.

The *100Pth normal percentile* is

$$t_P = \mu + z_P \sigma$$

Here z_P is the 100Pth standard normal percentile and is tabled below. The *median* (50th percentile) of the normal distribution is $t_{0.50} = \mu$, since $z_{0.50} = 0$. Some standard percentiles are

100P%	0.1	1	2.5	5	10	50	90	97.5	99
z_P:	−3.090	−2.326	−1.960	−1.645	−1.282	0	1.282	1.960	2.326

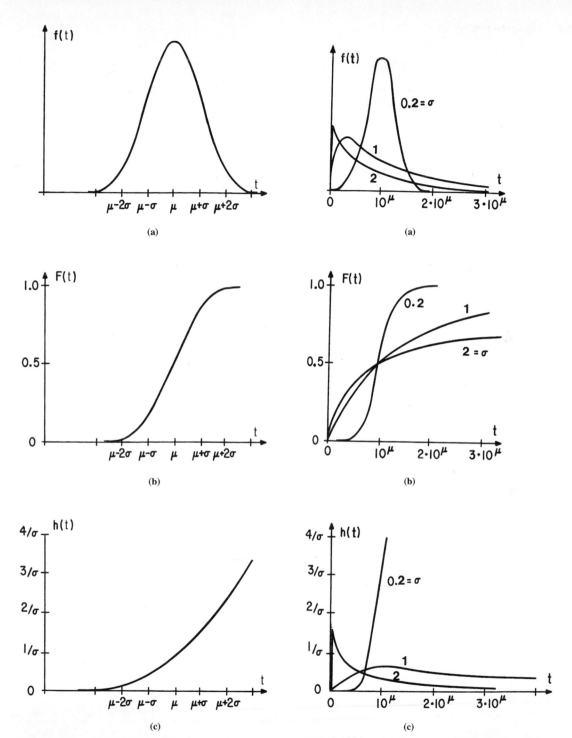

FIGURE 26.4 (*a*) Normal probability density; (*b*) normal cumulative distribution; (*c*) normal hazard function.

FIGURE 26.5 (*a*) Lognormal probability densities; (*b*) lognormal cumulative distributions; (*c*) lognormal hazard functions.

Median transformer life is $t_{0.50} = 6250$ h, and the 10th percentile is $t_{0.10} = 6250 + (-1.282)2600 = 2920$ h.

The *normal hazard function* appears in Fig. 26.4c, which shows that the normal distribution has an *increasing failure rate* (wearout) with age. A key issue was whether transformer failure rate increases with age. Since it does, older units should be replaced first. The increasing failure rate of the normal distribution indicates that older units are more failure-prone.

26.4 LOGNORMAL DISTRIBUTION

The *lognormal distribution* is used for many types of life data, for example, metal fatigue, semiconductor life, and electrical insulation life. The lognormal and normal distributions are related, a fact used for lognormal analysis of data with methods for a normal distribution.

The *lognormal probability density* is

$$f(t) = [(2\pi)^{1/2}t\sigma]^{-1} \exp\left[\frac{-[\ln(t) - \mu]^2}{2\sigma^2}\right], \qquad t > 0$$

Here, μ is called the *log mean* and may have any value; it is the mean of the *log* of life—not of life. σ is called the *log standard deviation* or *sigma* and must be positive; it is the standard deviation of the *log* of life—not of life. μ and σ are not "times" like t; instead, they are unitless pure numbers. The value of σ determines the shape of the distribution, and the value of μ determines its 50 percent point and spread. These parameters are estimated from data as shown in Chap. 27. Here $\ln(\ )$ denotes the base e logarithm. Some authors use the common (base 10) logarithm.[7,9] Figure 26.5a shows probability densities, which have a variety of shapes.

The *lognormal cumulative distribution function* for the population fraction failing by age t is

$$F(t) = \Phi\left[\frac{\ln(t) - \mu}{\sigma}\right], \qquad t > 0$$

Figure 26.5b shows lognormal cumulative distribution functions. $\Phi[\ ]$ appears in Table A.4 (App. A).

Locomotive Control Example.   The life (in thousand miles) of a population of electronic controls for locomotives was approximated by a lognormal distribution where $\mu = 5.149$ and $\sigma = 0.737$. The population fraction failing on an 80 thousand-mile warranty is $F(80) = \Phi\{[\ln(80)-5.149]/0.737\} = \Phi[-1.04] = 0.26$. This percentage was too high, and the control was redesigned.

The *100Pth lognormal percentile* is

$$t_p = \exp[\mu + z_p\sigma]$$

Here z_p is the 100Pth standard normal percentile, tabulated in Sec. 26.3. The *median* (50th percentile) is $t_{0.50} = \exp[\mu]$. For the locomotive control, $t_{0.50} = \exp[5.149] = 172$ thousand mi, regarded as a typical life. The 1 percent life is $t_{0.01} = \exp[5.149 + (-2.326)0.737] = 31$ thousand mi.

Lognormal hazard functions appear in Fig. 26.5c. For $\sigma \cong 1.0$, $h(t)$ is roughly constant. For $\sigma \leq 0.4$, $h(t)$ increases and is much like that of a normal distribution. For $\sigma \geq 1.5$, $h(t)$ decreases. This flexibility makes the lognormal distribution popular and suitable for many products. The lognormal hazard function has a property seldom seen in products. It is zero at time zero, increases to a maximum, and then (in the far upper tail for most σ values in practice) decreases to zero with increasing age. However, in the lower tail and middle, the lognormal distribution does fit life data on many products. For the locomotive control, $\sigma = 0.737$. So the behavior of its $h(t)$ is midway between the increasing and roughly constant hazard functions in Fig. 26.5c.

The relationship between the lognormal and normal distributions helps one understand the lognormal distribution in terms of the simpler normal distribution. The (base e) log of a variable with a lognormal distribution with parameters μ and σ has a normal distribution with mean μ and standard deviation σ. Thus the analysis methods for the normal distribution can be used for the logs of lognormal data.

26.5   WEIBULL DISTRIBUTION

The Weibull distribution is often used for product life because it flexibly describes increasing and decreasing failure rates. It may be suitable for a "weakest link" product; i.e., the product consists of many parts with comparable life distributions and the product fails with the first part failure. For example, the life of a capacitor or power cable is determined by the shortest-lived portion of its dielectric.

The *Weibull probability density function* is

$$f(t) = \left(\frac{\beta}{\alpha^\beta} \right) t^{\beta-1} \exp\left[-\left(\frac{t}{\alpha} \right)^\beta \right], \qquad t > 0$$

The *shape parameter* β and the *scale parameter* α are positive. α is called the *characteristic life*, as it is always 63.2th percentile. α has the same units as t, for example, hours, months, and cycles. β is a unitless pure number, usually in the range of 0.5 to 5. These parameters are estimated from data as shown in Chap. 27. The Weibull probability densities in Fig. 26.6a show that β determines the shape of the distribution and α determines the spread. For $\beta = 1$, the Weibull distribution is the exponential distribution. For much life data, the Weibull distribution is more suitable than the exponential, normal, and lognormal distributions.

The *Weibull cumulative distribution function* for the population fraction failing by age t is

$$F(t) = 1 - \exp\left[-\left(\frac{t}{\alpha} \right)^\beta \right], \qquad t > 0$$

Figure 26.6b shows Weibull cumulative distribution functions.

Winding Example.   The life of generator field windings was approximated with a Weibull distribution with $\alpha = 13$ years and $\beta = 2$. The population fraction of windings failing on a 2-year warranty is $F(2.0) = 1 - \exp[-(2.0/13)^2] = 0.023$ or 2.3 percent.

The *Weibull reliability function* for the population fraction surviving beyond age t is

$$R(t) = \exp\left[-\left(\frac{t}{\alpha} \right)^\beta \right], \qquad t > 0$$

For the windings, the population reliability for 2 years is $R(2.0) = \exp[-(2.0/13)^2] = 0.977$ or 97.7 percent.

The *100Pth Weibull percentile* is

$$t_P = \alpha \cdot [-\ln(1-P)]^{1/\beta}$$

Here $\ln(\ )$ is the natural logarithm. For example, $t_{0.632} \cong \alpha$ for any Weibull distribution. This may be seen in Fig. 26.6b. For the windings, $t_{0.632} = 13[-\ln(1-0.632)]^{1/2} = 13$ years, the characteristic life. The 10th percentile is $t_{0.10} = 13[-\ln(1-0.10)]^{1/2} = 4.2$ years.

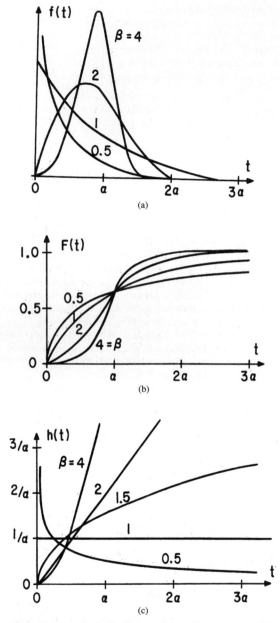

FIGURE 26.6   (*a*) Weibull probability densities; (*b*) Weibull cumulative distributions; (*c*) Weibull hazard functions.

The *Weibull hazard function* is

$$h(t) = \left(\frac{\beta}{\alpha^\beta} \right) t^{\beta - 1}, \qquad t > 0$$

Figure 26.6c shows Weibull hazard functions. $h(t)$ increases for $\beta > 1$ and decreases for $\beta < 1$. For $\beta = 1$ (the exponential distribution), the failure rate is constant. With increasing or decreasing failure rates, the Weibull distribution flexibly describes product life. For the windings, $\beta = 2$, and their population failure rate increases with age—wearout behavior. This tells utilities that preventive replacement of old windings will avoid costly service failures. When to replace depends on costs of windings and service failures.

26.6 POISSON DISTRIBUTION

The Poisson distribution is used for the *number* of occurrences of some event within some observed time, area, volume, etc. For example, it has been used to describe the yearly number of soldiers of a Prussian regiment kicked to death by horses, the number of flaws in a length of wire or computer tape, the number of failures of a repairable product over a certain period, and many other phenomena. It is appropriate if (1) the occurrences occur independently of each other over time (area, volume, etc.), (2) the chance of an occurrence is the same for each point in time (area, volume, etc.), and (3) the potential number of occurrences is unlimited.

The *Poisson probability* of y occurrences is

$$f(y) = \left(\frac{1}{y!} \right) (\lambda t)^y e^{-\lambda t}, \qquad y = 0, 1, 2, \ldots$$

Here t is the amount of exposure or observation; it may be a time, length, area, volume, etc. For example, for a power line, t is the product of length and time in thousand-ft-years. The *occurrence rate* λ must be positive; it is the expected number of occurrences per unit time, length, area, volume, etc. Figure 26.7 depicts Poisson probability functions.

Power Line.   For a power line, the yearly number of failures is assumed to have a Poisson distribution with $\lambda = 0.0256$ failures per year per thousand feet. For $t = 515.8$ thousand feet of line, the probability of no failures in a year is $f(0) = (1/0!)(0.0256 \times 515.8)^0 \exp(-0.0256 \times 515.8) = \exp(-13.2) = 1.8 \times 10^{-6}$, which is tiny.

The *Poisson cumulative distribution function* for the probability of y or fewer occurrences is

$$F(y) = \sum_{i=0}^{y} \left(\frac{1}{i!} \right) (\lambda t)^i e^{-\lambda t}$$

The Thorndike chart in Fig. 26.8 provides $F(y)$ as follows. Enter the chart on the horizontal axis (abscissa) at the value $\mu = \lambda t$. Go up to the curve labeled y. Then go horizontally to the vertical scale (ordinate) to read $F(y)$. For the power line, $\lambda t = 13.2$, and the probability of 15 or fewer failures is $F(15) = 0.75$ from the chart. $F(y)$ is tabulated in Table A.1 (App.A) and in most textbooks.

The *Poisson mean* (or *expected*) number Y of occurrences over time t is

$$\mu = \lambda t$$

For the power line, the expected (mean) number of failures in a year is $\mu = 0.0256(515.8) = 13.2$ failures, which is useful in maintenance planning.

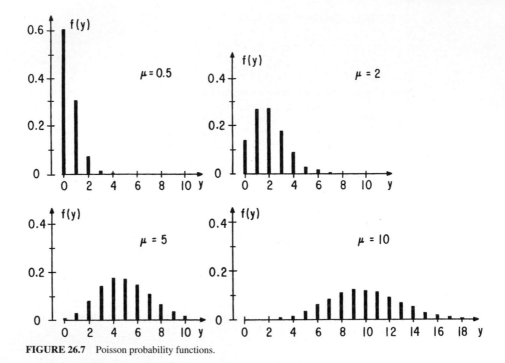

FIGURE 26.7   Poisson probability functions.

The *Poisson standard deviation* of the number Y of occurrences is

$$\sigma(Y) = (\lambda t)^{1/2}$$

For the power line, $\sigma(Y) = [13.2]^{1/2} = 3.63$ failures in a year.

A *normal approximation* to the Poisson $F(y)$ is

$$F(y) \cong \Phi\left[\frac{(y + 0.5 - \lambda t)}{(\lambda t)^{1/2}}\right]$$

Here $\Phi[\ ]$ is the standard normal cumulative distribution function, tabulated in Table A.4 (App. A). This approximation is satisfactory for most practical purposes if $\lambda t \geq 10$. For the power line, the approximate probability of 15 or fewer failures in a year is $F(15) \cong \Phi[(15+0.5-13.2)/3.63] = \Phi[0.63] = 0.74$. The exact probability is 0.75.

Demonstration testing commonly involves the Poisson distribution. Repairable hardware "demonstrates" required reliability if test units run a specified total time t with y or fewer failures. Units that fail are repaired and kept on test. A manufacturer designs the hardware to achieve a λ that assures passing the test with a desired high probability $100(1 - \alpha)$ percent. The hardware can fail the test with $100\,\alpha$ percent probability, called the *producer's risk*.

Electronic System Example.   Electronic systems were required to run a total time $t = 10,000$ h with $y = 2$ or fewer failures. For the electronic system, the producer's risk was to be 10 percent.

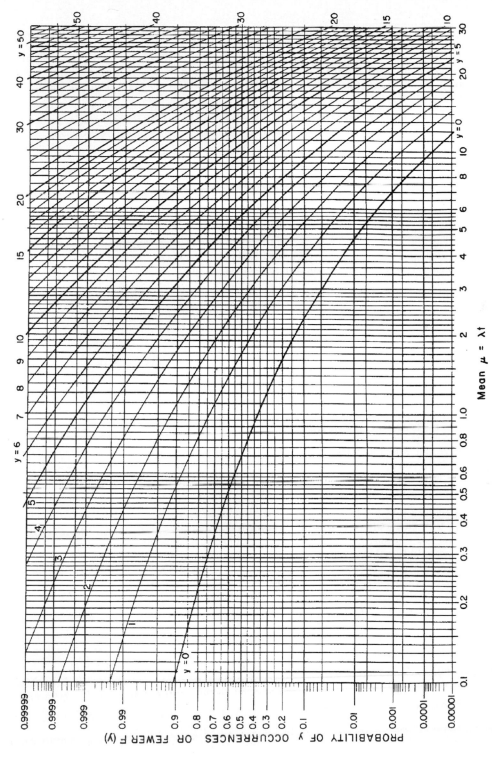

FIGURE 26.8 Poisson cumulative probabilities (Thorndike chart). *(From H. F. Dodge and H. G. Romig, Sampling Inspection Tables, Wiley, New York, Copyright 1959, Bell Telephone Labs, Inc., reprinted by permission.)*

To obtain the desired design λ, one must find $\mu = \lambda t$ such that the Poisson probability $F_{\lambda t}(t) = 1 - \alpha$. To do this, enter Fig. 26.8 on the vertical axis at $1 - \alpha$, go horizontally to the curve for y or fewer failures ($y = 2$ here), and then go down to the horizontal axis to read the appropriate μ value. Then the desired design failure rate is $\lambda = \mu/t$. For the electronic system, $1 - \alpha = 1 - 0.10 = 0.90$, $\mu = 1.15$, and the desired $\lambda = 1.15/10,000 = 0.115$ failures per thousand hours.

Relationship of Poisson and Exponential Distributions.   For a repairable product, suppose that times between failures are statistically independent and have an exponential distribution with failure rate λ. Then the *number* y of failures in a total running time t over any number of units has a Poisson distribution with mean λt.

26.7 BINOMIAL DISTRIBUTION

The *binomial distribution* is used as a model for the number of sample units that fall in a specified category. For example, it is used for the number of defective units in random samples from shipments and production, the number of units that fail on warranty, and the number of one-shot devices (used once) that work properly.

Its assumptions are (a) each sample unit has the same chance p of being in the category and (b) the outcomes of the n sample units are statistically independent.

The *binomial probability* of getting y category units in a sample of n units is

$$f(y) = \frac{n!}{y!(n-y)!} \, p^y(1-p)^{n-y}, \qquad y = 0,1,2,\ldots,n$$

where p is the population proportion in the category ($0 \leq p \leq 1$). Figure 26.9 depicts binomial probability functions.

In reliability work, if the category is "failure" of a device, the proportion p is the *failure probability,* sometimes incorrectly called the *failure rate*. The proportion p can be expressed as a percentage and differs from the Poisson failure rate λ, which has the dimensions of failures per unit time. If the category is "successful operation" of a device, the proportion p is called the *reliability* of the device.

Control Example.   A locomotive power control under development was assumed to fail on warranty with probability $p = 0.156$. $n = 96$ such controls were field-tested, and $y = 15$ failures occurred on warranty. The corresponding probability is $f(15) = 96![15!(96-15)!]^{-1} (0.156)^{15} \times (1-0.156)^{96-15} = 0.111$.

The *binomial cumulative distribution function* for the probability of y or fewer of the n sample items being in the category is

$$F(y) = \sum_{i=0}^{y} \frac{n!}{i!(n-i)!} \, p^i(1-p)^{n-i}, \qquad y = 0,1,2,\ldots,n$$

$F(y)$ is widely tabulated. For example, the probability of 15 or fewer warranty failures of the 96 controls is $F(15) = 0.571$ from a binomial table.

A *normal approximation* is

$$F(y) \cong \Phi\left\{ \frac{(y+0.5-np)}{[np(1-p)]^{1/2}} \right\}$$

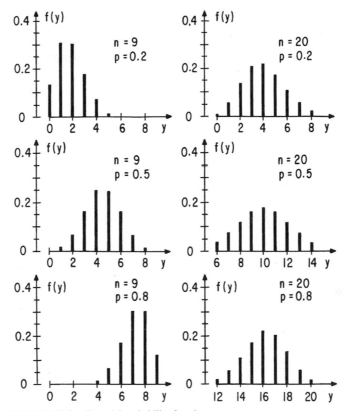

FIGURE 26.9   Binomial probability functions.

Here $\Phi[\ ]$ is the standard normal cumulative distribution function. This approximation is usually satisfactory if $np \geq 10$ and $n(1-p) \geq 10$. For example, for the controls, $F(15) \cong \Phi\{(15+0.5-96\times0.156)/[96\times0.156(1-0.156)]^{1/2}\} = 0.556$. Similarly, the approximate probability of 15 failures is $f(15) = F(15)-F(14) \cong 0.556 - \Phi\{(14+0.5-96\times0.156)/[96\times0.156\times(1-0.156)]^{1/2}\} = 0.112$ (0.111 exact). Here $96(0.156) = 15 > 10$ and $96(1 - 0.156) = 81 > 10$.

The *mean* of the number Y of sample items in the category is $\mu = np$. This is the number n of sample units times the population proportion p in the category. For example, the mean number of failures in samples of 96 locomotive controls is $\mu = 96 \times 0.156 = 15.0$ failures.

Acceptance sampling plans based on the binomial distribution appear in MIL-STD-105 and in quality-control books; for example, see Grant and Leavenworth[2] and Schilling.[11] An acceptance sampling plan specifies the number n of sample units and the acceptable number y of defective units in the sample. If the sample contains y or fewer defective units, the production (or lot) passes; otherwise, it fails. A plan had $n = 20$ and $y = 1$. If the production (or lot) has a proportion defective of $p = 0.01$, the chance it passes inspection is $F(1) = f(0)+f(1) = [20!/0!(20-0)!]0.01^0$ $0.99^{20}+[20!/1!(20-1)!] 0.01^1 0.99^{19} = 0.983$, which could be read from a binomial table. The chance of passing as a function of p is called the *operating characteristic* (OC) *curve* of the plan (n,y). The OC curve for $n = 20$ and $y = 1$ appears in Fig. 26.10.

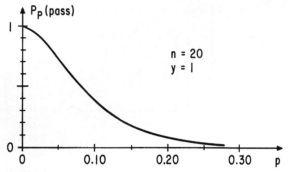

FIGURE 26.10   Operating-characteristic curve of an acceptance sampling plan.

26.8   THE SERIES-SYSTEM MODEL FOR PRODUCTS WITH MULTIPLE CAUSES OF FAILURE

Overview.   Many products fail from more than one cause. For example, any part in a toaster may fail and cause the toaster to fail. Also, humans may die from accidents, various diseases, etc. The series-system model is a model for the relationship between the product life distribution and those of its parts. This section presents the series-system model, the product rule for reliability, the addition "law" for failure rates, and the resulting distribution when some failure modes are eliminated. Graphical analyses of data with multiple causes of failure appear in Chap. 27.

Series Systems and the Product Rule.   Suppose that a product has a potential time to failure from each of M causes (also called *competing risks* or *failure modes*). Such a product is called a *series system* if its life is the smallest of those M potential times to failure. That is, the first part failure produces system failure. Let $R(t)$ denote the population reliability function of systems, and let $R_1(t), ..., R_M(t)$ denote the reliability functions of the M causes (each in the absence of all other causes). It is assumed that the M potential times to failure of a system are *statistically independent*. Such systems are said to have *independent competing risks* or to be *series systems* with independent causes of failure. For such systems, it can be shown that

$$R(t) = R_1(t)R_2(t) \cdots R_M(t)$$

This key result is the *product rule* for reliability of series systems (with independent components).

Three-way Bulb.   By engineering definition, a three-way light bulb is assumed to have both filaments on and to fail if either filament fails. Filament 1 (2) has a normal life distribution with a mean of 1500 (1200) h and a standard deviation of 300 (240) h. Filament reliability functions are depicted as straight lines on normal probability paper in Fig. 26.11. The life distribution of such bulbs was needed, in particular, the median life. Filament lives are assumed independent; so the bulb reliability is $R(t) = \{1 - \Phi[(t-1500)/300]\} \times \{1 - \Phi[(t-1200)/240]\}$; here $\Phi[\ ]$ is the standard normal cumulative distribution. For example, $R(1200) = \{1 - \Phi[(1200-1500)/300]\} \times \{1 - \Phi[(1200-1200)/240]\} = 0.421$. $R(t)$ is plotted in Fig. 26.11 and is not quite a straight line (not a normal distribution). The median life is obtained by solving $R(t_{0.50}) = 0.50$ to get $t_{0.50} = 1160$ h; this also can be obtained from the plot.

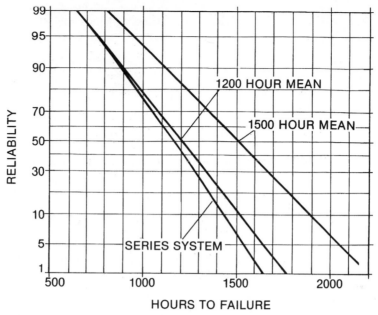

FIGURE 26.11 Reliability function of bulbs with two filaments.

Addition Law for Failure Rates. Denote the hazard function of the population of systems by $h(t)$ and those for the failure causes by $h_1(t)$, ..., $h_M(t)$. Then it can be shown that

$$h(t) = h_1(t) + h_2(t) + \cdots + h_M(t)$$

This is called the *addition law for failure rates* for *independent* failure modes (or competing risks). This "law" (actually a model) is depicted in Fig. 26.12, which shows the hazard functions of the two components of a series system and the system hazard function.

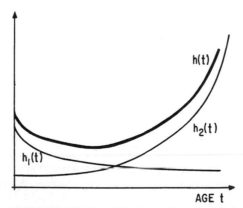

FIGURE 26.12 Hazard functions of a series system and its two components.

A pronounced increase in the failure rate of a product may occur at some age. This may indicate that a new failure cause with an increasing failure rate is becoming dominant at that age, as in Fig. 26.12.

Exponential Causes.   Suppose that M independent causes have exponential life distributions with failure rates $\lambda_1, ..., \lambda_M$. Then series systems consisting of such components have an exponential life distribution with a constant failure rate

$$\lambda = \lambda_1 + \cdots + \lambda_M$$

MIL-HDBK-217[5] uses this relationship, which is often *incorrectly* used for reliability analysis of systems with components that do not have constant failure rates. Then the previous equation for $h(t)$ is correct.

Freight Train.   A high-priority freight train was required to have its three locomotives all complete a one-day run. If a locomotive failed and delayed the train, the railroad had to pay a large penalty. The railroad needed to know the reliability of such trains. Time to failure for such locomotives has an exponential distribution with $\lambda_0 = 0.023$ failures per day. A series system, the train has an exponential distribution of time to delay with $\lambda = 0.023 + 0.023 + 0.023 = 0.069$ delays per day. Train reliability on a one-day run is $R(1) = \exp(-0.069 \times 1) = 0.933$.

Elimination of Failure Modes.   Often it is important to know how elimination of some causes of failure will improve product life. Suppose that cause 1 is eliminated (this may be a *collection of causes*). Then $R_1(t) = 1$, $h_1(t) = 0$, and the life distribution for the remaining causes has

$$R^*(t) = R_2(t) \times \cdots \times R_M(t), \qquad h^*(t) = h_2(t) + \cdots + h_M(t)$$

For example, if the 1500-h filament were replaced by one with essentially unlimited life, such bulbs would have the life distribution of the 1200-h filament.

Series Systems with Dependence.   Some series-system products contain parts with statistically dependent lifetimes. For example, adjoining segments of a cable may have positively correlated lives; that is, have similar lives. Models for dependent part lives are complicated; David and Moeschberger[1] and Hutchinson and Lai[3] survey multivariate distributions with dependence.

REFERENCES

1. David, H. A.; and Moeschberger, M. L., *The Theory of Competing Risks,* Griffin's Statistical Monograph No. 39, Methuen, London, 1979.

2. Grant, E. L.; and Leavenworth, R. S., *Statistical Quality Control,* 5th ed., McGraw-Hill, New York, 1980.

3. Hutchinson, T. P.; and Lai, C. D., *The Engineering Statistician's Guide to Continuous Bivariate Distributions,* Rumsby Scientific Publishing (P.O. Box 76, Rundle Mall, Adelaide, South Australia 5000), 1991.

4. Jensen, F.; and Petersen, N. E., *Burn-in: An Engineering Approach to the Design and Analysis of Burn-in Procedures,* Wiley, New York [(800) 879-4539], 1982.

5. MIL-HDBK-217, *Reliability Prediction of Electronic Equipment* [purchase from Naval Publications and Forms Center, 5801 Tabor Avenue, Philadelphia, PA 19120, (215) 697-3321].

6. Nelson, Wayne, *How to Analyze Data with Simple Plots,* Vol. 1 of the *ASQC Basic References in Quality Control: Statistical Techniques,* American Society for Quality Control [611 E. Wisconsin Ave., Milwaukee, WI 53201, (800) 248-1946], 1979.

7. Nelson, Wayne, *Applied Life Data Analysis,* Wiley, New York [(800)879-4539], 1982.

8. Nelson, Wayne, *How to Analyze Reliability Data,* Vol. 6 of *Basic References in Quality Control: Statistical Techniques,* American Society for Quality Control [611 E. Wisconsin Ave., Milwaukee, WI 53201, (800) 248-1946], 1983.

9. Nelson, Wayne, *Accelerated Testing: Statistical Models, Test Plans, and Data Analyses,* Wiley, New York [(800) 879-4539], 1990.

10. O'Connor, P. D. T., *Practical Reliability Engineering,* 3d ed., Wiley, New York [(800) 879-4539], 1991.

11. Schilling, E. G., *Acceptance Sampling in Quality Control,* Marcel Dekker, New York [(800) 228-1160], 1982.

12. Tobias, P. A.; and Trindade, D. C., *Applied Reliability,* 2d ed., Van Nostrand-Reinhold, New York [(212) 254-3232], 1995.

CHAPTER 27
GRAPHICAL ANALYSES OF RELIABILITY DATA

Wayne Nelson
PRIVATE RELIABILITY CONSULTANT
SCHENECTADY, NEW YORK

Necip Doganaksoy
GENERAL ELECTRIC RESEARCH AND DEVELOPMENT
SCHENECTADY, NEW YORK

27.1   INTRODUCTION

Purpose.   This chapter presents simple and informative graphical analyses of product reliability data. Section 27.2 shows how to construct and interpret probability plots of life-test data that are complete (all units are failed) and singly censored (unfailed units have a single common running time). Section 27.3 explains how to construct and interpret hazard plots of multiply censored life from actual service and from life tests (where unfailed units have differing running times). Section 27.4 presents a new method for plotting and interpreting product *repair* data where each unit can have recurring failures and repairs.

Background.   To use Secs. 27.2 and 27.3 on plotting of life data, readers need acquaintance with the life distributions and concepts presented in Chap. 26. To best use Sec. 27.4, readers would benefit from a previous basic statistics course, but still can use much of that section.

Random Sample.   Most statistical methodology rests on the assumption that the data come from a truly random sample from the population of interest. Roughly speaking, this means that every population unit has an equal chance of being in the sample. This is not always so. Clearly, development units in a lab test are not a random sample of production units in actual use. Lab tests measure *test reliability,* which only approximates *field reliability.* The usefulness of this approximation requires engineering judgment.

27.2   PROBABILITY PLOTS FOR LIFE DATA

27.2.1   Introduction

Background.   The life distribution of products is important to many manufacturers and customers. Interest centers on the percentage failing during warranty or design life, the nominal life, percentiles, and the failure rate (increasing or decreasing with the age of the population?). Probability plotting can be used with life data if every sample unit runs to failure; such data are called *complete*. However, most life data contain running times on unfailed units, and such data require special plotting methods. In some tests, the sample units start on test together, and the data are analyzed before all units fail. Then all failure times are below the common running time of the unfailed units (also called *suspensions*). Such data are called *singly censored*; they may be analyzed with probability plotting as explained below. In many tests and in the field, the units start at different times. Then the failure and running times are intermixed. Such data are called *multiply censored*; they may be analyzed with hazard plotting, which is like probability plotting and is explained in Sec. 27.3. When a failure is discovered by inspection, the exact failure time is not known. One knows only the previous inspection time, and that the failure occurred between the two. Such interval data can be analyzed with the methods below by assuming that each failure time is the midpoint of its interval. More exact methods for interval data are given by Nelson.[8]

Probability Paper.   On probability paper for a distribution (normal, lognormal, and Weibull), the data and cumulative probability scales are constructed so any such cumulative distribution function (see Chap. 26) plots as a straight line on the paper. To analyze sample data, plot the sample cumulative distribution function on such paper as described below. Then draw a straight line through the data. The line is an estimate of the population cumulative distribution function. Details on how to make and interpret such probability plots follow.

Cord Example.   Electric cords for irons are flexed by a test machine until failure. The test simulates actual use, but highly accelerated. Each week, 12 cords go on the machine and run for a week. After a week, the unfailed cords come off test to make room for a new sample of cords. Table 27.1 shows singly censored data on (1) the standard "old" cord and (2) a new, cheaper cord.

TABLE 27.1   Ironing-Cord Data (+ Unfailed)

Old			New		
Hours	Rank i	$\dfrac{100i}{n+1}$	Hours	Rank i	$\dfrac{100i}{n+1}$
57.5	1	4	72.4	1	7.7
77.8	2	8	78.6	2	15.4
88.0	3	12	81.2	3	23.1
96.9	4	16	94.0	4	30.8
98.4	5	20	120.1	5	38.5
100.3	6	24	126.3	6	46.2
100.8	7	28	127.2	7	53.8
102.1	8	32	128.7	8	61.5
103.3	9	36	141.9	9	69.2
103.4	10	40	164.1+	10	X
105.3	11	44	164.1+	11	X
105.4	12	48	164.1+	12	X
122.6	13	52			
139.3	14	56			
143.9	15	60			
148.0	16	64			
151.3	17	68			
161.1+	18	X			
161.2+	19	X			
161.2+	20	X			
162.4+	21	X			
162.7+	22	X			
163.1+	23	X			
176.8+	24	X			

Λ + marks the running time on each unfailed cord. The basic question is how do the lives of the two types of cords compare on test?

27.2.2 How to Construct a Probability Plot

Steps.   To construct a probability plot, do the following:

1. Order the n failure and running times from smallest to largest as in Table 27.1.
2. Assign the rank of 1 to the shortest time, the rank 2 to the second shortest, etc., and the rank n to the longest time, as in Table 27.1.
3. Calculate a probability plotting position F_i for each failure from its rank i and the number n of sample units as

$$F_i = \frac{100(i-0.5)}{n}, \qquad i = 1,\ldots,n$$

or as

$$F_i' = \frac{100i}{(n+1)}, \qquad i = 1,\ldots,n$$

Plotting positions are shown in Table 27.1. Nonfailures are not assigned plotting positions.

4. Choose a probability paper. The TEAM catalog[15] offers a wide selection of probability papers for various distributions. Label the data scale to span the data. Then plot each failure time against its plotting position. The cord data are plotted on normal probability paper in Fig. 27.1.

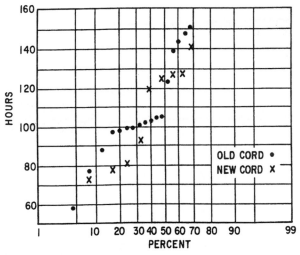

FIGURE 27.1   Normal plot of cord-life data.

Nonfailures are not plotted. Only the early failure times are observed, and they estimate the lower part of the cumulative life distribution, usually of greatest interest.

5. Assess whether the plotted data follow a straight line and if there are any peculiar data points. The two cord plots are relatively straight. Thus a normal distribution adequately describes (or fits) the data.

6. Draw a straight line through the plotted data. Such a line is an estimate of the cumulative distribution function. Figure 27.1 lacks the line; the plotted data serve in place of a line. Also, one can connect the dots with straight line segments or use a smooth curve. Then the fit is not the paper's theoretical distribution.

7. Obtain the desired information from the plot as described next. Various computer programs make such plots; for examples, see Abernethy,[1] Kececioglu,[4] and Nelson[11] (Chapter 5).

27.2.3   How to Interpret a Probability Plot

A probability plot yields estimates of distribution parameters, percentiles, and the percentage failing on warranty or during design life. Also, it indicates how well a theoretical distribution fits the data, and it reveals peculiarities in the data.

Percentiles.   As follows, estimate a population percentile. Enter the probability scale at that percentage. Go to the fitted line and then go to the corresponding point on the time scale to read the percentile estimate. For example, for the old cord, the estimate of the 50th percentile is 114 h, interpolated between data points.

Percent Failing.   As follows, estimate the population percentage failing by a given age (such as warranty or design life). Enter the plot on the time scale at the given age, go to the fitted line, and then go to the corresponding point on the probability scale to read the estimate. For example, for the new cord, the estimate of the population percentage failing by 80 h is 20 percent.

Distribution Check.   If the data tend to follow a straight line, then the chosen distribution describes the data adequately. The normal fit is adequate for the cord data. Thus the cords have an increasing failure rate (as described in Chap. 26). A curved plot indicates that the distribution does not fit the data. This assessment is subjective; different people may draw somewhat different conclusions from the same data. Plots of small samples have an erratic appearance and do not follow a straight line well, particularly in the tails of the plot. Section 27.3 shows a Weibull plot and how to interpret it. Nelson[6-8,11] and Abernethy[1] give many examples of different plots and their interpretations.

Peculiarities.   An observation that is clearly out of line with the rest of the data is called an "outlier" and may be suspected as bad data or due to a special cause. The cause should be sought, since that information may help improve the product or the data collection. Only pronounced discrepancies should be interpreted as suspect data or a wrong distribution. Inexperienced analysts tend to overinterpret plots and expect points to fall close to a straight line.

Comparison.   The two cord samples are plotted on normal paper in Fig. 27.1 after the data were plotted on several papers. It yields a reasonably straight plot, and the distribution was familiar to the responsible quality-control engineers. Plots for the two cords roughly coincide, whatever paper is used. Thus the life of new cord is comparable to that of standard old cord for engineering purposes. The new cord data show a gap between 105 and 140 h, roughly over the weekend of the test. No reason for this gap was found, but it does not affect the conclusion. Straight lines through the two samples would estimate the distributions, but they are not needed to answer the basic question.

Extrapolation.   Sometimes one estimates the lower or upper tail of a distribution from a singly censored sample by extending a straight line below or above the plotted points. The accuracy of such extrapolation depends on how well the theoretical distribution describes the true one in the extrapolated tail.

27.3 *HAZARD PLOTTING OF MULTIPLY CENSORED DATA*

Hazard and probability plots are widely used to analyze field and life-test data on products consisting of electronic and mechanical parts and ranging from microelectronics through heavy industrial equipment. This section* presents hazard plots, developed by Nelson[6] to estimate a life distribution from multiply censored life data. Such plots do not apply to failures found on inspection, since the failure occurred earlier at an unknown time. Computer programs for fitting distributions to life data and making data plots are surveyed by Nelson[11] (Chap. 5).

Appliance Component.   Data that illustrate hazard plotting appear in Table 27.2, which shows the cycles (number of times used) to failure of a snubber component of a toaster in a development program. Each nonfailure time has a + to indicate that the failure time of the unfailed snubber is beyond. Failure times are unmarked. Engineering wanted an estimate of the percentage failing on warranty (nominally 500 cycles) and an estimate of median life.

TABLE 27.2   Toaster Snubber Data and Hazard Calculations

Cycles	Reverse rank k	Hazard 100/k	Cumulative hazard	Cumulative probability	Cycles	Reverse rank k	Hazard 100/k	Cumulative hazard	Cumulative probability
45+	54				608+	27			
47	53	1.9	1.9	1.8	608+	26			
73	52	1.9	3.8	3.7	608+	25			
136+	51				608+	24			
136+	50				608+	23			
136+	49				608	22	4.6	24.8	22.0
136+	48				608+	21			
136+	47				608+	20			
145	46	2.2	6.0	5.8	630	19	5.3	30.1	26.0
190+	45				670	18	5.6	35.7	30.0
190+	44				670	17	5.9	41.6	34.0
281+	43				731+	16			
311	42	2.4	8.4	8.1	838	15	6.7	48.3	38.3
417+	41				964	14	7.1	55.4	42.5
485+	40				964	13	7.7	63.1	46.8
485+	39				1164+	12			
490	38	2.6	11.0	10.4	1164+	11			
569+	37				1164+	10			
571+	36				1164+	9			
571	35	2.9	13.9	13.0	1164+	8			
575	34	2.9	16.8	15.5	1164+	7			
608+	33				1164+	6			
608+	32				1198+	5			
608+	31				1198	4	25.0	88.1	58.6
608+	30				1300+	3			
608	29	3.4	20.2	18.0	1300+	2			
608+	28				1300+	1			

*This section is based on *How to Analyze Reliability Data* by Wayne Nelson. Copyright American Society for Quality Control, Inc., 1983, Milwaukee, Wisc. Reprinted by permission.

27.3.1 Steps to Construct a Hazard Plot

1. Order the n times from smallest to largest as shown in Table 27.2 without regard to which are failure or nonfailure times. Label the times with reverse ranks; that is, label the first time with n, the second with $n-1$, ..., and the nth with 1. There are $n = 54$ components in the example.

2. Calculate a hazard value for each *failure* as $100/k$, where k is its reverse rank, as shown in Table 27.2. For example, the failure at 145 cycles has reverse rank 46, and its hazard value is $100/46 = 2.2$ percent.

3. Calculate the cumulative hazard value for each *failure* as the sum of its hazard value and the cumulative hazard value of the preceding failure. For example, for the failure at 145 cycles, its cumulative hazard value of 6.0 is its hazard value 2.2 plus the previous cumulative hazard value, 3.8. Cumulative hazard values appear in Table 27.2. They have no physical meaning and may exceed 100 percent.

4. Choose a hazard paper. There are hazard papers for the exponential, Weibull, extreme value, normal, and lognormal distributions, offered in the TEAM catalog.[15] The distribution is often chosen from engineering knowledge of the product. To plot data on probability paper, convert each cumulative hazard value H into a cumulative probability plotting position F (a percentage) using $F = 100[1-\exp(-H/100)]$.

5. On the vertical axis of the hazard paper, mark a time scale that brackets the data. For the component data, normal hazard paper was chosen, and marked from 0 to 1200 cycles as shown in Fig. 27.2.

6. On the paper, plot each failure time vertically against its cumulative hazard value on the horizontal axis as shown in Fig. 27.2. Nonfailure times are not plotted; hazard and cumulative hazard values are not calculated for them. However, the nonfailure times do determine the proper plotting positions of the failure times through the reverse ranks.

7. If the plot of failure times is roughly straight, the theoretical distribution adequately fits the data. By eye, fit a straight line through the data points. Also, one can just use the plotted points without a line to get information.

Various computer programs do similar calculations and plot multiply censored data; for examples, see Abernethy,[1] Kececioglu,[4] and Nelson[11] (Chapter 5). Equivalent mean rank plotting positions for probability paper are given by Abernethy[1] and O'Connor.[14]

Fitting.   The line estimates the cumulative percentage failing (read from the horizontal probability scale at the top of the grid) as a function of age. The straight line, as explained below, yields information on the life distribution. If the plot is curved, plot the data on another hazard paper. If no hazard paper yields a straight enough plot, draw a smooth curve through the plotted data. Then, as described below, use the curve in the same way as a straight line to estimate percentiles and failure probabilities.

The Basic Assumption.   Hazard plotting is valid if the life distribution of units censored at a given age is the same as the life distribution of units that run beyond that age. For example, this assumption is not satisfied if unfailed units are removed from service when they look like they are about to fail.

27.3.2 How to Interpret a Hazard Plot

The probability and data scales on a hazard paper are exactly the same as those on the corresponding probability paper. Thus, a hazard plot is interpreted the same way as a probability plot,

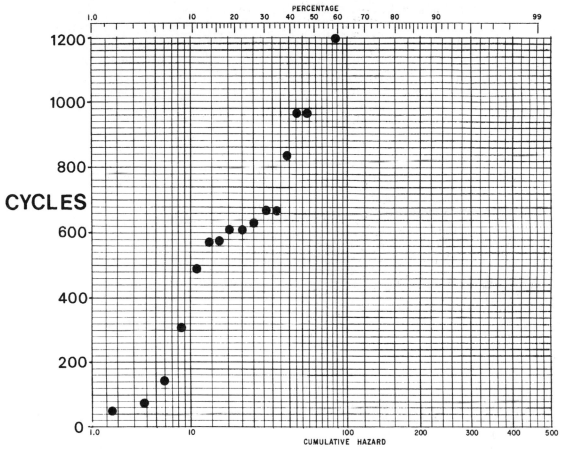

FIGURE 27.2 Normal hazard plot of snubber data.

and those scales on hazard paper are used like those on probability paper. The cumulative hazard scale is only an aid for plotting multiply censored data.

Estimate of the Percentage Failing. The population percentage failing by a given age is estimated from the fitted line or curve as follows. Enter the plot on the time scale at the given age, go to the fitted line (or data), and then go to the corresponding point on the probability scale to read the percentage. For example, the estimate of the percentage of snubbers failing by 500 cycles (typical use in a one-year warranty) is 12 percent; this answers a basic question. Clearly a better snubber is needed.

Percentile Estimate. To estimate a percentile, enter the plot on the probability scale at the given percentage, go to the fitted line (or data), and then go to the corresponding point on the time scale to read the percentile. For example, the estimate of the 50th percentile, nominal snubber life, is 1000 cycles (two years of typical use).

Normal Parameters.   As follows, graphically estimate the population mean μ and standard deviation σ. The μ estimate is the 50th percentile, 1000 cycles for the snubbers. The σ estimate is the difference of the 50th and 16th percentiles, $1000 - 700 = 300$ cycles for the snubbers.

27.3.3 Life Distribution with Failure Modes Eliminated

Hazard plotting also provides an estimate of the life distribution that would result if certain failure modes were eliminated by proposed design changes. It is costly and time-consuming to change a design and collect and analyze data to determine the value of design changes. Instead, this can be done using past data, as in the following example. It is assumed that the cause of each failure is identified.

The Method.   Suppose that a proposed design change of the toaster would eliminate mode 11 (snubber failures) and leave other failure modes unchanged. Past data (including mode 11 and other failure modes) are given in Table 27.3. To predict the resulting life distribution, hazard cal-

TABLE 27.3   Toaster Failure Modes and Hazard Calculations with Mode 11 Eliminated

Cycles	Failure mode	Cumulative hazard	Cycles	Failure mode	Cumulative hazard
45	1	1.85	608+		
47	11		608+		
73	11		608+		
136+			608+		
136	6	3.85	608+		
136+			608	11	
136+			608+		
136+			608+		
145	11		630	11	
190+			670	11	
190+			670	11	
281	12	6.18	731+		
311	11		838	11	
417	12	8.62	964	11	
485+			964	11	
485+			1164+		
490	11		1164+		
569	1	11.32	1164+		
571+			1164+		
571	11		1164+		
575	11		1164+		
608+			1164+		
608+			1198	9	31.32
608+			1198	11	
608+			1300+		
608	11		1300+		
608+			1300+		

culations for the life distribution without mode 11 are shown in Table 27.3. Each toaster failure with a mode 11 is treated as censored, since the new design would have run that long without failure. The failure times for the remaining modes are plotted against their cumulative hazard values as shown in Fig. 27.3 on Weibull paper. Chapter 26 presents the Weibull distribution. About 10

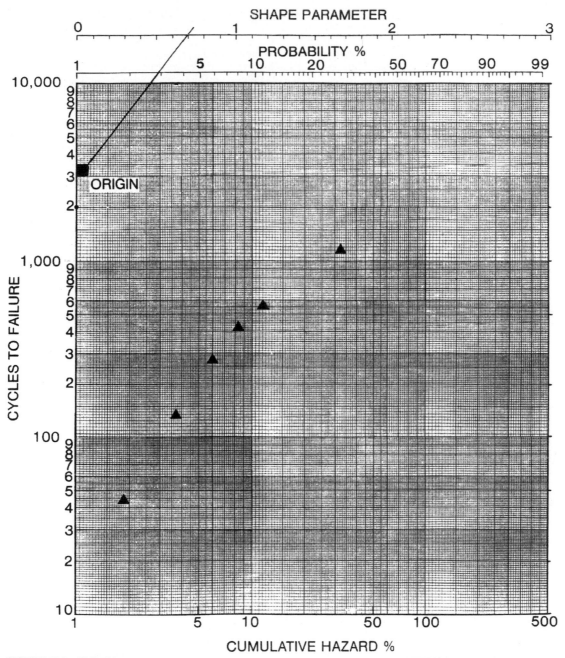

FIGURE 27.3   Weibull hazard plot of toaster life without mode 11.

percent would fail on warranty (500 cycles) with mode 11 eliminated. As 10 percent was too high, further improvement was needed.

Failure-Rate Behavior.   Often it is useful to know how the product failure rate depends on age. A failure rate that increases with age usually indicates that failures are due to wearout. A failure rate that decreases with age usually indicates that early failures are due to manufacturing or design defects.

Weibull Shape.   For data plotted on Weibull hazard paper, assess the nature of the failure rate as follows. A Weibull failure rate increases (decreases) if the shape parameter is greater (less) than 1. To estimate the Weibull shape parameter, draw a straight line parallel to the plotted data, so it passes through the "origin" of the Weibull hazard paper and through the shape parameter scale, as in Fig. 27.3. The value on that shape scale is the estimate; it is 0.7 in Fig. 27.3, indicating a decreasing failure rate (usually resulting from design or manufacturing defects). Also, the estimate of the Weibull characteristic life is the 63d percentile. For Fig. 27.3, this estimate is roughly 3000 cycles, using a line through the last two data points.

Assumptions.   The hazard plotting method above is based on four assumptions: (1) each unit has a potential failure time for each failure mode, (2) the observed time to failure for a unit is the smallest of its potential times to failure, (3) potential times to failure for different failure modes are statistically independent, and (4) the mode of each failure is identified. Thus the product is regarded as a series system (discussed in Chap. 26).

27.3.4   Life Distribution of a Failure Mode

Information is sometimes desired on the distribution of time to failure for a particular failure mode. An estimate of its distribution provides information on the nature of the failure mode and on the effect of design changes on that mode.

The Method.   An example of the method involves the toaster data with a mix of failure modes in Table 27.3. Hazard calculations for mode 11 (the snubber) are shown in Table 27.2. In these calculations, each failure time for another mode is treated as a censoring time for mode 11; that is, as if those units were removed from test before they failed by mode 11. The failure times for mode 11 (the snubber) are plotted against their cumulative hazard values in Fig. 27.2.

27.4   *GRAPHICAL ANALYSIS OF REPAIR DATA*

27.4.1   Introduction

Purpose.   Section 27.4* presents a simple, informative plot of reliability data on numbers and costs of repairs on a sample of repairable systems. The plot provides a nonparametric estimate of the population mean cumulative (1) *cost* or (2) *number* of repairs per system versus system age. The plot is used to

1. Evaluate whether the repair (or cost) rate of a population increases or decreases with age. This information helps with decisions on system retirement and burn-in.
2. Predict future numbers and costs of repairs.
3. Compare samples from different populations (from different designs, production periods, operating environments, maintenance policies, etc.).
4. Reveal unexpected information and insight, an advantage of plots.

---

*Sections 27.4.1 through 27.4.3 are reprinted in part from Nelson[10] with permission of the American Society for Quality Control.

The plot also applies to recurrence data on disease episodes, factory simulation data, borrower credit histories, and many other histories.

Repair Data. Figure 27.4 displays an artificial but typical sample of repair data on six systems. For a system, the data are (1) the cost of each repair and its corresponding age in months and (2)

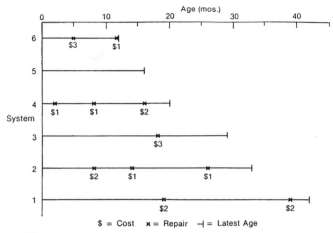

FIGURE 27.4 Display of system repair histories, artificial data.

the system's latest age. For example, the data on system 6 are (5, \$3), (12, \$1), and (12, +). The + indicates its latest age, called the "censoring" age. This can be the age when the system is retired or its current age if it is still in service. Censoring ages of sample systems differ and complicate the data analysis, requiring the analysis here.

MCF. For the following, imagine a population of system cumulative histories without censoring as depicted in Fig. 27.5. By age *t*, each system accumulates a total repair cost. So there is a

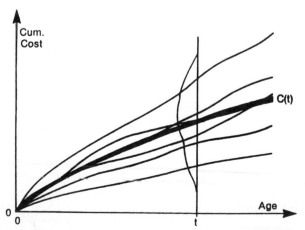

FIGURE 27.5 Uncensored population cumulative cost histories and MCF $C(t)$.

population distribution of cumulative costs at age *t*, as depicted. The distribution mean $C(t)$ is the population *mean cumulative function* (MCF) for *cost*. Assume that it has a derivative

$$c(t) = \frac{dC(t)}{dt}$$

where $c(t)$ is the mean rate at which repair costs accumulate per system at age *t*, expressed, say, in dollars per month per system.

Repairs.   Instead of costs, one may cumulate the *number* of repairs. Similarly, there is a discrete distribution of the cumulative number of repairs at age *t*, since different systems accumulate 0, 1, 2, ... repairs by age *t*, as depicted in Fig. 27.6. The mean $M(t)$ of this distribution is called the *pop-*

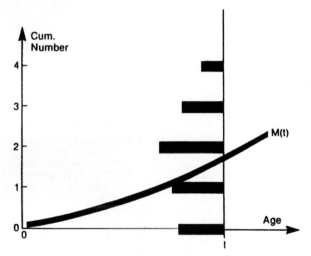

FIGURE 27.6   Population distribution of the cumulative number of repairs.

ulation mean cumulative function (MCF) for the *number* of repairs. $M(t) = C(t)$ when each repair costs one dollar. The derivative

$$m(t) = \frac{dM(t)}{dt}$$

is assumed to exist; it is called the *recurrence rate,* the *instantaneous repair rate,* and the *intensity function.* It has the dimensions of repairs per system per unit time. $m(t)$ differs from the "failure rate" (*hazard rate* of Chap. 26) of a life distribution for a nonrepaired item. Ascher and Feingold carefully distinguish the recurrence rate of repaired systems from the hazard rate of non-repaired items.[2]

27.4.2 How to Plot a Sample Mean Cumulative Function

Purpose.   This section shows how to calculate and plot a sample MCF for cost or number of repairs.

Plot.   The following steps yield a nonparametric estimate $C^*(t)$ for $C(t)$ and $M^*(t)$ for $M(t)$ from the N sample systems. $M^*(t)$ is similar in spirit to, but different from, the nonparametric estimate of the cumulative hazard function of a life distribution in Sec. 27.3.

1. List all repair and censoring ages in order from smallest to largest as in column (1) of Table 27.4. With each repair age, write the cost. With each censoring age, write "+." If a repair age of a system equals its censoring age, put the repair age first. If two or more systems have a common age, list them separately in random order.

TABLE 27.4   Artificial Data and MCF Calculations

Event	(1) (Age, cost)	(2) No. I in service	(3) Mean cost, $	(4) $C^*(t)$ $ MCF	(5) Mean repairs	(6) $M^*(t)$ No. MCF
1	(2, $1)	6	$1/6 = 0.17	0.17	1/6 = .17	.17
2	(5, $3)	6	$3/6 = 0.50	0.67	1/6 = .17	.34
3	(8, $1)	6	$1/6 = 0.17	0.84	1/6 = .17	.51
4	(8, $2)	6	$2/6 = 0.33	1.17	1/6 = .17	.68
5	(12, $1)	6	$1/6 = 0.17	1.34	1/6 = .17	.85
6	(12, +)	5				
7	(14, $1)	5	$1/5 = 0.20	1.54	1/5 = .20	1.05
8	(16, +)	4				
9	(16, $2)	4	$2/4 = 0.50	2.04	1/4 = .25	1.30
10	(18, $3)	4	$3/4 = 0.75	2.79	1/4 = .25	1.55
11	(19, $2)	4	$2/4 = 0.50	3.29	1/4 = .25	1.80
12	(20, +)	3				
13	(26, $1)	3	$1/3 = 0.33	3.62	1/3 = .33	2.13
14	(29, +)	2				
15	(33, +)	1				
16	(39, $2)	1	$2/1 = 2.00	5.62	1/1 = 1.00	3.13
17	(42, +)	0				

2. For each sample age, write the number I of systems then in service in column 2. If the first age is a censoring age, write $I = N-1$; otherwise, write $I = N$. Proceed down column 2 writing the same I value for each successive repair age. At each censoring age, reduce the I value by 1. For the last age, $I = 0$.

3. For each repair, calculate the *mean repair cost* as the cost (column 1) divided by I (column 2), the number of systems that went through that repair age. For example, for the repair at age 18, this is $3/4 = $0.75, which appears in column 3. For a censoring age, there is no mean repair cost. However, the censoring ages determine the I values and thus the mean repair costs.

4. In column 4, calculate the sample mean cumulative repair cost $C^*(t)$ for each repair as follows. For the first repair, it is its mean repair cost, $0.17 in Table 27.4. For each successive repair, it is the corresponding mean repair cost (column 3) plus the preceding cumulative cost (column 4). For example, at age 18, it is $0.75+$2.04 = $2.79. Censoring ages have no cumulative cost.

5. On linear graph paper, plot each mean cumulative repair cost (column 4) against its age (column 1) as in Fig. 27.7. This yields the plot of the estimate $C^*(t)$, the sample MCF for cost. Censoring times are not plotted.

Section 27.4.3 explains how to interpret and use such a plot.

Number.   Columns 5 and 6 (in Table 27.4) show the calculation of the MCF for repairs $M^*(t)$. For each repair, use 1 (one repair) in column 5 in place of the repair cost.

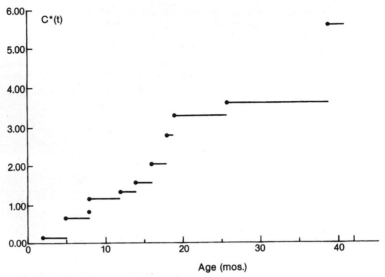

FIGURE 27.7   Artificial sample MCF $C^*(t)$.

Staircase.   Flat portions of a sample MCF may be plotted, if desired, yielding a staircase, as in Fig. 27.7. The sample MCF extends only to the last censoring time (tick marked on the time axis at 42 months). The true MCF of a large population is regarded as a smooth curve, imagined passing through the plotted points. Such a curve can be fitted to the plot by eye or mathematical means.

Assumptions.   The MCF estimate employs three assumptions:

1. The population MCF $M(t)$ exists (is finite) over ages of interest.
2. The sample systems are a truly random sample from the population. This is not so, for example, if prototype systems are lab-tested to predict performance of production systems in actual use.
3. Censoring of system histories is random. That is, the cost (or repair) histories of all N sample systems are statistically independent of their censoring ages. This is so if the N systems are randomly assigned to the N censoring times. Censoring is nonrandom if some systems under severe use are retired early because they had costly repair histories. Also, censoring is non-random if old production units (with long histories) have higher early repair rates and recent production units (with short histories) have lower early repair rates. Then $M^*(t)$ is biased.

Using these assumptions, Nelson shows that $C^*(t)$ is an unbiased estimate for $C(t)$, and $M^*(t)$ for $M(t)$.[10,13]

Parametric Models.   Most models and analyses for cost and repair data are parametric and involve more assumptions, which are often dubious, as discussed by Nelson.[10,13] Engelhardt,[3] Tobias and Trindade,[16] and Ascher and Feingold[2] discuss such models, analyses, and assumptions in detail for data on a *single system,* a situation rare in practice. For example, Nelson presents the assumptions and data analyses for the simplest parametric model, the Poisson process.[8,9]

Types of Repairs.   A data analysis may include all types of repairs, selected types of repairs, or a single type of repair. Also, scheduled and unscheduled maintenance costs and other service costs may be included if desired, as well as purchase price and (negative) scrap or trade-in value. The choice of what to include depends on the application and the desired information. In practice, one may analyze such data any number of times using different choices of the types of repairs or costs. Different choices may yield engineering, management, or accounting information.

Extensions.   Nelson[10,13] briefly describes how the MCF estimate and confidence limits extend to situations where

- The data contain various types of repairs.
- The censoring is more complex; for example, a patient may leave and reenter a medical study any number of times, leaving gaps in his history.
- Sample units have different sizes which need to be taken into account in the analyses; for instance, units are power cables of different lengths.
- Some "costs" are negative; for example, scrap value of a retired system.
- A variable used in place of "cost" varies continuously over time, such as (1) the energy consumed by a heat pump or (2) the actual system cumulative uptime as a function of the scheduled cumulative uptime, which yields availability information.

27.4.3 How to Use an MCF Plot

Purpose.   This section shows how to interpret and use an MCF plot.

Rate Behavior.   The derivative of an MCF (imagined or fitted) is the cost (or repair) rate. An increasing (decreasing) derivative means that average monthly repair costs increase (decrease) as systems age. The behavior of the rate is used to determine burn-in and retirement policies, as described by Nelson.[10] Figure 27.7 has a relatively constant derivative. Figure 27.8 has a complex derivative.

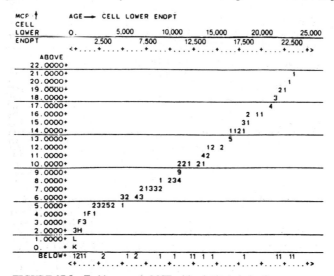

FIGURE 27.8   Turbine sample MCF with a bathtub derivative.

Turbines.   Figure 27.8 shows the sample MCF for repairs of 21 turbines versus operating hours. The plotted numbers are the numbers of repairs in that space on the plot. Also, $A = 10$, $B = 11$, etc. The bottom line of the plot (labeled BELOW), displays the 21 censoring ages. This plot shows a "bathtub" repair rate (that is, derivative). The initial repair rate (derivative) is high—typical of start-up with "infant mortality." After 3000 h, the rate is lower and gradually increases; that is, the plot is slightly convex upward after 3000 h, which is typical of wearout behavior.

Burn-in.   Some products are subjected to a factory burn-in. Units are run and repaired until the (instantaneous) population repair rate decreases to some desired value m'. An estimate of the suitable length t' of burn-in is obtained from the sample MCF as shown in Fig. 27.9. A straight line segment with slope m' is moved parallel to itself until it is tangent to the MCF. The corresponding age at the tangent point is the desired t' as shown in Fig. 27.9.

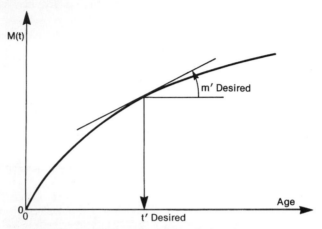

FIGURE 27.9   Method for estimating burn-in time t'.

Blood Analyzers.   Figure 27.10 shows the sample MCF from data on blood analyzers during burn-in. The purpose was to assess the effect of burn-in and determine a suitable burn-in time to achieve 1 repair per 10,000 actuations. The plot shows important features. First, data before 2000 actuations are not plotted. Second, the repairs accumulate linearly between 2000 and 5000 actuations—a constant repair rate. Third, beyond 5000 actuations, the repairs accumulate linearly but at a lower (constant) rate. The change of slope at 5000 actuations needed explanation. When questioned, the test engineers acknowledged that they had used different test procedures and definitions of failure before and after 5000 actuations. Subsequent data analyses employed a consistent

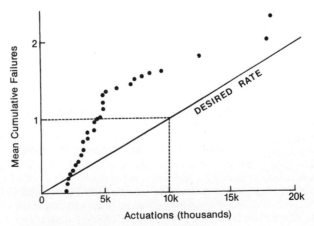

FIGURE 27.10   Blood analyzer sample MCF.

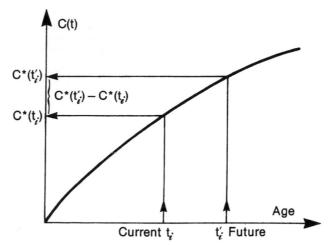

FIGURE 27.11   Method for predicting future cost.

definition of failure. The constant failure rate before and after 5000 cycles also suggests that burnin beyond 2000 actuations does not reduce the repair rate, which is important unexpected information, a key benefit of plots.

Prediction.   Such a plot can be used to predict the total number or cost of repairs of a fleet of systems in a future period as follows. For system i, suppose that its ages at the start and end of the period are t_i and t_i'. Use the sample MCF to obtain $C^*(t_i)$ and $C^*(t_i')$. Their difference $C^*(t_i') - C^*(t_i)$ predicts the expected incremental future repair cost. This difference appears graphically in Fig. 27.11. The sum of such estimates for all systems predicts the total fleet repair cost in the future period. Of course, t_i and t_i' differ from one system to another. Also, systems that go into service during the future period can be included in the calculation; for them, $t_i = 0$. If the age t_i' of a system exceeds the greatest (censoring) age for which $C^*(t)$ has been estimated, then fit a curve (by eye or by mathematics) to the plot and extrapolate the fitted curve to greater ages as needed. Prediction of the *number* of future repairs similarly employs $M^*(t)$.

Other Applications.   Nelson[10,13] and others[2,3,16] present various plots and their interpretations.

27.4.4   Confidence Limits for the Mean Cumulative Function

Purpose.   This section presents confidence limits for a mean cumulative function (MCF) for censored repair data. The limits help one assess whether a sample MCF is accurate enough for practical purposes. The limits are illustrated with the following application.

Braking Grids.   Figure 27.12 and Table 27.5 display repair data on two samples of locomotives, each with six braking grids. Locomotive engineers wanted to know whether the grid repair rate increases with population age. If so, preventive replacement would be considered. Also, they wanted to compare the sample-1 MCF with the sample-2 MCF of later production (Section 27.4.5). In Table 27.5, each line contains the data on a repair or the censoring age (length of history) of a locomotive. For a braking grid repair, (1) the first data value is the locomotive *ID*, (2) the second value is the locomotive *age* in days when repaired, and (3) the third value is the *number* of such repairs (usually 1) for that locomotive at that age. For example, the first line shows that locomotive S1-01 at 462 days had one grid replacement. For a censoring age (current age in service of a locomotive), (1) the first value is the locomotive *ID*, (2) the second value is the locomotive's current *age* in days,

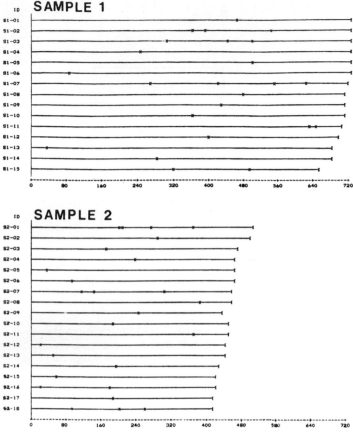

FIGURE 27.12   Display of grid repair ages.

and (3) the third value is a *code* −0. The data must contain the current (or retirement) age for each locomotive. For example, the last line shows that locomotive S1-15 was at 657 days, its current length of history. The data must contain a "length of history" for each locomotive.

Motivation.   Figure 27.5 motivates the variance of the MCF estimate $M^*(t)$ at age t. Suppose instead that the N cumulative history functions in that figure are a random sample from an infinite population, and they are uncensored at age t, where the vertical distribution of cumulative cost appears. For example, the braking grid histories in sample 1 are all uncensored up to 657 days. Before this age, the estimate $M^*(t)$ of the population mean $M(t)$ at age t is the sample average $\bar{Y}$ of the N observed cumulative costs $Y_1, Y_2, ..., Y_N$ at age t. By the basic central limit theorem, for large N, the sampling distribution of the estimate $M^*(t) = \bar{Y}$ here is approximately normal with a true mean of $M(t)$ and true variance $V[M^*(t)] = \sigma_t^2/N$, where σ_t^2 is the variance of the population cumulative cost distribution at age t. Here the estimate for $V[M^*(t)]$ is $v[M^*(t)] = s^2/N$, and s^2 is the usual (unbiased) sample variance of $Y_1, Y_2, ..., Y_N$. Thus two-sided normal approximate $C\%$ confidence limits for $M(t)$ are

$$\bar{Y} \pm K_C\left(\frac{s^2}{N}\right)^{1/2} = M^*(t) \pm K_C\{v[M^*(t)]\}^{1/2}$$

TABLE 27.5   Grid Repair Data (-0 End of History)

	Sample 1			Sample 2	
ID	Days	Value	ID	Days	Value
S1-01	462	1	S2-01	203	1
S1-01	730	−0	S2-01	211	1
S1-02	364	1	S2-01	277	1
S1-02	391	1	S2-01	373	1
S1-02	548	1	S2-01	511	−0
S1-02	724	−0	S2-02	293	1
S1-03	302	1	S2-02	503	−0
S1-03	444	1	S2-03	173	1
S1-03	500	1	S2-03	470	−0
S1-03	730	−0	S2-04	242	1
S1-04	250	1	S2-04	464	−0
S1-04	730	−0	S2-05	39	1
S1-05	500	1	S2-05	464	−0
S1-05	724	−0	S2-06	91	1
S1-06	88	1	S2-06	462	−0
S1-06	724	−0	S2-07	119	1
S1-07	272	1	S2-07	148	1
S1-07	421	1	S2-07	306	1
S1-07	552	1	S2-07	461	−0
S1-07	625	1	S2-08	382	1
S1-07	719	−0	S2-08	460	−0
S1-08	481	1	S2-09	250	1
S1-08	710	−0	S2-09	434	−0
S1-09	431	1	S2-10	192	1
S1-09	710	−0	S2-10	448	−0
S1-10	367	1	S2-11	369	1
S1-10	710	−0	S2-11	448	−0
S1-11	635	1	S2-12	22	1
S1-11	650	1	S2-12	447	−0
S1-11	708	−0	S2-13	54	1
S1-12	402	1	S2-13	441	−0
S1-12	700	−0	S2-14	194	1
S1-13	33	1	S2-14	432	−0
S1-13	687	−0	S2-15	61	1
S1-14	287	1	S2-15	419	−0
S1-14	687	−0	S2-16	19	1
S1-15	317	1	S2-16	185	1
S1-15	498	1	S2-16	419	−0
S1-15	657	−0	S2-17	187	1
			S2-17	416	−0
			S2-18	93	1
			S2-18	205	1
			S2-18	264	1
			S2-18	415	−0

Here the two-sided factor K_C is the $(100+C)/2$ standard normal percentile. For example, $K_{95} = 1.960$ and $K_{99} = 2.576$. These confidence limits apply to the braking grid data up to 657 days. Here $v[M^*(t)] = s^2/N$ is the estimate of the variance of $M^*(t)$ for *uncensored* history functions. The square root of this variance estimates the standard deviation of the sampling distribution of $M^*(t)$ and is called the *standard error* of $M^*(t)$. Nelson's[13] theory provides the true variance

$V[M^*(t)]$ and Nelson's unbiased estimate $v[M^*(t)]$ of it when some sample history functions are *censored* before age t. Nelson's variance in the confidence limits above yields approximate normal confidence limits for $M(t)$.

Plot. Figure 27.13 shows a plot of the MCF and 95 percent confidence limits for sample 1. The sample MCF increases little up to 240 days and linearly thereafter. So thereafter grid replacements occur at a constant rate. This suggests that preventive replacement of unfailed grids has no benefit after 240 days.

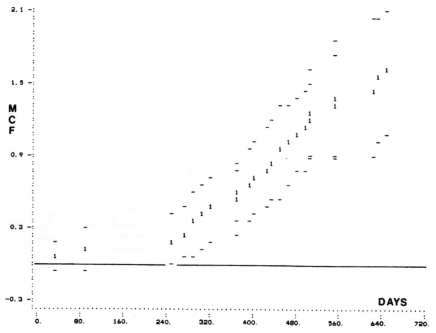

FIGURE 27.13 Grid sample 1 MCF and 95 percent confidence limits.

Programs. Calculation of Nelson's[13] confidence limits is complex and requires a special computer program. Nelson and Doganaksoy's FORTRAN program MCFLIM,[12] which does the calculations and plots may be purchased from Wayne Nelson, 739 Huntingdon Drive, Schenectady, NY 12309-2917, (518) 346-5138. The SAS Institute in a 1996 release will provide Nelson's confidence limits and plots in the SAS QC Software with the RELIABILITY procedure; contact Dr. Gordon Johnston of SAS at (919) 677-8000, extension 6679.

27.4.5 Comparison of Sample MCFs

Purpose. This section describes how to compare sample MCFs. This includes comparing two sample MCFs for a statistically significant (i.e., convincing) difference by means of a computer program.

Comparisons. MCF plots can be used to compare populations, which differ with respect to design, operation, maintenance, environment, application, etc. If population MCFs differ, one gains information on how to improve such systems. If one MCF is below another, as in Fig. 27.14*a*, repair costs (or rates) for that population are lower. If such functions cross, as in Fig. 27.14*b*, then one must decide whether low cost (rate) is more desirable at early or later ages. Of

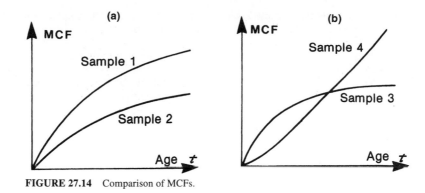

FIGURE 27.14   Comparison of MCFs.

course, the difference between two sample MCFs must be convincing (statistically significant) and also big enough for practical purposes (practically significant). Statistical confidence limits help one judge if observed differences are convincing.

Difference.   To compare two populations, one can envision the difference $M_1(t) - M_2(t)$ of their MCFs as a function of age t. This population difference is regarded as a smooth curve as a function of age t. For example, if Fig. 27.14a depicted population MCFs, their difference would be zero at age 0 and then would be positive and increase with age. Similarly, if Fig. 27.14b depicted population MCFs, their difference is zero at age 0, increases with age to a maximum, then decreases to zero, and continues to decrease, becoming more negative with age.

Grids.   Figure 27.15 shows the MCF for repairs and 95 percent confidence limits for sample 2 of locomotive grids. The sample-2 MCF increases earlier than the sample-1 MCF (Fig. 27.13).

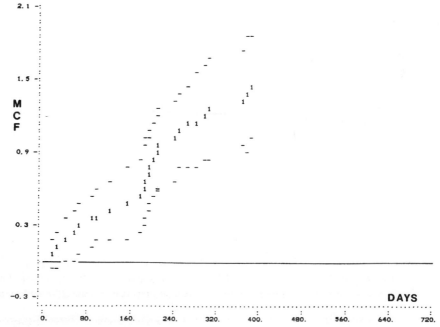

FIGURE 27.15   Grid sample 2 MCF and 95 percent confidence limits.

This suggests that sample 2 was not performing as well as sample 1. Whether this difference is convincing is assessed with a formal comparison below.

Limits.   Figure 27.16 is a plot of the difference of the MCF estimates of the two braking grid samples versus age in days. The difference is negative over the sample age range; that is, sample 1 has a smaller MCF. The plot includes approximate 95 percent confidence limits for the true difference. In the middle of the age range, the intervals do not enclose zero. There the sample-1 MCF is statistically significantly below the sample-2 MCF. With this convincing evidence, the engineers investigated and found that the grid supplier had changed the die used to make the grids between samples 1 and 2. The engineers remedied the die.

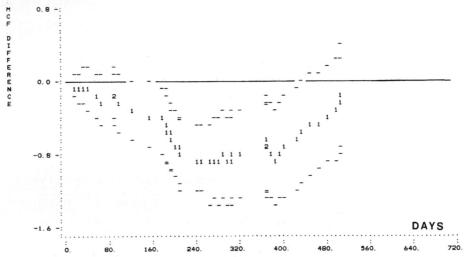

FIGURE 27.16   Difference of the grid MCFs and 95 percent limits.

Programs.   The calculations for Figure 27.16 are complex and require a computer program. Nelson and Doganaksoy's FORTRAN program MCFDIFF,[12] which does the calculations and plots, may be purchased from Wayne Nelson, 739 Huntingdon Drive, Schenectady, NY 12309-2917, (518) 346-5138. The SAS Institute in a 1996 release will provide these calculations and plot in the SAS QC Software with the RELIABILITY procedure; contact Dr. Gordon Johnston of SAS at (919) 677-8000, extension 6679.

REFERENCES

1. Abernethy, R. B., *The New Weibull Reliability Handbook,* 1993, $69.00. [Order from Dr. R. B. Abernethy, 536 Oyster Road, North Palm Beach, FL 33408-4328, (407) 842-4082.]

2. Ascher, H.; and Feingold, H., *Repairable Systems Reliability,* Marcel Dekker, New York [(800) 228-1160], 1984.

3. Engelhardt, M., "Models and Analyses for the Reliability of a Single Repairable System," in *Recent Advances in Life-Testing and Reliability,* N. Balakrishnan, ed., CRC Press, Boca Raton, Fla., 1993.

4. Kececioglu, D., *ReliaSoft's WEIBULL Program,* 1994. [Order from Dr. D. Kececioglu, 7340 N. La Oesta Ave., Tucson, AZ 85704-3119, (602) 621-6120.]

5. Lee, E. T., *Statistical Methods for Survival Data Analysis,* 2d ed., Wiley Interscience, New York [(800) 879-4539], 1992.

6. Nelson, Wayne, "Theory and Application of Hazard Plotting for Censored Failure Data," *Technometrics* **14,** 1972, pp. 945–966.

7. Nelson, Wayne, *How to Analyze Data with Simple Plots,* American Society for Quality Control [611 E. Wisconsin Ave., Milwaukee, WI 53201, (800) 248-1946], 1979.

8. Nelson, Wayne, *Applied Life Data Analysis,* Wiley, New York [(800) 879-4539], 1982.

9. Nelson, Wayne, *How to Analyze Reliability Data,* American Society for Quality Control [611 E. Wisconsin Ave., Milwaukee, WI 53201, (800) 248-1946], 1983.

10. Nelson, Wayne, "Graphical Analysis of System Repair Data," *Journal of Quality Technology,* **20,** 1988, pp. 24–35.

11. Nelson, Wayne, *Accelerated Testing: Statistical Models, Test Plans, and Data Analyses,* Wiley, New York [(800) 879-4539], 1990.

12. Nelson, Wayne; and Doganaksoy, Necip, *Documentation for MCFLIM and MCFDIFF—Programs for Recurrence Data Analysis,* 1994 [available from Wayne Nelson, 739 Huntingdon Drive, Schenectady, NY 12309-2917, (518) 346-5138].

13. Nelson, Wayne, "Confidence Limits for Recurrence Data—Applied to Cost or Number of Repairs," *Technometrics,* **37,** May 1995, pp. 147–157.

14. O'Connor, P. D. T., *Practical Reliability Engineering,* 3d ed., Wiley, New York [(800) 879-4539], 1991.

15. TEAM, *Catalog of Graph Papers,* TEAM [Box 25, Tamworth, NH 03886, (603) 323-8843], 1995.

16. Tobias, P. A.; and Trindade, D. C., *Applied Reliability,* 2d ed., Van Nostrand-Reinhold, New York [(212) 254-3232], 1995.

APPENDIXES

TABLE A.1   Summation of Terms of Poisson's Exponential Binomial Limit*
1,000 × probability of c or fewer occurrences of event that has average number of occurrences equal to c′ or np′

c′ or np′	0	1	2	3	4	5	6	7	8	9
0.02	980	1,000								
0.04	961	999	1,000							
0.06	942	998	1,000							
0.08	923	997	1,000							
0.10	905	995	1,000							
0.15	861	990	999	1,000						
0.20	819	982	999	1,000						
0.25	779	974	998	1,000						
0.30	741	963	996	1,000						
0.35	705	951	994	1,000						
0.40	670	938	992	999	1,000					
0.45	638	925	989	999	1,000					
0.50	607	910	986	998	1,000					
0.55	577	894	982	998	1,000					
0.60	549	878	977	997	1,000					
0.65	522	861	972	996	999	1,000				
0.70	497	844	966	994	999	1,000				
0.75	472	827	959	993	999	1,000				
0.80	449	809	953	991	999	1,000				
0.85	427	791	945	989	998	1,000				
0.90	407	772	937	987	998	1,000				
0.95	387	754	929	984	997	1,000				
1.00	368	736	920	981	996	999	1,000			
1.1	333	699	900	974	995	999	1,000			
1.2	301	663	879	966	992	998	1,000			
1.3	273	627	857	957	989	998	1,000			
1.4	247	592	833	946	986	997	999	1,000		
1.5	223	558	809	934	981	996	999	1,000		
1.6	202	525	783	921	976	994	999	1,000		
1.7	183	493	757	907	970	992	998	1,000		
1.8	165	463	731	891	964	990	997	999	1,000	
1.9	150	434	704	875	956	987	997	999	1,000	
2.0	135	406	677	857	947	983	995	999	1,000	

*From *Statistical Quality Control*, 3d ed., by Eugene L. Grant. Copyright, 1964. McGraw-Hill Book Company. Used by permission.

TABLE A.1   Summation of Terms of Poisson's Exponential Binomial Limit
(*Continued*)

c' or np' \ c	0	1	2	3	4	5	6	7	8	9
2.2	111	355	623	819	928	975	993	998	1,000	
2.4	091	308	570	779	904	964	988	997	999	1,000
2.6	074	267	518	736	877	951	983	995	999	1,000
2.8	061	231	469	692	848	935	976	992	998	999
3.0	050	199	423	647	815	916	966	988	996	999
3.2	041	171	380	603	781	895	955	983	994	998
3.4	033	147	340	558	744	871	942	977	992	997
3.6	027	126	303	515	706	844	927	969	988	996
3.8	022	107	269	473	668	816	909	960	984	994
4.0	018	092	238	433	629	785	889	949	979	992
4.2	015	078	210	395	590	753	867	936	972	989
4.4	012	066	185	359	551	720	844	921	964	985
4.6	010	056	163	326	513	686	818	905	955	980
4.8	008	048	143	294	476	651	791	887	944	975
5.0	007	040	125	265	440	616	762	867	932	968
5.2	006	034	109	238	406	581	732	845	918	960
5.4	005	029	095	213	373	546	702	822	903	951
5.6	004	024	082	191	342	512	670	797	886	941
5.8	003	021	072	170	313	478	638	771	867	929
6.0	002	017	062	151	285	446	606	744	847	916

c' or np'	10	11	12	13	14	15	16
2.8	1,000						
3.0	1,000						
3.2	1,000						
3.4	999	1,000					
3.6	999	1,000					
3.8	998	999	1,000				
4.0	997	999	1,000				
4.2	996	999	1,000				
4.4	994	998	999	1,000			
4.6	992	997	999	1,000			
4.8	990	996	999	1,000			
5.0	986	995	998	999	1,000		
5.2	982	993	997	999	1,000		
5.4	977	990	996	999	1,000		
5.6	972	988	995	998	999	1,000	
5.8	965	984	993	997	999	1,000	
6.0	957	980	991	996	999	999	1,000

TABLE A.1   Summation of Terms of Poisson's Exponential Binomial Limit
(*Continued*)

c' or np' \ c	0	1	2	3	4	5	6	7	8	9
6.2	002	015	054	134	259	414	574	716	826	902
6.4	002	012	046	119	235	384	542	687	803	886
6.6	001	010	040	105	213	355	511	658	780	869
6.8	001	009	034	093	192	327	480	628	755	850
7.0	001	007	030	082	173	301	450	599	729	830
7.2	001	006	025	072	156	276	420	569	703	810
7.4	001	005	022	063	140	253	392	539	676	788
7.6	001	004	019	055	125	231	365	510	648	765
7.8	000	004	016	048	112	210	338	481	620	741
8.0	000	003	014	042	100	191	313	453	593	717
8.5	000	002	009	030	074	150	256	386	523	653
9.0	000	001	006	021	055	116	207	324	456	587
9.5	000	001	004	015	040	089	165	269	392	522
10.0	000	000	003	010	029	067	130	220	333	458

	10	11	12	13	14	15	16	17	18	19
6.2	949	975	989	995	998	999	1,000			
6.4	939	969	986	994	997	999	1,000			
6.6	927	963	982	992	997	999	999	1,000		
6.8	915	955	978	990	996	998	999	1,000		
7.0	901	947	973	987	994	998	999	1,000		
7.2	887	937	967	984	993	997	999	999	1,000	
7.4	871	926	961	980	991	996	998	999	1,000	
7.6	854	915	954	976	989	995	998	999	1,000	
7.8	835	902	945	971	986	993	997	999	1,000	
8.0	816	888	936	966	983	992	996	998	999	1,000
8.5	763	849	909	949	973	986	993	997	999	999
9.0	706	803	876	926	959	978	989	995	998	999
9.5	645	752	836	898	940	967	982	991	996	998
10.0	583	697	792	864	917	951	973	986	993	997

	20	21	22
8.5	1,000		
9.0	1,000		
9.5	999	1,000	
10.0	998	999	1,000

TABLE A.1   Summation of Terms of Poisson's Exponential Binomial Limit
(*Continued*)

c' or np' \\ c	0	1	2	3	4	5	6	7	8	9
10.5	000	000	002	007	021	050	102	179	279	397
11.0	000	000	001	005	015	038	079	143	232	341
11.5	000	000	001	003	011	028	060	114	191	289
12.0	000	000	001	002	008	020	046	090	155	242
12.5	000	000	000	002	005	015	035	070	125	201
13.0	000	000	000	001	004	011	026	054	100	166
13.5	000	000	000	001	003	008	019	041	079	135
14.0	000	000	000	000	002	006	014	032	062	109
14.5	000	000	000	000	001	004	010	024	048	088
15.0	000	000	000	000	001	003	008	018	037	070

	10	11	12	13	14	15	16	17	18	19
10.5	521	639	742	825	888	932	960	978	988	994
11.0	460	579	689	781	854	907	944	968	982	991
11.5	402	520	633	733	815	878	924	954	974	986
12.0	347	462	576	682	772	844	899	937	963	979
12.5	297	406	519	628	725	806	869	916	948	969
13.0	252	353	463	573	675	764	835	890	930	957
13.5	211	304	409	518	623	718	798	861	908	942
14.0	176	260	358	464	570	669	756	827	883	923
14.5	145	220	311	413	518	619	711	790	853	901
15.0	118	185	268	363	466	568	664	749	819	875

	20	21	22	23	24	25	26	27	28	29
10.5	997	999	999	1,000						
11.0	995	998	999	1,000						
11.5	992	996	998	999	1,000					
12.0	988	994	997	999	999	1,000				
12.5	983	991	995	998	999	999	1,000			
13.0	975	986	992	996	998	999	1,000			
13.5	965	980	989	994	997	998	999	1,000		
14.0	952	971	983	991	995	997	999	999	1,000	
14.5	936	960	976	986	992	996	998	999	999	1,000
15.0	917	947	967	981	989	994	997	998	999	1,000

TABLE A.1   Summation of Terms of Poisson's Exponential Binomial Limit
(*Continued*)

c' or np' \ c	4	5	6	7	8	9	10	11	12	13
16	000	001	004	010	022	043	077	127	193	275
17	000	001	002	005	013	026	049	085	135	201
18	000	000	001	003	007	015	030	055	092	143
19	000	000	001	002	004	009	018	035	061	098
20	000	000	000	001	002	005	011	021	039	066
21	000	000	000	000	001	003	006	013	025	043
22	000	000	000	000	001	002	004	008	015	028
23	000	000	000	000	000	001	002	004	009	017
24	000	000	000	000	000	000	001	003	005	011
25	000	000	000	000	000	000	001	001	003	006

	14	15	16	17	18	19	20	21	22	23
16	368	467	566	659	742	812	868	911	942	963
17	281	371	468	564	655	736	805	861	905	937
18	208	287	375	469	562	651	731	799	855	899
19	150	215	292	378	469	561	647	725	793	849
20	105	157	221	297	381	470	559	644	721	787
21	072	111	163	227	302	384	471	558	640	716
22	048	077	117	169	232	306	387	472	556	637
23	031	052	082	123	175	238	310	389	472	555
24	020	034	056	087	128	180	243	314	392	473
25	012	022	038	060	092	134	185	247	318	394

	24	25	26	27	28	29	30	31	32	33
16	978	987	993	996	998	999	999	1,000		
17	959	975	985	991	995	997	999	999	1,000	
18	932	955	972	983	990	994	997	998	999	1,000
19	893	927	951	969	980	988	993	996	998	999
20	843	888	922	948	966	978	987	992	995	997
21	782	838	883	917	944	963	976	985	991	994
22	712	777	832	877	913	940	959	973	983	989
23	635	708	772	827	873	908	936	956	971	981
24	554	632	704	768	823	868	904	932	953	960
25	473	553	629	700	763	818	863	900	929	950

	34	35	36	37	38	39	40	41	42	43
19	999	1,000								
20	999	999	1,000							
21	997	998	999	999	1,000					
22	994	996	998	999	999	1,000				
23	988	993	996	997	999	999	1,000			
24	979	987	992	995	997	998	999	999	1,000	
25	966	978	985	991	994	997	998	999	999	1,000

TABLE A.2   Confidence Limits for a Proportion (One-sided), $n \leq 30$

(For confidence limits for $n > 30$, see Charts B.1 to B.4)

If the observed proportion is r/n, enter the table with n and r for an upper one-sided limit. For a lower one-sided limit, enter the table with n and $n - r$ and subtract the table entry from 1.

r	90%	95%	99%	r	90%	95%	99%	r	90%	95%	99%
	$n = 2$				$n = 3$				$n = 4$		
0	0.684	0.776	0.900	0	0.536	0.632	0.785−	0	0.438	0.527	0.684
1	0.949	0.975−	0.995−	1	0.804	0.865−	0.941	1	0.680	0.751	0.859
				2	0.965+	0.983	0.997	2	0.857	0.902	0.958
								3	0.974	0.987	0.997
	$n = 5$				$n = 6$				$n = 7$		
0	0.369	0.451	0.602	0	0.319	0.393	0.536	0	0.280	0.348	0.482
1	0.584	0.657	0.778	1	0.510	0.582	0.706	1	0.453	0.521	0.643
2	0.753	0.811	0.894	2	0.667	0.729	0.827	2	0.596	0.659	0.764
3	0.888	0.924	0.967	3	0.799	0.847	0.915+	3	0.721	0.775−	0.858
4	0.979	0.990	0.998	4	0.907	0.937	0.973	4	0.830	0.871	0.929
				5	0.983	0.991	0.998	5	0.921	0.947	0.977
								6	0.985+	0.993	0.999
	$n = 8$				$n = 9$				$n = 10$		
0	0.250	0.312	0.438	0	0.226	0.283	0.401	0	0.206	0.259	0.369
1	0.406	0.471	0.590	1	0.368	0.429	0.544	1	0.337	0.394	0.504
2	0.538	0.600	0.707	2	0.490	0.550	0.656	2	0.450	0.507	0.612
3	0.655+	0.711	0.802	3	0.599	0.655+	0.750	3	0.552	0.607	0.703
4	0.760	0.807	0.879	4	0.699	0.749	0.829	4	0.646	0.696	0.782
5	0.853	0.889	0.939	5	0.790	0.831	0.895−	5	0.733	0.778	0.850
6	0.931	0.954	0.980	6	0.871	0.902	0.947	6	0.812	0.850	0.907
7	0.987	0.994	0.999	7	0.939	0.959	0.983	7	0.884	0.913	0.952
				8	0.988	0.994	0.999	8	0.945+	0.963	0.984
								9	0.990	0.995−	0.999
	$n = 11$				$n = 12$				$n = 13$		
0	0.189	0.238	0.342	0	0.175−	0.221	0.319	0	0.162	0.206	0.298
1	0.310	0.364	0.470	1	0.287	0.339	0.440	1	0.268	0.316	0.413
2	0.415+	0.470	0.572	2	0.386	0.438	0.537	2	0.360	0.410	0.506
3	0.511	0.564	0.660	3	0.475+	0.527	0.622	3	0.444	0.495−	0.588
4	0.599	0.650	0.738	4	0.559	0.609	0.698	4	0.523	0.573	0.661
5	0.682	0.729	0.806	5	0.638	0.685−	0.765+	5	0.598	0.645+	0.727
6	0.759	0.800	0.866	6	0.712	0.755−	0.825+	6	0.669	0.713	0.787
7	0.831	0.865−	0.916	7	0.781	0.819	0.879	7	0.736	0.776	0.841
8	0.895+	0.921	0.957	8	0.846	0.877	0.924	8	0.799	0.834	0.889
9	0.951	0.967	0.986	9	0.904	0.928	0.961	9	0.858	0.887	0.931
10	0.990	0.995+	0.999	10	0.955−	0.970	0.987	10	0.912	0.934	0.964
				11	0.991	0.996	0.999	11	0.958	0.972	0.988
								12	0.992	0.996	0.999
	$n = 14$				$n = 15$				$n = 16$		
0	0.152	0.193	0.280	0	0.142	0.181	0.264	0	0.134	0.171	0.250
1	0.251	0.297	0.389	1	0.236	0.279	0.368	1	0.222	0.264	0.349
2	0.337	0.385+	0.478	2	0.317	0.363	0.453	2	0.300	0.344	0.430
3	0.417	0.466	0.557	3	0.393	0.440	0.529	3	0.371	0.417	0.503
4	0.492	0.540	0.627	4	0.464	0.511	0.597	4	0.439	0.484	0.569
5	0.563	0.610	0.692	5	0.532	0.577	0.660	5	0.504	0.548	0.630

This table was computed by Robert S. Gardner and is reproduced, by permission of the author and the publisher, from E. L. Crow, F. A. Davis, and M. W. Maxfield, *Statistics Manual*, China Lake, Calif., U.S. Naval Ordinance Test Station, NAVORD Report 3369 (NOTS 948), pp. 262–265.

TABLE A.2   Confidence Limits for a Proportion (One-sided), $n \leq 30$ (*Continued*)

r	90%	95%	99%	r	90%	95%	99%	r	90%	95%	99%
	$n = 14$ (Continued)				$n = 15$ (Continued)				$n = 16$ (Continued)		
6	0.631	0.675−	0.751	6	0.596	0.640	0.718	6	0.565+	0.609	0.687
7	0.695+	0.736	0.805+	7	0.658	0.700	0.771	7	0.625−	0.667	0.739
8	0.757	0.794	0.854	8	0.718	0.756	0.821	8	0.682	0.721	0.788
9	0.815−	0.847	0.898	9	0.774	0.809	0.865+	9	0.737	0.773	0.834
10	0.869	0.896	0.936	10	0.828	0.858	0.906	10	0.790	0.822	0.875−
11	0.919	0.939	0.967	11	0.878	0.903	0.941	11	0.839	0.868	0.912
12	0.961	0.974	0.989	12	0.924	0.943	0.969	12	0.886	0.910	0.945−
13	0.993	0.996	0.999	13	0.964	0.976	0.990	13	0.929	0.947	0.971
				14	0.993	0.997	0.999	14	0.966	0.977	0.990
								15	0.993	0.997	0.999
	$n = 17$				$n = 18$				$n = 19$		
0	0.127	0.162	0.237	0	0.120	0.153	0.226	0	0.114	0.146	0.215+
1	0.210	0.250	0.332	1	0.199	0.238	0.316	1	0.190	0.226	0.302
2	0.284	0.326	0.410	2	0.269	0.310	0.391	2	0.257	0.296	0.374
3	0.352	0.396	0.480	3	0.334	0.377	0.458	3	0.319	0.359	0.439
4	0.416	0.461	0.543	4	0.396	0.439	0.520	4	0.378	0.419	0.498
5	0.478	0.522	0.603	5	0.455+	0.498	0.577	5	0.434	0.476	0.554
6	0.537	0.580	0.658	6	0.512	0.554	0.631	6	0.489	0.530	0.606
7	0.594	0.636	0.709	7	0.567	0.608	0.681	7	0.541	0.582	0.655+
8	0.650	0.689	0.758	8	0.620	0.659	0.729	8	0.592	0.632	0.702
9	0.703	0.740	0.803	9	0.671	0.709	0.774	9	0.642	0.680	0.746
10	0.754	0.788	0.845−	10	0.721	0.756	0.816	10	0.690	0.726	0.788
11	0.803	0.834	0.883	11	0.769	0.801	0.855−	11	0.737	0.770	0.827
12	0.849	0.876	0.918	12	0.815−	0.844	0.890	12	0.782	0.812	0.863
13	0.893	0.915+	0.948	13	0.858	0.884	0.923	13	0.825−	0.853	0.897
14	0.933	0.950	0.973	14	0.899	0.920	0.951	14	0.866	0.890	0.927
15	0.968	0.979	0.991	15	0.937	0.953	0.975−	15	0.905−	0.925−	0.954
16	0.994	0.997	0.999	16	0.970	0.980	0.992	16	0.941	0.956	0.976
				17	0.994	0.997	0.999	17	0.972	0.981	0.992
								18	0.994	0.997	0.999
	$n = 20$				$n = 21$				$n = 22$		
0	0.109	0.139	0.206	0	0.104	0.133	0.197	0	0.099	0.127	0.189
1	0.181	0.216	0.289	1	0.173	0.207	0.277	1	0.166	0.198	0.266
2	0.245−	0.283	0.358	2	0.234	0.271	0.344	2	0.224	0.259	0.330
3	0.304	0.344	0.421	3	0.291	0.329	0.404	3	0.279	0.316	0.389
4	0.361	0.401	0.478	4	0.345+	0.384	0.460	4	0.331	0.369	0.443
5	0.415−	0.456	0.532	5	0.397	0.437	0.512	5	0.581	0.420	0.493
6	0.467	0.508	0.583	6	0.448	0.487	0.561	6	0.430	0.468	0.541
7	0.518	0.558	0.631	7	0.497	0.536	0.608	7	0.477	0.515+	0.587
8	0.567	0.606	0.677	8	0.544	0.583	0.653	8	0.523	0.561	0.630
9	0.615+	0.653	0.720	9	0.590	0.628	0.695+	9	0.568	0.605−	0.672
10	0.662	0.698	0.761	10	0.636	0.672	0.736	10	0.611	0.647	0.712
11	0.707	0.741	0.800	11	0.679	0.714	0.774	11	0.654	0.689	0.750
12	0.751	0.783	0.837	12	0.722	0.755+	0.811	12	0.695+	0.729	0.786
13	0.793	0.823	0.871	13	0.764	0.794	0.845+	13	0.736	0.767	0.821
14	0.834	0.860	0.902	14	0.804	0.832	0.878	14	0.775+	0.804	0.853
15	0.873	0.896	0.931	15	0.842	0.868	0.908	15	0.813	0.840	0.884
16	0.910	0.929	0.956	16	0.879	0.901	0.935−	16	0.850	0.874	0.912
17	0.944	0.958	0.977	17	0.914	0.932	0.959	17	0.885+	0.906	0.938
18	0.973	0.982	0.992	18	0.946	0.960	0.978	18	0.918	0.935+	0.961
19	0.995−	0.997	0.999	19	0.974	0.983	0.993	19	0.949	0.962	0.979
				20	0.995−	0.988	1.000	20	0.976	0.984	0.993
								21	0.995+	0.998	1.000

TABLE A.2   Confidence Limits for a Proportion (One-sided), $n \leq 30$ (*Continued*)

r	90%	95%	99%	r	90%	95%	99%	r	90%	95%	99%
	n = 23				n = 24				n = 25		
0	0.095+	0.122	0.181	0	0.091	0.117	0.175−	0	0.088	0.113	0.168
1	0.159	0.190	0.256	1	0.153	0.183	0.246	1	0.147	0.176	0.237
2	0.215+	0.249	0.318	2	0.207	0.240	0.307	2	0.199	0.231	0.296
3	0.268	0.304	0.374	3	0.258	0.292	0.361	3	0.248	0.282	0.349
4	0.318	0.355−	0.427	4	0.306	0.342	0.412	4	0.295−	0.330	0.398
5	0.366	0.404	0.476	5	0.352	0.389	0.460	5	0.340	0.375+	0.444
6	0.413	0.451	0.522	6	0.398	0.435−	0.505−	6	0.383	0.420	0.488
7	0.459	0.496	0.567	7	0.442	0.479	0.548	7	0.426	0.462	0.531
8	0.503	0.540	0.609	8	0.484	0.521	0.590	8	0.467	0.504	0.571
9	0.546	0.583	0.650	9	0.526	0.563	0.630	9	0.508	0.544	0.610
10	0.589	0.625−	0.689	10	0.567	0.603	0.668	10	0.548	0.583	0.648
11	0.630	0.665−	0.727	11	0.608	0.642	0.705−	11	0.587	0.621	0.684
12	0.670	0.704	0.763	12	0.647	0.681	0.740	12	0.625−	0.659	0.719
13	0.710	0.742	0.797	13	0.685+	0.718	0.774	13	0.662	0.695−	0.752
14	0.748	0.778	0.829	14	0.723	0.754	0.806	14	0.699	0.730	0.784
15	0.786	0.814	0.860	15	0.759	0.788	0.837	15	0.735−	0.764	0.815+
16	0.822	0.848	0.889	16	0.795+	0.822	0.867	16	0.770	0.798	0.845+
17	0.857	0.880	0.916	17	0.830	0.854	0.894	17	0.804	0.830	0.873
18	0.890	0.910	0.941	18	0.863	0.885+	0.920	18	0.837	0.861	0.899
19	0.922	0.938	0.962	19	0.895+	0.914	0.943	19	0.869	0.890	0.923
20	0.951	0.963	0.980	20	0.925+	0.941	0.964	20	0.899	0.918	0.946
21	0.977	0.984	0.993	21	0.953	0.965+	0.981	21	0.928	0.943	0.966
22	0.995+	0.998	1.000	22	0.978	0.985−	0.994	22	0.955+	0.966	0.982
				23	0.996	0.998	1.000	23	0.979	0.986	0.994
								24	0.996	0.998	1.000
	n = 26				n = 27				n = 28		
0	0.085−	0.109	0.162	0	0.082	0.105+	0.157	0	0.079	0.101	0.152
1	0.142	0.170	0.229	1	0.137	0.164	0.222	1	0.132	0.159	0.215−
2	0.192	0.223	0.286	2	0.185+	0.215+	0.277	2	0.179	0.208	0.268
3	0.239	0.272	0.337	3	0.231	0.263	0.326	3	0.223	0.254	0.316
4	0.284	0.318	0.385−	4	0.275−	0.308	0.373	4	0.265+	0.298	0.361
5	0.328	0.363	0.430	5	0.317	0.351	0.417	5	0.306	0.339	0.404
6	0.370	0.405+	0.473	6	0.358	0.392	0.458	6	0.346	0.380	0.445−
7	0.411	0.447	0.514	7	0.397	0.432	0.498	7	0.385−	0.419	0.484
8	0.451	0.487	0.554	8	0.436	0.471	0.537	8	0.422	0.457	0.521
9	0.491	0.526	0.592	9	0.475−	0.509	0.574	9	0.459	0.494	0.558
10	0.529	0.564	0.628	10	0.512	0.547	0.610	10	0.496	0.530	0.593
11	0.567	0.602	0.664	11	0.549	0.583	0.645+	11	0.532	0.565−	0.627
12	0.604	0.638	0.698	12	0.585−	0.618	0.679	12	0.567	0.600	0.660
13	0.641	0.673	0.731	13	0.620	0.653	0.711	13	0.601	0.634	0.692
14	0.676	0.708	0.763	14	0.655+	0.687	0.743	14	0.635+	0.667	0.723
15	0.711	0.742	0.794	15	0.689	0.720	0.773	15	0.669	0.699	0.753
16	0.746	0.774	0.823	16	0.723	0.752	0.802	16	0.701	0.731	0.782
17	0.779	0.806	0.851	17	0.756	0.783	0.831	17	0.733	0.762	0.810
18	0.812	0.837	0.878	18	0.788	0.814	0.857	18	0.765−	0.792	0.837
19	0.843	0.866	0.903	19	0.819	0.843	0.883	19	0.796	0.821	0.863
20	0.874	0.894	0.927	20	0.849	0.871	0.907	20	0.826	0.849	0.888
21	0.903	0.921	0.948	21	0.879	0.899	0.930	21	0.855+	0.876	0.911
22	0.931	0.946	0.967	22	0.907	0.924	0.950	22	0.883	0.902	0.932
23	0.957	0.968	0.983	23	0.934	0.948	0.968	23	0.911	0.927	0.952
24	0.979	0.986	0.994	24	0.958	0.969	0.983	24	0.936	0.950	0.969
25	0.996	0.998	1.000	25	0.980	0.987	0.994	25	0.960	0.970	0.984
				26	0.996	0.998	1.000	26	0.981	0.987	0.995−
								27	0.996	0.998	1.000

TABLE A.2 Confidence Limits for a Proportion (One-sided), n ≤ 30 (*Continued*)

r	90%	95%	99%	r	90%	95%	99%
		$n = 29$				$n = 30$	
0	0.076	0.098	0.147	0	0.074	0.095+	0.142
1	0.128	0.153	0.208	1	0.124	0.149	0.202
2	0.173	0.202	0.260	2	0.168	0.195+	0.252
3	0.216	0.246	0.307	3	0.209	0.239	0.298
4	0.257	0.288	0.350	4	0.249	0.280	0.340
5	0.297	0.329	0.392	5	0.287	0.319	0.381
6	0.335−	0.368	0.432	6	0.325−	0.357	0.420
7	0.372	0.406	0.470	7	0.361	0.394	0.457
8	0.409	0.443	0.507	8	0.397	0.430	0.493
9	0.445+	0.479	0.542	9	0.432	0.465+	0.527
10	0.481	0.514	0.577	10	0.466	0.499	0.561
11	0.515+	0.549	0.610	11	0.500	0.533	0.594
12	0.550	0.583	0.643	12	0.533	0.566	0.626
13	0.583	0.616	0.674	13	0.566	0.598	0.657
14	0.616	0.648	0.705−	14	0.599	0.630	0.687
15	0.649	0.680	0.734	15	0.630	0.661	0.716
16	0.681	0.711	0.763	16	0.662	0.692	0.744
17	0.712	0.741	0.791	17	0.692	0.721	0.772
18	0.743	0.771	0.818	18	0.723	0.750	0.799
19	0.774	0.800	0.843	19	0.752	0.779	0.824
20	0.803	0.828	0.868	20	0.782	0.807	0.849
21	0.832	0.855−	0.892	21	0.810	0.834	0.873
22	0.860	0.881	0.914	22	0.838	0.860	0.896
23	0.888	0.906	0.935−	23	0.865+	0.885+	0.917
24	0.914	0.930	0.954	24	0.891	0.909	0.937
25	0.938	0.951	0.970	25	0.917	0.932	0.955+
26	0.961	0.971	0.985	26	0.941	0.953	0.972
27	0.982	0.988	0.995−	27	0.963	0.972	0.985+
28	0.996	0.998	1.000	28	0.982	0.988	0.995+
				29	0.996	0.998	1.000

TABLE A.3   Confidence Limits for a Proportion (Two-sided), $n \leq 30$*

(For confidence limits for n >30, see Charts B.1 to B.4)
Upper limits are in boldface. The observed proportion in a random sample is r/n.

$n = 1$

r	90%		95%		99%	
0	0	**0.900**	0	**0.950**	0	**0.990**
1	0.100	**1**	0.050	**1**	0.010	**1**

$n = 2$

r	90%		95%		99%	
0	0	**0.684**	0	**0.776**	0	**0.900**
1	0.051	**0.949**	0.025+	**0.975−**	0.005+	**0.995−**
2	0.316	**1**	0.224	**1**	0.100	**1**

$n = 3$

r	90%		95%		99%	
0	0	**0.536**	0	**0.632**	0	**0.785−**
1	0.035−	**0.804**	0.017	**0.865−**	0.003	**0.941**
2	0.196	**0.965+**	0.135−	**0.983**	0.059	**0.997**
3	0.464	**1**	0.368	**1**	0.215+	**1**

$n = 4$

r	90%		95%		99%	
0	0	**0.500**	0	**0.527**	0	**0.684**
1	0.026	**0.680**	0.013	**0.751**	0.003	**0.859**
2	0.143	**0.857**	0.098	**0.902**	0.042	**0.958**
3	0.320	**0.974**	0.249	**0.987**	0.141	**0.997**
4	0.500	**1**	0.473	**1**	0.316	**1**

$n = 5$

r	90%		95%		99%	
0	0	**0.379**	0	**0.500**	0	**0.602**
1	0.021	**0.621**	0.010	**0.657**	0.002	**0.778**
2	0.112	**0.753**	0.076	**0.811**	0.033	**0.894**
3	0.247	**0.888**	0.189	**0.924**	0.106	**0.967**
4	0.379	**0.979**	0.343	**0.990**	0.222	**0.998**
5	0.621	**1**	0.500	**1**	0.398	**1**

$n = 6$

r	90%		95%		99%	
0	0	**0.345−**	0	**0.402**	0	**0.536**
1	0.017	**0.542**	0.009	**0.598**	0.002	**0.706**
2	0.093	**0.667**	0.063	**0.729**	0.027	**0.827**
3	0.201	**0.799**	0.153	**0.847**	0.085−	**0.915+**
4	0.333	**0.907**	0.271	**0.937**	0.173	**0.973**
5	0.458	**0.983**	0.402	**0.991**	0.294	**0.998**
6	0.655+	**1**	0.598	**1**	0.464	**1**

$n = 7$

r	90%		95%		99%	
0	0	**0.316**	0	**0.377**	0	**0.500**
1	0.015	**0.500**	0.007	**0.554**	0.001	**0.643**
2	0.079	**0.684**	0.053	**0.659**	0.023	**0.764**
3	0.170	**0.721**	0.129	**0.775**	0.071	**0.858**
4	0.279	**0.830**	0.225+	**0.871**	0.142	**0.929**
5	0.316	**0.921**	0.341	**0.947**	0.236	**0.977**
6	0.500	**0.985+**	0.446	**0.993**	0.357	**0.999**
7	0.684	**1**	0.623	**1**	0.500	**1**

$n = 8$

r	90%		95%		99%	
0	0	**0.255−**	0	**0.315+**	0	**0.451**
1	0.013	**0.418**	0.006	**0.500**	0.001	**0.590**
2	0.069	**0.582**	0.046	**0.685−**	0.020	**0.707**
3	0.147	**0.745+**	0.111	**0.711**	0.061	**0.802**
4	0.240	**0.760**	0.193	**0.807**	0.121	**0.879**
5	0.255−	**0.853**	0.289	**0.889**	0.198	**0.939**
6	0.418	**0.931**	0.315+	**0.954**	0.293	**0.980**
7	0.582	**0.987**	0.500	**0.994**	0.410	**0.999**
8	0.745+	**1**	0.685−	**1**	0.549	**1**

$n = 9$

r	90%		95%		99%	
0	0	**0.232**	0	**0.289**	0	**0.402**
1	0.012	**0.391**	0.006	**0.443**	0.001	**0.598**
2	0.061	**0.515+**	0.041	**0.558**	0.017	**0.656**
3	0.129	**0.610**	0.098	**0.711**	0.053	**0.750**
4	0.210	**0.768**	0.169	**0.749**	0.105+	**0.829**
5	0.232	**0.790**	0.251	**0.831**	0.171	**0.895−**
6	0.390	**0.871**	0.289	**0.902**	0.250	**0.947**
7	0.485−	**0.939**	0.442	**0.959**	0.344	**0.983**
8	0.609	**0.988**	0.557	**0.994**	0.402	**0.999**
9	0.768	**1**	0.711	**1**	0.598	**1**

$n = 10$

r	90%		95%		99%	
0	0	**0.222**	0	**0.267**	0	**0.376**
1	0.010	**0.352**	0.005+	**0.397**	0.001	**0.512**
2	0.055−	**0.500**	0.037	**0.603**	0.016	**0.624**
3	0.116	**0.648**	0.087	**0.619**	0.048	**0.703**
4	0.188	**0.659**	0.150	**0.733**	0.093	**0.782**
5	0.222	**0.778**	0.222	**0.778**	0.150	**0.850**
6	0.341	**0.812**	0.267	**0.850**	0.218	**0.907**
7	0.352	**0.884**	0.381	**0.913**	0.297	**0.952**
8	0.500	**0.945+**	0.397	**0.963**	0.376	**0.984**
9	0.648	**0.990**	0.603	**0.995−**	0.488	**0.999**
10	0.778	**1**	0.733	**1**	0.624	**1**

$n = 11$

r	90%		95%		99%	
0	0	**0.197**	0	**0.250**	0	**0.359**
1	0.010	**0.315+**	0.005−	**0.369**	0.001	**0.509**
2	0.061	**0.423**	0.033	**0.500**	0.014	**0.593**
3	0.105−	**0.577**	0.079	**0.631**	0.043	**0.660**
4	0.169	**0.685+**	0.135−	**0.667**	0.084	**0.738**
5	0.197	**0.698**	0.200	**0.750**	0.134	**0.806**
6	0.302	**0.803**	0.250	**0.800**	0.194	**0.866**
7	0.315+	**0.831**	0.333	**0.865−**	0.262	**0.916**
8	0.423	**0.895+**	0.369	**0.921**	0.340	**0.957**
9	0.577	**0.951**	0.500	**0.967**	0.407	**0.986**
10	0.685−	**0.990**	0.631	**0.995+**	0.500	**0.999**
11	0.803	**1**	0.750	**1**	0.641	**1**

$n = 12$

r	90%		95%		99%	
0	0	**0.184**	0	**0.236**	0	**0.321**
1	0.009	**0.294**	0.004	**0.346**	0.001	**0.445+**
2	0.045+	**0.398**	0.030	**0.450**	0.013	**0.555−**
3	0.096	**0.500**	0.072	**0.550**	0.039	**0.679**
4	0.154	**0.602**	0.123	**0.654**	0.076	**0.698**
5	0.184	**0.706**	0.181	**0.706**	0.121	**0.765+**
6	0.271	**0.729**	0.236	**0.764**	0.175−	**0.825+**
7	0.294	**0.816**	0.294	**0.819**	0.235−	**0.879**
8	0.398	**0.846**	0.346	**0.877**	0.302	**0.924**
9	0.500	**0.904**	0.450	**0.928**	0.321	**0.961**
10	0.602	**0.955−**	0.550	**0.970**	0.445+	**0.987**
11	0.706	**0.991**	0.654	**0.996**	0.555−	**0.999**
12	0.816	**1**	0.764	**1**	0.679	**1**

*This table was calculated by Edwin L. Crow, Eleanor G. Crow, and Robert S. Gardner and is reproduced, by permission of the authors and the publisher, from E. L. Crow, F. A. Davis, and M. W. Maxfield, *Statistics Manual*, China Lake, Calif., U.S. Naval Ordnance Test Station, NAVORD Reoprt 3369 (NOTS 948), 1955, pp. 257–261.

TABLE A.3   Confidence Limits for a Proportion (Two-sided), $n \leq 30$ (*Continued*)

n = 13

r	90%		95%		99%	
0	0	0.173	0	0.225+	0	0.302
1	0.008	0.276	0.004	0.327	0.001	0.429
2	0.042	0.379	0.028	0.434	0.012	0.523
3	0.088	0.470	0.066	0.520	0.036	0.594
4	0.142	0.545−	0.113	0.587	0.069	0.698
5	0.173	0.621	0.166	0.673	0.111	0.727
6	0.246	0.724	0.224	0.740	0.159	0.787
7	0.276	0.754	0.260	0.776	0.213	0.841
8	0.379	0.827	0.327	0.834	0.273	0.889
9	0.455+	0.858	0.413	0.887	0.302	0.931
10	0.530	0.912	0.480	0.934	0.406	0.964
11	0.621	0.958	0.566	0.972	0.477	0.988
12	0.724	0.992	0.673	0.996	0.571	0.999
13	0.827	1	0.775−	1	0.698	1

n = 14

r	90%		95%		99%	
0	0	0.163	0	0.207	0	0.286
1	0.007	0.261	0.004	0.312	0.001	0.392
2	0.039	0.365+	0.026	0.389	0.011	0.500
3	0.081	0.422	0.061	0.500	0.033	0.608
4	0.131	0.578	0.104	0.611	0.064	0.636
5	0.163	0.594	0.153	0.629	0.102	0.714
6	0.224	0.645+	0.206	0.688	0.146	0.751
7	0.261	0.739	0.207	0.793	0.195−	0.805+
8	0.355−	0.776	0.312	0.794	0.249	0.854
9	0.406	0.837	0.371	0.847	0.286	0.898
10	0.422	0.869	0.389	0.896	0.364	0.936
11	0.578	0.919	0.500	0.939	0.392	0.967
12	0.635−	0.961	0.611	0.974	0.500	0.989
13	0.739	0.993	0.688	0.996	0.608	0.999
14	0.837	1	0.793	1	0.714	1

n = 15

r	90%		95%		99%	
0	0	0.154	0	0.191	0	0.273
1	0.007	0.247	0.003	0.302	0.001	0.373
2	0.036	0.326	0.024	0.369	0.010	0.461
3	0.076	0.400	0.057	0.448	0.031	0.539
4	0.122	0.500	0.097	0.552	0.059	0.627
5	0.154	0.600	0.142	0.631	0.094	0.672
6	0.205+	0.674	0.191	0.668	0.135−	0.727
7	0.247	0.675−	0.192	0.706	0.179	0.771
8	0.325+	0.753	0.294	0.808	0.229	0.821
9	0.326	0.795−	0.332	0.809	0.273	0.865+
10	0.400	0.846	0.369	0.838	0.328	0.906
11	0.500	0.878	0.448	0.903	0.373	0.941
12	0.600	0.924	0.552	0.943	0.461	0.969
13	0.674	0.964	0.631	0.976	0.539	0.990
14	0.753	0.993	0.698	0.997	0.627	0.999
15	0.846	1	0.809	1	0.727	1

n = 16

r	90%		95%		99%	
0	0	0.147	0	0.179	0	0.264
1	0.007	0.235+	0.003	0.273	0.001	0.357
2	0.034	0.305+	0.023	0.352	0.010	0.451
3	0.071	0.381	0.053	0.429	0.029	0.525−
4	0.114	0.450	0.090	0.500	0.055+	0.579
5	0.147	0.550	0.132	0.571	0.088	0.643
6	0.189	0.619	0.178	0.648	0.125+	0.705−
7	0.235+	0.695−	0.179	0.727	0.166	0.739
8	0.299	0.701	0.272	0.728	0.212	0.788
9	0.305+	0.765−	0.273	0.821	0.261	0.834
10	0.381	0.811	0.352	0.822	0.295+	0.875−
11	0.450	0.853	0.429	0.868	0.357	0.912
12	0.550	0.886	0.500	0.910	0.421	0.945−
13	0.619	0.929	0.571	0.947	0.475+	0.971
14	0.695−	0.966	0.648	0.977	0.549	0.990
15	0.765−	0.993	0.727	0.997	0.643	0.999
16	0.853	1	0.821	1	0.736	1

n = 17

r	90%		95%		99%	
0	0	0.140	0	0.167	0	0.243
1	0.006	0.225+	0.003	0.254	0.001	0.346
2	0.032	0.290	0.021	0.337	0.009	0.413
3	0.067	0.364	0.050	0.417	0.027	0.500
4	0.107	0.432	0.085−	0.489	0.052	0.587
5	0.140	0.500	0.124	0.544	0.082	0.620
6	0.175+	0.568	0.166	0.594	0.117	0.662
7	0.225+	0.636	0.167	0.663	0.155+	0.757
8	0.277	0.710	0.253	0.746	0.197	0.758
9	0.290	0.723	0.254	0.747	0.242	0.803
10	0.364	0.775−	0.338	0.833	0.243	0.845
11	0.432	0.825−	0.406	0.834	0.338	0.883
12	0.500	0.860	0.456	0.876	0.380	0.918
13	0.568	0.893	0.511	0.915+	0.413	0.948
14	0.636	0.933	0.583	0.950	0.500	0.973
15	0.710	0.968	0.663	0.979	0.587	0.991
16	0.775−	0.994	0.746	0.997	0.654	0.999
17	0.860	1	0.833	1	0.757	1

n = 18

r	90%		95%		99%	
0	0	0.135−	0	0.157	0	0.228
1	0.006	0.216	0.003	0.242	0.001	0.318
2	0.030	0.277	0.020	0.325−	0.008	0.397
3	0.063	0.349	0.047	0.381	0.025+	0.466
4	0.101	0.419	0.080	0.444	0.049	0.534
5	0.135−	0.482	0.116	0.556	0.077	0.603
6	0.163	0.536	0.156	0.619	0.110	0.682
7	0.216	0.584	0.157	0.625+	0.145+	0.686
8	0.257	0.651	0.236	0.675+	0.184	0.772
9	0.277	0.723	0.242	0.758	0.226	0.774
10	0.349	0.743	0.325−	0.764	0.228	0.816
11	0.416	0.784	0.375−	0.843	0.314	0.855−
12	0.464	0.837	0.381	0.844	0.318	0.890
13	0.518	0.865+	0.444	0.884	0.397	0.923
14	0.581	0.899	0.556	0.920	0.466	0.951
15	0.651	0.937	0.619	0.953	0.534	0.975−
16	0.723	0.970	0.675+	0.980	0.603	0.992
17	0.784	0.994	0.758	0.997	0.682	0.999
18	0.865+	1	0.843	1	0.772	1

TABLE A.3 Confidence Limits for a Proportion (Two-sided), $n \le 30$ (*Continued*)

n = 19

r	90%		95%		99%	
0	0	0.130	0	0.150	0	0.218
1	0.006	0.209	0.003	0.232	0.001	0.305+
2	0.028	0.265+	0.019	0.316	0.008	0.383
3	0.059	0.337	0.044	0.365-	0.024	0.455+
4	0.095+	0.387	0.075+	0.426	0.046	0.515+
5	0.130	0.440	0.110	0.500	0.073	0.564
6	0.151	0.560	0.147	0.574	0.103	0.617
7	0.209	0.613	0.150	0.635+	0.137	0.695-
8	0.238	0.614	0.222	0.655+	0.173	0.707
9	0.265+	0.663	0.232	0.688	0.212	0.782
10	0.337	0.735-	0.312	0.768	0.218	0.788
11	0.386	0.762	0.345-	0.778	0.293	0.827
12	0.387	0.791	0.365-	0.850	0.305+	0.863
13	0.440	0.849	0.426	0.853	0.383	0.897
14	0.560	0.870	0.500	0.890	0.436	0.927
15	0.613	0.905-	0.574	0.925-	0.485-	0.954
16	0.668	0.941	0.635+	0.956	0.545-	0.976
17	0.735-	0.972	0.684	0.981	0.617	0.992
18	0.791	0.994	0.768	0.997	0.695-	0.999
19	0.870	1	0.850	1	0.782	1

n = 20

r	90%		95%		99%	
0	0	0.126	0	0.143	0	0.209
1	0.005+	0.203	0.003	0.222	0.001	0.293
2	0.027	0.255-	0.018	0.294	0.008	0.375-
3	0.056	0.328	0.042	0.351	0.023	0.424
4	0.090	0.367	0.071	0.411	0.044	0.500
5	0.126	0.422	0.104	0.467	0.069	0.576
6	0.141	0.500	0.140	0.533	0.098	0.601
7	0.201	0.578	0.143	0.589	0.129	0.637
8	0.221	0.633	0.209	0.649	0.163	0.707
9	0.255-	0.642	0.222	0.706	0.200	0.726
10	0.325	0.675+	0.293	0.707	0.209	0.791
11	0.358	0.745+	0.294	0.778	0.274	0.800
12	0.367	0.779	0.351	0.791	0.293	0.837
13	0.422	0.799	0.411	0.857	0.363	0.871
14	0.500	0.859	0.467	0.860	0.399	0.902
15	0.578	0.874	0.533	0.896	0.424	0.931
16	0.633	0.910	0.589	0.929	0.500	0.956
17	0.672	0.944	0.649	0.958	0.576	0.977
18	0.745+	0.973	0.706	0.982	0.625+	0.992
19	0.797	0.995-	0.778	0.997	0.707	0.999
20	0.874	1	0.857	1	0.791	1

n = 21

r	90%		95%		99%	
0	0	0.123	0	0.137	0	0.201
1	0.005+	0.197	0.002	0.213	0.000	0.283
2	0.026	0.245-	0.017	0.277	0.007	0.347
3	0.054	0.307	0.040	0.338	0.022	0.409
4	0.086	0.353	0.068	0.398	0.041	0.466
5	0.121	0.407	0.099	0.455+	0.065+	0.534
6	0.130	0.458	0.132	0.506	0.092	0.591
7	0.191	0.542	0.137	0.551	0.122	0.653
8	0.192	0.593	0.197	0.602	0.155-	0.661
9	0.245-	0.647	0.213	0.662	0.189	0.717
10	0.306	0.693	0.276	0.723	0.201	0.743
11	0.307	0.694	0.277	0.724	0.257	0.799
12	0.353	0.755+	0.338	0.787	0.283	0.811
13	0.407	0.808	0.398	0.803	0.339	0.845+
14	0.458	0.809	0.449	0.863	0.347	0.878
15	0.542	0.870	0.494	0.868	0.409	0.908
16	0.593	0.879	0.545-	0.901	0.466	0.935-
17	0.647	0.914	0.602	0.932	0.534	0.959
18	0.693	0.946	0.662	0.960	0.591	0.978
19	0.755+	0.974	0.723	0.983	0.653	0.993
20	0.808	0.995-	0.787	0.998	0.717	1.000
21	0.877	1	0.863	1	0.799	1

n = 22

r	90%		95%		99%	
0	0	0.116	0	0.132	0	0.194
1	0.005-	0.182	0.002	0.205+	0.000	0.273
2	0.024	0.236	0.016	0.264	0.007	0.334
3	0.051	0.289	0.038	0.326	0.021	0.396
4	0.082	0.340	0.065-	0.389	0.039	0.454
5	0.115-	0.393	0.094	0.424	0.062	0.505-
6	0.116	0.444	0.126	0.500	0.088	0.550
7	0.181	0.500	0.132	0.576	0.116	0.604
8	0.182	0.556	0.187	0.582	0.147	0.666
9	0.236	0.607	0.205+	0.617	0.179	0.682
10	0.289	0.660	0.260	0.674	0.194	0.727
11	0.290	0.710	0.264	0.736	0.242	0.758
12	0.340	0.711	0.326	0.740	0.273	0.806
13	0.393	0.764	0.383	0.795-	0.318	0.821
14	0.444	0.818	0.418	0.813	0.334	0.853
15	0.500	0.819	0.424	0.868	0.396	0.884
16	0.556	0.884	0.500	0.874	0.450	0.912
17	0.607	0.885+	0.576	0.906	0.495+	0.938
18	0.660	0.918	0.611	0.935+	0.546	0.961
19	0.711	0.949	0.674	0.962	0.604	0.979
20	0.764	0.976	0.736	0.984	0.666	0.993
21	0.818	0.995+	0.795-	0.998	0.727	1.000
22	0.884	1	0.868	1	0.806	1

TABLE A.3   Confidence Limits for a Proportion (Two-sided), $n \le 30$ (*Continued*)

$n = 23$

r	90%		95%		99%	
0	0	0.111	0	0.127	0	0.187
1	0.005−	0.174	0.002	0.198	0.000	0.365+
2	0.023	0.228	0.016	0.255	0.007	0.323
3	0.049	0.274	0.037	0.317	0.020	0.386
4	0.078	0.328	0.062	0.361	0.038	0.429
5	0.110	0.381	0.090	0.409	0.059	0.500
6	0.111	0.431	0.120	0.457	0.084	0.571
7	0.173	0.479	0.127	0.543	0.111	0.580
8	0.174	0.522	0.178	0.591	0.140	0.616
9	0.228	0.569	0.198	0.639	0.171	0.677
10	0.273	0.619	0.247	0.640	0.187	0.702
11	0.274	0.672	0.255−	0.683	0.229	0.735−
12	0.328	0.726	0.317	0.745+	0.265+	0.771
13	0.381	0.727	0.360	0.753	0.298	0.813
14	0.431	0.772	0.361	0.802	0.323	0.829
15	0.478	0.826	0.409	0.822	0.384	0.860
16	0.521	0.827	0.457	0.873	0.420	0.889
17	0.569	0.889	0.543	0.880	0.429	0.916
18	0.619	0.890	0.591	0.910	0.500	0.941
19	0.672	0.922	0.639	0.938	0.571	0.962
20	0.726	0.951	0.683	0.963	0.614	0.980
21	0.772	0.977	0.745+	0.984	0.677	0.993
22	0.826	0.995+	0.802	0.998	0.735−	1.000
23	0.889	1	0.873	1	0.813	1

$n = 24$

r	90%		95%		99%	
0	0	0.105+	0	0.122	0	0.181
1	0.004	0.165+	0.002	0.191	0.000	0.259
2	0.022	0.221	0.015+	0.246	0.006	0.313
3	0.047	0.264	0.035−	0.308	0.019	0.364
4	0.075−	0.317	0.059	0.347	0.036	0.416
5	0.105−	0.370	0.086	0.396	0.057	0.464
6	0.105+	0.423	0.115−	0.443	0.080	0.536
7	0.165−	0.448	0.122	0.500	0.106	0.584
8	0.165+	0.552	0.169	0.557	0.133	0.636
9	0.221	0.553	0.191	0.604	0.163	0.638
10	0.259	0.587	0.234	0.653	0.181	0.687
11	0.264	0.630	0.246	0.661	0.216	0.720
12	0.317	0.683	0.308	0.692	0.257	0.743
13	0.370	0.736	0.339	0.754	0.280	0.784
14	0.413	0.741	0.347	0.766	0.313	0.819
15	0.447	0.779	0.396	0.809	0.362	0.837
16	0.448	0.835−	0.443	0.831	0.364	0.867
17	0.552	0.835+	0.500	0.878	0.416	0.894
18	0.577	0.895−	0.557	0.885+	0.464	0.920
19	0.630	0.895+	0.604	0.914	0.536	0.943
20	0.683	0.925+	0.653	0.941	0.584	0.964
21	0.736	0.953	0.692	0.965+	0.636	0.981
22	0.779	0.978	0.754	0.985−	0.687	0.994
23	0.835−	0.996	0.809	0.998	0.741	1.000
24	0.895−	1	0.878	1	0.819	1

$n = 25$

r	90%		95%		99%	
0	0	0.102	0	0.118	0	0.175+
1	0.004	0.159	0.002	0.185+	0.000	0.246
2	0.021	0.214	0.014	0.238	0.006	0.305−
3	0.045−	0.255−	0.034	0.303	0.018	0.352
4	0.072	0.307	0.057	0.336	0.034	0.403
5	0.101	0.362	0.082	0.384	0.054	0.451
6	0.102	0.390	0.110	0.431	0.077	0.500
7	0.158	0.432	0.118	0.475−	0.101	0.549
8	0.159	0.500	0.161	0.525+	0.127	0.597
9	0.214	0.568	0.185+	0.569	0.155+	0.648
10	0.246	0.610	0.222	0.616	0.175+	0.658
11	0.255	0.611	0.238	0.664	0.205+	0.695+
12	0.307	0.640	0.296	0.683	0.245+	0.754
13	0.360	0.693	0.317	0.704	0.246	0.755−
14	0.389	0.745+	0.336	0.762	0.305−	0.795−
15	0.390	0.754	0.384	0.778	0.342	0.825−
16	0.432	0.786	0.431	0.815−	0.352	0.845−
17	0.500	0.841	0.475−	0.839	0.403	0.873
18	0.568	0.842	0.525+	0.882	0.451	0.899
19	0.610	0.898	0.569	0.890	0.500	0.923
20	0.638	0.899	0.616	0.918	0.549	0.946
21	0.693	0.928	0.664	0.943	0.597	0.966
22	0.745+	0.955+	0.697	0.966	0.648	0.982
23	0.786	0.979	0.762	0.986	0.695+	0.994
24	0.841	0.996	0.815−	0.998	0.751	1.000
25	0.898	1	0.882	1	0.825−	1

$n = 26$

r	90%		95%		99%	
0	0	0.098	0	0.114	0	0.170
1	0.004	0.152	0.002	0.180	0.000	0.235−
2	0.021	0.209	0.014	0.230	0.006	0.298
3	0.043	0.247	0.032	0.283	0.017	0.342
4	0.069	0.299	0.054	0.325+	0.033	0.393
5	0.097	0.343	0.079	0.374	0.052	0.442
6	0.098	0.377	0.106	0.421	0.073	0.487
7	0.151	0.419	0.114	0.465−	0.097	0.526
8	0.152	0.460	0.154	0.506	0.122	0.562
9	0.209	0.540	0.180	0.542	0.149	0.607
10	0.233	0.581	0.212	0.579	0.170	0.658
11	0.247	0.623	0.230	0.626	0.195−	0.678
12	0.299	0.657	0.282	0.675−	0.234	0.702
13	0.342	0.658	0.283	0.717	0.235−	0.765+
14	0.343	0.701	0.325+	0.718	0.298	0.766
15	0.377	0.753	0.374	0.770	0.322	0.805+
16	0.419	0.767	0.421	0.788	0.342	0.830
17	0.460	0.791	0.458	0.820	0.393	0.851
18	0.540	0.848	0.494	0.846	0.438	0.878
19	0.581	0.849	0.535−	0.886	0.474	0.903
20	0.623	0.902	0.579	0.894	0.513	0.927
21	0.657	0.903	0.626	0.921	0.558	0.948
22	0.701	0.931	0.675−	0.946	0.607	0.967
23	0.753	0.957	0.717	0.968	0.658	0.983
24	0.791	0.979	0.770	0.986	0.702	0.994
25	0.848	0.996	0.820	0.998	0.765+	1.000
26	0.902	1	0.886	1	0.830	1

TABLE A.3   Confidence Limits for a Proportion (Two-sided), $n \le 30$ (*Continued*)

r	90%		95%		99%		r	90%		95%		99%	
	n = 27							**n = 28**					
0	0	0.093	0	0.110	0	0.166	0	0	0.090	0	0.106	0	0.162
1	0.004	0.146	0.002	0.175−	0.000	0.225−	1	0.004	0.140	0.002	0.170	0.000	0.218
2	0.020	0.204	0.013	0.223	0.006	0.297	2	0.019	0.201	0.013	0.217	0.005+	0.273
3	0.042	0.239	0.031	0.270	0.017	0.332	3	0.040	0.232	0.030	0.259	0.016	0.323
4	0.066	0.291	0.052	0.316	0.032	0.384	4	0.064	0.284	0.050	0.307	0.031	0.365−
5	0.093	0.327	0.076	0.364	0.050	0.419	5	0.089	0.312	0.073	0.357	0.048	0.408
6	0.094	0.365+	0.101	0.415−	0.070	0.461	6	0.090	0.355−	0.098	0.384	0.068	0.449
7	0.145+	0.407	0.110	0.437	0.098	0.539	7	0.139	0.396	0.106	0.424	0.089	0.500
8	0.146	0.447	0.148	0.500	0.117	0.581	8	0.140	0.435+	0.142	0.463	0.112	0.551
9	0.204	0.500	0.175	0.563	0.143	0.587	9	0.197	0.473	0.170	0.537	0.137	0.592
10	0.221	0.553	0.202	0.570	0.166	0.617	10	0.208	0.527	0.192	0.576	0.162	0.635+
11	0.239	0.593	0.223	0.598	0.185−	0.668	11	0.232	0.565−	0.217	0.616	0.175+	0.636
12	0.291	0.635−	0.269	0.636	0.224	0.702	12	0.284	0.604	0.258	0.619	0.214	0.677
13	0.326	0.673	0.270	0.684	0.225−	0.716	13	0.310	0.645+	0.259	0.645+	0.218	0.727
14	0.327	0.674	0.316	0.730	0.284	0.775+	14	0.312	0.688	0.307	0.693	0.272	0.728
15	0.365+	0.709	0.364	0.731	0.298	0.776	15	0.335−	0.690	0.355−	0.741	0.273	0.782
16	0.407	0.761	0.402	0.777	0.332	0.815+	16	0.396	0.716	0.381	0.742	0.323	0.786
17	0.447	0.779	0.430	0.798	0.383	0.834	17	0.435+	0.768	0.384	0.783	0.364	0.825−
18	0.500	0.796	0.437	0.825+	0.413	0.857	18	0.473	0.792	0.424	0.808	0.365−	0.838
19	0.553	0.854	0.500	0.852	0.419	0.883	19	0.527	0.803	0.463	0.830	0.408	0.863
20	0.593	0.855−	0.563	0.890	0.461	0.907	20	0.565−	0.860	0.537	0.858	0.449	0.888
21	0.635−	0.906	0.585+	0.899	0.539	0.930	21	0.604	0.861	0.576	0.894	0.500	0.911
22	0.673	0.907	0.636	0.924	0.581	0.950	22	0.645+	0.910	0.616	0.902	0.551	0.932
23	0.709	0.934	0.684	0.948	0.616	0.968	23	0.688	0.911	0.643	0.927	0.592	0.952
24	0.761	0.958	0.730	0.969	0.668	0.983	24	0.716	0.936	0.693	0.950	0.635+	0.969
25	0.796	0.980	0.777	0.987	0.703	0.994	25	0.768	0.960	0.741	0.970	0.677	0.984
26	0.854	0.996	0.825+	0.998	0.775+	1.000	26	0.799	0.981	0.783	0.987	0.727	0.995−
27	0.907	1	0.890	1	0.834	1	27	0.860	0.996	0.830	0.998	0.782	1.000
							28	0.910	1	0.894	1	0.838	1
	n = 29							**n = 30**					
0	0	0.087	0	0.103	0	0.160	0	0	0.084	0	0.100	0	0.152
1	0.004	0.135−	0.002	0.166	0.000	0.211	1	0.004	0.130	0.002	0.163	0.000	0.206
2	0.018	0.190	0.012	0.211	0.005+	0.263	2	0.018	0.183	0.012	0.205+	0.005+	0.256
3	0.039	0.225−	0.029	0.251	0.015+	0.316	3	0.037	0.219	0.028	0.244	0.015−	0.310
4	0.062	0.279	0.049	0.299	0.030	0.354	4	0.059	0.266	0.047	0.292	0.028	0.345−
5	0.086	0.303	0.070	0.340	0.046	0.397	5	0.083	0.295−	0.068	0.325−	0.045−	0.388
6	0.087	0.345−	0.094	0.374	0.065+	0.438	6	0.084	0.336	0.091	0.364	0.063	0.430
7	0.134	0.385+	0.103	0.413	0.086	0.477	7	0.129	0.376	0.100	0.403	0.083	0.469
8	0.135−	0.425−	0.136	0.451	0.108	0.523	8	0.130	0.416	0.131	0.440	0.104	0.505+
9	0.189	0.463	0.166	0.500	0.132	0.562	9	0.182	0.455+	0.163	0.476	0.127	0.538
10	0.190	0.500	0.184	0.549	0.157	0.603	10	0.183	0.492	0.175+	0.524	0.151	0.570
11	0.225−	0.537	0.211	0.587	0.165+	0.646	11	0.219	0.524	0.205+	0.560	0.152	0.612
12	0.276	0.575+	0.247	0.626	0.206	0.654	12	0.265−	0.554	0.236	0.597	0.198	0.655+
13	0.294	0.615−	0.251	0.660	0.211	0.684	13	0.266	0.584	0.244	0.636	0.206	0.671
14	0.308	0.655+	0.299	0.661	0.260	0.737	14	0.295−	0.624	0.292	0.675+	0.294	0.692
15	0.345−	0.697	0.339	0.701	0.263	0.740	15	0.336	0.664	0.324	0.701	0.256	0.744
16	0.385+	0.706	0.340	0.749	0.316	0.789	16	0.376	0.705+	0.325−	0.708	0.308	0.751
17	0.425−	0.724	0.374	0.753	0.346	0.794	17	0.416	0.734	0.364	0.756	0.329	0.794
18	0.463	0.775+	0.413	0.789	0.354	0.835−	18	0.446	0.735+	0.403	0.764	0.345−	0.802
19	0.500	0.810	0.451	0.816	0.397	0.843	19	0.476	0.781	0.440	0.795−	0.388	0.848
20	0.537	0.811	0.500	0.834	0.438	0.868	20	0.508	0.817	0.476	0.825−	0.430	0.849
21	0.575+	0.865+	0.549	0.864	0.477	0.892	21	0.545−	0.818	0.524	0.837	0.462	0.873
22	0.615−	0.866	0.587	0.897	0.523	0.914	22	0.584	0.870	0.560	0.869	0.495−	0.896
23	0.655+	0.913	0.626	0.906	0.562	0.935−	23	0.624	0.871	0.597	0.900	0.531	0.917
24	0.697	0.914	0.660	0.930	0.603	0.954	24	0.664	0.916	0.636	0.909	0.570	0.937
25	0.721	0.938	0.701	0.951	0.646	0.970	25	0.705+	0.917	0.675+	0.932	0.612	0.955+
26	0.775+	0.961	0.749	0.971	0.684	0.985−	26	0.734	0.941	0.708	0.953	0.655+	0.972
27	0.810	0.982	0.789	0.988	0.737	0.995−	27	0.781	0.963	0.756	0.972	0.090	0.985+
28	0.865+	0.996	0.834	0.998	0.789	1.000	28	0.817	0.982	0.795−	0.988	0.744	0.995−
29	0.913	1	0.897	1	0.840	1	29	0.870	0.996	0.837	0.998	0.794	1.000
							30	0.916	1	0.900	1	0.848	1

TABLE A.4   Areas under the Normal Curve*

Proportion of total area under the curve that is under the portion of the curve from $-\infty$ *to* $(X_i - \bar{X}')/\sigma'$. *(X_i represents any desired value of the variable X.)*

$\dfrac{X_i - \bar{X}'}{\sigma'}$	0.09	0.08	0.07	0.06	0.05	0.04	0.03	0.02	0.01	0.00
−3.5	0.00017	0.00017	0.00018	0.00019	0.00019	0.00020	0.00021	0.00022	0.00022	0.00023
−3.4	0.00024	0.00025	0.00026	0.00027	0.00028	0.00029	0.00030	0.00031	0.00033	0.00034
−3.3	0.00035	0.00036	0.00038	0.00039	0.00040	0.00042	0.00043	0.00045	0.00047	0.00048
−3.2	0.00050	0.00052	0.00054	0.00056	0.00058	0.00060	0.00062	0.00064	0.00066	0.00069
−3.1	0.00071	0.00074	0.00076	0.00079	0.00082	0.00085	0.00087	0.00090	0.00094	0.00097
−3.0	0.00100	0.00104	0.00107	0.00111	0.00114	0.00118	0.00122	0.00126	0.00131	0.00135
−2.9	0.0014	0.0014	0.0015	0.0015	0.0016	0.0016	0.0017	0.0017	0.0018	0.0019
−2.8	0.0019	0.0020	0.0021	0.0021	0.0022	0.0023	0.0023	0.0024	0.0025	0.0026
−2.7	0.0026	0.0027	0.0028	0.0029	0.0030	0.0031	0.0032	0.0033	0.0034	0.0035
−2.6	0.0036	0.0037	0.0038	0.0039	0.0040	0.0041	0.0043	0.0044	0.0045	0.0047
−2.5	0.0048	0.0049	0.0051	0.0052	0.0054	0.0055	0.0057	0.0059	0.0060	0.0062
−2.4	0.0064	0.0066	0.0068	0.0069	0.0071	0.0073	0.0075	0.0078	0.0080	0.0082
−2.3	0.0084	0.0087	0.0089	0.0091	0.0094	0.0096	0.0099	0.0102	0.0104	0.0107
−2.2	0.0110	0.0113	0.0116	0.0119	0.0122	0.0125	0.0129	0.0132	0.0136	0.0139
−2.1	0.0143	0.0146	0.0150	0.0154	0.0158	0.0162	0.0166	0.0170	0.0174	0.0179
−2.0	0.0183	0.0188	0.0192	0.0197	0.0202	0.0207	0.0212	0.0217	0.0222	0.0228
−1.9	0.0233	0.0239	0.0244	0.0250	0.0256	0.0262	0.0268	0.0274	0.0281	0.0287
−1.8	0.0294	0.0301	0.0307	0.0314	0.0322	0.0329	0.0336	0.0344	0.0351	0.0359
−1.7	0.0367	0.0375	0.0384	0.0392	0.0401	0.0409	0.0418	0.0427	0.0436	0.0446
−1.6	0.0455	0.0465	0.0475	0.0485	0.0495	0.0505	0.0516	0.0526	0.0537	0.0548
−1.5	0.0559	0.0571	0.0582	0.0594	0.0606	0.0618	0.0630	0.0643	0.0655	0.0668
−1.4	0.0681	0.0694	0.0708	0.0721	0.0735	0.0749	0.0764	0.0778	0.0793	0.0808
−1.3	0.0823	0.0838	0.0853	0.0869	0.0885	0.0901	0.0918	0.0934	0.0951	0.0968
−1.2	0.0985	0.1003	0.1020	0.1038	0.1057	0.1075	0.1093	0.1112	0.1131	0.1151
−1.1	0.1170	0.1190	0.1210	0.1230	0.1251	0.1271	0.1292	0.1314	0.1335	0.1357
−1.0	0.1379	0.1401	0.1423	0.1446	0.1469	0.1492	0.1515	0.1539	0.1562	0.1587
−0.9	0.1611	0.1635	0.1660	0.1685	0.1711	0.1736	0.1762	0.1788	0.1814	0.1841
−0.8	0.1867	0.1894	0.1922	0.1949	0.1977	0.2005	0.2033	0.2061	0.2090	0.2119
−0.7	0.2148	0.2177	0.2207	0.2236	0.2266	0.2297	0.2327	0.2358	0.2389	0.2420
−0.6	0.2451	0.2483	0.2514	0.2546	0.2578	0.2611	0.2643	0.2676	0.2709	0.2743
−0.5	0.2776	0.2810	0.2843	0.2877	0.2912	0.2946	0.2981	0.3015	0.3050	0.3085
−0.4	0.3121	0.3156	0.3192	0.3228	0.3264	0.3300	0.3336	0.3372	0.3409	0.3446
−0.3	0.3483	0.3520	0.3557	0.3594	0.3632	0.3669	0.3707	0.3745	0.3783	0.3821
−0.2	0.3859	0.3897	0.3936	0.3974	0.4013	0.4052	0.4090	0.4129	0.4168	0.4207
−0.1	0.4247	0.4286	0.4325	0.4364	0.4404	0.4443	0.4483	0.4522	0.4562	0.4602
−0.0	0.4641	0.4681	0.4721	0.4761	0.4801	0.4840	0.4880	0.4920	0.4960	0.5000

*From *Statistical Quality Control*, 3d ed., by Eugene L. Grant. Copyright 1964. McGraw-Hill Book Company. Used by permission.

TABLE A.4   Areas under the Normal Curve (*Continued*)

$\dfrac{X_i-\bar{X}'}{\sigma'}$	0.00	0.01	0.02	0.03	0.04	0.05	0.06	0.07	0.08	0.09
+0.0	0.5000	0.5040	0.5080	0.5120	0.5160	0.5199	0.5239	0.5279	0.5319	0.5359
+0.1	0.5398	0.5438	0.5478	0.5517	0.5557	0.5596	0.5636	0.5675	0.5714	0.5753
+0.2	0.5793	0.5832	0.5871	0.5910	0.5948	0.5987	0.6026	0.6064	0.6103	0.6141
+0.3	0.6179	0.6217	0.6255	0.6293	0.6331	0.6368	0.6406	0.6443	0.6480	0.6517
+0.4	0.6554	0.6591	0.6628	0.6664	0.6700	0.6736	0.6772	0.6808	0.6844	0.6879
+0.5	0.6915	0.6950	0.6985	0.7019	0.7054	0.7088	0.7123	0.7157	0.7190	0.7224
+0.6	0.7257	0.7291	0.7324	0.7357	0.7389	0.7422	0.7454	0.7486	0.7517	0.7549
+0.7	0.7580	0.7611	0.7642	0.7673	0.7704	0.7734	0.7764	0.7794	0.7823	0.7852
+0.8	0.7881	0.7910	0.7939	0.7967	0.7995	0.8023	0.8051	0.8079	0.8106	0.8133
+0.9	0.8159	0.8186	0.8212	0.8238	0.8264	0.8289	0.8315	0.8340	0.8365	0.8389
+1.0	0.8413	0.8438	0.8461	0.8485	0.8508	0.8531	0.8554	0.8577	0.8599	0.8621
+1.1	0.8643	0.8665	0.8686	0.8708	0.8729	0.8749	0.8770	0.8790	0.8810	0.8830
+1.2	0.8849	0.8869	0.8888	0.8907	0.8925	0.8944	0.8962	0.8980	0.8997	0.9015
+1.3	0.9032	0.9049	0.9066	0.9082	0.9099	0.9115	0.9131	0.9147	0.9162	0.9177
+1.4	0.9192	0.9207	0.9222	0.9236	0.9251	0.9265	0.9279	0.9292	0.9306	0.9319
+1.5	0.9332	0.9345	0.9357	0.9370	0.9382	0.9394	0.9406	0.9418	0.9429	0.9441
+1.6	0.9452	0.9463	0.9474	0.9484	0.9495	0.9505	0.9515	0.9525	0.9535	0.9545
+1.7	0.9554	0.9564	0.9573	0.9582	0.9591	0.9599	0.9608	0.9616	0.9625	0.9633
+1.8	0.9641	0.9649	0.9656	0.9664	0.9671	0.9678	0.9686	0.9693	0.9699	0.9706
+1.9	0.9713	0.9719	0.9726	0.9732	0.9738	0.9744	0.9750	0.9756	0.9761	0.9767
+2.0	0.9773	0.9778	0.9783	0.9788	0.9793	0.9798	0.9803	0.9808	0.9812	0.9817
+2.1	0.9821	0.9826	0.9830	0.9834	0.9838	0.9842	0.9846	0.9850	0.9854	0.9857
+2.2	0.9861	0.9864	0.9868	0.9871	0.9875	0.9878	0.9881	0.9884	0.9887	0.9890
+2.3	0.9893	0.9896	0.9898	0.9901	0.9904	0.9906	0.9909	0.9911	0.9913	0.9916
+2.4	0.9918	0.9920	0.9922	0.9925	0.9927	0.9929	0.9931	0.9932	0.9934	0.9936
+2.5	0.9938	0.9940	0.9941	0.9943	0.9945	0.9946	0.9948	0.9949	0.9951	0.9952
+2.6	0.9953	0.9955	0.9956	0.9957	0.9959	0.9960	0.9961	0.9962	0.9963	0.9964
+2.7	0.9965	0.9966	0.9967	0.9968	0.9969	0.9970	0.9971	0.9972	0.9973	0.9974
+2.8	0.9974	0.9975	0.9976	0.9977	0.9977	0.9978	0.9979	0.9979	0.9980	0.9981
+2.9	0.9981	0.9982	0.9983	0.9983	0.9984	0.9984	0.9985	0.9985	0.9986	0.9986
+3.0	0.99865	0.99869	0.99874	0.99878	0.99882	0.99886	0.99889	0.99893	0.99896	0.99900
+3.1	0.99903	0.99906	0.99910	0.99913	0.99915	0.99918	0.99921	0.99924	0.99926	0.99929
+3.2	0.99931	0.99934	0.99936	0.99938	0.99940	0.99942	0.99944	0.99946	0.99948	0.99950
+3.3	0.99952	0.99953	0.99955	0.99957	0.99958	0.99960	0.99961	0.99962	0.99964	0.99965
+3.4	0.99966	0.99967	0.99969	0.99970	0.99971	0.99972	0.99973	0.99974	0.99975	0.99976
+3.5	0.99977	0.99978	0.99978	0.99979	0.99980	0.99981	0.99981	0.99982	0.99983	0.99983

TABLE A.5   Percentiles of the t Distribution*

df	$t_{.60}$	$t_{.70}$	$t_{.80}$	$t_{.90}$	$t_{.95}$	$t_{.975}$	$t_{.99}$	$t_{.995}$
1	.325	.727	1.376	3.078	6.314	12.706	31.821	63.657
2	.289	.617	1.061	1.886	2.920	4.303	6.965	9.925
3	.277	.584	.978	1.638	2.353	3.182	4.541	5.841
4	.271	.569	.941	1.533	2.132	2.776	3.747	4.604
5	.267	.559	.920	1.476	2.015	2.571	3.365	4.032
6	.265	.553	.906	1.440	1.943	2.447	3.143	3.707
7	.263	.549	.896	1.415	1.895	2.365	2.998	3.499
8	.262	.546	.889	1.397	1.860	2.306	2.896	3.355
9	.261	.543	.883	1.383	1.833	2.262	2.821	3.250
10	.260	.542	.879	1.372	1.812	2.228	2.764	3.169
11	.260	.540	.876	1.363	1.796	2.201	2.718	3.106
12	.259	.539	.873	1.356	1.782	2.179	2.681	3.055
13	.259	.538	.870	1.350	1.771	2.160	2.650	3.012
14	.258	.537	.868	1.345	1.761	2.145	2.624	2.977
15	.258	.536	.866	1.341	1.753	2.131	2.602	2.947
16	.258	.535	.865	1.337	1.746	2.120	2.583	2.921
17	.257	.534	.863	1.333	1.740	2.110	2.567	2.898
18	.257	.534	.862	1.330	1.734	2.101	2.552	2.878
19	.257	.533	.861	1.328	1.729	2.093	2.539	2.861
20	.257	.533	.860	1.325	1.725	2.086	2.528	2.845
21	.257	.532	.859	1.323	1.721	2.080	2.518	2.831
22	.256	.532	.858	1.321	1.717	2.074	2.508	2.819
23	.256	.532	.858	1.319	1.714	2.069	2.500	2.807
24	.256	.531	.857	1.318	1.711	2.064	2.492	2.797
25	.256	.531	.856	1.316	1.708	2.060	2.485	2.787
26	.256	.531	.856	1.315	1.706	2.056	2.479	2.779
27	.256	.531	.855	1.314	1.703	2.052	2.473	2.771
28	.256	.530	.855	1.313	1.701	2.048	2.467	2.763
29	.256	.530	.854	1.311	1.699	2.045	2.462	2.756
30	.256	.530	.854	1.310	1.697	2.042	2.457	2.750
40	.255	.529	.851	1.303	1.684	2.021	2.423	2.704
60	.254	.527	.848	1.296	1.671	2.000	2.390	2.660
120	.254	.526	.845	1.289	1.658	1.980	2.358	2.617
∞	.253	.524	.842	1.282	1.645	1.960	2.326	2.576
df	$-t_{.40}$	$-t_{.30}$	$-t_{.20}$	$-t_{.10}$	$-t_{.05}$	$-t_{.025}$	$-t_{.01}$	$-t_{.005}$

When the table is read from the foot, the tabled values are to be prefixed with a negative sign. Interpolation should be performed using the reciprocals of the degrees of freedom.

*The data of this table are taken from Table III of Fischer and Yates: *Statistical Tables for Biological, Agricultural and Medical Research*, published by Longman Group U.K., Ltd., London (previously published by Oliver & Boyd, Ltd., Edinburgh and by permission of the author and publishers. From *Introduction to Statistical Analysis*, 2d ed., by W. J. Dixon and F. J. Massey, Jr. Copyright, 1957. McGraw-Hill Book Company. Used by permission.

TABLE A.6 Percentiles of the x^2 Distribution*

df	Per Cent									
	.5	1	2.5	5	10	90	95	97.5	99	99.5
1	.000039	.00016	.00098	.0039	.0158	2.71	3.84	5.02	6.63	7.88
2	.0100	.0201	.0506	.1026	.2107	4.61	5.99	7.38	9.21	10.60
3	.0717	.115	.216	.352	.584	6.25	7.81	9.35	11.34	12.84
4	.207	.297	.484	.711	1.064	7.78	9.49	11.14	13.28	14.86
5	.412	.554	.831	1.15	1.61	9.24	11.07	12.83	15.09	16.75
6	.676	.872	1.24	1.64	2.20	10.64	12.59	14.45	16.81	18.55
7	.989	1.24	1.69	2.17	2.83	12.02	14.07	16.01	18.48	20.28
8	1.34	1.65	2.18	2.73	3.49	13.36	15.51	17.53	20.09	21.96
9	1.73	2.09	2.70	3.33	4.17	14.68	16.92	19.02	21.67	23.59
10	2.16	2.56	3.25	3.94	4.87	15.99	18.31	20.48	23.21	25.19
11	2.60	3.05	3.82	4.57	5.58	17.28	19.68	21.92	24.73	26.76
12	3.07	3.57	4.40	5.23	6.30	18.55	21.03	23.34	26.22	28.30
13	3.57	4.11	5.01	5.89	7.04	19.81	22.36	24.74	27.69	29.82
14	4.07	4.66	5.63	6.57	7.79	21.06	23.68	26.12	29.14	31.32
15	4.60	5.23	6.26	7.26	8.55	22.31	25.00	27.49	30.58	32.80
16	5.14	5.81	6.91	7.96	9.31	23.54	26.30	28.85	32.00	34.27
18	6.26	7.01	8.23	9.39	10.86	25.99	28.87	31.53	34.81	37.16
20	7.43	8.26	9.59	10.85	12.44	28.41	31.41	34.17	37.57	40.00
24	9.89	10.86	12.40	13.85	15.66	33.20	36.42	39.36	42.98	45.56
30	13.79	14.95	16.79	18.49	20.60	40.26	43.77	46.98	50.89	53.67
40	20.71	22.16	24.43	26.51	29.05	51.81	55.76	59.34	63.69	66.77
60	35.53	37.48	40.48	43.19	46.46	74.40	79.08	83.30	88.38	91.95
120	83.85	86.92	91.58	95.70	100.62	140.23	146.57	152.21	158.95	163.64

For large values of degrees of freedom the approximate formula

$$x_\alpha{}^2 = n \left(1 - \frac{2}{9n} + z_\alpha \sqrt{\frac{2}{9n}} \right)^3$$

where z_α is the normal deviate and n is the number of degrees of freedom, may be used. For example $x_{.99}{}^2 = 60[1 - .00370 + 2.326(.06086)]^3 = 60(1.1379)^3 = 88.4$ for the 99th percentile for 60 degrees of freedom.

* From *Introduction to Statistical Analysis*, 2d ed., by W. J. Dixon and F. J. Massey, Jr. Copyright, 1957. McGraw-Hill Book Company. Used by permission.

TABLE A.7 Percentiles of the $F(v_1, v_2)$ Distribution with Degrees of Freedom v_1 for the Numerator and v_2 for the Denominator*

v_2	Cum. Prop.	1	2	3	4	5	6	7	8	9	10	11	12	Cum. Prop.
1	.0005	$.0^{6}62$	$.0^{3}50$	$.0^{2}38$	$.0^{2}94$	.016	.022	.027	.032	.036	.039	.042	.045	.0005
	.001	$.0^{5}25$	$.0^{2}10$	$.0^{2}60$	.013	.021	.028	.034	.039	.044	.048	.051	.054	.001
	.005	$.0^{4}62$	$.0^{2}51$	.018	.032	.044	.054	.062	.068	.073	.078	.082	.085	.005
	.010	$.0^{3}25$	.010	.029	.047	.062	.073	.082	.089	.095	.100	.104	.107	.010
	.025	$.0^{2}15$	.026	.057	.082	.100	.113	.124	.132	.139	.144	.149	.153	.025
	.05	$.0^{2}62$	.054	.099	.130	.151	.167	.179	.188	.195	.201	.207	.211	.05
	.10	.025	.117	.181	.220	.246	.265	.279	.289	.298	.304	.310	.315	.10
	.25	.172	.389	.494	.553	.591	.617	.637	.650	.661	.670	.680	.684	.25
	.50	1.00	1.50	1.71	1.82	1.89	1.94	1.98	2.00	2.03	2.04	2.05	2.07	.50
	.75	5.83	7.50	8.20	8.58	8.82	8.98	9.10	9.19	9.26	9.32	9.36	9.41	.75
	.90	39.9	49.5	53.6	55.8	57.2	58.2	58.9	59.4	59.9	60.2	60.5	60.7	.90
	.95	161	200	216	225	230	234	237	239	241	242	243	244	.95
	.975	648	800	864	900	922	937	948	957	963	969	973	977	.975
	.99	405^{1}	500^{1}	540^{1}	562^{1}	576^{1}	586^{1}	593^{1}	598^{1}	602^{1}	606^{1}	608^{1}	611^{1}	.99
	.995	162^{2}	200^{2}	216^{2}	225^{2}	231^{2}	234^{2}	237^{2}	239^{2}	241^{2}	242^{2}	243^{2}	244^{2}	.995
	.999	406^{3}	500^{3}	540^{3}	562^{3}	576^{3}	586^{3}	593^{3}	598^{3}	602^{3}	606^{3}	609^{3}	611^{3}	.999
	.9995	162^{4}	200^{4}	216^{4}	225^{4}	231^{4}	234^{4}	237^{4}	239^{4}	241^{4}	242^{4}	243^{4}	244^{4}	.9995
2	.0005	$.0^{6}50$	$.0^{3}50$	$.0^{2}42$	.011	.020	.029	.037	.044	.050	.056	.061	.065	.0005
	.001	$.0^{5}20$	$.0^{2}10$	$.0^{2}68$	.016	.027	.037	.046	.054	.061	.067	.072	.077	.001
	.005	$.0^{4}50$	$.0^{2}50$	.020	.038	.055	.069	.081	.091	.099	.106	.112	.118	.005
	.01	$.0^{3}20$	.010	.032	.056	.075	.092	.105	.116	.125	.132	.139	.144	.01
	.025	$.0^{2}13$	.026	.062	.094	.119	.138	.153	.165	.175	.183	.190	.196	.025
	.05	$.0^{2}50$	.053	.105	.144	.173	.194	.211	.224	.235	.244	.251	.257	.05
	.10	.020	.111	.183	.231	.265	.289	.307	.321	.333	.342	.350	.356	.10
	.25	.133	.333	.439	.500	.540	.568	.588	.604	.616	.626	.633	.641	.25
	.50	.667	1.00	1.13	1.21	1.25	1.28	1.30	1.32	1.33	1.34	1.35	1.36	.50
	.75	2.57	3.00	3.15	3.23	3.28	3.31	3.34	3.35	3.37	3.38	3.39	3.39	.75
	.90	8.53	9.00	9.16	9.24	9.29	9.33	9.35	9.37	9.38	9.39	9.40	9.41	.90
	.95	18.5	19.0	19.2	19.2	19.3	19.3	19.4	19.4	19.4	19.4	19.4	19.4	.95
	.975	38.5	39.0	39.2	39.2	39.3	39.3	39.4	39.4	39.4	39.4	39.4	39.4	.975
	.99	98.5	99.0	99.2	99.2	99.3	99.3	99.4	99.4	99.4	99.4	99.4	99.4	.99
	.995	198	199	199	199	199	199	199	199	199	199	199	199	.995
	.999	998	999	999	999	999	999	999	999	999	999	999	999	.999
	.9995	200^{1}	200^{1}	200^{1}	200^{1}	200^{1}	200^{1}	200^{1}	200^{1}	200^{1}	200^{1}	200^{1}	200^{1}	.9995
3	.0005	$.0^{6}46$	$.0^{3}50$	$.0^{2}44$	.012	.023	.033	.043	.052	.060	.067	.074	.079	.0005
	.001	$.0^{5}19$	$.0^{2}10$	$.0^{2}71$	.018	.030	.042	.053	.063	.072	.079	.086	.093	.001
	.005	$.0^{4}46$	$.0^{2}50$	.021	.041	.060	.077	.092	.104	.115	.124	.132	.138	.005
	.01	$.0^{3}19$	.010	.034	.060	.083	.102	.118	.132	.143	.153	.161	.168	.01
	.025	$.0^{2}12$	.026	.065	.100	.129	.152	.170	.185	.197	.207	.216	.224	.025
	.05	$.0^{2}46$	.052	.108	.152	.185	.210	.230	.246	.259	.270	.279	.287	.05
	.10	.019	.109	.185	.239	.276	.304	.325	.342	.356	.367	.376	.384	.10
	.25	.122	.317	.424	.489	.531	.561	.582	.600	.613	.624	.633	.641	.25
	.50	.585	.881	1.00	1.06	1.10	1.13	1.15	1.16	1.17	1.18	1.19	1.20	.50
	.75	2.02	2.28	2.36	2.39	2.41	2.42	2.43	2.44	2.44	2.44	2.45	2.45	.75
	.90	5.54	5.46	5.39	5.34	5.31	5.28	5.27	5.25	5.24	5.23	5.22	5.22	.90
	.95	10.1	9.55	9.28	9.12	9.01	8.94	8.89	8.85	8.81	8.79	8.76	8.74	.95
	.975	17.4	16.0	15.4	15.1	14.9	14.7	14.6	14.5	14.5	14.4	14.4	14.3	.975
	.99	34.1	30.8	29.5	28.7	28.2	27.9	27.7	27.5	27.3	27.2	27.1	27.1	.99
	.995	55.6	49.8	47.5	46.2	45.4	44.8	44.4	44.1	43.9	43.7	43.5	43.4	.995
	.999	167	149	141	137	135	133	132	131	130	129	129	128	.999
	.9995	266	237	225	218	214	211	209	208	207	206	204	204	.9995

Read $.0^{3}56$ as .00056, 200^{1} as 2000, 162^{4} as 1620000, etc.

*From *Introduction to Statistical Analysis*, 2d ed., by W. J. Dixon and F. J. Massey, Jr. Copyright, 1957. McGraw-Hill Book Company. Used by permission.

TABLE A.7 Percentiles of the $F(v_1, v_2)$ Distribution with Degrees of Freedom v_1 for the Numerator and v_2 for the Denominator (*Continued*)

Cum. Prop.	15	20	24	30	40	50	60	100	120	200	500	∞	Cum. Prop.	v_2
.0005	.051	.058	062	.066	.069	.072	.074	.077	.078	.080	.081	.083	.0005	**1**
.001	.060	.067	.071	.075	.079	.082	.084	.087	.088	.089	.091	.092	.001	
.005	.093	.101	.105	.109	.113	.116	.118	.121	.122	.124	.126	.127	.005	
.01	.115	.124	.128	.132	.137	.139	.141	.145	.146	.148	.150	.151	.01	
.025	.161	.170	.175	.180	.184	.187	.189	.193	.194	.196	.198	.199	.025	
.05	.220	.230	.235	.240	.245	.248	.250	.254	.255	.257	.259	.261	.05	
.10	.325	.336	.342	.347	.353	.356	.358	.362	.364	.366	.368	.370	.10	
.25	.698	.712	.719	.727	.734	.738	.741	.747	.749	.752	.754	.756	.25	
.50	2.09	2.12	2.13	2.15	2.16	2.17	2.17	2.18	2.18	2.19	2.19	2.20	.50	
.75	9.49	9.58	9.63	9.67	9.71	9.74	9.76	9.78	9.80	9.82	9.84	9.85	.75	
.90	61.2	61.7	62.0	62.3	62.5	62.7	62.8	63.0	63.1	63.2	63.3	63.3	.90	
.95	246	248	249	250	251	252	252	253	253	254	254	254	.95	
.975	985	993	997	100^1	101^1	101^1	101^1	101^1	101^1	102^1	102^1	102^1	.975	
.99	616^1	621^1	623^1	626^1	629^1	630^1	631^1	633^1	634^1	635^1	636^1	637^1	.99	
.995	246^2	248^2	249^2	250^2	251^2	252^2	253^2	253^2	254^2	254^2	254^2	255^2	.995	
.999	616^3	621^3	623^3	626^3	629^3	630^3	631^3	633^3	634^3	635^3	636^3	637^3	.999	
.9995	246^4	248^4	249^4	250^4	251^4	252^4	252^4	253^4	253^4	253^4	254^4	254^4	.9995	
.0005	.076	.088	.094	.101	.108	.113	.116	.122	.124	.127	.130	.132	.0005	**2**
.001	.088	.100	.107	.114	.121	.126	.129	.135	.137	.140	.143	.145	.001	
.005	.130	.143	.150	.157	.165	.169	.173	.179	.181	.184	.187	.189	.005	
.01	.157	.171	.178	.186	.193	.198	.201	.207	.209	.212	.215	.217	.01	
.025	.210	.224	.232	.239	.247	.251	.255	.261	.263	.266	.269	.271	.025	
.05	.272	.286	.294	.302	.309	.314	.317	.324	.326	.329	.332	.334	.05	
.10	.371	.386	.394	.402	.410	.415	.418	.424	.426	.429	.433	.434	.10	
.25	.657	.672	.680	.689	.697	.702	.705	.711	.713	.716	.719	.721	.25	
.50	1.38	1.39	1.40	1.41	1.42	1.42	1.43	1.43	1.43	1.44	1.44	1.44	.50	
.75	3.41	3.43	3.43	3.44	3.45	3.45	3.46	3.47	3.47	3.48	3.48	3.48	.75	
.90	9.42	9.44	9.45	9.46	9.47	9.47	9.47	9.48	9.48	9.49	9.49	9.49	.90	
.95	19.4	19.4	19.5	19.5	19.5	19.5	19.5	19.5	19.5	19.5	19.5	19.5	.95	
.975	39.4	39.4	39.5	39.5	39.5	39.5	39.5	39.5	39.5	39.5	39.5	39.5	.975	
.99	99.4	99.4	99.5	99.5	99.5	99.5	99.5	99.5	99.5	99.5	99.5	99.5	.99	
.995	199	199	199	199	199	199	199	199	199	199	199	200	.995	
.999	999	999	999	999	999	999	999	999	999	999	999	999	.999	
.9995	200^1	200^1	200^1	200^1	200^1	200^1	200^1	200^1	200^1	200^1	200^1	200^1	.9995	
.0005	.093	.109	.117	.127	.136	.143	.147	.156	.158	.162	.166	.169	.0005	**3**
.001	.107	.123	.132	.142	.152	.158	.162	.171	.173	.177	.181	.184	.001	
.005	.154	.172	.181	.191	.201	.207	.211	.220	.222	.227	.231	.234	.005	
.01	.185	.203	.212	.222	.232	.238	.242	.251	.253	.258	.262	.264	.01	
.025	.241	.259	.269	.279	.289	.295	.299	.308	.310	.314	.318	.321	.025	
.05	.304	.323	.332	.342	.352	.358	.363	.370	.373	.377	.382	.384	.05	
.10	.402	.420	.430	.439	.449	.455	.459	.467	.469	.474	.476	.480	.10	
.25	.658	.675	.684	.693	.702	.708	.711	.719	.721	.724	.728	.730	.25	
.50	1.21	1.23	1.23	1.24	1.25	1.25	1.25	1.26	1.26	1.26	1.27	1.27	.50	
.75	2.46	2.46	2.46	2.47	2.47	2.47	2.47	2.47	2.47	2.47	2.47	2.47	.75	
.90	5.20	5.18	5.18	5.17	5.16	5.15	5.15	5.14	5.14	5.14	5.14	5.13	.90	
.95	8.70	8.66	8.63	8.62	8.59	8.58	8.57	8.55	8.55	8.54	8.53	8.53	.95	
.975	14.3	14.2	14.1	14.1	14.0	14.0	14.0	14.0	13.9	13.9	13.9	13.9	.975	
.99	26.9	26.7	26.6	26.5	26.4	26.3	26.3	26.2	26.2	26.2	26.1	26.1	.99	
.995	43.1	42.8	42.6	42.5	42.3	42.2	42.1	42.0	42.0	41.9	41.9	41.8	.995	
.999	127	126	126	125	125	125	124	124	124	124	124	123	.999	
.9995	203	201	200	199	199	198	198	197	197	197	196	196	.9995	

TABLE A.7   Percentiles of the $F(v_1, v_2)$ Distribution with Degrees of Freedom v_1 for the Numerator and v_2 for the Denominator (*Continued*)

v_2	Cum. Prop.	1	2	3	4	5	6	7	8	9	10	11	12	Cum. Prop.
4	.0005	$.0^644$	$.0^350$	$.0^246$	.013	.024	.036	.047	.057	.066	.075	.082	.089	.0005
	.001	$.0^518$	$.0^210$	$.0^273$	.019	.032	.046	.058	.069	.079	.089	.097	.104	.001
	.005	$.0^444$	$.0^250$	.022	.043	.064	.083	.100	.114	.126	.137	.145	.153	.005
	.01	$.0^318$	.010	.035	.063	.088	.109	.127	.143	.156	.167	.176	.185	.01
	.025	$.0^211$	.026	.066	.104	.135	.161	.181	.198	.212	.224	.234	.243	.025
	.05	$.0^244$	.052	.110	.157	.193	.221	.243	.261	.275	.288	.298	.307	.05
	.10	.018	.108	.187	.243	.284	.314	.338	.356	.371	.384	.394	.403	.10
	.25	.117	.309	.418	.484	.528	.560	.583	.601	.615	.627	.637	.645	.25
	.50	.549	.828	.941	1.00	1.04	1.06	1.08	1.09	1.10	1.11	1.12	1.13	.50
	.75	1.81	2.00	2.05	2.06	2.07	2.08	2.08	2.08	2.08	2.08	2.08	2.08	.75
	.90	4.54	4.32	4.19	4.11	4.05	4.01	3.98	3.95	3.94	3.92	3.91	3.90	.90
	.95	7.71	6.94	6.59	6.39	6.26	6.16	6.09	6.04	6.00	5.96	5.94	5.91	.95
	.975	12.2	10.6	9.98	9.60	9.36	9.20	9.07	8.98	8.90	8.84	8.79	8.75	.975
	.99	21.2	18.0	16.7	16.0	15.5	15.2	15.0	14.8	14.7	14.5	14.4	14.4	.99
	.995	31.3	26.3	24.3	23.2	22.5	22.0	21.6	21.4	21.1	21.0	20.8	20.7	.995
	.999	74.1	61.2	56.2	53.4	51.7	50.5	49.7	49.0	48.5	48.0	47.7	47.4	.999
	.9995	106	87.4	80.1	76.1	73.6	71.9	70.6	69.7	68.9	68.3	67.8	67.4	.9995
5	.0005	$.0^643$	$.0^350$	$.0^247$	.014	.025	.038	.050	.061	.070	.081	.089	.096	.0005
	.001	$.0^517$	$.0^210$	$.0^275$	.019	.034	.048	.062	.074	.085	.095	.104	.112	.001
	.005	$.0^443$	$.0^250$	.022	.045	.067	.087	.105	.120	.134	.146	.156	.165	.005
	.01	$.0^317$	.010	.035	.064	.091	.114	.134	.151	.165	.177	.188	.197	.01
	.025	$.0^211$	.025	.067	.107	.140	.167	.189	.208	.223	.236	.248	.257	.025
	.05	$.0^243$	.052	.111	.160	.198	.228	.252	.271	.287	.301	.313	.322	.05
	.10	.017	.108	.188	.247	.290	.322	.347	.367	.383	.397	.408	.418	.10
	.25	.113	.305	.415	.483	.528	.560	.584	.604	.618	.631	.641	.650	.25
	.50	.528	.799	.907	.965	1.00	1.02	1.04	1.05	1.06	1.07	1.08	1.09	.50
	.75	1.69	1.85	1.88	1.89	1.89	1.89	1.89	1.89	1.89	1.89	1.89	1.89	.75
	.90	4.06	3.78	3.62	3.52	3.45	3.40	3.37	3.34	3.32	3.30	3.28	3.27	.90
	.95	6.61	5.79	5.41	5.19	5.05	4.95	4.88	4.82	4.77	4.74	4.71	4.68	.95
	.975	10.0	8.43	7.76	7.39	7.15	6.98	6.85	6.76	6.68	6.62	6.57	6.52	.975
	.99	16.3	13.3	12.1	11.4	11.0	10.7	10.5	10.3	10.2	10.1	9.96	9.89	.99
	.995	22.8	18.3	16.5	15.6	14.9	14.5	14.2	14.0	13.8	13.6	13.5	13.4	.995
	.999	47.2	37.1	33.2	31.1	29.7	28.8	28.2	27.6	27.2	26.9	26.6	26.4	.999
	.9995	63.6	49.8	44.4	41.5	39.7	38.5	37.6	36.9	36.4	35.9	35.6	35.2	.9995
6	.0005	$.0^643$	$.0^350$	$.0^247$	.014	.026	.039	.052	.064	.075	.085	.094	.103	.0005
	.001	$.0^517$	$.0^210$	$.0^275$	.020	.035	.050	.064	.078	.090	.101	.111	.119	.001
	.005	$.0^443$	$.0^250$	.022	.045	.069	.090	.109	.126	.140	.153	.164	.174	.005
	.01	$.0^317$	.010	.036	.066	.094	.118	.139	.157	.172	.186	.197	.207	.01
	.025	$.0^211$	.025	.068	.109	.143	.172	.195	.215	.231	.246	.258	.268	.025
	.05	$.0^243$	.052	.112	.162	.202	.233	.259	.279	.296	.311	.324	.334	.05
	.10	.017	.107	.189	.249	.294	.327	.354	.375	.392	.406	.418	.429	.10
	.25	.111	.302	.413	.481	.524	.561	.586	.606	.622	.635	.645	.654	.25
	.50	.515	.780	.886	.942	.977	1.00	1.02	1.03	1.04	1.05	1.05	1.06	.50
	.75	1.62	1.76	1.78	1.79	1.79	1.78	1.78	1.78	1.77	1.77	1.77	1.77	.75
	.90	3.78	3.46	3.29	3.18	3.11	3.05	3.01	2.98	2.96	2.94	2.92	2.90	.90
	.95	5.99	5.14	4.76	4.53	4.39	4.28	4.21	4.15	4.10	4.06	4.03	4.00	.95
	.975	8.81	7.26	6.60	6.23	5.99	5.82	5.70	5.60	5.52	5.46	5.41	5.37	.975
	.99	13.7	10.9	9.78	9.15	8.75	8.47	8.26	8.10	7.98	7.87	7.79	7.72	.99
	.995	18.6	14.5	12.9	12.0	11.5	11.1	10.8	10.6	10.4	10.2	10.1	10.0	.995
	.999	35.5	27.0	23.7	21.9	20.8	20.0	19.5	19.0	18.7	18.4	18.2	18.0	.999
	.9995	46.1	34.8	30.4	28.1	26.6	25.6	24.9	24.3	23.9	23.5	23.2	23.0	.9995

TABLE A.7 Percentiles of the $F(v_1, v_2)$ Distribution with Degrees of Freedom v_1 for the Numerator and v_2 for the Denominator (*Continued*)

Cum. Prop.	15	20	24	30	40	50	60	100	120	200	500	∞	Cum. Prop.	v_2
.0005	.105	.125	.135	.147	.159	.166	.172	.183	.186	.191	.196	.200	.0005	**4**
.001	.121	.141	.152	.163	.176	.183	.188	.200	.202	.208	.213	.217	.001	
.005	.172	.193	.204	.216	.229	.237	.242	.253	.255	.260	.266	.269	.005	
.01	.204	.226	.237	.249	.261	.269	.274	.285	.287	.293	.298	.301	.01	
.025	.263	.284	.296	.308	.320	.327	.332	.342	.346	.351	.356	.359	.025	
.05	.327	.349	.360	.372	.384	.391	.396	.407	.409	.413	.418	.422	.05	
.10	.424	.445	.456	.467	.478	.485	.490	.500	.502	.508	.510	.514	.10	
.25	.664	.683	.692	.702	.712	.718	.722	.731	.733	.737	.740	.743	.25	
.50	1.14	1.15	1.16	1.16	1.17	1.18	1.18	1.18	1.18	1.19	1.19	1.19	.50	
.75	2.08	2.08	2.08	2.08	2.08	2.08	2.08	2.08	2.08	2.08	2.08	2.08	.75	
.90	3.87	3.84	3.83	3.82	3.80	3.80	3.79	3.78	3.78	3.77	3.76	3.76	.90	
.95	5.86	5.80	5.77	5.75	5.72	5.70	5.69	5.66	5.66	5.65	5.64	5.63	.95	
.975	8.66	8.56	8.51	8.46	8.41	8.38	8.36	8.32	8.31	8.29	8.27	8.26	.975	
.99	14.2	14.0	13.9	13.8	13.7	13.7	13.7	13.6	13.6	13.5	13.5	13.5	.99	
.995	20.4	20.2	20.0	19.9	19.8	19.7	19.6	19.5	19.5	19.4	19.4	19.3	.995	
.999	46.8	46.1	45.8	45.4	45.1	44.9	44.7	44.5	44.4	44.3	44.1	44.0	.999	
.9995	66.5	65.5	65.1	64.6	64.1	63.8	63.6	63.2	63.1	62.9	62.7	62.6	.9995	
.0005	.115	.137	.150	.163	.177	.186	.192	.205	.209	.216	.222	.226	.0005	**5**
.001	.132	.155	.167	.181	.195	.204	.210	.223	.227	.233	.239	.244	.001	
.005	.186	.210	.223	.237	.251	.260	.266	.279	.282	.288	.294	.299	.005	
.01	.219	.244	.257	.270	.285	.293	.299	.312	.315	.322	.328	.331	.01	
.025	.280	.304	.317	.330	.344	.353	.359	.370	.374	.380	.386	.390	.025	
.05	.345	.369	.382	.395	.408	.417	.422	.432	.437	.442	.448	.452	.05	
.10	.440	.463	.476	.488	.501	.508	.514	.524	.527	.532	.538	.541	.10	
.25	.669	.690	.700	.711	.722	.728	.732	.741	.743	.748	.752	.755	.25	
.50	1.10	1.11	1.12	1.12	1.13	1.13	1.14	1.14	1.14	1.15	1.15	1.15	.50	
.75	1.89	1.88	1.88	1.88	1.88	1.88	1.87	1.87	1.87	1.87	1.87	1.87	.75	
.90	3.24	3.21	3.19	3.17	3.16	3.15	3.14	3.13	3.12	3.12	3.11	3.10	.90	
.95	4.62	4.56	4.53	4.50	4.46	4.44	4.43	4.41	4.40	4.39	4.37	4.36	.95	
.975	6.43	6.33	6.28	6.23	6.18	6.14	6.12	6.08	6.07	6.05	6.03	6.02	.975	
.99	9.72	9.55	9.47	9.38	9.29	9.24	9.20	9.13	9.11	9.08	9.04	9.02	.99	
.995	13.1	12.9	12.8	12.7	12.5	12.5	12.4	12.3	12.3	12.2	12.2	12.1	.995	
.999	25.9	25.4	25.1	24.9	24.6	24.4	24.3	24.1	24.1	23.9	23.8	23.8	.999	
.9995	34.6	33.9	33.5	33.1	32.7	32.5	32.3	32.1	32.0	31.8	31.7	31.6	.9995	
.0005	.123	.148	.162	.177	.193	.203	.210	.225	.229	.236	.244	.249	.0005	**6**
.001	.141	.166	.180	.195	.211	.222	.229	.243	.247	.255	.262	.267	.001	
.005	.197	.224	.238	.253	.269	.279	.286	.301	.304	.312	.318	.324	.005	
.01	.232	.258	.273	.288	.304	.313	.321	.334	.338	.346	.352	.357	.01	
.025	.293	.320	.334	.349	.364	.375	.381	.394	.398	.405	.412	.415	.025	
.05	.358	.385	.399	.413	.428	.437	.444	.457	.460	.467	.472	.476	.05	
.10	.453	.478	.491	.505	.519	.526	.533	.546	.548	.556	.559	.564	.10	
.25	.675	.696	.707	.718	.729	.736	.741	.751	.753	.758	.762	.765	.25	
.50	1.07	1.08	1.09	1.10	1.10	1.11	1.11	1.11	1.11	1.12	1.12	1.12	.50	
.75	1.76	1.76	1.75	1.75	1.75	1.75	1.74	1.74	1.74	1.74	1.74	1.74	.75	
.90	2.87	2.84	2.82	2.80	2.78	2.77	2.76	2.75	2.74	2.73	2.73	2.72	.90	
.95	3.94	3.87	3.84	3.81	3.77	3.75	3.74	3.71	3.70	3.69	3.68	3.67	.95	
.975	5.27	5.17	5.12	5.07	5.01	4.98	4.96	4.92	4.90	4.88	4.86	4.85	.975	
.99	7.56	7.40	7.31	7.23	7.14	7.09	7.06	6.99	6.97	6.93	6.90	6.88	.99	
.995	9.81	9.59	9.47	9.36	9.24	9.17	9.12	9.03	9.00	8.95	8.91	8.88	.995	
.999	17.6	17.1	16.9	16.7	16.4	16.3	16.2	16.0	16.0	15.9	15.8	15.7	.999	
.9995	22.4	21.9	21.7	21.4	21.1	20.9	20.7	20.5	20.4	20.3	20.2	20.1	.9995	

TABLE A.7 Percentiles of the $F(v_1, v_2)$ Distribution with Degrees of Freedom v_1 for the Numerator and v_2 for the Denominator (*Continued*)

v_2	Cum. Prop.	1	2	3	4	5	6	7	8	9	10	11	12	Cum. Prop.
7	.0005	$.0^642$	$.0^350$	$.0^248$	.014	.027	.040	.053	.066	.078	.088	.099	.108	.0005
	.001	$.0^517$	$.0^210$	$.0^276$	.020	.035	.051	.067	.081	.093	.105	.115	.125	.001
	.005	$.0^442$	$.0^250$	.023	.046	.070	.093	.113	.130	.145	.159	.171	.181	.005
	.01	$.0^317$	.010	.036	.067	.096	.121	.143	.162	.178	.192	.205	.216	.01
	.025	$.0^210$	.025	.068	.110	.146	.176	.200	.221	.238	.253	.266	.277	.025
	.05	$.0^242$	.052	.113	.164	.205	.238	.264	.286	.304	.319	.332	.343	.05
	.10	.017	.107	.190	.251	.297	.332	.359	.381	.399	.414	.427	.438	.10
	.25	.110	.300	.412	.481	.528	.562	.588	.608	.624	.637	.649	.658	.25
	.50	.506	.767	.871	.926	.960	.983	1.00	1.01	1.02	1.03	1.04	1.04	.50
	.75	1.57	1.70	1.72	1.72	1.71	1.71	1.70	1.70	1.69	1.69	1.69	1.68	.75
	.90	3.59	3.26	3.07	2.96	2.88	2.83	2.78	2.75	2.72	2.70	2.68	2.67	.90
	.95	5.59	4.74	4.35	4.12	3.97	3.87	3.79	3.73	3.68	3.64	3.60	3.57	.95
	.975	8.07	6.54	5.89	5.52	5.29	5.12	4.99	4.90	4.82	4.76	4.71	4.67	.975
	.99	12.2	9.55	8.45	7.85	7.46	7.19	6.99	6.84	6.72	6.62	6.54	6.47	.99
	.995	16.2	12.4	10.9	10.0	9.52	9.16	8.89	8.68	8.51	8.38	8.27	8.18	.995
	.999	29.2	21.7	18.8	17.2	16.2	15.5	15.0	14.6	14.3	14.1	13.9	13.7	.999
	.9995	37.0	27.2	23.5	21.4	20.2	19.3	18.7	18.2	17.8	17.5	17.2	17.0	.9995
8	.0005	$.0^642$	$.0^350$	$.0^248$	.014	.027	.041	.055	.068	.081	.092	.102	.112	.0005
	.001	$.0^517$	$.0^210$	$.0^276$	.020	.036	.053	.068	.083	.096	.109	.120	.130	.001
	.005	$.0^442$	$.0^250$	.027	.047	.072	.095	.115	.133	.149	.164	.176	.187	.005
	.01	$.0^317$	.010	.036	.068	.097	.123	.146	.166	.183	.198	.211	.222	.01
	.025	$.0^210$	.025	.069	.111	.148	.179	.204	.226	.244	.259	.273	.285	.025
	.05	$.0^242$	.052	.113	.166	.208	.241	.268	.291	.310	.326	.339	.351	.05
	.10	.017	.107	.190	.253	.299	.335	.363	.386	.405	.421	.435	.445	.10
	.25	.109	.298	.411	.481	.529	.563	.589	.610	.627	.640	.654	.661	.25
	.50	.499	.757	.860	.915	.948	.971	.988	1.00	1.01	1.02	1.02	1.03	.50
	.75	1.54	1.66	1.67	1.66	1.66	1.65	1.64	1.64	1.64	1.63	1.63	1.62	.75
	.90	3.46	3.11	2.92	2.81	2.73	2.67	2.62	2.59	2.56	2.54	2.52	2.50	.90
	.95	5.32	4.46	4.07	3.84	3.69	3.58	3.50	3.44	3.39	3.35	3.31	3.28	.95
	.975	7.57	6.06	5.42	5.05	4.82	4.65	4.53	4.43	4.36	4.30	4.24	4.20	.975
	.99	11.3	8.65	7.59	7.01	6.63	6.37	6.18	6.03	5.91	5.81	5.73	5.67	.99
	.995	14.7	11.0	9.60	8.81	8.30	7.95	7.69	7.50	7.34	7.21	7.10	7.01	.995
	.999	25.4	18.5	15.8	14.4	13.5	12.9	12.4	12.0	11.8	11.5	11.4	11.2	.999
	.9995	31.6	22.8	19.4	17.6	16.4	15.7	15.1	14.6	14.3	14.0	13.8	13.6	.9995
9	.0005	$.0^641$	$.0^350$	$.0^248$	.015	.027	.042	.056	.070	.083	.094	.105	.115	.0005
	.001	$.0^517$	$.0^210$	$.0^277$	.021	.037	.054	.070	.085	.099	.112	.123	.134	.001
	.005	$.0^442$	$.0^250$	.023	.047	.073	.096	.117	.136	.153	.168	.181	.192	.005
	.01	$.0^317$	.010	.037	.068	.098	.125	.149	.169	.187	.202	.216	.228	.01
	.025	$.0^210$	.025	.069	.112	.150	.181	.207	.230	.248	.265	.279	.291	.025
	.05	$.0^240$	.052	.113	.167	.210	.244	.272	.296	.315	.331	.345	.358	.05
	.10	.017	.107	.191	.254	.302	.338	.367	.390	.410	.426	.441	.452	.10
	.25	.108	.297	.410	.480	.529	.564	.591	.612	.629	.643	.654	.664	.25
	.50	.494	.749	.852	.906	.939	.962	.978	.990	1.00	1.01	1.01	1.02	.50
	.75	1.51	1.62	1.63	1.63	1.62	1.61	1.60	1.60	1.59	1.59	1.58	1.58	.75
	.90	3.36	3.01	2.81	2.69	2.61	2.55	2.51	2.47	2.44	2.42	2.40	2.38	.90
	.95	5.12	4.26	3.86	3.63	3.48	3.37	3.29	3.23	3.18	3.14	3.10	3.07	.95
	.975	7.21	5.71	5.08	4.72	4.48	4.32	4.20	4.10	4.03	3.96	3.91	3.87	.975
	.99	10.6	8.02	6.99	6.42	6.06	5.80	5.61	5.47	5.35	5.26	5.18	5.11	.99
	.995	13.6	10.1	8.72	7.96	7.47	7.13	6.88	6.69	6.54	6.42	6.31	6.23	.995
	.999	22.9	16.4	13.9	12.6	11.7	11.1	10.7	10.4	10.1	9.89	9.71	9.57	.999
	.9995	28.0	19.9	16.8	15.1	14.1	13.3	12.8	12.4	12.1	11.8	11.6	11.4	.9995

TABLE A.7   Percentiles of the $F(v_1, v_2)$ Distribution with Degrees of Freedom v_1 for the Numerator and v_2 for the Denominator (*Continued*)

Cum. Prop. \ v_1	15	20	24	30	40	50	60	100	120	200	500	∞	Cum. Prop.	v_2
.0005	.130	.157	.172	.188	.206	.217	.225	.242	.246	.255	.263	.268	.0005	7
.001	.148	.176	.191	.208	.225	.237	.245	.261	.266	.274	.282	.288	.001	
.005	.206	.235	.251	.267	.285	.296	.304	.319	.324	.332	.340	.345	.005	
.01	.241	.270	.286	.303	.320	.331	.339	.355	.358	.366	.373	.379	.01	
.025	.304	.333	.348	.364	.381	.392	.399	.413	.418	.426	.433	.437	.025	
.05	.369	.398	.413	.428	.445	.455	.461	.476	.479	.485	.493	.498	.05	
.10	.463	.491	.504	.519	.534	.543	.550	.562	.566	.571	.578	.582	.10	
.25	.679	.702	.713	.725	.737	.745	.749	.760	.762	.767	.772	.775	.25	
.50	1.05	1.07	1.07	1.08	1.08	1.09	1.09	1.10	1.10	1.10	1.10	1.10	.50	
.75	1.68	1.67	1.67	1.66	1.66	1.66	1.65	1.65	1.65	1.65	1.65	1.65	.75	
.90	2.63	2.59	2.58	2.56	2.54	2.52	2.51	2.50	2.49	2.48	2.48	2.47	.90	
.95	3.51	3.44	3.41	3.38	3.34	3.32	3.30	3.27	3.27	3.25	3.24	3.23	.95	
.975	4.57	4.47	4.42	4.36	4.31	4.28	4.25	4.21	4.20	4.18	4.16	4.14	.975	
.99	6.31	6.16	6.07	5.99	5.91	5.86	5.82	5.75	5.74	5.70	5.67	5.65	.99	
.995	7.97	7.75	7.65	7.53	7.42	7.35	7.31	7.22	7.19	7.15	7.10	7.08	.995	
.999	13.3	12.9	12.7	12.5	12.3	12.2	12.1	11.9	11.9	11.8	11.7	11.7	.999	
.9995	16.5	16.0	15.7	15.5	15.2	15.1	15.0	14.7	14.7	14.6	14.5	14.4	.9995	
.0005	.136	.164	.181	.198	.218	.230	.239	.257	.262	.271	.281	.287	.0005	8
.001	.155	.184	.200	.218	.238	.250	.259	.277	.282	.292	.300	.306	.001	
.005	.214	.244	.261	.279	.299	.311	.319	.337	.341	.351	.358	.364	.005	
.01	.250	.281	.297	.315	.334	.346	.354	.372	.376	.385	.392	.398	.01	
.025	.313	.343	.360	.377	.395	.407	.415	.431	.435	.442	.450	.456	.025	
.05	.379	.409	.425	.441	.459	.469	.477	.493	.496	.505	.510	.516	.05	
.10	.472	.500	.515	.531	.547	.556	.563	.578	.581	.588	.595	.599	.10	
.25	.684	.707	.718	.730	.743	.751	.756	.767	.769	.775	.780	.783	.25	
.50	1.04	1.05	1.06	1.07	1.07	1.07	1.08	1.08	1.08	1.09	1.09	1.09	.50	
.75	1.62	1.61	1.60	1.60	1.59	1.59	1.59	1.58	1.58	1.58	1.58	1.58	.75	
.90	2.46	2.42	2.40	2.38	2.36	2.35	2.34	2.32	2.32	2.31	2.30	2.29	.90	
.95	3.22	3.15	3.12	3.08	3.04	3.02	3.01	2.97	2.97	2.95	2.94	2.93	.95	
.975	4.10	4.00	3.95	3.89	3.84	3.81	3.78	3.74	3.73	3.70	3.68	3.67	.975	
.99	5.52	5.36	5.28	5.20	5.12	5.07	5.03	4.96	4.95	4.91	4.88	4.86	.99	
.995	6.81	6.61	6.50	6.40	6.29	6.22	6.18	6.09	6.06	6.02	5.98	5.95	.995	
.999	10.8	10.5	10.3	10.1	9.92	9.80	9.73	9.57	9.54	9.46	9.39	9.34	.999	
.9995	13.1	12.7	12.5	12.2	12.0	11.8	11.8	11.6	11.5	11.4	11.4	11.3	.9995	
.0005	.141	.171	.188	.207	.228	.242	.251	.270	.276	.287	.297	.303	.0005	9
.001	.160	.191	.208	.228	.249	.262	.271	.291	.296	.307	.316	.323	.001	
.005	.220	.253	.271	.290	.310	.324	.332	.351	.356	.366	.376	.382	.005	
.01	.257	.289	.307	.326	.346	.358	.368	.386	.391	.400	.410	.415	.01	
.025	.320	.352	.370	.388	.408	.420	.428	.446	.450	.459	.467	.473	.025	
.05	.386	.418	.435	.452	.471	.483	.490	.508	.510	.518	.526	.532	.05	
.10	.479	.509	.525	.541	.558	.568	.575	.588	.594	.602	.610	.613	.10	
.25	.687	.711	.723	.736	.749	.757	.762	.773	.776	.782	.787	.791	.25	
.50	1.03	1.04	1.05	1.05	1.06	1.06	1.07	1.07	1.07	1.08	1.08	1.08	.50	
.75	1.57	1.56	1.56	1.55	1.55	1.54	1.54	1.53	1.53	1.53	1.53	1.53	.75	
.90	2.34	2.30	2.28	2.25	2.23	2.22	2.21	2.19	2.18	2.17	2.17	2.16	.90	
.95	3.01	2.94	2.90	2.86	2.83	2.80	2.79	2.76	2.75	2.73	2.72	2.71	.95	
.975	3.77	3.67	3.61	3.56	3.51	3.47	3.45	3.40	3.39	3.37	3.35	3.33	.975	
.99	4.96	4.81	4.73	4.65	4.57	4.52	4.48	4.42	4.40	4.36	4.33	4.31	.99	
.995	6.03	5.83	5.73	5.62	5.52	5.45	5.41	5.32	5.30	5.26	5.21	5.19	.995	
.999	9.24	8.90	8.72	8.55	8.37	8.26	8.19	8.04	8.00	7.93	7.86	7.81	.999	
.9995	11.0	10.6	10.4	10.2	9.94	9.80	9.71	9.53	9.49	9.40	9.32	9.26	.9995	

TABLE A.7   Percentiles of the $F(v_1, v_2)$ Distribution with Degrees of Freedom v_1 for the Numerator and v_2 for the Denominator (*Continued*)

v_2	Cum. Prop.	1	2	3	4	5	6	7	8	9	10	11	12	Cum. Prop.
10	.0005	$.0^641$	$.0^350$	$.0^249$	.015	.028	.043	.057	.071	.085	.097	.108	.119	.0005
	.001	$.0^517$	$.0^210$	$.0^277$	.021	.037	.054	.071	.087	.101	.114	.126	.137	.001
	.005	$.0^441$	$.0^250$	.023	.048	.073	.098	.119	.139	.156	.171	.185	.197	.005
	.01	$.0^317$	.010	.037	.069	.100	.127	.151	.172	.190	.206	.220	.233	.01
	.025	$.0^210$	.025	.069	.113	.151	.183	.210	.233	.252	.269	.283	.296	.025
	.05	$.0^241$	.052	.114	.168	.211	.246	.275	.299	.319	.336	.351	.363	.05
	.10	.017	.106	.191	.255	.303	.340	.370	.394	.414	.430	.444	.457	.10
	.25	.107	.296	.409	.480	.529	.565	.592	.613	.631	.645	.657	.667	.25
	.50	.490	.743	.845	.899	.932	.954	.971	.983	.992	1.00	1.01	1.01	.50
	.75	1.49	1.60	1.60	1.59	1.59	1.58	1.57	1.56	1.56	1.55	1.55	1.54	.75
	.90	3.28	2.92	2.73	2.61	2.52	2.46	2.41	2.38	2.35	2.32	2.30	2.28	.90
	.95	4.96	4.10	3.71	3.48	3.33	3.22	3.14	3.07	3.02	2.98	2.94	2.91	.95
	.975	6.94	5.46	4.83	4.47	4.24	4.07	3.95	3.85	3.78	3.72	3.66	3.62	.975
	.99	10.0	7.56	6.55	5.99	5.64	5.39	5.20	5.06	4.94	4.85	4.77	4.71	.99
	.995	12.8	9.43	8.08	7.34	6.87	6.54	6.30	6.12	5.97	5.85	5.75	5.66	.995
	.999	21.0	14.9	12.6	11.3	10.5	9.92	9.52	9.20	8.96	8.75	8.58	8.44	.999
	.9995	25.5	17.9	15.0	13.4	12.4	11.8	11.3	10.9	10.6	10.3	10.1	9.93	.9995
11	.0005	$.0^641$	$.0^350$	$.0^249$	.015	.028	.043	.058	.072	.086	.099	.111	.121	.0005
	.001	$.0^516$	$.0^210$	$.0^278$	.021	.038	.055	.072	.088	.103	.116	.129	.140	.001
	.005	$.0^440$	$.0^250$	.023	.048	.074	.099	.121	.141	.158	.174	.188	.200	.005
	.01	$.0^316$	.010	.037	.069	.100	.128	.153	.175	.193	.210	.224	.237	.01
	.025	$.0^210$	.025	.069	.114	.152	.185	.212	.236	.256	.273	.288	.301	.025
	.05	$.0^241$	.052	.114	.168	.212	.248	.278	.302	.323	.340	.355	.368	.05
	.10	.017	.106	.192	.256	.305	.342	.373	.397	.417	.435	.448	.461	.10
	.25	.107	.295	.408	.481	.529	.565	.592	.614	.633	.645	.658	.667	.25
	.50	.486	.739	.840	.893	.926	.948	.964	.977	.986	.994	1.00	1.01	.50
	.75	1.47	1.58	1.58	1.57	1.56	1.55	1.54	1.53	1.53	1.52	1.52	1.51	.75
	.90	3.23	2.86	2.66	2.54	2.45	2.39	2.34	2.30	2.27	2.25	2.23	2.21	.90
	.95	4.84	3.98	3.59	3.36	3.20	3.09	3.01	2.95	2.90	2.85	2.82	2.79	.95
	.975	6.72	5.26	4.63	4.28	4.04	3.88	3.76	3.66	3.59	3.53	3.47	3.43	.975
	.99	9.65	7.21	6.22	5.67	5.32	5.07	4.89	4.74	4.63	4.54	4.46	4.40	.99
	.995	12.2	8.91	7.60	6.88	6.42	6.10	5.86	5.68	5.54	5.42	5.32	5.24	.995
	.999	19.7	13.8	11.6	10.3	9.58	9.05	8.66	8.35	8.12	7.92	7.76	7.62	.999
	.9995	23.6	16.4	13.6	12.2	11.2	10.6	10.1	9.76	9.48	9.24	9.04	8.88	.9995
12	.0005	$.0^641$	$.0^350$	$.0^249$	.015	.028	.044	.058	.073	.087	.101	.113	.124	.0005
	.001	$.0^516$	$.0^210$	$.0^278$	.021	.038	.056	.073	.089	.104	.118	.131	.143	.001
	.005	$.0^439$	$.0^250$	.023	.048	.075	.100	.122	.143	.161	.177	.191	.204	.005
	.01	$.0^316$	.010	.037	.070	.101	.130	.155	.176	.196	.212	.227	.241	.01
	.025	$.0^210$	.025	.070	.114	.153	.186	.214	.238	.259	.276	.292	.305	.025
	.05	$.0^241$	.052	.114	.169	.214	.250	.280	.305	.325	.343	.358	.372	.05
	.10	.016	.106	.192	.257	.306	.344	.375	.400	.420	.438	.452	.466	.10
	.25	.106	.295	.408	.480	.530	.566	.594	.616	.633	.649	.662	.671	.25
	.50	.484	.735	.835	.888	.921	.943	.959	.972	.981	.989	.995	1.00	.50
	.75	1.46	1.56	1.56	1.55	1.54	1.53	1.52	1.51	1.51	1.50	1.50	1.49	.75
	.90	3.18	2.81	2.61	2.48	2.39	2.33	2.28	2.24	2.21	2.19	2.17	2.15	.90
	.95	4.75	3.89	3.49	3.26	3.11	3.00	2.91	2.85	2.80	2.75	2.72	2.69	.95
	.975	6.55	5.10	4.47	4.12	3.89	3.73	3.61	3.51	3.44	3.37	3.32	3.28	.975
	.99	9.33	6.93	5.95	5.41	5.06	4.82	4.64	4.50	4.39	4.30	4.22	4.16	.99
	.995	11.8	8.51	7.23	6.52	6.07	5.76	5.52	5.35	5.20	5.09	4.99	4.91	.995
	.999	18.6	13.0	10.8	9.63	8.89	8.38	8.00	7.71	7.48	7.29	7.14	7.01	.999
	.9995	22.2	15.3	12.7	11.2	10.4	9.74	9.28	8.94	8.66	8.43	8.24	8.08	.9995

TABLE A.7   Percentiles of the $F(v_1, v_2)$ Distribution with Degrees of Freedom v_1 for the Numerator and v_2 for the Denominator (*Continued*)

Cum. Prop. \ v_1	15	20	24	30	40	50	60	100	120	200	500	∞	Cum. Prop.	v_2
.0005	.145	.177	.195	.215	.238	.251	.262	.282	.288	.299	.311	.319	.0005	10
.001	.164	.197	.216	.236	.258	.272	.282	.303	.309	.321	.331	.338	.001	
.005	.226	.260	.279	.299	.321	.334	.344	.365	.370	.380	.391	.397	.005	
.01	.263	.297	.316	.336	.357	.370	.380	.400	.405	.415	.424	.431	.01	
.025	.327	.360	.379	.398	.419	.431	.441	.459	.464	.474	.483	.488	.025	
.05	.393	.426	.444	.462	.481	.493	.502	.518	.523	.532	.541	.546	.05	
.10	.486	.516	.532	.549	.567	.578	.586	.602	.605	.614	.621	.625	.10	
.25	.691	.714	.727	.740	.754	.762	.767	.779	.782	.788	.793	.797	.25	
.50	1.02	1.03	1.04	1.05	1.05	1.06	1.06	1.06	1.06	1.07	1.07	1.07	.50	
.75	1.53	1.52	1.52	1.51	1.51	1.50	1.50	1.49	1.49	1.49	1.48	1.48	.75	
.90	2.24	2.20	2.18	2.16	2.13	2.12	2.11	2.09	2.08	2.07	2.06	2.06	.90	
.95	2.85	2.77	2.74	2.70	2.66	2.64	2.62	2.59	2.58	2.56	2.55	2.54	.95	
.975	3.52	3.42	3.37	3.31	3.26	3.22	3.20	3.15	3.14	3.12	3.09	3.08	.975	
.99	4.56	4.41	4.33	4.25	4.17	4.12	4.08	4.01	4.00	3.96	3.93	3.91	.99	
.995	5.47	5.27	5.17	5.07	4.97	4.90	4.86	4.77	4.75	4.71	4.67	4.64	.995	
.999	8.13	7.80	7.64	7.47	7.30	7.19	7.12	6.98	6.94	6.87	6.81	6.76	.999	
.9995	9.56	9.16	8.96	8.75	8.54	8.42	8.33	8.16	8.12	8.04	7.96	7.90	.9995	
.0005	.148	.182	.201	.222	.246	.261	.271	.293	.299	.312	.324	.331	.0005	11
.001	.168	.202	.222	.243	.266	.282	.292	.313	.320	.332	.343	.353	.001	
.005	.231	.266	.286	.308	.330	.345	.355	.376	.382	.394	.403	.412	.005	
.01	.268	.304	.324	.344	.366	.380	.391	.412	.417	.427	.439	.444	.01	
.025	.332	.368	.386	.407	.429	.442	.450	.472	.476	.485	.495	.503	.025	
.05	.398	.433	.452	.469	.490	.503	.513	.529	.535	.543	.552	.559	.05	
.10	.490	.524	.541	.559	.578	.588	.595	.614	.617	.625	.633	.637	.10	
.25	.694	.719	.730	.744	.758	.767	.773	.780	.788	.794	.799	.803	.25	
.50	1.02	1.03	1.03	1.04	1.05	1.05	1.05	1.06	1.06	1.06	1.06	1.06	.50	
.75	1.50	1.49	1.49	1.48	1.47	1.47	1.47	1.46	1.46	1.46	1.45	1.45	.75	
.90	2.17	2.12	2.10	2.08	2.05	2.04	2.03	2.00	2.00	1.99	1.98	1.97	.90	
.95	2.72	2.65	2.61	2.57	2.53	2.51	2.49	2.46	2.45	2.43	2.42	2.40	.95	
.975	3.33	3.23	3.17	3.12	3.06	3.03	3.00	2.96	2.94	2.92	2.90	2.88	.975	
.99	4.25	4.10	4.02	3.94	3.86	3.81	3.78	3.71	3.69	3.66	3.62	3.60	.99	
.995	5.05	4.86	4.76	4.65	4.55	4.49	4.45	4.36	4.34	4.29	4.25	4.23	.995	
.999	7.32	7.01	6.85	6.68	6.52	6.41	6.35	6.21	6.17	6.10	6.04	6.00	.999	
.9995	8.52	8.14	7.94	7.75	7.55	7.43	7.35	7.18	7.14	7.06	6.98	6.93	.9995	
.0005	.152	.186	.206	.228	.253	.269	.280	.305	.311	.323	.337	.345	.0005	12
.001	.172	.207	.228	.250	.275	.291	.302	.326	.332	.344	.357	.365	.001	
.005	.235	.272	.292	.315	.339	.355	.365	.388	.393	.405	.417	.424	.005	
.01	.273	.310	.330	.352	.375	.391	.401	.422	.428	.441	.450	.458	.01	
.025	.337	.374	.394	.416	.437	.450	.461	.481	.487	.498	.508	.514	.025	
.05	.404	.439	.458	.478	.499	.513	.522	.541	.545	.556	.565	.571	.05	
.10	.496	.528	.546	.564	.583	.595	.604	.621	.625	.633	.641	.647	.10	
.25	.695	.721	.734	.748	.762	.771	.777	.789	.792	.799	.804	.808	.25	
.50	1.01	1.02	1.03	1.03	1.04	1.04	1.05	1.05	1.05	1.05	1.06	1.06	.50	
.75	1.48	1.47	1.46	1.45	1.45	1.44	1.44	1.43	1.43	1.43	1.42	1.42	.75	
.90	2.11	2.06	2.04	2.01	1.99	1.97	1.96	1.94	1.93	1.92	1.91	1.90	.90	
.95	2.62	2.54	2.51	2.47	2.43	2.40	2.38	2.35	2.34	2.32	2.31	2.30	.95	
.975	3.18	3.07	3.02	2.96	2.91	2.87	2.85	2.80	2.79	2.76	2.74	2.72	.975	
.99	4.01	3.86	3.78	3.70	3.62	3.57	3.54	3.47	3.45	3.41	3.38	3.36	.99	
.995	4.72	4.53	4.43	4.33	4.23	4.17	4.12	4.04	4.01	3.97	3.93	3.90	.995	
.999	6.71	6.40	6.25	6.09	5.93	5.83	5.76	5.63	5.59	5.52	5.46	5.42	.999	
.9995	7.74	7.37	7.18	7.00	6.80	6.68	6.61	6.45	6.41	6.33	6.25	6.20	.9995	

TABLE A.7   Percentiles of the $F(v_1, v_2)$ Distribution with Degrees of Freedom v_1 for the Numerator and v_2 for the Denominator (*Continued*)

v_2	Cum. Prop.	1	2	3	4	5	6	7	8	9	10	11	12	Cum. Prop.
15	.0005	$.0^641$	$.0^350$	$.0^249$	.015	.029	.045	.061	.076	.091	.105	.117	.129	.0005
	.001	$.0^516$	$.0^210$	$.0^279$	.021	.039	.057	.075	.092	.108	.123	.137	.149	.001
	.005	$.0^439$	$.0^250$	.023	.049	.076	.102	.125	.147	.166	.183	.198	.212	.005
	.01	$.0^316$	.010	.037	.070	.103	.132	.158	.181	.202	.219	.235	.249	.01
	.025	$.0^210$	.025	.070	.116	.156	.190	.219	.244	.265	.284	.300	.315	.025
	.05	$.0^241$	.051	.115	.170	.216	.254	.285	.311	.333	.351	.368	.382	.05
	.10	.016	.106	.192	.258	.309	.348	.380	.406	.427	.446	.461	.475	.10
	.25	.105	.293	.407	.480	.531	.568	.596	.618	.637	.652	.667	.676	.25
	.50	.478	.726	.826	.878	.911	.933	.948	.960	.970	.977	.984	.989	.50
	.75	1.43	1.52	1.52	1.51	1.49	1.48	1.47	1.46	1.46	1.45	1.44	1.44	.75
	.90	3.07	2.70	2.49	2.36	2.27	2.21	2.16	2.12	2.09	2.06	2.04	2.02	.90
	.95	4.54	3.68	3.29	3.06	2.90	2.79	2.71	2.64	2.59	2.54	2.51	2.48	.95
	.975	6.20	4.76	4.15	3.80	3.58	3.41	3.29	3.20	3.12	3.06	3.01	2.96	.975
	.99	8.68	6.36	5.42	4.89	4.56	4.32	4.14	4.00	3.89	3.80	3.73	3.67	.99
	.995	10.8	7.70	6.48	5.80	5.37	5.07	4.85	4.67	4.54	4.42	4.33	4.25	.995
	.999	16.6	11.3	9.34	8.25	7.57	7.09	6.74	6.47	6.26	6.08	5.93	5.81	.999
	.9995	19.5	13.2	10.8	9.48	8.66	8.10	7.68	7.36	7.11	6.91	6.75	6.60	.9995
20	.0005	$.0^640$	$.0^350$	$.0^250$	.015	.029	.046	.063	.079	.094	.109	.123	.136	.0005
	.001	$.0^516$	$.0^210$	$.0^279$	.022	.039	.058	.077	.095	.112	.128	.143	.156	.001
	.005	$.0^439$	$.0^250$	.023	.050	.077	.104	.129	.151	.171	.190	.206	.221	.005
	.01	$.0^316$	.010	.037	.071	.105	.135	.162	.187	.208	.227	.244	.259	.01
	.025	$.0^210$	.025	.071	.117	.158	.193	.224	.250	.273	.292	.310	.325	.025
	.05	$.0^240$	.051	.115	.172	.219	.258	.290	.318	.340	.360	.377	.393	.05
	.10	.016	.106	.193	.260	.312	.353	.385	.412	.435	.454	.472	.485	.10
	.25	.104	.292	.407	.480	.531	.569	.598	.622	.641	.656	.671	.681	.25
	.50	.472	.718	.816	.868	.900	.922	.938	.950	.959	.966	.972	.977	.50
	.75	1.40	1.49	1.48	1.47	1.45	1.44	1.43	1.42	1.41	1.40	1.39	1.39	.75
	.90	2.97	2.59	2.38	2.25	2.16	2.09	2.04	2.00	1.96	1.94	1.91	1.89	.90
	.95	4.35	3.49	3.10	2.87	2.71	2.60	2.51	2.45	2.39	2.35	2.31	2.28	.95
	.975	5.87	4.46	3.86	3.51	3.29	3.13	3.01	2.91	2.84	2.77	2.72	2.68	.975
	.99	8.10	5.85	4.94	4.43	4.10	3.87	3.70	3.56	3.46	3.37	3.29	3.23	.99
	.995	9.94	6.99	5.82	5.17	4.76	4.47	4.26	4.09	3.96	3.85	3.76	3.68	.995
	.999	14.8	9.95	8.10	7.10	6.46	6.02	5.69	5.44	5.24	5.08	4.94	4.82	.999
	.9995	17.2	11.4	9.20	8.02	7.28	6.76	6.38	6.08	5.85	5.66	5.51	5.38	.9995
24	.0005	$.0^640$	$.0^350$	$.0^250$	.015	.030	.046	.064	.080	.096	.112	.126	.139	.0005
	.001	$.0^516$	$.0^210$	$.0^279$	.022	.040	.059	.079	.097	.115	.131	.146	.160	.001
	.005	$.0^440$	$.0^250$	.023	.050	.078	.106	.131	.154	.175	.193	.210	.226	.005
	.01	$.0^316$	.010	.038	.072	.106	.137	.165	.189	.211	.231	.249	.264	.01
	.025	$.0^210$	.025	.071	.117	.159	.195	.227	.253	.277	.297	.315	.331	.025
	.05	$.0^240$	.051	.116	.173	.221	.260	.293	.321	.345	.365	.383	.399	.05
	.10	.016	.106	.193	.261	.313	.355	.388	.416	.439	.459	.476	.491	.10
	.25	.104	.291	.406	.480	.532	.570	.600	.623	.643	.659	.671	.684	.25
	.50	.469	.714	.812	.863	.895	.917	.932	.944	.953	.961	.967	.972	.50
	.75	1.39	1.47	1.46	1.44	1.43	1.41	1.40	1.39	1.38	1.38	1.37	1.36	.75
	.90	2.93	2.54	2.33	2.19	2.10	2.04	1.98	1.94	1.91	1.88	1.85	1.83	.90
	.95	4.26	3.40	3.01	2.78	2.62	2.51	2.42	2.36	2.30	2.25	2.21	2.18	.95
	.975	5.72	4.32	3.72	3.38	3.15	2.99	2.87	2.78	2.70	2.64	2.59	2.54	.975
	.99	7.82	5.61	4.72	4.22	3.90	3.67	3.50	3.36	3.26	3.17	3.09	3.03	.99
	.995	9.55	6.66	5.52	4.89	4.49	4.20	3.99	3.83	3.69	3.59	3.50	3.42	.995
	.999	14.0	9.34	7.55	6.59	5.98	5.55	5.23	4.99	4.80	4.64	4.50	4.39	.999
	.9995	16.2	10.6	8.52	7.39	6.68	6.18	5.82	5.54	5.31	5.13	4.98	4.85	.9995

TABLE A.7   Percentiles of the $F(v_1, v_2)$ Distribution with Degrees of Freedom v_1 for the Numerator and v_2 for the Denominator (*Continued*)

Cum. Prop. \ v_1	15	20	24	30	40	50	60	100	120	200	500	∞	Cum. Prop.	v_2
.0005	.159	.197	.220	.244	.272	.290	.303	.330	.339	.353	.368	.377	.0005	15
.001	.181	.219	.242	.266	.294	.313	.325	.352	.360	.375	.388	.398	.001	
.005	.246	.286	.308	.333	.360	.377	.389	.415	.422	.435	.448	.457	.005	
.01	.284	.324	.346	.370	.397	.413	.425	.450	.456	.469	.483	.490	.01	
.025	.349	.389	.410	.433	.458	.474	.485	.508	.514	.526	.538	.546	.025	
.05	.416	.454	.474	.496	.519	.535	.545	.565	.571	.581	.592	.600	.05	
.10	.507	.542	.561	.581	.602	.614	.624	.641	.647	.658	.667	.672	.10	
.25	.701	.728	.742	.757	.772	.782	.788	.802	.805	.812	.818	.822	.25	
.50	1.00	1.01	1.02	1.02	1.03	1.03	1.03	1.04	1.04	1.04	1.04	1.05	.50	
.75	1.43	1.41	1.41	1.40	1.39	1.39	1.38	1.38	1.37	1.37	1.36	1.36	.75	
.90	1.97	1.92	1.90	1.87	1.85	1.83	1.82	1.79	1.79	1.77	1.76	1.76	.90	
.95	2.40	2.33	2.39	2.25	2.20	2.18	2.16	2.12	2.11	2.10	2.08	2.07	.95	
.975	2.86	2.76	2.70	2.64	2.59	2.55	2.52	2.47	2.46	2.44	2.41	2.40	.975	
.99	3.52	3.37	3.29	3.21	3.13	3.08	3.05	2.98	2.96	2.92	2.89	2.87	.99	
.995	4.07	3.88	3.79	3.69	3.59	3.52	3.48	3.39	3.37	3.33	3.29	3.26	.995	
.999	5.54	5.25	5.10	4.95	4.80	4.70	4.64	4.51	4.47	4.41	4.35	4.31	.999	
.9995	6.27	5.93	5.75	5.58	5.40	5.29	5.21	5.06	5.02	4.94	4.87	4.83	.9995	
.0005	.169	.211	.235	.263	.295	.316	.331	.364	.375	.391	.408	.422	.0005	20
.001	.191	.233	.258	.286	.318	.339	.354	.386	.395	.413	.429	.441	.001	
.005	.258	.301	.327	.354	.385	.405	.419	.448	.457	.474	.490	.500	.005	
.01	.297	.340	.365	.392	.422	.441	.455	.483	.491	.508	.521	.532	.01	
.025	.363	.406	.430	.456	.484	.503	.514	.541	.548	.562	.575	.585	.025	
.05	.430	.471	.493	.518	.544	.562	.572	.595	.603	.617	.629	.637	.05	
.10	.520	.557	.578	.600	.623	.637	.648	.671	.675	.685	.694	.704	.10	
.25	.708	.736	.751	.767	.784	.794	.801	.816	.820	.827	.835	.840	.25	
.50	.989	1.00	1.01	1.01	1.02	1.02	1.02	1.03	1.03	1.03	1.03	1.03	.50	
.75	1.37	1.36	1.35	1.34	1.33	1.33	1.32	1.31	1.31	1.30	1.30	1.29	.75	
.00	1.84	1.79	1.77	1.74	1.71	1.69	1.68	1.65	1.64	1.63	1.62	1.61	.90	
.95	2.20	2.12	2.08	2.04	1.99	1.97	1.95	1.91	1.90	1.88	1.86	1.84	.95	
.975	2.57	2.46	2.41	2.35	2.29	2.25	2.22	2.17	2.16	2.13	2.10	2.09	.975	
.99	3.09	2.94	2.86	2.78	2.69	2.64	2.61	2.54	2.52	2.48	2.44	2.42	.99	
.995	3.50	3.32	3.22	3.12	3.02	2.96	2.92	2.83	2.81	2.76	2.72	2.69	.995	
.999	4.56	4.29	4.15	4.01	3.86	3.77	3.70	3.58	3.54	3.48	3.42	3.38	.999	
.9995	5.07	4.75	4.58	4.42	4.24	4.15	4.07	3.93	3.90	3.82	3.75	3.70	.9995	
.0005	.174	.218	.244	.274	.309	.331	.349	.384	.395	.416	.434	.449	.0005	24
.001	.196	.241	.268	.298	.332	.354	.371	.405	.417	.437	.455	.469	.001	
.005	.264	.310	.337	.367	.400	.422	.437	.469	.479	.498	.515	.527	.005	
.01	.304	.350	.376	.405	.437	.459	.473	.505	.513	.529	.546	.558	.01	
.025	.370	.415	.441	.468	.498	.518	.531	.562	.568	.585	.599	.610	.025	
.05	.437	.480	.504	.530	.558	.575	.588	.613	.622	.637	.649	.659	.05	
.10	.527	.566	.588	.611	.635	.651	.662	.685	.691	.704	.715	.723	.10	
.25	.712	.741	.757	.773	.791	.802	.809	.825	.829	.837	.844	.850	.25	
.50	.983	.994	1.00	1.01	1.01	1.02	1.02	1.02	1.02	1.02	1.03	1.03	.50	
.75	1.35	1.33	1.32	1.31	1.30	1.29	1.29	1.28	1.28	1.27	1.27	1.26	.75	
.90	1.78	1.73	1.70	1.67	1.64	1.62	1.61	1.58	1.57	1.56	1.54	1.53	.90	
.95	2.11	2.03	1.98	1.94	1.89	1.86	1.84	1.80	1.79	1.77	1.75	1.73	.95	
.975	2.44	2.33	2.27	2.21	2.15	2.11	2.08	2.02	2.01	1.98	1.95	1.94	.975	
.99	2.89	2.74	2.66	2.58	2.49	2.44	2.40	2.33	2.31	2.27	2.24	2.21	.99	
.995	3.25	3.06	2.97	2.87	2.77	2.70	2.66	2.57	2.55	2.50	2.46	2.43	.995	
.999	4.14	3.87	3.74	3.59	3.45	3.35	3.29	3.16	3.14	3.07	3.01	2.97	.999	
.9995	4.55	4.25	4.09	3.93	3.76	3.66	3.59	3.44	3.41	3.33	3.27	3.22	.9995	

TABLE A.7 Percentiles of the $F(v_1, v_2)$ Distribution with Degrees of Freedom v_1 for the Numerator and v_2 for the Denominator (*Continued*)

v_2	Cum. Prop	1	2	3	4	5	6	7	8	9	10	11	12	Cum. Prop.
30	.0005	$.0^640$	$.0^350$	$.0^250$	.015	.030	.047	.065	.082	.098	.114	.129	.143	.0005
	.001	$.0^516$	$.0^210$	$.0^280$	.022	.040	.060	.080	.099	.117	.134	.150	.164	.001
	.005	$.0^440$	$.0^250$	.024	.050	.079	.107	.133	.156	.178	.197	.215	.231	.005
	.01	$.0^316$	.010	.038	.072	.107	.138	.167	.192	.215	.235	.254	.270	.01
	.025	$.0^210$	.025	.071	.118	.161	.197	.229	.257	.281	.302	.321	.337	.025
	.05	$.0^240$	.051	.116	.174	.222	.263	.296	.325	.349	.370	.389	.406	.05
	.10	.016	.106	.193	.262	.315	.357	.391	.420	.443	.464	.481	.497	.10
	.25	.103	.290	.406	.480	.532	.571	.601	.625	.645	.661	.676	.688	.25
	.50	.466	.709	.807	.858	.890	.912	.927	.939	.948	.955	.961	.966	.50
	.75	1.38	1.45	1.44	1.42	1.41	1.39	1.38	1.37	1.36	1.35	1.35	1.34	.75
	.90	2.88	2.49	2.28	2.14	2.05	1.98	1.93	1.88	1.85	1.82	1.79	1.77	.90
	.95	4.17	3.32	2.92	2.69	2.53	2.42	2.33	2.27	2.21	2.16	2.13	2.09	.95
	.975	5.57	4.18	3.59	3.25	3.03	2.87	2.75	2.65	2.57	2.51	2.46	2.41	.975
	.99	7.56	5.39	4.51	4.02	3.70	3.47	3.30	3.17	3.07	2.98	2.91	2.84	.99
	.995	9.18	6.35	5.24	4.62	4.23	3.95	3.74	3.58	3.45	3.34	3.25	3.18	.995
	.999	13.3	8.77	7.05	6.12	5.53	5.12	4.82	4.58	4.39	4.24	4.11	4.00	.999
	.9995	15.2	9.90	7.90	6.82	6.14	5.66	5.31	5.04	4.82	4.65	4.51	4.38	.9995
40	.0005	$.0^640$	$.0^350$	$.0^250$	.016	.030	.048	.066	.084	.100	.117	.132	.147	.0005
	.001	$.0^516$	$.0^210$	$.0^280$	.022	.042	.061	.081	.101	.119	.137	.153	.169	.001
	.005	$.0^440$	$.0^250$	.024	.051	.080	.108	.135	.159	.181	.201	.220	.237	.005
	.01	$.0^316$	.010	.038	.073	.108	.140	.169	.195	.219	.240	.259	.276	.01
	.025	$.0^399$	.025	.071	.119	.162	.199	.232	.260	.285	.307	.327	.344	.025
	.05	$.0^240$	.051	.116	.175	.224	.265	.299	.329	.354	.376	.395	.412	.05
	.10	.016	.106	.194	.263	.317	.360	.394	.424	.448	.469	.488	.504	.10
	.25	.103	.290	.405	.480	.533	.572	.603	.627	.647	.664	.680	.691	.25
	.50	.463	.705	.802	.854	.885	.907	.922	.934	.943	.950	.956	.961	.50
	.75	1.36	1.44	1.42	1.40	1.39	1.37	1.36	1.35	1.34	1.33	1.32	1.31	.75
	.90	2.84	2.44	2.23	2.09	2.00	1.93	1.87	1.83	1.79	1.76	1.73	1.71	.90
	.95	4.08	3.23	2.84	2.61	2.45	2.34	2.25	2.18	2.12	2.08	2.04	2.00	.95
	.975	5.42	4.05	3.46	3.13	2.90	2.74	2.62	2.53	2.45	2.39	2.33	2.29	.975
	.99	7.31	5.18	4.31	3.83	3.51	3.29	3.12	2.99	2.89	2.80	2.73	2.66	.99
	.995	8.83	6.07	4.98	4.37	3.99	3.71	3.51	3.35	3.22	3.12	3.03	2.95	.995
	.999	12.6	8.25	6.60	5.70	5.13	4.73	4.44	4.21	4.02	3.87	3.75	3.64	.999
	.9995	14.4	9.25	7.33	6.30	5.64	5.19	4.85	4.59	4.38	4.21	4.07	3.95	.9995
60	.0005	$.0^640$	$.0^350$	$.0^251$	.016	.031	.048	.067	.085	.103	.120	.136	.152	.0005
	.001	$.0^516$	$.0^210$	$.0^280$	.022	.041	.062	.083	.103	.122	.140	.157	.174	.001
	.005	$.0^440$	$.0^250$	.024	.051	.081	.110	.137	.162	.185	.206	.225	.243	.005
	.01	$.0^316$	.010	.038	.073	.109	.142	.172	.199	.223	.245	.265	.283	.01
	.025	$.0^399$	.025	.071	.120	.163	.202	.235	.264	.290	.313	.333	.351	.025
	.05	$.0^240$	.051	.116	.176	.226	.267	.303	.333	.359	.382	.402	.419	.05
	.10	.016	.106	.194	.264	.318	.362	.398	.428	.453	.475	.493	.510	.10
	.25	.102	.289	.405	.480	.534	.573	.604	.629	.650	.667	.680	.695	.25
	.50	.461	.701	.798	.849	.880	.901	.917	.928	.937	.945	.951	.956	.50
	.75	1.35	1.42	1.41	1.38	1.37	1.35	1.33	1.32	1.31	1.30	1.29	1.29	.75
	.90	2.79	2.39	2.18	2.04	1.95	1.87	1.82	1.77	1.74	1.71	1.68	1.66	.90
	.95	4.00	3.15	2.76	2.53	2.37	2.25	2.17	2.10	2.04	1.99	1.95	1.92	.95
	.975	5.29	3.93	3.34	3.01	2.79	2.63	2.51	2.41	2.33	2.27	2.22	2.17	.975
	.99	7.08	4.98	4.13	3.65	3.34	3.12	2.95	2.82	2.72	2.63	2.56	2.50	.99
	.995	8.49	5.80	4.73	4.14	3.76	3.49	3.29	3.13	3.01	2.90	2.82	2.74	.995
	.999	12.0	7.76	6.17	5.31	4.76	4.37	4.09	3.87	3.69	3.54	3.43	3.31	.999
	.9995	13.6	8.65	6.81	5.82	5.20	4.76	4.44	4.18	3.98	3.82	3.69	3.57	.9995

TABLE A.7 Percentiles of the $F(v_1, v_2)$ Distribution with Degrees of Freedom v_1 for the Numerator and v_2 for the Denominator (*Continued*)

Cum. Prop.	v_1 15	20	24	30	40	50	60	100	120	200	500	∞	Cum. Prop.	v_2
.0005	.179	.226	.254	.287	.325	.350	.369	.410	.420	.444	.467	.483	.0005	30
.001	.202	.250	.278	.311	.348	.373	.391	.431	.442	.465	.488	.503	.001	
.005	.271	.320	.349	.381	.416	.441	.457	.495	.504	.524	.543	.559	.005	
.01	.311	.360	.388	.419	.454	.476	.493	.529	.538	.559	.575	.590	.01	
.025	.378	.426	.453	.482	.515	.535	.551	.585	.592	.610	.625	.639	.025	
.05	.445	.490	.516	.543	.573	.592	.606	.637	.644	.658	.676	.685	.05	
.10	.534	.575	.598	.623	.649	.667	.678	.704	.710	.725	.735	.746	.10	
.25	.716	.746	.763	.780	.798	.810	.818	.835	.839	.848	.856	.862	.25	
.50	.978	.989	.994	1.00	1.01	1.01	1.01	1.02	1.02	1.02	1.02	1.02	.50	
.75	1.32	1.30	1.29	1.28	1.27	1.26	1.26	1.25	1.24	1.24	1.23	1.23	.75	
.90	1.72	1.67	1.64	1.61	1.57	1.55	1.54	1.51	1.50	1.48	1.47	1.46	.90	
.95	2.01	1.93	1.89	1.84	1.79	1.76	1.74	1.70	1.68	1.66	1.64	1.62	.95	
.975	2.31	2.20	2.14	2.07	2.01	1.97	1.94	1.88	1.87	1.84	1.81	1.79	.975	
.99	2.70	2.55	2.47	2.39	2.30	2.25	2.21	2.13	2.11	2.07	2.03	2.01	.99	
.995	3.01	2.82	2.73	2.63	2.52	2.46	2.42	2.32	2.30	2.25	2.21	2.18	.995	
.999	3.75	3.49	3.36	3.22	3.07	2.98	2.92	2.79	2.76	2.69	2.63	2.59	.999	
.9995	4.10	3.80	3.65	3.48	3.32	3.23	3.15	3.00	2.97	2.89	2.82	2.78	.9995	
.0005	.185	.236	.266	.301	.343	.373	.393	.441	.453	.480	.504	.525	.0005	40
.001	.209	.259	.290	.326	.367	.396	.415	.461	.473	.500	.524	.545	.001	
.005	.279	.331	.362	.396	.436	.463	.481	.524	.534	.559	.581	.599	.005	
.01	.319	.371	.401	.435	.473	.498	.516	.556	.567	.592	.613	.628	.01	
.025	.387	.437	.466	.498	.533	.556	.573	.610	.620	.641	.662	.674	.025	
.05	.454	.502	.529	.558	.591	.613	.627	.658	.669	.685	.704	.717	.05	
.10	.542	.585	.609	.636	.664	.683	.696	.724	.731	.747	.762	.772	.10	
.25	.720	.752	.769	.787	.806	.819	.828	.846	.851	.861	.870	.877	.25	
.50	.972	.983	.989	.994	1.00	1.00	1.01	1.01	1.01	1.01	1.02	1.02	.50	
.75	1.30	1.28	1.26	1.25	1.24	1.23	1.22	1.21	1.21	1.20	1.19	1.19	.75	
.90	1.66	1.61	1.57	1.54	1.51	1.48	1.47	1.43	1.42	1.41	1.39	1.38	.90	
.95	1.92	1.84	1.79	1.74	1.69	1.66	1.64	1.59	1.58	1.55	1.53	1.51	.95	
.975	2.18	2.07	2.01	1.94	1.88	1.83	1.80	1.74	1.72	1.69	1.66	1.64	.975	
.99	2.52	2.37	2.29	2.20	2.11	2.06	2.02	1.94	1.92	1.87	1.83	1.80	.99	
.995	2.78	2.60	2.50	2.40	2.30	2.23	2.18	2.09	2.06	2.01	1.96	1.93	.995	
.999	3.40	3.15	3.01	2.87	2.73	2.64	2.57	2.44	2.41	2.34	2.28	2.23	.999	
.9995	3.68	3.39	3.24	3.08	2.92	2.82	2.74	2.60	2.57	2.49	2.41	2.37	.9995	
.0005	.192	.246	.278	.318	.365	.398	.421	.478	.493	.527	.561	.585	.0005	60
.001	.216	.270	.304	.343	.389	.421	.444	.497	.512	.545	.579	.602	.001	
.005	.287	.343	.376	.414	.458	.488	.510	.559	.572	.602	.633	.652	.005	
.01	.328	.383	.416	.453	.495	.524	.545	.592	.604	.633	.658	.679	.01	
.025	.396	.450	.481	.515	.555	.581	.600	.641	.654	.680	.704	.720	.025	
.05	.463	.514	.543	.575	.611	.633	.652	.690	.700	.719	.746	.759	.05	
.10	.550	.596	.622	.650	.682	.703	.717	.750	.758	.776	.793	.806	.10	
.25	.725	.758	.776	.796	.816	.830	.840	.860	.865	.877	.888	.896	.25	
.50	.967	.978	.983	.989	.994	.998	1.00	1.00	1.01	1.01	1.01	1.01	.50	
.75	1.27	1.25	1.24	1.22	1.21	1.20	1.19	1.17	1.17	1.16	1.15	1.15	.75	
.90	1.60	1.54	1.51	1.48	1.44	1.41	1.40	1.36	1.35	1.33	1.31	1.29	.90	
.95	1.84	1.75	1.70	1.65	1.59	1.56	1.53	1.48	1.47	1.44	1.41	1.39	.95	
.975	2.06	1.94	1.88	1.82	1.74	1.70	1.67	1.60	1.58	1.54	1.51	1.48	.975	
.99	2.35	2.20	2.12	2.03	1.94	1.88	1.84	1.75	1.73	1.68	1.63	1.60	.99	
.995	2.57	2.39	2.29	2.19	2.08	2.01	1.96	1.86	1.83	1.78	1.73	1.69	.995	
.999	3.08	2.83	2.69	2.56	2.41	2.31	2.25	2.11	2.09	2.01	1.93	1.89	.999	
.9995	3.30	3.02	2.87	2.71	2.55	2.45	2.38	2.23	2.19	2.11	2.03	1.98	.9995	

TABLE A.7   Percentiles of the $F(\nu_1, \nu_2)$ Distribution with Degrees of Freedom ν_1 for the Numerator and ν_2 for the Denominator (*Continued*)

ν_2	Cum. Prop.	1	2	3	4	5	6	7	8	9	10	11	12	Cum. Prop.
120	.0005	$.0^640$	$.0^350$	$.0^251$	.016	.031	.049	.067	.087	.105	.123	.140	.156	.0005
	.001	$.0^516$	$.0^210$	$.0^281$	.023	.042	.063	.084	.105	.125	.144	.162	.179	.001
	.005	$.0^439$	$.0^250$	.024	.051	.081	.111	.139	.165	.189	.211	.230	.249	.005
	.01	$.0^316$	.010	.038	.074	.110	.143	.174	.202	.227	.250	.271	.290	.01
	.025	$.0^399$	.025	.072	.120	.165	.204	.238	.268	.295	.318	.340	.359	.025
	.05	$.0^239$	.051	.117	.177	.227	.270	.306	.337	.364	.388	.408	.427	.05
	.10	.016	.105	.194	.265	.320	.365	.401	.432	.458	.480	.500	.518	.10
	.25	.102	.288	.405	.481	.534	.574	.606	.631	.652	.670	.685	.699	.25
	.50	.458	.697	.793	.844	.875	.896	.912	.923	.932	.939	.945	.950	.50
	.75	1.34	1.40	1.39	1.37	1.35	1.33	1.31	1.30	1.29	1.28	1.27	1.26	.75
	.90	2.75	2.35	2.13	1.99	1.90	1.82	1.77	1.72	1.68	1.65	1.62	1.60	.90
	.95	3.92	3.07	2.68	2.45	2.29	2.18	2.09	2.02	1.96	1.91	1.87	1.83	.95
	.975	5.15	3.80	3.23	2.89	2.67	2.52	2.39	2.30	2.22	2.16	2.10	2.05	.975
	.99	6.85	4.79	3.95	3.48	3.17	2.96	2.79	2.66	2.56	2.47	2.40	2.34	.99
	.995	8.18	5.54	4.50	3.92	3.55	3.28	3.09	2.93	2.81	2.71	2.62	2.54	.995
	.999	11.4	7.32	5.79	4.95	4.42	4.04	3.77	3.55	3.38	3.24	3.12	3.02	.999
	.9995	12.8	8.10	6.34	5.39	4.79	4.37	4.07	3.82	3.63	3.47	3.34	3.22	.9995
∞	.0005	$.0^639$	$.0^350$	$.0^251$	.016	.032	.050	.069	.088	.108	.127	.144	.161	.0005
	.001	$.0^516$	$.0^210$	$.0^281$	.023	.042	.063	.085	.107	.128	.148	.167	.185	.001
	.005	$.0^439$	$.0^250$	.024	.052	.082	.113	.141	.168	.193	.216	.236	.256	.005
	.01	$.0^316$	.010	.038	.074	.111	.145	.177	.206	.232	.256	.278	.298	.01
	.025	$.0^398$	.025	.072	.121	.166	.206	.241	.272	.300	.325	.347	.367	.025
	.05	$.0^239$	.051	.117	.178	.229	.273	.310	.342	.369	.394	.417	.436	.05
	.10	.016	.105	.195	.266	.322	.367	.405	.436	.463	.487	.508	.525	.10
	.25	.102	.288	.404	.481	.535	.576	.608	.634	.655	.674	.690	.703	.25
	.50	.455	.693	.789	.839	.870	.891	.907	.918	.927	.934	.939	.945	.50
	.75	1.32	1.39	1.37	1.35	1.33	1.31	1.29	1.28	1.27	1.25	1.24	1.24	.75
	.90	2.71	2.30	2.08	1.94	1.85	1.77	1.72	1.67	1.63	1.60	1.57	1.55	.90
	.95	3.84	3.00	2.60	2.37	2.21	2.10	2.01	1.94	1.88	1.83	1.79	1.75	.95
	.975	5.02	3.69	3.12	2.79	2.57	2.41	2.29	2.19	2.11	2.05	1.99	1.94	.975
	.99	6.63	4.61	3.78	3.32	3.02	2.80	2.64	2.51	2.41	2.32	2.25	2.18	.99
	.995	7.88	5.30	4.28	3.72	3.35	3.09	2.90	2.74	2.62	2.52	2.43	2.36	.995
	.999	10.8	6.91	5.42	4.62	4.10	3.74	3.47	3.27	3.10	2.96	2.84	2.74	.999
	.9995	12.1	7.60	5.91	5.00	4.42	4.02	3.72	3.48	3.30	3.14	3.02	2.90	.9995

For sample sizes larger than, say, 30, a fairly good approximation to the F distribution percentiles can be obtained from

$$\log_{10} F_\alpha(\nu_1, \nu_2) \approx \left(\frac{a}{\sqrt{h - b}} \right) - cg$$

where $h = 2\nu_1\nu_2/(\nu_1 + \nu_2)$, $g = (\nu_2 - \nu_1)/\nu_1\nu_2$, and a, b, c are functions of α given below:

α	.50	.75	.90	.95	.975	.99	.995	.999	.9995
a	0	0.5859	1.1131	1.4287	1.7023	2.0206	2.2373	2.6841	2.8580
b	—	0.58	0.77	0.95	1.14	1.40	1.61	2.09	2.30
c	0.290	0.355	0.527	0.681	0.846	1.073	1.250	1.672	1.857

TABLE A.7   Percentiles of the $F(v_1, v_2)$ Distribution with Degrees of Freedom v_1 for the Numerator and v_2 for the Denominator (*Continued*)

Cum. Prop.	15	20	24	30	40	50	60	100	120	200	500	∞	Cum. Prop.	v_2
.0005	.199	.256	.293	.338	.390	.429	.458	.524	.543	.578	.614	.676	.0005	120
.001	.223	.282	.319	.363	.415	.453	.480	.542	.568	.595	.631	.691	.001	
.005	.297	.356	.393	.434	.484	.520	.545	.605	.623	.661	.702	.733	.005	
.01	.338	.397	.433	.474	.522	.556	.579	.636	.652	.688	.725	.755	.0	
.025	.406	.464	.498	.536	.580	.611	.633	.684	.698	.729	.762	.789	.025	
.05	.473	.527	.559	.594	.634	.661	.682	.727	.740	.767	.785	.819	.05	
.10	.560	.609	.636	.667	.702	.726	.742	.781	.791	.815	.838	.855	.10	
.25	.730	.765	.784	.805	.828	.843	.853	.877	.884	.897	.911	.923	.25	
.50	.961	.972	.978	.983	.989	.992	.994	1.00	1.00	1.00	1.01	1.01	.50	
.75	1.24	1.22	1.21	1.19	1.18	1.17	1.16	1.14	1.13	1.12	1.11	1.10	.75	
.90	1.55	1.48	1.45	1.41	1.37	1.34	1.32	1.27	1.26	1.24	1.21	1.19	.90	
.95	1.75	1.66	1.61	1.55	1.50	1.46	1.43	1.37	1.35	1.32	1.28	1.25	.95	
.975	1.95	1.82	1.76	1.69	1.61	1.56	1.53	1.45	1.43	1.39	1.34	1.31	.975	
.99	2.19	2.03	1.95	1.86	1.76	1.70	1.66	1.56	1.53	1.48	1.42	1.38	.99	
.995	2.37	2.19	2.09	1.98	1.87	1.80	1.75	1.64	1.61	1.54	1.48	1.43	.995	
.999	2.78	2.53	2.40	2.26	2.11	2.02	1.95	1.82	1.76	1.70	1.62	1.54	.999	
.9995	2.96	2.67	2.53	2.38	2.21	2.11	2.01	1.88	1.84	1.75	1.67	1.60	.9995	
.0005	.207	.270	.311	.360	.422	.469	.505	.599	.624	.704	.804	1.00	.0005	∞
.001	.232	.296	.338	.386	.448	.493	.527	.617	.649	.719	.819	1.00	.001	
.005	.307	.372	.412	.460	.518	.559	.592	.671	.699	.762	.843	1.00	.005	
.01	.349	.413	.452	.499	.554	.595	.625	.699	.724	.782	.858	1.00	.01	
.025	.418	.480	.517	.560	.611	.645	.675	.741	.763	.813	.878	1.00	.025	
.05	.484	.543	.577	.617	.663	.694	.720	.781	.797	.840	.896	1.00	.05	
.10	.570	.622	.652	.687	.726	.752	.774	.826	.838	.877	.919	1.00	.10	
.25	.736	.773	.793	.816	.842	.860	.872	.901	.910	.932	.957	1.00	.25	
.50	.956	.967	.972	.978	.983	.987	.989	.993	.994	.997	.999	1.00	.50	
.75	1.22	1.19	1.18	1.16	1.14	1.13	1.12	1.09	1.08	1.07	1.04	1.00	.75	
.90	1.49	1.42	1.38	1.34	1.30	1.26	1.24	1.18	1.17	1.13	1.08	1.00	.90	
.95	1.67	1.57	1.52	1.46	1.39	1.35	1.32	1.24	1.22	1.17	1.11	1.00	.95	
.975	1.83	1.71	1.64	1.57	1.48	1.43	1.39	1.30	1.27	1.21	1.13	1.00	.975	
.99	2.04	1.88	1.79	1.70	1.59	1.52	1.47	1.36	1.32	1.25	1.15	1.00	.99	
.995	2.19	2.00	1.90	1.79	1.67	1.59	1.53	1.40	1.36	1.28	1.17	1.00	.995	
.999	2.51	2.27	2.13	1.99	1.84	1.73	1.66	1.49	1.45	1.34	1.21	1.00	.999	
.9995	2.65	2.37	2.22	2.07	1.91	1.79	1.71	1.53	1.48	1.36	1.22	1.00	.9995	

The values given in this table are abstracted with permission from the following sources:

1. All values for v_1, v_2 equal to 50, 100, 200, 500 are from A. Hald, *Statistical Tables and Formulas*, John Wiley & Sons, Inc., New York, 1952.

2. For cumulative proportions .5, .75, .9, .95, .975, .99, .995 most of the values are from M. Merrington and C. M. Thompson, *Biometrika*, vol. 33 (1943), p. 73.

3. For cumulative proportions .999 the values are from C. Colcord and L. S. Deming, *Sankhyā*, vol. 2 (1936), p. 423.

4. For cum. prop. = $\alpha < .5$ the values are the reciprocals of values for $1 - \alpha$ (with v_1 and v_2 interchanged). The values in Merrington and Thompson and in Colcord and Deming are to five significant figures, and it is hoped (but not expected) that the reciprocals are correct as given. The values in Hald are to three significant figures, and the reciprocals are probably accurate within one to two digits in the third significant figure except for those values very close to unity, where they may be off four to five digits in the third significant figure.

5. Gaps remaining in the table after using the above sources were filled in by interpolation.

$$\alpha = \frac{(v_1/v_2)^{\frac{1}{2}v_1}}{\beta(\frac{1}{2}v_1, \frac{1}{2}v_2)} \int_{-\infty}^{F_\alpha} F^{\frac{1}{2}v_1-1} \left(1 + \frac{v_1 F}{v_2}\right)^{-(v_1+v_2)/2} dF$$

TABLE A.8 Tolerance Factors for Normal Distribution (One-sided)*

Factors K such that the probability is γ that at least a proportion 1 − α of the distribution will be less than $\overline{X} + Ks$ (or greater than $\overline{X} − Ks$), where $\overline{X}$ and s are estimates of the mean and the standard deviation computed from a sample of size n.

| | γ = 0.75 | | | | | γ = 0.90 | | | | | γ = 0.95 | | | | | γ = 0.99 | | | | |
n \ α	0.25	0.10	0.05	0.01	0.001	0.25	0.10	0.05	0.01	0.001	0.25	0.10	0.05	0.01	0.001	0.25	0.10	0.05	0.01	0.001
3	1.464	2.501	3.152	4.396	5.805	2.602	4.258	5.310	7.340	9.651	3.804	6.158	7.655	10.552	13.857					
4	1.256	2.134	2.680	3.726	4.910	1.972	3.187	3.957	5.437	7.128	2.619	4.163	5.145	7.042	9.215					
5	1.152	1.961	2.463	3.421	4.507	1.698	2.742	3.400	4.666	6.112	2.149	3.407	4.202	5.741	7.501					
6	1.087	1.860	2.336	3.243	4.273	1.540	2.494	3.091	4.242	5.556	1.895	3.006	3.707	5.062	6.612	2.849	4.408	5.409	7.334	9.540
7	1.043	1.791	2.250	3.126	4.118	1.435	2.333	2.894	3.972	5.201	1.732	2.755	3.399	4.641	6.061	2.490	3.856	4.730	6.411	8.348
8	1.010	1.740	2.190	3.042	4.008	1.360	2.219	2.755	3.783	4.955	1.617	2.582	3.188	4.353	5.686	2.252	3.496	4.287	5.811	7.566
9	0.984	1.702	2.141	2.977	3.924	1.302	2.133	2.649	3.641	4.772	1.532	2.454	3.031	4.143	5.414	2.085	3.242	3.971	5.389	7.014
10	0.964	1.671	2.103	2.927	3.858	1.257	2.065	2.568	3.532	4.629	1.465	2.355	2.911	3.981	5.203	1.954	3.048	3.739	5.075	6.603
11	0.947	1.646	2.073	2.885	3.804	1.219	2.012	2.503	3.444	4.515	1.411	2.275	2.815	3.852	5.036	1.854	2.897	3.557	4.828	6.284
12	0.933	1.624	2.048	2.851	3.760	1.188	1.966	2.448	3.371	4.420	1.366	2.210	2.736	3.747	4.900	1.771	2.773	3.410	4.633	6.032
13	0.919	1.606	2.026	2.822	3.722	1.162	1.928	2.403	3.310	4.341	1.329	2.155	2.670	3.659	4.787	1.702	2.677	3.290	4.472	5.826
14	0.909	1.591	2.007	2.796	3.690	1.139	1.895	2.363	3.257	4.274	1.296	2.108	2.614	3.585	4.690	1.645	2.592	3.189	4.336	5.651
15	0.899	1.577	1.991	2.776	3.661	1.119	1.866	2.329	3.212	4.215	1.268	2.068	2.566	3.520	4.607	1.596	2.521	3.102	4.224	5.507
16	0.891	1.566	1.977	2.756	3.637	1.101	1.842	2.299	3.172	4.164	1.242	2.032	2.523	3.463	4.534	1.553	2.458	3.028	4.124	5.374
17	0.883	1.554	1.964	2.739	3.615	1.085	1.820	2.272	3.136	4.118	1.220	2.001	2.486	3.415	4.471	1.514	2.405	2.962	4.038	5.268
18	0.876	1.544	1.951	2.723	3.595	1.071	1.800	2.249	3.106	4.078	1.200	1.974	2.453	3.370	4.415	1.481	2.357	2.906	3.961	5.167
19	0.870	1.536	1.942	2.710	3.577	1.058	1.781	2.228	3.078	4.041	1.183	1.949	2.423	3.331	4.364	1.450	2.315	2.855	3.893	5.078
20	0.865	1.528	1.933	2.697	3.561	1.046	1.765	2.208	3.052	4.009	1.167	1.926	2.396	3.295	4.319	1.424	2.275	2.807	3.832	5.003
21	0.859	1.520	1.923	2.686	3.545	1.035	1.750	2.190	3.028	3.979	1.152	1.905	2.371	3.262	4.276	1.397	2.241	2.768	3.776	4.932
22	0.854	1.514	1.916	2.675	3.532	1.025	1.736	2.174	3.007	3.952	1.138	1.887	2.350	3.233	4.238	1.376	2.208	2.729	3.727	4.866
23	0.849	1.508	1.907	2.665	3.520	1.016	1.724	2.159	2.987	3.927	1.126	1.869	2.329	3.206	4.204	1.355	2.179	2.693	3.680	4.806
24	0.845	1.502	1.901	2.656	3.509	1.007	1.712	2.145	2.969	3.904	1.114	1.853	2.309	3.181	4.171	1.336	2.154	2.663	3.638	4.755
25	0.842	1.496	1.895	2.647	3.497	0.999	1.702	2.132	2.952	3.882	1.103	1.838	2.292	3.158	4.143	1.319	2.129	2.632	3.601	4.706
30	0.825	1.475	1.869	2.613	3.454	0.966	1.657	2.080	2.884	3.794	1.059	1.778	2.220	3.064	4.022	1.249	2.029	2.516	3.446	4.508
35	0.812	1.458	1.849	2.588	3.421	0.942	1.623	2.041	2.833	3.730	1.025	1.732	2.166	2.994	3.934	1.195	1.957	2.431	3.334	4.364
40	0.803	1.445	1.834	2.568	3.395	0.923	1.598	2.010	2.793	3.679	0.999	1.697	2.126	2.941	3.866	1.154	1.902	2.365	3.250	4.255
45	0.795	1.435	1.821	2.552	3.375	0.908	1.577	1.986	2.762	3.638	0.978	1.669	2.092	2.897	3.811	1.122	1.857	2.313	3.181	4.168
50	0.788	1.426	1.811	2.538	3.358	0.894	1.560	1.965	2.735	3.604	0.961	1.646	2.065	2.863	3.766	1.096	1.821	2.296	3.124	4.096

* Reproduced from "Tables for One-sided Statistical Tolerance Limits," by Gerald J. Lieberman, *Industrial Quality Control*, vol. XIV, no. 10, p. 8, April, 1958, with the permission of the author and journal.

TABLE A.9 Tolerance Factors for Normal Distributions (Two-sided)*

N \ P	γ = 0.75					γ = 0.90					γ = 0.95					γ = 0.99				
	0.75	0.90	0.95	0.99	0.999	0.75	0.90	0.95	0.99	0.999	0.75	0.90	0.95	0.99	0.999	0.75	0.90	0.95	0.99	0.999
2	4.498	6.301	7.414	9.531	11.920	11.407	15.978	18.800	24.167	30.227	22.858	32.019	37.674	48.430	60.573	114.363	160.193	188.491	242.300	303.054
3	2.501	3.538	4.187	5.431	6.844	4.132	5.847	6.919	8.974	11.309	5.922	8.380	9.916	12.861	16.208	13.378	18.930	22.401	29.055	36.616
4	2.035	2.892	3.431	4.471	5.657	2.932	4.166	4.943	6.440	8.149	3.779	5.369	6.370	8.299	10.502	6.614	9.398	11.150	14.527	18.383
5	1.825	2.599	3.088	4.033	5.117	2.454	3.494	4.152	5.423	6.879	3.002	4.275	5.079	6.634	8.415	4.643	6.612	7.855	10.260	13.015
6	1.704	2.429	2.889	3.779	4.802	2.196	3.131	3.723	4.870	6.188	2.604	3.712	4.414	5.775	7.337	3.743	5.337	6.345	8.301	10.548
7	1.624	2.318	2.757	3.611	4.593	2.034	2.902	3.452	4.521	5.750	2.361	3.369	4.007	5.248	6.676	3.233	4.613	5.488	7.187	9.142
8	1.568	2.238	2.663	3.491	4.444	1.921	2.743	3.264	4.278	5.446	2.197	3.136	3.732	4.891	6.226	2.905	4.147	4.936	6.468	8.234
9	1.525	2.178	2.593	3.400	4.330	1.839	2.626	3.125	4.098	5.220	2.078	2.967	3.532	4.631	5.899	2.677	3.822	4.550	5.966	7.600
10	1.492	2.131	2.537	3.328	4.241	1.775	2.535	3.018	3.959	5.046	1.987	2.839	3.379	4.433	5.649	2.508	3.582	4.265	5.594	7.129
11	1.465	2.093	2.493	3.271	4.169	1.724	2.463	2.933	3.849	4.906	1.916	2.737	3.259	4.277	5.452	2.378	3.397	4.045	5.308	6.766
12	1.443	2.062	2.456	3.223	4.110	1.683	2.404	2.863	3.758	4.792	1.858	2.655	3.162	4.150	5.291	2.274	3.250	3.870	5.079	6.477
13	1.425	2.036	2.424	3.183	4.059	1.648	2.355	2.805	3.682	4.697	1.810	2.587	3.081	4.044	5.158	2.190	3.130	3.727	4.893	6.240
14	1.409	2.013	2.398	3.148	4.016	1.619	2.314	2.756	3.618	4.615	1.770	2.529	3.012	3.955	5.045	2.120	3.029	3.608	4.737	6.043
15	1.395	1.994	2.375	3.118	3.979	1.594	2.278	2.713	3.562	4.545	1.735	2.480	2.954	3.878	4.949	2.060	2.945	3.507	4.605	5.876
16	1.383	1.977	2.355	3.092	3.946	1.572	2.246	2.676	3.514	4.484	1.705	2.437	2.903	3.812	4.865	2.009	2.872	3.421	4.492	5.732
17	1.372	1.962	2.337	3.069	3.917	1.552	2.219	2.643	3.471	4.430	1.679	2.400	2.858	3.754	4.791	1.965	2.808	3.345	4.393	5.607
18	1.363	1.948	2.321	3.048	3.891	1.535	2.194	2.614	3.433	4.382	1.655	2.366	2.819	3.702	4.725	1.926	2.753	3.279	4.307	5.497
19	1.355	1.936	2.307	3.030	3.867	1.520	2.172	2.588	3.399	4.339	1.635	2.337	2.784	3.656	4.667	1.891	2.703	3.221	4.230	5.399
20	1.347	1.925	2.294	3.013	3.846	1.506	2.152	2.564	3.368	4.300	1.616	2.310	2.752	3.615	4.614	1.860	2.659	3.168	4.161	5.312
21	1.340	1.915	2.282	2.998	3.827	1.493	2.135	2.543	3.340	4.264	1.599	2.286	2.723	3.577	4.567	1.833	2.620	3.121	4.100	5.234
22	1.334	1.906	2.271	2.984	3.809	1.482	2.118	2.524	3.315	4.232	1.584	2.264	2.697	3.543	4.523	1.808	2.584	3.078	4.044	5.163
23	1.328	1.898	2.261	2.971	3.793	1.471	2.103	2.506	3.292	4.203	1.570	2.244	2.673	3.512	4.484	1.785	2.551	3.040	3.993	5.098
24	1.322	1.891	2.252	2.950	3.778	1.462	2.089	2.480	3.270	4.176	1.557	2.225	2.651	3.483	4.447	1.764	2.522	3.004	3.947	5.039
25	1.317	1.883	2.244	2.948	3.764	1.453	2.077	2.474	3.251	4.151	1.545	2.208	2.631	3.457	4.413	1.745	2.494	2.972	3.904	4.985
26	1.313	1.877	2.236	2.938	3.751	1.444	2.065	2.460	3.232	4.127	1.534	2.193	2.612	3.432	4.382	1.727	2.460	2.941	3.865	4.935
27	1.309	1.871	2.229	2.929	3.740	1.437	2.054	2.447	3.215	4.106	1.523	2.178	2.595	3.409	4.353	1.711	2.446	2.914	3.828	4.888
30	1.297	1.855	2.210	2.904	3.708	1.417	2.025	2.413	3.170	4.049	1.497	2.140	2.549	3.350	4.278	1.668	2.385	2.841	3.733	4.768
35	1.283	1.834	2.185	2.871	3.667	1.390	1.988	2.368	3.112	3.974	1.462	2.090	2.490	3.272	4.179	1.613	2.306	2.748	3.611	4.611
40	1.271	1.818	2.166	2.846	3.635	1.370	1.959	2.334	3.066	3.917	1.435	2.052	2.445	3.213	4.104	1.571	2.247	2.677	3.518	4.493
45	1.262	1.805	2.150	2.826	3.609	1.354	1.935	2.306	3.030	3.871	1.414	2.021	2.408	3.165	4.042	1.539	2.200	2.621	3.444	4.399
50	1.255	1.794	2.138	2.809	3.588	1.340	1.916	2.284	3.001	3.833	1.396	1.996	2.379	3.126	3.993	1.512	2.162	2.576	3.385	4.323
55	1.249	1.785	2.127	2.795	3.571	1.329	1.901	2.265	2.976	3.801	1.382	1.976	2.354	3.094	3.951	1.490	2.130	2.538	3.335	4.260
60	1.243	1.778	2.118	2.784	3.556	1.320	1.887	2.248	2.955	3.774	1.369	1.958	2.333	3.066	3.916	1.471	2.103	2.506	3.293	4.206

*From *Introduction to Statistical Analysis*, 2d ed., by W. J. Dixon and F. J. Massey, Jr., Copyright, 1957. McGraw-Hill Book Company. Used by permission.

TABLE A.9 Tolerance Factors for Normal Distributions (Two-sided)* *(Continued)*

N	γ = 0.75					γ = 0.90					γ = 0.95					γ = 0.99				
P	0.75	0.90	0.95	0.99	0.999	0.75	0.90	0.95	0.99	0.999	0.75	0.90	0.95	0.99	0.999	0.75	0.90	0.95	0.99	0.999
65	1.239	1.771	2.110	2.773	3.543	1.312	1.875	2.235	2.937	3.751	1.359	1.943	2.315	3.042	3.886	1.455	2.080	2.478	3.257	4.160
70	1.235	1.765	2.104	2.764	3.531	1.304	1.865	2.222	2.920	3.730	1.349	1.929	2.299	3.021	3.859	1.440	2.060	2.454	3.225	4.120
75	1.231	1.760	2.098	2.757	3.521	1.298	1.856	2.211	2.906	3.712	1.341	1.917	2.285	3.002	3.835	1.428	2.042	2.433	3.197	4.084
80	1.228	1.756	2.092	2.749	3.512	1.292	1.848	2.202	2.894	3.696	1.334	1.907	2.272	2.986	3.814	1.417	2.026	2.414	3.173	4.053
85	1.225	1.752	2.087	2.743	3.504	1.287	1.841	2.193	2.882	3.682	1.327	1.897	2.261	2.971	3.795	1.407	2.012	2.397	3.150	4.024
90	1.223	1.748	2.083	2.737	3.497	1.283	1.834	2.185	2.872	3.669	1.321	1.889	2.251	2.958	3.778	1.398	1.999	2.382	3.130	3.999
95	1.220	1.745	2.079	2.732	3.490	1.278	1.828	2.178	2.863	3.657	1.315	1.881	2.241	2.945	3.763	1.390	1.987	2.368	3.112	3.976
100	1.218	1.742	2.075	2.727	3.484	1.275	1.822	2.172	2.854	3.646	1.311	1.874	2.233	2.934	3.748	1.383	1.977	2.355	3.096	3.954
110	1.214	1.736	2.069	2.719	3.473	1.268	1.813	2.160	2.839	3.626	1.302	1.861	2.218	2.915	3.723	1.369	1.958	2.333	3.066	3.917
120	1.211	1.732	2.063	2.712	3.464	1.262	1.804	2.150	2.826	3.610	1.294	1.850	2.205	2.898	3.702	1.358	1.942	2.314	3.041	3.885
130	1.208	1.728	2.059	2.705	3.456	1.257	1.797	2.141	2.814	3.595	1.288	1.841	2.194	2.883	3.683	1.349	1.928	2.298	3.019	3.857
140	1.206	1.724	2.054	2.700	3.449	1.252	1.791	2.134	2.804	3.582	1.282	1.833	2.184	2.870	3.666	1.340	1.916	2.283	3.000	3.833
150	1.204	1.721	2.051	2.695	3.443	1.248	1.785	2.127	2.795	3.571	1.277	1.825	2.175	2.859	3.652	1.332	1.905	2.270	2.983	3.811
160	1.202	1.718	2.047	2.691	3.437	1.245	1.780	2.121	2.787	3.561	1.272	1.819	2.167	2.848	3.638	1.326	1.896	2.259	2.968	3.792
170	1.200	1.716	2.044	2.687	3.432	1.242	1.775	2.116	2.780	3.552	1.268	1.813	2.160	2.839	3.627	1.320	1.887	2.248	2.955	3.774
180	1.198	1.713	2.042	2.683	3.427	1.239	1.771	2.111	2.774	3.543	1.264	1.808	2.154	2.831	3.616	1.314	1.879	2.239	2.942	3.759
190	1.197	1.711	2.039	2.680	3.423	1.236	1.767	2.106	2.768	3.536	1.261	1.803	2.148	2.823	3.606	1.309	1.872	2.230	2.931	3.744
200	1.195	1.709	2.037	2.677	3.419	1.234	1.764	2.102	2.762	3.529	1.258	1.798	2.143	2.816	3.597	1.304	1.865	2.222	2.921	3.731
250	1.190	1.702	2.028	2.665	3.404	1.224	1.750	2.085	2.740	3.501	1.245	1.780	2.121	2.788	3.561	1.286	1.839	2.191	2.880	3.678
300	1.186	1.696	2.021	2.656	3.393	1.217	1.740	2.073	2.725	3.481	1.236	1.767	2.106	2.767	3.535	1.273	1.820	2.169	2.850	3.641
400	1.181	1.688	2.012	2.644	3.378	1.207	1.726	2.057	2.703	3.452	1.223	1.749	2.084	2.739	3.499	1.255	1.794	2.138	2.809	3.589
500	1.177	1.683	2.006	2.636	3.368	1.201	1.717	2.046	2.689	3.434	1.215	1.737	2.070	2.721	3.475	1.243	1.777	2.117	2.783	3.555
600	1.175	1.680	2.002	2.631	3.360	1.196	1.710	2.038	2.678	3.421	1.209	1.729	2.060	2.707	3.458	1.234	1.764	2.102	2.763	3.530
700	1.173	1.677	1.998	2.626	3.355	1.192	1.705	2.032	2.670	3.411	1.204	1.722	2.052	2.697	3.445	1.227	1.755	2.091	2.748	3.511
800	1.171	1.675	1.996	2.623	3.350	1.189	1.701	2.027	2.663	3.402	1.201	1.717	2.046	2.688	3.434	1.222	1.747	2.082	2.736	3.495
900	1.170	1.673	1.993	2.620	3.347	1.187	1.697	2.023	2.658	3.396	1.198	1.712	2.040	2.682	3.426	1.218	1.741	2.075	2.726	3.483
1000	1.169	1.671	1.992	2.617	3.344	1.185	1.695	2.019	2.654	3.390	1.195	1.709	2.036	2.676	3.418	1.214	1.736	2.068	2.718	3.472
∞	1.150	1.645	1.960	2.576	3.291	1.150	1.645	1.960	2.576	3.291	1.150	1.645	1.960	2.576	3.291	1.150	1.645	1.960	2.576	3.291

APPENDIX B
CHARTS

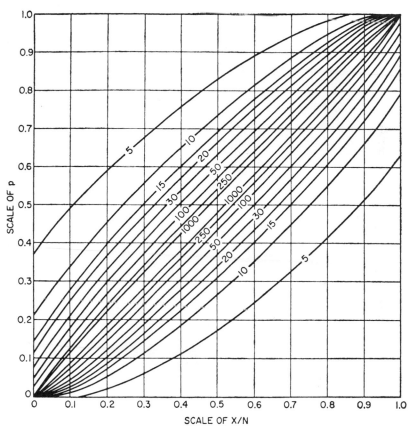

CHART B.1 Confidence belts for proportions (confidence coefficient 0.80). (*From Introduction to Statistical Analysis, 2d ed., by W. J. Dixon and F. J. Massey, Jr. Copyright 1957. McGraw-Hill Book Company. Used by permission.*)

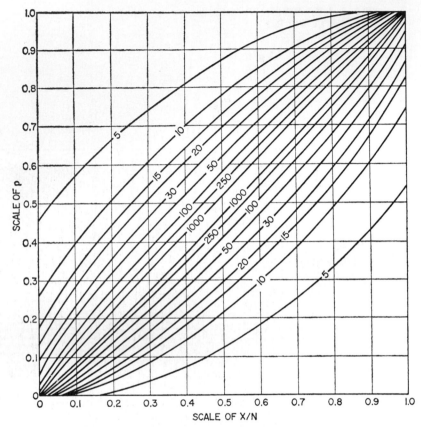

CHART B.2 Confidence belts for proportions (confidence coefficient 0.90). (*From Introduction to Statistical Analysis, 2d ed., by W. J. Dixon and F. J. Massey, Jr. Copyright 1957. McGraw-Hill Book Company. Used by permission.*)

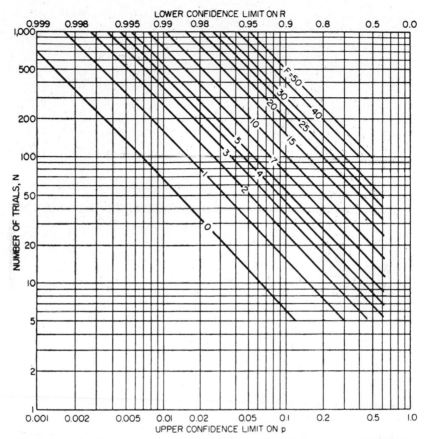

CHART B.3 Confidence limits on reliability. Upper confidence limit on unreliability (1 minus lower confidence limit on reliability), number of trials N, observed failures F, confidence coefficient $\gamma = 0.50$. (*David K. Lloyd and Myron Lipow, Reliability: Management, Methods and Mathematics.* © 1962, by permission of Prentice-Hall, Inc., Englewood Cliffs, New Jersey.)

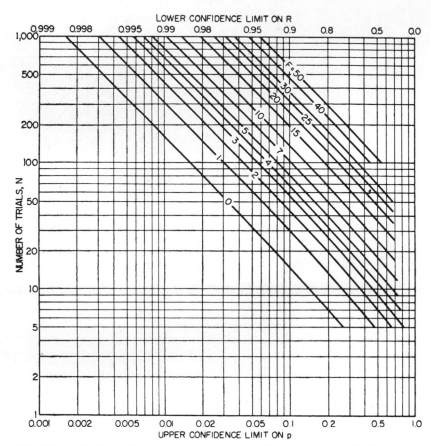

CHART B.4   Confidence limits on reliability. Upper confidence limit on unreliability (1 minus lower confidence limit on reliability), number of trials N, observed failures F, confidence coefficient $\gamma = 0.80$. (*David K. Lloyd and Myron Lipow, Reliability: Management, Methods and Mathematics. © 1962, by permission of Prentice-Hall, Inc., Englewood Cliffs, New Jersey.*)

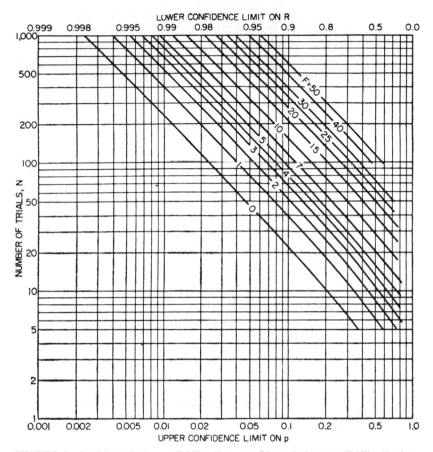

CHART B.5   Confidence limits on reliability. Upper confidence limit on unreliability (1 minus lower confidence limit on reliability), number of trials *N*, observed failures *F*, confidence coefficient $\gamma = 0.90$. (*David K. Lloyd and Myron Lipow, Reliability: Management, Methods and Mathematics.* © *1962, by permission of Prentice-Hall, Inc., Englewood Cliffs, New Jersey.*)

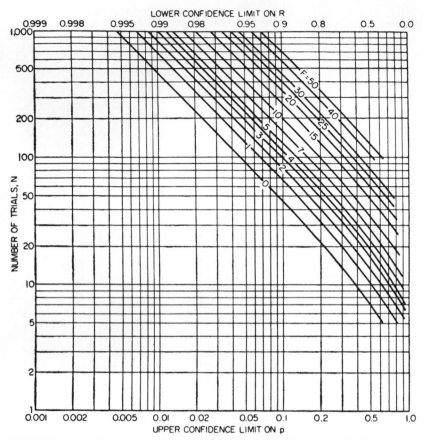

CHART B.6   Confidence limits on reliability. Upper confidence limit on unreliability (1 minus lower confidence limit on reliability), number of trials *N*, observed failures *F*, confidence coefficient γ = 0.95. (*David K. Lloyd and Myron Lipow, Reliability: Management, Methods and Mathematics. © 1962, by permission of Prentice-Hall, Inc., Englewood Cliffs, New Jersey.*)

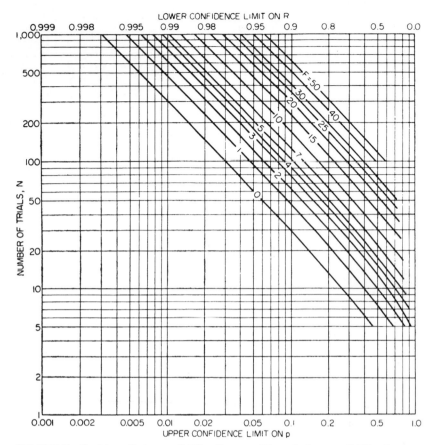

CHART B.7 Confidence limits on reliability. Upper confidence limit on unreliability (1 minus lower confidence limit on reliability), number of trials N, observed failures F, confidence coefficient $\gamma = 0.99$. (*David K. Lloyd and Myron Lipow, Reliability: Management, Methods and Mathematics.* © 1962, by permission of Prentice-Hall, Inc., Englewood Cliffs, New Jersey.)

APPENDIX C
RELIABILITY STANDARDS AND SPECIFICATIONS

Arthur A. McGill
LOCKHEED MARTIN MISSILES AND SPACE COMPANY
SUNNYVALE, CALIF.

The purpose of this appendix is to provide brief descriptions of some of the more commonly referenced reliability and maintainability documents and to assist in their selection and use. Military standards (MIL-STDs) generally impose requirements and are "what to do" documents. Military Handbooks (MIL-HDBKs) are generally "how to do it" documents. Many of the documents are extensive and complete, others are brief and of limited value. Since these documents are not always easy to obtain, some suggested sources are provided.

Internet contains web sites for many of the sources of the referenced specifications and documentation. Since addresses and telephone numbers change frequently, the Internet, coupled with the sources mentioned, and a "search engine" can provide current information on how to contact these organizations. As the information on the Internet expands, this source will become a significant tool.

SOURCES OF INFORMATION

Copies of military specifications, standards, and handbooks may be ordered from

DOSSP—Customer Service
700 Robbins Ave.
Building 4D
Philadelphia, PA 19111-5094
Phone (215) 697-2667, -2179

Department of Defense Directives and Instructions may be ordered from

Defense Technical Information Center
DTIC-FDAC
8725 John J. Kingman Raod, Suite 0944
Fort Belvoir, VA 22060-6218
Phone 1-800-225-3842

Many Rome Laboratory (formerly Rome Air Development Center RADC) technical reports may be obtained from

National Technical Information Service (NTIS)
5285 Port Royal Road
Springfield, VA 22161-2171
Phone (703) 487-4650

Reliability Analysis Center
P.O. Box 4700
Rome, NY 13442-4700
Phone (315) 337-0900

Institute of Environmental Sciences
940 East Northwest Highway
Mount Prospect, IL 60056
Phone (708) 255-1561

Copies of ISO-9000 (ANSI/ASQC Q90 Series) may be obtained from

The American Society for Quality Control (ASQC)
Technical Services Department
611 W. Wisconsin Ave., P.O. Box 3005
Milwaukee, WI 53201-3005
Phone (414) 272-8575 or 1-800-248-1946

American National Standards Institute (ANSI)
11 W. 42nd St.
New York, NY 10036
Phone (foreign/international) (212) 642-4995; (domestic) (212) 642-4900; (fax) (212) 302-1286 or (212) 398-0023. For ISO information: (212) 642-4953 or (212) 642-4993

GENERAL CATEGORIES OF INTEREST

Design

MIL-HDBK-251, MIL-HDBK-338, MIL-STD-454
MIL-STD-785, MIL-STD-1556, MIL-STD-1591
MIL-STD-1670, ISO 9000

Electrostatic Discharge

MIL-HDBK-263
DoD-STD-1686

Environmental Stress Screening

MIL-HDBK-344, MIL-STD-810, MIL-STD-1670
MIL-STD-2164
Environmental Stress Screening Guidelines for Assemblies published by Institute of Environmental Sciences, Mount Prospect, Ill.

FMECA

MIL-STD-1629, RADC-TR-83-72

Human Factors

      MIL-STD-1472, MIL-H-46855

Logistics

      MIL-STD-1388

Maintainability

      MIL-STD-470, MIL-STD-472, MIL-STD-721
      MIL-STD-741, RADC-TR-83-29

Quality

      ISO 9000 (ANSI/ASQC Q90 Series)

Reliability (Prediction)

      MIL-HDBK-217, MIL-STD-756, LC-78-2
      NPRD-91, RADC-TR-73-248, RADC-TR-75-22
      RADC-TR-83-91

Reliability (General)

      MIL-HDBK-338, MIL-STD-1543, MIL-STD-2155
      DoD-STD-4245.7, RADC-TR-83-29
      Rome Laboratory Reliability Engineer's Toolkit

Reliability Growth

      MIL-HDBK-189

Safety

      MIL-STD-882, MIL-STD-1574

Sampling

      MIL-STD-105, MIL-STD-690

Software

      MIL-STD-2167, DoD-STD-2168, MIL-S-52779

Testing

MIL-STD-781, MIL-HDBK-781, MIL-STD-785
MIL-STD-790, MIL-STD-810, MIL-STD-883
MIL-STD-2074, MIL-STD-2165

DESCRIPTIONS OF SELECTED STANDARDS AND SPECIFICATIONS

ISO 9000 (Ansi/ASQC Q90 Series)

ISO 8402 quality systems—vocabulary. Defines many of the common quality terms.

ISO 9000 quality management and quality-assurance standards—guidelines for selection and use. This standard assists in selection of the other standards and clarifies their interrelationships.

ISO 9001 quality systems—model for quality assurance in design and development, production, installation and servicing. This standard is a "what should be considered" not a "how to" document. This is the most stringent of the 9001 to 9003 series and covers suppliers in the titled phases.

ISO 9002 quality systems—model for quality assurance in production and installation. Supplier activities in titled phases.

ISO 9003 quality systems—model for quality assurance in final inspection and test. Supplier activities in titled phases.

ISO 9004 quality management and quality system elements—guidelines. These are internal guidelines for a company including selection of suppliers.

MIL-STD-105, Sampling Procedures and Tables for Inspection by Attributes

This document addresses the subjects of sampling plans, lot size, inspection levels, AQLs, classification of defects, multiple sampling, normal, tightened, and reduced sampling. For equipments where the sequential method of testing, based on operating time, may not be appropriate, this document, based on the success ratio, may be used. It includes numerous tables showing accept/reject levels and operating characteristic curves for sampling plans. It would help to have sampling theory well in hand to understand the applicability and limitations of this document.

MIL-HDBK-189, Reliability Growth Management

This document is designed for both managers and analysts covering everything from simple fundamentals to detailed technical analysis. Included are concepts and principles of reliability growth, advantages of managing reliability growth, and guidelines and procedures to be used to manage reliability growth. It allows the development of a plan that will aid in developing a final system that meets requirements and lowers the life-cycle cost of the fielded system. The document includes sections on benefits, concepts, engineering analysis, and growth models. It contains lots of plots and curves!

MIL-HDBK-217 F, Notice 1, Reliability Prediction of Electronic Equipment

This handbook includes two basic methods for reliability prediction of electronic equipment. The first is a simple method called the parts-count reliability prediction technique using primarily the number of parts of each category with consideration of part quality, environments encountered, and maturity of the production process. The second method is the part-stress analysis prediction technique employing complex models using detailed stress analysis information as well as environment, quality application, maximum ratings, complexity, temperature, construction, and a number of other application-related factors. The simple method is beneficial in early tradeoff studies and situations where the detailed circuit design is unknown. The complex method requires detailed study and analysis which is available when the circuit design has been defined. Samples of each type of calculation are provided. A bibliography on reliability prediction is included.

MIL-HDBK-251, *Reliability/Design Thermal Applications*

This document details approaches to thermal design, methods for the determination of thermal requirements, selection of cooling methods, natural methods of cooling, thermal design for forced air, and liquid-cooled, vaporization, and special (heat pipes) cooling systems. Topics covered are the standard hardware program thermal design, installation requirements, thermal evaluation, improving existing designs, and thermal characteristics of parts. Stress analysis methods are emphasized. Many graphs and nomographs are included. There is an excellent bibliography included. This is an excellent handbook.

MIL-HDBK-263, *An Electrostatic Discharge Control Handbook for Protection of Electrical and Electronic Parts, Assemblies and Equipment*

This document excludes information on electrically initiated explosive devices but includes definitions, causes and effects (including failure mechanisms), charge sources, list and category of electrostatic sensitive devices by part type, testing, application information, considerations, protective networks, and a bibliography. This is a useful, enlightening document on a problem that is very difficult to demonstrate to management and manufacturing personnel. It is a problem that is difficult to control, difficult to assess and pervasive. This document details the problem, determination of causes and identification by failure diagnosis and shows methods both in design and control to minimize its occurrence.

MIL-HDBK-338, *Electronic Reliability Design Handbook*

This document is virtually a text on reliability. Currently a two-volume set, it discusses the entire subject, heavily emphasizing the reasons for reliability discipline. It includes general information, referenced documents, definitions, reliability theory, component reliability design considerations, application guidelines, specification control during acquisition, logistic support (storage), failure reporting and analysis, reliability-maintainability (RM) theory, reliability specification allocation and prediction, reliability engineering design guidelines, reliability data collection and analysis, demonstration and growth, software reliability, systems reliability engineering, production and use (deployment) RM, and RM management considerations. This document is must reading for the reliability professional who has need of detailed explanations and theory.

MIL-HDBK-344, *Environmental Stress Screening (ESS) of Electronic Equipment*

The techniques for planning and evaluating ESS programs are presented. Data include screening strengths for vibration and temperature. Failure free acceptance test derivations, remaining defect density vs. failure rate and other topics are addressed. This is essential for creation and understanding of ESS.

MIL-STD-454N, *Standard General Requirements for Electronic Equipment*

This document establishes the technical baseline for design and construction of electronic equipment for the Department of Defense (DoD). It addresses 75 requirements such as brazing, substitutability, reliability, resistors, and casting. It gives numerous references on the subjects addressed but in itself is not rigorous or extensive in its treatment of subjects. It has some good illustrations for inspection attributes of selected processes such as soldering.

MIL-STD-470B, *Maintainability Program Requirements for Systems and Equipment*

This document includes application requirements, tailorable maintainability program tasks, and an appendix with an application matrix and guidance and rationale for task selection. The topics covered are program surveillance and control, design and analysis, modeling, allocations, predictions, FMEA, maintainability, and design criteria. Each task item includes a purpose, task description, and details to be specified.

MIL-HDBK-472, Notice 1, *Maintainability Prediction*

This document is to familiarize project managers and design engineers with maintainability prediction procedures. It provides the analytic foundation and application details of five prediction methods. Each procedure details applicability, point of application, basic parameters of measure, information required, correlation, and cautionary notes. This document includes equations and data analysis sheets.

MIL-STD-690B, Notice 3, *Failure Rate Sampling Plans and Procedures*

This document provides samples of life-test records, failure-rate sampling plans, failure-rate tables at 60 and 90 percent confidence levels, and a reliability nomograph. Tables are provided that allow the determination of the probability of qualification of a lot.

MIL-STD-721C, Notice 1, *Definition of Terms for Reliability and Maintainability*

This is a list of terms and definitions thereof.

MIL-STD-741A, Notice 2, *Maintainability Verification/Demonstration/Evaluation*

Procedures and test methods for MIL-HDBK-781.

MIL-STD-756B, Notice 1, *Reliability Modeling and Prediction*

This document establishes uniform procedures and ground rules for generating mission reliability models for electrical, mechanical, and ordinance equipment. It details the methods for determining service use (life cycle), creation of the reliability block diagram, and construction of the mathematical model for computing the item reliability. Some simple explanations on the applicability and suitability of the various prediction sources and methods are included.

MIL-STD-781D, *Reliability Design Qualification and Production Acceptance Tests: Exponential Distribution*

> This document covers the requirements and provides details for reliability development and growth testing, reliability qualification testing, production reliability acceptance tests, and environmental stress screening. These tasks are selectively applied in DoD-contracted procurements, requests for proposals, etc.

MIL-HDBK-781, *Reliability Test Methods, Plans, and Environments for Engineering Development, Qualification, and Production*

> Huge document (373 pages) on test methods, plans, development and growth models including the Duane Method, AMSAA model, ESS monitoring methods, MTBF assurance tests, and combined environment test conditions. Includes typical mission profiles for a number of typical applications. Supports the military standard.

MIL-STD-785B, Notice 2, *Reliability Program for Systems and Equipment, Development and Production*

> This document provides general requirements and specific tasks for reliability programs. It is of great importance for reliability program planning. It has task descriptions for basic application requirements, including sections on program surveillance and control, design and evaluation, and development and production testing. An appendix for application guidance for implementation of reliability program requirements is also included. The subsections are in the form of purpose, task description, and details to be specified by the procuring activity. This is a program management document, not a detailed "what to do document."

MIL-STD-790E, Notice 1, *Reliability Assurance Program for Electronic Parts Specifications*

> This document establishes the criteria for a reliability assurance program which is to be met by the manufacturer qualifying electronic parts to the specification. Typical topics covered are document submission, organizational structure, test facilities, and failure analysis reports. This document is 12 pages long.

MIL-STD-810E, Notice 1, *Environmental Test Methods and Engineering Guidelines*

> This document is designed to provide a more careful assessing of the environments to which items are exposed during their life as well as detailing test methods. Included in the numerous types of tests detailed are purpose, environmental effects, guidelines for determining test procedures and test conditions, references, apparatus, preparation for test, procedures, and information to be recorded. Numerous curves on environments are included, e.g., two-wheeled trailer–transverse axis frequency/power spectral-density curve.

MIL-STD-882C, *System Safety Program Requirements*

> This document provides requirements for developing and implementing a system safety program to identify the hazards of a system and to impose design requirements and management controls

to prevent mishaps by eliminating hazards or reducing risks. Twenty-two tasks are defined in the areas of program management and control and design and evaluation. Typical tasks are system safety program plan, preliminary hazard analysis, and software hazard analysis. An appendix is provided to give some rationale and methods for satisfying the requirements previously detailed.

MIL-STD-883D, *Test Methods and Procedures for Microelectronics*

This document establishes uniform methods and procedures for testing microelectronic devices. Basic environmental tests, physical (mechanical), and electrical tests (digital or linear) are specified. Also covered are test procedures for failure analysis, limit testing, wafer lot acceptance, and destructive physical analysis. This document gives extensive treatment to the subject of visual defects.

MIL-STD-965A, Notice 3, *Parts Control Program*

This document describes two procedures covering the submission, review, and approval of program parts selection lists (PPSLs). Typical topics covered are PPSL approval, meetings, parts control board, and Military Parts Control Advisory Group.

MIL-STD-1388-1A, Notice 3, *Logistics Support Analysis*

This document details logistic support analysis (LSA) guidelines and requirements. Tasks detail the purpose, task description, task input, and task output. Typical tasks are program planning and control; development of early LSA strategy; planning, program, and design reviews; mission hardware, software, and support; system standardization; early fielding analysis; and supportability assessment.

MIL-STD-1472D, Notice 2, *Human Engineering Design Criteria for Military Systems, Equipment and Facilities*

This document presents human-engineering principles, design criteria, and practices to integrate humans (their requirements) into systems and facilities. This is desired to achieve effectiveness, simplicity, efficiency, reliability, and safety of the system operation, training, and maintenance. This document contains interesting and useful information on items with which humans commonly interface, including data and illustrations on visual fields, controls and displays (manual, visual, and audio), physical dimensions and strengths of humans, anthropometry (DoD-HDBK 743, *Anthropometry of US Military Personnel,* is a referenced document), ground workspace design requirements, environments, design for maintainability, design for remote handling, and hazards and safety considerations. This document contains extensive figures and tables on human parameters.

MIL-STD-1543B, *Reliability Program Requirements for Space and Missile Systems*

This document covers topics such as design for reliability; FMECA; reliability analysis; modeling and prediction; discrepancy and failure reporting; maximum preacceptance operation; effects of testing, storage, and shelf life; and packaging, transportation, handling, and maintainability. It gives application guidance and an appendix for FMEA for space and launch vehicle systems. This document is mainly a "what" and definition document.

MIL-STD-1556B, *Government/Industry Data Exchange Program (GIDEP)*

This document defines the requirements for participation in the GIDEP program, which includes the engineering, failure experience, reliability-maintainability (RMDI), and metrology data interchanges. It is intended to be applied to prime contractors and major subcontractors (who are users of parts) for the government. The RMDI contains failure-rate, failure-mode, and replacement-rate data on parts, components, assemblies, subsystems, and materials based on field-performance information and reliability test of equipment, subsystems, and systems. This data interchange also contains reports on theories, methods, techniques, and procedures related to reliability and maintainability practices.

MIL-STD-1574A, Notice 1, *System Safety Program for Space and Missile Systems*

This document is a tailored application of MIL-STD-882A for space, missile, and related systems. It defines the management and technical requirements for system safety from concept to the end of the life cycle. It includes numerous system safety definitions such as accident, credible conditions, and system safety engineer. Includes the requirements for a system safety program plan such as associate contractor responses, subcontractor responses, software safety analysis, and a list of publications, specifications, and standards.

MIL-STD-1591A, *Command, Control and Communications (C3) System & Component Fault Diagnosis, Subsystems, Analysis/Synthesis of (On-Aircraft, Fault Diagnosis)*

This diagnosis provides criteria for conducting trade studies to determine optimal design for an aircraft fault diagnosis and isolation system, i.e., on-board built-in test system. It provides a cost model and maintainability workforce model.

MIL-STD-1629A, Notice 2, *Procedures for Performing a Failure Mode, Effects, and Criticality Analysis*

This document shows how to perform an FMECA. It details the modeling method and functional block diagrams, and defines severity classification and criticality numbers. It provides sample formats for an FMEA, criticality analysis, FMEA and criticality analysis maintainability information sheet, and damage mode and effects analysis sheet. Examples are provided.

MIL-STD-1670A, *Environmental Criteria and Guidelines for Air-Launched Weapons*

This document provides acquisition managers with guidelines for realistic test environments for air-launched weapons representing factory-to-target sequences. It covers environmental test criteria. Some good curves are presented, e.g., railroad vibration environments, jet/turboprop and prop cargo area vibration, and worldwide humidity environments. It does not cover nuclear environments, electromagnetic, or laser effect environments.

DoD-STD-1686B, *Electrostatic Discharge Control Program for Protection of Electrical and Electronic Parts, Assemblies and Equipment*

This document excludes information on electrically initiated explosive devices but covers the establishment and implementation of ESD control programs for design, test, inspection, servicing, manufacturing, processing, assembling, installation, packaging, labeling, or other handling of electrical and electronic equipment. Refer to DoD-HDBK-263 for "how to" information.

MIL-STD-2074, *Failure Classification for Reliability Testing*

This document contains criteria for classification of failures during reliability testing. This classification into relevant or nonrelevant categories allows the proper generation of MTBF reports.

MIL-STD-2155, *Failure Reporting, Analysis and Corrective Action System*

Establishes uniform requirements and criteria for failure reporting, analysis, and corrective action system to implement MIL-STD-785.

MIL-STD-2164 (*EC*), *Environmental Stress Screening Process for Electronic Equipment*

Provides requirements for environmental stress testing (ESS) of electronic equipment. Provides environmental test conditions, duration of exposure to the environments, procedures, and operation actions when defects are precipitated. Specifies test documentation.

MIL-STD-2165A, *Testability Program for Systems and Equipments*

Prescribes approach to testability planning, diagnostic concepts and testability concepts, testability design, and program reviews.

MIL-STD-2167A, *Defense System Software Development*

Establishes uniform requirements for software development. Includes data item descriptions for analyses and data to be submitted but does not provide detail on "how to do it."

DoD-STD-2168, *Defense System Software Quality Program*

Establishes the software quality program for defense systems.

DoD 4245.7-M, *Transition From Development to Production*

This document provides templates (diagrams) showing critical categories and considerations for design, test, production, facilities, logistics, and management during the transition from development to production. It details areas of risk in each category of concern, provides an outline for reducing the risk of the transition, and presents examples of analysis and control methods. Timelines indicating the general time each discipline will need to be implemented are provided. This document includes considerations for software design as well as computer-aided design. This is a "what to do," not a detailed "how to do it" manual.

MIL-H-46855B, Notice 2, *Human Engineering Requirements for Military Systems Equipment and Facilities*

This document details requirements and tasks to be applied during development and acquisition program phases to improve the human interface with equipment and software. Its use should allow achievement of effective and economical utilization of human resources. Topics covered are analysis functions, including human performance parameters, equipment capabilities, and task environments; design, test, and evaluation; and analyses to be performed such as work load analysis, dynamic simulation, and data requirements. Some selected terms are defined such as critical human factors. An application matrix is provided detailing program phases when each task is appropriate. Refer to MIL-STD-1472 for task details.

MIL-STD-52779A, *Software Quality Assurance Program Requirements*

This document covers software quality-assurance (QA) requirements to the extent that it says "go do it," but it isn't a "how to" document. Topics addressed are software QA programs, tools and technical methodologies, computer program design, work certification, documentation, documentation library control, reviews, configuration management, testing, corrective action, and subcontractor control. Other documents would provide insight into what is needed rather than what topics should be covered in a proposal.

LC-78-2, *Storage Reliability Analysis Summary Report*

This document (Vol. 1, *Electrical & Electronic Devices*; Vol. 2, *Electromechanical Devices*; Vol. 3, *Hydraulic and Pneumatic Devices*; Vol. 4, *Ordinance Devices*; Vol. 5, *Optical and Electrooptical Devices*) summarizes analyses on the nonoperating reliability of missile material. This document details the failure mechanisms observed on many part types, generally those itemized in MIL-HDBK-217B. Reliability models are included to allow reliability calculation.

NPRD-91, *Nonelectronic Parts Reliability Data,* 1991

This document provides failure-rate and failure-mode information for mechanical, electromechanical, electrical, pneumatic, hydraulic, and rotating parts. The assumption that the failures of nonelectronic parts follow the exponential distribution has been made because of the virtual absence of data containing individual times or cycles to failure. Generic failure-rate tables include data on environment, applications (military or commercial), failure rate, number of records, number failed, and operating hours. A 60 percent confidence interval is used.

RADC-TR-73-248, *Dormancy and Power On-Off Cycling Effects on Electronic (AD-768 619) Equipment and Part Reliability*

This document is the result of two 12-month programs by Martin Marietta. The first was to collect, study, and analyze reliability information and data on dormant military electronic equipment and parts and develop current dormant failure rates, factors, and prediction techniques. The second study was to provide similar information on power on/off cycling. Over 276 billion part-hours of dormancy information was gathered and 118 billion part-cycles of power on/off information was gathered. The power on/off cycling data resulted in limited success.

RADC-TR-75-22, *Nonelectronic Reliability Notebook*

This document contains sections on failure rates and analytical methods. The analytical section addresses applicable statistical methods, reliability prediction, demonstration, and specification. The analytical methods are in a cookbook format, and examples are provided. The section on prediction methods includes the methods of Lipson and Kececioglu in applying strength-stress interference methods. The demonstration test section contains instructions for using most of the standard methods. The statistical methods section describes methods for fitting failure distributions, point and interval estimation, tests for outliers, and tests for increasing hazard rates. The prediction tables list environment, failure rate (90 percent confidence limit), environment, and application factors for those environments.

RADC-TR-83-29, *Reliability, Maintainability, and Life Cycle Cost Effects of Commercial Off-the-Shelf Equipment*

This document provides the results of a study to determine the effects of using commercial electronic equipment in a military environment. Some terms such as militarized or best commercial practice are clarified. A computer program listing is included to perform a life-cycle cost study. Results of an industry survey performed as part of the study are included.

RADC-TR-83-72, *The Evolution and Practical Applications of Failure Modes and Effects Analysis*

This document gives a broad, general background in techniques available for failure effects and analysis. Sixteen techniques such as tabular FMEA, matrix FMEA, sneak circuit analysis, fault-tree analysis, and hardware-software interface analysis are discussed. This is a good working document.

RADC-TR-85-91, *Impact of Nonoperating Periods on Equipment Reliability*

This document is designed to provide the nonoperating reliability prediction equivalent of MIL-HDBK-217. The models were derived using empirical data analysis. Details on data sources are provided. Definitions of dormancy, storage, equipment power on/off cycles, etc. are provided and are especially helpful since their definitions are often misunderstood. Nonoperating failure-rate models analogous to MIL-HDBK-217-D were developed with examples of calculations. General models included base-failure rates, temperature factors, temperature coefficients, quality factors, and on/off cycling factors. This document is especially useful for failure mechanisms of dormancy and their implications. These considerations can provide solutions to many of the dormancy problems and give insight into critical areas where resources may be expended to improve the product reliability.

Rome Laboratory Reliability Engineer's Toolkit, April 1993

An application-oriented guide for the practicing reliability engineer. This book has sections on requirements, source selection, design, analysis, and testing. It is a superior reference book with many references and "how to" information.

Environmental Stress Screening Guidelines for Assemblies

This is an excellent "how to" guide for environmental stress screening. Provided by the Institute of Environmental Sciences.

INDEX

About the Editors

W. GRANT IRESON (deceased) was professor emeritus of industrial engineering at Stanford University. He was author and editor of *Factory Planning and Plant Layout, Handbook of Industrial Engineering and Management, Principles of Engineering Economy*, and the first edition of McGraw-Hill's *Handbook of Reliability Engineering and Management*.

CLYDE F. COOMBS, JR. has held a variety of technical and managerial positions in manufacturing, quality assurance, and marketing with Hewlett-Packard in California. He is the editor of the fourth edition of *Printed Circuits Handbook,* the second edition of *Electronics Instrument Handbook*, and the first edition of the *Handbook of Reliability Engineering and Management*, all published by McGraw-Hill.

RICHARD Y. MOSS is corporate reliability engineering manager for Hewlett-Packard. In addition to numerous papers presented at ASQC, IEEE, EOS/ESD Association Symposia, he is a contributor to McGraw-Hill's *Electronic Measurements and Instrumentation*, and the first edition of the *Handbook of Reliability Engineering and Management*.